# Archimedes

New Studies in the History and Philosophy of Science and Technology

Volume 69

**Archimedes** has three fundamental goals: to further the integration of the histories of science and technology with one another; to investigate the technical, social and practical histories of specific developments in science and technology; and finally, where possible and desirable, to bring the histories of science and technology into closer contact with the philosophy of science.

The series is interested in receiving book proposals that treat the history of any of the sciences, ranging from biology through physics, all aspects of the history of technology, broadly construed, as well as historically-engaged philosophy of science or technology. Taken as a whole, Archimedes will be of interest to historians, philosophers, and scientists, as well as to those in business and industry who seek to understand how science and industry have come to be so strongly linked.

*Submission / Instructions for Authors and Editors*: The series editors aim to make a first decision within one month of submission. In case of a positive first decision the work will be provisionally contracted: the final decision about publication will depend upon the result of the anonymous peer-review of the complete manuscript. The series editors aim to have the work peer-reviewed within 3 months after submission of the complete manuscript.

The series editors discourage the submission of manuscripts that contain reprints of previously published material and of manuscripts that are below 150 printed pages (75,000 words). For inquiries and submission of proposals prospective authors can contact one of the editors:


Please find on the top right side of our webpage a link to our *Book Proposal Form.*

Agathe Keller • Karine Chemla
Editors

# Shaping the Sciences of the Ancient and Medieval World

## Textual Criticism, Critical Editions and Translations of Scholarly Texts in History

Springer

*Editors*
Agathe Keller
Faculté Sociétés et Humanités / UFR IHSS /
Laboratoire SPHère
CNRS - Université Paris Cité
Paris, France

Karine Chemla
Faculté Sociétés et Humanités / UFR IHSS /
Laboratoire SPHère
CNRS - Université Paris Cité
Paris, France

ISSN 1385-0180 ISSN 2215-0064 (electronic)
Archimedes
ISBN 978-3-031-49616-5 ISBN 978-3-031-49617-2 (eBook)
https://doi.org/10.1007/978-3-031-49617-2



This Springer imprint is published by the registered company Springer Nature Switzerland AG
The registered company address is: Gewerbestrasse 11, 6330 Cham, Switzerland

Pour Pierre Chaigneau (1987–2022)

et

Pour Pauline Schmitt-Vandenbroeck

Ces pages qui cherchent à comprendre
comment le présent reste tissé des mots du passé

# Acknowledgements

This book is the outcome of two successive conferences that were prepared in the context of the ERC project 'Mathematical Sciences in the Ancient World (SAW)' (see http://sawerc.hypotheses.org/conferences/conference-shaping-the-sciences-of-the-ancient-world and http://sawerc.hypotheses.org/workshops/workshop-history-of-critical-editions). It is our pleasure to thank all those who participated in these meetings for their contribution and their feedback. It goes without saying that the funding we received from the European Research Council under the European Union's Seventh Framework Programme (FP7/2007–2013)/ERC Grant Agreement No. 269804 was instrumental in allowing us to develop the research whose outcomes are presented in the book. In the process of turning the original articles into a book, several colleagues played a key part: we have pleasure in expressing our deepest gratitude to Megan Forster, Jeremy Gray, Richard Kennedy, Virginie Maouchi, Karen Margolis, Christine Mousset, Elise Moeng, and Martin Sauvage.

# Contents

# Chapter 1
# Shaping the Sciences of the Ancient and Medieval World: An Introduction

**Karine Chemla and Agathe Keller**

**Abstract** This introductory chapter gives an overview of the research questions raised in the book as much for historians of science as for anyone working with, or producing editions of, ancient scholarly texts. It highlights the benefits that flow from a worldwide history of textual criticism and editions as well as from a focus on texts dealing with science—two key options that are taken in this book. Following the book's scheme, the introduction first concentrates on ancient editorial practices, in particular examining their potential impact on modern editions. We then highlight how, through time, perceptions changed concerning what a text is, and how this influenced in turn how scholarly texts were made accessible. We offer an analysis of the ways historical, political and social contexts shaped editions and translations of ancient scientific works and documents, using the case studies offered in this book, before turning to an analysis of the specifics of editions and translations that bear on scholarly documents rather than on literary or religious sources. Finally, this introduction looks at how some elements specific to texts dealing with science—such as diagrams and numbers—have been edited and the specific work that has been done editing mathematical and astronomical texts of the past. All these threads help us reflect on how editorial practices have heavily mediated the way we have access to ancient sources dealing with science. The scholarship displayed here lays the foundation for further studies on the history of critical editions. It also raises questions for those who make scholarly translations and critical editions today.

K. Chemla (✉)
Faculté Sociétés et Humanités / UFR IHSS / Laboratoire SPHère,
CNRS - Université Paris Cité, Paris, France

Radcliffe Institute, Harvard University, Cambridge, MA, USA
e-mail: chemla@univ-paris-diderot.fr

A. Keller
Faculté Sociétés et Humanités / UFR IHSS / Laboratoire SPHère,
CNRS - Université Paris Cité, Paris, France
e-mail: kellera@univ-paris-diderot.fr



A. Keller, K. Chemla (eds.), *Shaping the Sciences of the Ancient and Medieval World*, Archimedes 69, https://doi.org/10.1007/978-3-031-49617-2_1

## 1.1 Outlining the Project of the Book

### *1.1.1 Historical Remarks About Textual Criticism*

In 1996, shortly before his untimely death, the historian of mathematics in ancient Greece Wilbur Knorr (1945–1997) published an article that bore a striking title: 'The Wrong Text of Euclid: On Heiberg's Text and its Alternatives'. The title was referring to the critical edition of one of the major works of Greek antiquity—Euclid's *Elements*—by Johan Ludvig Heiberg (1854–1928), the philologist who brought out editions of virtually all the ancient Greek mathematical texts.[1] To this day, Heiberg's critical edition of the *Elements* (as well as most of his other editions) remains the authoritative reference. It has served as the basis for all the modern translations of the *Elements*, from the classic English translation published by Thomas Heath (1861–1940) in the first decades of the twentieth century (Heath 1908, second edition: 1926, reprinted as Heath 1956) to the most recent Italian translation (Acerbi 2007), including the French translation published at the end of the twentieth century (Vitrac 1990–2001). In other words, virtually all the historical works on Euclid's *Elements* that have appeared since the publication of this critical edition have built upon it. Yet, Knorr (1996) expresses doubts about the capacity of this critical edition to represent the oldest state of Euclid's *Elements* that could be reached with philological work. One could not explain more clearly how important it is for modern scholarship to reflect critically upon the critical editions of ancient writings that lie at the basis of today's historical and philosophical work. This is precisely the goal of this book.

The thrust of Knorr's argument is disturbing: In his view, because of the way Heiberg worked on his critical edition, he could concretely restore only a state of the text posterior to—and incorporating elements of—Theon of Alexandria's fourth-century recension of the *Elements*. However, Knorr continues, in theory, the available documents allow us to shed light on states of Euclid's *Elements* prior to Theon's work, and hence closer to Euclid's original text. The details of Knorr's analysis are important: For Knorr, Heiberg had *a priori* ideas about the way Euclid had written the *Elements*. In particular, in Heiberg's view, Euclid avoided logical gaps in proofs. This a-priori assumption led Heiberg to discard medieval Arabic and Latin manuscripts of the *Elements*, on the sole ground that they did not fit with his expectations concerning this Greek work. Indeed, they had logical gaps that the earliest extant Greek witnesses (some of which were more recent than these medieval editions) did not have. Accordingly, Heiberg based his editorial work on these Greek witnesses, thereby carving his own assumptions into the corpus at the basis of his philological work, and, accordingly, into the critical edition that we now all use in our research

[1] Heiberg's edition of the *Elements* (1883–1888) was published as part of that of Euclid's complete works by Heiberg and Heinrich Menge (1838–1904) (Heiberg and Menge 1883–1916). (Knorr 2001), published posthumously, returns to the same issue.

on the *Elements*. How can research on Euclid's proofs rely on Heiberg's edition? This is the key issue that Knorr's critical remarks raise.

By contrast, Knorr puts forward the thesis that these Latin and Arabic manuscripts themselves might be considered more faithful witnesses to a state of Euclid's *Elements* prior to Theon's edition and commentary than the Greek editions used by Heiberg, which represented a state of the text dependent on this later editorial work. Knorr further points out that this was precisely the hypothesis defended by the specialist of Arabic studies and history of mathematics Martin Klamroth (1855–1890), who during the same decades suggested basing a critical edition of Euclid's *Elements* on the Arabic witnesses from the Middle Ages. The key point is that, as we will see, this case illustrates the general rule with respect to critical editions of ancient works rather than a marginal phenomenon.

In fact, in the wake of Knorr's seminal article, doubts have also been raised about Heiberg's critical edition of the works of another major scholar of Greek antiquity, Archimedes. Heiberg's assumptions about Archimedes as a mathematician contrasted sharply with his hypotheses about the Euclid of the *Elements*. At variance with the idea that in the *Elements* the original text avoided logical gaps, in Archimedes' case Heiberg advanced the thesis that the mathematician did not bother with obvious arguments. This assumption led Heiberg to suspect a significant number of passages in the received text as 'interpolations', which he accordingly placed between square brackets in both the Greek text and the Latin translation (Chemla 1999). This holds true for Heiberg's first edition of *Archimedis Opera Omnia* (Heiberg 1880–1881) as well as for the second edition (Heiberg 1910–1915), which he published after new Archimedean writings had resurfaced (Netz 2004: 2).[2] As has been the case with Euclid's *Elements*, subsequent translations have drawn on Heiberg's critical editions. However, these translations dealt with Heiberg's suspicions of interpolations in quite different ways, as can be shown from taking a quick look at an example. We have chosen for this the second proposition of Archimedes' *On the Sphere and the Cylinder*. In his first edition, for this proposition, Heiberg (1880 (vol. 1): 14, Greek text) puts between square brackets two expressions, as he does in the translation into Latin (Heiberg 1880 (vol. 1): 14, Latin text). Consequently, in the translation, the bracketed expressions feature as having a status similar to the additions that Heiberg introduces into the text for the sake of explanation. In the second edition, the number of passages suspect of being interpolations increased (Heiberg 1910 (vol. 1): 12, Greek text). If we keep the example of the proposition 2, Heiberg added square brackets around another suspected term. Heath (1897), which offers a free English translation in which Archimedes' text is heavily modernized, relies on Heiberg's first edition: in the aforementioned second proposition, the passages suspect of being interpolations do not feature. In the twentieth century, two French translations of Archimedes' works appeared (Ver Eecke 1921; Mugler

[2] Decorps's chapter in this book outlines a history of editions of Greek mathematical texts of antiquity. She points out that Friedrich Hultsch's critical edition of Pappus' *Collection* (1876–1878) likewise considered several passages as spurious 'on the grounds of an ideal representation of what a Greek mathematical treatise should be' (see note 91).

1970–1972). For the same proposition 2, Paul Ver Eecke (1921: 8) silently translates the larger passage bracketed by Heiberg, but not the other two, without making his choices explicit, let alone what motivated them. However, in a footnote inserted in proposition 10, Ver Eecke makes clear that he does not translate a long passage suspected by Heiberg to be spurious (Ver Eecke 1921: 20). Mugler (1970 (vol. 1): 13–14) does for proposition 2 what he does everywhere: he reproduces Heiberg's brackets in the Greek text but translates the whole text into French without any indication—including the passages suspected of being interpolations. Finally, in the most recent English translation, Netz (2004: 43–45) translates the Greek text of proposition 2 according to Heiberg's second edition, including its square brackets, and he discusses their appropriateness as well as other editorial issues in the 'Textual Comments'. There, for proposition 2 of *On the Sphere and the Cylinder*, Netz expresses doubts about the validity of one of Heiberg's bracketing choices. Moreover, he argues in favor of another one, and introduces a new suspicion of interpolation that Heiberg had not pointed out. Netz's comment about the last and most important of Heiberg's bracketed passages (step 16 in the 2004 segmentation of the text) is worth reading for our purpose:

> Step 16 belongs to an important class: pieces of text which may be authentic (and then must shape accordingly our understanding of Archimedes' practices) or may be interpolated. How to tell? Only by our general understanding of Archimedes' practice—an understanding which is itself dependent upon such textual decisions! Heiberg imagined a purist, minimalist Archimedes. In this, he may have been right: my sense, too, is that Step 16 is by a later scholiast. But we should keep our minds open.

One could hardly better formulate how critical editions and translations shape texts of the past according to the image editors and translators had of them, and therefore surreptitiously communicate these actors' assumptions to the readers. With respect to Greek mathematical texts of antiquity, we still need to assess fully the consequences of Heiberg's editorial practices and assumptions for subsequent historiography.[3] More basically and more broadly, the above observations raise complex questions about the critical editions of ancient texts available today. We will not solve these specific questions here. Rather, this book aims to address general issues that these case studies point out and that can be summarized as follows: As we have already underlined, our work on the history of ancient science depends in an essential way on critical editions that were prepared in modern times in contexts sometimes far removed from our own. The problem is that, on the one hand, we cannot redo these critical editions for each new investigation into the texts that we undertake. On the other hand, the extant editions cannot be used lightly and indiscriminately. This dilemma raises a simple question: How then can we equip ourselves with the appropriate critical tools to use these editions today? The examples of Heiberg's editions of Euclid's *Elements* and Archimedes' works highlight how the authors of editions draw on documents that were in turn produced in contexts different from their own. These authors put into play criteria to select the source material

[3] For the case of Archimedes, Netz (2012) offers a reflection on this issue.

to be used to prepare their editions. They carry out textual criticism according to methods and values that depend not only on their personal choices, but also on the context in which they operate and the goals they assign to their editorial work. Highlighting these criteria, methods and values and analyzing them might enable us to use the results of these philological endeavors with the necessary critical distance. Our project for this book derives from these remarks. It appeared to us that an essential manner for the historian of science in the ancient world to acquire a critical distance of this kind was to develop a historical approach to the modern critical editions that we use daily in our research. This is one of the main aims pursued in this book.

### *1.1.2 Critical Editions and the Erasure of Clues About Practices*

The full deployment of our project requires that analysis such as Knorr's be first expanded. Indeed, Knorr focused primarily on Heiberg's editorial work on the *discursive part* of the *Elements*. His remarks thus have a crucial impact in estimating, for example, the extent to which we can rely on Heiberg's edition to conduct a historical investigation about Euclid's mathematical ideas or about his proofs. More recent work by Saito (2006) has further suggested that Heiberg's edition could also be an obstacle for us to work on Euclid's mathematical *practices*, more specifically in this case, his practices with diagrams. For Saito (2006) has shown that the geometrical diagrams of Heiberg's edition diverged quite dramatically from those of the ancient editions of Euclid's *Elements* on which Heiberg's philological work rested (see Fig. 1.1). These diagrams thus also call for critical analysis. Saito and Sidoli (2012) have actually established that in this case Heiberg seems to have ignored the evidence about diagrams found in the manuscripts and that he simply included into his edition the diagrams from an early nineteenth century edition of the *Elements* by Ernst Ferdinand August (1795–1870) (August 1826–1829).

It is true that some facets of Euclid's practice with diagrams, like those studied by Reviel Netz (1999), can be approached using clues that are found in the discursive part of the text of the *Elements*. Arguably, these facets might, at least partly, be captured using Heiberg's edition. However, Saito highlights other facets of that practice whose study depends in an essential way on material features of the diagrams themselves. The 'overspecification' of the diagrams is a case in point (Saito 2006: 82). By this term, Saito refers to the fact that ancient editions feature diagrams that are more specific than the proposition requires. Let us illustrate this phenomenon with the example of Proposition I.4 of the *Elements*, which is precisely the topic of Fig. 1.1. This proposition concerns two equal triangles and Fig. 1.1 brings together the diagrams for it included in the main ancient editions as well as in Heiberg's critical edition (Saito 2006: 100). Heiberg's diagram depicts these triangles as generic, whereas in Codex P the two triangles are equilateral and in Codex

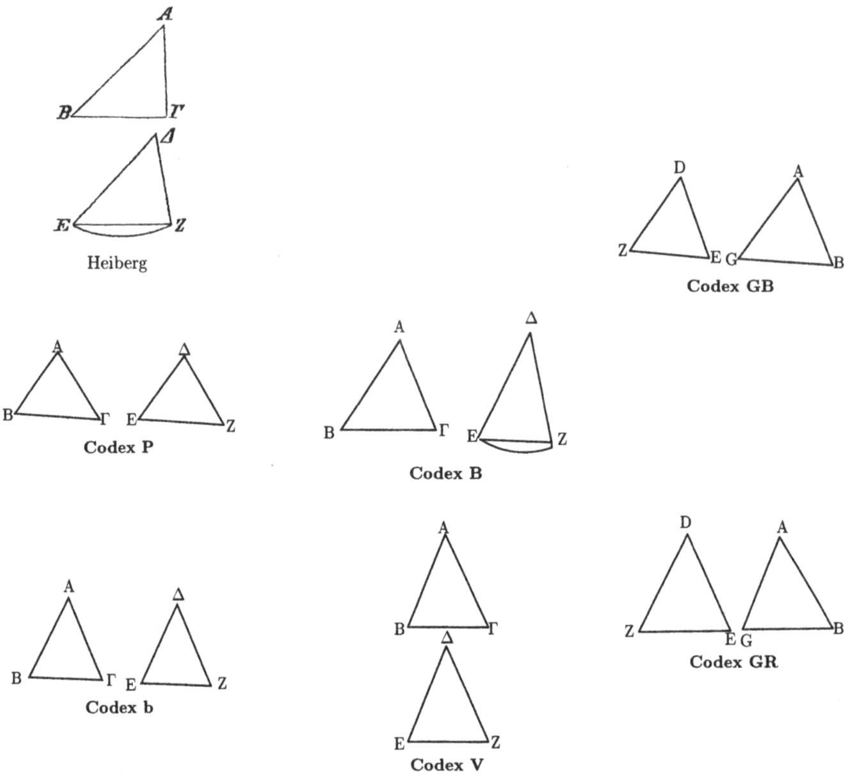


**Fig. 1.1** This figure is taken from (Saito 2006: 100). For Proposition 4 of Book I of the *Elements*, it shows the diagram found in Heiberg's edition (top left) as well as a set of diagrams contained in ancient editions. Under Heiberg's diagram, to the left, the reader can see the diagrams of the various Greek editions on which Heiberg relies. **Codex P, Vaticanus graecus 190,** was copied between 830 and 850; **Codex B, Bodleianus Dorvillianus 301**, was copied in 888; **Codex b, Codex Bononiensis, Bologna,** biblioteca comunale, 18-19, dates from the eleventh century, and **Codex V,** Vindobonensis phil.gr. 31, from the twelfth century. To the right, we reproduce the diagrams of two ancient Latin editions that Saito also provided: **GB: Bruges 521** and **GR, Vat. Rossiano 579,** which are two fourteenth-century copies of the Latin translation by Gerard of Cremona (1114–1187) (© Saito)

b and V they are isosceles. Are these witnesses outliers whose testimony should be disregarded, or do they reflect diagrams of a type Euclid used in his mathematical practice? The key point is that the oldest extant papyri also show diagrams that are overspecified, thus inviting us not to discard right away the hypothesis that the singular diagrams found in ancient editions might reflect those used by Euclid.

Figure 1.2a reproduces one of these papyri, Oxyrhynchus I 29, edited by Bernard P. Grenfell, and Arthur S. Hunt (1898). The editors identified this papyrus as containing the enunciation and diagram of Proposition 5 of Book II of the *Elements*. Clearly, the diagram borne by the papyrus shows, on the left hand side, the square dealt with in Proposition II.5 cut into four identical squares. However, the text of the proposition

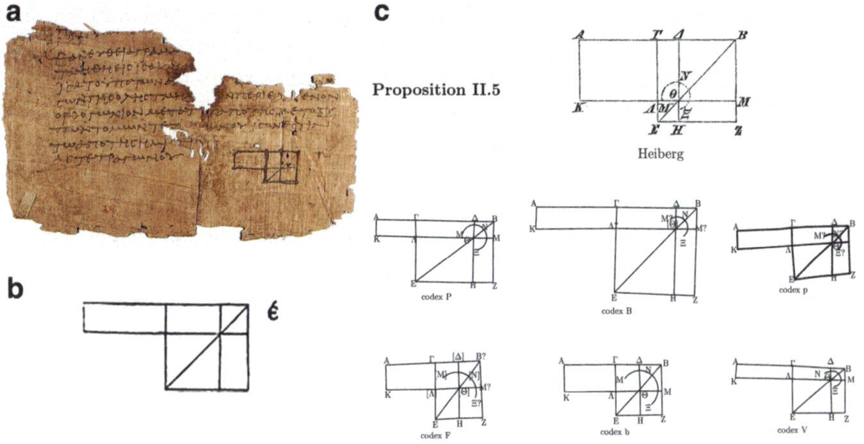


**Fig. 1.2** (**a**) Papyrus Oxyrhynchus I 29, University of Pennsylvania. Downloaded from https://en.wikipedia.org/wiki/Papyrus_Oxyrhynchus_29#/media/File:P._Oxy._I_29.jpg (accessed February 16, 2021, ©Wikimedia commons). According to Grenfell and Hunt (1898: 58), who provided the first edition of this papyrus, it shows 'the enunciation, with diagrams' of Proposition 5 from Book II of the *Elements,* and it was copied between the end of the third century and the beginning of the fourth century. Fowler (1987: 210–212) considers the copy was made between 75 and 125. (**b**) The edition of the diagram provided by Grenfell and Hunt (1898: 58). (**c**) Saito (2011: 47) reproduces Heiberg's diagram (on top) and (below) those of the manuscripts Heiberg used in his critical edition (© Saito)

and its proof more generally put into play a division of this square into two different squares and two rectangles. The diagram of the papyrus is thus overspecified, since it transforms shapes that in the general case are different from each other into four identical squares. Grenfell and Hunt's 1898 edition draws the diagram as shown in Fig. 1.2b. Like Heiberg's philological practice for diagrams, their edition thus erases the overspecification of the papyrus' diagram and substitutes it for a diagram that is closer to modern standards regarding the way the figure relates to the text of the proposition. The same remark holds true for Heiberg's treatment of the diagram for the related proposition in the *Elements*, which Fig. 1.2c reproduces along with the diagrams of the main ancient editions (Saito 2011: 47). Heiberg's diagram is lettered like those in the ancient editions of the *Elements*. However, as far as the shapes are concerned, he discarded the evidence of the overspecified diagrams to redraw a diagram similar to the witnesses that were closer to a modern practice of diagrams. In other words, Heiberg drew the diagrams according to his own modern representation.

Recently, the study of Euclid's diagrammatic practices has become a hot topic in the history and philosophy of mathematics. These remarks imply that Heiberg's edition cannot without any further reflection serve as a basis for an approach to Euclid's practice of diagrams. Knorr's analysis thus needs to be extended beyond the discursive part of the *Elements*. In fact, as Netz (2012) further showed, the same conclusions hold true for Heiberg's edition of Archimedes' writings. We have seen that they also apply to Grenfell and Hunt's edition of the Oxyrhynchus papyrus.

Two remarks are important here. To begin with, the editorial practices for diagrams that we have described have an impact on the historiography. Indeed, in redrawing the diagrams of ancient Greek mathematical texts in this way, Grenfell and Hunt as well as Heiberg do not merely ‘modernize’ them. They further increase their distance from the diagrams found in other ancient mathematical sources, e.g., in ancient Chinese mathematical texts that have come down to us. Let us illustrate this point with *The Gnomon of the Zhou [Dynasty]* (*Zhoubi* 周髀, completed in the first century BCE or CE)—this is the oldest extant Chinese mathematical classic handed down with ancient commentaries through the written tradition—and, more precisely, with diagrams used by the third century commentator Zhao Shuang 趙爽. Figure 1.3a, b, c shows these diagrams as they are found in the earliest extant edition of this work, published by Bao Huanzhi 鮑澣之 in 1213. The other ancient editions share the same diagrammatic features as those illustrated by Fig. 1.3. Zhao Shuang refers to these diagrams to discuss the correctness of algorithms that are associated with right triangles. The use of square units in the diagrams indicates clearly that the figures are drawn for the right triangle whose three sides are, respectively, 3, 4 and 5. Zhao Shuang’s text also mentions these particular dimensions. Using Saito’s

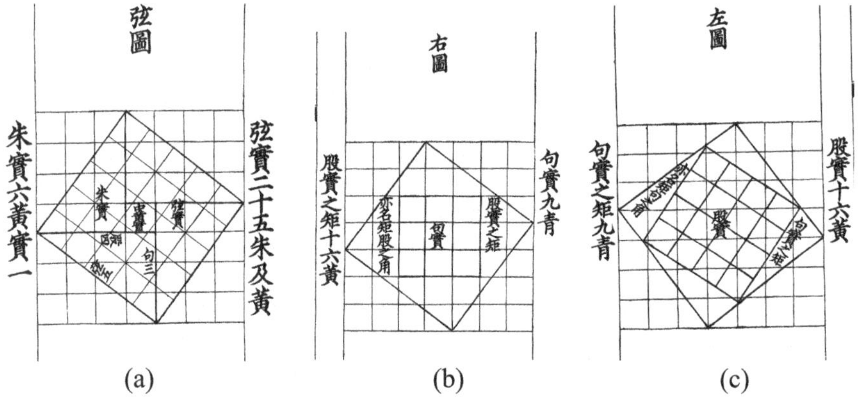


**Fig. 1.3** Bao Huanzhi’s 鮑澣之1213 edition of *The Gnomon of the Zhou [Dynasty]* (*Zhoubi* 周髀, first century BCE or CE) (Volume *Zhoubi*, in (Shanghai Tushuguan and Beijing Daxue Tushuguan 1980: 3b–4b)). The three diagrams open a section of Zhao Shuang’s 趙爽 commentary (third century) titled: ‘Figures of the base (*gou*) and of the height (*gu*), of the square and of the circle.’ Translation of the textual indications on the diagram: (**a**) The two characters at the top: ‘Figure of the hypotenuse’. Then, from top to bottom, from right to left: ‘The square (*shi*) of the hypotenuse, twenty-five, is vermillion and yellow.// The square of the hypotenuse//The base is three.//Central yellow area (*shi*).//(in horizontal characters) The height is four.//Vermillion area (*shi*)//(slantwise) The hypotenuse is five.//The vermillion areas are six. The yellow area is one.//’ (**b**) The two characters at the top: ‘Right Figure’. Then, from top to bottom, from right to left: ‘The square of the base, nine, is blue-green.//The gnomon of the square of the height//The square of the base//Is also called the angle of the height as gnomon//The gnomon of the square of the height, sixteen, is yellow.’ (**c**) The two characters at the top: ‘Left Figure’ Then, from top to bottom, from right to left: ‘The square of the height, sixteen, is yellow.//The gnomon of the square of the base//The square of the height //Is also called the angle of the base as gnomon//The gnomon of the square of the base, nine, is blue-green’

words, these diagrams are overspecified, in the same way as those found in ancient Greek mathematical documents.[4] This remark underlines a similarity between practices with diagrams to which ancient Greek and ancient Chinese sources attest. However—and this is the key point—the similarity can be seen only if we rely on ancient editions. It is hidden if we compare Chinese sources with Grenfell and Hunt's as well as Heiberg's critical editions.

This is where the second remark comes into play. Indeed, the editorial practices concerning diagrams that we have highlighted in Grenfell and Hunt's edition as well as in Heiberg's are by no means an exception. Modern editions of ancient scientific texts exhibit phenomena of this kind much more broadly. We can illustrate this remark using precisely the way modern critical editions have dealt with the diagrams contained in Zhao Shuang's commentary on *The Gnomon of the Zhou [Dynasty]* that we have mentioned above. Let us take, as an example, the first modern critical edition of *The Gnomon of the Zhou [Dynasty]*, which Qian Baocong 錢寶琮 published in 1963. As we have seen, Fig. 1.3a, b, c reproduces the diagrams as they occur in the earliest extant edition from 1213, whereas, as has already been mentioned, the diagrams found in the other extant ancient editions all share the features that this 1213 edition exhibits. In addition to showing that the shapes displayed bear unit squares, these ancient editions all contain diagrams with the same textual indications. These indications refer to colors (yellow, vermillion, blue-green), to shapes (e.g., gnomon), to places (e.g., center) and to specific dimensions, which echo those shown using the unit squares. Finally, the three figures seem to constitute a set of fundamental figures, from which the correction of all the algorithms Zhao Shuang gives about the right triangle can be established (Chemla 2005). Figure 1.4 reproduces the diagrams that Qian (1963: 15–16) drew for his critical edition in order to feature Zhao Shuang's diagrams. Clearly, the diagrams in the 1963 edition delete salient features of the witnesses. For instance, Qian's diagrams do not make use of unit squares and, in correlation with this point, the textual indications in the modern diagrams do not refer to any particular value. Moreover, Qian replaces the set of three diagrams, as shown in all the ancient editions, by five diagrams, thereby modifying the nature of the relationship between the diagrams and the text. We argue that these changes delete key clues about ancient actors' practice with diagrams, in exactly the same way as Grenfell and Hunt as well as Heiberg did for Greek mathematical texts of Antiquity. Moreover, Qian's substitution of diagrams has had a clear impact on the later transmission of the text of *The Gnomon of the Zhou [Dynasty]*. Indeed, in the same way as Heiberg's diagrams were reproduced in all the translations based on his edition, subsequent critical editions of *The Gnomon of the Zhou [Dynasty]* also used the diagrams that Qian (1963) inserted into the text and that differed radically from those of the ancient editions on which he drew.[5]

[4] Chemla (2005) argues that Zhao uses these diagrams as paradigms.

[5] See, e.g., for the *Elements*, (Vitrac 1990, volume I: 200–202) and, for *The Gnomon of the Zhou [Dynasty]*, the critical edition by Guo Shuchun 郭書春 and Liu Dun 劉鈍 (1998: 2). Note that, in

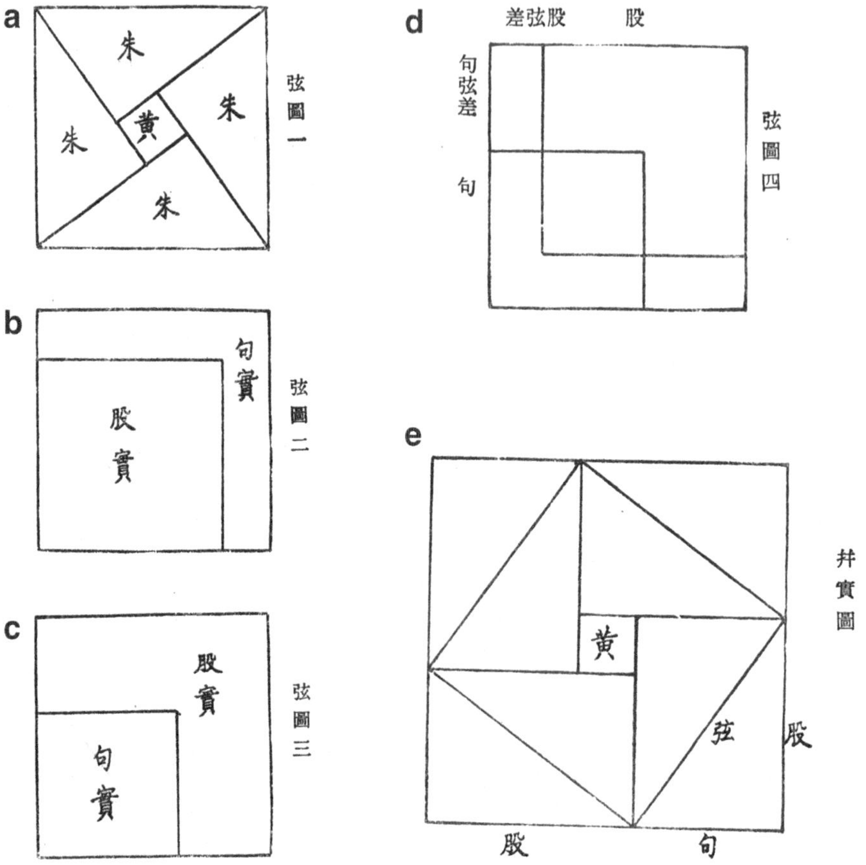


**Fig. 1.4** Critical edition by Qian Baocong 錢寶琮 (1963: 15–16) of Zhao Shuang's diagrams. Translation of the textual indications on the diagrams: (**a**) From top to bottom, from right to left: 'Figure of the hypotenuse 1: Vermillion/Vermillion/Yellow/Vermillion/Vermillion' (**b**) From top to bottom, from right to left: 'Figure of the hypotenuse 2: Square of the base/Square of the height.' (**c**) From top to bottom, from right to left: 'Figure of the hypotenuse 3: Square of the height/Square of the base.' (**d**) From top to bottom, from right to left: 'Figure of the hypotenuse 4: Height/Difference between the hypotenuse and the height/Difference between the base and the hypotenuse/Base.' (**e**) From top to bottom, from right to left: 'Figure of the square of the sum (of the base and the height): Height/Hypotenuse/Base/Yellow/Height'

One might argue that Fig. 1.3b, c as displayed in the ancient editions were erroneous and hence that in his critical edition, Qian tried to restore correct figures, as they might have been drawn before mistakes were introduced in the course of the written transmission. Li Jimin (1990: 371) clearly thought along these lines, since he too suggested replacing the diagrams in the ancient editions with correct ones. However, his way of restoring the same diagrams, which is reproduced in Fig. 1.5,

the latter edition, the vertical presentation of the sequence of diagrams given by Qian was transformed into a horizontal one.

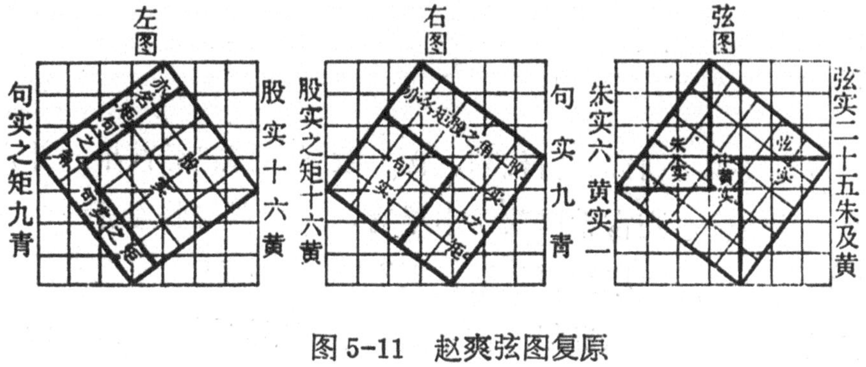


**Fig. 1.5** Li Jimin's 李繼閔 (1990: 371) reconstruction of Zhao Shuang's left and right diagrams. Translation of the textual indications on the diagram: To the left, on top: 'Left Figure' Then, from top to bottom, from right to left: 'The square of the height, sixteen, is yellow.//The gnomon of the square of the base//The square of the height //Is also called the angle of the base as gnomon//The gnomon of the square of the base, nine, is blue-green.' To the right, on top: 'Right Figure' Then, from top to bottom, from right to left: 'The square of the base, nine, is blue-green.//The gnomon of the square of the height//The square of the base//Is also called the angle of the height as gnomon// The gnomon of the square of the height, sixteen, is yellow'

clearly follows principles different from Qian's. Li Jimin's diagrams might seem to us closer to those in the ancient editions than Qian's, and we might be tempted to conclude that they are thus more faithful to those that once occurred in the original text. However, we should not jump to conclusions too quickly here: Chemla (2004) argues that the *way* in which the diagrams in the ancient editions are erroneous gives clues on the nature of the original diagrams that are essential for a historical inquiry into the practices with diagrams—clues that are precisely erased in Li Jimin's edition.

This remark illustrates clearly the dilemma that an editor faces: each solution for the diagrams has its merits and its drawbacks. More importantly, exactly as noted above concerning the discursive part of a text, we see here the latitude that editors have in their negotiation between the different criteria by which they could abide in their shaping of diagrams. Here too, the solutions they adopt depend, among other things, on the context in which they operate and the goals they assign to the edition, as much as on their personal assumptions regarding the edited text. The examples concerning diagrams given above highlight what is at issue in the variety of editorial practices that have been put into play in the making of critical editions. These practices have led to quite different ways of presenting editions of the same text to the reader. Depending on the editorial choices, historical work about diagrammatic practices can rely more or less on the editions and thus unfold more or less fully.

This book is predicated upon the conviction that a historical approach to modern editorial work can enhance our understanding of the features of ancient texts to which critical editions have applied (often tacitly) changes. Such a historical approach can further help assess the impact of these changes on the historical work based on these editions. Indeed, as previous historians have sometimes noted, other elements and features of ancient texts underwent reshaping in the course of editorial endeavors. We need to identify them. This book intends to shed light on changes of this kind that have received less attention and on their bearing upon the historiography. However, before we set out to tackle these issues, let us explain why we need, once again, to broaden our perspective.

### *1.1.3 Historicizing the Modernization of Ancient Texts in Editions and Translations*

Modernizing diagrams, albeit in different ways, suppresses clues with which we could have addressed the issue of ancient actors' mathematical practices with their figures. However, discourse and diagrams are not the only parts of ancient scientific texts that have undergone modernization—or, more generally, changes—in the successive editions, thereby making it more difficult and even sometimes impossible to use these editions to describe ancient actors' knowledge and practices. The same conclusion has been drawn about another facet of ancient scientific texts whose importance for the history of science can hardly be denied, that is, the numbers and quantities that they contain. Another historian of Greek mathematics who also passed away all too early, David Fowler (1937–2004), drew our attention to this issue, offering remarks that will prove useful in defining more precisely the project of this book. Writing about the notation and uses of fractions in early Greek mathematical texts, Fowler (1992: 134) observed:

> Almost all of our written evidence about Greek culture has passed via Egypt, and almost all of it has been later rewritten, from the ninth century AD onwards, in a modernised Byzantine script. Numerical material in these Byzantine manuscripts is liable to have been modernised and uniformised in what might then have been considered to be unimportant ways—this applies, in particular, to the treatment of numbers and fractions (One needs only to look at modern editions and translations, even by the most scrupulous of scholars, to see similar processes at work today.)

Two remarks are essential for us in Fowler's observations. To begin with, Fowler notes that the Greek manuscripts on which modern philologists have relied to produce their critical editions are for the most part posterior to the transliteration of texts from antiquity into the minuscule script, which took place in the Byzantine world starting between the end of the eighth and the ninth centuries. Fowler thus raises the question of the transformations that the notation of numbers and quantities might have undergone in the latter context as well as that of the impact of these changes on subsequent editions and more broadly on the work of historians of science.

This first remark highlights a general and essential issue: a historical and critical approach to modern editions of scientific texts, for the development of which this book pleads, must also take into consideration the contexts and circumstances in which the source material on which the philological work is based was produced as well as the possible changes to the original features that these old materials may have already incorporated. Research of this kind would not only provide crucial tools to exercise our critical acumen, but it would also allow us to examine how the modern philologists whom we study have dealt with the same question. This constitutes an important facet in the description and the contextualization of their philological practice. Several chapters in this book offer reflections on these issues, notably the two specific case studies presented in Part I.

Fowler points out a concrete example. He emphasizes that changes in the notation of divisions in a few papyri and in later Byzantine manuscripts have led historians to conclude that the concept of fraction that we commonly use today already existed in ancient Greece (Fowler 1992: 137). However, for him, such was not the case, and this gives a distorted view of numbers and arithmetic in ancient Greece. In particular, this creates a gap between hieratic and Greek mathematical documents, where in fact ancient documents point to a great continuity with respect to the nature and the concept of fraction used.[6] Exactly as we had seen above for the diagrams displayed in the critical editions, which create the illusion of a greater distance between Greek and Chinese ancient mathematical texts and a smaller distance between Greek ancient texts and modern mathematics texts, we see here the shaping of a divide between ancient Greek and hieratic texts and that of a similarity between fractions in Greek texts and modern fractions.

In fact, Fowler continues, the same type of modernization recurs in modern translations of ancient source material. This is the second remark which Fowler's quotation above highlights and which is equally important to widen adequately the perspective adopted in this book. Indeed, Fowler notes that, just as ancient editions do, modern editions as well as translations—both crucial tools for historians of science—tacitly modernize in ways that look innocuous but in effect have significant consequences. Fowler's argument relies on two documents (Fowler 1992: 138–140 and 140–141, respectively). Figure 1.6 reproduces the plate with which Fowler illustrates the first document. To the left (Fig. 1.6a), the plate displays part of the papyrus Hibeh I 27 written in *ca.* 300 BCE. The first line shows the notation of a fraction as a sequence—the juxtaposition means a sum—of what we would call 'unit fractions'. Each of these 'unit fractions' is not written as a pair of a numerator and a denominator, but as a number topped by a stroke.[7] These two features indicate that the notation of said quantity in the Greek papyrus is essentially the same as the way of writing fractions attested by hieratic mathematical texts. The edition of the papyrus published by Bernard P. Grenfell, and Arthur S. Hunt (1906: 146) is faithful to the notation as it appears in the original document (Fig. 1.6b). However, beneath the photo of the papyrus and the 1906 edition, Fowler reproduces the English translation given by Grenfell and Hunt (1906: 152)

[6] Fowler (1992) upholds this thesis. In an earlier publication, Knorr (1982) had also highlighted this point.

[7] Fowler and Turner (1983) provide a systematic description of the notation of integers and fractions in the papyrus Hibeh I 27.

**a**

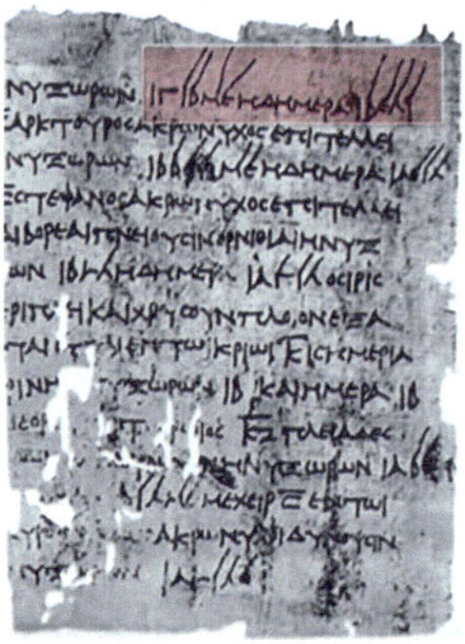

**b**

Fr. (*b*), 2nd hand. Col. iv. PLATE VIII.

[ἡ] νὺξ ὡρῶν ιγί β′ μ́ ε′, ἡ δʹ ἡμέρα ιβ′ έ Χ′ q́.
[ι]ς Ἀρκτοῦρος ἀκρώνυχος ἐπιτέλλει,
[ἡ] νὺξ ὡρῶν ιββ′ ί έ μ́ έ, ἡ δʹ ἡμέρα ιαθ′ ί Χ′.
[κ]ς Στέφανος ἀκρώνυχος ἐπιτέλλει
[κ]αὶ βορέαι πνείουσιν ὀρνιθίαι, ἡ νὺξ
[ὡρ]ῶν ιβ∠Χ′, ἡ δʹ ἡμέρα ιαγ′ ί Χ′. Ὀσῖρις
[π]εριπλεῖ καὶ χρυσοῦν πλοῖον ἐξά-
[γε]ται. Τῦβι ⟨ε⟩ ἐν τῶι Κριῶι. κ ἰσημερία
[ἐα]ρινή, [ἡ] νὺξ ὡρῶν ιβ καὶ ἡμέρα ιβ,
[κ]αὶ ἑορ[τ]ὴ Φιτωρώιος. κζ Πλειάδες
[ἀκ]ρώνυχ[οι] δύνου[σ]ιν, ἡ νὺξ ὡρῶν ιαβ′ ς′ q́,
[ἡ] δʹ ἡμέρα [ι]βί Χ′ μ́ έ. Μεχεὶρ ς ἐν τῶι
[Τ]αύρωι. Ὑάδες ἀκρώνυχοι δύνουσιν,
[ἡ] νὺξ ὡρῶν ια∠ί Χ′ έ,

55. l. ί έ for ί β. 57. ί έ corr. from έ. 65. ς′ corr. 68. l. Χ′ q́ for Χ′ έ.

**c**

55–205. '(Choiak 1st:) . . . The night is $13\frac{4}{45}$ hours, the day $10\frac{41}{45}$. 16th, Arcturus rises in the evening. The night is $12\frac{34}{45}$ hours, the day $11\frac{11}{45}$. 26th, Corona rises in the evening, and the north winds blow which bring the birds. The night is $12\frac{8}{15}$ hours and the day $11\frac{7}{15}$. Osiris circumnavigates, and the golden boat is brought out.

'Tubi 5th, the sun enters Aries. 20th, spring equinox. The night is 12 hours and the day 12 hours. Feast of Phitorois. 27th, Pleiades set in the evening. The night is $11\frac{38}{45}$ hours, the day $12\frac{7}{45}$.

**Fig. 1.6** Fowler's first document: Papyrus Hibeh I 27, edited with translations and notes by Bernard P. Grenfell, and Arthur S. Hunt in 1906. These three pieces ((**a**): Facsimile; (**b**): Transcription; (**c**): Translation) are reproduced from Grenfell and Hunt (1906: Plate VIII, 146, and 152, resp.). They were brought together in (Fowler 1992: 139)

(Fig. 1.6c). Just as noted above about the edition of the diagram occurring on a papyrus, the translation by Grenfell and Hunt transforms the fractions of the original document into a completely different concept: a single fraction with a numerator and a denominator. Interestingly, this transformation also precisely echoes the troubling problem in the historiography of fractions that Fowler exposes: an ancient notation, which indicated the use of a specific concept of fraction, is replaced with another—more modern—notation that might induce the reader to assume that the papyrus makes use of a modern notion of fraction. Note that in this case, modernization takes place in the translation. As several examples analyzed in this book suggest, modern editorial work on ancient source material has sometimes been inseparable from the production of translations into modern languages, to the point that sometimes the editorial work takes the form of a translation. This remark will appear to be all the more significant for our book as we adopt a world-wide view on the problems presented above, and at this global level, translations into languages foreign to that of the base text often incorporate editorial work. As a result, the historical inquiry presented in this book has taken both modern editions and translations of ancient scholarly texts into its focus.

### *1.1.4 Our project*

Clearly, ancient and modern editions, translations and, more generally, publications of texts from the past are crucial moments in the multiple processes by which ancient and medieval works as well as other types of documents are made available to us. As we have recalled above, since the 1980s, an awareness has gradually emerged that historians of science ought to scrutinize the changes undergone by ancient scholarly texts, when said texts are presented or represented in editions and translations that are used in historical work. However, these reflections have remained scattered and punctual. We now need to adopt a more systematic approach to this issue. This book aims to take a step in this direction. More importantly, we cannot just expose the problematic character of what until recently was often taken unquestioningly as a direct access to source-texts. It is true that, if pressed, nobody would claim that Heiberg's edition allows us to read Euclid's text of the *Elements*, in an immediate and transparent way, even though, in practice, scholars have behaved as if they believed this. In fact, we also need to analyze *how* editions and translations have tacitly transformed the sources on which they relied, and how these changes left their imprint on the historiography of science, when it was based on these editions and translations.

In a sense, this book belongs to a recently renewed approach to histories of text criticism, philology, and translations, including new perspectives on histories of the book and of critical editions.[8] In this context, it is nevertheless characterized by two key features.

To begin with, we adopt a worldwide perspective on the issues addressed. In contrast, whether we think of Knorr, Fowler or Saito, the first forays into the topic under consideration typically focused on Greek geometrical texts of antiquity. The fact of taking a wider perspective brings to the fore general questions that might otherwise have been overlooked.[9]

For instance, from a world-wide viewpoint, it becomes crystal-clear that scholarly texts of the past were edited and translated in ways that, in particular, made them understandable in new environments and comparable with texts produced in other contexts. This phenomenon is all the more conspicuous if we think that from the eighteenth century onwards, editions and translations of Chinese and Sanskrit scholarly texts were produced in Europe, while since the seventeenth century, Persian and Arabic texts had been translated into Sanskrit, and Latin editions of Greek texts of antiquity translated into Chinese. To the issue of anachronism, which appears to characterize the aforementioned treatments of diagrams and numbers, we thus need to add what Kim Plofker (2021) has referred to as 'anachorism,' that is, in our terms, the problem of overlooking that the texts dealt with were produced not

[8] See notably (Judet de La Combe 1990), (Cerquiglini 1999), (Chartier 2021).

[9] Here again, we are not alone in striving to broaden the discipline in this way. See, notably, (Suarez and Woudhuysen 2013), (Pollock et al. 2015), (Grafton and Most 2016) as well as the contributions to the journal *Philological Encounters*, for instance (Dayeh et al. 2018) and (Pecchia et al. 2021).

only in different times, but also in places and scholarly cultures far removed from those of the editors and the translators. When the evidence has been reshaped, which features of this reshaping can be associated with the fact that texts circulated in places and times in which their original languages, their textual genres, and also the practice of mathematics to which they adhered were unfamiliar? How did editions and translations tacitly make ancient "exotic" texts either more "exotic" or, conversely, comparable in bits and pieces to texts that would have been familiar to the readers? It is precisely on such issues that a world-wide perspective sheds interesting light.

From another angle, poring over sources from different parts of the world raises another key question, that of the methods by which the texts attested by these sources were edited and translated. As a first approximation, the world-wide perspective we adopt suggests distinguishing between two types of situation—which are the two poles of a spectrum of possibilities.

Sometimes, ancient works were edited by scholars whose working language was intimately related to the language of the original works in question, and who were using methods that had been fashioned to deal with sources in cognate textual traditions. We can think of the example of the Chinese mathematical work *The Gnomon of the Zhou [Dynasty]*—which we mentioned above—as it was edited by Dai Zhen 戴震 (1724–1777) in the context of the preparation of the great encyclopedic compilation the *Complete Library of the Four Branches* (四庫全書*Siku quanshu*) at the end of the eighteenth century. Dai Zhen had access to several ancient editions of the work—which he compared to establish the text—as well as editorial tools and methods that had been developed in the context of the movement of 'evidential research' (*Kaozheng* 考證).[10]

In contrast, Edouard Biot's French translation of *The Gnomon of the Zhou [Dynasty]*, which was the first ever translation of the work into a European language, could only rely on the single edition, which he found in the collection of the Royal Library.[11] Moreover, for his approach to the text, which Biot endeavored to render literally, he could rely only on the ancient commentaries with which the work has been handed down as well as on the first Chinese dictionaries in foreign languages published, and, to begin with, that published by Chrétien-Louis-Joseph De Guignes (1759–1845) in 1813.[12] This example illustrates the second type of situation for a translation.

As shown in, e.g., Chap. 10, in which Cooper studies how Sumerian texts were edited in the past, this book also exemplifies situations of this kind not only for

[10] See this book, Chap. 4, to which we return below. (Chu 2010: 147–151) outlines more generally Dai Zhen's restoring of the Chinese mathematical 'canon' in the context of his work for the *Complete Library of the Four Branches*.

[11] Biot had access to *The Gnomon of the Zhou [Dynasty]* by means of the Ming edition included in the *Jindai mishu* 津逮祕書 (Biot 1841: 596–597). On this edition, see (Chemla 2020: 286). (Biot 1841) gives a translation of the text, and is complemented by (Biot 1842), which relies on astronomical data and computation to analyze *The Gnomon of the Zhou [Dynasty]* further.

[12] On Biot's literal approach, see (Chemla 2021: 49–57). (Martija-Ochoa 2001–2002) has analyzed Biot's translation more broadly.

translations but also for editions. Indeed, sometimes, editorial work was carried out adopting assumptions that had been shaped for sources produced by different scriptural acts and attesting to texts that derived from different conceptions of authorship. The backdrop for the example is this: In nineteenth century Europe, philological techniques had been devised, in particular in the context of editing ancient Greek and Latin sources, or biblical Hebrew. For sources of this kind, a philological method imposed itself, which consisted of organizing written evidence using a stemma and then of focusing on only part of the sources in relation to the structure of the stemma. Note that, just as we have seen above for Heiberg's edition of Euclid's *Elements*, the method led to carving an assumption about the history of the sources in the corpus on the basis of which editorial work was performed. In this context, in the first decades of Assyriology, some European scholars applied this philological method directly when dealing with cuneiform sources, despite the fact that these sources were the outcome of wholly different stories and processes. Was it appropriate to believe that this method was suited for all types of ancient documents worldwide? Chapter 10 explains why there are reasons to doubt the validity of the operation for contexts in which the sources available derive from scriptural acts other than copying.[13] Which assumptions about the sources and which related editorial practices were more generally transferred from one context to another, and what consequences did this have on the editions and translations produced? These are thus other key questions that a world-wide perspective highlights as promising.

For us, this global approach to the issues addressed is all the more needed that, from the outset, the history of ancient and medieval science has adopted an international perspective, comparing writings produced in different parts of the world through the editions and translations available. It is thus essential to examine in a critical way the material foundations on the basis of which these comparisons were carried out. What is at issue here is the historical study of the fashioning of an estrangement or, conversely, of a homogeneity of texts of science.

The latter remarks bring us to the second key feature that characterizes our book in the wider context of an interest for the histories of philological endeavors: we focus on scholarly texts, with a special emphasis on documents that attest to activities in mathematics, astral sciences, and medicine. One might doubt that this fact has any bearing on the questions on which we focus. However, it clearly does. One example will suffice to illustrate this point. For the German philologist Georg Friedrich Wilhelm Thibaut (1848–1914), who prepared, with the Indian scholar Sudhākara Dvivedin (1855-*ca.* 1910), a critical edition of Varāhamihira's *Pañcasiddhāntikā* (a sixth century work on astronomy),

> texts of purely mathematical or astronomical contents may, without great disadvantages, be submitted to a much rougher and bolder treatment than texts of other kinds. What interests

[13] On the genesis of the philological method associated with the name of Karl Lachmann (1793–1851), see (Timpanaro 2005). Interestingly, the history of the production of sources that Timpanaro assumes in his reflections shows a clear bias due to the type of works to which philological endeavor was first applied in Europe. Indeed, sources are assumed to have been produced by copying, and copyists are assumed either to have made mistakes or to have produced copies following specific types of scenario.

> us in these works, is almost exclusively their matter, not either their general style or the particular words employed, and the peculiar nature of the subject often enables us to restore with nearly absolute certainty the general meaning of passages the single words of which are past trustworthy emendation.[14]

Thibaut's declaration illustrates one way in which the editing of mathematical or astronomical texts was done in a specific fashion, compared to other types of texts. In this book, we will be interested in understanding how, in the nineteenth century, and sometimes well into the twentieth century, different editors and translators perceived scientific documents as specific and separate in their study from other scholarly and literary texts. We also question how that perception may have affected their editing practices.

Clearly, scientific documents raise specific editorial issues. Scientific practices sometimes put into play textual practices that are not purely discursive, like interacting with diagrams and images, and carrying out computations. Some of these practices leave specific traces in the texts in the form of non-discursive components, like drawings and tables, while others—no less specific, like computations carried out materially—leave only clues. The fact of attending to the edition of scientific texts thus commonly requires that editors deal with several kinds of specific non-discursive elements, and all the more so that, as we have pointed out, all these textual facets have historically been subject to many editorial manipulations. How they have attended to this task, and also how they have dealt with traces and clues are questions that are central for us, in particular because of the potential impact of the result on the historiography of science. Indeed, as has already been emphasized, in the past decades, non-discursive elements of scientific practices and texts have become a key issue in the history and philosophy of science. This has led to question how these elements have been passed down, edited, and translated in what until then was often taken unquestioningly as a direct access to source-texts. Addressing this issue in a systematic way can certainly benefit these discussions.

However, the example from Thibaut quoted above shows that these issues are also worth addressing with respect to the discursive parts of these texts, which might have been perceived as less specific. Indeed, for Thibaut, because of the nature of the subjects treated in texts of this kind, the discursive part of a scholarly work allows for a specific type of editing. This is in line with the argument that what counts for scientific texts is their content not their form, thus inducing specific text criticism, modes of translation and editing for them. In fact, the two examples of Heiberg's editions of Euclid's *Elements* and Archimedes's works that we have sketched above show two strikingly different illustrations of how in the editions of scientific texts, the discursive parts were molded in relation to the philologist's assumptions about these two practitioners' mathematical activities. Here too, thus, Heiberg's philological practice for these discursive parts was intimately correlated with his perception of these texts as scientific writings. However, the assumptions

[14] The assertions, taken from Thibaut's preface (p. v) to (Thibaut and Dvivedin 1889), were quoted in (Keller 2012: 265).

he adopted and how they were brought to bear on his editorial practice differed for both cases, and in both cases, they differed from those in Thibaut's edition. Both philologists valued authenticity and faithfulness. However, they understood these values differently and they also translated them into different editorial practices. The general issues of the values prized by editors and translators and also of how they shaped their practices in response to their respective values appear as meaningful for our project.

Much is thus at issue for historians and philosophers of mathematics in deconstructing the appearance of immediacy and transparency that readers focusing essentially on the contents sometimes attach to editions and scholarly translations.

The specificity of scientific works and documents in these respects should nevertheless not obscure the fact that, to a certain extent, practitioners of editorial work and translation have applied to such texts operations that they would have applied to any other text. Seen from this angle, we may still benefit from considering the ancient and modern production of editions and translations from the broader perspective of a more general history of texts, translations and books. Conversely, precisely because of the singularity of their subject matter, editions and translations of scholarly works and documents could allow us to better perceive transformations undergone in the course of philological and translation work that would be difficult to apprehend for other types of writing. This is a conclusion that can be drawn from Part I of this book, which is devoted to ancient editorial practices.

## 1.2 Ancient and Modern Actors and Institutions at Work in the Manufacturing of Sources for Ancient and Medieval Texts

As we have argued in Sect. 1.1.3, inquiring into editions and translations of ancient and medieval scholarly texts implies that we begin with a reflection on editorial practices before early modern and modern times. Indeed, we have seen that the documents upon which early modern and modern actors relied for their philological endeavors already incorporated the results of operations carried out by ancient actors in the context of their editorial activities and that these operations had significant consequences on the historiography of science. We have encountered above the impact, on modern scholarship, of the edition of the *Elements* made by Theon of Alexandria, who operated in a context upon which historical work has already been devoted. We have also encountered the potential impact, on the inscription of numbers contained in our sources, of the transliteration of ancient Greek sources from majuscule to minuscule script, which began in the Byzantine world at the end of the eighth century. To deal with this issue more broadly, in this book, we have concentrated on ancient editors from other parts of the world.

### *1.2.1 Ancient Editorial and Cross-Linguistic Practices*

Interestingly, these other ancient actors broaden our views on the types of editorial intervention that were carried out, whether texts were worked upon, copied or translated. These actors also give us clues as to how we can detect such interventions in the documents that came down to us. This is what Piotr Michalowski argues in Chap. 2 of the book. In fact, he holds a radical view. Indeed, for him, the oldest written documents we know—that is, the first accounting texts from the fourth millennium BCE, which attest to the birth of writing—are already edited texts. The crucial remark is that these texts are quite uniform and thus seem to reflect the intention to shape an organized accounting system, using standardized writings.

In Michalowski's view, the same holds true for another basic type of text that was crucial in elementary scribal education: lexical lists. These texts were inventories of nouns in Sumerian, probably composed at the same time as the accounting documents mentioned above. However, the first material testimonies of these texts are the standardized lists that are associated with the expansion of the Ur III state (ca. 2112–2004 BCE), its languages and institutions—schools in particular—throughout much of the Middle East. Piotr Michalowski's chapter argues that the lists borne by these earliest surviving tablets should also be considered as already shaped by editorial practices. To highlight operations ancient actors carried out as they were editing these lists, Piotr Michalowski concentrates on a widespread professions list, which travelled all over Mesopotamia, from South to North and beyond, notably to Ebla in today's Syria. For the unique palace archive of Ebla reveals the manner in which the lists were copied. Some of the professions lists were copied from tablets coming directly from southern Babylonia, while others were produced from copies. At times, features of these copies testify to the scribes' intention to preserve formal textual properties of the original. Indeed, the first copies respected the norm of nineteen lines per column, while the copies of copies did not. The former copies added a double line after each nineteenth line, showing again that the scribes—although adopting a new format—wanted to keep traces of the old ones as well. Michalowski interprets these often imperceptible acts as clues of philological activity. Indeed, they reveal that ancient actors not only reflected on how to preserve the original but also created textual acts with this aim in mind. On the other hand, Ebla's palace archive provides evidence that scribes also adapted these lists. This is another facet of their editorial work. Here language is important. In southern Babylonia we know that texts were read in Sumerian while in the north they were probably read in semitic languages. Michalowski shows that the migration of lists involved all sorts of 'interlanguage procedures': lists were sometimes translated or gave birth to new lexical lists in regional languages. These operations were also editorial in nature: they aimed to re-actualize the list to make it readable for new audiences. In some cases, the antiquated professions list was adapted by listing regional professions. In others, the professions list was made into a bilingual list, serving then as a translating tool or an indicator on how to pronounce some Sumerian words.

In short, Piotr Michalowski argues that these ancient tablets testify to processes of standardization and acculturation, while at the same time providing evidence of ancient actors' reflection on how to preserve features of the original. All these acts have left clues that only indirectly tell us how texts were shaped and reshaped so that new audiences could read them. The argument highlights how challenging it is, for ancient contexts of this kind, to characterize what past editorial work consisted of. This is true because this work bears on textual aspects that we do not always consider important, such as the diagrammatic features of texts seen above. And yet features of this kind might yield crucial pieces of evidence for the historiography of science (Chemla 2020). This is also true because we only have traces of this work. What are the operations observable today that testify to the fact that in the past actors reflected on how to preserve and transmit a given text? This is a key question for our endeavor. To this question, Sheldon Pollock can give another kind of answer, because of the nature of the writings on which he relies. Indeed, he focuses on different types of commentaries on ancient Sanskrit works, which enable him to deploy a contextual approach to modes of edition in a wide range of sources in this corpus.

In comparison with the pieces of evidence discussed by Michalowski, Sanskrit commentaries attest to other types of additions to a base text, which meant to offer their readers a new form of approach to the text. Each of them reflects modalities that actors shaped in order to transmit the base text. Moreover, commentaries quote the root text, and hence they are compositions in which editions are carried out and conceptions of original texts are discussed. In this context, Pollock sheds light on another phenomenon worth contemplating. For his contextual approach allows him to establish that, depending on the genre, ancient commentators wanted to preserve and transmit different aspects of a Sanskrit text. Indeed, Pollock first underlines how little information we obtain directly. Nevertheless, he highlights that commentators put into play, in their discussions, key editorial notions such as that of 'interpolation' (*prakṣipta*, *kṣepaka*). This is the case of Haradatta Miśra, a ninth century commentator of the work of the grammarian Pāṇini, who uses precisely a term to qualify the 'original' text—the term *sāṃpradāyika* ('traditional' or 'original') being employed in opposition with that of 'interpolated'. The use of such notions allows historians to approach actors' varying representations of the genuine base text.

Moreover, Pollock sets forth clues indicating the principles by which actors determined what was interpolated, and which variant represented the original composition. The essential point for us is that the criteria used by classical, medieval and early modern editor-commentators of Sanskrit works, as they opposed original and interpolated texts, reveal two main types of editorial approaches. Indeed, Pollock suggests distinguishing between commentaries that gave pride of place to content and those that mainly relied on stylistic features, to decide over issues of interpolation. Pollock further suggests that commentators adopted one or the other approach in relation to the types of works commented upon.

For him, commentaries to knowledge forms (*vidyasthāna*s) belong to the first category, as do scriptures—an intermediate textual form between scholarly texts and poetry. Pollock illustrates the case of the edition of scriptures using the example of Buddhist texts and analyzing how the editors aim to establish what for them is 'word of the Buddha' (*buddhavacana*). Commentators on these scriptures, Pollock shows, judge the issue of the authenticity of a received text through a discussion of the quality of the truth enunciated, rather than through historical or textual considerations. Thus, in contrast to the examples given above by Michalowski, what counts in the transmission of the text in such cases has to do with its knowledge claims, rather than with formal elements. This stands also in contrast with the commentaries on epic literature and court poetry of the second millennium, for which Pollock argues that a diversity of criteria becomes explicit. For example, Vallabhedava, a tenth century commentator/editor of poetry, compares diverging readings from various recensions and gives insights into his values as an editor when he puts forward arguments to support judgments that a given reading can be considered as 'correct', 'authoritative', 'false', 'unmetrical', 'interpolated' or 'beautiful'. The yardstick by which elements of a text would be considered worth preserving would, in this case, be less the evaluation of their claims than the consideration of their conformity with what was expected from a specific genre of text.

As Pollock further underlines, in their editorial practice, commentators were aware of characteristic features of the preservation and transmission of texts in South Asia, since they might have taken into account the key role played by orality. In particular, they show an awareness of how oral instantiations may have contaminated the original wording in specific ways, as when literary texts were affected by memorization and performance. Assumptions of this kind seem to have left an impact on the concepts with which they approached textual variations. For instance, Pollock suggests that the term *pāṭha*, 'variant', should be understood as also taking into account variants in the recitation of a text. In other words, for medieval Sanskrit commentators, the notion of corruption did not only refer to an accident occurring when copying manuscripts by tracing words on a given medium but extended to the transformations that might occur in oral transmission. To put it in more general terms, the materiality and modalities of reproduction of the word and their impact on transmission do have a bearing on the philological concepts that ancient actors form to describe the events affecting the various witnesses on which they rely to produce an edition.

To summarize what this foray into ancient sources tells us, we see that we detect editorial work being performed as soon as we perceive ancient actors striving to preserve features of a text they reproduce. Should we straightforwardly consider that these actors conceived of said text as a master text? A conclusion of this kind might derive from a projection of an understanding of philological work that became dominant in nineteenth century Europe. However, we will see below that ancient sources have compelled editors to widen their understanding of editorial practices and accordingly their approach to the sources to which these were applied.

The reflections mentioned above have moreover shown that the way in which editorial work is performed depends on ancient—and modern—actors' representation of the source to be preserved and, in case variants are considered, the principles by which they are judged. It also depends on how these actors imagined the transmission process, and on the values they upheld. Adopting this perspective has already required that we diversify our own representations of editorial acts. Clearly, in these processes, scholarly texts may have been treated differently from other types of documents. How were texts dealing with mathematics, astral sciences, and medicine edited? This issue lies at the center of our interests in this book.

Our earlier considerations have highlighted how and why the contexts in which editions and translations were produced are important elements for our understanding of these textual productions. In particular, it appears as essential to note that editorial work is carried out—and editions are handed down—by different types of institutions.

Contexts of the ancient world are notably difficult to address with precision, even though this general statement might need to be nuanced depending on the regions of the world dealt with and on the ways in which access to the ancient sources was fashioned. However, when we turn to early modern and modern actors—as we do in the subsequent part of the book—, things change radically, and in these cases, we can observe in greater detail the historical and social contexts within which a person—or a group of people—could find it meaningful to embark in editions and translations of ancient scientific texts. We can moreover shed light on motivations that presided over the launching of editorial work, the types of actors who embarked on these endeavors and their social backgrounds, as well as the technologies they put into play. These are precisely the issues dealt with in Part II of the book, which concentrates on the historical and social contexts of early modern and modern work on ancient scientific texts.

### 1.2.2 What Was at Issue When Returning to Ancient Texts in Early Modern and Modern Times?

In a first case study, Han Qi examines the reasons why, and the circumstances in which, a significant interest in one of the oldest mathematical texts from antiquity, *The Gnomon of the Zhou [Dynasty]* (*Zhoubi* 周髀), grew in seventeenth- and eighteenth-century China. Why did early modern actors find this work important to the point that the book became one of the major works for the historiography of mathematics in China in subsequent centuries? Han Qi argues that this resulted from opportunity and specific political contexts—a combination that we will see recurring in several other cases. The value that through this process came to be attached to *The Gnomon of the Zhou [Dynasty]* finds its clearest expression in the fact that an edition of the book was given pride of place in the imperially-commissioned encyclopedic work *Complete Library of the Four Branches* (*Siku quanshu* 四庫全書) at the end of the eighteenth century. Han analyzes the different factors that entered into the increasing value attached to the book. Moreover, he sheds light on the process through which this edition came into being.

What circumstances allowed seventeenth- and eighteenth-century scholars to develop an interest in *The Gnomon of the Zhou [Dynasty]*? This ancient mathematical treatise—which included elements of mathematical knowledge required for cosmography and the calendar—was the only mathematical work of antiquity that had been edited by Ming scholars at the beginning of the seventeenth century. In contrast with any other ancient mathematical work, it was thus more readily available at the time. Han shows that what first sparked a new interest for *The Gnomon of the Zhou [Dynasty]* was the translation into Chinese, in 1607, of part of Clavius' (1538–1612) Latin commentary on Euclid's *Elements*, and more broadly the translation and dissemination of writings of 'European' mathematics, as early modern actors in China perceived them. The translation of the *Elements* was carried out by the Jesuit Matteo Ricci (1552–1610) and the Chinese convert Xu Guangqi 徐光啓 (1562–1633). Xu understood the potential of 'Western knowledge' to reform the imperial calendar which was a very political affair. This development triggered a reaction by scholars who refused what they called 'Western knowledge'. In this context, Xu used reference to *The Gnomon of the Zhou [Dynasty]* as a strategic move to convince his Chinese opponents that knowledge in China and knowledge brought from the West were identical, and that hence there was nothing wrong in using the latter. Reference to this ancient Chinese work was thus instrumental to legitimize the study of 'Western knowledge'. This first set of circumstances thus illustrates what was at issue for early modern actors in returning to mathematical works of antiquity. Interestingly, this occurs in a context in which various struggles for power hinged on the belief that the bodies of knowledge introduced from Europe and translated by Jesuit missionaries were 'Western'.

More than fifty years later, the Manchus having conquered China and ruling the Middle Kingdom, court politics entered the stage more decisively. This derived from the complex strategy of the Manchu emperor Kangxi (康熙 reign: 1662–1711), faced with Han scholars questioning his legitimacy as a Chinese emperor. To avoid Han scholars' opposition to 'Western science' being used in court institutions, and to show off his knowledge of ancient 'Chinese science', Kangxi, among other moves, referred to *The Gnomon of the Zhou [Dynasty]* to put forward the thesis of a 'Chinese origin of western knowledge *xixue zhongyuan* 西學中源', especially that of calendars and of right-triangles. The most important practitioner of mathematics and astral sciences at the time in China, Mei Wending (梅文鼎, 1633–1721), embraced the emperor's thesis and subsequently drew on *The Gnomon of the Zhou [Dynasty]* to further expand this view. Therefore, it was a complex situation at court that gave voice to specific discourses on the origins of mathematics and astronomy which in turn triggered interest in an ancient Chinese text of Antiquity.

The presence at the time of many Jesuits at Kangxi's court gave this phenomenon an international dimension. Indeed, European correspondents of the Jesuits—among whom Leibniz—were curious of the history of geometry in China. In particular, they were interested in knowledge and proofs of the 'Pythagorean Theorem' and this is how *The Gnomon of the Zhou [Dynasty]* entered a more global conversation. In return, this interest would resonate back to China, and have a new echo. Leibniz' questions on the nature and antiquity of 'Chinese mathematics' would find their way to Kangxi's court. Accordingly, in *The Fine Essence of Mathematical Principles* (*Shuli jingyun* 數理精蘊)—the synthetic work on mathematics that had been commissioned by Kangxi and completed in 1722 by imperial compilers among whom Mei Wendings's grandson, Mei Juecheng (梅瑴成, 1681–1763)—, the Chinese origin of astronomy and geometry is presented as being endorsed by Jesuits at Kangxi's court. Whatever the case may be, J. F. Foucquet (1665–1741), a Jesuit at Kangxi's court, did bring back to Europe precisely the Ming annotated edition of the text thanks to which *The Gnomon of the Zhou [Dynasty]* was available in China at the time. This sheds light on the processes through which evidence for ancient scholarly activity outside Europe was being gradually gathered in Europe, creating the conditions for editions and translations of works such as *The Gnomon of the Zhou [Dynasty]* to be carried out there. It was precisely on the basis of a reprint of this same Ming edition—which he found in the collections of the Bibliothèque Royale—that, about a century later, Edouard Biot would carry out his translation of the work into French. Biot's introduction to his publication shows his awareness of the fact that in China, *The Gnomon of the Zhou [Dynasty]* was considered 'the fundamental basis of the mathematical and astronomical knowledge of all peoples' (Biot 1841: 597). In particular, at the Bibliothèque Royale, Biot also had access to *The Fine Essence of Mathematical Principles*, in which this view was expounded. He treated the latter theory with irony, but he had his own antiquarian agenda for embarking in a translation of *The Gnomon of the Zhou [Dynasty]*. He seeks out the earliest extant books that testify to the knowledge in mathematics and astronomy available 'in the Orient' in ancient times—which he would use notably to assess the relative value of peoples. From this perspective, given the antiquity that Biot lends to *The Gnomon of the Zhou [Dynasty]* (for him, the book goes back, at least partly, to the eleventh century BCE), to his eyes the work is unique, thereby justifying a translation (Biot 1841: 593–594, 598). Developments of this kind—e.g., the constitution of libraries with 'Oriental' sources, the emergence of training in 'Oriental' languages in Europe—thus illustrate the circumstances in which European publication projects provided access to an international past of sciences.

If we go back in time and to China, the 1722 work on mathematics that was mentioned above—*The Fine Essence of Mathematical Principles*—had the aim of offering a synthesis of mathematical knowledge East and West. The fact that, in it, *The Gnomon of the Zhou [Dynasty]* was given a central place expressed, in the very structure of the work, that this ancient book was the foundation of all mathematical sciences. The claim was instrumental in the fact that half a century later, in 1772, Dai Zhen (1724–1777) would undertake a crucial critical edition of the work—along with that of all Chinese mathematical canons used as textbooks in the Imperial University—for publication in the encyclopedic work *Complete Library of the Four Branches*. Thanks to this editorial work, for the first time since the thirteenth century, the entire corpus of Chinese canonical works in the field of mathematical sciences was re-constituted, hence becoming available for historical inquiry (Chu 2010). More generally, the imperially-commissioned *Complete Library of the Four Branches* offered editions of all the Chinese writings of the past that, in the eyes of the imperial institutions, were deemed worth preserving—including works brought from Europe and translated into Chinese. The enterprise of the *Complete Library* thereby fashioned a Chinese written heritage, on the editions of which most present-day editions of the same works depend. What is more, in line with the theory that had emerged during Kangxi's reign, in the section that contained astronomical works, *The Gnomon of the Zhou [Dynasty]* was given the highest status, as the first book. Texts speak not only through their words, but also through their positioning in a textual structure.

Han's contribution illustrates how, in fact, various types of context need to be taken into account to understand the different meanings early modern actors attach to a mathematical work of antiquity like *The Gnomon of the Zhou [Dynasty]*. As a result, the late eighteenth century imperially commissioned re-edition of *The Gnomon of the Zhou [Dynasty]* is not a merely scholarly event: it echoes complex political debates that spanned two centuries and involved international networks of scholars. It was shaped in the midst of a global debate on China's scientific past, and in return then, once re-edited, contributed to shaping discourses on China's mathematical past. The same type of intricate relations between international networks and projects of a political nature, albeit of a different kind, can be perceived in the case of the *Bibliotheca Indica* series of the Asiatic Society of Bengal, to which Alessandro Graheli devotes his chapter.

Graheli focuses on the *editio princeps* of a seminal work of dialectics, epistemology, logic and metaphysics that Jayanārāyaṇa Tarkapañcānna (1806–1872) published within the *Bibliotheca Indica* in 1864–1865: the commentary by Vātsyāyana Pākṣilasvāmin (ca. fifth century CE) called the *Nyāyabhāṣya*. This *editio princeps* is a case in point to show how the politics of scholarly publications in South Asian languages in the second half of the nineteenth century in Bengal shaped, in a significant manner, the printing of ancient Sanskrit works.

The *Bibliotheca Indica* was initially created by James Prinsep (1799–1840), the then secretary of the Asiatic Society of Bengal. Its establishment was prompted by a change in policy in Great Britain which favored English and more generally European texts for education in India over attempts to make bridges with scholarly texts from South Asia. To counter such policies, Prinsep encouraged the publication of important South Asian works of the past. At first then, in part at least, the aim of the series was to publish counter-manuals for the education of English and Indian employees of the East India Company. In other words, as in Han's case study, politics—here, specifically, the educational policies of a British colony—played a role in shaping the choice of texts published by the *Bibliotheca Indica.*

However, from the beginning, these publications seem also to have been motivated by an antiquarian and scholarly interest. Although, progressively, the latter overtook the whole purpose of the collection, it nevertheless owed its creation to a mix of factors: a reaction to the colonial power's policies and the logic of scholarly endeavor, which were the initial goal of the Society. The aims of the institution funding the *Bibliotheca Indica* seem to have been an even more important factor in determining the kinds of works it specifically published. Indeed, other editorial projects in Calcutta at the time were undertaken in educational institutions such as the Fort William College (which trained British employees of the East India

Company) and at the Calcutta Sanskrit College (which trained British-friendly pandits—traditional scholars of Sanskrit): these projects focused exclusively on textbooks. By contrast, as a systematic planned endeavor, the *Bibliotheca Indica* published much more—and more widely—concerning Sanskrit lore.

Graheli shows how the progressive shift in the targeted readership of the *Bibliotheca Indica* series was inscribed in the way the selected works were published. In the beginning, in 1847, the series required that a translation be made for each edited text so that it was accessible to any reader fluent in English. By 1853, in place of a translation an analysis could be supplied in the form of an introduction. In other words, at first, the text had to be made accessible to those unfamiliar with oriental languages, but later, the texts seemed to be made for scholars who could read Sanskrit.

The edition of the *Nyāyabhāṣya* was also published in the *Bibliotheca Indica* for more specific reasons. Interest in *nyāya,* and in particular in the oldest possible texts of this philosophical school dealing with dialectics, epistemology, logic and metaphysics, had been sparked, in Europe, by a memoir issued in 1824 by the British Indologist H. T. Colebrooke (1765–1837). Colebrooke (1824) reported in particular on Sanskrit authors' description of a form of syllogism, which elicited the question of whether there existed a peculiar Hindu or Eastern logic, and also of whether *nyāya* had been influenced or had affected developments in Greek philosophy (Ganeri 1996). These issues inspired further work on Sanskrit texts dealing with logic. In Calcutta at the time there was an important school of 'new *nyāya*' (*navanyāya*). The interest in history shifted the focus of publications on *nyāya* during the nineteenth century from textbooks to teach 'the new *nyāya*' at the Calcutta Sanskrit College to an antiquarian search for the 'oldest text' of *nyāya,* which was fostered in the Asiatic Society of Bengal. It is precisely this antiquarian pursuit that contributed to rescuing a work, the *Nyāyabhāṣya,* that had been neglected with the development of 'the new *nyāya*' from fifteenth century onwards in Bengal. Consequently, the *editio princeps* of the *Nyāyabhāṣya* was based on a single recension of a quite marginal text. In other words, as in the case studied by Han, it is, in part, international conversations that contributed to attaching new values to an ancient work and hence to the production of a modern edition for it.

Just as in the case discussed by Han, the foregoing remarks illustrate the variety of contexts that need to be taken into account to explain modern actors' interest in an old and almost forgotten work. Graheli goes one step further and examines the production of the edition itself. In the first place, he focuses on the modern edition *qua* material object. As D. McKenzie (1999) has long argued, the material features of books are not innocuous. Graheli argues that what such features convey in the case of this edition of the *Nyāyabhāṣya* is quite meaningful, thereby inviting us to take them into consideration in our general reflection on editions.

Indeed, in this case, to begin with, just as for any other edited work, the *Nyāyabhāṣya* underwent a major transformation, when, instead of being made available in manuscript form, it was printed. Printing may seem to be just another technology for reproducing a text. However, in nineteenth century South Asia, this technique of reproduction attached a modern and above all foreign flavor to past texts. Along with printing, the script used to produce the *Nyāyabhāṣya* was also changed for the edition. In scholarly circles in Bengal, Bengali script was used to transcribe Sanskrit. However, Devanagarī rather than Bengali was used to print Sanskrit in the *Bibliotheca Indica* editions. This is still the case today, attesting to the enduring impact of the choices made in the printing *milieu* of Calcutta in the late eighteenth and early nineteenth century on indological publications to this day. Furthermore, the font used was a trademark of the Baptist Mission Press—of which Graheli tells the story—which was responsible for printing the *Bibliotheca Indica* publications. It became a standard font type with an enduring impact on twentieth century Devanagarī typography[15]. In other words, the *Bibliotheca Indica* project highlights the historical shaping of standards of publications for critical editions.

The printing format represented another silent and yet meaningful modification that the *Nyāyabhāṣya* underwent: South Asian manuscripts were habitually written in landscape format—the shape of the traditional palm-leaves. Publications of the *Bibliotheca Indica*, however, were in book format—a choice most probably made for cost purposes. As a result, texts issued in the shape of a book might have had an appearance similar to other texts from elsewhere in the world—this is one facet of the manufacture of homogeneity that was discussed above—, hence becoming more apt to be compared with them.

Finally, the *Bibliotheca Indica* books, bought by subscriptions, were published in different fascicles, bounded together once the last fascicle came out; provisional cardboard covers were used for the first fascicles while the last one had a cover which englobed all the others. In other words, the texts of the *Bibliotheca Indica* series were published as books in progress.

These remarks raise a key question, that of understanding the consequences of publishing texts in this way: What happens when a text hand-written in Bengali script on a landscape format manuscript becomes a set of fascicles printed in Devanagarī, later bound between two covers and turned into a book? How does this transformation shape the perception various readers have of it? What could appear as anecdotal changes in a modernization process that make ancient texts available to new publics might affect in important ways the way these texts are received. In the case under consideration, altogether these transformations were formulating a discourse. For some actors at least, these new material features were tacitly recounting how the work had been uprooted from its primary contexts of emergence and use and now bore marks of foreignness. This case in point illustrates more generally what is at issue in examining the refashioning of a work or a document that takes place in the production of an edition.

In addition to examining the material reshaping of the work that occurred with its inclusion into the *Bibliotheca Indica* series, Graheli attends to the textual technologies put into play to prepare the edition. Indeed, for the *Nyāyabhāṣya* Jayanārāyaṇa created a new kind of edition which used practices and values inherited from

[15] For the political implications behind this choice, see (Singh 2018: 472–474), which also writes a larger international history of the printing of Devanagarī in the twentieth century.

Sanskrit textual critical traditions as well as others coming from the Asiatic Society *milieu*. For instance, the edition included editorial colophons after each section of the text, of a type standard in Sanskrit commentaries. It also systematically mentioned variant readings (*pathāntara*), as in the elaboration of a critical edition following the Asiatic Society style. In other words, critical works published in the *Bibliotheca Indica* series embody new, syncretic, conceptions of what a published text should be. How, more generally, were editions carried out in contexts in which different traditions of editorial scholarship were concurrently practiced? Attending to this issue, which is raised by the case of the *Nyāyabhāṣya*, might help us better perceive meaningful differences between ways of editing ancient works worldwide.

Han and Graheli's chapters likewise illustrate, in two completely different contexts, how scientific texts of the past that had become marginal, or even forgotten, all of a sudden turned out to be the object of a revived interest and how, accordingly, they were the object of modern editions—in Han's case, in the context of state institutions, and, in Graheli's case, in that of foreign institutions. By contrast, Karin Preisendanz's contribution deals with a wholly different case, for she discusses the modern editions of an ancient medical work that was still widely read and used by practitioners of medicine, among others, in nineteenth century India. Indeed, she focuses on *Caraka's compendium* (*Carakasaṃhitā*), one of the canonical ancient texts of *ayurveda*—the ancient scholarly discipline of medicine in South Asia. In relation to the active and persisting interest in the work, Preisendanz has identified more than fifty different nineteenth and twentieth century printed editions of the compendium—not to mention manuscripts owned privately—produced by types of actors and for kinds of uses wholly different from what Han and Graheli describe. Editions of *Caraka's compendium* are a far cry from the single and prestigious editions that were the outcome of the processes analyzed by Han and Graheli, offering to us another facet of the meaning and forms of the re-creation of ancient texts for all those who value them. As in Graheli's study, Preisendanz's corpus sheds light on how the different aims motivating each of these publications—which were influenced by different kinds of sponsors and fueled by international stakes related to the status of scholarly knowledge in colonial India—shaped in multiple but decisive ways the published editions and translations of *Caraka's compendium*.

The fact that this work, originally composed in Sanskrit, was part of a live dynamic tradition is strikingly embodied by the annotations in diverse languages and the complex textual apparatus that surround its publications. Indeed, the targeted readers of these editions were not supposed to be trained scholars in Sanskrit. As a result, most editions of *Caraka's compendium* did not include the sole text in Sanskrit, but involved commentaries, footnotes and translations. The text was vernacularized in many different ways, through scripts, translations and commentaries in the languages of the place where they were edited (e.g., in Marathi, Bengali, Tibetan, Hindi, etc.). From a different perspective, the languages of these publications also reflect the social milieu of the editors. Some editions of *Caraka's compendium* were made by traditional Sanskrit scholars. This is the case for the *editio princeps*, prepared by Gangadhar Ray in 1868, which for the first time turned *Caraka's compendium* into a printed book. However, the subsequent editions and translations were for the most part made not by philologists or pandits, but by practicing physicians of *ayurveda*. This would have a deep impact not only on the kinds of edition they would publish, but also on the concrete production and the availability of these works. Indeed, by its very nature, this case study highlights another general issue about editions and translations: publishing, printing and selling printed matter requires many different resources, including the financial support of sponsors, a press, bookstores, etc. In fact, a great variety of networks entered into each printing of *Caraka's compendium*. Physician-editors were sometimes also press owners. More generally, to publish, editors relied on intricate relations between students and teachers within close family ties, related to clinics, pharmacies, bookstores and printing presses. These facilities were often on the same streets. Sometimes these networks were pan-Indian, for instance involving editions relating to medical schools with actors circulating between Calcutta and Jaipur.

Preisendanz further argues that the type of funding available was also decisive in shaping printed texts. When sponsors made the printing of works possible in the first place, they also left their imprint on the kind of books published and the language chosen for them. For instance, Maharajas and lineages of wealthy scholars could promote Sanskrit editions or prestigious vernacular translations/commentaries. In other cases, prints were published in fascicles with advanced subscriptions or payments. Fascicle after fascicle, they could display their prestigious patrons but also, sometimes, their lack of decisive sponsorship. Preisendanz nevertheless shows that cheap vernacular publications were much more successful than prestigious scholarly Sanskrit ones—which were, just as in the *Bibliotheca Indica* series, printed in Devanagarī script. Therefore, not only the material conditions of publication—as previously underlined by Graheli—but also the networks to which editors belonged—whether they were respected pandits or local practicing physicians—all participated in the making of the printed edition as well as in its popularity.

Why was printing *Caraka's compendium* important? As previously, these publications also reflected wider international issues. Just as in the cases treated by Han and Graheli, Preisendanz demonstrates that for editors, printing was part of a modernist project which involved integrating the Ayurvedic medical tradition into a world conversation and a world history. Accordingly, for the promotion of their agenda the editors took advantage of the new print media introduced by the colonial rulers. In their view, printing texts allowed them to promote knowledge they felt was despised by colonial rule, and to disseminate what they perceived to be neglected foundational works of science. [16] In other words, here too, editing ancient texts—and in particular fashioning the text in a certain way—is in and of itself a political statement.

In line with these concerns, the prefaces of these editions reflect how *ayurveda* slowly became part of a nationalist discourse. The issue for the editors, then, was

[16] Preisendanz notes that to this day, the edition and publication of ancient works are deemed important and relevant to readers and editors in India. The sheer number of published books runs contrary to the feelings expressed by editors that *Caraka's compendium* would be a forgotten, neglected, text.

not only to valorize what seemed to be a downplayed tradition, but also to prove that *ayurveda* was at the origin of all other medical traditions. As in the case described by Han, the point was to establish not only that European allopathic medicine originated from *ayurveda*, but also that it was the origin of all the medical traditions that were its direct competitor, namely 'homeopathy' or the Persian medical tradition known as *Unani*. Editions of scientific texts, therefore, belong to a discourse on the history of science that is worth documenting. Furthermore, we can study the impact of such a discourse on the way these editions are carried out: editions influence the writing of such histories, and, in return, these histories participate in giving shape to these editions.

Beyond the content of their prefaces, editions of *Caraka's compendium* embody—and thus assert—these historical and genealogical points of view. In particular, some editors were 'integrationists' forging different types of hybrids out of their knowledge of *ayurveda* and what they perceived to be 'foreign knowledge'. The synthetic character of the editions extended to the textual features of the books produced. Indeed, some editions mimicked scholarly Western editions, noting variations of manuscripts in footnotes without, however, describing the manuscripts themselves. Another sign of this integration and accommodation of 'foreign knowledge' is the appearance—in the editions themselves—of illustrations, and notably of anatomical drawings. There seems to have been no such images earlier. The labels of these new images involve, to use Michalowski's expression, 'interlanguage procedures'. Indeed, these labels also display linguistically how these images establish dialogs between several scholarly worlds. For instance, the eminent scholar Shankar Daji Pade's (1876–1909) Marathi translation and edition of *Caraka's compendium* published in 1907 has anatomical drawings. They are labelled with numbers, and include names in Sanskrit, Marathi, English and Latin. To take another example illustrating both kinds of integration—that is, integration between bodies of knowledge as well as between textual practices—, Khemraj Shrikrishnada was one of those scholars who were at the same time merchants, publishers and press owners responsible for vernacular editions of the *Caraka's Compendium*. His edition with Mihirchandra's Hindi commentary, published in 1898, contains multilingual illustrations, notably of parts of the body not usually studied in *ayurveda*, such as the brain or the nervous system. Such illustrations sometimes included English terms transliterated in Devanagarī. However, these anatomical drawings co-existed with mythological representations of diseases as demons, and the illustrations of the divine origin of *ayurveda*. Despite the similarity between the textual devices employed, the use of illustrations in these two editions may be interpreted as formulating opposite meanings: if Shankar Daji Pade displays how he is constructing a new hybrid knowledge, Khemraj Shrikrishnada and Mihirchandra's point with their choice of illustrations might be to prove that European medicine was a later development of *ayurveda*.

In brief, as the case of *Caraka's compendium* shows, publications of editions embody not only the social positions of the editors, but also the way they want to situate themselves within the world of what was perceived as 'European knowledge' with its critical editions and footnotes. Editions and within them a specific kind of object—the illustrations—reflected also the ways that *ayurveda* was being thought and adapted. These editions with illustrations and translations more broadly exhibit their editor's goal, and speak of the histories of science they wish to be integrated in: editions and translations are performative acts. More largely such a corpus offers rich evidence for the issue of how past texts were accommodated for present uses.

Taking a world-wide view on editions and translations of ancient scholarly texts in early modern and modern times hence sheds lights on how international networks, political considerations and all sorts of material constraints have had an impact in inspiring interest in ancient texts and shaping the outcomes of these endeavors. But there is more. The anatomical illustrations evoked in Preisendanz's contribution invite another range of questions. Indeed, they display the diversity of factors that enter more generally into the shaping of specific elements making up editions and translations. More importantly, fashioned as they are, these elements will have a bearing upon how subsequent readers interpret the ancient scientific works to which these elements belong. As we have suggested in Sect. 1.1 of this introduction, this is a key general issue. How have editions and translations reshaped the various components specific to ancient scientific documents? How have these transformations left their imprint on interpretations and also, more broadly, on the historiography of science? These are the important questions that the book addresses next. Accordingly, in what follows, in contrast with a global, macroscopic point of view, we turn to micro-case studies focusing on how editions and translations showcase the components of ancient scientific sources and how this approach enables us to historicize the treatment of these specific elements. This forms Part III of the book.

### 1.2.3 Shaping Specific Features of Scientific Texts

The first two chapters of Part III return to the issue of the edition and, more broadly, of the representation of numbers and quantities, which we touched upon in Sect. 1.1.3. They do so, by concentrating on cuneiform documents. Indeed, modern editions of sources from Mesopotamia reveal important facets of the treatment of numbers and quantities, as well as, more generally, of the shaping of editions and translations.[17] However, to understand these aspects, we must first add a few words about these documents themselves.

Mesopotamian clay-tablets are witnesses of scientific knowledge and practice in ways that differ from most of the other sources mentioned in this book. First, these tablets are excavated objects, which, for the most part, were found close to the places where they had been inscribed and used, rather than texts transmitted through a continuous process of copying and willful preservation in libraries and archives.

[17] In the three annexes appended to (Chemla et al. 2022), the reader will find information on the conventions used in this volume to deal with numbers, measurement units and measurement values in, respectively, cuneiform, Chinese and Sanskrit sources.

Second, these tablets are written in scripts that can be read in Sumerian or in Akkadian. A little explanation is in order here, since much is at issue in this point. Sumerian was inscribed using a logographic writing (Sumerograms), while Akkadian can be written using a syllabary, that is, in a phonetic fashion. However—and this is a crucial point for us—, sometimes the inscription of Akkadian does not make use of the Akkadian syllabary, but of Sumerian characters, which are accordingly read in Akkadian.

These two features—the very nature of these sources and the language in which they were composed (presumably, since it is not always easy to determine the actual language lying behind the script)—have led modern actors to shape, for their editions, specific kinds of textual device. Indeed, the nature of the material evidence and the possible disjunction between the script and the language are correlated with the fact that an edition of cuneiform source material consists standardly of several of the following components: a photograph, a hand-drawn copy, and several types of representation of the text inscribed on the tablet with a modern Latin alphabet. The latter include: transliterations (which encode graphemes and thus represent the written form of the signs) and transcriptions (which encode phonemes and thus focus on how the text was pronounced).[18] In addition, these Latin-alphabet representations of cuneiform texts are often accompanied by translations and followed by a commentary which provides an interpretation of the sources' content. Such editions therefore represent the witnesses with textual *dispositifs* that mirror the sources in different ways, depending on the editor's choice. The case of clay tablets displays immediately why editions do not provide a direct access to the documents they rely on. It also clearly sheds light on the artificial character of the text of these editorial works *qua* text. What is more, this example highlights right away various aspects of the historicity of the text of editions: the printed output depends, for example, on the material resources available to reproduce the tablet (drawings, photographs in black and white or in color, 3D images, etc.).

Mathematical tablets give noteworthy illustrations of how editors' decisions fashion the edition of a text—in particular with respect to the languages in which sources were composed and read. Indeed, Proust provides stunning examples of how different editors gave wholly different transcriptions of the same Old Babylonian mathematical cuneiform tablets (2000 BCE-1600 BCE), thereby shaping different editions for the same source.

With the example of Tablet YBC 4710#1, she shows how the different transcriptions given by Neugebauer and Thureau-Dangin derive from the fact that they understood differently the function of Sumerograms used to write down the text, in

[18] Proust notes that transliterations and transcriptions are historical constructions. She uses here the definitions given by the Cuneiform Digital Library Initiative (CDLI), which itself relies on the definitions of these categories given at the twenty-first congress of Orientalists in Paris in 1948. For instance, Neugebauer's 'transcription' corresponds to what is now for the CDLI a 'transliteration'. In practice, Assyriologists have also used other modes (hybrids of both notably) of representation of the tablet in Latin script. For the sake of simplicity, in what follows we will stick to the CDLI definitions.

ways that have a significant impact for the history of mathematics. Indeed, for Thureau-Dangin, when Sumerograms were used in an Akkadian context, they were merely employed to write the Akkadian language. Accordingly, Thureau-Dangin transcribes them in Akkadian. For instance, his transcription of the Sumerogram meaning 'field'—which he transliterates as *a-ša*—is *eqlum*. By contrast, for Neugebauer, Sumerograms in Akkadian texts were to be understood as mathematical symbols, and should thus be preserved as Sumerograms in the transcribed text—note that Neugebauer thereby gives another meaning to what a "transcription" is. The kinds of texts used for an edition are a flexible matter that is adapted to the scientific work carried out with them.

These diverging choices seem to derive from a fundamental disagreement between the two Assyriologists about how they understand the function of a mathematical inscription, and this entails deep historiographic consequences: contrary to Thureau-Dangin, for whom a mathematical text simply writes down oral utterances, Neugebauer's practice might reflect the assumption that mathematical work is primarily written and that the inscription represents the traces of this engagement with the written word. From the viewpoint of the editions, in a spectacular way, a same witness then will give rise to the publication of different transcriptions and translations. Furthermore, obviously, if signs are interpreted as being mathematical symbols, this has bearing on their relevance for the history of mathematics. The tablet containing them would then be eligible as a source to be included in the rich and somewhat prestigious field of the history of mathematical symbolism.

Let us leave here the issue of the language corresponding to the written word—and the different functions that accordingly different editors lend to the inscription—to focus on our main topic in the first two chapters of Part III: the edition and translation of signs denoting numbers and quantities. The observation of these signs sheds light on another phenomenon. Indeed, we might believe that transliterations straightforwardly reflect the written word. However, Proust shows that even signs of this kind highlight that interpretation has an impact on transliteration.[19] The tablet that we have just mentioned illustrates this remark: the same sign (for instance, ) might be thought, by some, as being an un-pronounced graphical indicator and be accordingly transliterated in superscript, whereas it might be interpreted by others as a surface unit (in our example, this is, for instance, the case for the young Neugebauer), or as a lexical word meaning 'field' (this is the option chosen by Friberg). Here, thus, as above, but for different reasons, a same witness is given three distinct transliterations and widely diverging translations in relation to three editors' different interpretations. This example clearly illustrates that the way numbers are transcribed, transliterated and translated does not only depend on the language in which editors think texts were composed (although this factors in,

[19]As Proust also notes, these remarks further extend to the treatment of computations. After all, transliterations cannot always convey the fact that the addition of quantities of goods to which some tablets attest could have been obtained by bringing together and regrouping the numerical signs representing the addends. The use of Arabic numerals in transliterations might even at times encourage artificial interpretations of numerical relations.

as we have seen above), but it also rests on how the function of the numerical sign, and the information given by a quantitative sign are understood. These remarks extend more generally to numerical and quantitative signs: a sign can elicit different transliterations, transcriptions and translations. The reader of an edition and a translation will therefore have an access to the source material that is heavily loaded by the editor's choices.

On the other hand, the signs used to write down numbers also illustrate a converse phenomenon, which highlights the complexity of the issue. Proust makes this point in her historical remarks about the interpretation of sexagesimal place-value notations. The analysis is developed in two steps, both of which are important to us.

To begin with, Proust argues that the interpretation put forward by Neugebauer in a seminal article published in 1932–1933 deeply shaped the historiography of the sexagesimal place-value numeration system, in particular because his choices were inscribed in his translations of the sources. The crucial point is this: Although Neugebauer observes that, in Old-Babylonian mathematical tablets, many notations using a sexagesimal place-value numeration system are inscribed without any indication of an order of magnitude, in his translations he specifies the orders of magnitude, as he understands them. This is precisely the operation through which he inserts his interpretation—which is that ancient actors tacitly associated an order of magnitude with the numerical signs—into his translation of the original text. This historiographic practice durably shaped the understanding subsequent historians—and more generally readers—will have of these texts and their numerical component. However, Proust argues, one can interpret these numerical signs as they feature in the sources, without assuming that ancient actors added a layer of meaning that they did not write down: it suffices to imagine that ancient actors employed numerals written using the sexagesimal place-value notation only to compute and not to express quantities. In this case, although Neugebauer and Proust transliterate the same signs in comparable ways, given that their interpretations of these signs differ, they translate in quite different ways. In line with the remarks above, this example again clearly illustrates the potential impact of decisions taken in translations on the historiography of mathematical sciences.

But there is more, and this is where the converse phenomenon appears. Indeed, Proust attempts to identify the reasons behind Neugebauer's choice in this respect. Much as Heiberg editing Archimedes homogenized a text to avoid its 'interpolations', Proust suggests that Neugebauer's translation practice seems to have aimed to render in a uniform way all sexagesimal place-value notations found in cuneiform sources. Astronomical tablets written in the last centuries BCE,—that is, 2000 years after the mathematical tablets mentioned above—attest to the fact that at the time, actors apparently used the sexagesimal place-value notations as indicating orders of magnitude. Neugebauer seems to have assumed that similar notations in earlier mathematical texts were of the same nature as the latter, and he thus translated them accordingly. It is important to note that this amounted to assuming that over 2000 years the sexagesimal place-value system has not changed in meaning or use. What is more, translations tacitly conveyed this assumption of immutability to the reader. The historiographic impact of this assumption for the history of mathematics in Mesopotamia has been tremendous: the historical evolutions of the sexagesimal place-value notation have been overlooked. However, if we follow Proust's interpretation, it appears that throughout these two millennia, there were deep transformations in the use of this notation, and these changes need to be attended to. The additional theoretical point for our purpose is this: we have observed above that sometimes the same sign in a document was the object of diverging interpretations and hence different transliterations. We now see that the same numerical sign might have wrongly been assumed to have the same meaning, when different uses suggest that in different contexts it needed to be interpreted differently. Thus, use and also context might leave no trace on the sign itself, and yet they need to be attended to when carrying out editions and translations.

We will come back below to how the will to homogenize sources could tacitly engrave an assumption of uniformity into the edition of mathematical and astronomical sources themselves. For now, let us note that a decision to homogenize, in the translations, a numeration system found in sources shaped in crucial ways how non-Assyriologists could reflect on numerical signs.

What we have just seen suggests that cuneiform sources attest to a clear diversity in the ways of inscribing and using numerical signs. However, in order for this diversity to be attended to, it must be distinguished from and rid of another diversity, which stems from the diverging interpretations of editors and translators. This is where a historical reflection such as the one whose development this book intends to encourage might be helpful. In Proust's view, the diversity in representing numerical and quantitative signs springs from a lack of collective reflection, which is related to a deeper historical phenomenon. Historically, in the field of Assyriology, the scientific dimensions of the sources have not been a central concern—which shows a striking contrast with the field of classical studies in ways that call for further reflection. This phenomenon is reflected by the fact that the editorial conventions presiding over editorial work did not mention anything concerning numerical and quantitative signs. For instance, the conventions defined by Ignace J. Gelb at the twenty-first congress of Orientalists in Paris in 1948 are silent on this topic. At the time, for Orientalists, the most significant editions were those of literary and religious texts. Accordingly, editorial conventions focus on elements deemed relevant for the edition of these genres. On the other hand, historians of mathematics and astronomy concentrated only on what they thought relevant for their scientific analysis, neglecting what they considered to be less important elements of the context. Thus, in his conventions Neugebauer omits the complicated transcriptions of measurement values, probably because he believed they had no mathematical interest. More generally, in this no man's land, each author devised his or her own editorial conventions, focusing on the numerical signs and the features that he or she perceived as important. These remarks reveal a crucial phenomenon, touched upon also by Glenn Most in the postface of this book: they show the impact of the modern divide between disciplines on the editions of different types of source material and even on the different parts of a document.

For Proust, a study of the diversity artificially generated by the practices of modern actors should pay heed to an essential issue. Despite the variety of practices adopted by twentieth-century editors, their editorial conventions for the transcription, transliteration and translation of cuneiform numbers, measurement values and quantities generally share a common problem: they do not distinguish between the different types of signs contained in cuneiform sources. Indeed, ancient actors seem to have shaped distinct sets of signs for different uses—an issue of major importance for the history of science. Some of these signs incorporate both numerical and metrological information, while others are used specifically to count discrete items and yet others to carry out computations. The differences between such notations were quite stable in time, which explains why it is crucial to reflect them in transliteration and translation. However, editions often do not give us access to them in a faithful fashion, which severely limits the understanding that non-Assyriologists could acquire of these practices with numerical notations. The next chapter, by Cécile Michel, continues this line of analysis, by examining, from a historical viewpoint, the various problems elicited by modern transliterations and translations of the expression of quantities in the context of the edition of cuneiform source material. Michel thereby allows us to understand in greater detail the nature of the distortions that since the beginning of the twentieth century, various editors have applied to quantitative information. For this, she concentrates on tablets produced in contexts different from those evoked by Proust—that is, in economic and business milieus. These tablets abound in expressions of quantities—including measurement values—rather than numbers expressed using the sexagesimal place-value notation. Michel studies specifically how twentieth-century editors have transliterated the original notations, her aim being not only to show the profound transformations undergone by the sources even in transliteration, but also to explore the historiographic consequences of the various solutions adopted.

The main thrust of her contribution is to highlight several ways in which, in the original documents, the expression of quantities has a structure in the cuneiform script that the transliterations and also the translations have replaced with another. An example given by Michel will suffice to explain what is at issue. A tablet inscribed in the first centuries of the second millennium BCE in Mari attests to the fact that the scribe used two different systems to write numbers. He first employs the sexagesimal system called system S to express the number resulting from a count of men and women. Then he uses a second numeration system, apparently decimal,[20] to express the sum of the numbers listed. In the transliteration of the tablet given by Jean Bottéro (1914–2007), the first result of the count of men is given as '60(+)16 *awîlû*(meš)', whereas a transliteration following the conventions of the CDLI would indicate that the 60, the 10 and the 6 are written using, respectively, three types of signs in a uniform way (one for 60, one for 10, and one for 6, which is repeated six times). In other words, the transliteration introduces a kind of opposition between the first sign (transliterated as 60) and the last two (transliterated together as 16), which structures the notation in a way different from what the source displays.

[20] (Proust 2002) and (Chambon 2012) consider that this system is centesimal. See also (Chemla in dialogue with Keller and Proust 2022: 31, 42). What follows is based on these publications.

Moreover, the transliteration renders the graphemes using the decimal place-value notation commonly employed today. Exactly the same remark holds true for the transliteration of the sum, in which the number of persons reads '1 ME 29', which means '1 hundred 29 men', while the use of the CDLI conventions gives '1 me 2(u) 9(diš)'.

The key point is this: by transliterating the separate counts of people in each line using a decimal place-value system, Bottéro makes the related numerical signs appear closer to those expressing the sum than they actually are. The transliteration thus fails to highlight the crucial change of numeration system that has taken place between the moment when the scribe determines the numbers to be added and the moment when the sum is computed. However, when we place this tablet in the context of contemporary Mari tablets, this detail appears essential: we now know that in this context, additions were carried out with a specific numeration system, into which the expression of the numbers to be added was first converted, before the sum was computed. This practice leaves a clue in the way the sum is expressed in the tablet under discussion. Consequently, the tablet bears witness to this practice, which can be perceived only if the transliteration faithfully reflects the numerical signs as they actually occur. The example illustrates what was at issue when ancient actors used different numerical signs to inscribe different quantities: the differences bespeak the distinct natures and meanings of the quantities expressed. In the case under consideration, some quantities derive from an operation of counting while the sum is yielded by a computation. What is more, against this backdrop, we see how misleading it is to add a sign '+' between the sign transliterated as 60 and those transliterated as 16.

The other examples discussed by Michel show that these two issues—refashioning the structure of the expression as found in the original document and thereby deleting clues about the operations that produced this expression—are quite general. With the various systems of transliteration used for the expression of quantities, the structure and nature of the original signs were regularly misrepresented. Michel further argues that translations were also the site where major transformations took place.

Like in the cases studied by Fowler concerning the notation of numbers in ancient Greek texts, Michel notes how the specific signs for fractions that can be found in administrative texts—often respected in transliteration—might disappear in translations if authors, for instance translate a transaction involving 2/3 *mana* and 4 *gin* as '44 *gin* of silver'. Notice that, in this case, the fraction occurs in an unexpected position in the expression of the measurement value. Indeed, as is commonly the case in these documents, the quantity is expressed using several measurement units in an order of decreasing magnitude. It is noteworthy that the fraction does not occur at the end of the expression, as a part of the smallest of all the measurement units used, but at the beginning. Why is that so, and what does this tell us about the practice of measurement or about that of computation? To our knowledge, these questions are not yet fully settled. It is clear that to address them, historians must rely on as faithful a representation of the original expression as possible.

In addition, Michel shows that, quite standardly it seems, translations convert the expression of all measurement values into a decimal number of the highest or the lowest measuring unit. François Thureau-Dangin (1872–1944) for instance, in an 1894 publication, converts all capacity measurement values contained in a document into quantities of *sila* (akkadian QA)—a measurement unit that is deemed to be equivalent to the liter. Thureau-Dangin does not give a transliteration of said tablet. Cecile Michel does, and she transliterates one of the quantities in question according to CDLI conventions as '3 (*bariga*) 2 (*ban*$_2$) 4 (*sila*$_3$) *kaš ús*'. Such a transliteration indicates that there are three *bariga* signs, 2 *ban*$_2$ signs and four *sila*$_3$ signs. Thus here, for each component of the measurement value, both numbers and measurement units are expressed with a same sign. Thureau-Dangin gives directly a translation of this expression as 204 QA (*silas*). In other words, what the original text formulates as an expansion with respect to a sequence of measurement units, whose relationship one to the next varies from one pair to the next—1 *bariga* is 6 *ban* and 1 *ban* is 10 *sila*—, becomes a number of *silas* expressed using a modern decimal place-value numeration system. What is more, the translation separates the number and the measurement unit in ways that do not reflect the original document. Michel suggests that Thureau-Dangin may have done so to make the text easy to read for those interested in economic history: Editions and translations in this case seem to have been shaped to enable readers to conveniently retrieve the quantitative data they contained, rather than to be faithful to an original. In this respect, Thureau-Dangin's practice evokes what we have said about Thibaut's conception of the editorial work in the case of scientific sources: for such documents, in editor-actors' view, what matters appears to be the mere content, irrespective of the form it takes in the sources. However, whether converting into *sila*, or into the highest unit of the capacity system, the *gur*, such translations erase what was written on the tablet and suppress clues regarding how ancient actors computed with such measurement values. Moreover, consequently, by giving the illusion of a direct access to the source, translations also made possible assumptions on quantitative signs and numbers that had nothing to do with what actually was in the original text. In this way, such editions and translations subsequently provided a representation of source material on the basis of which was shaped a history of numerical notations and quantitative signs that could assume a uniform use of the sexagesimal place value notation or the absence of fractions.

Michel identifies a third type of transformation that was applied to the expressions of quantity in translation. Indeed, these expressions often make use of negative formulations. This is, for instance, the case of '1 *lá* 1/4 *gín*', which refers to a quantity by indicating that it is one *gin* from which ¼ *gin* was subtracted. Michel notes that Larsen and Møller translate '3/4 shekels'—in other words, '3/4 *gin*'—as do many Assyriologists in similar cases. Again, when do such quantities occur? Are they the results of measurement or of computation? What clues does their expression give about these questions? Questions of this kind can only be address on the basis of transliteration and translation that reproduce the main features of the original expression.

Michel's chapter puts forward a last remark that raises important issues for the reflection on philological work in mathematics and that interestingly returns to the remarks we have made on diagrams, in Sect. 1.1.2. Indeed, she notes that the ways in which an '8' or a '9' are written on different tablets might present variations that, in practice, were not considered worth mentioning in transliterations. For instance, in a transliteration following the CDLI norms, these numerical signs might be rendered as '8(*diš*)' or '9(*diš*)', that is, by indicating that the signs are formed by the repetition of the sign '*diš*' as many times as there are units. However, the way in which the 'diš' signs are arranged in the document is not indicated. Indeed, there is a standard way of placing them, in three rows, in the upper two of which three *diš* are placed, but Assyriologists also encounter non-standard ways. Why are there variations in the arrangement? What do they tell us about the scribes? For Michel, observing these variations systematically would be a meaningful task, not least because it might open the way to a social history of numbers. However, to this day, these differences between the signs have been overlooked in transliterations: the signs were treated as the same, and hence concretely uniformized. Michel argues that this practice hides the diversity of actors, which implies that historians will have to return to the original documents anew, when and if they intend to address this issue.

What Proust and Michel have shown holds true much more generally: the expression of numbers and quantities—that is, mathematical objects that can be found in many different kinds of texts—has historically been subject to several sorts of rewritings in modern editions and translations. Their shapes were thereby often deeply transformed. They have been sometimes standardized, and sometimes modernized to meet the needs of imagined readers. These transformations have resulted in the erasure of the various contexts to which the notations of numerical and quantitative signs were attached, no matter how complex the editorial devices made to give access to original sources were. The coherence of different sets of notations as well as the various uses of different numerical and quantitative signs to which our sources attest then become inaccessible to the reader. These conclusions call for the deployment of a broader critical reflection on modern editions and translations of numbers and quantities in ancient documents. They also call for a systematic analysis of how editions and translations depend on a given historiography of numbers and of how, conversely, they also inform the historiographies that rely on them.

Reviel Netz's contribution continues this analysis of how specific elements of editions and translations were shaped, this time focusing on diagrams. His chapter deals with the history of the editions of mathematical diagrams in Greek geometrical texts from antiquity to modern critical editions, notably those published by Heiberg and Friedrich Hultsch (1833–1906). However, the issues Netz's chapter addresses differ from those dealt with for cuneiform documents for one fundamental reason: the witnesses available for the Greek works from antiquity considered by Netz are of a different nature and have a different history than the source material available in cuneiform. Accordingly, the history of editions and translations is also quite different. In particular, in this case the historian is confronted with what, by comparison, appears as an almost continuous series of attempts to recreate these works of the past for the present.

At first sight, Netz's study is in line with the historiography evoked at the beginning of this introduction. As we have seen in Sect. 1.1.2, recent work has established that there are key differences between the diagrams preserved in the manuscripts and those found in the critical editions of the late nineteenth and early twentieth century. Some of these differences—relating to the supposed generality and the visual accuracy of these diagrams—have been the object of a collective effort in recent decades (De Young 2005, 2012, 2020; Lee 2018, 2020; Netz 1999, 2012, 2013, 2020; Saito 2006, 2011, 2019; Sidoli and Saito 2012). However, Netz here, precisely by relying on the possibility of observing editorial work over the long term, aims to go beyond this and raise new questions: Noting the wide gap between manuscripts and modern editions with respect to diagrams, he seeks to establish how this divide occurred in history. The point is to understand the processes that produced the diagrams that feature in the editions commonly used today and thereby to interpret the meaning of these changes.

To this effect, Netz develops a historical inquiry into the transformations undergone by the diagrams in Greek geometrical texts from antiquity. More precisely, he focuses on diagrams in the context of proofs proceeding by *reductio ad absurdum*, which Netz identifies as particularly revealing of the processes he is interested in. To begin with, like any other diagram, these diagrams generally show striking differences when we compare their occurrences in the various manuscripts and standard printed editions. However, in their case, the differences take a specific form: ancient diagrams emphasize the absurdity of the situation assumed in the proof, while modern editions reduce the absurdity. For instance, where, in manuscripts of Archimedes' work *Spiral lines*, Proposition 13 has a diagram depicting a tangent line touching a spiral in two distinct points with a broken line, Heiberg's 1913 edition restores the drawing of a straight line. Such transformations naturally suffer exceptions. Thus, in the case of Proposition 35 of Book IV of Apollonius's work *Conics*—where the issue is the representation of the intersection of conic sections and the question is raised whether the curves are tangent or not at these points—Heiberg's 1893 edition remains quite faithful to the curves of the manuscript tradition (normalizing it only 'slightly') while Halley's 1710 edition dramatizes the absurdity found in the manuscript diagrams. We return to the contrast between these two editions in Sect. 1.2.4.

In this context, Netz's question is thus to identify when and why these breaks in the traditions of drawing diagram took place. Going beyond just noting the differences existing between the various occurrences of the same diagrams, his argument makes clear the benefits that can be drawn from writing a history of the diagrams and their transmissions in manuscript and print.

Looking at ancient sources, from ancient Greek papyrii to medieval Byzantine manuscripts of works by Archimedes, Euclid and Apollonius, Netz argues first that, as a rule, diagrams there show very little variation. As a result, the diagram attached to a given proposition in the shared archetype can in general be reconstructed. This diagram is characterized notably by two features. First, it is overspecified—which we have discussed in Sect. 1.1.2. Second, it is schematic: the ancient figure represented the important traits of the situation under consideration without apparently giving any importance to it being iconic—that is, without attempting to imitate the object under consideration. Similarly, Netz shows that Heiberg's editions contain diagrams often in keeping with those that appeared in early modern and modern editions, and quite different from the restored archetypes. Clearly, thus, these results allow us to measure in which respect a reader using today's modern editions deals with diagrams utterly different from the original.

Netz goes one step further, when he asks how we can account for the amount of diagram variation we find between the ancient witnesses and the modern figures. He identifies the moment when, and the context in which, the ancient diagrams were simply replaced by the versions we find in modern editions. More precisely, Netz suggests that the break did not take place in the shift from manuscript to print cultures but that it was largely—though not always—carried out by humanist *milieus*. Indeed, these humanists felt the diagrams handed down in medieval manuscripts were corrupt, since these diagrams seemed absurd with respect to their own sense of what a correct diagram should look like. Striving more generally to retrieve the 'Greek original' behind what was perceived as the corruption of the 'Middle Ages', humanists thus discarded the extant manuscript evidence to replace it with diagrams that looked more reasonable to them and thus closer to what they imagined Greek scholars of antiquity would have drawn. Ironically, while they attempted to be faithful to the original, their editions betrayed the ancient texts, introducing more modern diagrams in their restoration. For instance, Memus' (1537) *editio princeps* of the *Conics* seems to be the place where a reduction of the absurdity of *Conics* diagrams took place. These early modern diagrams left their imprint on Heiberg's edition of the *Conics*. However, in this case, Heiberg seems to have also taken into account the diagrams of the single twelfth-century manuscript on which, in his eyes, all available manuscripts of the *Conics* eventually depended, thus eventually giving diagrams that present a mixture of both Memus' and the manuscript's. Commandino's (1565) edition of Archimedes' *Floating Bodies* and the Basel printed edition of Archimedes' *Spiral lines* in 1544 would be other such moments where the transformation towards a reduction of diagram's absurdity would have taken place—here maybe ultimately influencing Heiberg's edition of these texts. Concerning *Spiral lines*, Netz further argues that the initial transformation may have taken place in certain manuscripts, notably William of Moerbeke's (d. 1286) translation of it into Latin, before showing up in certain Renaissance printed editions.

Netz's inquiry is an example of what a long-term reflection on the processes of copying, and making critical editions can contribute to the history of Greek geometrical texts of antiquity. Taking into account the complex processes through which ancient texts were given to read to new readerships enables him to go beyond the mere observation that modern editions hide ancient diagrammatic practices. Netz retraces the genealogy of these new types of diagrams, their history, and the forces at play in promoting a replacement of evidence by newly drawn figures. He identifies the ancient actors and the issues they attached to the diagrams, explaining *how* and *why* these actors produced the diagrams we are used to reading as belonging to the sources, although these diagrams might be, in some cases, quite different from the original ones. This of course has consequences. The kind of history set forth by Netz shows that, in certain cases, humanist-produced 'modern' diagrams consequently helped in shaping what was understood as the 'modernity' of 'the Greeks'.

Netz's historical analysis of diagrams thus clearly points out some of the new insights that historians might more generally derive from the fulfilment of the following critical tasks: attending to the different transformations elements of a text might have undergone in various ancient testimonies as well as later editions and translations to then account for why some manuscripts present these elements in ways that differ considerably from others. Tasks of this kind are meaningful not only for diagrams, but also for any other element of a scholarly text. Yet, a history of the transformations of specific elements of texts such as numbers and diagrams still remains to be written, as underlined by Most in the postface.

The foregoing explorations have shown that the expression of quantities as well as the diagrams encountered in the sources are sometimes reproduced in editions and translations in ways that depend more on the representations forged by editors and translators than on the attempt to faithfully mirror those of witnesses. These remarks clearly highlight facets of the editorial and translation work specific to scientific sources. Cooper's contribution, for his part, sheds light on another more general and more fundamental feature of editions which has not, to our knowledge, been the subject of reflection, despite the fact that it has undergone major changes: the textual devices that actor-editors shaped on the basis of their reflections on how best to represent the witnesses. This issue draws our attention to another facet of the production of editions, already touched upon by Graheli: editors operate in the context of philological cultures, which they contribute to shape in relation to the specificities of the source material they deal with. Here again, due to the nature of the cuneiform witnesses, Old-Babylonian documents will appear as offering precious pieces of evidence to approach this facet of the editorial work, which turns out to be of capital importance.

To deal with the phenomenon under consideration, Cooper has at hand a very telling kind of source: tablets attesting to Sumerian literary compositions. These compositions, whose editions Cooper—like Netz—examines from a historical perspective, were among the works taught in Old-Babylonian schools. Other works included royal panegyrics, epics, and mythological tales that were learned after word lists, advanced sign lists, metrology and mathematics. Pupils may have learned to write these texts, first producing excerpts found in single-column tablets, and then with larger portions in multi-column tablets. Consequently, most of the witnesses used for the editions of literary compositions of this kind were the innumerable discarded tablets of school education, dating from 1825 to 1725 BCE. In fact, quite concretely, no 'master copy' of a Sumerian literary composition exists. Nor do we know anything about authorship. As we now understand the nature of the evidence, each school, each archeological layer, enables the reconstruction of what could have been for a given teacher, in a given place and time, a correct rendering of the composition. As a result, recorded variants cannot be interpreted in the same way as they are when editing single author texts: variants do not arise by copying a document that was itself a copy of another document and so on. However, this awareness emerged precisely through a long struggle with the issue of how to represent the extant witnesses.

Cooper's contribution considers the problems raised as Assyriologists strove to produce editions of these peculiar works with such specific witnesses, examining, to begin with, how editors understood the task to fulfil. First, editors tried to reconstruct a stable original text. Early scholars, trained in classics, imported editing techniques devised for ancient Greek and Latin sources, or biblical Hebrew, and attempted to apply such philological cultures to texts with different stories and, above all, with different types of witnesses. Consequently, published editions first reflected their editor's desperate struggle in reconstructing an eclectic best text. For instance, S. N. Kramer's (1897–1990) edition concentrated on the wording of the composition, line by line. The critical apparatus, displayed in his first 1937 publication, ordered tablets according, first, to the first line of text contained in a tablet and then—for tablets with the same first line—to their length, the manuscript sigla allowing the editor to list them in alphabetical order. Footnotes indicated when a tablet did not contain a given passage. The reader had to reconstruct from a set of dispersed footnotes and the sigla what text a given witness contained. This way of proceeding was standard until the mid-seventies of the twentieth century. Cooper notes that such editions appeared essentially as line-by-line texts 'in a forest of footnotes.' The work itself seemed to become meaningful only if and when it was put in visual relation with a facing page translation, enabling the reader to grasp that the Sumerian composition was indeed a piece of literature. Translation and edition were thus deeply intertwined. In the mid-seventies, the impossibility of grasping such a fragmented text led to the creation of a new kind of textual *dispositif.* Editors experimented with ways of making editions that could visually display witnesses and their variants, still aiming to reconstruct a standard best text. In this vein, in 1974, D. Edzard introduced a new textual device inspired by an edition of 'scores', providing line after line all the contents of all the witnesses, furnishing full-fledged data on the text contained in each tablet. This had the advantage, over the previous style of edition, of being very clear as to what exactly the text contained in each tablet was—the reader was not required to reconstruct it any more.

These changes were not of pure form. As is often the case, new textual representations led to seeing the documents in a different light and to questioning the goal that should be ascribed to editorial work. Indeed, new features of the sources were progressively deemed meaningful for editions. Previously unrepresented, they were now included in publications: editors concentrated on different types of visual display and various kinds of *varia.* Editors also began paying attention to provenance to organize the complexity of the source material, which allowed them to identify the existence of groups of texts. From another perspective, to make their editions they could make use of new text-editing tools. Thus, in Civil's 1994 edition, recorded variants did not only concern text and content but also the kind of signs used. The listing of variants was coded in such a way that it enabled him, as he was in the editing process, to make changes from the main text to the *varia* section quite easily, with the help of his computerized text editor. Indeed, editing practices changed as automation and large data hashing were increasingly made possible by the progressive use of computers in the second half of the twentieth and the beginning of the twenty-first centuries. Such changes went hand in hand with what became a new perception of the witnesses and their *varia*. New editions indeed are now shaped to display what is considered to be different layers of transmission of the text. Thus, Delnero's 2012 study and edition rest on 740 witnesses. In his edition, groups of similar variants are understood as testimonies of a transmission taking place in a same school or place. The edition has the shape of a score which, in conformity with Edzard's style, records each witness' content. An eclectic best text is provided, and variants from this text feature in boldface. The tablet sigla, using a system of subscripts and upper-case letters, indicates in a left-hand column the provenance of a tablet as well as the nature and the state of the source. Indeed, some witnesses might be broken fragments of tablets while others could be extant double-columned ones. This edition also separates the standard school compositions of the second millennia from the scarcer sources that have sometimes come down from the first millennium.

Such an edition then not only attempts to take into account the multifarious versions to which a widely diffused text can give rise when it has no fixed author, but it also enables a visualization of both the material state of the sources and the complex history of the text. At a glance, different states of the text become accessible in the published edition. Cooper's chapter is a testimony of the editorial creativity of editors faced with specific sources, new tools and new editorial goals. They gave birth to new kinds of editions, and, above all, new ways of representing witnesses and their *varia*. Here, these editions are precisely correlated with the specific nature of the documents dealt with, while reflecting the material environment in which editors operate. In a sense, we thus see the circumstances in which and the processes through which new philological cultures emerge.

Assyriologists began editing their texts through the importation of philological practices that had been shaped for other kinds of source material. Gradually, they adopted new goals and mobilized new techniques for their editions. Their sources compelled them to pay attention to the very diverse conceptions of text and authorship to which various documents testify. These sources also led Assyriologists to question classic conceptions—such as that the 'Ur' text—and the general validity of some of the classic tools for making editions—such as the stemma. Could the goals and techniques Assyriologists devised now inspire new ways of editing documents produced in other contexts and other times? The question is worth considering, and it illustrates the benefits that may derive from approaching the history of editorial work from a world-wide perspective.

In fact, even though for different reasons, both Netz and Cooper show why editions sometimes are not just about recovering an original text, or elements of the original text. Such a goal might sometimes be out of reach, sometimes meaningless or even hide essential information for the history of science. Rather, editions could display different states for which a text and its elements are documented. They would then testify to these states rather than expose the ever-elusive search for an original.

With these remarks, we conclude our observation of how inscriptions specific to scientific texts were dealt with in editions and translations as well as how, in relation to the nature of the documents available, editors and translators shaped artificial textual forms to re-present the source material and the features that they deemed important. Clearly, these elements present major variations depending on the editions dealt with. How can these analytical tools help us approach the differences between different editions of the same work and understand their impact? This is the issue to which Part IV of the book is devoted.

### *1.2.4 Publishing Ancient Mathematical and Astronomical texts: Comparative Perspectives*

Comparing the ways in which different types of ancient and modern actors have constructed editions or elaborated translations of a same mathematical or astronomical work can more generally help us better perceive the correlation between, on the one hand, the specific purposes for which, and the specific readers for whom, editions and translations were made and, on the other, the texts produced. Reflecting upon the differences between editions and translations made on the basis of the same documents may in particular allow us to increase our critical awareness with respect to texts that purport to make the past available for the present. What is more, a comparison of this kind can help us further identify key specific features of the philological cultures in the context of which these editions and translations were prepared and how these cultures left their hallmark on the works produced.

A first chapter, in which Micheline Decorps outlines a long-term history of editions and translations of Apollonius' *Conics*, takes us back to an issue that was discussed above when evoking Pollock's contribution—the diverging criteria with which editors with different scholarly backgrounds make choices between variants—, but this time for a mathematical work. In the cases presented by Decorps, like in those studied by Pollock, a contrast can be drawn between the variants adopted for the text edited on the basis of mathematical concerns and those made to ensure textual faithfulness to the witnesses that were deemed important. Decorps illustrates such an opposition quite strikingly, for instance, with the difference between the edition and translation into Latin of Apollonius's *Conics* made by a mathematician who knew Greek and Arabic, Edmund Halley (1656–1742), and the critical edition of the same text made in 1891–1893 by Heiberg, the philologist evoked at the beginning of this introduction as well as in the previous section for his editions of Euclid and Archimedes' works.

The notion of 'best text' embodied by Halley's *editio princeps* of the *Conics*, in 1710, was that of a clear mathematical exposition that would adequately display and articulate the logic of the reasoning, leaving space so that mathematical commentaries transmitted with the text—those of Eutocius (480–540) and Pappus—could be conveniently perused. These commentaries included Eutocius' annotations composed in the context of his sixth-century edition of the first four books of the *Conics*. It also included the auxiliary propositions ('lemmas') needed for some proofs in the *Conics* and collected in Pappus' fourth-century *Collection*. Halley's Latin translation similarly aimed to be a tool for subsequent mathematicians interested in a mathematical reading. Mathematical meaning was thus here the chief guiding principle for the establishment of the text, even if it was for reasons differing from those of Thibaut. Heiberg's edition of the *Conics*, on the other hand, embodies both the kind of choices for the text that are motivated by beliefs with respect to the nature of the text *qua* text, as well as by the kind of conservatism that reflects a certain kind of fetishism towards the ancient documents chosen as the basis of editorial work. To the former can be attached the decision to consider as interpolated all the auxiliary propositions that had parallels in Pappus' *Collection*. As for the latter, it is best exemplified by the preservation of bizarre syntaxes and singular meanings (*hapax*). However, it should not escape us that the fetishism is exercised here on the basis of a corpus of documents for which decisions of inclusion and exclusion are based on judgments heavily loaded with personal assessments.

Indeed, the contrast between Halley and Heiberg can also be grasped at the level of their treatment and their textual representation of the sources. Halley's seamless text is composed as a marquetry of variant readings drawn from all possible sources, treated on an equal footing, without allowing the reader to keep track of their origins. By contrast, in line with the modern methods of textual criticism that he applied to mathematical texts of Antiquity, Heiberg's philological work starts with a classification of the witnesses that expresses the historical and textual relationships between them. The point of this first operation is to assess their relative value as witnesses and to eliminate, from the corpus on which editorial work would concentrate, those that are deemed derivative. The shaping of the evidence to be considered thus brings into play wholly different textual technologies. As a result, in Heiberg's work, not all variant readings are granted the same historical value. What is more, the textual technology of the footnotes is put into play to fulfil the goal of allowing the reader to get an overview of the witnesses taken into consideration and to judge the choices made. The different philological cultures in the context of which the two editions were produced are hence reflected in the textual *dispositifs* of the editions.

Although, as Decorps underlines, Halley and Heiberg illustrate, in a sense, two opposite ways of making an ancient text available to a given public, they are nevertheless united by the same wish: that of presenting, each in his own way, a coherent text. However, this goal derives from a major assumption, which in turn guides how witnesses are handled: both editors assume that the witnesses reflect what originally was a coherent work. Decorps highlights how the search for consistency would lead both Halley and Heiberg to select a reading that was deemed to be the 'best'—the 'best' being, according to the case, decided either by linguistic or by mathematical criteria, and hence eventually producing different texts. By contrast, Decorps-Federspiel's editorial work and French translation, published at the beginning of the twenty-first century, led them to question this assumption.

What these two editors suggest, instead, is that the internal linguistic and textual variety to which the witnesses of the *Conics* attest have a meaning. What is more, for them, this variety can be correlated with the actual history of the production and transmission of Apollonius' *Conics*. As a result, homogenizing the language and more generally the text, wherever a witness allows it, as previous editors have done, erases clues that offer precious historical evidence about these ancient processes. In other words, for Decorps, what lies behind the variety to which the witnesses of the *Conics* testify are the different versions which circulated, co-existed, and were sometimes regrouped and (re-)homogenized. In fact, such a history underlines how, to begin with, a work takes shape, and how it is then reshaped according to the different assumptions of what editors think the work should be: linguistically homogeneous, logically structured, or having specific elements of a certain kind, according to what is demanded of a mathematical text.

Indeed, Apollonius composed the *Conics* on the basis of a compilation/synthesis of other texts on the topic, which he then significantly developed. He sent out several drafts of the successive books, writing the parts that eventually made up the work in stages. We presume Apollonius's work circulated in schools where it was studied and commented upon; his compositions were thus worked upon. As in the case described by Cooper, works that were the objects of study make for particular—and particularly modified—witnesses. Like Theon for Euclid's *Elements*, the *Conics* was transmitted through Eutocius' re-edition of the first four books, to which Eutocius attached his commentaries. The *Conics* thus underwent a first radical transformation when at least its first books became a unitary work. Decorps argues that Halley's mathematical edition continues a tradition of mathematical edition of the *Conics* inaugurated by Eutocius. However, the ancient commentator describes how he has put together several recensions of the work on which his editorial work relied. According to his own testimony, he notably collected from them all the different demonstrations he could find for the various propositions in the text, and selected those that would go in the main part of the text for their simplicity and clarity, while others would be put in the margins. Eutocius then himself had started this work of homogenizing the first four books of the *Conics* according to mathematical criteria.

Crucial to Decorps and Federspiel's edition is that Apollonius's *Conics* was translated into Arabic by a pool of practitioners of mathematics, which included the ninth-century scholar Thābit ibn Qurra. However, for the first four books that they edited and translated, neither Halley nor Heiberg took the Arabic witnesses into consideration. They thus neglected key testimonies of the work. Such an example shows how the editors' conceptions of the sources shape the evidence that is deemed meaningful for the editorial work. The preface of the Arabic translation makes clear that these scholars had a copy of Eutocius's commentary, which they claim to have followed for the first four books, and another manuscript of the seven first books, which they used for the translations of books V to VII. This Arabic translation of Apollonius's *Conics* is striking precisely because its style is not homogenous and also because it is partly similar to and partly divergent from the extant Greek witnesses. For Decorps and Federspiel, both the lack of homogeneity and the discrepancies, far from being a stumbling block, offer precisely means to examine how the witnesses shed light on the processes of composition and transmission of the *Conics*. In other words, instead of aiming to restore a single text, the editors' goal is now rather to unfold before the readers' eyes the processes of composition and transmission reflected by the witnesses. Different contexts, different *Conics*!

The comparison carried out by Decorps highlights how scholars with different disciplinary backgrounds and operating in different times understood editorial work differently. The comparison drawn in Agathe Keller's chapter adds another dimension, since she compares translations and editions produced by two nineteenth-century actors working in two different contexts. We have already encountered them, since they are the Indologist H. T. Colebrooke (see Sect. 1.2.2) and the *pandit*, librarian and teacher at the Government Sanskrit College in Benares, S. Dvivedin (1855-*ca.*1910), whose collaboration with Thibaut was discussed in Sect. 1.1.4. Colebrooke and Dvivedin are of interest to us here insofar as they shaped quite different books out of a same work. Indeed, Keller centers on the scholarly translation and the edition of Chapter 12 of a Sanskrit astronomical treatise: the *Theoretical treatise on the true Brahmā [school of astral science] (Brāhmasphuṭasiddhānta)* completed in 628 by Brahmagupta, to which both scholars, respectively, devoted a publication. What Keller examines in particular are the tacit textual operations that molded the publications of this treatise: the typographical structure of the publications, their sections and paratexts serve a vision of the work which influences the way it and its parts are understood.

In Sect. 1.2.2, we have encountered H. T. Colebrooke as the author of an 1824 memoir on Hindu authors' logical works. We meet again with him here, as the author of an English translation of mathematical documents. Colebrooke's first operation was to take two chapters of Brahmagupta's astronomical treatise as separate tracts and thus to detach them from the other astronomical chapters with which they composed the treatise. This was a radical move, which gave artificial autonomy to chapters that Brahmagupta had integrated in a treatise. This transformation further artificially associated two chapters that might not have been thought of as closely related to one another by the original author: Chapter 12 called 'chapter on mathematics' (*gaṇitādhyāya*) and Chapter 18 called 'chapter on the pulverizer' (*kuṭṭakādhyāya*). What both disciplines ('mathematics' and 'pulverizer') were for Brahmagupta, and how they were to be related are still in need of elucidation, maybe precisely because Colebrooke and others in his wake have attempted to make the contents of these chapters familiar to the modern reader, associating the first chapter to 'arithmetic' and the second to 'algebra.' More, Colebrooke took the contents of Chapter 12 to necessarily cover all the topics included in the definition of 'mathematics' given as a preliminary by Brahmagupta. This involved reducing this chapter to two parts, 'operations' and 'practices', and gathering all that did not fit into this structure at the end of the text as 'supplements.' Colebrooke's translation further highlights what he sees as an internal structural similarity between arithmetic and algebra in Brahmagupta's treatise. He also draws parallels between this 'arithmetic' embodied in Chapter 12 and what might have been for his English reader antiquated but familiar subdivisions of what Colebrooke identified as 'logistics' or 'practical mathematics'. These two facets of Colebrooke's interpretation of the *Theoretical treatise on the true Brahmā [school]* derive from the fact that he puts these two chapters in relation with what was for him the most important text of Sanskrit mathematics and astronomy: Bhāskara II's *Crown of Theoretical Astronomical Treatises* (*Siddhāntaśiromāṇi*) from the twelfth-century—and, more precisely, what he considered as two autonomous chapters: the arithmetical *Līlāvatī* and the *Algebra* (*Bījagaṇita*). Both chapters—which were likewise extracted from what was seen as their astronomical environment—were also translated in Colebrooke's 1817 publication.[21] By projecting the opposition between the twelfth-century *Līlāvatī* and *Bījagaṇita* onto the two chapters extracted from the seventh-century *Theoretical treatise on the true Brahmā [school of astral science]*, H. T. Colebrooke inscribes, in the translation itself, his perception that the two works reflect an unchanging tradition. In other words, the translation creates a structure that conveys to the reader what Colebrooke understands as an ahistorical tradition of mathematics, embodied both in Brahmagupta's work (in which it is seen in an older form) and in Bhāskara II's (in which it was best expounded). This remark holds true more generally: Colebrooke's point of view on the contents of the mathematical chapters as devised by Brahmagupta, as well as his perception of the history of mathematics in Sanskrit sources, with its canonical authors, both shape many details of his published translation.

The translator's approach to and understanding of the text shapes not only the different parts of the published work, but also how the treatise relates to its commentaries. Indeed, Pṛthūdaka's (fl. 860) commentary on Brahmagupta's *Theoretical treatise on the true Brahmā [school of astral science]* was an essential tool for Colebrooke, notably to access the contents of Brahmagupta's mathematical rules. However, Colebrooke operates another major transformation when he refashions both Brahmagupta's treatise and Pṛthūdaka's commentary as he finds them in his sources. What the manuscripts presented as a running commentary, embedding the treatise in the form of successive quotations of the sections commented upon, becomes, in the translation, a marginal and fragmented composition, featuring in bits and pieces in the footnotes to what now is given pride of place as the main text: Brahmagupta's composition. As in the translations studied by Michel in which numbers were shaped for historians of economics, Colebrooke's translation might have been composed to make ancient 'exotic' works comparable to other books, more familiar to his readers. The reduction of Pṛthūdhaka's commentary into fragments placed in footnotes might have had the aim notably of making its text

[21] The status of Bhāskara II's *Līlāvatī* and *Algebra* in relation to his astronomical *Crown of Theoretical Astronomical Treatises* is still under investigation. In the wake of Colebrooke's publication, which follows some Sanskrit commentators, they were often thought of as chapters extracted from the astronomical treatise, and having had a more or less autonomous life. Recent studies of the inner coherence of these texts have suggested that these chapters might have been from the beginning three independent treatises (Ramasubramanian et al. 2019: xxi sqq). An edition of all three works taking into account the different strata of editorial work on the texts is thus in need.

accessible to readers for whom some of its exegeses might have seemed at least partly incomprehensible—for instance, when he developed grammatical analyses of the original. It might have intended to enable comparisons with other mathematical texts in other languages, using portions that would have made the comparative work easier. This had important consequences on how Brahmagupta's mathematics would be perceived by nineteenth-century readers of Colebrooke's translation.

Furthermore, the translator's personal interest also comes into play, having a hand in the way the translation is shaped. For instance, Colebrooke is intent to be as close as possible to the reconstruction of computational practices offered in the commentaries. He carefully reproduces tabular displays of numbers that such commentaries describe.

Many of Colebrooke's choices differ from those made by Dvivedin, the author of the *editio princeps* of Brahmagupta's Sanskrit treatise almost seventy years later. To begin with, Dvivedin integrates—in continuity perhaps with a Sanskrit tradition in which he was trained as a *pandit*—Pṛthūdaka's commentary within his own Sanskrit commentary. What is more, he edits the whole astronomical treatise. Lastly, in contrast with the interpretation promoted by Colebrooke, Dvivedin does not underline the structural similarity between Brahmagupta's treatments of arithmetic and algebra. Thus, Colebrooke's English translation and Dvivedin's Sanskrit edition offer two different kinds of 'source' texts. Clearly, a translation like an edition embodies points of views on a composition, neither of them providing a direct access to a given work.

However, one might argue that the translation and the edition both display, each in its own way, the adherence of Bragmagupta's *Theoretical treatise on the true Brahmā [school of astral science]* to a broader Sanskrit context. Indeed, Dvivedin and Colebrooke converge when they shape a history of mathematics which takes Brahmagupta as an older testimony of what would have been a homogeneous 'Hindu' tradition in mathematics and astronomy; a tradition essentially defined by the works of Bhāskara II. In Dvivedin's commentary on Chapter 12 of the *Theoretical treatise on the true Brahmā [school of astral science],* for instance, each of Brahmagupta's rules is likened to what can be found in Bhāskara II's compositions.

From the viewpoint of the textual *dispositif* of the edition, Dvivedin edits the work in new ways that are reminiscent of the nineteenth century actors discussed in Graheli and Preisendanz's contributions. His editions are hybrids which incorporate principles inherited from the printing medium of the book and German nineteenth-century philology, on the one hand, and, on the other, modes of re-edition notably through commentary, familiar to Sanskrit scholarly milieus, as described in Pollock's contribution. Thus, just as what we have seen with Heiberg's editions, Dvivedin uses philological tools such as the critical footnote, to display the results of the comparison between different manuscripts—a practice which he may have learned from the German philologist whom we have encountered above, Georg Thibault, with whom he had closely worked. Further, although he opts for the edition carried out in the context of a commentary that embeds the treatise and other commentaries, his position as a modernist pandit might best be visually embodied by his use of mathematical symbolism in his Sanskrit commentary. The latter feature is in line with the fact that Dvivedin focusses on mathematical ideas and does not reproduce elements of the computational practices which can be found in Pṛthūdaka's commentary. He might also have inherited from Thibault the idea that we have mentioned above, according to which scientific texts require a specific kind of edition that focusses on the contents rather than on the style.[22] However, Dvivedin's edition of Brahmagupta's treatise also pays attention to a feature of the text that the English translation leaves unattended: its textual form as a set of verses. In this respect, Dvivedin's commentary shapes a linguistically coherent text, notably by homogenizing its versification. The comparison between his works and Colebrooke's thus interestingly highlights different aspects on which, in editions and translations, homogenizing is carried out.

This example further illustrates how multifarious are the kinds of textual operations implicitly carried out by editors and translators as they publish ancient texts. In the same way as what was shown in Preisendanz's study of the labels of anatomical drawings in modern editions of *Caraka's compendium*, more or less visible editorial acts, such as the harmonization of verses and the choice of mathematical symbolism in commentaries, combine to embody a point of view on the edited source, to assert the social position of the editor, and to put the editorial work in dialog with different editorial practices. Where Colebrooke's English translation can be seen as a work fashioned for transmitting something foreign and maybe exotic to his European readers, Dvivedin's edition was transmitting what he perceived as his heritage, updated for a new Sanskrit reading public: The editor's and the translator's aims shaped their publications. This, then, influenced the way the texts were read and understood. But then again, these works were read in ways that might have been different from the initial project of those who had them published. For instance, readers of Colebrooke's translation took it as a quasi-direct access to a primary source.[23]

More largely, mathematical and astronomical texts in Sanskrit were subsequently read as testifying to a logical coherent system, best embodied by Bhāskara II's works. The fabrication of a 'Hindu' tradition as a coherent mathematical system was inherent to the way the texts were shaped in translation and in edition, and this belief, to which both the editor and the translator were committed, was furthermore tacitly transmitted to those who accessed the past through these channels. This might account for why and how this view has been so enduring since.

Colebrooke's and Dvivedin's editorial choices also left their imprint on many other elements of the text, in ways that shaped the historiography of mathematics in India. Until today, reflections on mathematical proofs in Sanskrit commentaries lament that they are often fragmented and irregular—a view no doubt reinforced, if not inherited, from Colebrooke's piecemeal rendering of commentaries. Colebrooke and Dvivedin were also influential with respect to the type of text and editorial

[22] See Sect. 1.1.4.

[23] (Charette 2012), (Smadja 2016).

practices that they shaped for the purpose of giving Sanskrit mathematical texts to read: throughout the twentieth century, Sanskrit mathematical and astronomical texts have often been edited and translated in very similar manners. A first enduring practice is that the Sanskrit text is generally translated into English with a commentary which focuses on its mathematical content. What is more, mathematical and astronomical components of the same treatise are often published and studied separately from one another.

As noted above, the case study dealt with in Keller's chapter compares a translation and an edition of the same text—the 'mathematical' part of a seventh-century treatise—which were carried out by scholars situated in two different scholarly worlds. Colebrooke was a British Indologist who had been in contact with *pandits* in India, whereas Dvivedin was a 'modernist pandit', both rooted in a Sanskrit tradition and learned in works of mathematics that had been introduced into India from the British Isles. Knowledge of mathematics from other traditions played out in the way the translation and the editing of the Sanskrit mathematical text were carried out, albeit in different ways. The case study that Zhu Yiwen and Zheng Cheng present in the next chapter deals with editions of *Mathematical Book in Nine Chapters* (*Shushu jiuzhang* 數書九章, hereafter *Mathematical Book*, 1247) by Qin Jiushao 秦九韶 (1208–1268) that were prepared by Chinese scholars in the eighteenth and nineteenth centuries, and it differs from the previous case study in at least two key respects. Firstly, although mathematical knowledge that had been introduced from Europe into China since the beginning of the seventeenth century certainly had an impact on these editions, the way in which it played out was quite different. Incidentally, this remark highlights the benefits that can be derived from adopting a world-wide outlook on our topic, as we do in this book. Secondly, these editions were making available to a larger number of readers a thirteenth-century work, parts of which had in the meantime become incomprehensible to Chinese readers, just as had several other mathematical works from about the same period. As a result, copies of *Mathematical Book* had become scarce and difficult to access. The decisive fact is that, shortly before the editions under study were undertaken, Chinese scholars made a breakthrough, and managed to partly recover the meaning of those of the works that were still available in China, precisely thanks to the mathematical knowledge that had been introduced from Europe by the mediation of some Jesuits. In other words, in this case, the relationship between the eighteenth- and nineteenth-century actors involved in editing and the work whose edition they prepared was the product of quite specific historical circumstances.

To make these points clearer, we need to give elements of information, which brings us back to one of the contexts discussed by Han Qi in relation to *The Gnomon of the Zhou [Dynasty]* (see Sect. 1.2.2). Indeed, in our discussion of this other case study, we mentioned Mei Wending's contribution to the recovery of ancient Chinese works on mathematical sciences and the role of his grandson Mei Juecheng in the compilation of the 1722 imperially-commissioned encyclopedia *The Fine Essence of Mathematical Principles*. They are also key figures in the case study offered by Zhu and Zheng. The fact that large parts of thirteenth-century mathematical works could no longer be understood since at least the sixteenth century was partly due to this: these works gave pride of place to a mathematical practice with calculating rods, and numerals related to them—which feature in thirteenth-century works. In particular, they testify to the use of these calculating rods to write key mathematical concepts and to carry out complex mathematical algorithms. Among these, there was an innovative algorithm for solving any type of algebraic equation (known today as the Ruffini-Horner algorithm) and an algorithm associated with what is called today the 'Chinese remainder theorem'—precisely, two algorithms that were central in Qin Jiushao's *Mathematical Book*. However, at the time when editorial work on ancient mathematical texts began, the practice with calculating rods was abandoned, notably because, for computations, since the fifteenth century at the latest, calculating rods generally had been replaced with the abacus. Partly in correlation with this change, the notations and algorithms associated with calculating rods were gradually forgotten. The works in which they played a central role became accordingly incomprehensible. In his effort to restore ancient mathematical knowledge to which Chinese writings attested, Mei Wending devoted a work to the ancient use of calculating rods. Then, in 1745, his grandson Mei Juecheng caused a sensation with the publication of *Pearls Lost in the Red River* (赤水遺珍, 1745): in this work, using knowledge of a type of algebra that had been introduced from Europe, Mei Juecheng managed to crack some of the lost meaning of thirteenth-century mathematical works, establishing that these works were precisely dealing with a kind of algebra similar to that imported from Europe. Indeed, he established that the notations with calculating rods that had become unintelligible were simply notations for polynomials and algebraic equations. This discovery made a strong impact on Chinese scholars who then turned to the analysis of the relationship between 'Western science' and 'Chinese science' in new terms. It also elicited an active movement of search and interpretation of ancient Chinese mathematical works. It is at this juncture that the modern editions of Qin Jiushao's *Mathematical Book* were prepared.

As Zhu and Zheng explain, the first modern edition was carried out for the imperially commissioned *Complete Library of the Four Branches*, also evoked in Han's chapter, which was completed in 1781. This holds true for *Mathematical Book* as well as—as we explained in Sect. 1.2.2—for all the other Chinese mathematical works composed before the fourteenth century, with the exceptions of *The Gnomon of the Zhou [Dynasty]* and another auxiliary work used as a textbook in the seventh century Imperial University.[24] Therefore, these late eighteenth-century editions were all produced in the context of a major enterprise, through which the imperial institutions aimed to shape the national heritage according to their views. In particu-

[24] Note that this assertion is true for China, but not for Korea or for Japan, where some of these works were the objects of earlier editions. Moreover, this statement implies that we do not consider as 'editions' works based on thirteenth-century works like Gu Yingxiang's 顧應祥 (1483–1565) works based on Li Ye's *Measuring the circle on the Sea-Mirror* (*Ceyuan haijing* 測圓海鏡, 1248) and Wu Jing's 吳敬 *Great Compendium of The Nine Chapters on Mathematical Methods with Analogies* (*Jiuzhang suanfa bilei daquan* 九章算法比類大全, 1450). On these works, see, respectively, (Guo Shirong 2015) and (Zhou Xiaohan 2018).

lar, in the 'mathematics' section of the *Complete Library*—which had been carefully designed (Chu 2010: 148–157)—the comparison between China and 'the West' was part of the subtext: works translated from European sources were placed alongside Chinese works. However, Chinese works were always positioned at the forefront. Moreover, the editors again took up the thesis of the 'Chinese origin of Western knowledge,' notably in relation to *Mathematical Book*: Zhu and Zheng emphasize how the 'summary' placed before its edition states explicitly that the algebra associated with 'the West' originated precisely from Qin's work.

Like several other mathematical books for which the *Complete Library* offered an edition, Qin Jiushao's work had also been selected for inclusion in the previous gigantic imperially-commissioned undertaking of this kind, at the beginning of the fifteenth century: the *Great Compendium of the Yongle Era* (1505–1508).[25] In such cases, the editors relied primarily—and in Qin's case, exclusively—on this former edition. Zhu and Zheng explain why this was a challenging task. What is clear for our reflection is the impact that, in this context, official projects of this type have had in saving such mathematical works from oblivion. In ways reminiscent of Dvivedin's work, this edition of Qin Jiushao's *Mathematical Book* bears the mark of a long-lasting interest in China for shaping the written heritage, notably the mathematical heritage. What is also clear is that, when the edition of *Mathematical Book* was prepared for the *Complete Library*, large sections of the work remained incomprehensible. Since the base-text is lost, we cannot examine how exactly this played out in the way the editing was carried out. However, Zhu and Zheng can still identify key features of the base-text that were not reproduced in the edition of the *Complete Library.* In fact, unlike other thirteenth-century mathematical books that simply disappeared from China—and were later discovered in Korea and in Japan—, Qin's *Mathematical Book* was copied at the beginning of the seventeenth century. Zhu and Zheng consider that these copies derived from the edition that had been used for the *Great Compendium of the Yongle Era*. The copy made by Zhao Qimei 趙琦美 (1563–1624) in 1616 is still extant and can thus give us insights into how the *Complete Library* edition was made.

Regardless of the defects of the edition, the publication contributed to draw Chinese scholars' attention to Qin's work, at a time when restoring a national heritage in mathematics and situating it with respect to works translated from European writings appeared as valuable endeavors. Not only copies of this edition, but also copies of Zhao Qimei's manuscript began circulating among scholars involved in both historical and mathematical work in China at the time. Their notes and their interpretations attest to their engagement with passages of the book that had long been obscure. They made clear the need for a new edition.

These were the circumstances in which, in 1842, a more commercial edition appeared. It had been prepared by Song Jingchang 宋景昌 (fl. 1840), a scholar who belonged to these circles and who, for his composition, relied on exegetical and

[25] On the way in which mathematical and astronomical works were edited, drawing on that earlier project, see (Chu 2010: 148–149; 160–162 (Tables 3 and 4)).

editorial work carried out by his peers and his master. Zhu and Zheng show that both the base of the edition and the editorial techniques distinguish the 1842 publication from that of the *Complete Library*. For instance, by contrast with the latter, Song's edition took two witnesses into account, and he further added editorial notes to the text, which testified to his textual collation and his attitude towards his sources. What is more, Song certainly understood the mathematical ideas at issue better than his predecessors.

However, Zhu and Zheng focus on another difference between the two editions, which brings an additional dimension to our discussion. Qin's *Mathematical Book* features many computational diagrams, in which numbers were represented using numerals made with calculating rods (also known as 'rod-numerals'). At the end of the eighteenth century, the way in which numbers were represented with rod-numerals was already well understood. If we judge from Zhao Qimei's copy, however, the edition of Qin's work that was included in the *Great Compendium of the Yongle Era* presented in this respect a remarkable feature, which, to our knowledge, is specific to *Mathematical Book*: in some of these computational diagrams, the representations of numbers are regularly linked by different types of lines. In a recent work, Zheng and Zhu (2010) drew on Zhao Qimei's copy to establish that these lines represented in fact a symbolic notation of the operations applied to the numbers thereby joined. More precisely, they showed that, in these diagrams, different operations were written using different types of line. In other words, Zhao Qimei's copy contains a symbolic and dynamic inscription of a computational procedure. This notation was not understood at the time the editions were prepared. On the basis of the 1616 manuscript, it appears that the *Complete Library* edition of the work neither completely erases this feature of the base-text, nor does it seem to have reproduced its base-text faithfully in this respect. In other words, the eighteenth-century edition still contains hints of this feature of the original text, without offering the readers a solid base to interpret the lines. By contrast, in Song Jingchang's 1842 edition, these lines have simply disappeared, and with them, the possibility for readers to understand what was at issue in Qin Jiushao's computational diagrams. Zhu and Zheng surmise that the copy of Zhao Qimei's manuscript used by Song Jingchang might not have reproduced the lines of its base-text. Perhaps also, Song Jingchang's versatility in mathematics led him to decide that the lines that still featured in the *Complete Library* edition were meaningless. Whatever the reason, Song's edition neither shows nor mentions these lines. As a consequence, exegetes all interpreted these 'computational diagrams' as static illustrations of computations carried out elsewhere. A key element of the original book—its use of a symbolic representation of the dynamics of a procedure—was lost in the shaping of successive editions of the text.

What matters for the reflection that this book aims to develop is this: the editions of ancient works often erase all kinds of clues that their sources still contain on mathematical practices. To put it differently, editions and translations are molded by what editors know and what they pay attention to as much as by what they are ignorant of and what they consider irrelevant. As a result, readers using these editorial works are effectively deprived of essential information that could have drawn their attention to the authors' mathematical practices. We can reformulate in these terms the remarks that we have put forward in Sect. 1.1, about how diagrams and the expression of numbers were dealt with in nineteenth- and early twentieth-century editions and translations. Through a comparison between an eighteenth- and a nineteenth-century edition of *Mathematical Book*, Zhu and Zheng show that the same conclusion applies to the specific inscriptions that Qin Jiushao used to write down his computational procedures. However, in this case, two elements played a specific role: the fact that the tradition in which *Mathematical Book* made sense was discontinued together with the fact that the modern editions were shaped in a context in which actors perceived a form of competition between different traditions.

Zhu and Zheng further suggest that the transformation that took place in these editions had a major influence on the modern reception of Qin Jiushao's *Mathematical Book*: in correlation with how modern editions have distorted these notations, subsequent historiographies have given more weight to Qin's mathematical achievements than to his writing practices. Indeed, today *Mathematical Book* is famous for its procedures to solve simultaneous linear congruences and its mode of resolution of equations of higher degree. This is correlated with the fact that, after its publication, Song Jingchang's 1842 edition of the work became the most accessible edition. In particular, it was taken up by nineteenth-century European scholars, notably Alexander Wylie (1815–1857), who recognized in *Mathematical Book*—probably with the help of Li Shanlan 李善蘭 (1810–1882), with whom he cooperated on translations of mathematical books into Chinese—the 'Chinese remainder theorem', which he compared with the 'Hindu Cuttaca' he knew from H.T. Colebrooke's translations of Brahmagupta and Bhāskara II evoked in Keller's chapter. Subsequently, *Mathematical Book* acquired an important status in the writing of a history of mathematics in China.

Zhu and Zheng reflect that in some way, the re-shaping of the diagrams of Qin Jiushao's original book is not unlike the reading of Qin Jiushao's reasonings, at the end of the nineteenth and beginning of the twentieth century, with a modern algebraical notation. Indeed, the tools that editors use to interpret as well as those they assume that ancient actors used to practice mathematics shape the way texts are edited. In return these tools change deeply how these texts are understood by subsequent readers, and thus influence the way the histories of mathematics that rely on these editions are written.

Similar conclusions emerge from the next chapter, in which Matthieu Ossendrijver analyzes the pioneering efforts to make sense of and edit astronomical material in cuneiform script from the first millennium BCE that began to surface in the 1870s from excavations in the Middle East. Before the emergence of these tablets, the only existing evidence of astral science activity in Mesopotamia came from scattered references in ancient Greek and Roman sources. In this case, too, thus, source material produced in the context of traditions that had been discontinued was suddenly available. Almost nothing was known about these traditions: neither the technical vocabulary used and type of approach followed, nor the phenomena that had interested the ancient actors, nor even the types of text used by the practitioners. In addition, the sources discovered were not 'stand-alone works': They were tablets, which, moreover, were often broken into disjointed pieces. Finally, in contrast with the Chinese work discussed above, in this case, the source material was not claimed by modern local actors seeking to retrieve their own tradition.

Nevertheless, as early as 1881, a first publication concerning these documents appeared. From that point onwards, and until 1955, when Otto Neugebauer (1899–1990) published his seminal *Astronomical Cuneiform Texts*, Ossendrijver examines in a diachronic fashion how practices of edition and translation were gradually shaped and how they interacted with the progressive work of interpretation of these sources. The first publication was authored by two Jesuit priests, and more generally, until Otto Neugebauer entered the field, most editors and translators were catholic priests. This illustrates from yet another perspective a point that we have regularly emphasized: studying editions and translations of ancient source material involves paying attention to the social identity of editors and translators and, accordingly, to the corresponding significance they attach to this type of work—a direction of inquiry that Ossendrijver explores.

The 1881 article was the result of cooperation between two Jesuit authors, in a context in which astronomy and issues of chronology were meaningful in the Jesuit order. Beyond the fact that they were both Jesuits, Johann Nepomuk Strassmaier (1846–1920) and Joseph Epping (1835–1894) had quite different disciplinary backgrounds: Strassmaier had competences in Assyriology whereas Epping was particularly versed in astronomy. Ossendrijver points out how, over the decades they spent editing and translating these sources, the alliance of the two disciplinary backgrounds proved essential. Further, Ossendrijver highlights how editors and translators mobilized these two skills in different ways and in different combinations. To begin the work of interpretation, knowledge in mathematics and astronomy was crucial. Deciphering started with tablets that were purely numerical tables, and relied on an analysis of the structure and nature of the sets of numbers recorded. An approach of this kind meant that numbers were central not only to the interpretation, but also to the edition, in which 'errors' were rectified to fit the pattern identified. Ossendrijver pursues this line of inquiry, examining diachronically how, at different stages, the various editors and translators made use of their scientific knowledge and how this had an impact on the edition and the translation carried out.

In order for Epping to deploy this approach, Strassmaier had to copy the source material from the British Museum—which, for the two authors, remained the single location in which they found their sources. This dimension of the editorial work highlights another general issue, on which Ossendrijver dwells: that of how editors and translators have access to the source material. This issue involves all kinds of consideration, including—as Ossendrijver points out—the way the documents are catalogued and identified in collections, which, in the case of dozens of thousands of tablets and bits thereof, is not a minor concern. For all sorts of reason, direct access to a collection was not always an available option. Further, in the case of the cuneiform documents dealing with astral sciences, producing hand-copies of the sources required knowledge in Assyriology which the copies in turn passed on in part to their users. As a result, for decades Strassmaier's copies of the documents were reproduced to serve as a basis for the work of several subsequent Assyriologists' works—to begin with, for those of the Jesuit Franz Xaver Kugler (1862–1929) and the catholic priest Johann Baptist Schaumberger (1855–1955). This explains that these reproductions were found in several Assyriologists' archives, and that they were employed even after the technology of photography began being used as a means of access to the source material. Indeed, slowly the photograph became an essential tool to carry out the edition. Eventually, it was published within it. However, Strassmaier's copies continued to be reproduced as part of later editions of tablets. This remark brings us to another general issue that this case study highlights: that of the shaping of types of text for editions and translations.

Ossendrijver distinguishes different axes along which we can contrast different editions of the source material from a textual viewpoint. As Cooper's contribution showed for the case of literary texts, editors provided access to astronomical tablets through different settings. Part of these settings included diagrammatic representations of the text. Depending on the case, these pictorial representations used copies, drawings, and/or photographs. To represent, this time discursively, the text deciphered on the tablets, types of transliterations were progressively honed, evolving with the better understanding of the technical terms, the underlying approaches in the astral sciences and the languages written using cuneiform script. From the beginning, specific types of translations were introduced—with the label 'factual translations.' They aimed to make the mathematical and astronomical component of a tablet explicit rather than to be faithful to the text, especially when the early editors were not always sure of the readings of a given tablet. For this aspect, too, the interpenetration of the Assyriological and the astral science approaches to the tablets is manifest. Scholars shaped editions and translations, recycling practices common in other domains of Assyriology, while innovating in relation to the demands specific to their source material.

Another facet of the editorial enterprises appears to be revealing: it concerns the scope of the source material under consideration. To begin with, Ossendrijver shows, the edition bore on parts of a tablet that could be interpreted, like a couple of columns of numbers. With the expansion of the text understood, or the discovery that distinct bits were originally part of the same tablet, greater portions of a tablet were edited. In fact, the astronomical tables that can be found among cuneiform astronomical tablets were the first to be edited. Several textual artefacts were introduced to convey the meaning of the table *qua* table. Columns were given labels, the labelling being organized in such a way as to indicate simultaneously the place of a column in the table and tablet, and the astronomical component computed. The latter label was used in the commentary to the edition, as a symbol in the writing of formulas and equations to represent and analyze the algorithms to which the tables displayed testified. Distinct types of fonts as well as other types of signs were introduced to distinguish between the material on the tablet and the material that could be restored thanks to the structural analysis of the table. For the edition of tables, diagrammatic elements such as lines and double lines were also employed to make apparent their inner structure. All these textual additions display what we have referred to as the artificial character of the text of an edition.

On the basis of textual and exegetical work of this kind, relationships progressively appeared between tablets. These links delineated corpora of tablets, which then became the purpose of the editions. As Ossendrijver shows, editions and translations of astronomical texts were progressively made according to the idea of exposing what was thought to be a system of Babylonian astronomy. We thus see the correlation that can be established between how editors understand the ancient actors' scientific practice and how they then conceive their editorial projects. This new goal required a systematic scheme of edition for the entire corpus, and notably the use of uniform symbols which might present more systematically the mathematical and astronomical components of the system as a whole. Put differently, symbols previously had been introduced to explain a tablet and prepare the mathematical commentary that would go with its edition. The system of symbols gradually developed and structured scholarly discussions about the sources. They eventually stabilized and formed the backbone of the edition of a corpus.

Moreover, the project of editing and translating corpora was also articulated around the shaping of a typology of texts reflecting the different facets of the system: tabular texts were distinguished and edited separately from procedure texts, these being separated from astronomical diaries and astrological texts, each type of text also requiring its specific tools for editing. Neugebauer's publications, and notably his edition of tables and procedure texts in *Astronomical Cuneiform Texts* (1955), constitute the apex of such a project. In this classic work, the editions of different types of texts are scattered across text genres: Tables are always edited as a whole, and transliterations are provided 'in an idealized layout with added ruling that reflect the mathematical structures of the column'. The procedure texts are edited procedure by procedure rather than tablet by tablet. Finally, the translations, as well, are more intent in conveying to the reader the meaning of the texts in modern terms than in respecting, for instance, variations in terminology that could have been identified.

Such a way of editing, Ossendrijver further argues, rests on different editorial cultures, elements of which would be used to create a new specific edition. We have highlighted its dependence upon some features of editorial practices in Assyriology. Neugebauer was further inspired by the editions of Greek and Latin texts made by Heiberg, Hultsch and Tannery, whom we have encountered above: editorial practices do not only circulate within Assyriology, but also across the various branches of history of science. However, these editorial traditions contributed to the forging of a textual *dispositif* that was idiosyncratic for the cuneiform tablets dealing with astral sciences, with its system of photographs, drawings, transliterations, editions, translations and analysis. The device was understood as the process necessary to make texts accessible 'easily' and 'in their best available form'. For the editor and translator, this complex manufacture was providing, as best as it could, a first-hand source to the reader. Thus, Ossendrijver's contribution highlights how the editorial and translational practices reflect editors' perceptions of the mathematical tools used by the ancient actors under consideration, and even more, their perceptions of whole ancient mathematical and astronomical theories. While this systematic way describes especially Neugebauer's *Astronomical Cuneiform Texts*, Ossendrijver documents also how Neugebauer uses processes that had been explored by his predecessors. The artificial text of an edition is the outcome of a cumulative process that is intimately related with the advancement of the history of science. In this case, it was also shaped through a continuous tradition of editorial efforts that has been applied to the same tablets.

This cumulative and layered historical work which, in the end, produces a standard frame for critical editions seems to summarize, or totalize, the succession of case studies of this book. Gathered together, these contributions give us a glimpse into the great diversity of texts, goals and the multifarious creations that presided over the edition, publication and translation of ancient texts.

## 1.3 Conclusion

At the end of this exploration, a few general questions arise naturally: how can we situate our historical foray into the editions and translations of ancient and medieval scientific texts within the broader fields of the history of text criticism and that of translation? And also, what can such undertakings in the history of science contribute to a general history of these topics?

These are some of the issues considered by Glenn Most in the postface with which he concludes this book. However, to these queries, he adds another, which seems innocuous, but points out important issues with respect to our subject: why, does he ask, have we so few historical studies focused on the text criticism, critical edition and translation of ancient scientific texts?

In examining this question, Most develops a reflection on the way in which texts of science have been perceived with respect to texts on which the history of philology has concentrated, that is, in his own words, 'central literary and especially poetic texts.' In particular, in Most's view, if historians of philology and translation have shown no interest in past editions and translations of ancient scientific texts, this might derive from widespread assumptions about science and accordingly about scientific works. This remark invites us to ponder how images of and values attached to science have played out more globally in our history. Of Most's reflection, several points come out that draw our attention.

To begin with, the hypothesis that what matters for practitioners of science is what a text says, and not how this text is formulated might have led historians of philology to assume that the editing of scientific texts has followed aberrant practices and could not contribute to a reflection on the general history of editing. There is some truth to this hypothesis: we have mentioned above editions produced by editors who shared precisely this belief about scientific texts and prepared editions of ancient texts accordingly. However, our book also examines editions that were carried out with entirely different assumptions regarding works of this kind. In other words, the history of the editions of scientific texts attests to a large spectrum of editing practices that, precisely because of its variety, deserve to be taken into consideration.

What is more, as we emphasized at the beginning of this introduction, we cannot, with each new piece of research, redo the critical editions of the scientific works on which we need to draw. In this respect, a historical understanding of how a given edition was carried out seems crucial. This historical awareness provides tools which help us to find ways of using this edition, despite the fact that it does not fulfil present-day requirements. This is why, for historians working on ancient documents that attest to scholarly activities, there are still many topics that need further exploration. These include the changing purposes and modes of editions and translations, the practices of ancient philologies (and their modes of edition) world-wide, the history of shapes of scientific texts, the international networks of publishers and scientists in the nineteenth and twentieth centuries, the politics and economics of publications as well as the way technologies have shaped editions and translations, to name a few. These topics, obviously, do not only concern historians of ancient science.

In Most's view, the lack of interest by historians of philology for editions of scientific texts may stem from another assumption they have made about these texts: these historians might have perceived these texts as having been less important—from a cultural and social viewpoint—than 'other canonical or scholarly literary texts.' As a result, they might have assumed that the editing practices on the former texts were derivative with respect to those that bore on the latter. To put it differently, considering the history of the editions of scientific texts was deemed as bringing nothing new to a history of philology.

Is this true? For sure, as we have emphasized above, several examples on which the book focusses show editors of scientific texts emulating philological practices that had been designed for other types of texts. For example, some of them made use of critical apparatus and of stemma, which, as they were well aware, had been introduced in other contexts. Indeed, more generally, the way in which editions of scientific texts might have been the by-product of philological techniques elaborated to deal with other types of sources is worthy of further explorations.

However, this phenomenon does not exhaust the range of questions that such editions call for. As Most emphasizes, scientific texts are specific because they commonly contain, e.g., expressions of numbers and quantities as well as diagrams. Editors of such works must thus elaborate solutions for problems that were not central for the edition of other types of text and that philologists have hence not addressed in a systematic way. Accordingly, the historical approach to editions of scientific texts that this book contributes to has compelled us to introduce new analytical tools. These tools allow us to describe more accurately how technical non-discursive elements were dealt with in early-modern and modern editions. Such studies and the related analytical tools certainly constitute a contribution to a general history of philology

We could actually argue that this book goes further, and opens perspectives that could be of interest to all historians working on texts. Indeed, editions and translations of scientific documents are also texts in their own right. The studies published here shed light on the specific kinds of text that editors created to convey both a representation of an ancient work that is faithful to its witnesses—'faithfulness' being an epistemological value that was understood in different ways depending on the context—and these editors' understanding of the work. Such texts are often surprising, non-conventional, precisely because they deal with documents that reflect very peculiar technical content textualized in a very peculiar way. Indeed, the textual invention of editors of cuneiform astronomical tablets or the multilinguistic labels of drawings in some editions of Ayurvedic works are some striking examples of how scientific writings, and accordingly their editions and translations, are also creative modes of textuality that should be of interest to all.

In fact, the present book brings another input to a general history of text criticism, in relation precisely to the fact that it concentrates on texts of science. This point appeared clearly when we evoked above how Chinese actors in the seventeenth and eighteenth centuries reacted to the introduction of knowledge brought from Europe. The search for ancient Chinese books of science and the effort to interpret and edit them that the encounter between Jesuits, Chinese scholars and high-level officials elicited were motivated by the project of shaping a place for China in a world narrative of history of science. More generally, we have examined how the symbolic value attached to science played out in various ways on how editions of ancient texts were constructed. From this viewpoint, the history of editing ancient scientific texts sheds light on the values that different social groups associate with the past and invites us to a wider reflection on the various meanings that editions and translations take for actors. We might refer to this type of concern as 'the politics of philology', a subject dear to S. Pollock.[26] Perhaps no better than with scientific texts can we measure the meanings of such an expression and its effects on editions and translations. For example, crucially, in editions and translations, past scientific texts have been subject to a philology aiming to modernize them. In case studies presented in this book, we have examined how the modernizing of scientific content in ancient texts could be an efficient way of constructing editions in which some texts belonged to a contemporary scholarly conversation while at the same time, others, which were not subject to such a modernization, could appear as ancient and/or exotic. Scientific texts are indeed a crucial place where the politics of modernization encounters philology.

This wider political dimension of philology might be indeed one of the key lessons to be taken from this book and a major way in which a historical approach to the philology of ancient scientific texts might be of interest for a general history of philology.

Last but not least, these historical considerations also question the kind of philological studies, critical editions and scholarly translations we want to compose today or in the near future. Such questions are raised by the historical study of scholarly technical texts but are addressed to philologists and translators at large. Indeed, the studies contained in this book have invited us to reconsider some of the usual objects of critical editions, reminding us of how notions of 'original texts',

[26] (Pollock 2009), (Pollock 2016).

'best texts', or 'variants' could be multifarious, and at times misleading. How then should we carry out critical editions today? Which bearing might the reflections presented in this book have on our editions of scientific texts? Studies on how specific parts of scientific texts were dealt with in former editions convince us that we need to find new ways of editing configurations of noted signs as well as diagrams, to begin with, through historicizing how notations for numbers and quantities and diagrams were transformed. The changing social and cultural function of a same text also raises questions for the future editor. For instance, how should we edit works that fulfilled different functions in different contexts and time periods? Can we, and should we, edit them as being at the same time literary, pedagogical, and scientific texts? We could go on listing many other such questions that historical investigations raise to editors of today.

In a way, we seem to be historically extending the 'maximally inclusive' editions that Pollock described for ancient Sanskrit editors of poetry, involving the transmission (and discussion) of all readings. Pollock notes that such editions could be a fruitful example of a philological pluralism which would non-dogmatically take into account all the known versions of a 'true' text. And indeed, digital humanities provide us with a wealth of new tools to host this chorus of multiple voices. But the stories in this volume also indicate that a good edition remains one that makes choices. Cooper's contribution highlights how we should beware of the information overload such maximal editions could contain, encouraging us to create visual devices which would enable an edition to both contain large volumes of information and remain readable.

Our awareness of the blind spots of past editorial enterprises mingled with technological euphoria make us hope for more. More than a record (history and analysis) of the various states of a text and of its different meanings, digital editions bear new and greater fantasies. One of these is the ambition to produce editions which could also stand the scrutiny of future questions. Can we imagine an 'open' editorial device that would enable us to question ancient sources with new criteria and new inquiries without requesting the whole editorial project to be re-done?

In our quest to critically and historically examine reconstructions of the unreachable 'Ur' text, we have undoubtedly forged a new ideal. This editorial utopia can be drawn from the methodological questions we raise as we want to undertake new publication projects: in print as well as digitally can we, with sufficient economy, display the multifarious information and stories retrieved around a document, its travels and metamorphoses, while still leaving the door open to new interpretations? Can we further do so, highlighting the specificities of the text as a scientific text, notably with its peculiar non-discursive elements? The past teaches us that in the future we will probably invent creative *dispositifs* to publish new editions. Maybe, somewhat like this book of studies made with multiple hands and expertise, can we dream of (or would it be a nightmare?) an international community of critical editors, Wikipedia style, perpetually enhancing the critical edition of a text, with new points of view, new discussions and new stories about its past.

## References


Acerbi, Fabio. 2007. *Euclide, Tutte le Opere, Il Pensiero Occidentale*. Bompiani.

August, Ernst Ferdinand. 1826–1829. *Eukleidou stoicheia. Euclidis Elementa*. Berolini: Impensis T. Trautweinii.

Qian Baocong 錢寶琮. 1963. *Suanjing shishu* 算經十書 (*Qian Baocong jiaodian* 錢寶琮校點) (Critical punctuated edition of *The Ten Classics of Mathematics*). Beijing 北京: Zhonghua shuju 中華書局.

Biot, Edouard. 1841 (juin). Traduction et examen d'un ancien ouvrage chinois intitulé *TCHEOU-PEI*, littéralement "Style ou signal dans une circonférence"; par M. Edouard BIOT. *Journal Asiatique* 3e série, 11: 593–639.

———. 1842 (février). Note supplémentaire à la traduction du *Tcheou-pei*. *Journal Asiatique* 3e série, 13: 198–202.

Cerquiglini, Bernard. 1999. *In Praise of the Variant: a Critical History of Philology*. Baltimore: John Hopkins University Press. Originally published *as Éloge de la variante: Histoire critique de la philologie*. Paris: Éditions du Seuil, 1989.

Chambon, Grégory. 2012. Notations de nombres et pratique de calcul en Mésopotamie: Réflexions sur le système centésimal de position. *Revue d'histoire des mathématiques* 18: 5–36.

Charette, François. 2012. The Logical Greek *Versus* the Imaginative Oriental: On the Historiography of 'Non-Western' Mathematics During the Period 1820–1920. In *The History of Mathematical Proof in Ancient Traditions*, ed. K. Chemla, 274–293. Cambridge: Cambridge University Press.

Chartier, Roger. 2021. *Éditer et traduire. Mobilité et matérialité des textes (XVIe–XVIIIe siècle)*. Paris: EHESS, Gallimard et Seuil, Hautes Etudes.

Chemla, Karine. 1999. Commentaires, éditions et autres textes seconds: quel enjeu pour l'histoire des mathématiques ? Réflexions inspirées par la note de Reviel Netz. *Revue d'histoire des mathématiques* 5: 127–148.

———. 2004. *Editing the earliest extant mathematical figures from China*. Paper presented at the Conference "Diagrams and Images criticism in mathematical textual traditions," Pisa (Italy), 11-2004.

———. 2005. Geometrical figures and generality in ancient China and beyond: Liu Hui and Zhao Shuang, Plato and Thabit ibn Qurra. *Science in context* 18: 123–166.

———. 2020. On the Diagrammaticity of Ancient Texts and its Importance for the History of Science, Based on the Example of the Early Chinese Mathematical Text *The Gnomon of the Zhou*. *EASTS (East Asian Science, Technology and Society)* 14: 279–308. https://doi.org/10.1215/18752160-8538529.

———. 2021. From Reading Rules to Reading Algorithms: Textual Anachronisms in the History of Mathematics and their Effect on Interpretation. In *Anachronisms in the History of Mathematics. Essays on the Historical Interpretation of Mathematical Texts*, ed. Niccolò Guicciardini, 42–82. Cambridge, UK: Cambridge University Press.

——— (in dialogue with Agathe Keller and Christine Proust). 2022. Cultures of computation and quantification in the ancient world: An introduction. In *Cultures of computation and quantification in the ancient world. Numbers, measurements, and operations in documents from Mesopotamia, China and South Asia*, eds. Karine Chemla, Agathe Keller, and Christine Proust, 1–140. Cham: Springer Nature.

Chemla, Karine, Agathe Keller, and Christine Proust, eds. 2022. *Cultures of computation and quantification in the ancient world. Numbers, measurements, and operations in documents from Mesopotamia, China and South Asia*. Cham: Springer Nature.

Chu, Pingyi (祝平一 Zhu Pingyi). 2010. Scientific Texts in Contest, 1600–1800. In *Looking at it from Asia: The processes that shaped the sources of history of science*, ed. Florence Bretelle-Establet, 141–166. Dordrecht: Springer.

Colebrooke, Henry Thomas. 1824. On the Philosophy of the Hindus [Part II]. *Transactions of the Royal Asiatic Society of Great Britain and Ireland* 1: 92–118.

Dayeh, Islam, Umar Ryad, and Pascale Rabault-Feuerhahn. 2018. Nineteenth and Twentieth Century Philological Encounters. *Philological Encounters* 3 (1–2): 1–2. https://doi.org/10.1163/24519197-12340044.

De Young, Gregg. 2005. Diagrams in the Arabic Euclidean tradition: a preliminary assessment. *Historia mathematica* 32: 129–179.

———. 2012. Mathematical diagrams from manuscript to print: examples from the Arabic Euclidean transmission. *Synthese* 186: 21–54.

———. 2020. On "Overspecification" in Medieval Mathematical Diagrams. In *Diagrammatic Representation and Inference*, eds. Peter Chapman Ahti-Veikko Pietarinen, Leonie Bosveld-de Smet, Valeria Giardino, James Corter, and Sven Linker, 3–9. Cham: Springer International Publishing.

Fowler, David. 1987. *The Mathematics of Plato's academy. A new reconstruction*. Oxford: Clarendon Press.

———. 1992. Logistic and fractions in early Greek mathematics. In *Histoire de Fractions, Fractions d'Histoire*, ed. Paul Benoit, Karine Chemla, and Jim Ritter, 133–147. Stuttgart: Birkhaüser.

Fowler, David H., and E.G. Turner. 1983. Hibeh Papyrus I 27: An Early Example of Greek Arithmetical Notation. *Historia Mathematica* 10: 344–359.

Ganeri, Jonardon. 1996. The Hindu syllogism: Nineteenth-century perceptions of Indian logical thought. *Philosophy East and West* 46: 1–16.

Grafton, Anthony, and Glenn W. Most, eds. 2016. *Canonical Texts and Scholarly Practices: A Global Comparative Approach*. Cambridge: Cambridge University Press.

Grenfell, Bernard P., and Arthur S. Hunt. 1898. *Oxyrhynchus Papyri. Part I: Edited with Translations and Notes*. London: Egypt Exploration Fund.

———. 1906. *Hibeh Papyri. Part I: Edited with Translations and Notes*. London: Egypt Exploration Fund.

Guo Shirong 郭世榮. 2015. Gu Yingxiang's Method of Solving Numerical Equations with the Abacus. In *A Delicate Balance: Global Perspectives on Innovation and Tradition in the History of Mathematics. A Festschrift in Honor of Joseph W. Dauben*, eds. David E. Rowe, and Horng Wann-Sheng, 343–359. Cham: Birkhäuser-Springer.

Guo Shuchun 郭書春, and Liu Dun 劉鈍. 1998. *Suanjing shishu* 算經十書 *(The Ten Canonical Texts of Mathematics). Guo Shuchun, Liu Dun dianjiao* 郭書春，劉鈍 點校 *(Punctuated critical edition by Guo Shuchun and Liu Dun)*. Shenyang 瀋陽: Liaoning jiaoyu chubanshe 遼寧教育出版社.

Heath, Thomas. 1897. *The Works of Archimedes*. London: Cambridge University Press.

———. 1908. *Euclid. The Thirteen Books of the Elements. Translated with introduction and commentary by Sir Thomas Heath*. Cambridge: Cambridge University Press.

———. 1926. *Euclid. The Thirteen Books of the Elements. Translated with introduction and commentary by Sir Thomas Heath*. 2nd ed. Cambridge: Cambridge University Press.

———. 1956. *The Thirteen Books of Euclid's Elements*. New York: Dover Publications.

Heiberg, Johan Ludvig. 1880–1881. *Archimedis Opera Omnia cum Commentariis Eutocii*. Leipzig: B. G. Teubner.

——————. 1883–1888. *Euclidis Elementa edidit et latine interpretatus I. L. Heiberg*, vol. I. Elementa I-IV (1883); II. Elementa V-IX (1884); III. Elementa X (1886); IV. Elementa XI-XIII (1885); V. Elementa XIV-XV, Scholia, Prolegomena critica (1888). Book series: *Euclidis Opera omnia*, eds. I. L. Heiberg, and H. Menge. Leipzig: B. G. Teubner.

—————. 1910–1915. *Archimedis Opera Omnia cum Commentariis Eutocii Iterum Edidit*. Leipzig: B. G. Teubner.

Heiberg, J.L., and H. Menge, eds. 1883–1916. *Euclidis Opera omnia*. Leipzig: Teubner.

Hultsch, Friedrich. 1876–1878. *Pappi Alexandrini Collectionis quae supersunt* (3 vols.). Berlin: Weidmann.

Judet de La Combe, P. 1990. Philologie classique et légitimité. Quelques questions sur un 'modèle'. In *Philologiques I. Contribution à l'histoire des disciplines littéraires en France et en Allemagne au XIXe siècle*, eds. M. Espagne and M. Werner, 23–42. Paris: Éditions de la Maison des Sciences de l'homme.

Keller, Agathe. 2012. Overlooking mathematical justifications in the Sanskrit tradition: the nuanced case of G. F. W. Thibaut. In *The History of Mathematical Proof in Ancient Traditions*, ed. Karine Chemla, 260–273. Cambridge, UK: Cambridge University Press.

Knorr, Wilbur R. 1982. Techniques of Fractions in the Ancient Egypt and Greece. *Historia Mathematica* 9: 133–141.

————. 1996. The Wrong Text of Euclid: On Heiberg's Text and Its Alternatives. *Centaurus: International Magazine of the History of Mathematics, Science, and Technology* 38: 208–276.

———. 2001. On Heiberg's Euclid. *Science in context* 14: 133–143.

Lee, Eunsoo. 2018. Let the Diagram Speak: Compass Arcs and Visual Auxiliaries in Printed Diagrams of Euclid's Elements. *Endeavour* 42: 78–98.

———. 2020. *Visual Agency in Euclid's Elements: A Study of the Transmission of Visual Knowledge*. Ph.D.: Stanford University.

Li Jimin 李繼閔. 1990. *Dongfang shuxue dianji Jiuzhang suanshu ji qi Liu Hui zhu yanjiu* 東方數學典籍九章算術及其劉徽注研究 *(Research on the Oriental mathematical Classic The Nine Chapters on Mathematical Procedures and on its Commentary by Liu Hui)*. Xi'an.: Shaanxi renmin jiaoyu chubanshe.

Martija-Ochoa, Isabelle. 2001–2002. 'Edouard et Jean-Baptiste Biot. L'astronomie chinoise en France au XIXe siècle'. Master Thesis, University Paris Diderot.

McKenzie, D.F. 1999. *Bibliography and the Sociology of Texts*. Cambridge: Cambridge University Press.

Mugler, Charles. 1970–1972. *Archimède. I. De la sphère et du cylindre, La mesure du cercle, Sur les conoides et sphéroïdes; II. Des spirales, De l'équilibre des figures planes, L'arénaire, La quadrature de la parabole; III. Des corps flottants, Stomachion, La méthode, Le livre des lemmes, Le problème des bœufs; IV. Commentaires d'Eutocius et fragments. Texte établi et traduit par Charles Mugler*. Paris: Les Belles Lettres.

Netz, Reviel. 1999. *The Shaping of Deduction in Greek Mathematics: A Study in Cognitive History*. West Nyack, NY, USA: Cambridge University Press.

———. 2004. *The works of Archimedes. Translation and commentary. Volume 1: The two books On the sphere and the cylinder*. Cambridge: Cambridge University Press.

———. 2012. The Texture of Archimedes' Arguments: through Heiberg's Veil. In *The History of Mathematical Proof in Ancient Traditions*, ed. Karine Chemla, 163–205. Cambridge: Cambridge University Press.

———. 2013. Authorial Presence in the Ancient Exact Sciences. In *Writing Science: Medical and Mathematical Authorship in Ancient Traditions*, ed. Markus Asper, 217–254. Berlin/Boston: De Gruyter.

———. 2020. Why Were Greek Mathematical Diagrams Schematic? *Nuncius* 35: 506–535.

Pecchia, Cristina, Johanna Buss, and Alaka A. Chudal. 2021. Print Cultures in the Making in 19th- and 20th-Century South Asia: Beyond Disciplinary Boundaries. *Philological Encounters* 6 (1–2): 1–14. https://doi.org/10.1163/24519197-bja10019.

Plofker, Kim. 2021. Anachronism and anachorism in the study of mathematics in India. In *Anachronisms in the History of Mathematics. Essays on the Historical Interpretation of Mathematical Texts*, ed. Niccolò Guicciardini, 83–104. Cambridge, UK: Cambridge University Press.

Pollock, Sheldon. 2009. Future Philology? The Fate of a Soft Science in a Hard World. In *The Fate of Disciplines*, eds. J. Chandler and A. I. Davidson. *Critical Inquiry* 35 (4): 931–961.

———. 2016. Philology and Freedom. *Philological Encounters* 1 (1-4): 4–30. https://doi.org/10.1163/24519197-00000012.

Pollock, Sheldon, Benjamin A. Elman, and Ku-ming Kevin Chang, eds. 2015. *World Philology*. Harvard: Harvard University Press.

Proust, Christine. 2002. Numération centésimale de position à Mari. In *Recueil d'études à la mémoire d'André Parrot, Florilegium Marianum 6 = Mémoires de Nabu 7*, eds. D. Charpin and J.-M. Durand, 513–516. Paris.

Ramasubramanian, K., Takao Hayashi, and Clemency Montelle. 2019. Introduction. In *Bhāskara Prabhā*, eds. K. Ramasubramanian, Takao Hayashi, and Clemency Montelle, XV-XXXIV. New Delhi/Singapore: Hindustan Book Agency and Springer Nature.

Saito, Ken. 2006. A preliminary study in the critical assessment of diagrams in Greek mathematical works. *SCIAMVS* 7: 81–144.

———. 2011 (April 3). The Diagrams of Book II and III of the *Elements* in Greek Manuscripts. In *Diagrams in Greek Mathematical Texts (May 8, 2008, last update April 3, 2011). Report version 2.03*, ed. Ken Saito, 39–80. Online publication: https://www.greekmath.org/diagrams/Diagrams_in_Greek_Mathematical_Texts_Report_Ver_2_03_20110403.pdf. Accessed 26 Apr 2023.

———. 2019. Re-examination of the different origins of the arithmetical books of Euclid's *Elements*. *Special issue: On Mathemata: Commenting on Greek and Arabic Mathematical Texts*. *Historia Mathematica* 47: 39–53.

Saito, Ken, and Nathan Sidoli. 2012. Diagrams and arguments in ancient Greek mathematics: Lessons drawn from comparisons of the manuscript diagrams with those in modern critical editions. In *The History of Mathematical Proof in Ancient Traditions*, ed. Karine Chemla, 135–162. Cambridge: Cambridge University Press.

Shanghai Tushuguan 上海圖書館, and Beijing Daxue Tushuguan 北京大學圖書館. 1980. *Songke suanjing liu zhong* 宋刻算經六種 *(Six mathematical Canons printed during the Song dynasty)*. Beijing: Wenwu chubanshe.

Sidoli, Nathan, and Ken Saito. 2012. Comparative analysis in Greek geometry. *Historia Mathematica* 39: 1–33.

Singh, Vaibhav. 2018. The Machine in the Colony: Technology, Politics, and the Typography of Devanagari in the Early Years of Mechanization. *Philological Encounters* 3: 469–495. https://doi.org/10.1163/24519197-12340051.

Smadja, Ivahn. 2016. On two conjectures that shaped the historiography of indeterminate analysis: Strachey and Chasles on Sanskrit sources. *Historia Mathematica* 43: 241–287. https://doi.org/10.1016/j.hm.2016.04.001.

Suarez, Michael F., and H.R. Woudhuysen, eds. 2013. *The Book: A Global History*. Oxford: Oxford University Press.

Thibaut, Georg, and Sudhākara Dvivedin. 1889. *The Pañchasiddhântikâ, the Astronomical Work of Varâha Mihira. The Text, Edited with an Original Commentary in Sanskrit and an English Translation and Introduction*. Benares: E. J. Lazarus and co, at the Medical Hall Press.

Timpanaro, Sebastiano. 2005. *The Genesis of Lachmann's Method*. Edited and translated by Glenn W. Most. Chicago/London: The University of Chicago Press.

Ver Eecke, Paul. 1921. *Les oeuvres complètes d'Archimède, traduites du grec en français avec une introduction et des notes*. Paris: Desclée, De Brouwer & cie.

Vitrac, Bernard. 1990–2001. *Euclide: Les Eléments. Vol. 1: Introduction générale (par Maurice Caveing) et Livres I à IV (1994). Vol. 2: Livres V à IX (1994). Vol. 3: Livre X (1998). Vol. 4: Livres XI à XIII (2001)*. Book series *Bibliothèque d'histoire des sciences*. Paris: Presses Universitaires de France.

Zheng, Cheng 鄭誠, and Yiwen Zhu 朱一文. 2010. *Shushu Jiuzhang liuchuan xin kao: Zhao Qimei chaoben chutan* 數書九章流傳新考: 趙琦美鈔本初探 (A new study on the textual transmission of *Shushu Jiuzhang*). *Ziran Kexueshi Yanjiu* 自然科學史研究 *(Studies in the History of Natural Sciences)* 29 (3): 319–328.

Zhou, Xiaohan 周霄漢. 2018. Elements of Continuity between Mathematical Writings from the Song-Yuan (13th–14th century) Dynasties and the Ming Dynasty (15th century): Comparing Yang Hui's *Mathematical Methods* (1261 CE) and Wu Jing's *Great Compendium* (1450 C.E.). PhD. Dissertation, University Paris Diderot.

# Part I
# Ancient Editorial and Cross-Linguistic Practices

# Chapter 2
# Before the Library of Babel: On Some Very Early Philologers

**Piotr Michalowski**

> *...but the Library will endure: illuminated, solitary, infinite, perfectly motionless, equipped with precious volumes, useless, incorruptible, secret. (Jorge Luis Borges, The Library of Babel)*

**Abstract** Writing began in ancient Mesopotamia more than five and a half thousand years ago. From the beginning, scribes, and teachers, alone and in groups, orally, and in writing, reflected on the texts they copied, composed, memorized, edited, and redacted. These anonymous people should be recognized as the first philologists known to us today. Cuneiform texts from the first millennium BCE preserve the works and names of those who composed, transmitted, reworked, and re-edited a rich textual tradition in at least two languages, Akkadian, and Sumerian, and much of this late Mesopotamian scholarly work has been well studied by modern academics. This paper addresses aspects of less well-known philological endeavours from early third millennium BCE Mesopotamia.

## 2.1 Introduction

Histories of Western textual criticism and redaction usually anchor their narratives in the eighteenth century, linking the foundations of the endeavour in Classical and Biblical research. The modern era of theory and practice in this area begins with Karl Lachmann (1793–1851), although selected predecessors such as Spinoza are

The research leading to these results has received funding from the European Research Council under the European Union's Seventh Framework Programme (FP7/2007–2013)/ERC Grant Agreement No. 269804.

P. Michalowski (✉)
Department of Middle East Studies, The University of Michigan, Ann Arbor, MI, USA
e-mail: piotrm@umich.edu



A. Keller, K. Chemla (eds.), *Shaping the Sciences of the Ancient and Medieval World*, Archimedes 69, https://doi.org/10.1007/978-3-031-49617-2_2

often brought into the discussion. Some might take the story much earlier with a perfunctory bow to the third and fourth century CE Alexandrian redactors of Homer and other early Greek poets, to the great librarians such as Zenodotus, Aristophanes, and Aristarchus. As such, this remains very much a Western European history, with a mythical past firmly founded in Hellenism. But as Sheldon Pollock (2009) reminds us, this is but part of a much larger story; there have been robust traditions of textual criticism on the Indian subcontinent, in the Jewish and Islamic traditions, in China and throughout the globe that are so often left out of our disciplinary histories. Much of this was remedied in a collection of impressive studies on *World Philology*,[1] published in 2015, but this seemingly comprehensive volume begins with ancient Greece and tells only half of the tale, silencing over two millennia of efforts of ancient Near Eastern philologists. Thus, to be fair, we should revise our histories to include the more than three millennia of ancient Near Eastern textual traditions, practiced by different civilizations in many languages and places: in Egypt, the Levant, in Hittite, Elamite, Hurrian and Urartian lands, and in Mesopotamia.

This earlier history is rich and complex; as a starting point, one may invoke the words that a man named Balasi, who worked at philology a thousand years before the librarians of Alexandria, once wrote in a letter to the Assyrian king Assurbanipal (ca. 668–627 BCE), under whose patronage large libraries were collected in his capital of Nineveh and who was himself literate (indeed, Balasi was his tutor):

> <The omen series> *šumma izbu* is difficult to interpret! The first time that I come before my king, my lord, I shall <personally> show, with this tablet, that I am sending to the king, my lord, how the omen is written. Really, [anyone] who has [not] had <the meaning> pointed out to him with a finger cannot possibly understand it! (Parpola 1993: 44)

Balasi was alerting the king against precipitous interpretation, urging caution and patience, but also asserting his interpretive power at court, but he could not know that was anticipating what a latter-day colleague, Calvert Watkins (1990: 25), would write so many years later, channelling his own masters in an essay decrying modern avoidance of very ancient textual scholarship:

> What, then, is philology? Let me conclude with the definition of philology that my teacher Roman Jakobson gave (who got it from his teacher, who got it from his): 'Philology is the art of reading slowly.'

## 2.2 Ancient Mesopotamian Writings

The Mesopotamian textual record, written in the Sumerian and Babylonian/Assyrian languages, spanned over three millennia of time with a prestige status-driven spread throughout most of the territory of Western Asia. The primary vehicle for the cuneiform logo-syllabic sign system was clay and secondarily stone. When dried or baked, clay tablets become virtually indestructible, subject only to the vicissitudes of human

[1] Pollock et al. (2015).

disruption of their habitat: some were purposely kept for reference or posterity; some were thrown out into streets or trash heaps; and many were preserved by chance in abandoned or violently destroyed houses, temples, or palaces. As a result, scholars today have access to hundreds of thousands of cuneiform tablets, plundered, robbed, or professionally excavated, and countless more lie covered by sand and earth in the mounds of Iraq, Syria, and neighbouring countries. Most of these texts are records, accounts, and letters that bear witness to private and state business, but a sizable portion consists of scholarly and pedagogical texts, including literary poetic narratives. The use of cuneiform died out some time during the first century or second CE, and except for some elements alluded to or paraphrased in Biblical, Hellenistic Greek, and other early writings, the literary and scholarly achievements of the Mesopotamian and other ancient Near Eastern literati disappeared, seemingly forever, covered up by dunes and wind.

From the middle of the second millennium BCE, cuneiform texts were collected in personal, temple, and palace libraries, composed on and copied from clay tablets, creating complex textually based traditions. Authorship was generally anonymous, albeit sometimes ascribed to divine or ancient human actors, but we know something about the often-difficult redactional activities of various scribe-scholars.[2] Not only did they create new editions from existing manuscripts that often varied in much detail, but also sifted through the textual repertoire, adding as well as deleting compositions from the accepted canon. This is well illustrated by the words of the scholar Akkulanu, written in the seventh century to the Assyrian king at Nineveh (Parpola 1993: 80): 'The king, my lord, should have a look; let them remove what is to be removed and add what is to be added.'

Throughout much of this textual history Mesopotamians eschewed all overt metatext but beginning sometime in the seventh or eighth century BCE, ancient intellectuals began to construe elaborate commentaries that can comfortably be described as philological work (Frahm 2011). There is some evidence that this tradition had a longer history that is currently not recoverable and there are hints that such efforts may have been pursued orally and in writing much earlier. Here, I would like to move chronologically further back, before the libraries of the late period, to a time when Mesopotamian texts occupied very different physical and semiotic spaces.

## 2.3 Early Mesopotamian Writing

Writing was developed for accounting purposes in southern Mesopotamia, in the area early on designated as Sumer some time before 3500 BCE,[3] but the first known texts consist of numerical and commodity word signs, with very little direct linguistic information and it took it another seven hundred years or so before it was

[2] See, for example, the works of a family of such scholars described by Finkel (1988). To situate the places mentioned in this chapter, see Annex A.

[3] The Sumerian native designation was kengir, *Šumeru* in Akkadian, which eventually resulted in the English designation Sumer and its similar equivalents in other modern languages.

sufficiently developed for composing written poetry, as far as we can presently discern from the surviving documentation. The fact that the proto-cuneiform recording system originated as an adjunct to accounting and not as a vehicle for the direct preservation of speech determined much of its early history. This was what Juri Lotman (1977) and his Tartu cohorts would have labelled a secondary modelling system; it was a cultural artefact that was homologous with natural language, but was not part of it; the two were parallel communicative devices.[4] The relationship between proto-cuneiform and language is a matter of some dispute and even among those who posit an underlying natural language there are disagreements between those who believe that this language would have been Sumerian and others who suspect that some other tongue may have been involved.

During a period of a few hundred years, the two parallel lines somehow came to cross, and although this process is poorly documented, one must assume that it was not a simple, linear process. It took over half a millennium before the transformed system, which we can now call writing, could carry complex linguistic messages. There are a few hundred tablets from Ur, ca. 2700 BCE, two or three hundred years younger than the proto-cuneiform texts, which were certainly in Sumerian (Lecompte and Verderame 2013). Other than the Ur texts and a few scattered documents from elsewhere, the various processes involved in the development of early Mesopotamian writing are hidden to us by the mists of time and by the chance nature of archaeological discovery.

Even if we acknowledge the considerable gaps in our knowledge, the archaeological record reveals a flowering of literature and scholastic writing in the Sumerian language, documented in Mesopotamian cities such as Shuruppak, Nippur and Tell Abu Salabikh (ancient name unknown), in Kish further to the north, but also in Syrian Mari and Ebla (and undoubtedly elsewhere) a few generations later, ca. 2400 BCE. While we do not fully understand any of the poems from this time, we can say with confidence that they all seem to be concerned with mythological issues, and never referred—as far as we can determine—to contemporary political events. The function of these literary clay tablets is difficult to determine. Were they the products of inspired poets, records of a vast oral tradition, or was this an entirely new form of literary discourse, grounded in writing; were they kept for public recitation or simply for teaching children how to write, as in later times? The poetry was inscribed in a manner presupposing that both writers and readers either knew much of the text already or understood the compositional and linguistics aspects of the narrative well enough to reconstruct it from a partial notation. Signs were often not written in the order they were read and much of the grammatical information was not set down. Most important, the Mesopotamian scholastic texts of this time radiated throughout Western Asia as part of a new intellectual *koiné*, perhaps the first identifiable instance of intellectual globalization in world history.

[4] This is a convenient oversimplification; the notions of primary and secondary modelling systems have been the subject of much critique; see, for example, Chang (2003).

Few literary texts from subsequent centuries have been recovered. Most of what we know about early Sumerian poetry comes down to us in copies of older works made by students in Old Babylonian learning contexts from the eighteenth century BCE.

Many of the Old Babylonian literary clay tablets are copies of works that were composed in the Ur III period (ca. 2110–2003), a short century of strong political and economic centralization and integration, although very few literary tablets from this period have been recovered to date and most of those remain unpublished.[5] The first two kings of the new polity, which we call the Ur III state, apparently adopted many of the decorum and self-representational strategies of an earlier political formation from the city-state of Lagash, inheriting a new way of utilizing scribal culture for political ends. Under the founder of the Ur dynasty, Ur-Namma (2110–2093 BCE), and then under his son and successor Shulgi (2092–2045 BCE), scribal masters adapted and redesigned most of the basic teaching tools such as word lists but also created new poetic pedagogical materials written in honour of the contemporary ruling house—royal hymns and stories about their mythical royal ancestors from the city of Uruk: Enmerkar, Lugalbanda, and Gilgamesh. I suggest that the primary impetus that accelerated this process was an ideological crisis precipitated by the battlefield death of Ur-Namma, a calamity that required major damage control. Such a royal demise would have been interpreted as a sure signal of divine anger and abandonment and may have precipitated vigorous questioning of the legitimacy of central authority. To bolster the regime, the process of reinvention of royal self-representation was likely reinvigorated, grounded in the concept of divine kingship and the concomitant reinvention of the written tradition, which was now partially reoriented to represent charismatic rule. As already noted, the roots of this political tradition can be traced back at least as far back as Gudea, the last great king of the earlier polity of the Lagash region, who reigned immediately before the rise of Ur-Namma's kingdom, but we know the elaboration only through the prism of later versions of Ur III poems.

The centralized, patrimonial state run from Ur required a well-regulated and well-trained bureaucracy that could be held accountable for all fiscal and organizational activities. Writing was the instrument by which the Crown exercised oversight and control, as documented by the roughly hundred thousand or so published administrative documents from the period. The hearts and minds of these literate servants had to be moulded through schooling that taught them writing skills but also indoctrinated them into the ideological aspirations of the new state (Michalowski 1991).

Literature and other schooling materials contributed to the socialization of lowly scribes, bureaucrats, and children of elites. Although we still know very little about schooling in Ur III times, we are in a better position to study how the pedagogical materials were refashioned and reinterpreted by schoolmasters for educational

[5] On the Ur III period, see the detailed survey by Sallaberger (1999); on the unpublished Ur III literary texts, see Rubio (2000).

purposes centuries later. A closer look at the archaeological context of these later copies, the majority from the city of Nippur, allows us to date most of them to the reigns of two eighteenth-century kings, Hammurabi (1792–1750 BCE) and his son Samsuiluna (1750–1712 BCE), altogether 53 years at the most; that is less than two generations. To be sure, there are other Sumerian literary texts from other cities, and some of them are either earlier or later, but much of what we have derives from the destruction and abandonment layers of cities that were defeated after a great rebellion against the Hammurabi's son (Michalowski 2011). This means that the literature we are considering constitutes but a snapshot of a long process and we cannot ascertain how it stands in the flow of time.

There is no information on the compositional process or on the social context of textual redaction in Ur III times. A Shulgi hymn and a few literary and lexical texts mention places designated with signs that translate literally as the 'House of Knowledge/Wisdom of the [Scribal Training Goddess] Nisaba',[6] seemingly recalling the Bayt al Ḥikma, 'The House of Wisdom', in ninth- to thirteenth-century CE Baghdad (Balty-Guesdon 1992), claiming royal patronage of poetic composition, but in all honesty, it remains uncertain if this was a physical space or a poetic metaphor for divine knowledge and the analogy with the Abbasid library or centre of learning may be spurious. Moreover, the eighteenth-century versions of Sumerian literature differ in writing conventions and even in substance from their Ur III originals. We have no direct evidence that would provide any information on the redactional work that took place between the collapse of the Ur III dynasty around 2004 and the recovered school texts from the eighteenth century BCE. There is a short interlude during the reign of the first post-Ur III sovereign—Ishbi-Erra of Isin (ca. 2019–1987)—when attempts were made to create a new form of royal literature focused on specific events in the life of the sovereign, primarily of a martial nature, but this new style was soon abandoned and his successors on the throne of Isin returned to the language of power and authority pioneered by the philologists of Ur and their Lagash predecessors (Michalowski 2005). The philologists who created, recreated, and redacted the texts that came into existence during this politically motivated flowering of literary and pedagogical endeavours will forever remain unknown, as shall their intellectual debates and inspirations.

During the next century or so, the received texts were sifted and redacted; some were altered, others were discarded. The very language of the texts was reformed, and the writing rules regularized in a somewhat artificial manner. This type of cultural recreation is hardly unexpected. Since I have referred to Juri Lotman, I shall proceed to cite a classic text by him and his colleague Boris Uspensky:

> The history of the destruction of texts, of the purging of texts from the reserves of the collective memory, proceeds alongside the history of the creation of new texts. Every new movement in art revokes the authority of the texts by which preceding epochs oriented themselves, by transferring them into the category of nontexts, texts of a different level, or by physically destroying them. Culture by its very essence is against forgetting. It overcomes forgetting, turning it into one of the mechanisms of memory. (Lotman and Uspensky 1978: 216)

[6] See, most recently, Klein and Sefati (2014: 90–91). The actual reading of the sign combination is unknown at the present time.

This form of redactional forgetting served to create an idealized 'Sumerian' past, as argued so well by Jerry Cooper (2010). The earliest known dated manuscript written in the classic Old Babylonian style that is best known from the eighteenth-century schooling establishments and which became the norm for almost two millennia is a tablet of unknown provenance housed in the British Museum that contains part of a hymn to the goddess Nungal (CT 52, 27; Michalowski 1995: 50). This composition is well attested among the pedagogical exercises from the mid-eighteenth century, where it was part of the so-called Decad, a group of ten poems used at the beginning of the second tier of instruction (Tinney 1999). This tablet is dated to the time of a dynasty (Manana) that was roughly contemporary with Sumuabum and Sumulael of Babylon; the latter reigned ca. 1880–1845, about one hundred years after the death of Ishbi-Erra. Visually and in content, this tablet is indistinguishable from the later eighteenth-century exercises and therefore one must conclude that at least some aspects of the early Old Babylonian redactional revolution had to take place sometime during this century-long gap in our documentation, before 1880 or so.

As things stand today, we are in the dark concerning the motivations, authorizations, and the very procedures of these redactional events; we can only speculate how the philologists who took part in this project acquired the power and prestige to adapt and revise received written knowledge and to recreate a new normalized form of the literary Sumerian language. Nor do we know how the results of these reforms were transmitted and further modified by succeeding generations. We do not have any library copies or master copy texts, nor do we know if the poems were communicated orally or on clay from master to master. Recently, it has become accepted that the surviving Old Babylonian Sumerian literary texts were mostly written by schoolchildren from teachers' dictation (Delnero 2012). In such a system, the repository of texts would have been human memory: teachers who knew the compositions by heart would dictate them to students who would in turn memorize them, sometimes in short segments, and then write them down. This was probably the case for many of the tablets we have, but not for all of them. The poems of the Decad and other texts from early in the instructional cursus have few variants, and indeed many of these can be explained by recourse to memorization errors. There are other texts, however, that have more variation and some of this seems to more likely resemble copying errors or purposeful change instigated by a need to understand the often-difficult Sumerian that was the subject of their studies.

Some textual variation is difficult to explain by any means. Imagine seven manuscripts of the same composition from three different cities. Differences in spelling or even of word choice are distributed in the following manner (Table 2.1).

It would be impossible to ascribe such differences to local preference or to common error. Clearly, our knowledge of ancient redactional and scribal conventions is inadequate for the task at hand and modern editors have yet to find a satisfactory solution to the editorial challenges of such problems.

**Table 2.1** Hypothetical distribution of variants in a text

| City | Text | Feature |
|---|---|---|
| Nippur | 1 | A |
| | 2 | A |
| | 3 | B |
| Ur | 1 | A |
| | 2 | B |
| Sippar | 1 | B |
| | 2 | C |

But even if we cannot trace the workings of the ancient philological processes that operated on poetry or literary prose, we are somewhat better off when it comes to the backbone of early Mesopotamian educational activity: the lexical texts, or word lists.[7] To understand these, we must turn to the very beginning of literacy in Sumer.

## 2.4 The Beginnings of Mesopotamian Philology

The earliest Mesopotamian texts known to us, from fourth millennium BCE Uruk, from ca. 3500–3000 BCE, provide the first evidence of human philological activity. The uniformity of the accounts, as well as pedagogical texts in the form of word lists, testify to the invisible hands of some person or persons responsible for the organization of the accounting systems and the teaching materials. Although there are apparently unpublished tablets that may be older than the excavated archaic tablets from Uruk, at present, the typology of the texts recognizes an older phase, named after the archaeological Level IV (ca. 3500–3250 BCE) and a younger one, named after Level III (ca. 3250–3000 BCE). However, almost all these the tablets were discovered in secondary trash deposits, and their association with archaeological levels is a convention based on scholarly reconstruction. There does not seem to be any sizable time gap between these two stages and most of the accounts and word lists derive from the younger one ('Uruk III').

The word lists were the reproductive organs of the system, so to speak, witnesses to the kinetic elements of the learning process that allowed this semiotic experiment to survive for more than one generation. One can imagine that a robust oral explanatory tradition evolved around this learning process, but all that has survived are the remnants of student exercises. It cannot be ruled out that some of better word list copies may have been teachers' models, but this can only be conjectured.

[7] The bulk of this essay was written before the appearance of Veldhuis (2014b), an authoritative history of the Mesopotamian lexical tradition that deals with all of this in a much more comprehensive manner.

Thirteen distinct relatively standardized lexical lists were used in leaning writing in Uruk times, although some fragmentary tablets suggest that there may have been other such compositions that were rarely used.[8] These consist of word lists that were by and large thematic, although some diverge into other semantic domains and the general meaning of one of them remains a subject of debate (*Word List C*); their modern names, '*Vessels and Garments*', '*Metal*', '*Wood*', '*Cities*', or '*Birds*', etc., describe their basic contents, but the one attested in the largest number of exemplars (over 200), is generally referred to as Lu A or '*Professions*' ($lu_2$ was the Sumerian word for 'man, person', and a classifier marking professional names). Only three of these are currently attested on tablets from the earlier Uruk IV period and they differ substantially from their later versions (Veldhuis 2014b: 32). The clear majority of the recovered Uruk pedagogical texts belong to the later Phase III and as a result it is impossible to know which ones were created in the later stage in the development of cuneiform and which were there already in Uruk IV but have yet to be found. It is indeed probable that the teaching materials were invented together with the first forms of proto-cuneiform or soon thereafter, thus ensuring the survival of this form of notation over more than one generation. Either way, we have before us a textual production that is not haphazard; it was well planned and organized, reflecting not only a strong hierarchical bureaucratic power but also some form of philological authority that dictated and controlled the parameters, diachronic as well as synchronic, of the accounts as well as of the learning materials. This is all the more interesting for our purposes, because there is some evidence to suggest that the earlier versions were reworked and standardized at the beginning of the Uruk III phase (Englund 1998: 103). All of this required authoritative activity by one or more persons; we have no alternative to celebrating them among the first philologists on earth.

The word lists were essentially inventories of nouns and would remain so for millennia; when verbs were added, they were cited in as verbal nouns. But the actual accounts focus not just on commodities but, most prominently, on their numbers, true to the numerical origins of writing itself.[9]

## 2.5 Third Millennium Developments

During the middle of the third millennium BCE, a time usually designated as 'Early Dynastic', cuneiform writing and Mesopotamian textual traditions spread throughout much of the Middle East. The Sumerian language instructional texts that travelled west and northwest were adapted for local instruction, but the structure of the

[8] On these lexical texts see Krispijn (1992), Englund (1998: 82–110), Wagensonner (2021) and Veldhuis (2014b: 32–59).

[9] Sheldon Pollock (2015: 3) laments the fall of philology in modern corporate universities: 'Indeed, say the students, what is the point of learning to read well when all you need to know is how to count.' The Assyriologist, sadly in agreement with Pollock, may admire the irony of how writing may have come full circle to its beginnings in numeracy.

**Table 2.2** General chronology of referenced texts

| | |
|---|---|
| 3500–3000 | Uruk IV/III |
| 2700 | Early dynastic I/II (Ur "archaic" texts) |
| 2575–2310 | Early dynastic III (Shuruppak and slightly later Abu Salabikh and Ebla tablets) |
| 2110–2003 | Ur III |
| 2019–1595 | Old Babylonian |

writing system was also modified tailored for the local Semitic tongues.[10] At present, Early Dynastic texts (see Table 2.2) are known from mainly from Mari, Urkesh, and most prominently from Ebla, but there can be little doubt the knowledge of cuneiform had spread throughout the whole area corresponding to present-day Syria and its environs, radiating from various nodes a writing system as well as whole compositions and literary languages (see, most recently, Biga 2021; for the general chronological scheme used here, see Table 2.2). Even though the documentation for this spread is fragmentary and uneven, the general outlines of this new intellectual *koiné* can be sketched out in broad terms.

How this knowledge was disseminated is far from clear to us today: it is possible that teachers from northern Babylonia instructed local students and future instructors, dictating, or writing model tablets from memory, but it can also be demonstrated that actual tablets were brought from Mesopotamia and then passed on together with additional local copies from centres along the way. It is important to observe that only a limited number of compositions were adopted outside of Mesopotamia proper: Sumerian word lists and magical charms (incantations) as well as a handful of literary narratives, written in Sumerian and in a Semitic language.

Although variations abound, the overall unity of pedagogical literary practices in Western Asia in the time around 2600–2400 BCE contrasts with the political and religious fragmentation that was characteristic of the period. In Southern Babylonia, relatively small polities, often labelled as city-states, were the norm. We know less of what was happening upstream from there; I.J. Gelb (1981) posited a cultural but not necessarily politically unified complex that he referred to as the Kish Civilization, but Piotr Steinkeller (2013) believes that this was also a supremely powerful larger political and administrative polity. In the expansive rain-fed steppes of Syria, powerful large urban centres such as Mari, Nagar, or Ebla controlled sizeable hinterlands with various degrees of authority, but smaller polities thrived as well. Sporadic tablet finds document the fact that literacy, albeit known to only small numbers of scribes and accountants, was widespread throughout the area. But there was an important fissure in what seems like pedagogical unity and the source of this divide lay in Mesopotamia. However one views the political divisions of the time, it is now

[10] The Sumerian language was an isolate, that is it has no known relatives. The structurally very different languages of early Mesopotamia and Syria such as Akkadian and Eblaite and related tongues belonged to the East Semitic branch of the Afro-Asiatic language family.

apparent that pedagogical texts were used differently in the north and in the south. As Veldhuis (2014a, b: 103–116) has recently argued, the southern texts, known mainly from Fara, were read in Sumerian, while in the north similar texts, or even identical ones, including lexical lists, were read in some form of East Semitic, with perhaps a mixture of the two traditions in border areas between the two such as Abu Salabikh. According to Veldhuis, six Semitic language compositions, attested mainly in Abu Salabikh and Ebla, which will be discussed below, provide evidence for the northern 'Kish' pedagogical repertoire, five of them are lexical,[11] but one is a poetic hymn to the sun god Shamash.

The difference between the southern and northern repertoires is perhaps best exemplified by the distribution of examples of a pair of instructional texts that originated in different parts of Mesopotamia and became standbys of mid-third millennium scribal training. The first, the most widely used early lexical text, was a list of professional names that is commonly referred to as *Ed Lu A*,[12] already mentioned above, first documented among the earliest Uruk exercises (*Lu A*) and is attested in both southern and northern third millennium Babylonia in the cities of Uruk, Ur, Shuruppak, Nippur, Girsu, Adab, and Abu Salabikh, in Iranian Susa, and in Syria at Ebla and possibly somewhat later at Nagar (Michalowski 2003: 1).[13] A separate list of professional names, commonly designated as *Ed Lu E*, was apparently created during this time to replace, or perhaps to augment the standard professions list (*Ed Lu A*) by providing occupational designations that were actually used at the time in place of the obsolete entries of the traditional composition; moreover, if Veldhuis (2014a) is correct, it was meant to be read in Semitic. The distribution of the known exemplars of this text is instructive: these have been found at Abu Salabikh, in the northern part of southern Babylonia, in northern Babylonian Kish and in Syria at Ebla and Urkesh (Buccellati 2003). This new list is not presently documented in southern Babylonia and therefore appears to have been a northern innovation that spread out from the Kish area towards the west and northwest. Indeed, this is the only component of the assumed northern Semitic-language repertoire that was actually discovered in the city of Kish, the purported site of origin of this tradition.

Although Kish, some 80 km south of Baghdad, was excavated three times beginning in 1912, its large size and long occupational history, which seems to begin at the end of the Uruk period, has made it difficult to find early textual materials and hardly any Early Dynastic tablets have been recovered to date (Yoffee 2013). But the muse of archaeology does play with dice and by pure chance, a complete mid-third millennium palace archive that had never been disturbed was discovered far away from Kish, at a mound named Tell Mardikh, ~55 km southwest of Aleppo, which we now know was the ancient city of Ebla. In 1975, an Italian team that had been working for years at the site discovered a collection of 1200 or so tablets,

[11] The professions list *Ed Lu E*, the *List of Geographical Names*, *Animals B* and *Practical Vocabulary A*.

[12] The Sumerian word lu (written with the sign $lu_2$) means 'man, person'.

[13] The Nagar exemplar may have to be dated somewhat later than the Ebla archives (Veldhuis 2014b: 63 Note 90).

provided us with the only full example of what such files must have looked like in most of the major cities at the time (Matthiae 2010). Most of the tablets were found undisturbed in a specially designed storage room adjacent to what seems to have been a royal audience space. The repository had wooden shelves on all four sides and the tablets lay on the floor where they had fallen when the ledges on which they rested had disintegrated. As a result, the Ebla palace collection represents the only complete relatively undisturbed third or second millennium tablet archive ever discovered. The contents are instructive: most of the tablets were accounts of various sorts but also included smaller numbers of diplomatic letters and state rituals. These texts obviously provided information for use by the royal administration, allowing officials easy access to review past performance or to plan for the future. But the royal archive also included scholarly works. Some of these came from the hands of scholars associated with the Crown, others were possibly deposited by proud elite students, their teachers or even by their parents.

## 2.6 The Word Lists of Ebla

The instructional tablets discovered in Sumerian places such as Shuruppak, Abu Salabikh, Adab, or Nippur consisted of both Sumerian literary narratives and of lexical texts. In Ebla, the word lists predominated. And while in general the texts are quite like the Mesopotamian counterparts, there are significant differences that bear witness to philological activity and provide clues to the way in which third millennium scribes indulged in analysis and reinterpretation of received textual tradition.

The Sumerian language, whatever its temporal and spatial parameters may have been, was certainly not spoken in northern Babylonia and Syria at the time. Rather, the local vernaculars consisted of various forms of the eastern branch of the Semitic language family; it is even likely that the intermingling of different languages and dialects, albeit related, produced varying forms of multilingualism and diglossia in a highly interrelated, if fragmented, political, and economic landscape (Rubio 2006a, 2006b). The introduction of Sumerian and northern Mesopotamian Semitic-based learning added more levels of complexity to the socio-linguistic matrix: a form of written and recitative multilingualism. Although it is the scholarly norm to refer to the written Semitic of Ebla as Eblaite, it is difficult to ascertain the level of familiarity between the written conventions and the languages of the street, nor can we discern the differences between written Eblaite, local vernaculars and the Semitic of northern Mesopotamia ('Kish'). One can also ascertain which Semitic language texts were composed at Ebla and which were imported. Some, such as the coronation rituals or copies of outgoing letters, were undoubtedly local, but the Semitic poetic narratives and the lexical materials that include Semitic words may have been redacted elsewhere and brought to Ebla in a form similar to or even identical to what we have at hand. One Sumerian monolingual exemplar of the *List of Animals* may be a physical import from northern Mesopotamia (Archi 1992: 6).

## 2.7 The *Standard Professions List*—A Case in Point

The history of *Lu A* or '*The Standard Professions List*', is indicative of a widespread interest in survivals from the early history of the cuneiform system (Veldhuis 2014b: 63–64). As already observed above, the earliest known exemplars of this list harken back to Uruk IV, to the time of the oldest currently known period of Mesopotamian writing, and although these are few in number, but there is enough of this early material to observe that the list was somewhat fluid at the time. By the time of Uruk III, this professions list was more standardized, albeit not without some remaining variation. By the middle of the third millennium, it was stabilized and became the most widely distributed and best attested of all the Sumerian word lists and even though most of the entries were anachronistic by this time, it took a few more centuries before it was completely discarded from the main educational curricula. But even after it was discarded from mainstream pedagogy in eighteenth-century Babylonia, scribes occasionally still copied it or used it for illustrative teaching purposes.

The four Ebla manuscripts of the main professions list provide clues as to the way such tablets came to be as well as on the role of such artefacts in this palace archive and this sheds some light on philology at Ebla. Four manuscripts of the list were found in the archive, each one originally containing the complete 129-line composition. Based on an analysis of several textual variants, Archi (1992: 3) observed that A and B were copied from different originals; he furthermore claims that C and D were in turn copies of B (Table 2.3). More important, however, is the obvious conclusion that A and B were copied, directly or indirectly, from tablets imported from Babylonia. The Ebla scribes accentuated this fact by keeping strictly to the 19 lines per column format that apparently characterized the source manuscripts and the copyists of C and D, who used a different layout, still noted the end of each column in their source by inserting a double line after every nineteenth line.

**Table 2.3** Ebla sources of *LU A* (Archi 1992: 2)

| Text | Square | Round | 19 cases per column | Double lines after every 19 lines |
|---|---|---|---|---|
| A | x | | x | |
| B | x | | x | |
| C | | x | | x |
| D | | x | | x |

A written by Azi, various other 'scholars' involved, but Azi is described as um-mi-a, 'master'
B Abu-Malik 'established the (text) of the tablet' (Sumerian or rather, Sumerogram, dub ki-ŋar. For the meaning of this, see, most recently, Fronzaroli (2012: 173))
C similar to B but written (dub mu-sar) by $bu_3$-da-BE
D no colophon

## 2.8 The New Semitic Language Philology

A variety of texts illustrate the new Semitic philology. Some bilingual lexical lists provided Semitic translations of Sumerian words and word signs (logograms) that were used in recording Semitic texts. Perhaps more significant are two tablets that have jointly been misleadingly designated as *The Ebla Sign List* (*ESL*; Archi 1987).[14] One of these is a list of 122 words or signs; the opening section is a selection from the first 56 lines of the great professional names list (*Lu A*), followed, seemingly randomly, by entries from various other lists (Arcari 1983; Veldhuis 2014b: 133–135). A semi-duplicate, written in a similar manner by another hand on a roundish tablet of the kind usually used for student practice, has the same list of signs, augmented to a total of 150, but also provides 'Semiticized' pronunciation guides to the Sumerian words.

Here are the opening nine lines of *Ed Lu A* compared with the first five lines of *ESL* (Table 2.4).

These are occupational designations that are not well understood at present; indeed, all of them were obsolete by the middle of the third millennium.[15] Note that both basic elements of *Lu A*: 1 are explained in the order they were written, which is out of order, and that Lines 3 through 7 are omitted; this is also the case for some other entries. The Semiticized pronunciation of individual signs contains the added nominative ending /u/ and mimation (*−u* + *m*), an ending added in word final position on singular nouns in early East Semitic, including Akkadian and Eblaite. Elena Arcari (1983: 167), who first observed the connection between the 'List' and the professions list, concluded that the opening section of the 'List' was 'una schedatura del segni che sono parte integrante dei

**Table 2.4** The opening lines of *Ed Lu A* and *ESL*

| *Ed Lu A* | *ESL A* | *ESL B* |
|---|---|---|
| 1. $nam_{2}$-$^{ŋiš}$šita (ŠITA.ŊIŠ.$NAM_{2}$) | šita | šita = *ti-iš-da-*$num_{2}$ |
| 2. $nam_{2}$-tuku | $nam_{2}$/umuš | umuš = $u_{2}$*-mu-šum* |
| 3. $nam_{2}$-di | tuku | tuku = [*tu*$^{?}$*-*$gu_{2}$$^{?}$*-u*]*m* |
| 4. $nam_{2}$-umuš | | |
| 5. $nam_{2}$-uru | | |
| 6. $nam_{2}$-$šeš_{2}$ | | |
| 7. $ukur_{2}$ | | |
| 8. $nam_{2}$-$uru_{4}$ | $uru_{4}$ | $uru_{4}$ = [$u_{2}$*-r*]$u_{12}$*-um* |
| 9. $suš_{x}$ | $suš_{x}$ | $suš_{x}$ = *su-šum* |

[14] Online edition: http://oracc.museum.upenn.edu/dcclt/Q000038 (last accessed April 20, 2022).

[15] The first sign in most lines is $nam_{2}$, which probably means 'occupation of/responsibility for' and became grammaticalized as a derivational prefix creating abstract nouns. However, the same sign was also used for the word umuš, 'intelligence, understanding, planning', and this is what a scribe chose here.

**Table 2.5** ESL 17–20

| *Ed Lu A* | *ESL* A | *ESL* B |
|---|---|---|
| 31. umbisaŋ.umun$_2$ | umbisaŋ | umbisaŋ |
| | | *ub-bi$_2$-sa-ga-im* |
| | kum | kum |
| | kam$_4$ | kam$_4$ |
| | | *ga-mu-um* |
| | gan | gan |
| | il | il |

lessemi presenti in ED LU A',[16] but a few years later, Civil and Rubio (1999: 265) referred to it as 'a sort of card index that enabled the scribes to read ED Lu A'. In a later study, Civil (2009, 64) observed that this includes evidence of textual exegesis. In addition to the Semiticized explanations, there are a few extra signs that are not in the original professions list. For example, compare the following (Table 2.5 *ESL* 16–20).

The associations provided here are based on semantic, graphemic, and phonetic similarity: the Sumerian word umbisaŋ was an early word for 'scholar', one that would soon become largely obsolete. The sign that follows, umun$_2$, which appears by itself earlier in these texts where its pronunciation is explained as *nu-u$_9$-mu-um*, seems to mean 'knowledge, expertise', or the like, according to much later lexical texts.[17]

Whether or not this derived exercise was composed at Ebla, it reveals an interpretive strategy that resulted in a new way of looking at Sumerian from a Semitic language point of view and in the reworking of standard lexical texts in a novel manner. It is possible that the inspiration for this kind of interlanguage procedure is to be sought in the contemporary northern Mesopotamian repertoire that was also used at Ebla, in lists such as *Lu E* that were seemingly written in Sumerian but were read aloud in Semitic.

The sign combination umbisaŋ.umun$_2$ of Line 31 of *Lu A* was commented in *ESL* by providing a Semiticized reading of the first sign; it was followed by signs that are not in the source document. The first, kum, is similar in shape to umbisaŋ, which in turn is followed by gam$_3$/kam$_4$, a kind of stick; in turn, it was associated with kum only by assonance, made explicit by the name *ga-mu-um*, followed by another phonological association, gan (possibly a kind of pot or pot stand), succeeded by il, again graphically like gan. One can only speculate on the social context that prompted the production of this commentary-like text. It could be a home or collective schooling environment or a gathering of scholars, or perhaps both. Either way, one could paraphrase this as:

[16] A record of the signs that were an integral part of the words used in *ED Lu A*.

[17] On this word, see Sjöberg (1973: 121–122).

> In *Lu A* line 13 the first sign is umbisaŋ. What sign do you think sounds much like this? – gan. Is there another? – $kam_4$, which is *gamum*, as is gan. And what is a sign that you would associate visually with the latter? –il.

A second example of the third millennium philological practices that comes to mind is exemplified by tablets from the Ebla archive that appear to be linked more closely with pedagogical pursuits. In addition to lexical texts imported from Mesopotamia, the archive revealed a family of manuscripts pertaining to an innovative native or regional lexical list, which is named $EŠ_2$.BAR.$KIN_5$ (Veldhuis 2014b: 135–138) from its opening line (the reading and meaning of these signs is unclear). The arrangement is by sign forms, on acrographic principles, and seems to represent a local or regional attempt to understand Sumerian words and Sumerograms; some versions are monolingual, but there is also a bilingual one (Pettinato 1981, 1982). This organizational norm contrasts with what is encountered in Babylonian lexical texts that were construed largely on semantic principles. There is much instability in the manuscript tradition of $EŠ_2$.BAR.$KIN_5$; for example, words are often listed in varying order, and there are also excerpts that contain one or a few more words with translations, some of which differ from those found on longer pieces (Bonechi 2007, 2008; Krecher 1983). Physically, the tablets differ from those copied from Mesopotamian models; as noted by Archi (1992: 17), 'the cases are rather smaller than in the copies of texts from Mesopotamian originals, and the signs closer'. For some, such variety may signal unmotivated messy disorder, but it is probable that it provides insight into fluid local teaching practices geared to immediate student needs and to language play that was more pronounced in conjunction with local texts as opposed to the staid imported tradition that may have been held down to ground by a sense of prestige and tradition. This complicated manuscript practice, one that was not rigidly standardized but allowed for much scribal and pedagogical latitude, is indicative of the kinds of intellectual innovations that were taking place in third millennium Syria.

The Ebla archives offer vivid witness to the cuneiform philological enterprise outside of Mesopotamia proper. The presently available documentation of these processes is incomplete but provides some intimation of philological practices at work in northern Babylonia and Syria. While it is possible that cuneiform literature and other study materials spread together with teachers who knew the texts by heart, it is equally likely that they brought actual exemplars with them; indeed, there is indirect evidence that this was the case. The manner of transmission is thus unknown and can only be surmised, but there is an often-cited year name that appears on some texts, including manuscripts of the great Mesopotamian professions list, that commemorates the occasion when 'the young scribes came up from Mari' and one or more were probably written by a scribe from Kish (Archi 1992: 20–21). It would be a mistake to make too much of such data, since we really have no idea what the journey of 'young scribes' entailed, how young they were and whether they brought knowledge with them or if they came to Ebla in search of enlightenment. The Kish reference is not as simple as may appear because there is some evidence suggesting that at least some of the Ebla mentions of this city reference a homophonous Syrian city and not the northern Babylonian metropolis.

## 2.9 Third Millennium Mathematical Exercises—Some Examples

The mid-third millennium pedagogical and scholarly texts are mainly lexical lists and poetic narratives, but there are also a small number of mathematical texts that prefigure the important role that numeracy would play in early second millennium education. The number of discovered Early Dynastic mathematical exercises is indeed small—seven from Fara (Friberg 2007: 147–153), one from Adab and four, possibly five, from Ebla—but they provide important insight into the philology of the time (Robson 2008: 31–33, 301–302). As explained by Friberg (2005: 1):

> ... all known mathematical exercises from the 3rd millennium BC are 'metro-mathematical,' in the sense that they are not simply concerned with relations between abstract numbers but rather with relations between measures for lengths, areas, volumes, capacities, etc.

The surviving Early Dynastic metro-mathematical texts are either demonstrations of calculations or student answers to questions that were provided orally by a teacher. Some show the procedure that was followed while others only provide the answers. As a result, there are no duplicates in the classic sense because the mathematical texts were not copied from master texts nor were they written down from memory.[18] These exercises have been discussed many times; here, I provide three related examples that are directly relevant to the matter at hand.[19] The first, *TSŠ* 81 (A), concerns the calculation of daily flour rations given the total number of workers and flour:

| | |
|---|---|
| *Obverse (problem)* | |
| 40 dumu šidim | 40 apprentice builders |
| $zi_3$ ba | wages paid in flour[20] |
| 2($ban_2$) šu ba-ti | 2($ban_2$) were received (by each) |
| *Reverse (solution)* | |
| 3 $lid_2$-ga $zi_3$[21] 1(bariga) 2($ban_2$) | 3 lidga (= 72 ban), 1 bariga (= 6 ban), and 2 ban of flour (= 80 ban) |

[18] *TSŠ* is an abbreviation of *Tablettes sumériennes de Šuruppak conservées au Musée de Stamboul* (Jestin 1937).

[19] See the references provided by Robson (2008: 301) and Friberg (2005, 2007: 414–415); note especially Guitel (1963), Høyrup (1982), and Melville (2002).

[20] On the translation 'wages' or 'allotments' paid to workers, rather than 'rations' see Prentice (2010: 91–95) and in more detail Steinkeller (2015: 26–30). Flour was rarely used for wages in this period; as noted by Visicato (1995: xxx), most occurrences seem to involve the military. The expression $zi_3$ ba occurs only twice in the Shuruppak documentary record.

[21] This time the student wrote the word for 'flour' by slightly tilting it to our left (they held tablets differently), in a manner technically designated much later as $zi_3$-*tenû*.

This can be paraphrased as 'Question: if 40 apprentice builders receive flour rations of 2 ban each, how much flour is needed (for the project)? Answer: 80. In the metrological system in use at the time, six ban made up one bariga and four bariga made up one lidga. We must infer how the student made the calculations: only the basic problem and the solution were written in an abbreviated manner. Little else is specified; normally, one would expect a daily allotment of 2 L per worker; 2 $ban_2$ consisted of 20 L so, a somewhat unusual number, since such wages or allotments ranged from 2 to 7 or 8, rarely 10 L a day.[22] It should be noted that flour rations are extremely rare in administrative texts from the time.

The inverse problem—how to calculate the number of workers involved, given a total of the disbursed commodity but this time noting the daily wage rate—is provided by two very similar exercises, also from Fara/Shuruppak (Melville 2002: 242–246, Friberg 2005: 21–22). In both cases the abbreviated problem was quite similar, but the way the texts were written was different.

The first of these, *TSŠ* 50 (B), was inscribed in two columns on the obverse while the reverse was left blank:

| | |
|---|---|
| *Column 1 (problem)* | |
| 1 še $guru_7$ | 1 'granary' of barley |
| $sila_3$ 7 | (daily wages of) 7 L (of barley) |
| 1 $lu_2$ šu ba-ti | was received per person |
| $lu_2$-bi | —its persons (how many)? |
| *Column 2 (solution)* | |
| 45($šar_2$), 42 ($ŋeš_2$), 51[23] | 164,571(daily wages = persons) |
| še 3 $sila_3$ šu $taka_4$ | 3 L of barley, 'sent off' |

The first column may be paraphrased as 'Given an amount of grain A and daily wages for each person of B, how many people were involved?' The second column provides the answer: 'C daily wages; left over D grain left over,' i.e., 'C men, with a remainder of D amount of grain'.

To understand the challenge of this metro-mathematical exercise, one must be aware of the complexities of the numerical and metrological systems involved, which are well explained by Friberg (2005: 20–21): in later texts, the 'granary' ($guru_7$)—normally a construction for long-term grain storage (Borrelli 2020: 47)--was a very large capacity measure, although apparently its precise volume varied, but it was one that is otherwise unattested in this period. Here, there were 1,152,000 sila (i.e., litres) in a 'granary'. If one divides this number by 7, the result matches with the number of workers in the solution, with 3 left over. This is how we would do this, but there are different proposals concerning the way the ancients went about this, given their metrology and procedures; the discussion on such matters lies outside the scope of the present essay.

[22] The Sumerian $sila_3$ capacity measure $sila_3$ was equivalent to 0.86 L or 0.91 quarts, conventionally translated as litre here.

[23] $= 45 \times 3600 + 42 \times 60 + 51$.

The second example is a similar mathematical exercise, likewise from Shuruppak (*TSŠ* 671 = C), albeit written by an adept who was not having a good day, which apparently provides an answer to a problem posed by a master that was either the same or very similar in nature. The mistakes of the youngster are revealing. Rather than separate the proposition and answer in two distinct columns on one side of the tablet, this student wrote them on the two sides of the object, but he apparently got them wrong and began on what would normally be the reverse and carried the 'question' onto the second side, but also got the calculations wrong:

| | |
|---|---|
| *Reverse (problem)* | |
| še sila$_3$ 7 guru$_7$ | 1 'granary' of grain |
| 1 lu$_2$ šu ba-ti | Each person received (daily wages of) 7 L |
| *Obverse (question and solution)* | |
| guruš | (How many) young workers? |
| 40 (šar$_2$) | 162,760 |
| 5(šar$_2$) | |
| 46(ŋeš$_2$)[24] | |

It would be out of place here to reprise all the mistakes made by this 'bungler', as Marvin Powell (1976: 432–433) referred to the student in his extensive treatment of the text, which is provided here only as an example of how similar these exercises were.

More important, in the present context, is the hypothetical nature of A, and even more so in B and C. During this period, the standard daily worker ration was 10 L of barley per person. In A this was given as 2 ban$_2$, which was equal to 20 L; either this was incorrect, or the implication was that this was for 2 days work. The details in B and C, however, are obviously non-standard, taking the students out of their comfort zones. In both the daily wages consisted of 7 rather than 10 L. The use of the term 'granary' in a manner unusual for the times, and the then rare notion of flour allotments, among other features, testifies to the speculative nature of these exercises, which could be understood to silently begin with 'supposing that', suggest that already in this period, if not much earlier, even mathematical exercises had at least one foot in the realm of fiction.

## 2.10 Conclusions

The scribal materials cited in this essay exemplify the workings of an ancient philological imagination that contrasts vividly with the static traditionalist perspective one gains from looking at the textual history of the standard word lists. It is sometimes noted that that this traditionalism reflects a worship of the past as a purist way of constructing latter day Mesopotamian intellectual identities.

[24] = 40 × 3600 + 5 × 3600 + 46 × 60.

There is undoubtedly some validity to this perspective, but the philological texts afford a corrective to this view. The actual texts may have been transmitted in unchanging manner over long periods, but the tablets in question are only the surface markers of a rich interpretive tradition that accompanied the manuscripts. These oral practices, which may be reflected in a very small way in the philological texts cited above, may have quite fluid and have provided the arena for perpetuating the received tradition by innovation by means of speculation and debate among scholars and educators. Some of this would have surfaced in pedagogical settings, but the real work of this type probably took place through face-to-face communion by learned folk, in small and larger circles, and emanated from there to teaching practice. Thus, in the third millennium Near East, there existed diverse forms of philological practice, prompted and inspired by curiosity and reverence for a textual tradition that was sometimes at odds with current writing conventions, but also by multilingual and diglossic environments and a need to penetrate obtuse anachronisms.

The lexical and mathematical examples illustrated here provide some insight into the intricacies of very early philological procedures. It is apparent that complex relationships between oral and written text, aural and visual perceptions of language and memory were vividly at play in the Early Dynastic period. Many literary narratives and scholarly exercises were registered in clay in a manner that required an understanding of a variety of philological conventions as individual signs were not always written in the order that they were to be read, only basic grammatical information was explicitly provided, and texts were sometimes abbreviated in various degrees. As was the case throughout much of the history of cuneiform, metatextual information was never committed to writing and therefore we have no direct information about such principles, but it is obvious that they were the products of conscious interpretive efforts. The Ebla, Shuruppak, and Lagash examples cited earlier provide clues to the textual exuberance and play that accompanied the early attempts to explore the written word as an autonomous semiotic universe and to test its relationships with spoken language and languages. As a result, the ancient editorial practices, which on the surface may seem haphazard, fall into better relief, appropriate to the kind of textuality that was in play. Perhaps most important, the search for insights into the deeper workings of language, manifest in both the lexical and mathematical traditions, point towards the discovery of textually based fiction, more obviously apparent in the new written contemporary mythological poetry.

The modern interpreter, however, who seeks comfortable order, may be hard pressed to find and apposite format for this kind of active textual performance that straddled the written and spoken word, skimming across sound and meaning. In trying to unravel these very ancient texts, I am reminded of José Ortega y Gasset's postulates for a 'New Philology', brought to my attention many years ago by Alton Becker: 'Every utterance is deficient—it says less than it wishes to say. Every utterance is exuberant—it conveys more than it plans.'[25]

[25] Ortega and Gasset (1959): 2); see Becker (2000: 5).

## References


Arcari, E. 1983. Sillabario di Ebla e ED LU A: rapporti intercorrenti tra le due liste. *Oriens Antiquus* 22: 167–178.

Archi, A. 1987. The "sign-list" from Ebla. *Eblaitica* 1: 91–113.

———. 1992. Transmission of the Mesopotamian lexical and literary texts from Ebla. In *Literature and literary language at Ebla, Quaderni di Semitistica* 18, ed. Pelio Fronzaroli, 1–39. Florence: Dipartimento di Linguistica Università di Firenze.

Balty-Guesdon, M.-G. 1992. Le Bayt al-ḥikma de Baghdad. *Arabica* 39: 131–150.

Becker, A.L. 2000. *Beyond translation: Essays toward a modern philology*. Ann Arbor: University of Michigan Press.

Biga, G. M. 2021. The diffusion of cuneiform writing in Syria in the third millennium BC. *Pasiphae* 15: 49–62.

Bonechi, M. 2007. Studies in the Ebla lexical lists, I. MEE 4 77, 83, 87. *Quaderni del Dipartimento di Linguistica – Università di Firenze* 17: 199–214.

———. 2008. Studies in the Ebla lexical lists, II: *MEE* 4 82, 84, 85, 86. *Studi Epigrafici e Linguistici* 25: 1–26.

Borrelli, N. 2020. Institutional grain storage and its control network in the Ur III province of *Ĝirsu/ Lagaš. In After the harvest: Storage practices and food processing in bronze age Mesopotamia, Subartu* 43, ed. N. Borrelli and G. Scazzosi, 45–63. Turnhout: Brepols.

Buccellati, G. 2003. A Lu E school tablet from the service quarter of the royal palace AP at Urkesh. *Journal of Cuneiform Studies* 55: 45–48.

Chang, H.L. 2003. Is language a primary modeling system? On Juri Lotman's concept of semiosphere. *Σημειωτκή-Sign Systems Studies* 31: 9–23.

Civil, M. 2009. The Mesopotamian lexical lists: Authors and commentators. In *Reconstruyendo el Pasado Remoto. Estudios sobre el Próximo Oriente Antiguo en homenaje a Jorge R. Silva Castillo*, ed. D.A. Barreyra Fracaroli and G. del Olmo Lete, 63–69. Sabadell: AUSA.

Civil, M., and G. Rubio. 1999. An Ebla incantation against insomnia and the Semiticization of Sumerian: Notes on ARET 5 8b and 9. *Orientalia NS* 68: 254–266.

Cooper, J.S. 2010. 'I have forgotten my burden of former days!' Forgetting the Sumerians in ancient Iraq. *Journal of the American Oriental Society* 130: 327–335.

Deimel, A. 1924. *Wirtschaftstexte aus Fara. Die Inschriften Von Fara III. Wissenschaftliche Veröffentlichungen der Orient-Gesellschaft 45*. Leipzig: J.C. Heinrichs.

Delnero, P. 2012. Memorization and the transmission of Sumerian literary compositions. *Journal of Near Eastern Studies* 71: 189–208.

Englund, R.K. 1998. Texts from the Late Uruk Period. In *J. Bauer, R.K. Englund, and M. Krebernik. Mesopotamien. Späturuk-Zeit und Frühdynastische Zeit, Orbis Biblicus et Orientalis* 160/1, ed. P. Attinger and M. Weffler, 15–233. Freiburg/Göttingen: Universitätsverlag Freiburg/ Vandenhoeck & Ruprecht.

Finkel, I.L. 1988. Adad-apal-iddina, Esagil-kīn-apli, and the series SA.GIG. In *A scientific humanist: Studies in memory of Abraham Sachs*, ed. E. Leichty, M. deJong Ellis, and P. Girardi, 143–159. Philadelphia: The Samuel Noah Kramer Fund.

Frahm, E. 2011. *Babylonian and Assyrian text commentaries: Origins of interpretation. Guides to the Mesopotamian textual record 5*. Münster: Ugarit-Verlag.

Friberg, J. 2005. On the alleged counting with sexagesimal place value numbers in mathematical cuneiform texts from the third millennium BC. *Cuneiform Digital Library Journal* 2005 (2): 1–23.

———. 2007. *A remarkable collection of Babylonian mathematical texts. Manuscripts in the Schøyen collection: Cuneiform texts I*. New York: Springer.

Fronzaroli, P. 2012. The Eblaic king's supplication to the gods of the night (TM. 75. G. 756+ 771+ 815). *Revue d'assyriologie et d'archéologie orientale* 106: 165–176.

Gelb, I.J. 1981. Ebla and the Kish civilization. In *La Lingua di Ebla*, ed. L. Cagni, 9–73. Naples: Istituto Universitario Orientale.

Guitel, G. 1963. Signification mathématique d'une tablette sumérienne. *Revue d'Assyriologie et d'Archéologie Orientale* 57: 145–150.

Høyrup, J. 1982. Investigations of an early Sumerian division problem c. 2500 B.C. *Historia Mathematica* 9: 19–36.

Jestin, R. 1937. *Tablettes sumériennes de Šuruppak conservées au Musée de Stamboul. Mémoires de l'Institut Francais d'Archéologie d'Istanbul 3*. Paris: de Boccard.

Klein, J., and Y. Sefati. 2014. The "stars (of) heaven" and cuneiform writing. In *He has opened Nisaba's house of learning: Studies in honor of Åke Waldemar Sjöberg on the occasion of his 89th birthday on august 1st 2013, Cuneiform monographs* 46, ed. L. Sassmannshausen, 85–102. Leiden: Brill.

Krecher, J. 1983. Eine unorthographische sumerische Wortliste aus Ebla. *Oriens Antiquus* 22: 179–189.

Krispijn, Th.J.H. 1992. The early Mesopotamian lexical lists and the dawn of linguistics. *Jaarbericht Ex Oriente Lux* 32: 12–22.

Lecompte, C., and L. Verderame. 2013. *Archaic Tablets and Fragments from Ur (ATFU) from L. Woolley's excavations at the Royal Cemetery. Nisaba 25*. Messina: Dipartimento di Civilità antiche e moderne.

Lotman, J.M. 1977. Primary and secondary communication-modeling systems. In *Soviet semiotics: An anthology*, ed. D.P. Lucid, 95–105. Baltimore: The Johns Hopkins University Press.

Lotman, Y.M., and B.A. Uspensky. 1978. On the semiotic mechanism of culture. *New Literary History* 9: 211–232.

Matthiae, P. 2010. *Ebla. La città del trono. Archeologia e storia*. Torino: Einaudi.

Melville, D.J. 2002. Ration computations at Fara: Multiplication or repeated addition? In *Under one sky: Astronomy and mathematics in the ancient near east, Alter orient und Altes testament* 297, ed. J.M. Steele and A. Imhausen, 237–252. Münster: Ugarit-Verlag.

———. 2014. The Mathland Mirror: On using mathematical texts as reflections of everyday life. In *Krieg und Frieden im Alten Vorderasien: 52e rencontre Assyriologique Internationale international congress of Assyriology and near eastern archaeology Münster, 17.–21. Juli 2006, Alter orient und Altes testament* 401, ed. H. Neumann, R. Dittmann, S. Paulus, G. Neumann, and A. Schuster-Brandis, 517–526. Münster: Ugarit-Verlag.

———. 2016. After Neugebauer: Recent developments in Mesopotamian mathematics. In *A Mathematician's journeys, Archimedes* 45, ed. A. Jones, C. Proust, and J. Steele, 237–263. Cham: Springer.

Michalowski, P. 1991. Charisma and control: On continuity and change in early Mesopotamian bureaucratic systems. In *The Organization of Power: Aspects of bureaucracy in the ancient near east*, ed. R.D. Biggs and M. Gibson, 2nd ed., 45–58. Chicago: University of Chicago Press.

———. 1995. Review of B. Alster and M. J. Geller, *Cuneiform texts from Babylonian tablets in the British museum part 58. Sumerian literary texts* (London, 1990). *Journal of Near Eastern Studies* 54: 49–51.

———. 2003. An early dynastic tablet of ED Lu a from tell Brak (Nagar). *Cuneiform Digital Library Journal* 2003 (3): 1–6.

———. 2005. Literary works from the court of king Ishbi-Erra. In *'An experienced scribe who neglects nothing': Ancient near eastern studies in honor of Jacob Klein*, ed. Y. Sefati et al., 199–212. Bethesda: CDL Press.

———. 2011. *The correspondence of the kings of Ur: An epistolary history of an ancient Mesopotamian kingdom. Mesopotamian civilizations 15*. Winona Lake: Eisenbrauns.

———. 2013. Networks of authority and power in Ur III times. In *From the 21st century B.C. to the 21st century A.D.: Proceedings of the international conference on neo-Sumerian studies held in Madrid 22–24 July 2010*, ed. S.J. Garfinkle and M. Molina, 169–205. Winona Lake: Eisenbrauns.

Ortega y Gasset, J. 1959. The difficulty of reading. *Diogenes* 7: 1–17. Trans. C. E. Parmenter.

Parpola, S. 1993. *Letters from Assyrian and Babylonian scholars. State archives of Assyria 10*. Helsinki: Helsinki University Press.

Pettinato, G. 1981. *Testi lessicali monolingui della biblioteca L. 2769. Materiali epigrafici di Ebla 3*. Naples: Istituto Universitario Orientale di Napoli.

———. 1982. *Testi lessicali bilingui della biblioteca L. 2769. Materiali epigrafici di Ebla 4*. Naples: Istituto Universitario Orientale di Napoli.

Pollock, S. 2009. Future philology? The fate of a soft science in a hard world. *Critical Inquiry* 35: 931–961.

———. 2015. Introduction. In *World philology*, ed. S. Pollock, B.A. Elman, and K.K. Chang, 1–24. Cambridge, MA: Harvard University Press.

Pollock, S., B.A. Elman, and K.K. Chang. 2015. *World philology*. Cambridge, MA: Harvard University Press.

Powell, M.A. 1976. The antecedents of old Babylonian place notation and the early history of Babylonian mathematics. *Historia Mathematica* 3: 417–439.

Prentice, R. 2010. *The exchange of goods and services in Pre-Sargonic Lagash. Alter Orient und Altes Testament 368*. Münster: Ugarit-Verlag.

Robson, E. 2008. *Mathematics in ancient Iraq: A social history*. Princeton: Princeton University Press.

Rubio, G. 2000. On the orthography of the Sumerian literary texts from the Ur III period. *Acta Sumerologica* 22: 203–225.

———. 2006a. Eblaite, Akkadian, and East Semitic. In *The Akkadian language in its Semitic context: Studies in the Akkadian of the third and second millennium BC, Publications de l'Institut historique-archéologique néerlandais de Stamboul* 106, ed. G. Deutscher and N.J.C. Kouwenberg, 110–139. Leiden: Nederlands Instituut voor het Nabije Oosten.

———. 2006b. Writing in another tongue: Alloglottography in the ancient near east. In *Margins of writing, origins of cultures, Oriental institute seminars* 2, ed. S.L. Sanders, 33–66. Chicago: The oriental Institute of the University of Chicago.

Sallaberger, W. 1999. Mesopotamien: Akkade–Zeit und Ur III–Zeit. In *W. Sallaberger and A. Westenholz, Mesopotamien. Akkade-Zeit und Ur III-Zeit, Orbis Biblicus et Orientalis* 160/3, ed. P. Attinger and M. Weffler, 119–390. Freiburg/Göttingen: Universitätsverlag Freiburg/ Vandenhoeck & Ruprecht.

Sjöberg, Å.W. 1973. Der Vater und sein missratener Sohn. *Journal of Cuneiform Studies* 25: 105–169.

Steinkeller, P. 2013. An archaic "prisoner plaque" from Kiš. *Revue d'Assyriologie et d'Archéologie Orientale* 107: 131–157.

———. 2015. Introduction. Labor in the early states: An early Mesopotamian perspective. In *Labor in the ancient world: A colloquium held at Hirschbach (Saxony), April 2005, International scholars conference on ancient near eastern economies* 5, ed. P. Steinkeller and M. Hudson, 1–35. Dresden: ISLET.

Tinney, S. 1999. On the curricular setting of Sumerian literature. *Iraq* 61: 159–172.

Veldhuis, Niek C. 2010. Guardians of tradition: Early dynastic lexical texts in old Babylonian copies. In *Your praise is sweet: A memorial volume for Jeremy black from students, colleagues and friends*, ed. H.D. Baker, E. Robson, and G. Zólyomi, 379–400. London: British Institute for the Study of Iraq.

Veldhuis, N.C. 2014a. The early dynastic Kiš tradition. In *He has opened Nisaba's house of learning: Studies in honor of Åke Waldemar Sjöberg on the occasion of his 89th birthday on August 1st 2013*, Cuneiform monographs 43, ed. L. Sassmannshausen, 241–259. Leiden: Brill.

———. 2014b. *History of the cuneiform lexical tradition. Guides to the Mesopotamian textual record 6*. Münster: Ugarit-Verlag.

Visicato, G. 1995. *The bureaucracy of Šuruppak: Administrative Centres, central offices, intermediate structures and hierarchies in the economic documentation of Fara. Abhandlungen zur Literatur Alt-Syrien-Palästinas 10*. Münster: Ugarit-Verlag.

Wagensonner, K. 2021. Early lexical lists revisited: Structures and classification as a mnemonic device. In *Language in the ancient near east: Proceedings of the 53e rencontre Assyriologique Internationale. Vol. I. Part 1, Babel und Bibel* 4/1, ed. L. Kogan, N. Koslova, S. Loesov, and S. Tishchenko, 285–310. Winona Lake: Eisenbrauns.

Watkins, C. 1990. What is philology? *Comparative Literature Studies* 27: 21–25.

Yoffee, Norman. 2013. Towards a biography of Kish: Notes on urbanism and comparison. In *Literature as politics, politics as literature: Essays on the ancient near east in honor of Peter Machinist*, ed. D.S. Vanderhooft and A. Winitzer, 527–544. Winona Lake: Eisenbrauns.

# Chapter 3
# A Theory of Philological Practice in Early Modern India

**Sheldon Pollock**

**Abstract** Premodern Indian philology in the sense of ecdotics and interpretation begins late in the scholarly tradition, at the end of the first millennium CE. Although the knowledge form was entirely unsystematized, a philological theory can be derived from commentarial practices. These are reviewed and synthesized across the principal genres, and the implicit theory of the text reconstructed.

## 3.1 Sanskrit Philology in Practice But Not in Theory

Although India was among the most densely textualized cultures of the premodern world, we know less about the concrete textual practices of that culture than perhaps any other, certainly for the pre-Islamic period. We know, for example, next to nothing about the slow transition from oral to manuscript culture beginning in the third century BCE (the transition was never discussed in the tradition) or about the interactions between oral culture, which persisted well into the early modern period, and literate culture (works produced by literate writers as late as the seventeenth century could be orally disseminated, sometimes virtually without variation).[1] We know little about the creation of written texts, whether and, if so, how, they were dictated

Note: The final draft of this essay was submitted in 2016, and (with one exception) could not take account of scholarship published after that date. The research leading to these results has received funding from the European Research Council under the European Union's Seventh Framework Programme (FP7/2007–2013)/ERC Grant Agreement No. 269804.

[1] See Pollock (2006).

S. Pollock (✉)
Arvind Raghunathan Professor Emeritus of Sanskrit and South Asian Studies, Columbia University. MESSAS, Columbia University, New York, NY, USA
e-mail: sp2356@columbia.edu



A. Keller, K. Chemla (eds.), *Shaping the Sciences of the Ancient and Medieval World*, Archimedes 69, https://doi.org/10.1007/978-3-031-49617-2_3

to scribes, for example, or inscribed by the author himself (and what difference this distinction might make). We know little about scribal culture itself, about the processes or social character of manuscript reproduction (whether by individuals or through what, in the case of northwest Buddhism, may have been professionalized scriptoriums), about the book market (which undoubtedly existed for secular books and, in a rather different way, for sacred works such as the Jain canon), about the practices of individual collectors or the organization or functioning of libraries (very few premodern manuscript catalogues have been preserved).[2] We have little idea how works once created were published, if that is even the appropriate word. All we do know is that publication, libraries, collectors, a market, established mechanisms of reproduction, scribal culture and all the rest did once exist.

This area of darkness extends to the very heart of the study of Indian textual culture, the discipline of philology. Although I am prepared to defend a conception of philology that embraces the widest possible number of its many senses—it should be understood, in the knowledge order of the contemporary university, as the discipline of making sense of texts and thereby includes everything from paleography, codicology, and textual criticism to the history, interpretation and comparison of all textualized language—I use the term on the present occasion to refer more narrowly to textual criticism: recension, emendation and other practices of text editing. Philology in that sense has neither a corresponding term in Sanskrit (or any other South Asian language) nor, consequently, was it ever the object of a corresponding form of discourse.

The absence of a cognitive category and disciplinary form of textual criticism is especially curious, given the development of other subdisciplines of philology and the general scope of systematization in Sanskrit intellectual history. In many ways, Sanskrit is the most philologized language in human history. The philological habit as such is fostered above all by a language's time–space distance, and Sanskrit—never a language of everyday life but instead (according to the dominant language ideology) the language of the gods—was maximally distant from the human world. Accordingly, grammar, phonology, metrics, lexicography and hermeneutics all attained astonishing refinement. We need think only of the rules devised by Pāṇini in the fifth or fourth century BCE that reduce to order the apparent chaos of phonological and morphological transformations, and contrast those with the quite random way phonology and morphology in Greek and Latin have been understood and taught for centuries. But philology in the sense of textual criticism not only arose late in the history of Sanskrit culture; it was never codified as a practice, notwithstanding the codification of all its subdisciplines. Indeed, in Sanskrit India, virtually every human practice that could be reduced to the descriptive–prescriptive rigors of *śāstra*, or work of systematic knowledge, was so reduced, but not philology.

[2]The fourteenth-century *Bṛhaṭṭipaṇikā* (*Great Annotation*) and the seventeenth-century *Kavīndrācāryasūcipatram* (*Index of Kavīndrācārya['s Library]*) are the only two known to me (for the former, for which I cannot locate a published version; see Tripathi (1975: 5)). On Buddhist text reproduction, see Schopen (2009), especially page 195.

It may not be surprising that people can have a conception of the parts of a thing without having a conception of the whole.[3] It may also be true that the textual practices of philology too thoroughly pervaded the Indian thought world for it even to be identified. At all events, the fact remains that we are left to gather the principles of Sanskrit philology from the raw evidence of the texts themselves; that is, from the practices of philological commentators. While my principal concern here is to give some coherent sense of those practices and to try to draw from them a more general theory, I also want to think about their emergence and consolidation. The lack of systematicity about those practices makes the first task difficult, and the relative lateness of their emergence presents something of an historical conundrum.

## 3.2 The Genre of Philological Commentary

What is striking about Sanskrit philology is not that it was embedded exclusively in commentary but that philological commentary arose so late, relatively speaking, in the history of Sanskrit textuality. Commentary as such is far older, of course, virtually coeval with the primary texts. But commentary of the sort we find in the early period is decidedly not philological. Thus, we have commentary on *śāstra*, especially the principal knowledge forms (*vidyāsthānas*) such as grammar, logic and hermeneutics, from the beginning of the Common Era, but here commentary is actually the form that the substantive conceptual development of the knowledge system took: one contributed to the system by writing commentary on the base text (the *sūtra*) or on one of its primary exegeses (the Sanskrit terms *vṛtti*, *bhāṣya*, *vārttika*, *ṭīkā* and so on all refer to different species of commentary or subcommentary). We occasionally meet with grammatical or semantic exegesis, but rarely, if ever, are questions of recension and emendation raised (techniques like *yogavibhāga*, or the splitting of *sūtras*, in the early grammatical tradition, are more interpretative than text-critical manoeuvers). By contrast, commentary on those textual forms where philology is central or shown to be central—namely epic literature, court poetry and (in a more hermeneutical sense of philology) scripture—is largely a phenomenon of the early second millennium and thus is centuries later (if not millennia later, in the case of Vedic scripture) than the texts to which such commentary is directed.

It stands to reason that philology in traditional India was focused in the first instance on works of literature (only secondarily on scripture). The tenth-century literary critic Bhaṭṭa Nāyaka introduced a typology, widely accepted by later thinkers, that sought to distinguish among three text genres: one where wording is predominant, a second where meaning is, and a third where wording and meaning are

[3] Recall Bruno Snell's old argument on the absence in archaic Greece of a conception of the human body as a totality, i.e., something other than 'a mere construct of independent parts variously put together' (Snell 1953: 6).

equal.[4] The first category comprises scripture, since the Veda's efficacy lies in *how* something is said, that is, in the particular sequence of its phonemes.[5] The second category is occupied by *śāstra*, or science broadly conceived, for here (at least for Indian thinkers, who, of course, knew nothing of what has come to be called the 'linguistic turn') what counts is *what* is said, not how it is said. The third category pertains to poetry, where wording and meaning share predominance. While this typology does nothing to explain why the philology of epic literature, court poetry and (to some extent) scripture should manifest a substantial presence first in the early centuries of the second millennium, it does help us understand why philology was directed toward expressive texts and why, therefore, the philology of scientific texts in traditional India should be as underdeveloped, relatively speaking, as it appears to me to be.

## 3.3 Scriptural Commentary, Buddhist and Other

Recent discussions of early Buddhist scriptural commentary and exegesis, which are a very early phenomenon, may be thought to be inconsistent with the picture I have just sketched of what constitutes the relevant pool of data for a history of Sanskrit philology.[6] Explaining why they are in fact consistent will help clarify some of the points I have just tried to make.

Buddhist commentators were essentially exegetes, not philologists. Vasubandhu's important *Vyākhyāyukti* (*Arguments for Exegesis*, ca. fifth century; extant only in Tibetan), which defends the Mahāyānasūtras (early centuries CE) against their older opponents, the Śrāvakas (Theravada, late centuries BCE), argues for the acceptance of these *sūtras* as *buddhavacana*, 'word of the Buddha', the key term for authenticity and authority. Vasubandhu has been taken to be making a general *philological* argument against the Śrāvakas, that even such texts as the *Mahāparinirvāṇa Sūtra* were recited differently among the different schools and even 'authorized editions' (those prepared by Mahākāśyapa at the first Buddhist council and other early disciples) were gradually affected by variations, signaled by divergent chapter divisions and the like and even different passages and chapters. And if 'authorized editions' had degenerated, it was unclear, argued Vasubhandu, what exactly constituted *buddhavacana*.[7]

[4] First cited in *Dhvanyāloka* of Ānandavardhana, with the *Locana* of Abhinavagupta, (Ānandavardhana 1940: 87)—Bhaṭṭa Nāyaka's work itself has vanished.

[5] *viśiṣṭānupūrvī*, to use the term of Kumārila, the seventh-century master of hermeneutics, author of the *Tantravārttika* (Kumārila 1970: Vol. 1, 155).

[6] The history of philological exegesis in Jain scriptural commentary complicates the picture I draw in what follows. Compare Pollock (2011: 426).

[7] On this text, see (Cabezón 1992), especially page 227. Translating *yang dag par bsdus pa'i gzhi bo* as 'authorized edition' adds, I am told, too much to the Tibetan, which seems only to be referring to texts 'crafted by *arhats* such as Mahākaśyapa on the basis of summaries' (Richard Nance,

This comment aside, however, Vasubandhu and subsequent Buddhist tradition did not consider *buddhavacana* to be an object of either general historical or particular philological knowledge, and did not assess it on the basis of the *sūtras*' language or textual form, let alone by invoking linguistic or textual archaism or particularity. Their basis for authenticity is instead the *doctrinal truth* of the *sūtras*. The Buddha was fully understood to have been a historical being, yet from the beginning of the tradition, his teaching was not restricted to his native tongue; on the contrary, there was a scriptural obligation to transmit it in any given local language (although one would think that precisely this obligation, and the concomitant prohibition against using Sanskrit, would have made the philological argument, in some sense of 'philology', even more compelling). For this reason, if for no other, *buddhavacana* was identifiable only by the quality of the truth it enunciated. Śāntideva's *Bodhiycaryāvatāra* (eighth century) with the (ca. ninth-century) commentary of Prajñākaramati implicitly rejects anything resembling a philological–historical method in its defense of Mahāyāna as *buddhavacana*.[8] A Mahāyāna text adduced by Prajñākaramati (the *Adhyāśayasaṃcodanasūtra*, not extant) moves far beyond philology to the outer edge of the hermeneutical with four criteria 'for recognizing any insight as being *buddhavacana*':

> It must be sensible and not nonsensical; it must be in accordance with reality and not at variance; it must remove the afflictions and not increase them; it must convey praise for the virtues of nirvana and not those of transmigration ... To whomever insight arises, Maitreya, or ever will arise according to these four criteria ... should be named a Buddha; he should be named a teacher and his teaching learned as dharma. Why? Because whatever is properly spoken, Maitreya, is all the Word of the Buddha. Any hateful person who maligns these insights, Maitreya, saying they were not spoken by the Buddha, or shows them disrespect, maligns all the insights spoken by the Buddha.[9]

---

personal communication.) Skilling (2000) has observed that the *Vyākhyāyukti* is concerned not just with philosophical interpretation: *sūtras* are explained according to 'the summarized meaning' (*sapiṇḍārtham*), 'the sense of the words' (*padārtha*, which can be polysemic, etymologically derived, and so on), or contextual 'sequence or connection' (*sānusaṃdhikaḥ*) (2000: 319).

[8] See *Bodhicaryāvatāra* 9.42–44 and commentary *ad loc.*, (Śāntideva 1960: 204–206).

[9] *pratibhānam arthopasaṃhitaṃ bhavati nānarthopasaṃhitam. dharmopasaṃhitam bhavati nādharmopasaṃhitam. kleśaprahāyakaṃ bhavati na kleśavivardhakam. nirvāṇaguṇānuśaṃsasaṃdarśakaṃ bhavati na saṃsāraguṇānuśaṃsasaṃdarśakaṃ ... yasya kasyacin maitreya etaiś caturbhiḥ pratibhāti pratibhāsyati vā tatra ... buddhasaṃjñotpādayitavyā. śāstṛsaṃjñāṃ kṛtvā sa dharmaḥ śrotavyaḥ. tat kasya hetoḥ. yat kiṃcin maitreya subhāṣitaṃ sarvaṃ tad buddhabhāṣitam. tatra maitreya ya imāni pratibhānāni pratikṣipet naitāni buddhabhāṣitānīti teṣu cāgauravam utpādayet pudgalavidveṣeṇa tena sarvaṃ buddhabhāṣitaṃ pratibhānaṃ pratikṣiptaṃ bhavati* (Prajñākaramati's commentary on Śāntideva's *Bodhiycaryāvatāra*, (Śāntideva. 1960: 205, lines 9–15). These four criteria respond to those of the Śrāvakas, who offer them in response to the charge that their own canon is beset with precisely the same defects of authenticity, contradiction, and the like for which they censure the Mahāyāna: 'Something that has been transmitted from teacher to pupil as the Word of the Buddha; that penetrates into the sense of a *sūtra* text (*sūtre avatarati*), is reflected in the Vinaya, and does not stand at cross purposes (*vilomayati*) with reality (*dharmatā*), must be considered *buddhavacana*, and nothing else' (Śāntideva 1960: 205, lines 1–3).

In such circumstances, where the authenticity (and hence authority) of a text is a function of its (*a priori*) truth—and not its truth a function of its (demonstrated) authenticity—philology has no role to play in the establishment or purification of a canon. It is therefore unsurprising that no philological apparatus was produced by Indian Buddhists until relatively late (with respect to Sanskrit grammar, for example, the mid-fifth century *Candravyākaraṇa*; with respect to Pali grammar, the twelfth-century *Saddanīti*).

It might accordingly seem sensible to contrast the Buddhists' 'content model' with the 'linguistic model' of the Vedas and to register, as a consequence, the importance of comprehension in the one case and accuracy in the other.[10] But the distinction, surprisingly, carries no consequential differences for a history of philology as edition and emendation. While the peculiar character of Vedic language was acknowledged from the beginning of systematic philological reflection (being marked in Pāṇini's grammar as *chandas*, 'The Metrical', i.e., the Veda, in contrast to *bhāṣā*, 'the spoken')[11] and was identified centuries later by Mīmāṃsā as key diagnostic of canonicity,[12] the unparalleled commitment—or ideology of commitment—to textual stability in the transmission of the Vedas completely excluded philological engagement except indeed at the level of content. Aside from the Vedic hermeneuts discussed below, none of the actual commentators on these texts (though they are vast and I cannot pretend to have examined them all) ever raises a question of textual variation.

The afterlife of these tendencies with regard to scripture across Hindu religious communities merits brief comment. Later apologists for the new scriptures (*āgama*) of communities devoted to the worship of Vishnu and Shiva that we begin to find from the early medieval period similarly held their texts to be in essential harmony with the Veda—indeed, in the eyes of some, as cognate with the Veda—and they were not in the least disturbed by, or even cognizant of, philological criteria. A good example of this indifference is offered by the defenders of the authority of the Pāñcarātra (Vaishnava) texts that began to be produced some time in the middle of the first millennium. While the tenth-century theologian Yāmunācārya argues that these texts were 'composed by the Supreme Being himself'[13] just like the Vedas (which he regards as created by God, unlike Mīmāṃsā, which holds them to be authorless), no attempt is made to justify this claim on the basis of philology. Notwithstanding what we might call a practical Vedicization in the creation of other Vaishnava texts of the period—the tenth-century *Bhāgavatapurāṇa*, for example, is remarkable for its archaizing language—Yāmunācārya offers no defense on the basis of language, nor accounts for the fact that while God created both the Vedas

---

[10] Davidson (1990: 296–297).

[11] 'Spoken', that is, not in everyday life but in educational and comparable contexts.

[12] 'That the Veda is an autonomous source of true knowledge is vouchsafed by its very form' (*tena vedasvatantratvaṃ rūpād evāvagamyate*, Kumārila, *Tantravārttika*, (Kumārila 1970: 166, line 2)).

[13] *paramapuruṣaviracita*, Yāmunācārya, *Āgamaprāmāṇya*, (Yāmunācārya 1976: 2); they are 'based on the veridical knowledge-experience' of God (*avitathasahajasarvasākṣātkāra* (Yāmunācārya 1976: 84)).

and the Pāñcarātra scriptures equally,[14] their linguistic characters are so radically divergent. Instead, like the Buddhists, he is concerned only with Pāñcarātra truth, the harmony of that truth with that of the Vedas, the refutation of apparently contrary Vedic or quasi-Vedic texts that deny that truth, and the exclusion of other sectarian scriptures from the realm of Vedic truth.[15] It is not until the seventeenth century that a kind of text-critical (if not language-focused) philology begins to challenge theology in the adjudication of scriptural claims, a point to which I return at the end of this essay.

Philology as recension and emendation in the world of Sanskrit, then, is essentially a set of practices found in commentaries on epic and court literature from the beginning of the second millennium, but for which we have no evidence of their ever having been systematized and theorized by the commentators themselves. It is on those practices that I shall concentrate in the remainder of this essay.

## 3.4 Sanskrit Textual Criticism

There exists no scholarly account of the origins of textual criticism in Sanskrit intellectual history; indeed, the very idea that textual criticism might have an origin seems to be unknown. If the contrast between philological and philosophical commentary is rarely drawn, at least we are finally beginning to get strong accounts of the long historical development of the latter[16]; for the former, scholarship is still in its infancy. As already noted, philological commentary, both for epic and secular poetry, is an invention of the late first millennium CE (almost certainly in Kashmir), and in the following centuries, we can clearly observe an intensification of textual criticism.

Text-critical terms like 'variant reading' (*pāṭha*) and 'interpolation' (*prakṣipta*, later *kṣepaka*) that come into common use in the second millennium are hardly to be found before this period. One of the earliest discussions I have located is in

[14] See, for example, Yāmunācārya's *Āgamaprāmāṇya* (Yāmunācārya 1976: 37, 41, 50).

[15] The Mīmāṃsā critic in Yāmunācārya's *Āgamaprāmāṇya* (Yāmunācārya 1976: 7) does represent Pāñcarātra scriptures (*āgama*) as *smṛti*—the gist of Vedic texts 'preserved in memory', not 'heard word for word' (*śruti*)—which would, of course, render the question of linguistic character less probative. But Yāmunācārya is not clear about this, and while he sometimes suggests that like the Vedic *smṛtis*, the Pāñcarātra *smṛtis* derive their validity from Vedic texts no longer extant, he elsewhere asserts that they were created by God and depend for their validity on God having himself 'perceived' dharma (see e.g., Yāmunācārya 1976: 91); they are not 'memories' of texts that have since disappeared, as is argued by Mīmāṃsā for the validity of the Vedic *smṛtis* (the Pāñcarātra *smṛtis* are said to constitute a 'summary' of the Vedas (*tadarthaṃ saṃkṣipya*) for devotees less competent in studying and retaining the vast Vedas themselves; (Yāmunācārya 1976: 102)). Ongoing work by Guy St. Amant on Kṣemarāja's eleventh-century commentary on a Śaiva Tantric text (especially his invented category of *aiśa*, 'God's idiolect'), will provide nuance to my reflections here.

[16] A careful study of the philosophical commentary is offered in Preisendanz (2008).

neither epic nor court literature commentary but in Vedic hermeneutics. The seventh-century scholar Kumārila, when discussing a purity rule in the law books (*smṛti*), observes that a particular word in the rule 'is not found to be used in the oldest *pāṭha*' of the text in question,' though no further detail is offered.[17] It is only an eleventh-century commentator who raises questions more pertinent to our concern here: 'It may be that someone interpolated the word and transmitted that *pāṭha*,' he argues. 'It certainly cannot be the *pāṭha* of the sages since it is not that of the bulk of learned scholars.'[18] This commentator's criteria for textual authenticity thus include, as we would expect, antiquity, but antiquity itself is determined by numbers and quality: the witnesses have to be credible and numerous.

What importantly complicates the question of textual variation in the case of the Veda is the concomitant one of orality. I intentionally left untranslated the word *pāṭha* in the above discussion, for whereas it later comes to be used to mean a 'variant reading' of a literate manuscript tradition, its sense for Vedic culture is different. There, literate procedures were decidedly denigrated for scriptural transmission; as Kumārila puts it, 'Something ceases to be regarded as revealed moral law (*dharma*) if it derives from the Veda that has been learned in ways contrary to reason, as, for example, through a written text'. The way that is not contrary to reason is the precise 'after-pronouncing [i.e., reproduction] of what has been pronounced by the mouth of the guru'.[19] And hence *pāṭh/pāṭha* must connote 'recite/recitation' as well, whereby Kumārila's usage in the passage cited earlier (*pāṭhe cirantane*) comes to mean 'original recitation'. Nor is there any reason to suppose that variants cannot be interpolated—or that Indians in the classical period did not believe that variants could not be interpolated—into oral traditions: here, numbers would seem to count.[20]

Such early rarities set aside, much of our best data regarding practices of recension, or the examination of manuscripts in order to determine the earliest available state of the text, come from commentators on the epics, who often—and for the

[17] The word is *lipta* (smeared): *na liptagrahaṇaṃ tatra pāṭhe 'sti tu cirantane* (*Tantravārttika* 1.3.3, (Kumārila 1970: 182)). Kumārila also uses the term *saṃyakpāṭha* (Kumārila 1970: 551), which (like *prakṣip-*) appears nowhere in his (fifth-century) predecessor Śabara, who does, however, know *pramādapāṭha* in the sense of 'erroneous transmission', e.g., of a whole species of text (the *arthavādas*) inserted into the Veda (*Bhāṣya* on *Pūrvamīmāṃsāsūtra* 1.2.8), (Kumārila 1970: 122); see also (Kumārila 1970: 183, 550, 1139).

[18] '*yady api kena cid ... liptapadaṃ prakṣipya paṭhyate tathāpy abhiyuktabahujanāpaṭhitatvān nāsau ārṣaḥ pāṭhaḥ*' (Someśvara Bhaṭṭa, *Nyāyasudhā*, (Someśvara 1909: 150)).

[19] Kumārila, *Tantravārttika*,1.3.3.43: '*yathaivānyāyavijñātād vedāl lekhyādipūrvakāt ... dharmajñānaṃ na saṃmatam*', on 1.3.7, (Kumārila 1970:123); the phrase *gurumukhoccāraṇānūccāraṇa-* is found frequently in early modern authors (e.g., *Śāstrasiddhāntaleśasaṃgraha*, (Appayya 1935: 53)). This denigration did not, of course, apply to the creation and criticism of *scholarly* texts. Kumārila often refers to Śabara's scribal mistakes, *pramādalikhita-*, etc.

[20] What is unclear for the passage under discussion is to what degree the lawbooks themselves, *smṛtis*, were transmitted orally at any time, let alone at the end of the first millennium, when Someśvara was writing.

most part confessedly—functioned as editors.[21] (I should note at once that there exists no term for 'editor', 'edition', or the like in any Indian language, outside of Persianate culture, before the modern period.)[22] Commentators on court poetry, such as the earliest among them from tenth-century Kashmir, including Vallabhadeva, seem often (though not always) to have been editors as well. While fully aware of variation in manuscript copies, such commentators nowhere describe collation of manuscripts, though something like that may be implied by the evidence, direct or indirect and usually vague, that they compared them.[23]

Epic commentators, on the other hand, often provide clearer testimony on the matter, especially with respect to the culturally foundational *Mahābhārata*, a work that was repeatedly edited and somehow—by a process as yet entirely unclear to us—published in the period 1000–1700. Thus, Nīlakaṇṭha Caturdhara, a Maharashtrian Brahman who worked in north India in the last quarter of the seventeenth century, gathered 'many manuscripts from different regions' and 'critically established the best readings'. He occasionally evaluates manuscripts according to age ('old,' 'recent,' 'damaged', 'good') and readings by their frequency ('common', 'occasional', 'rare'); frequently discusses variants of individual readings or in the sequence of verses, and even once admits failure ('Only Vyāsa himself [the putative author] knows the true reading here'), though it is only rarely he tells us where, let alone why, he edited the text the way he did.[24] Nīlakaṇṭha's explicit acknowledgment of the transregional dissemination of manuscripts, and his tacit recognition that these are all versions of the same text and must be compared with each other to attain the singular textual truth (which indubitably exists for him), are important markers of a theory of textuality in general as well as of the understanding of this particular text's mode of being. And far from being peculiar to Nīlakaṇṭha, these beliefs were shared by every editor who bothered to explain, or at least allude to, his editorial procedures.[25]

It seems very likely that this kind of transregional consciousness in editing was a phenomenon of the early second millennium, following a long period of the regionalization of recensions—no doubt affected, to some degree, by the regionalization

[21] The following section draws substantially on Pollock (2015).

[22] In north-Indian languages, *saṃpādaka, saṃpādana* ('put together') are recent neologisms; (*saṃ*)*śodhana* ('purification') is older but was never associated with a specific edition or editor.

[23] Dakṣiṇāvartanātha, a twelfth-century south Indian commentator on the court epic *Raghuvaṃśa* of Kālidāsa, tells us that he 'prepared his commentary after examining variants in manuscripts from various regions, adopting the right readings and rejecting the others'. See Unni (1987: 42).

[24] See Nīlakaṇṭha's commentary on *Mahābhārata Ādiparvan* v. 6; on *Harivaṃśa* 1.37.30 ('true reading', *pāṭhatattvam*); cited in Bhattacharya (1990: 220 n). (See also page 224 n. on transpositions). For a general account of Nīlakaṇṭha, see Minkowski (2005).

[25] The edition of Vidyāsāgara (eighteenth-century Bangladesh), his transregional collection of manuscripts, and his use of at least a dozen earlier commentaries (including Devabodha's by then eight-hundred-year-old *Jñānadīpikā*), are discussed in Pollock (2015). Note that not only were older commentators systematically studied (Nīlakaṇṭha follows 'the explanations of early teachers' (*prācāṃ gurūṇām anusṛtya vācām*), v. 6 of his introduction), but the very chronology of their succession was preserved in memory and understood to represent a meaningful order.

of writing systems that we first begin to notice with the rise of vernacular or regional literacy in the last quarter of the first millennium—that is observable across the history of Sanskrit literature. Indeed, epic commentary itself is a phenomenon of that period, finding its origins around the beginning of the second millennium—there is no evidence that the earliest (Devabodha) had a predecessor—and experiencing a dramatic upsurge after about the mid-thirteenth century. It is then that commentary on the *Vālmiki Rāmāyaṇa*, the second great Sanskrit epic, commences, with a south Indian commentator arguing for the need to establish the 'correct reading (*samyakpāṭha*) corrupted by scribes unskilled in the various scripts,' by 'examining multiple manuscripts from multiple regions'.[26]

Indian scholars were fully aware that the textual condition required clear principles that needed to be followed in text editing, but again, given the lack of programmatic statements, we can discover them only by sifting the texts of our commentators.[27] Vallabhadeva, to return to our tenth-century literary scholar, wrote basically word-for-word commentaries (*pañc[j]ikā*s), which required him to address very closely the textual state of a work, and he left us several dozen text-critical discussions on the various manuscripts that he compared.[28] These show that his criteria were multifarious: readings (or passages) could be judged as grammatically or contextually 'correct'/'reasonable'/'proper'/'right' or 'more correct'/'more reasonable'; 'authoritative', 'false', 'mistaken', 'corrupt', 'unmetrical' and 'ancient'; 'interpolated', 'in need of emendation', or 'obscene' and last but not least, 'lovely', 'beautiful' and 'more beautiful.'[29] At least once, he adduces paleographical or at least graphical criteria, as when he notes that a variant is a result of confusing two similar characters (*lipisārūpyamohāt*) and he rejects it on the grounds that it would seriously contradict the narrative.[30] Vallabhadeva often evaluates a reading on the familiar principle of difficulty and the antiquity such difficulty implies: 'This must be the ancient reading precisely because it is unfamiliar'. He sometimes combines principles of antiquity and aestheticism when asserting that 'the old reading in this verse is more beautiful'. But antiquity can be too ancient, as it were, if it produces a grammatical (or lexical or metrical or rhetorical) irregularity such as a Vedicism.

[26] On Uḍāḷi Varadarāja, see Raghavan (1941–1942).

[27] For an initial survey, see Colas (1999).

[28] On Vallabhadeva, see Goodall and Isaacson (2003). Vallabhadeva's terminology alone indicates comparison of exemplars, but he elsewhere also implies something like *recensio* when noting that a given verse is 'only infrequently transmitted' in manuscripts (*viralo 'sya ślokasya pāṭhaḥ*, on *Raghuvaṃśa* ad 18:17; cited by Goodall and Isaacson (2003: xxxi)).

[29] The Sanskrit terms are, respectively, *sādhu/yukta/samīcīna/samyak, sādhiyān/yuktatara, prāmānika, ayukta* or *apapāṭha, prāmādika, duṣṭa, asaṃbaddha, ārṣa/prācīna/jarat* (all modifying *pāṭha*), *prakṣipta śloka, śodhana, asabhya* (once, on *Kumārasaṃbhava* 3.41), and *sundara/ramya/ramyatara pāṭha*. See also (on *Meghadūta* 72) *anārya*, 'meritless,' or 'inferior' (from the point of view of grammatical correctness). Compare also Colas (1999: 35–36). The sources of such readings are rarely indicated, and then only vaguely ('an old manuscript', 'an eastern manuscript', and the like).

[30] *Meghadūta* 2 (*atīva viruddham*, a position that Mallinātha demolishes, while claiming that Vallabhadeva's reading is a conjecture (*kalpayanti*)).

Here and elsewhere, like other commentators, Vallabhadeva was on occasion prepared to suggest a revision in order to save his author from a supposed solecism, but he typically hesitates to actually alter the text and winds up transmitting the offending reading.[31]

In the matter of emendation—or, in this context, perhaps better 'correction', since the restoration of the original text was not invariably the goal—a tension manifests itself that will mark the whole long history of Sanskrit philology. On the one hand, as manuscripts show, some scribes and editors were highly attuned to text-critical problems and fully prepared to 'improve' the text, whether to remove a grammatical deviation or to correct a supposed aesthetic or logical fault.[32] On the other hand, some scholars explicitly rejected doing so, such as Mallinātha, who took care to assure readers that he was not transmitting anything not found in his manuscripts.[33] And generally, it seems, editor-commentators did seek to establish as coherent and authoritative a text as they could on the basis of received manuscript tradition (*āgata*) rather than conjecture (*kalpita*) (though the later proverb, 'We must explain that text as we find it', is not attested before the eighteenth century).[34] Yet even Mallinātha sometimes adopted a conjecture that his predecessors had only suggested while they themselves preserved the received text.[35]

The work of the literary theorist Mahima Bhaṭṭa (ca. 1000, Kashmir) is instructive here and for his general approach to textual criticism. In his treatise on poetics (and this is common in such works), the phrase 'correct reading' (*yuktaḥ pāṭhaḥ*) often connotes, not what a study of the manuscripts indicates to be correct, but what in the view of the critic the text *should be* if the passage is to avoid some putative fault and conform to aesthetic norms. Mahima Bhaṭṭa will often assert that a given

[31] Commentary on *Kumārasaṃbhava* 3.44, and compare 3.28; see also Goodall (2001). For Vallabhadeva's first principle, see *Kumārasaṃbhava* 1.46, *aprasiddhatvād ārṣaḥ pāṭhaḥ*, the Sanskrit version of the familiar maxim *lectio difficilior melior/potior est*; for the second, 2.26, cf. 2.37, *jaratpāṭho 'tra ramyataraḥ*.

[32] Here 'Die sprichwörtliche Aversion zwischen Dichtern und Philologen' (König 2013: 15) is apposite. For a perspective on this question of one twelfth-century Kashmir poet, see Pollock (2003: 112).

[33] 'I transmit nothing that is not found in the original' (*nāmūlaṃ likhyate kiṃcit*; some take this as a reference to the 'sources' of his exegesis), a statement repeated in the introductions to his commentaries on all the major *kāvya*s.

[34] *sthitasya* (or *sthiter*) *gatiś cint[anī]yā*. See Gerschheimer (2010) and Pollock (2011). Evidence of a more manipulative approach to texts can complicate this picture. Already in the seventh century, Kumārila could suggest that the author of the *Mahābhāṣya* himself changed the wording of a Vedic text on phonetics, turning the phrase 'corrupted *mantra*' into 'corrupted *language*', and thereby sought to enhance the importance of the study of grammar (*Tantravārttika* 1.3.24 v. 780; (Kumārila 1970: 268); that the charge is made only in a *pūrvapakṣa* does not alter the main point that textual manipulation was a conceptual possibility historically available to Kumārila). See also Arjunvarmadeva on the *Amaruśataka* (Amaru. 1916: 42) (he notes that 'others have introduced' a variant to remove a redundancy).

[35] Compare Mallinātha and Vallabhadeva on *Meghadūta* (v. 2.39 and v. 99 respectively).

verse of some great poet 'should more properly read' such and such.[36] He does, however, occasionally note that the normative reading is actually found in some manuscripts,[37] and at times he directly addresses an acknowledged text-critical problem. But if visions of normativity did not prompt Mahima Bhaṭṭa (or any other poetician I know of) to actually emend a text, this was not always the case, as we have just seen, with commentator-philologists. Here, for example, is Mahima Bhaṭṭa's discussion of a line in Kālidāsa's poem *Meghadūta*, which, while dealing with trivial syntactical problems, (about which Mahima was nonetheless much exercised), illuminates larger issues:

> Commentators, too, with their false air of learning, are often found to bring shame not only on themselves but on celebrated poets as well by their erroneous comments. For example, in the *Meghadūta* line *jātā manye śiśiramathitā padminīvānyarūpā* [v. 80 = 2. 16], the commentators are completely ignorant of the reading that offers the nominative phrase in conjunction with the verb 'to think' ('I think the girl must be ...') and, failing to understand the poet's true intention,[38] they reject the profound beauty of his own thought—and this, too, despite the parallel to the nominative phrase in conjunction with the verb 'to think' that is found elsewhere in Kālidāsa.[39] They thereby turn the nominative 'the girl ...' into the direct object, simply inventing the word *vā* in their confusion when it is actually the word *iva* transformed by vowel *sandhi*, and so proceed with their commentary. This adds no beauty whatever to the meaning. Nor is it possible to find this construction in any work of the great poet Kālidāsa, so as to convince us that he would use the word *vā* in the sense of *iva*—a veritable canker upon this poem, which is, in fact, a treasure-trove of aesthetic emotion.[40]

Notice the three criteria of text-critical judgment Mahima Bhaṭṭa uses: grammaticality, usage and beauty. The syntactic construction that he believes to be original is perfectly in keeping with Sanskrit grammar (and the putatively false reading can be explained by a failure to understand the euphonic combination of the original); it is

[36] Mahima's *Vyaktiviveka* (Mahima 1983: 234–235) (relating to poems of Bhavabhūti and Kālidāsa). The *yuktaḥ pāṭhaḥ* argument is made some three dozen times in Chap. 2 of the work (on literary 'faults'), and is also found in other later treatises on the subject, such as *Kāvyaprakāśa* 7.

[37] E.g., (Mahima 1983: 268): *kvāpy ayam api pāṭho dṛśyate.*

[38] I read with hesitation *–vākalitakavihevākāḥ* (for *–va kalitakavihevākāḥ*); for this sense of *hevāka* see Abhinavagupta's *Abhinavabhāratī* (Abhinavagupta 1992: volume 1, 35).

[39] He cites *Raghuvaṃśa* 1.78: *avaimi tadavajñānād yatnāpekṣo manorathaḥ*. Here again, in his own commentary, Vallabhadeva reads the accusative, *avaimi tadapadhyānād yatnāpekṣaṃ manoratham* (so, too, the southern tradition as represented by Aruṇagirinātha: *īpsitaṃ tadavadhyānād* [*tadavajñānād*, Mallinātha] *viddhi sārgalam ātmanaḥ*, 1. 76/79). But a mid-twelfth-century scholar from Kashmir quotes the line as *avaimi tadapadhyānād yatnāpekṣo mahodayaḥ*, precisely in the course of a discussion of the nominative construction with verbs of knowing, hearing, etc. (Someśvara Bhaṭṭa on the *Kāvyaprakāśa*, (Someśvara 1909: 141)).

[40] *vyākyātāro 'py alīkavidvanmānitayā prāyeṇāpavyākhyānair na kevalam ātmānaṃ yāvat tatrabhavato mahākavīn api hrepayanto dṛśyante. tad yathā ... ity atra pāṭham imam abuddhvaivākalitakavihevākāḥ parākṛtapratīticārutātiśayās te. avaimi ... ity ādau dṛṣṭām api vākyārthakarmatāṃ manyater apaśyanto bālāyāḥ karmatām asya manyamānāḥ svarasandhivaśād vikṛtam ivaśabdam eva bhramād vāśabdaṃ parikalpyāpavyākhyām ārabhante. na caivam arthasya vaicitrī kācit samunmiṣati. nāpi mahākaveḥ kālidāsasyānvayagatir iyaṃ kvacanāpi prabandhe 'vadhāritapūrvā yad ayam rasavidhāne kāvye vyādhim iva vāśabdam ivārthe prayuñjīteti.* (Mahima 1983: 485).

found elsewhere in the poet's work and the false reading nowhere; while the latter is, for this critic, inherently ugly. What Mahima Bhaṭṭa does not do here, however, is ground his argument by reference to actual manuscripts—yet he is not simply inventing an opponent. The reading Mahima Bhaṭṭa criticizes is, in fact, that of Vallabhadeva, perhaps a generation or two his senior and, if the latter, a typically conservative editor, did not emend it himself, some other commentator may well have done so.[41] In other instances, however, Mahima Bhaṭṭa seems rather to be inventing than discovering text-critical problems.[42]

The commitment to fidelity toward the received text that we see here, and that was clearly widespread, is corroborated in the treatment of interpolation. As a rule, commentators continue to transmit materials they consider to be interpolated (as they continue to transmit clearly corrupt readings) and do not expunge them. Indeed, they and scribes in general sometimes went out of their way to ensure that material they *knew* to be interpolated was preserved in their transmission. The quest for what is thus the maximally inclusive edition, as is evidenced in manuscript culture, persisted into the early print era.[43]

The question of interpolation seems to be one of the few where commentators on scientific treatises exercise text-critical judgment, for here it is precisely *what* is said rather than *how* it is said that can engage them philologically. One of the more telling cases is found in language analysis (*vyākaraṇa*). Beginning in the thirteenth century, with Haradatta Miśra in his commentary *Padamañjarī* (*Bouquet of Words*) on the *Kāśikāvṛtti* (*Benares Gloss*), a grammatical treatise of the early ninth century, we find a new attention to the integrity of the *sūtra* text of Pāṇini (fifth to fourth century BCE), the foundational work for the science. Haradatta often notes that the authors of the *Kāśikāvṛtti* inserted a given term into the *sūtras*, usually from another grammatical source, the later *Vārttika* (*Exegesis*) of Kātyāyana (third century BCE).

---

[41] *jātāṃ manye śiśiramathitāṃ padminīṃ vānyarūpām*. That Vallabha's text is unlikely to be original is suggested by the commentary of Dakṣiṇāvartanātha, who has the same reading as Mahima Bhaṭṭa and cites a parallel from the *Rāmāyaṇa* (arguing as he does elsewhere that Kālidāsa sought to recreate the *Rāmāyaṇa* narrative in the *Meghadūta*: *śrīrāmāyaṇavacanānusāreṇa kaveḥ pūrvokto rāmakathābhilāṣaḥ spaṣṭaḥ*; (Mahima 1983: 52)) that goes to vindicate the nominative (*Rāmāyaṇa* 5.14.30, *himahatanalinīva naṣṭaśobhā*). Note also that the nominative construction is the reading of Jinasena's adaptation in his *Pārśvanāthābhyudaya*, which dates to the mid-ninth century. S. K. De imprudently follows Mallinātha in his critical edition. Vallabhadeva's usual conservatism aside, he clearly inherited interpolated texts, as in the case of *Śiśupālavadha*; see Bronner and McCrea (2012).

[42] See, for example, *Vyaktiviveka*, (Mahima 1983: 485), where Mahima Bhaṭṭa asserts as original a reading for which there is no textual evidence.

[43] Bronner and McCrea (2012: 442–444), though, as they show, Mallinātha himself silently suppressed a famous passage in the *Śiśupālavadha* that he considered to be (and that is) an interpolation. The transmission of acknowledged interpolation is very frequent among *Rāmāyaṇa* commentators (see my notes on 2.89.19; 3.45.27, and 47.30 in Pollock (1986)); a well-known example from *dharmaśāstra* is Medhātithi on *Manu* 9.93 (he expresses doubt about the authenticity of a verse and yet transmits it anyway; see Lariviere (1989: 5)). On Arjunavarmadeva's identification and preservation of interpolations, see *Amaruśataka*, (Amaru 1916: 46–48; 54). A similar conservatism can be noticed among Alexandrian scholars.

Thus we frequently encounter statements such as 'Since this is found in the *Vārttika*, we can conclude that it has been interpolated into the *sūtra*'; 'this was interpolated into the *sūtras* by the author of the *Kāśikāvṛtti*' and 'present-day scholars have interpolated this into the *sūtra*'.[44] All the examples of this sort of editorial attention I have been able to locate date to the late medieval or early modern period.[45] The remarkably innovative seventeenth-century grammarian Bhaṭṭojī Dīkṣita offers a good number of text-critical remarks about the *sūtra* text of Pāṇini, Patañjali's *Mahābhāṣya* (*Great Commentary*) on it and other works, as, for example, when he notes with respect to a given reading, that 'according to older authorities it must have been interpolated by some scholar or other, since it is not available in the *Mahābhāṣya* and related texts. But it is, in fact, to be found in present-day manuscripts of both the *Mahābhāṣya* and the *Vārttika*.'[46] Such observations become commonplace by the eighteenth century, as in the work of the Thanjavur scholar Vāsudeva Dīkṣita.[47] It is in early modern grammatical literature, too, that scholars first begin to use the term *sāṃpradāyika* (*saṃpradāya* is attested in epic commentators in the sense of recension, here in the sense of a group of manuscripts related by script and region) in a text-critical context as the counterpositive to 'interpolated', and thus signifying 'traditional' or 'original'.[48] The criterion of *sāṃpradāyika* is also found in the literary realm, but there it does not have the probative force it has in the domain of text transmission; on the contrary, it suggests (to borrow a phrase from the Islamic tradition of *ijtihad*), a closing of the doors of interpretation.[49]

[44] *vārttike 'darśanāt sūtre prakṣiptam*, ad Pāṇini 4.2.2., 4.3.134, 4.4.17, 5.1.36, 5.2.10, 8.3.16, 8.3.116; *vṛttikṛtā tu sūtreṣu prakṣiptam*, ad 3.1.118, 3.3.122, 4.1.14, 4.1.167, 4.2.43, 5.2.102; *sa idānīntanaiḥ prakṣiptaḥ*, ad 1.2.65, 4.1.63. Several examples are discussed in Birwe (1958), who did not, however, comment on the frequency or innovative quality of Haraddata's observations.

[45] The sole exception known to me is Kaiyaṭa ad *Mahābhāṣya* 4.1.166.

[46] *atrohaśabdaḥ kaiścit prakṣipto bhāṣyādau tu na dṛśyata iti prāñcaḥ. Idānīntanapustakeṣu tu bhāṣyavārttikayor ūhaśabdo dṛśyata eva* (*Prauḍhamanoramā* ad Pāṇini 6.1.89).

[47] See *Bālamanoramā* ad Pāṇini 3.2.78, 4.1.54, etc. Jinendrabuddhi's *Tattvabodhinī*, another commentary on the *Siddhāntakaumudī*, is also much concerned with identifying interpolation.

[48] Thus *prakṣipto na tu sāṃpradāyika* in *Prauḍamanoramā* ad Pāṇini 4.1.176; *Tattvabodhinī* ad 7.3.19; *Bālamanoramā* (*asāṃpradāyika* ad 1.1.37). The term itself is found as early as *Pūrvamīmāṃsāsūtra* 1.2.8 (*tulyaṃ ca sāṃpradāyikam*), but there it refers to *extrinsic* features of Vedic text genres: calendrical restrictions on studying, oral transmission, the teacher–student relationship itself, and the like.

[49] The eighteenth-century scholar Harihara comments on a verse of the eighth-century dramatist Bhavabhūti: 'Kashmiri scholars conjecture a reading here and offer a completely non-traditional interpretation' (*kāśmīrāṇāṃ pāṭhāntaraṃ kalpayitvā vyākhyānam asāṃpradāyikam* (Bhavabhūti. 1999: 436); the *varia lectio* is nowhere unrecorded and would, in fact, produce an infrequent variation on the *upajāti* meter).

## 3.5 The Historical Moment

The data assembled above present two challenges to our understanding of Sanskrit philology: how to explain the moment of its historical emergence and how to understand the theory of textuality that underlies it. I deal with these in turn.

While Indian scholars clearly had some sense of textual variability prior to the early second millennium—they knew that readings could be corrupted, that interpolations could be inserted and that transmissions could be disrupted—it is only around 1000, rather dramatically and with increasing intensity thereafter, that text criticism became a self-conscious scholarly practice—something that cannot easily be explained away as a mere artifact of the survival of sources beginning in that period. This originary moment of text criticism may now be taken as a fact, but it is one we are far from being able to explain.

There was no sudden transformation in the material or intellectual or institutional context leading to the emergence of philological commentary; no new technical development or conceptual discovery or state-led innovation that lit the text-critical fuse. The introduction of paper early in the second millennium, for example, by no means displaced palm leaf and birch bark in book production. Thus its consequences in India were hardly comparable to the European and Islamicate experience, where, in providing a cheap alternative to parchment, paper vastly expanded communication practices.[50] The *khagaz raj*, or paper kingdom, of the Mughals marks a shift, but that is a seventeenth-century phenomenon. Nor can text criticism have been a response to some unprecedented estrangement from Sanskrit, for there had never occurred a moment of true rupture in Sanskrit cultural history. The arrival of new ruling groups from Central Asia beginning around 1000 certainly produced nothing of the sort, and in any case, the earliest stages of the philological revolution came several generations earlier.

The histories of other philological traditions suggests a range of causal factors. Philology developed in late imperial China—as 'evidential research studies' (*kaozheng xue*)—when scholars concluded that it was because of their failure to read the classics properly that the empire had been lost to barbarians (the Manchus) and they devised new philological principles to solve old problems of unintelligibility. Indians had no empire to lose except an imaginary one and if some felt even this was lost (to Mahmud of Ghazni in the eleventh century, the Delhi Sultanate in the fourteenth, or the Mughal Empire in the sixteenth), the event was rarely seen as a cataclysm and was never ascribed to some intellectual failure. The Arab scholars who invented the 'art of philology' (*ṣināʿat al-adab*) or 'the literary arts' (*al-ʿulūm al-adabiyya*) in early Abbasid lands were confronting the presence of an immensely important new text, the Qur'an, whose language they had, at all costs, to understand. New kinds of revelation did come to India in the course of the first millennium, whether Mahayana, Vaishnava, or Shaiva, but—as we have seen and, surprisingly, to my way of thinking—questions of their validity were typically not adjudicated on

[50] See O'Hanlon (2013), with some overstatement of the consequences.

the basis of their language but rather on the basis of the doctrine: authenticity of a text was a function of its truth, and not its truth a function of its authenticity as proven by philology. As for the Greeks, they only seem to have stumbled upon *ekdosis* and *diorthotes* because they stumbled upon, or rather into, the Alexandrian library. The very fact that multiple written sources of a single work thereby became available made the need for those philological acts of edition and correction both obvious and necessary for the first time. In India, by contrast, where orality long persisted as a cultural value, the very idea of accumulating multiple copies of manuscripts of the same text would probably have seemed outlandish, and so too the idea of seeking out and explaining textual variation—until, for some reason, they no longer did. It is improbable that manuscript culture in India could have reached critical mass only at the beginning of the second millennium so as to provide the foundations for philological reflection, but it is then for the first time that Indian scholars begin to behave as if it did.

However, we explain this text-critical transformation, its effects were new and far-reaching. More than ever before, and in some ways never before, Sanskrit texts came to be mediated by a philological apparatus, one that with growing sophistication emphasized the dynamic changeability of transmission, the need for purification and the systematicity of reading, and whose growing density and wide circulation bespoke, or seemed to bespeak, new pedagogical needs and perhaps even correlated with new reading publics, of the sort we begin to perceive in other aspects of early modern culture (a good example is the growing popularity of introductory manuals, especially for grammar, hermeneutics and logic, the Sanskrit trivium). New institutional forms of pedagogy consolidating in the early centuries of the second millennium, whether in temples or Brahman settlements, and supported by wealthy courts, undoubtedly also had an important role to play.[51] To the degree one is prepared to make something of it, the Indian date when philological commentary attains real cultural significance correlates with European and Chinese 'early modernity', if the story of that contested periodization is taken to start with the twelfth-century Renaissance and the Song (where the rapid expansion of print culture is especially pertinent) rather than with the date of 1500, which marks the beginning of global modernization, something quite different.[52] That said, a later 'early modernity', beginning in the seventeenth century, seems also to be signaled by other new kinds of philological attention. The ordering and wording of foundational texts of the various knowledge systems (grammar, for example, or logic) come under systematic scrutiny for the first time in the seventeenth or eighteenth century, and the validity of sectarian scriptures—overturning more than a

[51] I thank Andrew Ollett for this observation.

[52] For some parallels to the European case, and some counterarguments or at least hesitations about a supposed symmetry (and the 'Axial Age' theoretical model itself), see Pollock (2004); for Song modernity, Woodside (2006) and especially Cherniack (1994); for rethinking of the European twelfth century, see Bynum (1984), and for Byzantium, Browning (1992).

millennium of purely doctrinal adjudication—comes to be vigorously debated on the basis of recensions, quotations and other textual (if not linguistic) features.[53]

## 3.6 The Theory of the Text

The theory of textuality that underlies Sanskrit philology can be described both negatively and positively, by its exclusions and its inclusions.

A wide range of questions were never addressed by the philologists of premodern India. Take the effects of a still-living oral culture on manuscript transmission, a phenomenon that crucially distinguishes the Sanskrit case from the Greek and Latin, and that the philologists of India did not and perhaps could not address precisely because they were *inside* that culture. First, because tradition was an oral-literate hybrid, textual transmission—and this pertains to scientific as well as literary texts—often shows the consequences of memorization and performance.[54] The manuscripts of the *Śatakatrayam* (*The Three Hundreds*), for example, of the seventh-century (?) poet Bhartṛhari present countless variants that are neither scribal mistakes nor learned corrections but oral variants in what by any standard was nonetheless fundamentally a literate culture. Second, because the tradition was active and not dormant, text-critical models that make sense for, say, Catullus or Propertius (where the descent of manuscript begins with unique copies of the ninth century) do not work for Bhartṛhari.[55] Manuscripts of his work were produced by the thousands well into the nineteenth century, and while, to some degree, these can be reduced to regional recensions (largely defined by script), they were always on the move and interacting with manuscripts from other regions. 'Contamination' in this world is therefore not the exception but often the normal state of affairs (and a phenomenon that clearly needs a new name).

Thus, the text as embodiment of an authorial intention—however much a value explicitly acknowledged by the participants in the literary culture—was constantly and in some cases irremediably destabilized by the messy business of bringing works to life in an oral world, whether in the classroom (where the set text was

[53] On the poet and theologian Nīlakaṇṭha Dīkṣita of Madurai, see Fisher (2013); on disputes over the *Bhāgavatapurāṇa*, Minkowski (2010). Sixteenth-century Shaivas like Appayya Dīkṣita and Nīlakaṇṭha Dīkṣita, and their Vaishnava opponents, dispute the authenticity of each other's scriptures on the basis of larger philological arguments (that these scriptures are so long that they must contain interpolations; that their original recensions have vanished and what is left is inauthentic, etc.) and they sometimes attack their opponent's arguments by showing that the 'scriptural' citations adduced have been invented whole cloth.

[54] The issue has been raised by (Rath and Houben 2012: 23 and 35). How substantially oral exegesis transformed the text of scientific works as recent as the seventeenth century is shown by Gerschheimer (1996).

[55] There are numerous examples of major works extant in few or even single manuscripts (*Arthaśāstra, Śikṣāsamuccaya, Abhinavabhāratī, Rājataraṅgiṇī,* and so on), but these never became objects of text-critical attention until the modern era.

typically recited from memory) or in literary performance. There were some exceptions. The memory culture of the Vedic tradition ensured its invariant transmission, and this value seems to have been transferred to other quasi-sacred texts such as the sixteenth-century *Rāmcaritmānas*, a fully literate work transmitted with little variation despite, or perhaps precisely because, it was constantly presented in oral performance.

There was, however, a vast domain of questions the philologists of India could address, as we have seen, and if they never openly theorized their practices (a lack that can hardly be merely an artifact of the extant data), they provided materials enabling us to do so. Let me now lay out elements of such a theory of textuality.

Classical Indian philologists understood texts to be unitary creations embodying authorial intention, even for texts that we today consider paradigms of composite authorship, such as the epics and *purāṇas* (repositories of myth and legend). This intention, they felt, could be recovered by the judicious assessment of manuscript variants—one that was never, however, conceived of as genealogical—and they developed criteria of textual criticism in harmony with that fundamental principle. If variants could be adjudicated on the basis of antiquity, as we have seen, it was only because of the implicit conviction that the older the reading, the closer it brought us to that authorial original. When Indian philologists took cognizance of the problem of regional variation, as they did from an early period (the thirteenth-century *Rāmāyaṇa* commentator Uḍāli Varadarāja),[56] it was out of the same implicit conviction that a single text underlies variation, and variation therefore constitutes deviation—no '*éloge de la variante*' here. It is only because texts were viewed as coherent wholes that the notion of interpolation could ever have developed into the widespread text-critical principle that it became. Not only were they taken as wholes (so that higher-order criteria such as non-contradiction across the narrative could be invoked), but so was an author's complete oeuvre: thus the eleventh-century critic Mahima Bhaṭṭa could appeal to Kālidāsa's works in their entirety to rule out a given usage.[57]

Other principles of the philologists of India derived from other presuppositions, which sometimes worked in tension with those just discussed. Since any text in Sanskrit was a part of Sanskrit culture, it was expected to adhere to the rules that defined that culture. The conflict here between two values, authorial intentionality on the one hand and normativity (in grammar, metrics, rhetoric, and the rest) on the other, was rarely discussed as such, but the checkered history of emendation in service of those rules demonstrates its consequences. A similar problem, if in some ways more elusive because nowhere discussed in the tradition, concerns large-scale textual expansion, no doubt closely related to the scribal conservatism already noted. Editors clearly believed that fidelity was a virtue, omission a sin, and, in

[56] They never perceived, however, the textual isogloss, so to call it, that produced the north–south hyparchetypes found in the transmission histories of many works.

[57] By contrast, the authorial text could also, curiously, be viewed as static: there was no conception that a second edition of a work could be produced, though this almost certainly occurred from time to time (Pollock 2007: 54; Harrison 2007).

general, bigger texts—where bigger texts were available—were better than smaller texts. A tendency toward agglomeration can thus be found in almost every genre, epic (*Mahābhārata*), scientific treatise (*Yogaśāstra* of Patañjali), or court poem (*Śiśupālavadha*).[58] Other kinds of text-critical interventions, about which we know far less, are visible in, say, the re-edition of the *Nāṭyaśāstra* in medieval Kashmir, where the very structure of the work was altered to make way for new—and sometimes disruptive—material, such as the addition of a ninth aesthetic emotion (*rasa*) to the canonical eight.

Aside from such large axioms, and the tensions that could arise when these conflicted, we can identify a range of other more restricted principles at work in Sanskrit philology. Scribes then were as prone to make mistakes as we are now and 'copyist's error' (*lekhakapramāda*) was a small principle that found wide application. Finer discriminations of such error (haplography, dittography, and the like) were not made, but script confusion—as when southerners misread ligatures in the Śāradā script of Kashmir—could be invoked to emend a passage. Aesthetic criteria for text-critical judgment are in evidence, too, for belles lettres, where 'the more beautiful reading' was determined less on the basis of subjective taste than by the invocation of principles from the neighboring science of poetics. It is here, at the intersection of text criticism with other forms of knowledge— evaluation of evidence, modes of reasoning, forms of proof—which both shaped and were shaped by philology, that a deeper, altogether unexplored, realm of intellectual history opens before us.

## 3.7 A Dead or a Living History?

The attempt to reconstruct a theory of philological practice in premodern India—indeed, the empirical investigation itself—raises an obvious but difficult question, with which I close: Are such investigations and reconstructions merely a chapter in intellectual history, or do they have continuing relevance to the practice of philology today? In other words, do traditional Indian notions of textuality have any proper role to play in the present-day study of Indian texts?

Both the practice of philology in the wide sense and the philological study of Indian texts in the narrow concern a search for truth. The truth of the text, however, is not singular and unique. This is very clear in the case of meaning, which can usefully be seen as triadic: a work's meaning for its original audience; its meaning for the many subsequent generations of its traditional readers; its meaning for us today. These meanings, I have long argued, are *all are equally true*, however irreconcilable with each other they may be, since meaning will always be meaning for different readers and cannot be reduced to any one of them. We must strive, difficult though it will be, to hold these three truths in balance, for it is only in their combination that

[58] Consider, in the vernacular tradition, the case of the sixteenth-century poet Sūrdās, whose corpus grew over the century or two after his date from about 250 to 5000 poems (Bryant and Hawley 2015).

the notion of the 'true meaning' of the work makes any sense.[59] The question for us here is whether the same logic holds in the case of the text itself.

Just as we present-day scholars want to know, in a historicist sense, what a fifth-century author *meant,* so we want to know—and this is, of course, prior—what he *wrote*. This is the foundation of modern textual criticism. But just as we also want to know what tenth- or fifteenth-century readers thought the author meant, we also want to know what *they* thought the author wrote.[60] These two different text-critical goals, now typically associated with the names Lachmann and Bédier, need not be mutually exclusive, as they are always represented as being. For me, the point is not that we cannot know or should not bother trying to know the original text (or that in some cases there cannot have been an original), nor that all versions of a work can be 'just as good' in some aesthetic sense. It is rather that textual variation embodies variable human consciousness, and that we need to take all such variation seriously if we are to seriously understand the history of consciousness. Indian philologists themselves, at least, seemed to accommodate all such options, not incoherently but pluralistically, for example complementing their views of authorial intention with the desire for maximally complete texts. Such pluralism—again, to the degree we have the intelligence (and now the technology) to hold these competing claims in equipoise[61]—may be one important lesson that Indian philologists of the past can teach today's philologists of India. And that therefore makes their theory of philological practice essential for our own.

## References

### *Primary Sources*

Abhinavagupta. 1992. *Abhinavabhāratī*. In *Nāṭyaśāstra of Bharata*, ed. K. Krishnamoorthy, vol. Vol. I, 4th ed. Baroda: Oriental Institute.

Amaru. 1916. In *Amaruśataka, with the commentary of Arjunavarmadeva*, ed. Durgaprasada and Vasudeva Laxman Sastri, 3rd ed. Bombay: Nirnaya Sagar Press.

Ānandavardhana. 1940. *Dhvanyāloka, with the Locana of Abhinavagupta and the Bālapriyā of Rāmaṣāraka*, ed. Pattabhirama Shastri. Varanasi: Chowkhamba Sanskrit Series Office.

Appayya Dīkṣita. 1935. In *Śāstrasiddhāntalesasaṃgraha, with the commentary of Acyutānandakṛṣṇa Tīrtha, Vol. I*, ed. P.P.S. Sastri. Srirangam: Vani Vilas Press.

59 See further in Pollock (2014). Contrast nineteenth-century philology, for which 'the true meaning' of a text 'must have been one, and not many' (thus the American Sanskritist W. D. Whitney (1873: 125)).

60 An illustration of such competing claims is offered by Goethe's Weimar edition revisions and the earlier versions known to his readers (Hanneder 2009–2010: 8).

61 The fact that this approach—'radical' though it may be—is now becoming 'widely accepted as a legitimate approach to editing' is considered historically in Hult (2010: 37–50), especially pages 47 and 50.

Bhaṭṭojī Dīkṣita. 1929. In *Siddhāntakaumudī, with the commentary of Vāsudeva Dīkṣita*, ed. C. Sankararama Sastri, 2nd ed. Madras: Balamanorama Press.
———. 1939. In *Prauḍhamanoramā, with three commentaries*, ed. Gopal Sastri Nene. Varanasi: Jaya Krishna Das Hari Das Gupta.
Bhavabhūti. 1999. In *Mālatīmādhava, with the commentary of Harihara*, ed. François Grimal. Pondichery: Institut Français de Pondichery.
De, S. K., ed. 1970. *The Meghadūta of Kālidāsa*. 2nd revised edition, ed. V. Raghavan. New Delhi: Sahitya Akademi.
Kālidāsa. 1902. In *Meghadūta, with the commentary of Mallinātha*, ed. Kashinath Pandurang Parab. Bombay: Nirnaya Sagar Press.
Kinjawadekar, Ramchandrashastri, ed. 1929. *Mahābhārata, with the commentary of Nīlakaṇṭha, 6 volumes*. Pune: Chitrashala Press.
Kumārila Bhaṭṭa. 1970. *Tantravārttika*. In *Mīmāṃsādarśanam*, ed. Subbasastri, vol. 6 volumes, 2nd ed. Pune: Anandashrama Press.
Mahima Bhaṭṭa. 1983. In *Vyaktiviveka with the commentary of Ruyyaka*, ed. Rewaprasada Dwivedi, vol. 2 volumes. Varanasi: Chaukhambha Sanskrit Sansthan.
Mammaṭa. 1959. In *Kāvyaprakāśa, with the commentary of Someśvara Bhaṭṭa*, ed. Rasiklal C. Parikh. Jodhpur: Rajasthan Oriental Research Institute.
Narayana Murti, M.S., ed. 1980. *Vallabhadeva's Kommentar (Śāradā-Version) zum Kumārasambhava des Kālidāsa*. Wiesbaden: F. Steiner.
Patañjali. 1987. In *Mahābhāṣya, with the commentary of Kaiyaṭa*, ed. Guruprasad Shastri, vol. 7 volumes. Varanasi: Vani Vilas Press.
Śāntideva. 1960. In *Bodhicaryāvatāra, with the commentary of Prajñākaramati*, ed. P.L. Vaidya. Darbhanga: Mithila Institute.
Sastri, R. Anantakrishna, ed. 1921. *Kavīndrācāryasūcipatra*. Baroda: Oriental Institute.
Someśvara Bhaṭṭa. 1909. In *Nyāyasudhā*, ed. Mukunda Sastri. Varanasi: Chowkhamba.
Vāmana and Jayāditya. 1984. In *Kāśikā, with the commentaries of Jinendrabuddhi and Haradatta*, ed. Jayashankar Lal Tripathi and Sudhakar Malaviya. Varanasi: Tara Printing Works.
Yāmunācārya. 1976. In *Āgamaprāmāṇya*, ed. M. Narasimhachary. Baroda: Oriental Institute.

## Secondary Works

Bhattacharya, Ram Shankar. 1990. Use of manuscripts in textual criticism by our commentators. In *Sampādana ke siddhānta aur upādāna (Principles of editing and instrumentation)*, ed. V. Dvivedi et al., 219–229. Varanasi: Central Institute of Higher Tibetan Studies.
Birwe, Robert. 1958. *Variae Lectiones* in *Adhyāya* IV und V der *Aṣṭādhyāyī*. *Zeitschrift der Deutschen Morgenländischen Gesellschaft* 108: 133–154.
Bronner, Yigal, and Lawrence McCrea. 2012. To be or not to be *Śiśupāla*. *Journal of the American Oriental Society* 132: 427–455.
Browning, Robert. 1992. Byzantium and Homer. In *Homer's ancient readers*, ed. R. Lamberton and J. Keaney, 134–186. Princeton: Princeton University Press.
Bryant, Kenneth, and John S. Hawley, eds. and trans. 2015. *Sur's Ocean: Poems from the early tradition*. Cambridge, MA.: Harvard University Press.
Bynum, Caroline Walker. 1984. Did the twelfth century discover the individual? In *Jesus as mother*, ed. C.W. Bynum, 83–107. Berkeley: University of California Press.
Cabezón, José Ignacio. 1992. Vasubhandu's *Vyākhyāyukti* on the authenticity of the *Mahāyāna Sūtras*. In *Texts in context*, ed. J. Timm, 221–242. Albany: State University of New York Press.
Cherniack, Susan. 1994. Book culture and textual transmission in sung China. *Harvard Journal of Asiatic Studies* 54: 5–125.

Colas, Gérard. 1999. The criticism and transmission of texts in classical India. *Diogenes* 47 (2): 30–43.
Davidson, Ronald. 1990. An introduction to the standards of scriptural authority in Indian Buddhism. In *Chinese Buddhist apocrypha*, ed. R.E. Buswell, 291–325. Honolulu: University of Hawaii Press.
Fisher, Elaine. 2013. *A new public theology: Sanskrit and the religious landscape of early modern South India*'. Ph.D. diss. Columbia University.
Gerschheimer, Gerdi, ed. 1996. *La théorie de la signification chez Gadādhara: le Sāmānyakāṇḍa du Śaktivādavicāra*. Paris: Collège de France; Edition-Diffusion de Boccard.
———. 2010. *Sthitasya gatis cintaniya*? A propos de la "Glose concise" (*Laghuvrtti*) du *Ṣaḍdarśanasamuccaya* de Haribhadra. In *Ecrire et transmettre en Inde classique*, ed. G. Colas and G. Gerschheimer, 287–306. Pondichery: Ecole française d'Extrême-Orient.
Goodall, Dominic. 2001. *Bhūte 'āha' iti pramādāt*: Firm evidence for the direction of change where certain verses of the *Raghuvaṃśa* are variously transmitted. *Zeitschrift der Deutschen Morgenländischen Gesellschaft* 151: 103–124.
Goodall, Dominic, and Harunaga Isaacson, eds. 2003. *The Raghupañcikā of Vallabhadeva, being the earliest commentary on the Raghuvaṃśa of Kālidāsa, Vol. I*. Groningen: Egbert Forsten.
Hanneder, Jürgen. 2009–2010. Introduction. Wiener Zeitschrift für die Kunde Südasiens 52–53: 5–16.
Harrison, Paul. 2007. The case of the vanishing poet: New light on Śāntideva and the *Śikṣāsamuccaya*. In *Indica et Tibetica: Festschrift für Michael Hahn*, ed. K. Klaus and J.-U. Hartmann, 215–248. Vienna: Arbeitskreis für Tibetische und Buddhistische Studien, University of Vienna.
Hult, David F. 2010. Text editing: Past, present, and future. *Romanic Review* 101 (1–2): 37–50.
König, Christoph. 2013. Grenzen der Cyklisation. In *Friedrich Schlegel und die Philologie*, ed. U. Breuer, R. Bunia, and A. Erlinghagen, 15–43. Paderborn: Schöningh.
Lariviere, Richard W. 1989. *The Nāradasmṛti*. Vol. 2 volumes. Philadelphia: Department of South Asia Regional Studies, University of Pennsylvania.
Minkowski, Christopher. 2005. What makes a work 'traditional'? On the success of Nīlakaṇṭha's *Mahābhārata* commentary. In *Boundaries, dynamics and construction of traditions in South Asia*, ed. F. Squarcini, 1–28. Florence: Firenze University Press.
———. 2010. I'll wash your mouth out with my boot: A guide to philological argument in Mughal-Era Banaras. In *Epic and argument in Indian Sanskrit literary history*, ed. S. Pollock, 117–141. Delhi: Manohar.
O'Hanlon, Rosalind. 2013. Performance in a world of paper: Puranic histories and social communication in early modern India. *Past and Present* 219: 87–126.
Pollock, Sheldon. 1986. *Rāmāyaṇa of Vālmīki: Ayodhyā*. Princeton: Princeton University Press.
———. 2003. Sanskrit literature from the inside-out. In *Literary cultures in history: Reconstructions from South Asia*, ed. S. Pollock, 39–130. Berkeley: University of California Press.
———. 2004. The transformation of culture-power in Indo-Europe, 1000–1300. *Medieval Encounters* 10 (1–3): 247–278.
———. 2006. Literary culture and manuscript culture in precolonial India. In *History of the book and literary cultures*, ed. Simon Eliot et al., 77–94. London: British Library.
——— ed. and trans. 2007. *Rama's last act: The Uttararāmacarita of Bhavabhūti*. New York: New York University Press.
———. 2011. Indian philology and India's philology. *Journal Asiatique* 299 (1): 423–442.
———. 2014. Philology in three dimensions. *Postmedieval* 5 (4): 398–413.
———. 2015. What was philology in Sanskrit? In *World philology*, ed. S. Pollock, B. Elman, and K. Chang, 114–141. Cambridge, MA: Harvard University Press.
Preisendanz, Karin. 2008. Text, commentary, annotation: Some reflections on the philosophical genre. *Journal of Indian Philosophy* 36: 599–618.
Raghavan, V. 1941–1942. Uḍāli's commentary on the *Rāmāyaṇa*: The date and identity of the author and the discovery of his commentary. *Annals of Oriental Research* 6 (2): 1–8.

Rath, Saraju, and Jan E.M. Houben. 2012. Manuscript culture: Contours and parameters. In *Aspects of manuscript culture in South India*, ed. S. Rath, 1–53. Leiden: E.J. Brill.
Schopen, Gregory. 2009. On the absence of urtexts and otiose *ācāryas*: Buildings, books, and lay Buddhist ritual at Gilgit. In *Ecrire et transmettre en Inde classique*, ed. G. Colas and G. Gerschheimer, 189–219. Pondichéry: Ecole française d'Extrême-Orient.
Skilling, Peter. 2000. Vasubandhu and the *Vyākhyāyukti* literature. *Journal of the International Association of Buddhist Studies* 23 (2): 297–350.
Snell, Bruno. 1953. *The discovery of the mind: The Greek origins of European thought*. Trans. T.G. Rosenmeyer. Cambridge, MA: Harvard University Press.
Tripathi, Chandrabhal. 1975. *Catalogue of the Jaina manuscripts at Strasbourg*. Leiden: E.J. Brill.
Unni, N.P. 1987. *Meghasandeśa of Kālidāsa*. Delhi: Bharatiya Vidya Prakashan.
Whitney, W.D. 1873. The translation of the Veda. In *Oriental and linguistic studies.*, First series, ed. W.D. Whitney, 100–132. New York: Scribner's.
Woodside, Alexander. 2006. *Lost modernities*. Cambridge, MA: Harvard University Press.

# Part II
# What Was at Issue in Returning to Ancient Texts in Early Modern and Modern Times?

# Chapter 4
# Rethinking the Ancient Mathematical Text: Ming-Qing Scholars' Critical Reflections on *The Gnomon of the Zhou [Dynasty]*

**Qi Han**

**Abstract** European mathematics was introduced to China since the late sixteenth century. From then on, Chinese scholars used it as a tool to study traditional Chinese mathematics. They paid great attention to *The Gnomon of the Zhou [Dynasty],* the oldest Chinese mathematical book. This chapter first explains the views of Xu Guangqi (1562–1633) on *The Gnomon of the Zhou [Dynasty]*. Then it analyzes why the Kangxi Emperor (1654–1722) and the Qing mathematician Mei Wending (1633–1721) were interested in the study and elaboration of Chinese mathematical texts, especially *The Gnomon of the Zhou [Dynasty]*. The chapter further examines why the French Jesuits Dominique Parrenin (1665–1741) and Antoine Gaubil (1689–1759) were interested in the ancient history of Chinese mathematics and how they influenced Voltaire's views of Chinese history. It also discusses the influence of Ming-Qing scholars' views in the mid-eighteenth and early nineteenth century.

The research leading to these results has received funding from the European Research Council under the European Union's Seventh Framework Programme (FP7/2007–2013)/ERC Grant Agreement No. 269804. I am very grateful to the participants in the SAW workshop held in June 2013 and in December 2014. My sincere thanks to Karine Chemla for her kind and valuable comments. This paper was supported by the Palace Museum's open project in 2021, New Research on the Sino-Western Scientific Exchanges during the Kangxi and Qianlong Period. This subject has been funded by the Mercedes-Benz Star Wish Fund of China Youth Foundation.

Q. Han (✉)
Department of Chinese History and Culture, The Hong Kong Polytechnic University, Hong Kong, P.R. China
e-mail: qi.han@polyu.edu.hk



A. Keller, K. Chemla (eds.), *Shaping the Sciences of the Ancient and Medieval World*, Archimedes 69, https://doi.org/10.1007/978-3-031-49617-2_4

## 4.1 Introduction

During the late Ming and mid-Qing periods (ca. 1582–1800), European science and religion were introduced into China. In order to avoid opposition from some conservative Chinese scholars, and to promote the calendar reform and the spread of European science, other scholars expressed positive attitudes toward Western learning. Meanwhile, they used European science as a tool for interpreting ancient Chinese mathematical and astronomical texts, one of which, *The Gnomon of the Zhou [Dynasty]* (*Zhoubi suanjing*, 周髀算經), was the oldest of the *Ten Mathematical Classics* (*Suanjing shishu*, 算經十書).[1] Through the study of ancient mathematical classics, Chinese scholars tried to prove that they had their own established tradition which was identical with Western learning. As a case study, I will talk about how Ming–Qing scholars began to be interested in *The Gnomon of Zhou [Dynasty]*. Firstly, I will explain the views of Xu Guangqi (徐光啟, 1562–1633) on *The Gnomon of the Zhou [Dynasty]*. Then I will analyze why the Kangxi Emperor (康熙, 1654–1722) and the Qing mathematician Mei Wending (梅文鼎, 1633–1721) were interested in the study and elaboration of Chinese mathematical texts, especially *The Gnomon of Zhou [Dynasty]*, and why the French Jesuits of China mission were interested in the ancient history of Chinese mathematics. I shall also discuss the influence of their views in the mid-eighteenth and early nineteenth century.

## 4.2 Comments on *The Gnomon of the Zhou [Dynasty]* in Late Ming and Early Qing

It is commonly accepted that Chinese traditional science was lagging behind European science in the late sixteenth century when Western science was introduced to China by the European Jesuits. The transmission of European science aroused Chinese scholars' interest in their own scientific tradition and helped them to improve their interpretation of ancient Chinese mathematical texts.

Xu Guangqi, one of the 'three great pillars' of Catholicism in late Ming and one of the first scholars to come into contact with Western science, was regarded as an outstanding scholar, politician and scientist. In the course of the exchange between Chinese and Western science, he developed his own original ideas about how to avoid opposition to adopting European science from some conservative scholars. Because of his leading role in his academic circle, some of his views greatly influenced Qing scholars' understanding of Western learning. The views Xu Guangqi

[1] In 656, a mathematical school was established in the Imperial Academy. *Ten Mathematical Classics* was therefore selected by Li Chunfeng (李淳風, 602–670) and was used as textbook. Qian Baocong gives a critical edition of *The Gnomon of Zhou [Dynasty]* and some ancient commentaries on this classic; see Qian (1963) and Chemla (2013: 176–177). The first printed edition of *The Gnomon of Zhou [Dynasty]* was published in 1084.

advocated at the time of the calendar reform during the reign of Chongzhen (崇禎, 1627–1644) were always held up by later scholars as unchangeable theories (Han 2001: 360–379).

In collaboration with Matteo Ricci (1552–1610), Xu was the first person to translate part of *Euclidis Elementorum Libri* XV (Rome, 1574), written by the Jesuit mathematician Clavius (1538–1612), and he published its Chinese translation in 1607 (Engelfriet 1998: 2, 127–128). His colleague and close friend Li Zhizhao (李之藻, 1565–1630) also translated Clavius' *Epitome Arithmeticae Practicae* into Chinese. Xu wrote a preface for this book extolling the superiority of Western mathematics and denouncing ancient Chinese mathematical texts as 'tattered sandals' (Hart 2013: 190–191). However, after finishing his collaboration with Ricci on the *Methods and Principles of Measurement* (測量法義, ca. 1608), Xu wrote a book on right-angled triangles, *Explanations of Gougu* ("Right-angled triangles") (*Gougu yi*, 勾股義, ca. 1609), in which he mentioned *The Gnomon of the Zhou [Dynasty]*,[2] the oldest canon of ancient Chinese astronomy and mathematics. To some extent, Xu's objective in writing *Explanations of Gougu* was to recover meanings lost in ancient traditional mathematics, especially the principle for right-angled triangles in *The Gnomon of the Zhou [Dynasty]*, commonly called Pythagorean theorem.

In the *Explanations of Right-angled triangles*, Xu cited the dialogue between the Duke Zhou (周公, active eleventh century BCE) and a mathematician, Shang Gao (商高), as described in *The Gnomon of the Zhou [Dynasty]*. He pointed out the importance of *The Gnomon of the Zhou [Dynasty]* and said:

> The first chapter of *The Gnomon of the Zhou* [*Dynasty*] was the earliest ancestor of right-angled triangles in the *Nine Chapters*. Scholars like Zhen Luan [甄鸞, active sixth century], and Li Chunfeng reinterpreted and clarified it. It really ranks first among ancient mathematical books. (《周髀》首章，《九章》勾股之鼻祖，甄鸞、李淳風輩為之重釋，頗明悉，實為算術中古文第一。)

Xu stressed the importance of *The Gnomon of the Zhou [Dynasty]* in the application of measurement, especially in irrigation projects. However, Xu discarded its astronomical aspect and thought that the dialogue between Rong Fang (榮方) and Master Chen (陳子), which discussed the mathematical principles of the sun, moon, heaven and earth, was 'the most foolish in ancient times' (至於商高問答之後，所謂榮方問於陳子者，言日月天地之數，則千古大愚也。).[3] Xu Guangqi's ideas concerning Western science were accepted by a number of his contemporaries and followers. Some of his views were shared by scholars of his time. Chen Jinmo (陳藎謨, ca. 1600–ca. 1692) (Fung 2007: 209–262), a scholar from Zhejiang province, was interested in synthesizing Western and Chinese traditional mathematics. In his book *On Measurement* (*Duce*, 度測), Chen tried to use Western mathematics and instruments to interpret *The Gnomon of the Zhou [Dynasty]* in detail. He cited Xu

---

[2]The edition that Xu Guangqi used was probably the Wanli edition (1573–1620), which is still accessible in some libraries.

[3]This dialogue between Rong Fang and Chen Zi, on which *The Gnomon of Zhou [Dynasty]* was based, depicts their discussion about the heavens. See Cullen (1996: 74).

Guangqi's *Methods and Principles of Measurement* and *Explanations of Right-angled Triangles*, and claimed that *The Gnomon of the Zhou [Dynasty]* was 'the canon of right-angled triangles' and *Methods and Principles of Measurement* consisted of the commentaries of right-angled triangles (徐玄扈先生有《測量法義》、《勾股義》，是《周髀》者勾股之經，《法義》者勾股之疏傳也。). In addition, Chen thought that the 'right-angled triangles method was especially subtle and marvelous. It originated from *The Gnomon of the Zhou [Dynasty]*.' (勾股尤奇奧，其法肇見周髀。)[4] This stressed the important status of *The Gnomon of the Zhou [Dynasty]* in Chinese traditional mathematics.

## 4.3 The Kangxi Emperor's Test Using Two Terms from *The Gnomon of the Zhou [Dynasty]* and Early Qing Scholars' Reactions

Kangxi (reigned 1662–1722) was the second Manchu emperor to show great interest in Western science. In the 1660s, the anti-Christianity movement launched by Yang Guangxian (楊光先, 1597–1669) was a very serious political, cultural and scientific phenomenon which greatly influenced the course of Western science and religion in China. The German Jesuit Adam Schall von Bell (1592–1666) and his confrères from the order were arrested and some Chinese Christian astronomers were sentenced to death. This case, which concerned Schall's work on reforming the Chinese calendar, greatly impressed Kangxi when he was still very young and directly influenced him to study Western science. Shortly afterwards, he began studying with the Jesuits Lodovico Buglio (1606–1682), Gabriel de Magalhães (1610–1677) and Ferdinand Verbiest (1623–1688).

The year 1678 was crucial not only for the Sino-Portuguese relationship but also in the Chinese history of science. A Portuguese envoy was sent to Beijing with an African lion as a gift (Bertuccioli 1976: 223–240). After the lion arrived at the Imperial court, many Chinese and Manchu officials were summoned to watch it on the invitation of the Kangxi Emperor and were asked to compose a prose piece in honor of the lion. The Chinese pronunciation of *Shi* (獅, lion) sounds like *Shi* (師, master). On this special occasion, a Chinese scholar flattered the emperor by saying to him that it is necessary to learn from 'masters'. The Emperor agreed. This gathering became a symbol for the Kangxi Emperor's motivation to learn from wise men, both Chinese and Western.

This event coincided with the establishment of the Department for Erudite Literati (*Boxue hongci ke*, 博學鴻詞科). In the same year (1678), the Kangxi Emperor summoned one hundred and eighty-six Chinese scholars to Beijing and questioned them on two topics. One of these was the 'prose of *Xuanji yuheng*'

[4] See Chen Jinmo's preface to his *On Measurement* written in 1640, manuscript copy preserved at the Library of the Institute for the History of Natural Sciences, Chinese Academy of Sciences.

(*Xuanji yuheng* fu, 璿璣玉衡賦) (Han 1997: 14–15).[5] The prose pieces composed on that occasion can be still found in various collected writings of contemporary scholars of the time. The Emperor evidently intended to test these scholars' knowledge in ancient Chinese astronomy. This event undoubtedly stimulated their interest in mathematics and astronomy. What we should also highlight here is that the Belgian Jesuit Ferdinand Verbiest compiled the *Perpetual Calendar of Kangxi Reign* (*Kangxi yongnian lifa*, 康熙永年曆法) in the same year, 4 years after he constructed six astronomical instruments based on Tycho Brahe's design.[6]

Imperial support of Western learning aroused Chinese scholars' interest in Chinese astronomical texts. Before Mei Wending, the great mathematician of the Qing dynasty, arrived in Beijing in 1689, he heard the news from his friends who had participated in the Emperor's examination. He, thus, composed a lengthy piece, the 'prose of *Xuanji yuheng*',[7] depicting a brief history of Chinese traditional astronomy in an elegant literary fashion and showing off his wider knowledge of ancient Chinese science. This prose work attracted much attention from scholars and became widespread in academic circles at that time (Han 1997: 15–16).

Hearing about the Kangxi Emperor's test involving prose may have led to Mei Wending's deep concentration on *The Gnomon of the Zhou [Dynasty]*. When Mei Wending began researching it, the astronomical knowledge it contained was considered absurd. In 1692, he first expressed his opinion about the points of similarity between *The Gnomon of the Zhou [Dynasty]* and Western science.[8] In the following year, he wrote *Doubts on the Calendar* (*Lixue yiwen*, 曆學疑問, 1693) and elaborated on it in detail. Relying on Western learning, he argued that *Gaitian* (蓋天) theory was identical with *Huntian* (渾天) theory, which supported the sphericity of the earth.[9] This overthrew the traditional view of *Gaitian* theory, which assumed

[5]The compound term *xuanji yuheng* 璿璣玉衡 originated from the chapter "Canon of Yao" (Yaodian 堯典) in the *Classic of Documents* (Shangshu 尚書). The earliest extant commentaries, produced in the Western Han, commonly interpreted the expression as referring to stars near the north celestial pole. More specifically, *xuanji* was understood as the polar star or the first star of the Northern Dipper (Beidou 北斗), the nexus around which the whole Northern Dipper, what *yuheng* stood for, revolved. However, prominent commentators in the Eastern Han, such as Ma Rong 馬融 and Zheng Xuan 鄭玄, presented a very different interpretation: for them, the expression designated an astronomical instrument, the armillary sphere (*huntian yi* 渾天儀). In this perspective, according to Ma Rong, *xuanji* referred to the main structure of the sphere, whereas *yuheng* denoted the rotatable sighting tube. Up to the Qing dynasty, the term has been widely used as an alternative name for the sphere. I am very grateful to Dr. Guo Jinsong for the detailed explanation. The term *Xuanji yuheng* was also mentioned in commentaries of *The Gnomon of the Zhou [Dynasty]*.

[6]In Verbiest's memorial to the Kangxi Emperor in 1674, he associated these European style astronomical instruments with Chinese texts related to *Xuanji yuheng*. On Verbiest's memorial, see Han Qi and Wu Min (2006: 118). On Verbiest's astronomical work, see Golvers (1993).

[7]Mei Wending, *Jixuetang wenchao* (績學堂文鈔) (*Collected Writings of Mei Wending*) (*juan* 1, 3–10), edition published in the reign of Qianlong.

[8]'Who could expect that the words of Europeans are identical with those of *Zhoubi*?' (誰知歐邏言，乃與周髀通), in Mei Wending, *Jixuetang shichao* 績學堂詩鈔 (*Collected Poems of Mei Wending*) (*juan* 3, 20–21), edition published in Qianlong reign.

[9]On the *Gaitian* and *Huntian* theories, see Needham (1959: 210–219).

that the earth was flat. Following this, Mei Wending worked on *The Gnomon of the Zhou [Dynasty]* and wrote a book titled *The Supplementary Annotations to The Gnomon of the Zhou [Dynasty]* (*Zhoubi suanjing buzhu*, 周髀算經補注).[10]

## 4.4 Kangxi's Theory of the Chinese Origin of Western Learning and Mei Wending's Reaction Based on *The Gnomon of Zhou [Dynasty]*

After the case of the calendar reform, the Kangxi Emperor believed that Western science was greatly superior to Chinese science. He expressed this opinion in his discussions with his senior officials on many occasions. However, in 1703, he wrote an essay titled 'Imperially composed treatise on the derivation of triangles'(*Yuzhi sanjiaoxing tuisuan falun*, 御制三角形推算法論) and proposed the theory of the Chinese origin of Western learning (*Xixue zhongyuan*, 西學中源), which argued that Western astronomical and mathematical sciences originated in China. In this essay, the Kangxi Emperor explained why he studied Western science and went on to discuss the relationship between the Western and the Chinese traditional calendars:

> Those who argue that the ancient and modern methods are different absolutely do not understand the calendar. The calendar, originated in China, was transmitted to the Far West. (論者以古法今法之不同，深不知曆，曆原出自中國，傳及於極西。)[11]

During the 'calendar case' in the 1660s, European astronomy, which was regarded as the exotic knowledge of barbarians, faced strong opposition from some conservative Chinese scholars and Muslim astronomers at the Imperial Board of Astronomy. The reason why Kangxi proposed this theory is quite interesting. To some extent, his purpose was to mitigate the dispute between Western and Chinese learning, and to counteract the deeply negative effects of the anti-Christianity movement. On the other hand, it was also a good excuse for the Chinese to learn European mathematics and astronomy. Because of his advocacy and the subsequent propagation by some Chinese scholars, the theory of the Chinese origin of Western learning became a very important perspective which influenced the study of mathematics and calendrical science in the eighteenth and nineteenth centuries. This could explain why several studies and reprints of *The Gnomon of the Zhou [Dynasty]* appeared at that time.

Shortly after his concentrated tuition in mathematics by the Jesuits from 1688 to 1691 (Landry-Deron 1995), Kangxi began searching for some accomplished Chinese scholars who were well versed in mathematics. In late 1691, he sent

[10] See Mei Wending, *A Descriptive Catalogue of Wu'an's Mathematical Works* (勿庵曆算書目). Kangxi edition, now preserved in the Library of Tsinghua University (Beijing).

[11] Kangxi, *Collected Writings of the Kangxi Emperor* (御制文集), *Ji* 3, *Juan* 19. See Han (2014: 1223).

someone to check Mei Wending's ability to measure the shadows cast by the sun. However, Mei's performance did not please him (Han 2011: 4–5). But later, after more than 10 years of preparation, Mei Wending's mathematical ability was greatly admired by the Kangxi Emperor. On the recommendation of Li Guangdi (李光地, 1642–1718), a grand minister and his patron, Mei was granted an audience with the Kangxi Emperor in 1705. During their dialogue, the Emperor showed Mei his essay on triangles, which expounded the theory that European mathematics originated in China. The Kangxi Emperor undoubtedly wanted to show off his new knowledge of geometry using his recently printed essay (Han 2016: 1–9). The audience with the Emperor moved Mei greatly. A year later, in 1706, Mei praised the Emperor's essay on triangles highly and again mentioned the similarity between European astronomy and the theory contained in *The Gnomon of the Zhou [Dynasty]*.[12] To propagate the Emperor's idea, Mei Wending deliberately expanded his research on *The Gnomon of the Zhou [Dynasty]* and wrote *The Supplement to the Doubts on the Calenda*r (*Lixue yiwen bu*, 曆學疑問補), in which he extended Kangxi's argument that European astronomy and mathematics originated in China. His theory had a great influence on Chinese mathematicians in the latter half of the eighteenth century. From the description above, we can see the close relationship and mutual reaction between Mei Wending and the Kangxi Emperor with regard to *The Gnomon of the Zhou [Dynasty]*. Mei Wending's approach to *The Gnomon of the Zhou [Dynasty]* had a great influence on some official scholars and was especially admired by Li Guangdi, who regarded Mei as the forerunner of calendrical sciences.

## 4.5 The Kangxi Emperor's Interest in *The Gnomon of Zhou [Dynasty]* and Mei Juecheng's Role in the Compilation of *The Fine Essence of Mathematical Principles*

In addition to his interest in European mathematics, to enhance the impression that he was also a master of traditional Confucian learning, the Kangxi Emperor devoted much attention to traditional Chinese mathematical works such as those written by the prince scholar Zhu Zaiyu (朱載堉, 1536–1611),[13] the abacus expert Cheng Dawei (程大位, 1533–1606) and Fang Zhongtong (方中通, 1634–1698), who had learned European science from a Polish Jesuit, J. N. Smogulecki (穆尼閣 or彌閣, 1610–1656). In addition, by that time, *The Gnomon of the Zhou [Dynasty]* had already aroused the Kangxi Emperor's interest. In the year 1713, Yinzhi (胤祉, 1677–1732), the third son of the Kangxi Emperor, presented a memorial to Kangxi in which he mentioned that the Emperor had read *The Gnomon of the Zhou [Dynasty]*

[12] Mei Wending, *Jixuetang shichao* 績學堂詩鈔 (*Collected Poems of Mei Wending*) (*Juan* 4, 21).

[13] Zhu Zaiyu was famous for the study of musical theory in the late Ming period. He also wrote a book entitled *Illustrated Explanations of The Gnomon of Zhou [Dynasty]* (*Zhoubi tujie*, 周髀圖解).

and had written some commentaries on it.[14] Kangxi's interest in *The Gnomon of the Zhou [Dynasty]* was widely known in the academic sphere. In the same year as a major celebration for the Kangxi Emperor's sixtieth birthday, Huang Qianren (黃千人, 1694–1771), the grandson of the famous philosopher Huang Zongxi (黃宗羲, 1610–1695), composed a poem dedicated to the Emperor. In this poem, he mentioned that the Kangxi Emperor had made some comments on *The Gnomon of the Zhou [Dynasty]*.

The Kangxi Emperor's interest in *The Gnomon of the Zhou [Dynasty]* doubtless exercised a great influence on the imperial compilers of *The Fine Essence of Mathematical Principles* (*Shuli jingyun*, 數理精蘊), a major mathematical encyclopedia of the Qing dynasty, published in 1722.[15] In its opening chapter, the compilers inserted an *Annotation of The Gnomon of the Zhou [Dynasty]* (*Zhoubi jingjie*, 周髀經解), which emphasized that China possessed the most ancient tradition in mathematics and was thus the fount of mathematical knowledge. As one of the chief editors of *The Fine Essence Mathematical Principles*, Mei Juecheng (梅瑴成, 1681–1763), a grandson of Mei Wending, may have written the following paragraph:

> After the founding of our dynasty, people from far away admired our civilization and came to our country one after another. Among them are Adam Schall von Bell, Ferdinand Verbiest, Antoine Thomas and Claudio Filippo Grimaldi, who were successively in charge of calendrical sciences. They were also good at mathematics. Consequently, principles of measurement and number gradually became much more complete. However, when we asked them their origin, they all said they were transmitted from the land of China. (我朝定鼎以來，遠人慕化，至者漸多，有湯若望、南懷仁、安多、閔明我，相繼治理曆法，間明算學，而度數之理，漸加詳備。然詢其所自，皆云本中土所流傳。)[16]

This shows that the idea of the Chinese origin of Western learning had spread to all the fields of mathematical sciences.

Mei Juecheng, an important figure in the history of Chinese science, seems to have been less open-minded and more conservative than his grandfather. Taught by

[14] This could explain why the bronze movable type edition of *The Gnomon of Zhou [Dynasty]* was included in the *[Imperially authorized] Comprehensive Corpus of Illustrations and Books, from Ancient Times to the Present* (*Gujin tushu jicheng*, 古今圖書集成) which was compiled by Chen Menglei (陳夢雷, 1650–1741), whose patron was the third prince, Yinzhi. This edition was based on the Zhao Kaimei (趙開美) edition from the Wanli era (1573–1620).

[15] On the compilation of *The Fine Essence of Mathematical Principles*, see Han (1991).

[16] This excerpt is part of the preface of the *Zhoubi jing jie* (周髀經解), *The Fine Essence of Mathematical Principles*, Shangbian, *Juan* 1, p.10, Kangxi edition. Adam Schall von Bell, a German Jesuit from Cologne, was the first president of the Imperial Board of Astronomy and played a leading role in the translation and compilation of European astronomical books. Ferdinand Verbiest, a Jesuit from Belgium, was Schall's successor and the tutor of the Kangxi Emperor. He edited several books on European astronomical instruments, some of which were constructed in 1674 and are still preserved today in the Beijing Ancient Observatory. Antoine Thomas (1644–1709), a Jesuit from Namur, was also a mathematics tutor of the Kangxi Emperor. Claudio Filippo Grimaldi (1638–1712) was an Italian Jesuit who was sent to Europe as an envoy and had contact with the German philosopher Leibniz.

the Kangxi Emperor, he discovered that the *tianyuan* (天元), the traditional Chinese method of the Song-Yuan dynasties, was identical to the European algebra newly introduced by the Jesuits. This provided him with strong evidence for the theory of the Chinese origin of Western learning. In his book, *Pearls Lost in the Red Rive*r (赤水遺珍, 1745), he wrote:

> Later, I served at the Imperial court. I am indebted to the benevolent Emperor [Kangxi] for teaching me the 'method of borrowing radix and powers [Jiegenfang, 借根方].'[17] And His Majesty instructed, 'The Westerners name this book *a-er-re-ba-da* [i.e. algebra], which means "method from the East" [Dong lai fa, 東來法].' I read it respectfully. This method, mysterious and wonderful, is really the guide to mathematical methods. (後供奉內廷，蒙聖祖仁皇帝授以借根方法，且諭曰:西洋人名此書為阿爾熱八達，譯言東來法也。敬受而讀之，其法神妙，誠算法之指南。)[18]

Mei Juecheng also played a crucial role in the compilation of the *Ming History*, the official history of the Ming dynasty (1368–1644). 'Loss of rites rescued by barbarians' (*Li shi qiu ye*, 禮失求野) also became an official theory and was adopted in the 'Memoir on the Calendar' (*Li zhi*, 曆志) of the *Ming Histo*ry (published in 1739). In the second *juan* of the 'Memoir on the Calendar', an ancient path of transmission of Chinese knowledge to the West is traced:

> Westerners who come to China all call themselves Europeans; their calendrical science is similar to that of Muslims, only more accurate. If we search in previous dynasties, people from distant countries who discussed the calendar were from Western marches; we never heard that there were [astronomers] in the north, south, and east.…Yao had ordered the Xi and He younger brothers to disperse in the four directions. The second and third Xi brothers and the third He brother only went to lands east, south, and north. Only the second He brother is said to have gone west beyond the frontier.[19] How could that knowledge not have been transmitted far away to the West at that time? (西洋人之來中土者，皆自稱歐羅巴人，其曆法與回回同，而加精密。嘗考前代，遠國之人言曆法者多在西域，而東南北無聞。……蓋堯命羲、和仲叔分宅四方，羲仲、羲叔、和叔則以嵎夷、南交、朔方為限，獨和仲但曰'宅西'，而不限以地，豈非當時聲教之西被者遠哉?)[20]

By the end of the Zhou dynasty, the text continues, most of China's astronomical knowledge had already been transferred to the West. Western calendrical science thus has the same origin as the Muslim calendar. Many of its ideas stemmed from the ancient Chinese classic, *The Gnomon of the Zhou [Dynasty]*. This kind of description formed an official theory of diffusion of Chinese knowledge westward.

---

[17] On this method and the first translation of European algebra in China, see Han (2003): 105–114).

[18] *Pearls Lost in the Red River* 赤水遺珍, in *Collected Mathematical Writings of Mr. Mei* 宣城梅氏曆算叢書輯要，*Juan* 61, 8–9, Chengxuetang edition (1745).

[19] This refers to the first chapter of *Book of Documents* (*Shujing*, 書經).

[20] Zhang Tingyu ed., *Ming History* 明史, *Juan* 31, Beijing: Zhonghua Book Company, 1974, 544.

## 4.6 Was Pythagorean Theorem a Chinese Invention? French Jesuits and Their Interest in *The Gnomon of the Zhou [Dynasty]*

From the late sixteenth century onward, Jesuits wrote many letters home about China. This correspondence led a considerable number of European scholars to become interested in Chinese history and cultural tradition. Leibniz was one of the most active scholars who wished to establish close contact with China. He had been interested in China from 1666 at the latest. In 1689, he began corresponding with the Italian Jesuit C. F. Grimaldi (1638–1712) after they had met in Rome. On 19 July 1689, Leibniz asked Grimaldi some questions, in particular whether there had been geometrical demonstrations in ancient China (Widmaier 1990: 5). He also asked Bouvet similar questions and in 1697 he published the *Latest News from China* (*Novissima Sinica*) (Chemla 2012: Introduction, 3; Han 2014: 1225–1226).

In addition, some Jesuits from the China mission were also interested in geometrical demonstrations like Pythagorean theorem and *The Gnomon of the Zhou [Dynasty]*, as revealed in the correspondence between the French Jesuit Dominique Parrenin (1665–1741) and Dortous de Mairan (1678–1771), the permanent secretary of the French Royal Academy of Sciences. Parrenin arrived in Canton in 1698 and subsequently stayed in Beijing until his death in 1741. He had a gift for languages and a good command of Chinese and Manchu. He was commissioned to translate European anatomy into Manchu and was summoned to the court to translate European documents. He had a close relationship with the Kangxi Emperor and served as an interpreter when European envoys visited China.

On 11 August 1730, in his letter to de Mairan, Parrenin wrote:

> One sees past historians talk about the right-angled triangle as being something common and explained by the famous Duke of Zhou, who lived eleven hundred years before Jesus Christ. I have heard someone say to the late Emperor that this [i.e., the right-angled triangle] was one of the longest-known things in China, and some claim that the famous Yu used it. If this is the case, then Pythagoras was not the first who made this glorious discovery. (On voit des anciens historiens parler du triangle rectangle comme d'une chose commune & expliquée par le célébre *Tcheou-Kong*, qui vivoit onze cens ans avant Jesus-Christ. J'ai oui-dire au feu Empereur que c'étoit une des plus anciennes connoissances de la Chine, & il y en a qui prétendent que le fameux *Yu* s'en servit: si cela est, Pythagore n'est pas le premier qui ait eu la gloire de cette découverte.)[21]

Parrenin's information came from the Imperial court and was probably based on Mei Wending's commentaries of *The Gnomon of the Zhou [Dynasty]*. This indicated that the Kangxi Emperor's interest in *The Gnomon of the Zhou [Dynasty]* was not only widely shared by Chinese scholars, but also by the Jesuits.

---

[21] English translation from *Lettres édifiantes et curieuses écrites des missions étrangeres*. T.21, Paris, 1734, 108–109. I am very grateful to Dr. Jerome Kerlouegan for his kind help in the translation of some French texts.

On 20 September 1740, in another letter to Dortous de Mairan, Parrenin once again mentioned Chinese ancient mathematics and gave a more detailed description:

> I think I have already mentioned to you that they have known the right-angled triangle since olden times. According to the Kangxi Emperor, it is not possible to know from when this knowledge dates back. One reads that the predecessor of the famous Duke of Zhou, who lived about eleven centuries before Jesus Christ, told his disciple that with this instrument, one can make several observations, and that Yu had used it to measure heights. It is not said that Yu invented it, but that he used it.
>
> How did this knowledge later pass on to Pythagoras, who earned so much honour from it? Did he invent it (for it is not impossible to meet in the same fields of knowledge)? Or did he receive it from the Indians, who in turn had received it from the Chinese? This is pure guesswork. One cannot be certain until one unearths other monuments than those we have hitherto been able to see.
>
> (Il me semble que je vous ai déja parlé de l'ancienne connoissance qu'ils avoient du triangle rectangle, de laquelle, selon le témoignage de l'Empereur *Cang-hi*, on ne pouvoit assigner le commencement. On lit que le prédécesseur du fameux *Tcheou-cong*, qui vivoit environ onze siecles avant Jesus-Christ, disoit à son disciple, qu'avec cet instrument on pouvoit faire plusieurs observations, & que *Yu* s'en étoit servi pour mesurer les hauteurs. Il n'est pas dit que *Yu* en fût l'Inventeur, mais qu'il en avoit fait usage.
>
> Comment cette connoissance passa-t-elle dans la suite à Pythagore, auquel elle fit tant d'honneur? L'inventa-t-il; car il n'est pas impossible qu'on se rencontre dans les mêmes connoissances? ou bien l'avoit-il reçu des Indiens, & ceux-ci des Chinois? Pure conjecture; on ne peut rien assurer jusqu'à ce qu'on déterre d'autres monumens que ceux que nous avons pu voir jusqu'ici.)[22]

From the above citation, it is clear that the Kangxi Emperor knew something about the right-angled triangle (i.e. the so-called Pythagorean theorem).

Voltaire showed great interest in Jesuit writings on China and Parrenin's letters in *Lettres édifiantes*. He had a very high opinion of Parrenin, as revealed in his *Essais sur les mœurs et l'esprit des nations*. He said:

> One is surprised that this inventive people did not make any breakthrough in geometry beyond the elements. It makes no doubt that they knew about the elements several centuries before Euclid wrote them down when residing among the Greeks of Alexandria. In our days, the Kangxi Emperor has assured Father Parrenin, one of the most knowledgeable and sage missionaries to have approached this Prince, that Emperor Yu made use of the properties of the right-angled triangle to make a geographical chart of a province, more than three thousand nine hundred and sixty years ago. And Father Parrenin himself quotes a book written eleven hundred years before the Christian era in which it is said that the famous demonstration attributed in the West to Pythagoras long ranked among the best-known theorems.
>
> (On est étonné que ce peuple inventeur n'ait jamais percé dans la géométrie au-delà des élémens. Il est certain qu'ils connaissaient ces élémens plusieurs siècles avant qu'Euclide les eût rédigés chez les Grecs d'Alexandrie. L'Empereur Cang-hi assura, de nos jours, au

[22] English translation from *Lettres édifiantes et curieuses écrites des missions étrangeres*. T.26, Paris, 1743, 17–18.

père Parenin (sic), l'un des plus savans & des plus sages missionnaires qui aient approché de ce Prince, que l'empereur Yu s'était servi des propriétés du triangle rectangle, pour lever un plan géographique d'une province, il y a plus de trois mille neuf cent soixante années; & le père Parenin (sic) lui même cite un livre écrit onze cents ans avant notre ère, dans lequel il est dit que la fameuse démonstration, attribuée en occident à Pythagore, était depuis longtemps au rang des théorèmes les plus connus.) (Voltaire 1773: 358–359)

The book Parrenin cited was written in about 1100 BCE, very close to the era in which Duke Zhou lived. In addition, the book mentioned the famous Chinese demonstration similar to that of Pythagoras. This shows quite clearly that Parrenin had read *The Gnomon of the Zhou [Dynasty]*.

In addition to Parrenin, the French Jesuit J.-F. Foucquet (1665–1741) was also interested in *The Gnomon of the Zhou [Dynasty]*. Invited by the Kangxi Emperor to study the *Book of Changes*, he arrived in Beijing in 1711 and served at the court for more than 11 years. He translated European algebra and many astronomical texts into Chinese for the Kangxi Emperor (Witek 1982). He also collected many Chinese books, some of which he took back to Europe when he left China in 1722. They included a Ming dynasty edition of *The Gnomon of the Zhou [Dynasty]*.[23] When I checked this edition in the British Library in 1995, I discovered the existence of many commentaries from Foucquet himself, who cited a considerable number of Chinese book titles related to *The Gnomon of the Zhou [Dynasty]*.

Aside from Parrenin and Foucquet, another French Jesuit, Antoine Gaubil (1689–1759), also did some research on *The Gnomon of the Zhou [Dynasty]*. Gaubil tried to prove that China had an ancient tradition of astronomy and mathematics in order to refute some European scholars' negative views of ancient Chinese science (Han 1998: 43–59). He wrote numerous texts on Chinese chronology and the history of Chinese astronomy (Han 2012: 71–79). It was Gaubil who was responsible for widely publicizing the history of ancient Chinese astronomy in the eighteenth and nineteenth centuries.

## 4.7 *The Gnomon of the Zhou [Dynasty]* and Its Commentators in the Mid-eighteenth Century

By the early eighteenth century, Chinese scholars were aware of the possible contributions that European methods could make to traditional Chinese mathematical astronomy. Consequently, many Chinese scholars regretted that they could not understand the details of Western mathematics needed for astronomy in general and the determination of the calendar in particular. This also reflected the basic interest of Chinese scholars in Western science. Following on from the early Qing period, scholars of the Qianlong–Jiaqing period (1736–1820) were also very interested in Western learning and were thus conscious of many different advances in Western

[23]This Ming edition (Wanli era) was collated by Zhao Kaimei (趙開美) and published by Hu Zhenheng (胡震亨).

science. They applied Western sciences to the study of Chinese classics and traditional mathematical sciences, specifically to studies of the Chinese calendar and chronology, because Western sciences offered methods of calculation that were especially effective in the prediction of solar and lunar eclipses, the conjunction of the five planets and so forth. Consequently, their studies stimulated major advances in traditional learning.

With support from the Qianlong Emperor (1711–1799), a large-scale project was launched in 1772 to compile the *Complete Library of the Four Branches* (*Siku quanshu*, 四庫全書). Dai Zhen (戴震, 1724–1777), who was very famous for the historical scholarship characterized as 'evidential research' (*Kaozheng* 考證) in the Qianlong period,[24] was invited to join the compilation and was involved in editing and collating the *Ten Mathematical Classics*. He also tried to study traditional science based on Western science. He systematically collated the text of *The Gnomon of the Zhou [Dynasty]* using different versions held in the Imperial Library.[25] His edition was later printed with wooden movable type and was widely distributed throughout China. He also commented on *Xuanji* and *Yuheng*, the astronomical instruments in *The Gnomon of the Zhou [Dynasty]* (Qian Baocong 1958: 45–46). As a leading figure in *Kaozheng* scholarship, Dai Zhen's views exercised considerable influence on other scholars. To some extent, he played a crucial role in stimulating the study of traditional calendrical and mathematical sciences in the Qianlong period. His influence helps to explain further how official opinions of Western learning were transmitted from the court to the scholarly region of Jiangnan (the present-day Jiangsu, Zhejiang and Anhui provinces).[26]

During the compilation of the *Complete Library of the Four Branches*, many rare books were presented to the Imperial court by book collectors from all over China. One of the most valuable is the delicate handwritten copy of *The Gnomon of the Zhou [Dynasty]* based on the Song edition (1213). The Qianlong Emperor composed a poem about it in 1783 depicting the gifts and ability of his grandfather (the Kangxi Emperor) in mathematics,[27] especially in the knowledge of triangle. He lamented that he had missed the opportunity to learn from his grandfather because he was too young, and confessed that he was ashamed of his lack of ability in mathematics. He therefore encouraged his children to learn from his uncle Prince Zhuang (允祿, 1695–1767), who was the sixteenth son of the Kangxi Emperor and had learned mathematics from the Jesuits. The Qianlong Emperor's support for the academic project stimulated the flourishing of research on traditional mathematics.

---

[24] Eighteenth-century evidential research scholars placed emphasis on painstaking research, rigorous analysis and impartial evidence drawn from ancient artifacts and historical documents and texts. On *Kaozheng* scholarship, see Elman (2005: 191).

[25] Dai Zhen relied in fact on Hu Zhenheng's edition and consulted the edition of the *Great Compendium of the Yongle Era*.

[26] On the social context of the shift of scholarship from the court to the Jiangnan region, see Han (2015): 86–87).

[27] The Imperial copy is still preserved in the Palace Museum, Taipei.

## 4.8 An Qingqiao and His Research on *The Gnomon of the Zhou [Dynasty]*

The theory of the Chinese origin of Western learning proposed by the Kangxi Emperor and Mei Wending was repeated and cited by many scholars in the eighteenth century. However, in the late eighteenth and early nineteenth century, An Qingqiao (安清翹, 1756–1829), a mathematician from Yuanqu (垣曲) County, Shanxi Province, opposed this theory based on his research on *The Gnomon of the Zhou [Dynasty]*. He began by discussing the conflict between Western and Chinese mathematics, and continued:

> In late Ming, when mathematics became lax, [people from] the West Ocean took advantage of this weak point and entered [China]. At that time, those who observed [maintained] old methods rigidly adhered to preconceived ideas and could not adapt [to Western methods] in order to incorporate the strong points. However, those who talked about Western methods were also puzzled by Westerners. They always used new theories to slander people from the distant past. [They are like] water and fire, diametrically opposed to each other and disputing together endlessly. How regrettable this is! At that time, Xu Wending [徐文定, i.e. Xu Guangqi] ranked first among those who knew mathematics. However, he was almost led into confusion by Westerners. The so-called 'Melting the elegant Western calculations to cast them in the *Datong* (大統, a calendar of Ming dynasty) Mould' was mere phraseology. (明季當數學廢弛之際，西洋乘其虛而入，爾時守古法者拘於成見，不能變通以兼收西法之長，而講西法者又為西人所惑，每執新說以詆諆古人，水火交爭，聚訟不休，良可歎也。其時知數者首屬徐文定，然文定亦幾為西人所惑矣。所謂鎔西方之巧算，入大統之型模者，亦徒有其說耳。) (An Qingqiao 1811: *Juan* 4, 49).[28]

He also criticized Jiang Yong's (江永, 1681–1762) favorable attitude toward Western mathematics[29]:

> [Mr Mei, i.e., Mei Wending] was not willing to discard [abandon] ancient methods and to follow Western methods completely. That was really a judicious judgment. However Jiang Shengxiu [江慎修, i.e., Jiang Yong] slandered him. What was his motivation? Mei Yuru [梅玉汝, i.e., Mei Juecheng] said that Jiang was flattering the Westerners. Although his words were excessively unkind, they were indeed not without reason. ([梅勿庵先生]不肯棄古法盡從西法，誠為卓識。而江慎修又從而詆之，是誠何心哉?梅玉汝謂江氏諂附西人，雖其言過刻，要不為無因矣。) (An Qingqiao 1811: *Juan* 4, 50).

Although An Qingqiao personally supported Chinese mathematics, he opposed the theory of the Chinese origin of Western learning and made detailed criticisms of Mei Wending's opinion. He wrote:

> When Western methods entered China, people from the land of China mostly doubted them. By the time those methods prevailed, they all followed their [Western] theory. Mei Wu'an [梅勿庵, i.e., Mei Wending] knew both Chinese and Western methods and therefore proposed the theory that Western methods originated from the land of China. (西法始

[28] This section is entitled 'Mathematics should mainly rely on the Chinese method' (象數以中法為主).

[29] Jiang Yong, a native from Anhui province, was radical in his critique of Mei Wending for exalting indigenous natural studies. He recognized the advantages of European astronomy over native traditions. See Elman (2005: 256).

入中土，中土人多疑之。及其法盛行，則又悉從其說。梅勿庵兼綜中西，遂有西法傳自中土之論。) (An Qingqiao 1811: *Juan* 4, 51).

He went on to list Mei's arguments in four parts. Mei mentioned the story in the *Canon of Yao* (*Yaodian*, 堯典), a chapter of the *Book of Documents* (*Shujing*, 書經). He recounted that when the Zhou dynasty declined, mathematicians dispersed, taking instruments for calculation with them to far distant places. The Xi and He brothers were dispersed in all four directions.[30] Only the second He brother was sent to the west and ordered to make observations. This meant, Mei argued, that Westerners' knowledge of mathematical sciences could originate from ancient Chinese mathematicians. Mei's second argument was that the Western theory about cold and warm zones coincided with that contained in *The Gnomon of the Zhou [Dynasty]*. The European theory of the round earth was also illustrated in *The Gnomon of the Zhou [Dynasty]*. Mei's third argument was that the cosmological system introduced by M. Ricci was the same as the old method of *Gai-tian* theory.[31] Finally, Mei argued that '*Sanjiao baxian*' (三角八線) was known by ancient people. Westerners, however, could apply it more subtly.[32] Based on these four reasons, Mei concluded that the Western methods originated from China. However, An Qingqiao disagreed with these ideas and thought that the evidence in the *Canon of Yao* was not reliable proof of the Chinese transmission of knowledge to Western countries. In addition, he believed that *The Gnomon of the Zhou [Dynasty]* did not clearly explain the round earth. He stated:

> What *Zhoubi* and the Western method share is that they have the same principle. The basic principle [*li*, 理] does not differ because of [its origins from] the East or West Ocean. If one thinks that the Western method originated from *Zhoubi* just because they have the same principle, is that really right? (然則西法同于《周髀》者，乃其理同耳。理之至者，不以東海、西海而異，今因其理之同，而謂西法出於《周髀》，其果然否耶?) (An Qingqiao 1811: *Juan* 4, 51–52)[33]

This long essay was quite exceptional in the history of Chinese mathematics. To date, we have not found any similar discussion which boldly opposed the theory of the Chinese origin of Western learning, the official slogan propagated by the Kangxi Emperor.

---

[30] This story was mentioned earlier in the *Ming History*. Concerning this story, see Needham (1959: 186–188).

[31] Ricci's cosmological system was depicted in the *Illustrated Explanation of the Unified Principle of Huntian and Gaitian Theories* (渾蓋通憲圖說), translated from Clavius' *Astrolabium* (Rome, 1593) with the assistance of Li Zhizao. This title implies that *Gaitian* and *Huntian* theories could be united in a single principle.

[32] *Sanjiao baxian* was firstly introduced into China in the early eighteenth century. *Sanjiao* literally means 'trigonometry', while *baxian* means eight trigonometric lines related to the division of the circle.

[33] This section is entitled 'Western methods were not necessarily transmitted from the land of China'. (西法不必傳自中土).

## 4.9 Concluding Remarks

The study of *The Gnomon of the Zhou [Dynasty]* in Qing China represents an interesting episode in the historiography of Chinese science. One of the earliest ancient mathematical texts, this book was used by the Kangxi Emperor and Mei Wending, as well as other later Qing scholars, a number of French Jesuits and even Voltaire as a tool for comparing and establishing relationships/similarities between Chinese and European mathematics.

Xu Guangqi was the first to mention the similarity between the *Gougu* method in *The Gnomon of the Zhou [Dynasty]* and the Pythagorean theorem in Western mathematics. His "discovery" was followed by Chen Jinmo, Mei Wending and other mathematicians in the mid-eighteenth and the nineteenth centuries. The restoration of historical memory greatly stimulated the study of traditional Chinese mathematics.

Proposing the theory of the Chinese origin of Western learning made the Chinese literati aware that traditional Chinese science had certain strengths as compared with Western science. It was therefore worthwhile to propagate them to enhance Chinese self-confidence. This led to a great deal of new research by Qianlong–Jiaqing scholars in the field of mathematical science. Some mathematicians at that time devoted their lives to the study of mathematics and calendrical sciences. The new approach to traditional mathematics mingled with new Western methods that had recently been introduced into China. This reflected the new tendency of the times, which helped the Chinese to achieve a better understanding of their own traditional science.

## References

### *Primary Sources*

1734. *Lettres édifiantes et curieuses écrites des missions étrangères*. T.21, Paris.

An Qingqiao. 1811. *Shuxue wushu* 數學五書 (推步惟是) (*Five books of mathematics*), Shurentang edition.

Chen Jinmo, *Duce* 度測 (*On Measurement*), preface written in 1640, manuscript copy preserved at the Library of the Institute for the History of Natural Sciences, Chinese Academy of Sciences.

Han Qi (韓琦) and Wu Min (吳旻) (eds). 2006. Xichao chongzheng ji, Xichao ding'an wai san zhong <熙朝崇正集><熙朝定案>外三種. Beijing: Zhonghua Book Company.

Landry-Deron, I. 1995. *Les leçons de sciences occidentales de l'empereur de Chine Kangxi (1662–1722): Textes des Journaux des Pères Bouvet et Gerbillon*. Paris: Ecole des Hautes Etudes en Sciences Sociales.

Mei Juecheng (1745), *Chishui yizhen* 赤水遺珍 (*Pearls Lost in the Red River*), in *Xuancheng Mei shi lisuan congshu jiyao* 宣城梅氏曆算叢書輯要 (*Collected Mathematical Writings of Mr. Mei*), Chengxuetang edition.

Mei Wending, *Lixue yiwen bu* 曆學疑問補 (*The Supplement to the Doubts on the Calendar*), Jianjitang edition, 1723.

———, *Wu'an lisuan shumu* 勿庵曆算書目 (*A Descriptive Catalogue of Wu'an's Mathematical Works*). Kangxi edition.

———, *Jixuetang wenchao* 績學堂文鈔 (*Collected Writings of Mei Wending*), Qianlong edition.

———, *Jixuetang shichao* 績學堂詩鈔 (*Collected Poems of Mei Wending*), Qianlong edition.

Qian Baocong (錢寶琮). 1963. *Suanjing shishu* 算經十書 (*Critical punctuated edition of ten mathematical classics*). Beijing, Zhonghua Book Company.

Ricci, Matteo, *Hun gai tong xian tushuo* 渾蓋通憲圖說 (*Illustrated Explanation of the Unified Principle of Huntian and Gaitian Theories*), translated from Clavius' *Astrolabium* (Rome, 1593) with the assistance of Li Zhizao.

*Shuli jingyun* 數理精蘊 (*The Fine Essence of Mathematical Principles*), Kangxi edition.

Voltaire, *Essais sur les mœurs et l'esprit des nations*. Tome1, Neuchâtel, 1773, Chapitre premier. De la Chine, de son antiquité de ses forces, de ses loix, de ses usages & de ses sciences.

Widmaier, R., ed. 1990. *Leibniz korrespondiert mit China: der Briefwechsel mit den Jesuitenmissionaren (1689–1714)*. Frankfurt am Main: Vittorio Klostermann.

*Yuzhi sanjiaoxing tuisuan falun* 御制三角形推算法論 (*Imperially composed treatise on the derivation of triangles*). Kangxi edition.

Zhang Tingyu, ed. 1974. *Mingshi* 明史 *(Ming History)*. Beijing: Zhonghua Book Company.

*Zhoubi suanjing* 周髀算經 (*The Gnomon of the Zhou [Dynasty*]), Ming edition (Wanli era), collated by Zhao Kaimei (趙開美) and published by Hu Zhenheng (胡震亨).

Zhu Zaiyu, *Zhoubi tujie* 周髀圖解 (*Illustrated Explanations of The Gnomon of Zhou [Dynasty]*, Zhu's private edition of the Wanli era (1573–1620).

## Secondary Sources

Bertuccioli, G. 1976. A lion in Peking: Ludovicus Buglio and the embassy of Bento Pereira de Faria in 1678. *East and West* 26: 223–240.

Chemla, K., ed. 2012. *The History of Mathematical Proof in Ancient Traditions*. Cambridge, UK: Cambridge University Press.

———. 2013. Shedding some light on a possible origin of a concept of fraction in China: Division as a link between the newly discovered manuscripts and *The Gnomon of the Zhou [Dynasty]*. *Sudhoffs Archiv: Zeitschrift für Wissenschaftsgeschichte* 97 (2): 174–198.

Chen Cheng-Yih, J. (程貞一) and Wenren Jun (聞人軍). 2012. *Zhoubi suanjing yizhu* 周髀算經譯注 (*Annotated edition of The Gnomon of the Zhou [Dynasty]*). Shanghai: Shanghai Classics Publishing House.

Cullen, C. 1996. *Astronomy and mathematics in Ancient China: The Zhou bi suan jing*. Cambridge, UK: Cambridge University Press.

Elman, B. 2005. *On their Own Terms: Science in China, 1550–1900*. Cambridge, MA: Harvard University Press.

Engelfriet, P.M. 1998. *Euclid in China: The Genesis of the First Chinese Translation of Euclid's Elements, Books I–VI (Jihe Yuanben, Beijing, 1607) and its reception up to 1723*. Leiden: Brill.

Fung Kam Wing (馮錦榮). 2007. Chen Jinmo zhi shengping ji xixue yanjiu陳藎謨(1600?–1692?)之生平及西學研究 (Chen Jinmo: his life and the study of Western learning). *Mingqing shi jikan*明清史集刊 (*Bulletin of Ming-Qing Studies*) 9:209–262.

Golvers, N. 1993. *The Astronomia Europaea of Ferdinand Verbiest, S.J. (Dillingen, 1687): Text, translation, notes and commentaries*. Nettetal: Steyler Verlag.

Han Qi, 1991. *Kangxi shidai chuanru de xifang shuxue jiqi yingxiang* 康熙時代傳入的西方數學及其影響 *(The Transmission of Western Mathematics during the Kangxi Reign and Its Influence on Chinese Mathematics)*. Ph.D diss. Institute for the History of Natural Sciences, Chinese Academy of Sciences.

———. 1996. Junzhu he buyi zhijian: Li Guangdi zai Kangxi shidai de huodong jiqi dui kexue de yingxiang 君主和布衣之間:李光地在康熙時代的活動及其對科學的影響(Between the Emperor and mathematician: Li Guangdi's activity during the Kangxi Reign and its influence on science). *Qinghua xuebao* 清華學報 *(Tsing Hua Journal of Chinese Studies) (Hsinchu), New Series* 26 (4): 421–445. (in Chinese).

———. 1997. Patronage scientifique et carrière politique: Li Guangdi entre Kangxi et Mei Wending. *Etudes Chinoises* 16 (2): 7–37.

———. 1998. Sino-British scientific relations through Jesuits in the seventeenth and eighteenth centuries. In *La Chine entre amour et haine*, ed. M. Cartier, 43–59. Paris: Desclée de Brouwer.

———. 2001. Astronomy, Chinese and Western: The influence of Xu Guangqi's views in the early and mid-Qing. In *Statecraft and intellectual renewal in late Ming China: the cross-cultural synthesis of Xu Guangqi (1562–1633)*, eds. C. Jami, P. Engelfriet, and G. Blue, 360–379. Leiden: Brill.

———. 2003. Antoine Thomas, SJ, and his mathematical activities in China: A preliminary research through Chinese sources. In *The history of the relations between the Low Countries and China in the Qing Era (1644–1911)*, ed. W.F. Vande Walle, 105–114. Leuven: Leuven University Press.

———. 2011. Kexue, zhishi yu quanli: riying guance yu Kangxi zai lifa gaige zhong de zuoyong 科學、知識與權力——日影觀測與康熙在曆法改革中的作用 (Science, knowledge and power: Observations of the shadows of the sun and the Kangxi Emperor's role in the calendrical reform). *Ziran kexueshi yanjiu* 自然科學史研究 *(Studies in the History of Natural Sciences)* 30 (1): 1–18.

———. 2012. The Jesuits and their study of Chinese astronomy and chronology in the seventeenth and eighteenth centuries. In *Europe and China: Science and arts in the 17th and 18th centuries*, History of Mathematical Sciences: Portugal and East Asia IV, ed. Luís Manuel Roberto Saraiva, 71–79. Singapore: World Scientific Publishing Co Pte Ltd.

———. 2014. Knowledge and power: A social history of the transmission of mathematics between China and Europe during the Kangxi Reign (1662–1722). In *Proceedings of the International Congress of Mathematicians, Vol. IV*, ed. S.Y. Jang, Y.R. Kim, D.-W. Lee, and I. Yie, 1218–1229. Seoul: Kyung Moon Sa. Co. Ltd.

———. 2015. Chinese literati's attitudes toward Western science: Transition from the late Kangxi period to the mid-Qianlong period. *Historia Scientiarum* 24 (2): 76–87.

———. 2016. Kangxi di zhi zhi shu yu Xixue zhongyuan shuo xin lun: *Yuzhi sanjiaoxing tuisuan fa lun* de chengshu jiqi beijing 康熙帝之治術與"西學中源"說新論——<御制三角形推算法論>的成書及其背景 (Using knowledge as power: A new interpretation of the birth of the Kangxi Emperor's 'Imperially Composed Treatise on the Derivation of Triangles' and his theory of the Chinese origin of Western learning). *Ziran kexueshi yanjiu* 自然科學史研究 *(Studies in the History of Natural Sciences)* 35 (1): 1–9.

Hart, R. 2013. *Imagined civilizations: China, the West and their first encounter*. Baltimore, Md: Johns Hopkins University Press.

Needham, J. 1959. Mathematics and the Sciences of the Heavens and Earth, in Science and civilisation in China. vol. 3. Cambridge, UK: Cambridge University Press.

Qian Baocong (錢寶琮). 1958. Gaitian yuanliu kao 蓋天源流考 (A study of the origin and history of Gaitian theory). *Kexue shi jikan* 科學史集刊 (*Bulletin of History of Science*), 1: 29–46.

Witek, J.W. 1982. *Controversial ideas in China and in Europe: A Biography of J. F. Foucquet, S.J. (1665–1741)*. Rome: Institutum Historicum S.I.

# Chapter 5
# The Asiatic Society, the Bibliotheca Indica and Devanāgarī Printing in Bengal: The Historical Context of the *Editio Princeps* of the *Nyāyabhāṣya*

**Alessandro Graheli**

**Abstract** The *editio princeps* of the *Nyāyabhāsya* of Vātsyāyana Pākṣilasvāmin (ca. fifth century CE) was published by the Asiatic Society of Bengal in 1864–1865 and edited by the Nyāya scholar Jayanarayana Tarkapanchanana. The influential *Nyāyabhāṣya* is the textual root of the prominent Nyāya tradition of epistemology and logic, yet in the nineteenth century it was largely unknown and understudied even by Indian scholars. In his foreword, Jayanarayana predicted the possibility of saving the *Nyāyabhāṣya* from oblivion and reviving it by means of the pioneering movable type technology in Devanāgarī characters. His endeavor was certainly fulfilled, since his seminal edition has inspired several improved editions, translations and studies.

This contribution contains a thorough description of Jayanarayana's edition, analyzed in the broader context of the history of the Asiatic Society, of the Bibliotheca Indica Series, and of the early Devanāgarī press in British India. Particularly, Jayanarayana's foreword in Sanskrit verses and some key variant readings in his edited text of the *Nyāyabhāṣya* are here discussed in their historical context and matched with the information derived from the manuscript tradition. Although Jayanarayana's manuscript sources are barely mentioned in his edition, it is now possible to identify at least two of the three manuscripts used by him with a fair

The research leading to these results has received funding from the European Research Council under the European Union's Seventh Framework Programme (FP7/2007–2013)/ERC Grant Agreement No. 269804. This paper was written with the support of the Fonds zur Förderung der wissenschaftlichen Forschung (Austrian Science Fund), during FWF projects M1160 G-15 (2009–2011), P19328 (2011–2012), P24388 (2012–2015), and P28069 (2015–2019). The manuscript images used in this paper were generously supplied by the Sanskrit College of Kolkata, India, and by the Jawahar Lal Nehru Library, University of Kurukshetra, India.

A. Graheli (✉)
Faculty of Arts and Science, Department for the Study of Religion, University of Toronto, Toronto, ON, Canada
e-mail: alessandro.graheli@utoronto.ca



A. Keller, K. Chemla (eds.), *Shaping the Sciences of the Ancient and Medieval World*, Archimedes 69, https://doi.org/10.1007/978-3-031-49617-2_5

degree of certainty, after the thorough investigation on the textual transmission of the *Nyāyabhāṣya* done in Vienna in the course of several research projects. We now know how his three manuscripts represent the latest stage of the much richer transmission of the *Nyāyabhāṣya*. The present study also shows how a late manuscript from Kashmir in Śāradā script likely descended from this *editio princeps*, an hypothesis that opens a new scenario of horizontal transmission of readings.

More in general, the purpose of this paper is to provide the context for a better understanding of the editorial strategies of the Bibliotheca Indica, of its editors' motives, and of the impact of the early press on Sanskrit on Sanskrit scholarship.

## Abbreviations and Primary Sources


ATV Jayanārāyaṇa Tarkapañcānana, ed. (1849). *Ātmatattvaviveka of Udayanācārya*. Kolkata: Sanskrita Yantra.

C1$^{D}$ "Kolkata, Sanskrit College, ms. 201" (n.d.). Foolscap paper, Devanāgarī, 129 fols., 203 × 305 mm. The manuscript begins in the middle of NBh ad NS 1.1.2, missing the first five folios, and ends within NBh ad NS 5.2.17, missing the last folio.

C2$^{D}$ "Kolkata, Sanskrit College, ms. 202" (n.d.). Country-made paper, Devanāgarā, 24 fols., 356 × 178 mm. The manuscript abruptly ends within NBh ad NS 2.1.36.

C3$^{D}$ "Kolkata, Sanskrit College, ms. 332" (n.d.). Country-made paper, Devanāgarā, 121 fols., 210 × 75 mm, missing the first and last folios. The manuscript begins in the middle of the introduction to NBh ad NS 1.1.1 and ends abruptly within NBh ad NS 5.2.11.

Ku$^{S}$ "Kurukshetra, Jawahar Lal Nehru Library, Kurukshetra University, ms. 52903" (n.d.). Paper, Śāradā, 170 × 220, 141 fols., complete, Saṃvat 56 (1880 CE?).

NBh$^{Ga}$ Gangadhara Shastri, ed. (1896). *The Nyāyasūtras with Vātsyāyana's Bhāṣya and Extracts from the Nyāyavārttika and the Tātparyaṭīkā*. Vizianagram Sanskrit Series 9. E.J. Lazarus and Co.

NBh$^{Ja}$ Jayanarayana Tarkapanchanana, ed. (1864–1865). *Nyāyadarśana with the Commentary of Vātsyāyana*. Bibliotheca Indica 50. Calcutta: Asiatic Society of Bengal.

NBh$^{Tha}$ Anantalal Thakur, ed. (1997). *Gautamīyanyāyadarśana with Bhāṣya of Vātsyāyana*. Nyāyacaturgranthikā 1. New Delhi: Indian Council of Philosophical Research.

NKu E. B. Cowell, ed. (1864). *The Kusumànjali or Hindu Proof of the Existence of a Supreme Being*. Bibliotheca Indica. With the commentary of Hari Dàsa Bhaṭṭàchàrya. Edited and translated by E.B. Cowell, M.A., assisted by Paṇḍita Mahésa Chandra Nyàyaratna. Calcutta.

NSM$^{Kā}$ Kāśinātha Tarkapañcānana, ed. (1821). *A System of Logic Written in Sunscrit by the Venerable Sage Boodh, and Explained in a Sunscrit*

*Commentary by the Very Learned Viswonath Turkaluncar. Maharṣigautamakṛta nyāyadarśanaṃ mahāmahopadhyāya śrīviśvanātha tarkālaṃkārakṛta tadīya bhāśāparicchedaḥ*. Printed in Bengali characters at the Baptist Mission Press, Circular Road. Calcutta: Calcutta School-Book Society.

NSM$^{\mathrm{Ro}}$ Eduard Röer, ed. (1850). *Viśvanāthapañcānanakṛtasid-dhāntamuktāvalī sahitabhāśāpariccheda. Division of the Categories of the Nya'ya Philosophy*. Vol. 9. Bibliotheca Indica 32, 35. Edited in Devanagari characters and translated. Printed at the Baptist Mission Press, Circular Road. Calcutta: Asiatic Society of Bengal.

NSM$^{\mathrm{sc}}$ Committee of Public Instruction (1827). *The Bhasha Pariccheda and Siddhanta Muktavali. An Elementary Treatise on the Terms of Logic with its Commentary. By Viswanatha Panchanana Bhatta*. Published for the use of the Sanskrit College. Calcutta: Committee of Public Instruction, Education Press.

NSu$^{\mathrm{Ba}}$ J.R. Ballantyne, ed. (1850–1854). *The Aphorisms of the Nyàya Philosophy by Gautama, with Illustrative Extracts from the Commentary by Viśvanàtha in Sanskrit and English*. Printed for the use of the Benares College, by order of Govt. N. W. P., at the Presbyterian Mission Press, Rev. Jos. Warren, Superintendent. Vol. 1 1950, vol. 2 1953, vol. 3 1954. Allahabad: Benares College.

NSuV Mahacandra Siromani, ed. (1828). *Nyàya Sùtra Vritti, the Logical Aphorisms of Gotama, with a commentary by Viswanàtha Bhattàchàrya. Published under the authority of the General Committee of Public Instruction*. Devanagari title-page: śrīviśvanāthabhaṭṭācāryya — kṛtonyāyasātravṛttināmāyaṃ granthaḥ — śrāyuktakamiṭṭisāhebānām anujãayā — śrīmahācandraśiromaṇiśodhitaḥ — kalikātākhyarājadh ānyāṃ saṃskṛta — pāṭhaśālāyantrālaye mudritaḥ — pañcāśadadhikasap tadaśaśākavatsare — iṃlaṇḍīyāṣṭāviṃśatyadhikāṣṭādaśavatsare //. Education Press (Sanskrit College Press).

PTS Jayanārāyaṇa Tarkapañcānana (1867). *Padārthatattvasāra*. Kalikātā: Nūtanasaṃskṛtayantra.

ṚV Eduard Röer, ed. (1848–1849). *Ṛgveda*. Vol. 1. Bibliotheca Indica, Old Series 1–4. Printed at the Baptist Mission Press, J. Thomas. Calcutta: The Asiatic Society of Bengal.

SDS$^{\mathrm{Ja}}$ Jayanarayana Tarkapanchanana, ed. (1865). *Sarbbadarśanasaṃgraha. Jayanārāyaṇa Tarkapaṇcānana karttṛka Bāṅgalā bhāṣāya saṇkalita*. In Bengali characters. Kalikātā: Bidyāratna Yantra.

SDS$^{\mathrm{Vi}}$ Īśvaracandra Vidyāsāgara, ed. (1853). *Sarvadarśanasaṇgraha*. Bibliotheca Indica, Old Series 63. Printed by J. Thomas, Baptist Mission Press. Calcutta.

SVij Nawadvipa Chandra Goswami and Jayanārāyaṇa Tarkapañcānana, eds. (1864). *The S'ankara-Vijaya*. Bibliotheca Indica, New Series 46, 137, 138. Printed by Somaprakāśa Press, up to ch. 33, and by C.B. Lewis, the Baptist Mission Press. Calcutta: The Asiatic Society of Bengal.

VSu Jayanārāyaṇa Tarkapañcānana, ed. (1861). *The Vai'seshika Dars'ana. With the Commentaries of S'ankara Mis'ra, and Jayana'ra'yana Tarka Pancha'nana.* Bibliotheca Indica 18. Kolkata: The Asiatic Society of Bengal. Printed by C.B. Lewis, Baptist Mission Press.

## 5.1 Introduction

Nyāya is one of the major philosophical traditions of India, specialised in dialectics, metaphysics and epistemology. The oldest known Nyāya work is the *Nyāyasūtra*, an aphoristic treatise in five chapters ascribed by the tradition to Akṣapāda, also known as Gotama or Gautama. Its aphorisms were embedded in its earliest commentary, the *Nyāyabhāṣya* (henceforth NBh), composed by Pākṣilasvāmin Vātsyāyana not later than the fifth century CE. This commentary has in many ways shaped the later history of Nyāya, most importantly by preserving a manuscript tradition of the *Nyāyasūtra* contained in it. In fact, the transmission of independent *Nyāyasūtra* manuscripts is a later phenomenon, and such *Nyāyasūtra* texts seem to be the product of an extrapolation of the aphorisms from the NBh and later commentaries. In this sense, the NBh can be considered the very foundation of the Nyāya school.

In the course of the research project "Metaphysics and Epistemology of the Nyāya Tradition" based at the University of Vienna,[1] copies of about sixty manuscripts of the NBh were procured, and their textual evidence was collated and studied, resulting in a thorough description of the available testimonies, a full bibliography of the sources, the complete collation of the text of the extensive and foundational first part of the NBh, the formulation of a well-founded stemmatic hypothesis and finally, in a new critical edition supported by the *stemma codicum*.

Most of the existing printed editions of the NBh were also acquired, described, collated and analyzed in the course of the project. To date, at least 26 editions of the NBh have been published. Among these, only a few can be considered "critical" in the sense that they use readings from manuscript sources, most notably $\text{NBh}^{\text{Ja}}$, $\text{NBh}^{\text{Ga}}$, and $\text{NBh}^{\text{Tha}}$.

The topic of this paper is the historical context of the publication of $\text{NBh}^{\text{Ja}}$, the *editio princeps* of the NBh edited by Jayanārāyaṇa Tarkapañcānana in Kolkata, West Bengal, in 1865. The editor's relationship with the traditional lore, with the printing technology, and with British institutions will be discussed with detailed references to the early history of Devanāgarī printing, of the Asiatic Society and of the Bibliotheca Indica Series.

[1] Specifically, projects P17244, P19328 and P24388 granted by the FWF (Fonds zur Förderung der wissenschaftlichen Forschung).

## 5.2 The Novelty of an Ancient Work

Prior to NBh$^{\text{Ja}}$, a few Nyāya publications had already been printed, all in Kolkata and Varanasi. The *Siddhāntamuktāvalī*, also known as *Bhāṣapariccheda*, had been recently translated for the Bibliotheca Indica (NSM$^{\text{Ro}}$). In Rüer's words, (NSM$^{\text{Ro}}$ : xxiv), "[The *Bhāṣapariccheda*] is a text-book in the schools of Nyáya in Bengal. There is no Pandit of any repute who does not know it well, and many know the whole by heart." The Sanskrit text had previously appeared in print, both in Bengali (NSM$^{\text{Kā}}$) and in Devanāgarī characters (NSM$^{\text{sc}}$). In 1828, the Calcutta Sanskrit College had published a Devanāgarī edition of Viśvanātha's *Nyāyasūtravṛtti* (NSuV), which was later translated into English in 1850–1854 (NSu$^{\text{Ba}}$).

All these editions and translations were conceived as textbooks to be used in teaching and learning Navyanyāya, the "new Nyāya" flourished in eastern India after the fifteenth century. Both the *Siddhāntamuktāvalī* and the *Nyāyasūtravṛtti*, indeed, were well known works, which had long been taught and studied in the traditional curriculum. By contrast, the more ancient Nyāya works of Prācīnanyāya ("old Nyāya"), such as the NBh, were falling into oblivion. In the anonymous preface of NBh$^{\text{Ja}}$, Henry Thomas Colebrooke (1765–1837) is quoted as saying that "the Bha'shya 'is repeatedly cited by modern commentators as well as by writers of separate treatises,' but he [i.e., Colebrooke] had never met with a copy of the work; and in fact, although it is, after the Su'tras, the most ancient authority for the Naiya'ika system, few paṇdits know more of it than the name."

The concern about the threat of losing this ancient work forever and the hope of imparting new life to it by means of a printed edition were apparently shared by Jayanārāyaṇa as well, when he wrote: "This ancient Nyāyabhāṣya was rare, almost lost. This [Asiatic] Society has engaged me in editing and printing it."[2]

In 1864 E.B. Cowell, who was about to leave his position as principal of the Calcutta Sanskrit College, where Jayanārāyaṇa was teaching, explicitly affirmed the importance of an edition of the work (NKu: Preface, ix, footnote): "I hope that the Asiatic Society will publish the Nyàyabhàshya in the Bibliotheca Indica ... No work could be of greater importance for giving an insight into the older Nyàya." In fact, the first of the three fascicles of NBh$^{\text{Ja}}$ was issued soon after that, at the end of 1864, and NBh$^{\text{Ja}}$ was destined to be the earliest printed text in the genre of Prācīnanyāya.

Why was the NBh published in the Bibliotheca Indica Series? What type of audience did the editors of the series have in mind? Why were Indian pandits collaborating with the British establishment? Were there political motives behind the activities of the Asiatic Society? Such questions can be satisfactorily addressed only in the broader context of the early history of the Asiatic Society, of Devanāgarī printing in Bengal, and of the Bibliotheca Indica.

---

[2] *nyāyabhāṣyam idaṃ pūrvvaṃ viralaṃ luptavat sthitaṃ / mudraṇāśodhane 'syeyaṃ sā sabhā māṃ nyayojayat //6//* (NBh$^{\text{Ja}}$ : ii).

## 5.3 The Asiatic Society

### *5.3.1 Foundation, Mission and Membership*

The "Asiatick Society," later known as "Asiatic Society of Bengal," was established in Kolkata on 15 January 1784, during a meeting of "thirty gentlemen" who "represented the *élite* of the European community in Calcutta at the time" (Mitra et al., 1885, Part I, 2). The mission statement was originally formulated as the object of an "enquiry into the history and antiquities, arts, sciences and literature of Asia." In his inaugural address, Sir William Jones, the founder and first president (1784–1794), expressed the aim of the Society in these words: "If now it be asked, what are the intended objects of our enquires within these spacious limits, we answer, Man and Nature; whatever is performed by the one, or produced by the other."

In 1793, in his tenth anniversary discourse, William Jones stressed that the centre of their enquiries was "the history of India" (Jones et al., 1798, vi), and that "[t]his outline of human transactions," i.e., the "civil and natural history" (1798, xx) concerning "Egyptians, Indians, Goths," etc. (1798, iv), could be completed and strengthened only "by the help of Asiatic literature" (1798, vi). In this very same discourse Jones also promised some "remarks on Asiatic philosophy" for the following year's meeting (1798, xx). In February 1794, indeed, he spoke on "the philosophy of the Asiatics" (1798, 165–185), a speech in which he seemed impressed by the depth of philosophical Sanskrit works. In particular, he praised Śaṅkara's commentary of the Vedāntasūtra, which he had probably read.[3] He went on to assert that "until an accurate translation of it shall appear in some European language, the general history of philosophy must remain incomplete", and argued in favor of "one correct version of any celebrated Hindu book," which would have been "of greater value than all the dissertations or essays" (1798, 170–171). This conviction would later be reflected in the editorial choices of the Bibliotheca Indica.

Jones's vision was partially reflected, some seventy years later, in Jayanārāyaṇa's own description of the Society:

> This earth has been divided into four parts [i.e., continents] by the English scientists. That [part] which includes Bhāratavarṣa (i.e., India) is called Asia. The celebrated Asiatic Society, based in this city of Kolkata and shining with experts in several sciences, is committed to bring to light things that exist and have existed there [in Asia].[4]

Jones has been both glorified and criticized as one of the fathers of Orientalism. For instance, he was criticized by his contemporary James Mill (1773–1836), in *The History of British India*, for his "illusions about the Hindus" and for being worse

[3] Jones et al. (1798, 173) did not have "the leisure" to study Nyāya, but he confirmed that he had "frequently seen perfect syllogisms in the philosophical writings of the Bráhmens".

[4] *bhāgaiś caturbhir bhaktā bhūr iyam iṃlaṇḍapaṇḍitaiḥ / yadbhāge bhārataṃ varṣam āśiyā so 'bhidhīyate //1// tatra yāvanti vastūni tiṣṭhanti sthitavanti ca / atyantasvāntahārīṇi tatprakāśanatatparā //2// kalkyātākhye 'tra nagare vikhyātā yāśiyāsabhā / vidyotamānā vidvadbhir nānāvidyāviśāradaiḥ //3//* ( NBh$^{Ja}$ : ii).

than Rousseau in his "his rhapsodies on the happiness and virtues of savage lives" (quoted in Mukherjee, 1964, 37). He has been described as the first scholar who studied Sanskrit "for a non-utilitarian standpoint" (Master, 1946, 799), but has also been blamed as one of the early causes of the Orientalistic bias: "To rule and to learn, then to compare Orient with Occident: these were Jones's goals …" (Said, 1979, 78). For sure, he was a man of his times with his own contradictions, who "wrote devotional hymns to Hindu deities and launched modern linguistics by postulating the Indo-European link, yet he also judged in court that an Indian servant may be beaten to death with a cane …" (Hoerner, 1995, 215). For the purpose of this paper, it is important to stress how there were different motives behind Jones's judicial writings and his works on poetic or religious literature, as convincingly argued by Rocher (1993b): while Jones's publications on law were commissioned by the Government, his religious and philosophical studies were "Literary Delights … unrelated to governmental concerns and without governmental applications" (Rocher, 1993b, 241–242).

Jones died on 27 April 1794, and his successor at the Presidency of the Asiatic Society, John Shore, disclosed a list of "desiderata" left by him, mostly concerning research on Indian sciences and, more marginally, on Arabia, Persia, China and Tartary. The list of desiderata related to India included fourteen items, which would later be partly covered by the Bibliotheca Indica series (Jones et al., 1798, 192–193):

I The Ancient Geography of India &c. from the Purános.
II A Botanical Description of Indian Plants, from the Cóshas, &c.
III A Grammar of the Sanscrit Language, from Pánini, &c.
IV A Dictionary of the Sanscrit language, from thirty-two original Vocabularies and Niructi.
V On the Ancient Music of the Indians.
VI On the Medical Substances of India, and the Indian Art of Medicine.
VII On the Philosophy of the Ancient Indians.
VIII A Translation of the Véda.
IX On Ancient Indian Geometry, Astronomy, and Algebra.
X A Translation of the Purános.
XI Translations of the Mahábharat Rámáyan.
XII On the Indian Theatre, &c. &c. &c.
XIII On the Indian Constellations, with their Mythology, from the Purános.
XIV The History of India before the Mahommedan conquest, from the Sanscrit-Cashmír-Histories.

Jones had often expressed the purpose of the Society as "pure quest for knowledge," in statements such as:

> If the result of our observations had been totally different, we should nevertheless have published them, not indeed with equal pleasure, but with equal confidence; for *Truth is mighty*, and whatever be its consequences, *must always prevail* (Jones et al., 1798, iv).[5]

[5] Italics in the original.

On another occasion he said:

> It may be advisable at first, in order to prevent any difference of sentiment on particular points not immediately before us, to establish but one rule — namely, to have no rules at all. This only I mean, that, in the infancy of any society, there ought to be no confinement, no trouble, no expense, no unnecessary formality. … I recommend with earnestness — on no account to admit a new member who has not expressed a voluntary desire to become so; and in that case, you will not require, I suppose, any other qualification than a love of knowledge and a zeal for the promotion of it (Mitra et al., 1885, Part I, 5–6).

Until January 1829, membership of the Society was reserved for Europeans. At that point the secretary, H.H. Wilson, proposed some Indian new members, who were then elected. In that same year, Jacquemont (1844, 189) reported that "Five or six native Indians, members of the Asiatic Society, attended the meeting. They were addressed directly as Babou or Monsieur, and in the third person as 'gentlemen;' … they are all Hindus."[6]

Most members of the society were civil or military servants, or businessmen, so the activities of the Society were apparently sporadic and improvised (Mitra et al., 1885, Part I, 18). There is no record of government assistance in the early years, not even when the Society began to raise funds to meet the need for an independent building. "In 1796, when the idea of providing a suitable house was first mooted, funds had to be raised by subscription, and it was ruled that Ordinary members should pay a quarterly contribution of one gold mohur[7] each for the support of the Society" (Mitra et al., 1885, Part I, 8). In 1804, however, an application for land at the Chowringhee Street—Park Street junction was filed and the government granted it in 1805. The house was raised at the cost of 30,000 rupees, apparently the fruit of subscriptions and donations (Mitra et al., 1885, Part I, 22). The Asiatic Society moved to the new premises in 1808.[8]

In the succeeding decades the Asiatic Society gradually built up a collection of books, manuscripts, coins and other articles, and eventually a library and a museum. In the beginning the artifacts were mainly occasional gifts from members, but as the Society gained prestige, important collections were entrusted to it, such as the Mackenzie manuscript collection and the whole library of the college at Fort William.[9] By 1885 2507 manuscripts belonging to the Government of India, and altogether more than 8000 manuscripts, were preserved at the Asiatic Society library (Mitra et al., 1885, Part I, 28). The maintenance of these activities was no longer feasible solely through subscriptions, so the Society occasionally applied for

[6] "Cinq ou six natifs, membres de la Société asiatique, assistaint a la seance. On les appelle *Babou* ou Monsieur; et parlant d'eux, on dit *Gentlemen*; *a native Gentleman*. … ils sont tous Hindous."

[7] A gold coin. In a coinage of the Bengal Presidency, in 1790, it weighed 12.36 grams of pure gold and equalled 15 Rs. (Michael, 2010, 948 and following).

[8] An early lithographic image of the Asiatic Society building, ca. 1830, can be found in Losty (1990, 108).

[9] The library of Fort William, in turn, included older acquisitions and libraries, such as the collection previously belonging to Tipu Sultan of Hyderabad (Buchanan, 1805, 40).

Government grants, such as funding for a curator of the museum in 1837 (Mitra et al., 1885, Part I, 40).

### 5.3.2 Publishing Endeavors and the Challenge of Devanāgarī

Publications were contemplated by the Asiatic Society already in its early days. The first effort resulted in a periodical, first called *Asiatick Miscellany* and then *Asiatick Researches*. Although some lines in Devanāgarī script had already appeared in the *Asiatick Miscellany*, it was only in a rudimentary form and lacked a fully developed font technology. Intensive publishing in the Devanāgarī script would not have been possible without the development of a readable font and a viable printing technique. In that period Western scholars were beginning to fathom the depths of Sanskrit literature, and an effective font was becoming a pressing need. In January 1797, after completing his translation of the *Digest of Hindu Law*, Henry Thomas Colebrooke wrote to his father (Colebrooke, 1839, 27):

> I shall probably visit Calcutta this month, to arrange the mode of putting the work in to the press. ... Types have been lately cast in Calcutta, for printing the Sanscrit language in its appropriate character. This will be early followed by the publication of a dictionary of that language, which a friend of mine is preparing for the press; and this again may be accompanied by a grammar which I have some intention of giving to the public. When these helps to the study of Sanscrit have issued from the press, it is possible that it may be taken up at our universities in England, as Arabic is now studied there.

Incidentally, the choice of Devanāgarī as the "appropriate character" for Sanskrit, and more precisely the choice of a specific variation of Devanāgarī, was due to complex reasons that are beyond the scope of this chapter. Briefly, Sanskrit had been written in several regional scripts throughout the ages. In North India, however, the overwhelming majority of Sanskrit manuscripts happened to be in the Devanāgarī script. Although in Bengal Sanskrit was customarily written in Bengali script, the popularity of Devanāgarī was well established in scholarly circles, so it was the most obvious choice for the dissemination of printed Sanskrit texts.

Indic characters in South Indian scripts had already appeared in print in 1577 (see Naik, 1971, Vol. 1, 228 and Kesavan, 1985, 26 and following). North Indian scripts, however, presented technical obstacles in printing by movable types, especially because of the large number of ligatures. Except for rare instances of block-printing, the earliest specimen using a northern script made by efficient movable types appeared only at the end of the eighteenth century, first in Bengali and then in Devanāgarī.[10] The history of Devanāgarī printing can be traced back to Nathaniel Halhed, a writer in the East India Company service, who in 1778 published a

[10] See Naik (1971, Vol. 1, 242–261), Kesavan (1985, 181–184), and Graheli (2024).

Bengali grammar adopting the first known Bengali font.[11] In the introduction of his grammar, Halhed (1778, xxii) mentions some of the obstacles he encountered:

> That the Bengal letter is very difficult to be imitated in steel will readily be allowed by every person who shall examine the intricacies of the strokes, the unequal length and size of the characters, and the variety of their positions and combinations.
>
> It was no easy task to procure a writer accurate enough to prepare an alphabet of a similar and proportionate body throughout, and with the symmetrical exactness which is necessary to the regularity and neatness of a fount. Mr. Bolts (who is supposed to be well versed in this language) attempted to fabricate a set of types for it, with the assistance of the ablest artists in London. But as he has egregiously failed in executing even the easiest part, or primary alphabet, of which he has published a specimen, there is no reason to suppose that his project when completed ...

According to Halhed's account, it was Charles Wilkins (1749–1836) who deserved the credit for cutting the punches and, most importantly, for devising a new technique to accommodate the complex patterns of ascendants, descendants and ligatures of the script.

> The advice and even solicitation of the Governor General prevailed upon Mr. Wilkins, a gentleman who has been some years in the India Company's civil service in Bengal, to undertake a set of Bengal types. He did, and his success has exceeded every expectation. In a country so remote from all connexion with European artists, he has been obliged to charge himself with all the various occupations of the Metallurgist, the Engraver, the Founder and the Printer.

Once the main problems of the architecture of a Bengali font had been solved, the creation of a Devanāgarī counterpart was only a matter of time, given the structural similarity of the two scripts. In 1795 Wilkins began working at a Devanāgarī font, but as he was ready to print, unfortunately, his workshop was engulfed in flames and his efforts abruptly halted (Wilkins, 1808, xii):

> At the commencement of the year 1795, residing in the country, and having much leisure, I began to arrange my materials, and prepare them for publication. I cut letters in steel, made matrices and moulds, and cast from them a fount of types of the Dēvanāgari character, all with my own hands; and with the assistance of such mechanics as a country village could afford, I very speedily prepared all the other implements of printing in my own dwelling-house; for by the second of May of the same year, I had taken proofs of sixteen pages, differing but little from those now exhibited in the first two sheets ... when alas! The premises were discovered to be in flames ... the whole building was presently burnt to the ground. Greatest part of the punches and matrices was saved but types were ruined.

Wilkins travelled back to England soon afterward, taking the surviving Devanāgarī punches with him. There, in 1808, he eventually managed to use his font in the edition of his *Grammar of Sanskrit Language*.[12]

[11] A complete and compelling biography of Halhed can be found in Rocher (1983a).

[12] Meanwhile, his font had apparently been used for some publications in Tanjore, though it is not clear when exactly this occurred (Naik, 1971, Vol. 1, 262–264).

Wilkins is thus acknowledged as the inventor of the new font technology, but he was also assisted in his efforts by a local blacksmith (*karmakāra*), Panchanan.[13] Once Wilkins had left India for good, the know-how was left in Panchanan's hands.

## 5.4 The Baptist Mission and the Press

William Carey (1761–1834) was a Baptist missionary whose activities had an enormous impact in early typography in Bengal. In 1799, with Joshua Marshman, William Ward and others, he established the Baptist Mission in Serampore, where he soon managed to set up an efficient press and print several volumes in various Indic characters.[14]

By the time Carey had established the Baptist Mission in Serampore, Bengali typography was already gaining ground under the impulse generated by Wilkins and Halhed. Carey was teaching Bengali and Sanskrit at Fort William College (Buchanan, 1805, 127; Drewery, 1979, 125), where he also helped with the production of textbooks (Buchanan, 1805, 135). At some point he came in touch with Panchanan, who was at that time assisting Colebrooke at Fort William. Later, Colebrooke would claim the idea of ligatures cut in single glyphs, which he called "logographs", as his own (Rocher and Ludo, 2013, 72). In this period, apparently, Colebrooke directed the production of types and published books in Devanāgarī both with the Baptist Mission in Serampore and with the Baburam Sanskrit Press in Kolkata (2013, 44). Eventually, Carey managed to hire Panchanan, who moved to Serampore, where there was an urgent need for types to print texts in Bengali, Devanāgarī and other regional scripts for preaching purposes.[15]

> The first major work undertaken by Panchanan at the Serampore Mission was a fount of Devanāgarī types required primarily for printing the Sanskrit grammar then in preparation by William Carey. To assist him in this project Panchanan took on an apprentice of the same

[13] On opinions about the possibility that the engraver Joseph Shepherd helped Wilkins with the first fonts, and on the documented role of Panchanan, see Naik (1971, Vol. 1, ch. 9), Shaw (1981, 69–71), Ross (1999, 10–12), and Ogborn (2007, 242–243).

[14] There must have been good economical and political reason behind the choice of Serampore: "The Danish town of Serampore ... established as a Danish settlement as early as 1755 ... due to wars between England and Napoleon and his allies" was "the only non-English commercial centre which remained free," and "became the great emporium of European private trade in India" (Khan, 1961, 222).

[15] In his speech at Fort William College on September 29, 1804, Richard Colley Wellesley, the then Governor-General, may have been referring to Panchanan when he stated that "Great improvements have been introduced in the art of printing the Oriental character, by native artists" (Buchanan, 1805, 136). And from Buchanan (1805, 233)—"A New improved Devanagari type has been cast for the Shanscrit language, under the superintendence of Mr. William Carey"—it appears that Panchanan was assisting Carey already at Fort William. The account of Carey seducing Panchanan against Colebrooke's will is effectively refuted in Khan (1961, 249–250); in fact, there are accounts of Colebrooke being on good terms with Carey and collaborating with the Serampore Mission even after the alleged incident.

> caste and trade, Manohar, who was to become his son-in-law and eventually his successor as the master typefounder at the Mission. Manohar remained at the foundry for over forty years (Ross, 1999, 46).

In the words of one of Carey's biographers,

> When type was difficult or impossible to obtain, they located a former Indian blacksmith named Panchanan who had been instructed in the art of punch-cutting by Sir Charles Wilkins ... Panchanan trained others with the result that the Serampore press not only produced all its own type, in whatever tongue was required, but sold it as well (Drewery, 1979, 133).

The available evidence shows that Panchanan and his apprentice Manohar were the creators of the Devanāgarī fonts that would later be used by the Baptist Presses and other letterpresses in Kolkata, Benares, and elsewhere for most of the nineteenth century, with a major influence on the history of Devanāgarī typography until the present day.

> His apprentice, Mohonur, continued to make elegant fonts of type for many Eastern languages for more than forty years. Rev. James Kennedy saw him cutting the matrices and casting the type for the Bibles while he squatted before his favorite idol, under the auspices of which alone he would work. Serampore continued until 1860 to be the principal Oriental type-foundry of the East (Lanman, 1920, 230).

In 1817 the Calcutta Baptist Missionary Society was founded, and the Baptist Mission Press was established in Kolkata in 1818. This press soon started producing Bengali fonts of improved quality.[16] According to Ross (1999, 73),

> The publications of the Baptist Mission Press ... significantly improved upon the standard of typography normally exhibited in the earlier Serampore imprints. The type designs of the Calcutta Baptist missionaries, their choice of founts from other foundries, and their arrangement of type matter on the page all indicate a move towards a typography closer to Bengali chirography. This tendency, which recognized the limitations of the current typefounding and printing technology, was influential in shaping the developing Bengali printed character ...

In those same years, in England and in Germany, Colebrooke and August Wilhelm von Schlegel (1767–1845) were earnestly improving the Devanāgarī printing technique, and were developing independently cut fonts. From 1820 to 1837 they had frequent exchanges on the printing technology and constant interactions with the Asiatic Society of Bengal (Rocher and Ludo, 2013, particularly, letters 13 to 15). Colebrooke described "native editors and publishers" as "slovenly with the press, as with manuscripts," and was rather skeptical about the possibility of producing good Devanāgarī editions in India, without "European superintendence" (2013, 43–44). Yet, simultaneously, the quality of Devanāgarī typography at the Baptist Mission Press increased, together with the heightened standard of the Bengali fonts, probably a reason why this press was chosen for many Asiatic Society publications and most of the Devanāgarī ones. In 1866 the Baptist Mission Press of Kolkata was advertised as follows in Thacker's Directory (1866, 363):

[16] See Ross (1999, 61–72) for further details.

BAPTIST MISSION PRESS

No. 21, *Lower Circular Road, Town side. Superintendent*, Rev. C. B. Lewis (*Europe*); *officiating ditto*, Rev. J. Wenger.

*Printer*, W. Thomas

This Press is extensively engaged in printing Religious Works, in English and the Oriental languages, connected with the Mission; it also undertakes the execution of all descriptions of typographical Printing and Book-binding in a superior style.

The undermentioned periodicals are printed at this Press.

The Journal of the Asiatic Society, published at the Society's Premises, Park Street; and The Calcutta Christian Observer.

The first translations of oriental works were published through a collaboration of the Asiatic society with the Baptist Mission in Serampore, and in 1806 a series with the name "Bibliotheca Asiatica" was contemplated, although little came of it. In the first three decades of the century, indeed, publications of the Asiatic Society were rather sparse, a symptom of a period of relative inertia. In July 1829 Jacquemont (1844, 188) described some of the activities of the Society, in less than flattering terms: "... like the Royal Society in London and all the scientific and literary societies in Great Britain, it is composed of a small number of clever men who are not very well considered and of some rich and powerful, but not clever men who enjoy all the honours."[17]

In 1835, however, the Asiatic Society became a player in "the battle which had long raged between the Anglicists and the Orientalists, in regard to the language best adapted for the people" in British India, which "came to a close with the overthrow of the latter" (Mitra et al., 1885: part I, 57; Emmott, 1965: 163–164). The editorial activity on a number of oriental works was abruptly discontinued and the Asiatic Society offered to resume the projects to prevent those partial efforts from going to waste. This effort by the Asiatic Society eventually led to a government grant of 500 rupees per month for publishing "works on instruction in Eastern languages" (Mitra et al., 1885, Part I, 60). These key events took place during the tenure of James Prinsep (1799–1840) as secretary from 1832 to 1839, a period acknowledged as one of the most vibrant of the Asiatic Society (Mitra et al., 1885: part I, 80; Losty, 1990: 108). Incidentally, James Prinsep's elder brother, Henry Thoby Prinsep (1792–1878), was in 1835 the Secretary of the Government and a leader of the Orientalists (Emmott, 1965, 163).

## 5.5 The Bibliotheca Indica

Once Prinsep had left office, the editorial activities of the Asiatic Society were reduced for several years until a new plan for using the 500-rupees grant was devised in 1847:

[17] "... elle est composée, comme la Société royale de Londres et toutes les Sociétés scientifiques et littéraires de la Grande-Bretagne, d'homme habiles en petit nombre qui y sont assez peu considérés, et d'hommes riches ou puissants, mais inhabiles, qui y jouissent de tous les honneurs."

> The immediate question before the Council of 1847 was, how to utilize the grant, and a committee was appointed to devise means to carry out the Court's wishes regarding the publication of the Vedas. The plan approved by the Society was ... to start a monthly serial under the name of 'Bibliotheca Indica' and the editorship of a competent scholar, aided by a staff of Pandits (Mitra et al., 1885, Part I, 61).

Although a significant amount of Arabic and, later, Persian works were also published, the majority of Bibliotheca Indica publications are editions or translations from Sanskrit. Mitra et al. (1885, 65) summarize the genres represented in the Sanskrit titles of the series as follows:

> The Vedas are represented by twenty-five different works; the Puranas, by three; the philosophical schools, by the text-books of all the six leading systems and several commentaries; the Yotishah [*sic*], by three, two with translations. The law-books, the ritual of the Vedas, grammar, rhetoric, and other branches have also been represented by important works.

A complete list of the works published in the Bibliotheca Indica is difficult to compile, even from the catalogues of major libraries such as the British Library and the Library of Congress, or in catalogs of the Asiatic Society, such as Bibliotheca Indica. Catalogue of Works (1929). The evolution of the series remains hidden to some extent, without details about its early monthly issues and the later developments of the Old and New Series. The sheer number of titles is enormous and, especially in the first years, the maze of volumes, numbers, parts, fascicles and dates is hard to reconstruct. For this reasons, a list of all the published issues, with the essential details, up to the publication of $\mathrm{NBh}^{\mathrm{Ja}}$ , is provided in Tables 5.1, 5.2, and 5.3 at the end of this chapter.

There is a noticeable difference from previous editorial enterprises, such as those at Fort William and the Sanskrit College, that were stimulated by the need of text books in educational activities.[18] In the Bibliotheca Indica, at least in its first years, the antiquarian purpose is quite evident and the target readership was also, or perhaps mostly, composed of scholars overseas. The series was the first systematic, well-funded and planned effort to publish editions of Sanskrit works. The final result, made possible by the matured printing technology, was an unprecedented standard of elegantly presented and widely marketed literature. Jayanārāyaṇa might have had some of these achievements in mind when he wrote: "Books that are verging towards disappearance are being printed with earnest effort and shaped in a structured form (*kāyavyūha*)."[19]

The first issue of the Bibliotheca Indica was published in 1849. In the introduction to this first fascicle, the editor and at that time secretary of the Oriental Department of the Asiatic Society, E. Röer, sketched the editorial plan as follows (ṚV: Preface, i):

---

[18] A list of publications printed in 1801–1805 at Fort William College can be found in Buchanan (1805, 219–231).

[19] *ye granthā luptapadavīpānthās te yatprayatnataḥ / vidyante mudritāḥ santaḥ kāyavyūhaṃ vahanti ca //5//* ( $\mathrm{NBh}^{\mathrm{Ja}}$ : ii).

> The present work is the first of a series of Oriental publications, which, as the title intimates, will be printed under the patronage of the Hon. Court of Directors and the superintendence of the Asiatic Society. It is to appear in the form of a Journal, in monthly numbers of 80 to 100 pages. Every work is intended to be accompanied with an English translation, and if taken from Sanscrit literature, with its commentary also. The works of this collection may be of any Oriental language, provided they relate to India …

With few exceptions, fascicles consisted of about one hundred pages, ninety-six in most cases. The journal format with monthly issues (see Table 5.1), however, was discontinued after the first three years (1848–1850). The original plan of monthly fascicles apparently proved to be an overly ambitious target and demanded a change in the editorial policy. From 1851 onward, indeed, the month was no longer mentioned on the cover. Moreover, each published work was originally conceived as a numbered volume of the series, often divided into parts and issued in different fascicles. This led to a maze of numbers that was further simplified by omitting the volume number. These changes became visible and were finalized in 1853, as explained on the internal cover of one of the fascicles of the $SDS^{Vi}$ (fifth point):

> The following changes have been made in the plan and management of this publication, under a resolution of the Society adopted, at the recommendation of the Council, on the 6th of April last.[20]
>
> 1st. The publication of the series will be superintended by the Philological Committee.
>
> 2st. Proposed contribution to the series should be sent in, as at present, to the Council who, after reference to the Philological Committee,[21] will report on the proposition for the Society's sanction.
>
> 3rd. Editors of texts will not be called upon to supply translations of the works published, but an analysis of these will be required in the form of an Introduction.
>
> 4th. Contributions will be paid for, if required, at rates not exceeding 12 Rs. per sheet of 8 octavo pages. The exact amount of renumeration [*sic*] will be fixed by the Council in each special case. The renumeration allowed is to include all charges, as for procuring MS., Copying, Collating, &c.: proofs to be corrected by the editor.
>
> 5th. Each fasciculus to have a consecutive No. without reference to volumes; the paging to be continuous for each work and for each translation of a work, so that each work or translation shall form a distinct volume and have its separate title-page.
>
> 6th. The price to be reduced to 10 annas per No.

The original resolve of producing English translations along with the original Sanskrit texts may have been intended a reassurance to the prevalent Anglicist approach in the education of the Indians. Yet this resolve was modified in the third point of the new policy published in 1853 and quoted above, and indeed very few translations were actually ever published (see Tables 5.1, 5.2, and 5.3).

At the beginning of the 1960s the New Series of the Bibliotheca Indica was started, keeping the simplified system of numeration, i.e., omitting the month of publication and without the additional Roman number for volumes. Some of the delayed issues of the Old Series were still in production, while the New Series was printed with renewed impetus. $NBh^{Ja}$ was published in this New Series.

[20] I.e., 6 April 1852 or 1853.

[21] Probably described in the internal cover of some of the fascicles, see below.

Concerning the fourth point of the new policy, by 1865 the Bibliotheca Indica fascicles were marketed at the cost of ten annas each, or one shilling and eight pence. In the inside cover of $NBh^{Ja}$ its three fascicles are listed with a price of 1.75 Rs. (one rupee and twelve annas).

An hypothesis about the intended audience can be built by comparing these prices with the average wages in eighteenth-century India. In 1820, at the Benares Sanskrit College, the Rector was paid 200 rupees, while the Pandits teaching there were earning from 60 to 120 rupees a month (Nicholls, 2005, 44, 50). In 1884, at the same college, a European professor would earn 500 rupees, an Indian professor 250 rupees, and an assistant professor 100 rupees (Nicholls, 2005, 168).

In 1829, Jacquemont described a meeting of the Asiatic Society during which the request for financial help submitted by an unnamed Hungarian philologist was discussed and granted (Jacquemont, 1844, 188–189). The philologist's request itemized four rupees a month intended for the servant, and twenty for a "Lama" teacher.

In 1839, student allowances at the Benares Sanskrit College amounted to three rupees per month, though this was considered a regrettably low sum that was hopefully to be raised to five rupees. In any case, these allowances were only granted to a minority of the student population (Nicholls, 2005, 115–116). In 1844, however, student grants ranged from five to twenty rupees per month (Nicholls, 2005, 129–133).

O'Conor (1886, 43–48) provides a table of monthly wages for 1873, arranged by districts and types of workers; for instance, in Kolkata agricultural labourers were earning about ten rupees a month, and masons, carpenters and blacksmiths nine rupees.[22]

The price of the Bibliotheca Indica fascicles was obviously too high for a large part of the Indian population. In fact, in that period even an educated, yet poor person may still have thought it more reasonable to copy the book by hand rather than spend 1.75 rupees on it.

To sum up, the early Bibliotheca Indica fascicles were certainly targeting Western scholars, who were eager to read and study previously unpublished Sanskrit works, but were also accessible to relatively wealthy Indian intellectuals, such as the teachers employed at the Sanskrit College and other British-funded institutions.

---

[22] In Bengal, before 1800, as reported by Colebrooke, daily wages of unskilled and skilled labourers amounted, respectively, to one and two annas per day or 2.7 and 7 kg of rice daily, that is, two to four rupees a month (Bhattacharya and Roy Malakar, 1977, 65–66). In Bihar, in 1811–1813, monthly expenditure pro capita was less than one rupee for more than half of the population, while less than one per cent of the population had a monthly expenditure superior to four rupees. And in 1880, daily wages of less than two annas a day, i.e., about four rupees per month, could not "supply the wants of the number of persons ordinarily dependent on bread-winner." (Bhattacharya and Roy Malakar, 1977, 69). In 1874, according to Broadberry and Bishnupriya (2006, 14), daily wages were 1.79 silver grams (about five rupees a month) and 5.27 silver grams (about 16 rupees a month) for unskilled and skilled labourers, respectively, in Northern and Western India. This would amount to 2.5 and 7.5 kg of wheat grain. Wages were thus quite varied in different areas of India. For a better comparison, in 1864–1865, the rupee-pound ratio was 1:10, specifically in relation to silver rupees and golden sovereigns (Waterfield, 1866, 25–26).

## 5.6 The Editor of the NBh, Jayanārāyaṇa Tarkapañcānana

Not much is known about Jayanārāyaṇa Tarkapañcānana (1806–1872), apart from what can be gathered from his publications. Like most other Naiyāyikas, he probably adhered to the Śaiva faith, as indicated by the beginning of his commentary to the *Vaiśeṣikasūtra*, where he invokes the favors of Bhava, i.e., Śiva (VSu: 1).[23]

Jayanārāyaṇa had the habit of introducing or concluding his editions—VSu, NBh$^{\mathrm{Ja}}$ , and SVij—with autobiographical colophons in Sanskrit verses, in many ways following the customs of the manuscript tradition. In VSu: 374–376 the metrical colophon is placed at the end of the edition, as normally done in South Asian manuscripts, while in the latter two books it is found at the beginning, akin to a preface. Most of the following information is derived from these three colophons.

### *5.6.1 His Early Years*

Jayạnārāyaṇa was born in an illustrious family of brahmins of the Kāṇva lineage. He grew up in the Vaḍsya area of Kolkata (probably today's Barisha), west of Tollygunge.[24] His grandfather was Rāmacandra Tarkālaṅkāra, a famous Nyāya scholar and nephew of Śyāmasundara Vācaspati. He was the second son of Harīścandra Vidyāsāgara, from whom he received in his childhood a basic education in grammar, epics and morals, "just like food and clothes."[25]

[23] *so'yaṃ vaḥ sthirabhaktiyogasulabho bhūyād bhavo bhūtaye.*

[24] ... *ṭāligañjapratīcyām* ... *vaḍsyāsaṃjñābhiṣaṇgā* ... *//1//* (VSu: 474); *vaḍsyākhyaḥ prathitaḥ grāmas tatra vāsaḥ purā mama //9//* ( NBh$^{\mathrm{Ja}}$ : i).

[25] *śyāmasundaranāmāsau bhaktimān śyāmasundare / vācaspatir atikhyāto vācaspatir ivāparaḥ //2//* ... *tadbhrātṛjo rāmacandras tarkālaṇkārasaṃjñitaḥ //3//* ... *tatsutaḥ śrīhariścandrovidyāsāg arasaṃjñakaḥ / dharmmaśāstre purāṇe ca nipuṇo 'dyāpi vidyate //5// taddvitīyasutaḥ śrīmān jayanārāyaṇaḥ punaḥ / tarkapañcānanābhikhyāvikhyātas tarkasaṃśrayāt //6// vātsye vyākaraṇādīni dharmmaśāstraṇi ca kramāt / yas tātapādapadmasya sannidhānād adhītavān //7//* (VSu: 474); *naptāhaṃ rāmacandrasya tarkālaṇkārasaṃjñinaḥ / nyāyādinānāvidyāsu vikhyātasya mahātmanaḥ //7// putras tathā hariścandravidyāsāgaradhīmataḥ / smṛtau vyutpattim āptasya purāṇādau ca tattvataḥ //8//* ... *bālye vyākaraṇādīni dharmaśāstrāṇi kānicit / pituḥ sakāśāt prāpto 'ham annavastrādivat purā //10//* ( NBh$^{\mathrm{Ja}}$ : i–ii); *śyāmasundarasaṃjño'sau vācaspatipadāspadaṃ / pratīcyavaidikaśreṇīlasatkāṇvāyanānvayaḥ //10//* ... *bhrātṛvyas tasya dhīrendro rāmacandreti nāmadhṛk / tarkālaṇkāropanāmā tatra tarkam apāṭhayat //12// tasyāṇgajo hariścandravidyāsāga rasañjñakaḥ / purāṇe dharmmaśāstre ca nipuṇo'bhūn mahāmatiḥ //13// tasya dvitīyas tanayo jayanārāyaṇābhidhaḥ / tarkapañcānanopākhyāvikhyātas tarkasaṃśrayāt //14// kāvye vyākaraṇaṃ dharmmaśāstrañ ca smārttanirmmitaṃ / yo'dhītavān yatnaśālī kathitāt pitur ātmanaḥ //15//* (SVij: i–ii).

In his formative years Jayanārāyaṇa studied poetics under Rāmatoṣaṇa Vidyālaṅkāra,[26] and Vedānta, among other disciplines, under Nāthūrāma Śāstrī.[27]

Jayanārāyaṇa, as his Tarkapañcānana title suggests, was a renowned scholar of Nyāya (also called *tarka*). His Navyanyāya tutor was Jaganmohana Śarmā Tarkasiddhānta, a resident of Śālikā town, which could be present-day Shalika, about 50 km west of Kolkata.[28] Under him Jayanārāyaṇa studied the *Tattvacintāmaṇi*, the main work of the Navyanyāya school, as well as important commentaries such as the *Dīdhiti* and the *Gādādharī*.[29]

## 5.6.2 His Career at the Sanskrit College

After his education, at around 1840, Jayanārāyaṇa moved to the Kolkata area of Nārikelaḍānga and was employed at the nearby Calcutta Sanskrit College, where he taught philosophical disciplines to bright and illustrious students.[30]

The Calcutta Sanskrit College had been founded in 1824 and underwent major changes during Jayanārāyaṇa's tenure (Bose, 1953, i–ii).

> To begin with, it was an organised Sanskrit Tol with 55 stipendiary students and 8 professors who used to teach Nyaya, Smriti, Darshan, Vyakaran, Jyotisha and Ayurveda ... In 1851 the renowned Iswar Chandra Vidyasagar became the Principal of the college, and introduced the western method of teaching Sanskrit in place of the time-honoured method followed in Tols.

---

[26] *latāyāḥ prāṇatoṣaṇyāḥ kārakād rāmatoṣaṇāt / vidyālaṇkāravikhyātādyo 'laṇkāram athāgamat //8//* (VSu: 474); *rāmatoṣaṇasaṃjñasya vidyālaṇkāradhīmataḥ / chātro 'haṃ suprasiddhasyālaṇk āragranthapāṭhane //11//* ( NBh[Ja] : ii); *yenā'laṇkāraśāstrāṇi prāptāni rāmatoṣaṇāt / vidyālaṇkāravikhyātād āgamagranthakārakāt //16//* (SVij: ii).

[27] *tathā gurjjaradeśīyād yo nāthūrāmaśāstriṇaḥ / vāgdevyāḥ pauruṣād rūpād vedāntādīny adhītavān //12//* (VSu: 474); *vedāntādīni śāstrāṇi nāthūrāmasya śāstriṇaḥ / sakāśād āptavān asmi purā gurjjaravāsinaḥ //12//* ( NBh[Ja] : ii); *yena gurjjaradeśīyān nāthūrāmākhyaśāstriṇaḥ / vedāntādīny adhītāni vāgdevyāḥ pauruṣākṛteḥ //20//* (SVij: ii).

[28] *asti svastimad ūrjjivāryyanivahaṃ bhūṣāyamāṇaṃ bhuvaḥ / kalkyātānagaraṃ sudhābdhilaharī nyakkārisaudhojjvalam // pāregaṇgam amuṣya cāṇgam iva yā śākhāpurī śālikā / tatrāsīd badhavṛndavanditagaṇāmbodhir jaganmohanaḥ //9//* (VSu: 475); *tarkasiddhāntasaṃjñasya jaganmohanaśarmmaṇaḥ / śuśrūṣayāptavān asmi nyāyavidyāṃ mahāmateḥ //13//* ( NBh[Ja] : ii); *kalkyātānagarād ārāt śālikāgrāma uttamaḥ / gaṇgāpaścimakūlasthas tatrāsīd gotamopamaḥ //17// tarkasiddhāntopanāmā jaganmohananāmadhṛk /* (SVij: ii).

[29] *tasmād atisuvikhyātād adhītaṃ yena yatnataḥ //18// tarkaśāstraṃ tattvacintāmaṇiprabhṛtivistṛ taṃ / dīdhitipramukhañ cāpi gādādharyyādikan tathā //19//* (SVij: ii).

[30] *kalkyātārājadhānyantarvirājat pāṭhamandire / darśanākhyāpane rājñā yaḥ parīkṣya niyojitaḥ //13// yo nārikelaḍāṇgākhyasthale lohādhvasannidhau / nivasan sudhiyaḥ śiṣyān adhyāpayati santatam //14//* (VSu); *tathā maṭhe rājakīye nagare 'smin niyojitaḥ / darśanādhyāpane rājñā pañcaviṃśativatsarān //16//* ( NBh[Ja] : ii).

The purpose of the Calcutta Sanskrit College may also be inferred from the words of James R. Ballantyne (1813–1864), who stated the mission of the Benares Sanskrit College as its newly appointed principal in 1846 (Nicholls, 2005, 142):

> The object of such an institution ... ought to be this, to produce Pandits, not merely with Sanskrit learning equal to that which can be acquired in the native schools, but with minds so far tinctured with European habits of thought as shall render each of them in some degree a moral light among his countrymen. ... if we succeed in establishing in the Sanskrit College the standard of training which I propose, we shall have in the case of each pupil so trained a Brahman whose acquirement in Sanskrit learning must command respect ...

The targets of these two institutions were quite different from those of the pre-existing College at Fort William, which was opened in 1800 for Europeans and not for Indians, more specifically for "the improvement of the education of the young men destined to the civil service in India" (Buchanan, 1805, 16). These young Britishers were expected to leave for India at the age of fifteen or sixteen; there they also needed an acquaintance with the local languages and thus an education "of a mixed nature" (Buchanan, 1805, 20). The Sanskrit College, instead, was rather meant for the formation of British-friendly pandits. Curiously enough, it flourished right after the debate among Anglicists and Orientalists, despite the eventual success of the Anglicists and apparently in competition with "Anglicist" British institutions, such as the present Scottish Church College, which was founded in those years by the Scottish missionary Alexander Duff (1806–1878) (Emmott, 1965, 163).

An eminent alumnus of the Calcutta Sanskrit College was Īśvaracandra Vidyāsāgara (1820–1891), who studied there from 1828 until his graduation in 1841 (Roy, 1921, 4–17), and was later professor and principal of the college from 1851 to 1858. His career was intertwined with that of Jayanārāyaṇa, who began teaching at the College just before Īśvaracandra's graduation and flourished as professor under his headship. In this period, Īśvaracandra famously reformed the admission rules, opening the College to Kayasthas and other non-Brahmins. Together with Madanamohana Tarkālaṇkara, Īśvaracandra also founded the influential Sanskrit Press in 1847, where he introduced typographical innovations such as the use of Latin punctuation in Bengali texts (Ross, 1999, 129) and where Jayanārāyaṇa published his first book (ATV).

Maheśacandra Nyāyaratna (1836–1906), the principal of the College from 1876 to 1895, is also renowned for his educational reforms and had previously been Jayanārāyaṇa's pupil.

Another famous scholar associated with Jayanārāyaṇa is Edward Byles Cowell (1826–1903), principal of the College from 1858 to 1864 and later Sanskrit professor at Cambridge, "perhaps the *guru* of the greater number of English Sanskrit scholars" (Ridding, 1931, 461). Cowell held Jayanārāyaṇa in great esteem, so much so that he kept his portrait in his room at Cambridge (1931, 463). In the introduction of his edition of the *Nyāyakusumāñjalī* he praises him as follows (NKu: xiii):

> But I must not omit to express my many obligations to Paṇḍita Jayanáráyaṇa Tarkapanchánana and Táránátha Tarkavâchaspati, the Professors of Philosophy and Grammar, the two most learned Hindus I have met during my residence in India ...

Jayanārāyaṇa's name is also mentioned in the introduction to the translation of the *Bhāṣapariccheda*, published in the Bibliotheca Indica Series of the Asiatic Society. There Röer ( $NSM^{Ro}$ : xxvi) acknowledges his assistance as follows:

> The present edition has been carefully revised partly by comparison with Mahàdevabhaṭṭa's commentary, partly by a reference to the corrections in the copy of Jayanàràyana Tarkapanchànana, the professor of Nyàya in the Sanscrit College, which he had the kindness to lend me, and for which I offer my grateful acknowledgements.

Jayanārāyaṇa would later contribute to the Bibliotheca Indica with three scholarly Sanskrit editions of his own, the enduring evidence of his erudition and philological commitment.

### 5.6.3 His Works

Altogether, Jayanārāyaṇa Tarkapañcānana is known to have edited or authored the following works, three of which were published in the Bibliotheca Indica Series:

- editor of the *Ātmatattvaviveka* of Udayana (ATV);
- editor of the *Śaṇkaravijaya* of Anantānandagiri (SVij);
- editor of the *Nyāyabhāṣya* of Vātsyāyana ( $NBh^{Ja}$ );
- editor and translator, in Bengali, of the *Sarvadarśanasaṃgraha* of Mādhava ($SDS^{Ja}$) ;
- editor of the *Vaiśeṣikasūtra* of Kaṇāda and author of the *Vivṛti* commentary thereon (VSu);[31]
- author of the *Padārthatattvasāra*, an original Vaiśeṣika work in Sanskrit (PTS).

The editing of the *Nyāyabhāṣya* began in 1864 and culminated with the printing of the last of the three fascicles in the spring of 1865, as attested in the last stanza of the colophon. There the year is conveyed in word numerals, as routinely done in the manuscript tradition: "In the year 6 (*rasa*, 'the six flavours'), 8, 7 (*acala*, 'the seven mountains'), 1 (*bhū*, 'the earth') [i.e., 1786, to be read inversely] of the Śaka era, during the bright half of the Caitra month (early spring), I completed the printing of this Bhāṣya."[32]

In a note found in the introduction to (NKu: ix), Cowell recommended the preparation of a NBh edition after the discovery of manuscripts that Jayanārāyaṇa would have used in his edition. Cowell's introduction is dated March 1864, so assuming that Cowell was aware of Jayanārāyaṇa's activities it may have taken less than a year for Jayanārāyaṇa to complete the editing of $NBh^{Ja}$ .

[31] Jayanārāyaṇa's commentary became relatively influential. For instance it is translated along with Śaṅkaramiśra's *Upaskāra* in Sinha (1911).

[32] *rasāṣṭācalabhūmāne śākābde sauracaitrike / mudrām etasya bhāṣyasya sampannāṃ kṛtavān aham //17//* ( $NBh^{Ja}$ : ii).

Interestingly, both in this versified date and in the passage already quoted above—"The Asiatic Society has engaged me in editing and printing it"[33]—Jayanārāyaṇa refers to the actual process of printing, suggesting that he may have been involved not only in the editing process, but also in the typographical production of his NBh edition.

## 5.7 Description of the *Editio Princeps*

The size is 130 × 202 mm, in quarto. The full NBh text is printed in three fascicles, the first two of 96 pages each, the third of 106 pages. Each fascicle has a paperback cover in pink color. The three covers have all been preserved in the case of the Vienna copy.[34] The front cover of the first fascicle reads (see also Fig. 5.2 and Fig. 5.3 below).

> BIBLIOTHECA INDICA; | a | COLLECTION OF ORIENTAL WORKS | published under the superintendence of the | ASIATIC SOCIETY OF BENGAL. | New Series, No. 56. ||[lithographic image of the old Asiatic Society Building, Park Street, with the inscription "Museum Asiatic Society Calcutta" at the lower edge, 70 × 35 mm] | न्यायदर्शनम् | गोतममुनिप्रणीतम् | वात्स्यायनमुनिप्रणीतभाष्यसहितम् | श्रीयुक्त-जयनारायण-तर्कपञ्चाननेन | परिशोधितम् | THE NYA'YA DARS'ANA | OF GOTAMA, | with the | COMMENTARY OF VA'TSYA'YANA, | edited by | Paṇdita Jayan'ara'yaṇa Tarkapancha'nana. | fasciculus i. | calcutta: | printed by c. b. lewis, at the baptist mission press. | 1864.

The front cover of the first fascicle is followed by a title page on white paper, identical to the front cover save for the missing image and for the line "Nos. 50, 67, and 70," thus summarising the numbers of the issues of the three fascicles, and the date "1865" in the last line, i.e., the publication date of the third fascicle. The front cover of the second fascicle reads "New Series, No. 67" in line 6. The cover of the third fascicle reads "New Series, No. 70" at line six, and "1865" in the last line. Otherwise, the front covers are identical. The back sides of the three front covers contain a list of published and forthcoming works in the same series. The back cover of each fascicle lists further works, "in progress" and "in the old series."

C. B. Lewis, mentioned in the penultimate line of the front cover, was the superintendent of the Baptist Mission Press (Thacker's Directory, 1866, 363). He is likely to be Charles Bennett Lewis (1821–1891), who wrote a biography of the early

[33] *mudraṇāśodhane 'syeyaṃ sā sabhā māṃ nyayojayat* ( $NBh^{Ja}$ : ii).

[34] In many other Bibliotheca Indica editions such covers were dispensed with when the fascicles were bound into a single volume. In such cases the publication date of the first fascicle has been lost, since the date on the title-page corresponded to that of the last fascicle.

Baptist missionary John Thomas (Lewis, 1871). Lewis was an active missionary himself (see Lewis, 1869a,b) and contributed to the series as the editor of Christian scriptures translated into Sanskrit, Bengali and Persian, also published by the Baptist Mission Press.

After the title page, there is a brief English preface on a single page, followed by 17 verses in Sanskrit composed by the editor, the table of contents for all the three fascicles, and an errata corrige. Altogether this front-matter amounts to ten pages, unnumbered but marked by a growing pattern of small vertical and oblique strokes on the upper and outer edges of the page. It is followed by signatures in quarto, with page numbers starting from 1 in Devanāgarī digits. Each quarto is marked by a letter (B,C, ...) on the lower internal margin, with the internal folios of the quarto marked respectively by B2, C2, etc.

The NBh text ends on p. 297 with the regular formula found in NBh manuscripts, *iti vātsyāyanīye nyāyabhāṣye pañcamādhyāyasya dvitīyam āhnikaṃ samāptaś cāyaṃ pañcamo 'dhyāyaḥ*. This is further followed by the line *samāptañ cedaṃ śāstram*, and again by a colophon that reads *śubham astu / sambat 1921 / śakābdaḥ 1786 / 25 phālgunaḥ //*. This date corresponds to Wednesday March 22, 1865.[35]

The book is neatly and elegantly typeset, with a balanced use of white space, font sizes, indentations, page size, and margins. The *sūtra* and *bhāṣya* passages is contrasted both by means of *sūtra* numbers and by the larger, bold typeface used for the *sūtra* text, and are further specified by means of the abbreviations *sū.* and *bhā.* in the left margin. This last feature also occurs in VSu, and may have been introduced by Jayanārāyaṇa himself, since it is not found in other Bibliotheca Indica editions. The reader's orientation is further assisted by running headers with chapter and section numbers.

Selected variant readings from the three manuscripts used as sources are provided in the footnotes; in some cases the mere variant is mentioned, while in others, variants are diversely qualified with the words *iti pāṭhāntaram* ("variant reading"), *idam api pāṭhāntaram* ("there is also this variant reading"), *ity api kvacit pāṭhaḥ* ("this reading is also found elsewhere"). The editor frequently mentions variants or opinions found in the *Nyāyasūtravṛtti*, e.g., "In the opinion of the Vṛttikāra, this is not a *sūtra*, but rather the commentary" (*vṛttikāramātre nedaṃ sūtraṃ kin tu bhāṣyam* ( $\text{NBh}^{\text{Ja}}$ : 97)). Furthermore, there are rare cases in which the editor gives an alternative reading which seems to be his own conjecture, e.g. *prāptīccheti samīcinaḥ pāṭhaḥ*, "*prāptīcchā* is the correct reading" ( $\text{NBh}^{\text{Ja}}$ : 182).

The fascicles of the Bibliotheca Indica were sold with a provisional binding, by a string through holes 5 mm from the spine. The fascicles were meant to be bound together, with the title page provided by the publisher. This title page listed the numbers of the fascicles in the Series under the title and contained further general information about the collection. The date at the bottom of the title page was that of

[35] Calculated by Michio Yano's Pancanga, version 3.14 (http://www.cc.kyoto-su.ac.jp/~yanom/pancanga/).

the last fascicle.[36] By 1865, the Bibliotheca Indica volumes were marketed and disseminated both in India and in the West. For the first four years the cost of each fascicle was one rupee in India and two shillings in the UK. The price was reduced to 10 annas/one shilling and eight pence in 1853. From the internal cover of NBh$^{Ja}$ one can see that its three fascicles cost 1.75 rupees (one rupee and 12 annas). In the fascicles from 1881 onwards, NBh$^{Ja}$ no longer appears in the lists of published works provided on the back covers, so it may have been out of stock by then. Unfortunately, no information on the number of printed copies can be found. There is also no trace of reprints or revised versions in the Bibliotheca Indica.

In the case of the Vienna copy, the three fascicles are bound together in a single tome with a black and brown hardcover. There are two older stickers on the spine, with obsolete inventory or call numbers: "Bibl. Ind. N.S. 56 u.a." and "76185." On the front cover a more recent library label reads "Fachbibliothek — für Indologie — IND 18776–B — Ph I 6:280," the latter number being the one in current use. On the cover of the second fascicle, the words "Harr 25/1 1901" are written in pencil, showing that the copy was probably acquired by the Vienna library in 1901.[37]

The Leipzig copy has the call number Ind/W Gcc 2, in the Weller collection of the Sächsische Akademie. It was part of the Hermann Jacobi Collection in Bonn, and then of the Weller collection in Leipzig, as evidenced by the personal stamps of "Prof. H. Jacobi, Bonn a. Rh." and "Friedrich Weller, Lepzig C1, Stephanstra e 2" on the inside front cover, as well as from a pencil remark, "Prof. Weller, 15,—".

## 5.8 Jayanārāyaṇa's Sources

About a year before the publication of NBh$^{Ja}$ , Cowell wrote that "the only *Nyàyabhàshya* known to exist is that which bears the name of 'Vàtsyàyana.' By the kindness of Professor Griffith, I have procured a MS. of it from the Benares College Library …" In a note, he added: "it contains 123 leaves, and its colophon is *iti vātsyāyanīye nyāyabhāsye pañcamo'dhyāya*" . Moreover, he referred to a further Bengali manuscript (NKu: Preface, ix, footnote)): "I hope that the Asiatic Society will publish the Nyàyabhàshya in the Bibliotheca Indica, as a second MS. has been just procured from Nuddea. No work could be of greater importance for giving an insight into the older Nyàya."

---

[36] Gaskell (1972, 146–147) explains that "throughout the hand-press period the craft of binding books was normally separate from that of printing them. … Edition binding, in fact, was never normal … the economics of the trade were against it. The most important factor was that every book had to be bound separately by hand, and binding costs did not go down with quantity; it costs ten times as much to bind a thousand copies of a book as it costs to bind a hundred. … It was moreover to the provincial bookseller's advantage to buy books in sheets, for they were cheaper to transport than were bound copies."

[37] The abbreviation "Harr" probably refers to Harrassowitz Booksellers and Subscription Agents, Wiesbaden.

In NBh$^{Ja}$, three manuscripts are explicitly mentioned as the basis for the edition:

I. A very correct MS. belonging to the Benares College library.
II. A fair MS. in the Bengali character, belonging to a paṇdit of Nuddea.
III. An imperfect MS. belonging to the editor, Paṇdita Jayana'ra'yana Tarkapancha'anana.

The first two manuscripts are probably those mentioned by Cowell. The only manuscript of the NBh recorded in catalogues before 1864 is the one in Fitz-Edward Hall's catalogue (Hall, 1859, 20), succinctly described as "96 fols. 3400 ślokas Ben. Coll." The number of leaves does not match the manuscript mentioned by Cowell, but since no other NBh manuscript preserved in Benares with 96 folios is known to exist, and since known NBh manuscripts were quite rare at that time, this could be a mistake in Fitz-Edward Hall's list.

According to the collation work and the stemmatic analysis of the transmission of the NBh done in the Vienna projects, the first of the three must be C3$^{D}$, which matches also Cowell's remarks in NKu).[38] The fact that C3$^{D}$ is presently preserved at the Sanskrit College in Kolkata, where Jayanārāyaṇa was still teaching in 1864–1865, also points to this identification. If this assessment is correct, after Jayanārāyaṇa's editing work was completed the manuscript was never returned to the Benares College.

The second manuscript, in Bengali script, remains missing. None of the Bengali manuscripts collated during the Vienna projects, in fact, has variant readings suggesting a relation with NBh$^{Ja}$.

The third manuscript was presumably also in Devanāgarī script, since no specific script is mentioned in the preface of NBh$^{Ja}$. It may be or at least be related to C2$^{D}$, because the variants recorded in the footnotes of NBh$^{Ja}$, mostly labelled as "kvacit pāṭhaḥ," correspond to the readings found in this manuscript.

Among the manuscripts collated in the Vienna project there are two more manuscripts closely related to NBh$^{Ja}$, C1$^{D}$ and Ku$^{S}$.

C1$^{D}$ was probably a draft prepared in view of the printed edition. This manuscript has a vertical layout (Fig. 5.4), in contrast to the horizontal and more common *pothī* form. Through an analysis of its peculiarities it has been proven that it is an apograph of C3$^{D}$. A proof of the genealogical relation of C1$^{D}$ and NBh$^{Ja}$ is the sentence *anuvāde tv apunar uktaṃ śabdābhyāsād arthaviśeṣopapatteḥ*, in NBh *ad Nyāyasūtra* 5.2.14. This sentence is part of the commentary in all the other branches of the NBh transmission, while it is marked as *sūtra* 5.2.15 both in C1$^{D}$ and in NBh$^{Ja}$ (see Fig. 5.4, line 7, and Fig. 5.5, lines 14–15). In fact, *sūtra* numbers in Bengali script were systematically added in the margin of C1$^{D}$, perhaps as an indication for the typographer.

[38] An analysis of NBh$^{Ja}$ variants throughout the *Trisūtrī* part of the NBh, indeed, shows that NBh$^{Ja}$ is directly connected to C3$^{D}$, C2$^{D}$, C1$^{D}$ and Ku$^{S}$. A remarkable instance is also found in NBh *ad Nyāyasūtra* 4.1.61. There are fewer cases, however, in which NBh$^{Ja}$ reads together with Ku$^{S}$.

$\mathrm{Ku}^{\mathrm{S}}$ is a manuscript in Śāradā, the script traditionally used in Kashmir (Fig. 5.6). It is dated *saṃ 56*, which may be year 4956 of the Laukika or Saptarṣi era, i.e., 1880 CE. It is in a vertical lay-out, just like $\mathrm{C1}^{\mathrm{D}}$, though this is a common feature of Śāradā manuscripts and should not be independently taken as an indication of its relationship with $\mathrm{C1}^{\mathrm{D}}$ or $\mathrm{NBh}^{\mathrm{Ja}}$, with which $\mathrm{Ku}^{\mathrm{S}}$ shares a number of conjunctive readings. For instance, as in $\mathrm{C1}^{\mathrm{D}}$ and $\mathrm{NBh}^{\mathrm{Ja}}$, the *anuvāde ... -papatteḥ* passage is marked as *sūtra* 5.2.15 (see Fig. 5.6, line 10). $\mathrm{Ku}^{\mathrm{S}}$ also shares conjunctive readings exclusively with $\mathrm{NBh}^{\mathrm{Ja}}$, so the most plausible explanation is that the scribe of $\mathrm{Ku}^{\mathrm{S}}$ used $\mathrm{NBh}^{\mathrm{Ja}}$ as a source.

In Fig. 5.1 the genealogy of the transmission is shown. The manuscripts related to the *editio princeps* have been magnified in the diagram and are all in the lower-right part of the stemma, i.e., in the area of the *vulgata*.

## 5.9 Conclusions

A glance at the titles of the Old Series of the Bibliotheca Indica reveals an antiquarian interest of the editorial board in the early years. Originally, every Sanskrit edition was meant to be complemented by an English translation, although such translations did not always materialise and became less and less a priority over the years. The Old Series of the Bibliotheca Indica was replaced, in 1861, by a New Series with an increasing number of philosophical works from the Vaiśeṣika, Nyāya, Sāṇkhya, Mīmāṃsā and Vedānta traditions. Vedic works continued to be published, but in most cases they were completions of unfinished editions which had only partially being printed during the Old Series period.

Incidentally, editions of Arabic works were discontinued and those of Persian works began to appear. Editors of the series such as Röer, Ballantyne, Cowell, etc., were certainly not Hindu patriots, and there is hardly trace of an influence of Ram Mohan Roy's ideas on the activities of the Asiatic Society. Still, the corpus of Sanskrit literature built up in the Bibliotheca Indica might have helped the case of "certain upper class Hindus" who later found in Ram Mohan Roy's discourse "an almost impeccable legitimation for their communal political ends," as argued elsewhere (van Bijlert, 1996, 366).

Most of the editors of the Bibliotheca Indica, particularly in the case of philosophical works, were professors from the Calcutta and Benares Sanskrit Colleges, and we cannot rule out that these books were used for teaching purposes. Yet, remarkably, even if there were many philosophical texts in the New Series, these were still mostly ancient works, as seen in the contrast among Navyanyāya and Prācīnanyāya works. This kind of differentiation from the works studied in the indigenous traditional system was probably satisfying to the historical sentiments of the British intelligentsia, but the publication of rare ancient works must have also appealed to Indian pandits, who were probably quite happy to be given an occasion to unveil the ancient roots of their traditions. In any case, the result was a very productive spirit of cooperation among Indian and European scholars. One should not forget, however, that the collaboration of illustrious pandits in teaching and editing activities was also motivated by economic factors: after all, Indian professors and editors were handsomely paid, at least by the Indian standards of those times.

Scholars such as Jayanārāyaṇa were personally involved in the typographical activities. They may have seen the printing technique as an opportunity to preserve, revive and perpetuate the knowledge of their tradition. If this was the case, it must have been important for them to monitor the editorial passages closely with a view to an optimal result. Jayanārāyaṇa even had the peculiar habit of ending or beginning his editions with a Sanskrit colophon in verse, including biographical data, acknowledgements and respects to his teachers, and the date of completion expressed in the traditional format. This practice preserves a continuity from the traditional manuscript transmission to the novelty of the press, rather than revealing a rupture from the past.

By the time $\mathrm{NBh}^{\mathrm{Ja}}$ was published, the relatively young technology of Devanāgarī printing had already reached sufficient maturity. The choice of Devanāgarī was not an obvious one in Bengal, where Sanskrit texts in Bengal were traditionally written in Bengali script. The use of Devanāgarī is, in itself, a sign that such texts were intended for wider dissemination beyond Bengal. The Bibliotheca Indica editions, however, were obviously not meant for the Indian masses, as their prices clearly indicate. The Sanskrit Colleges themselves were off-limits for non-Brahmins, at least in the first decades of their existence. It must be therefore assumed that these enterprises were targeting the intellectual élite of the society.

The Asiatic Society survived on members' subscriptions for many decades, and the official funds granted for publications were at least partially used for purposes not intended by the Government, e.g., in the dispute between Anglicists and Orientalists, which was won by the former, yet followed by several new editions of Sanskrit texts published in the Bibliotheca Indica. At least some of the scholars involved in editing such texts must have been genuinely interested in the acquisition and dissemination of knowledge, as attested by the high standard of editions such as Jayanārāyaṇa's *editio princeps* of the *Nyāyabhāṣya*.

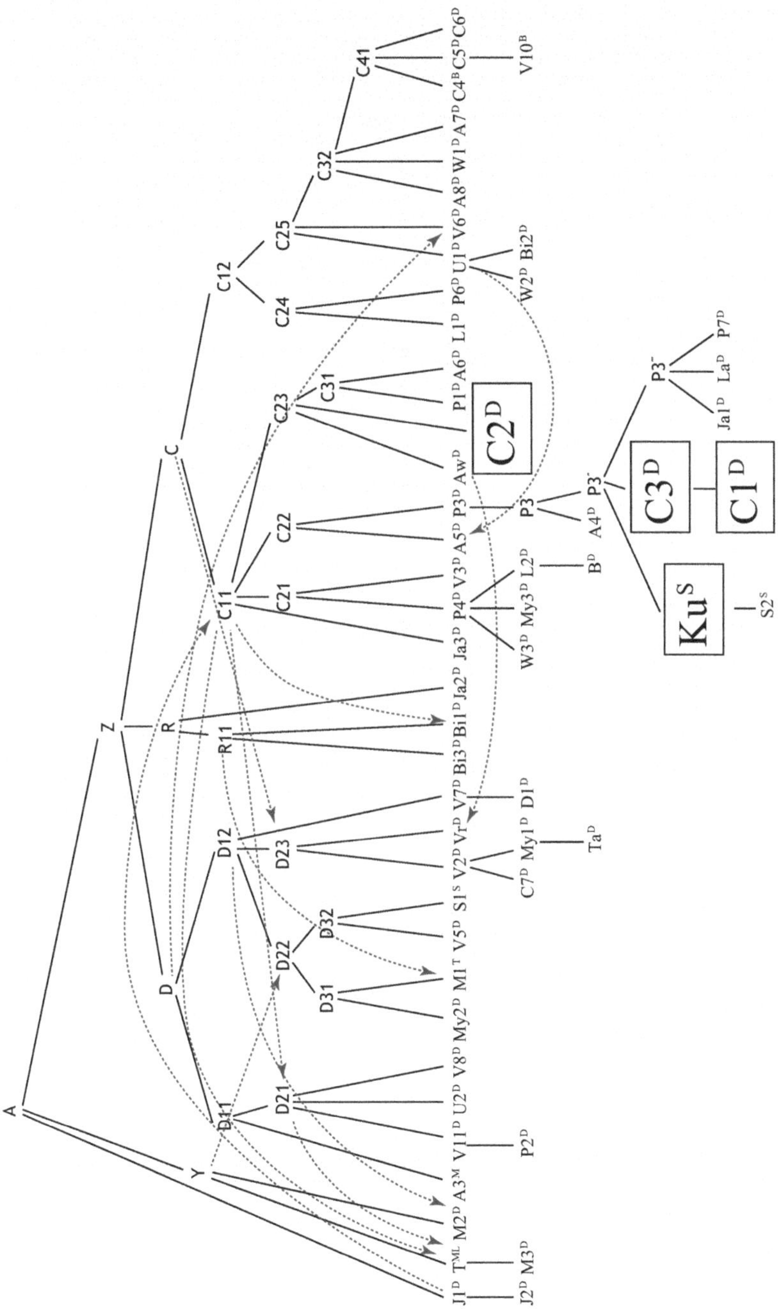

**Fig. 5.1** The manuscript transmission of the NBh

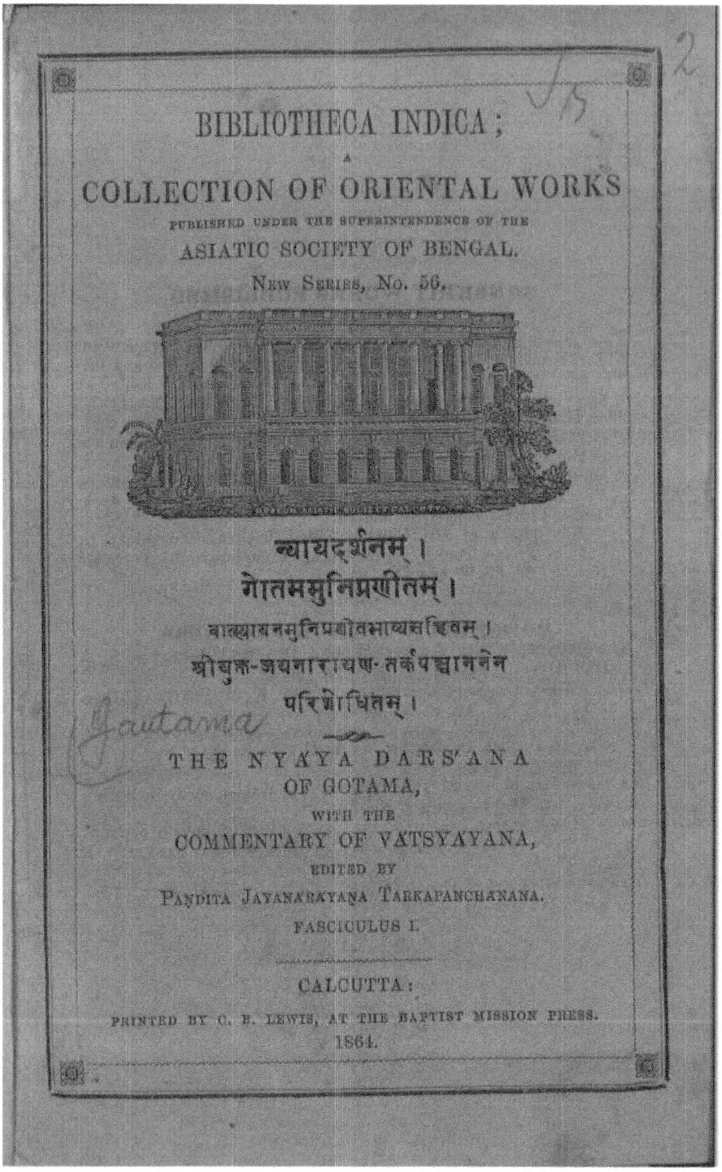

BIBLIOTHECA INDICA;

A

COLLECTION OF ORIENTAL WORKS

PUBLISHED UNDER THE SUPERINTENDENCE OF THE

ASIATIC SOCIETY OF BENGAL.

New Series, No. 56.

न्यायदर्शनम् ।

गौतममुनिप्रणीतम् ।

वात्स्यायनमुनिप्रणीतभाष्यसहितम् ।

श्रीयुक्त-जयनारायण-तर्कपञ्चाननेन

परिशोधितम् ।

THE NYÁYA DARS'ANA

OF GOTAMA,

WITH THE

COMMENTARY OF VÁTSYÁYANA,

EDITED BY

Paṇdita Jayanárayaṇa Tarkapanchánana.

FASCICULUS I.

CALCUTTA:

PRINTED BY C. B. LEWIS, AT THE BAPTIST MISSION PRESS.

1864.

**Fig. 5.2** Front cover of the *Editio Princeps*

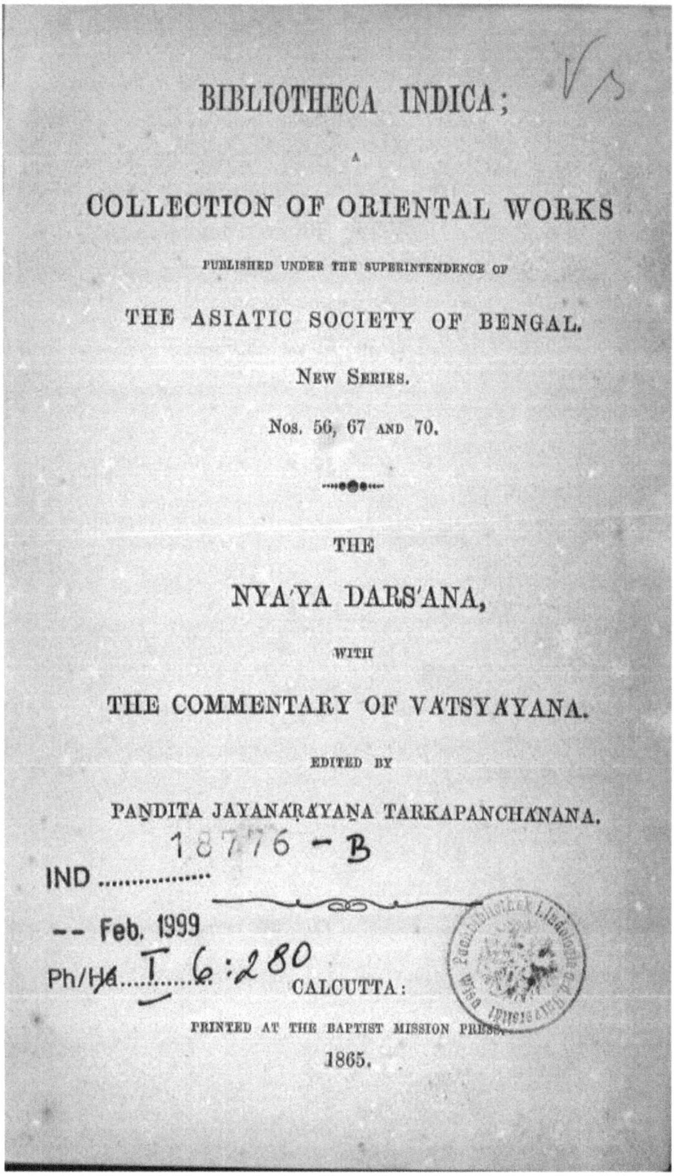

BIBLIOTHECA INDICA;

A

COLLECTION OF ORIENTAL WORKS

PUBLISHED UNDER THE SUPERINTENDENCE OF

THE ASIATIC SOCIETY OF BENGAL.

NEW SERIES.

NOS. 56, 67 AND 70.

THE

NYA'YA DARS'ANA,

WITH

THE COMMENTARY OF VA'TSYA'YANA.

EDITED BY

PAṆDITA JAYANA'RA'YAṆA TARKAPANCHA'NANA.



CALCUTTA:

PRINTED AT THE BAPTIST MISSION PRESS.

1865.

**Fig. 5.3** Title-page of the *Editio Princeps*

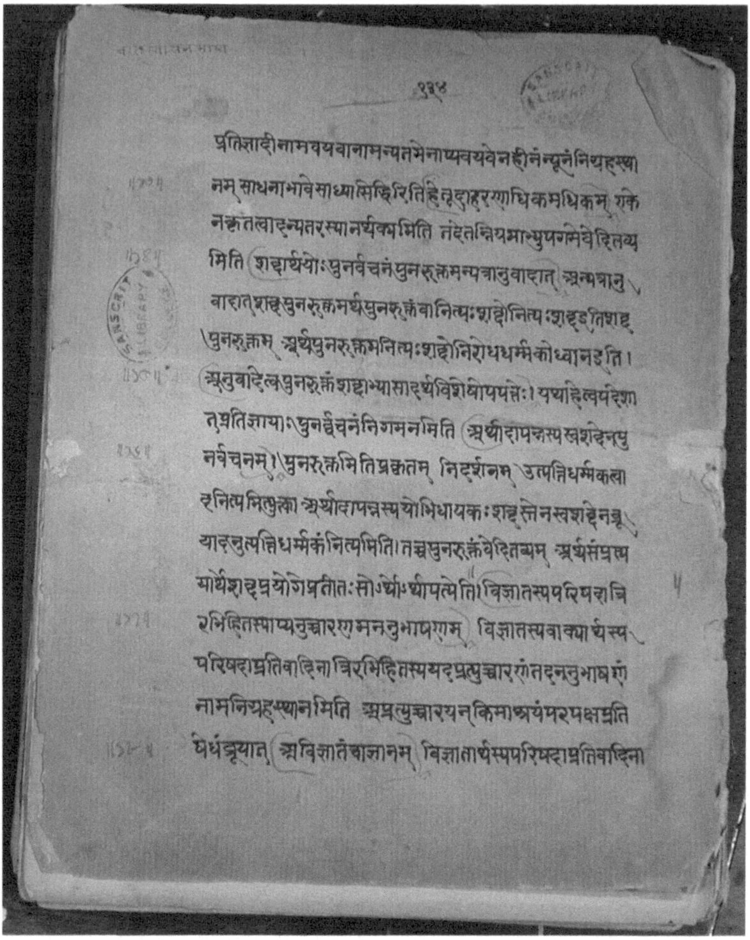

**Fig. 5.4** The Calcutta Manuscript: NBh *ad* 5.2.15

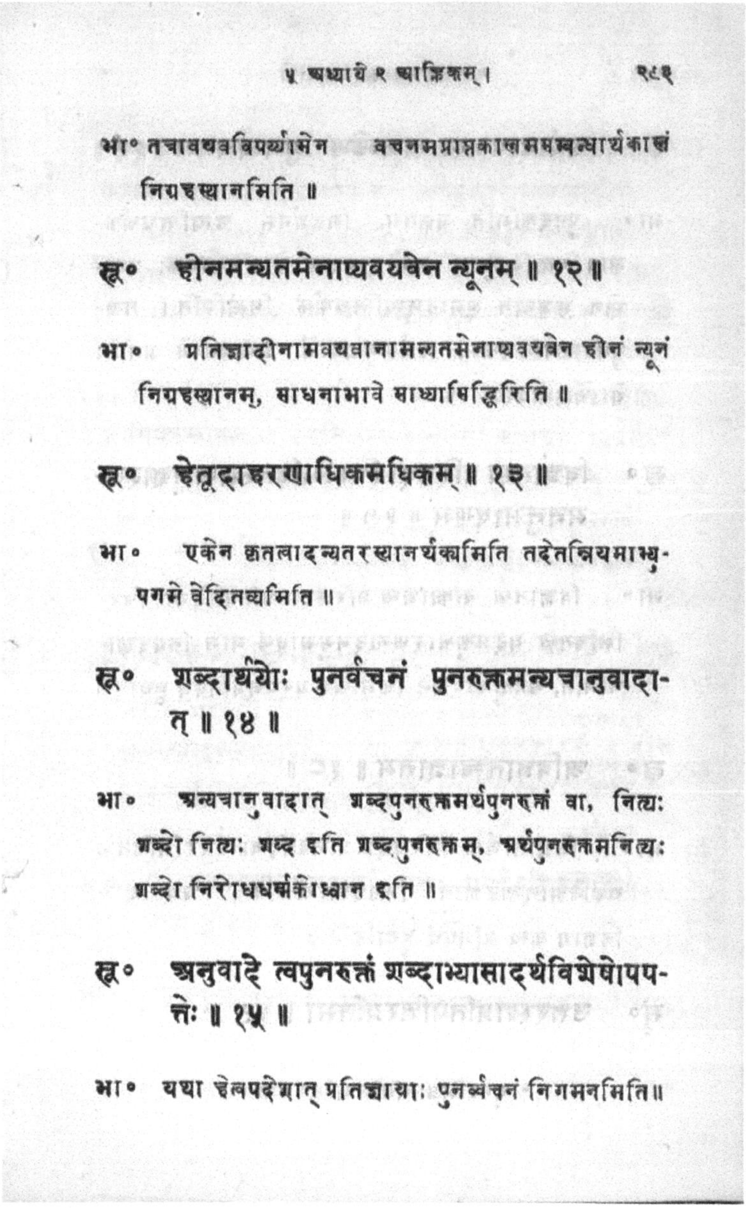

५ अध्याये २ आह्निकम्। २८९

भा॰ तत्रावयवविपर्य्यासेन वचनमप्राप्तकालमसम्बद्धार्थकालं निग्रहस्थानमिति ॥

सू॰ हीनमन्यतमेनाप्यवयवेन न्यूनम् ॥ १२ ॥

भा॰ प्रतिज्ञादीनामवयवानामन्यतमेनाप्यवयवेन हीनं न्यूनं निग्रहस्थानम्, साधनाभावे साध्यासिद्धिरिति ॥

सू॰ हेतूदाहरणाधिकमधिकम् ॥ १३ ॥

भा॰ एकेन कृतत्वादन्यतरस्यानर्थक्यमिति तदेतन्नियमाभ्युपगमे वेदितव्यमिति ॥

सू॰ शब्दार्थयोः पुनर्वचनं पुनरुक्तमन्यत्रानुवादा-
त् ॥ १४ ॥

भा॰ अन्यत्रानुवादात् शब्दपुनरुक्तमर्थपुनरुक्तं वा, नित्यः शब्दो नित्यः शब्द इति शब्दपुनरुक्तम्, अर्थपुनरुक्तमनित्यः शब्दो निरोधधर्म्मको ध्वान इति ॥

सू॰ अनुवादे त्वपुनरुक्तं शब्दाभ्यासादर्थविशेषोपप-
त्तेः ॥ १५ ॥

भा॰ यथा हेत्वपदेशात् प्रतिज्ञायाः पुनर्व्वचनं निगमनमिति॥

**Fig. 5.5** The *Editio Princeps*: NBh *ad* 5.2.15

**Fig. 5.6** The Kurukshetra Manuscript: NBh *ad* 5.2.15

**Table 5.1** Bibliotheca Indica, Old Series, 1848–1850 (Fasc.=Fasciculus, Rs.=rupees, Shill.=shillings, tr.=translated)

| No. | Month, Year | Vol. | Part, Fasc. | Pages | Editor | Title | Lang., Genre | Press | Typographer | Price |
|---|---|---|---|---|---|---|---|---|---|---|
| 1 | Jan., 1848 | I | Fasc. I | 108 | E. Röer | Ṛgveda | Sanskrit, Veda | Baptist Mission Press | J. Thomas | 1 Rs. |
| 2 | Feb., 1848 | I | Fasc. II | 96 | E. Röer | Ṛgveda | Sanskrit, Veda | Baptist Mission Press | J. Thomas | 1 Rs. |
| 3 | Mar., 1848 | I | Fasc. III | 84 | E. Röer | Ṛgveda | Sanskrit, Veda | Baptist Mission Press | J. Thomas | 1 Rs. |
| 4 | Apr., 1848 | I | Fasc. IV | 104 | E. Röer | Ṛgveda | Sanskrit, Veda | Baptist Mission Press | J. Thomas | 1 Rs. |
| 5 | May, 1848 | II | Part I, Fasc. I | 96 | E. Röer | Bṛhadāraṇyaka Up. | Sanskrit, Veda | Baptist Mission Press | J. Thomas | 1 Rs. |
| 6 | Jun., 1848 | II | Part I, Fasc. II | 96 | E. Röer | Bṛhadāraṇyaka Up. | Sanskrit, Veda | Baptist Mission Press | J. Thomas | 1 Rs. |
| 7 | Jul., 1848 | II | Part I, Fasc. III | 96 | E. Röer | Bṛhadāraṇyaka Up. | Sanskrit, Veda | Baptist Mission Press | J. Thomas | 1 Rs. |
| 8 | Aug., 1848 | II | Part I, Fasc. IV | 96 | E. Röer | Bṛhadāraṇyaka Up. | Sanskrit, Veda | Baptist Mission Press | J. Thomas | 1 Rs. |
| 9 | Sep., 1848 | II | Part I, Fasc. V | 96 | E. Röer | Bṛhadāraṇyaka Up. | Sanskrit, Veda | Baptist Mission Press | J. Thomas | 1 Rs. |

(continued)

**Table 5.1** (continued)

| No. | Month, Year | Vol. | Part, Fasc. | Pages | Editor | Title | Lang., Genre | Press | Typographer | Price |
|---|---|---|---|---|---|---|---|---|---|---|
| 10 | Oct., 1848 | II | Part II, Fasc. VI | 96 | E. Röer | Bṛhadāraṇyaka Up. | Sanskrit, Veda | Baptist Mission Press | J. Thomas | 1 Rs. |
| 11 | Nov., 1848 | II | Part II, Fasc. VII | 96 | E. Röer | Bṛhadāraṇyaka Up. | Sanskrit, Veda | Baptist Mission Press | J. Thomas | 1 Rs. |
| 12 | Dec., 1848 | II | Part II, Fasc. VIII | 96 | E. Röer | Bṛhadāraṇyaka Up. | Sanskrit, Veda | Baptist Mission Press | J. Thomas | 1 Rs. |
| 13 | Jan., 1849 | II | Part II, Fasc. IX | 96 | E. Röer | Bṛhadāraṇyaka Up. | Sanskrit, Veda | Baptist Mission Press | J. Thomas | 1 Rs. |
| 14 | Feb., 1849 | III | Fasc. I | 96 | E. Röer | Chandogya Up. | Sanskrit, Veda | Baptist Mission Press | J. Thomas | 1 Rs. |
| 15 | Mar., 1849 | III | Fasc. II | 96 | E. Röer | Chandogya Up. | Sanskrit, Veda | Baptist Mission Press | J. Thomas | 1 Rs. |
| 16 | Apr., 1849 | II | Part II, Fasc. X | 96 | E. Röer | Bṛhadāraṇyaka Up. | Sanskrit, Veda | Baptist Mission Press | J. Thomas | 1 Rs. |
| 17 | May, 1849 | III | Fasc. III | 96 | E. Röer | Chandogya Up. | Sanskrit, Veda | Baptist Mission Press | J. Thomas | 1 Rs. |
| 18 | Jun., 1849 | II | Part II, Fasc. XI | 128 | E. Röer | Bṛhadāraṇyaka Up. | Sanskrit, Veda | Baptist Mission Press | J. Thomas | 1 Rs. |

| | | | | | | | | | | |
|---|---|---|---|---|---|---|---|---|---|---|
| 19 | Jul., 1849 | IV | Fasc. I | 144 | Rajendralal Mitra | Kāmandakīyanītisāraḥ | Sanskrit, Dharmaśāstra | Baptist Mission Press | J. Thomas | 1 Rs. |
| 20 | Aug., 1849 | III | Fasc. IV | 96 | E. Röer | Chandogya Up. | Sanskrit, Veda | Baptist Mission Press | J. Thomas | 1 Rs. |
| 21 | Sep., 1849 | VI | Fasc. I | 100 | A. Sprenger | Two Works on Arabic Bibliography | Arabic | Baptist Mission Press | J. Thomas | 1 Rs. |
| 22 | Oct., 1849 | VII | Fasc. I | 136 | E. Röer | Taittirīya, Aitareya, Śvetāśvatara Up. | Sanskrit, Veda | Baptist Mission Press | J. Thomas | 1 Rs. |
| 23 | Nov., 1849 | III | Fasc. V | 96 | E. Röer | Chandogya Up. | Sanskrit, Veda | Baptist Mission Press | J. Thomas | 1 Rs. |
| 24 | Dec., 1849 | VIII | Fasc. I | 96 | E. Röer | Īśa, Kena, Kaṭha, Praśna, Muṇḍaka, Māṇḍukya Up. | Sanskrit, Veda | Baptist Mission Press | J. Thomas | 1 Rs. |
| 25 | Jan., 1850 | III | Fasc. VI | 156 | E. Röer | Chandogya Up. | Sanskrit, Veda | Baptist Mission Press | J. Thomas | 1 Rs. |
| 26 | Feb., 1850 | VIII | Fasc. II | 96 | E. Röer | Īśa, Kena, Kaṭha, Praśna, Muṇḍaka, Māṇḍukya Up. | Sanskrit, Veda | Baptist Mission Press | J. Thomas | 1 Rs. |
| 27 | Mar., 1850 | II | Part III, Fasc. I | 88 | E. Röer | Bṛhadāraṇyaka Up. (tr.) | English, Veda | Baptist Mission Press | J. Thomas | 1 Rs. |

(continued)

**Table 5.1** (continued)

| No. | Month, Year | Vol. | Part, Fasc. | Pages | Editor | Title | Lang., Genre | Press | Typographer | Price |
|---|---|---|---|---|---|---|---|---|---|---|
| 28 | Apr., 1850 | VIII | Fasc. III | 96 | E. Röer | Īśa, Kena, Kaṭha, Praśna, Muṇđaka, Māṇđukya Up. | Sanskrit, Veda | Baptist Mission Press | J. Thomas | 1 Rs. |
| 29 | May, 1850 | VIII | Fasc. IV | 96 | E. Röer | Īśa, Kena, Kaṭha, Praśna, Muṇđaka, Māṇđukya Up. | Sanskrit, Veda | Baptist Mission Press | J. Thomas | 1 Rs. |
| 30 | Jun., 1850 | VIII | Fasc. V | 96 | E. Röer | Īśa, Kena, Kaṭh. Praśn., Muṇđ., Māṇđ. Up. | Sanskrit, Veda | Baptist Mission Press | J. Thomas | 1 Rs. |
| 31 | Jul., 1850 | VIII | Fasc. VI | 120 | E. Röer | Īśa, Kena, Kaṭha, Praśna, Muṇđaka, Māṇđukya Up. | Sanskrit, Veda | Baptist Mission Press | J. Thomas | 1 Rs. |
| 32 | Aug., 1850 | IX | Fasc. I | 116 | E. Röer | Bhāṣapariccheda (tr.) | Sanskrit, Navyanyāya | Baptist Mission | J. Thomas | 1 Rs. |
| 33 | Sep., 1850 | VII | Fasc. II | 128 | E. Röer | Taittirīya, Aitareya, Śvetāśvatara Up. | Sanskrit, Veda | Baptist Mission Press | J. Thomas | 1 Rs. |
| 34 | Oct., 1850 | VII | Fasc. III | 128 | E. Röer | Taittirīya, Aitareya, Śvetāśvatara Up. | Sanskrit, Veda | Baptist Mission Press | J. Thomas | 1 Rs. |
| 35 | Nov., 1850 | IX | Fasc. II | 152 | E. Röer | Bhāṣapariccheda (tr.) | Sanskrit, Navyanyāya | Baptist Mission Press | J. Thomas | 1 Rs. |
| 36 | Dec., 1850 | X | Fasc. I | 124 | E. Röer, J.R. Ballantyne (tr.) | Sāhityadarpaṇa | English, Poetics | Baptist Mission Press | J. Thomas | 1 Rs. |

**Table 5.2** Bibliotheca Indica, Old Series, 1851–1865 (d. =pence, Fasc.=Fasciculus, Rs.=rupees, Shill.=shillings, tr.=translated)

| No. | Year | Vol. | Part, Fasc. | Pages | Editor/Author | Title | Lang., Genre | Press | Typographer | Price |
|---|---|---|---|---|---|---|---|---|---|---|
| 37 | 1851 | X | Fasc. II | 96 | E. Röer | Sāhityadarpaṇa | Sanskrit, Poetics | Baptist Mission | J. Thomas | 1 Rs. / 2 Shill. |
| 38 | 1851 | II | Part III, Fasc. II | 82 | E. Röer | Bṛhadāraṇyaka Up. (tr.) | English, Veda | Baptist Mission | J. Thomas | 1 Rs. / 2 Shill. |
| 39 | 1851 | XI | Part I, Fasc. I | 96 | E. Röer | Uttaranaiṣadhacarita | Sanskrit, Poetry | Baptist Mission | J. Thomas | 1 Rs. / 2 Shill. |
| 40 | 1851 | XI | Part I, Fasc. II | 96 | E. Röer | Uttaranaiṣadhacarita | Sanskrit, Poetry | Baptist Mission | J. Thomas | 1 Rs. / 2 Shill. |
| 41 | 1852 | XV | Fasc. I | 88 | E. Röer | Taitt., Ait., Śvet., Kena, Īśa Up. (tr.) | English, Veda | Bishop's College | T.J. M'Arthur | 1 Rs. / 2 Shill. |
| 42 | 1852 | XI | Part I, Fasc. III | 96 | E. Röer | Uttaranaiṣadhacarita | Sanskrit, Poetry | Baptist Mission | J. Thomas | 1 Rs. / 2 Shill. |
| 43 | 1852 | XII | | | M. Basherood-Deen et al. | Kitāb Alma'ārif | Arabic | | | |
| 44 | 1852 | | Fasc. I | | M. Sadeedood-deen Khan et al. | Soyúty's Itqán | Arabic | | | |
| 45 | 1852 | XI | Part I, Fasc. IV | 96 | E. Röer | Uttaranaiṣadhacarita | Sanskrit, Poetry | Baptist Mission | J. Thomas | 1 Rs. / 2 Shill. |
| 46 | 1852 | XI | Part I, Fasc. V | 96 | E. Röer | Uttaranaiṣadhacarita | Sanskrit, Poetry | Baptist Mission | J. Thomas | 1 Rs. / 2 Shill. |
| 47 | 1853 | XIV | Fasc. I | 116 | Rajendralal Mittra | Caitanyacandrodaya | Sanskrit, Drama | Baptist Mission | J. Thomas | 1 Rs. / 2 Shill. |
| 48 | 1853 | XIV | Fasc. II | 96 | Rajendralal Mittra | Caitanyacandrodaya | Sanskrit, Drama | Baptist Mission | J. Thomas | 1 Rs. / 2 Shill. |
| 49 | 1853 | | Fasc. II | | M. Sadeedood-deen Khan et al. | Soyúty's Itqán | Arabic | | | |

(continued)

**Table 5.2** (continued)

| No. | Year | Vol. | Part, Fasc. | Pages | Editor/Author | Title | Lang., Genre | Press | Typographer | Price |
|---|---|---|---|---|---|---|---|---|---|---|
| 49 | 1853 | XVI | Fasc. I | | A.M. Shustery, A. Sprenger | Sikandar-Namah of Nizamy | Persian | | | |
| 50 | 1853 | XV | Fasc. II | 96 | E. Röer | Taittirīya, Aitareya, etc., Up. (tr.) | English, Veda | Bishop's College | T.J. M'Arthur | 1 Rs. / 2 Shill. |
| 51 | 1853 | V | Fasc. I | 96 | Rajendralal Mitra | Lalitavistara | Sanskrit, Buddhism | Baptist Mission | J. Thomas | 1 Rs. / 2 Shill. |
| 52 | 1853 | XI | Part I, Fasc. VI | 96 | E. Röer | Uttaranaiṣadhacarita | Sanskrit, Poetry | Baptist Mission | J. Thomas | 1 Rs. / 2 Shill. |
| 53 | 1853 | X | Fasc. III | 96 | E. Röer | Sāhityadarpaṇa | Sanskrit, Poetics | Baptist Mission | J. Thomas | 1 Rs. / 2 Shill. |
| 54 | 1853 | X | Fasc. IV | 96 | E. Röer | Sāhityadarpaṇa | Sanskrit, Poetics | Baptist Mission | J. Thomas | 1 Rs. / 2 Shill. |
| 55 | 1853 | X | Fasc. V | 88 | E. Röer | Sāhityadarpaṇa | Sanskrit, Poetics | Baptist Mission | J. Thomas | 1 Rs. / 2 Shill. |
| 56 | 1853 | | Fasc. I | | W. Nassau Lees | The Fotooh al-Shaám | Arabic | Baptist Mission | J. Thomas | |
| 57 | 1853 | | Fasc. III | | M. Sadeedood-deen Khan et al. | Soyúty's Itqán | Arabic | | | |
| 58 | 1853 | | Fasc. I | | Mohammad Wajih et al. | A Dictionary of Technical Terms | Arabic | Bengal Military Orphan | F. Carberry | |
| 59 | 1853 | | Fasc. I | | W. Nassau Lees | The Conquest of Syria | Arabic | | | |
| 60 | 1853 | | Fasc. I | | A. Sprenger | Túsys list of Shy'ah books | Arabic | | | |
| 61 | 1853 | | Part I, Fasc. I | | A. Sprenger et al. | A Biographical Dictionary of Persons | Arabic | | | |
| 62 | 1853 | | Fasc. II | | W. Nassau Lees | The Fotooh al-Shaám | Arabic | | | |
| 63 | 1853 | | Fasc. I | 96 | Īśvaracandra Vidyāsāgara | Sarvadarśana Saṇgraha | Sanskrit, Doxography | Baptist Mission | J. Thomas | 10 annas / 1 shill. 8 d |

| | | | | | | | | | | |
|---|---|---|---|---|---|---|---|---|---|---|
| 64 | 1853 | | Fasc. I | | Rāmanārāyaṇa Vidyāratna | Brahmasūtra | Sanskrit, Vedānta | | | |
| 65 | 1853 | | Fasc. II | | Mohammad Wajih et al. | A Dictionary of Technical Terms | Arabic | Bengal Military Orphan | F. Carberry | |
| 66 | 1853 | | Fasc. II | | W. Nassau Lees | The Conquest of Syria | Arabic | | | |
| 67 | 1853 | XI | Part II, Fasc. VII | 96 | E. Röer | Uttaranaiṣadhacarita | Sanskrit, Poetry | Baptist Mission | J. Thomas | 10 annas / 1 shill. 8 d |
| 68 | 1853 | | Fasc. IV | | M. Sadeedood-deen Khan et al. | Soyúty's Itqán | Arabic | | | |
| 69 | 1853 | | Part I, Fasc. II | | A. Sprenger et al. | A Biographical Dictionary of Persons | Arabic | | | |
| 70 | 1853 | | Fasc. V | | M. Sadeedood-deen Khan et al. | Soyúty's Itqán | Arabic | | | |
| 71 | 1853 | | Fasc. II | | A. Sprenger | Túsys list of Shy'ah books | Arabic | | | |
| 72 | 1853 | XI | Part II, Fasc. VIII | 96 | E. Röer | Uttaranaiṣadhacarita | Sanskrit, Poetry | Baptist Mission | J. Thomas | 10 annas / 1 shill. 8 d |
| 73 | 1854 | V | Fasc. II | 96 | Rajendralal Mitra | Lalitavistara | Sanskrit, Buddhism | Baptist Mission | J. Thomas | 10 annas / 1 shill. 8 d |
| 74 | 1854 | | Fasc. VI | | M. Sadeedood-deen Khan et al. | Soyúty's Itqán | Arabic | | | |
| 75 | 1854 | | Part I, Fasc. III | | A. Sprenger et al. | A Biographical Dictionary of Persons | Arabic | | | |
| 76 | 1854 | | | | A. Sprenger | A Dictionary of Technical Terms | Arabic | Bengal Military Orphan | F. Carberry | |
| 77 | 1854 | | Fasc. VII | | M. Sadeedood-deen Khan et al. | Soyúty's Itqán | Arabic | | | |
| 78 | 1854 | | Fasc. I | 96 | Rajendralal Mittra | Chandogya Up. (tr.) | English, Veda | Baptist Mission | J. Thomas | |

(continued)

**Table 5.2** (continued)

| No. | Year | Vol. | Part, Fasc. | Pages | Editor/Author | Title | Lang., Genre | Press | Typographer | Price |
|---|---|---|---|---|---|---|---|---|---|---|
| 79 | 1854 | | Fasc. I | 108 | A.M. Fitz-Edward Hall, Bāpudeva Śāstrī | Sūryasiddhānta with Gūḍhārthaprakāśa | Sanskrit, Astronomy | Baptist Mission | J. Thomas | |
| 80 | 1854 | XIV | Fasc. III | 80 | Rajendralal Mittra | Caitanyacandrodaya | Sanskrit, Drama | Baptist Mission | J. Thomas | 10 annas / 1 shill. 8 d |
| 81 | 1854 | | Fasc. VIII | | M. Sadeedood-deen Khan et al. | Soyúty's Itqán | Arabic | | | |
| 82 | 1854 | | Fasc. III | | Mohammad Wajih et al. | A Dictionary of Technical Terms | Arabic | Bengal Military Orphan | F. Carberry | |
| 83 | 1854 | | Part I, Fasc. IV | | A. Sprenger et al. | A Biographical Dictionary of Persons | Arabic | | | |
| 84 | 1854 | | Fasc. III | | W. Nassau Lees | The Fotooh al-Shaám | Arabic | Baptist Mission | J. Thomas | |
| 85 | 1854 | | Fasc. IV | | W. Nassau Lees | The Fotooh al-Shaám | Arabic | Baptist Mission | J. Thomas | |
| 86 | 1854 | | Part I, Fasc. V | | A. Sprenger et al. | A Biographical Dictionary of Persons | Arabic | | | |
| 87 | 1854 | XI | Part II, Fasc. IX | 96 | E. Röer | Uttaranaiṣadhacarita | Sanskrit, Poetry | Baptist Mission | J. Thomas | 10 annas / 1 shill. 8 d |
| 88 | 1854 | | Fasc. IV | | A. Sprenger | A Dictionary of Technical Terms | | Bengal Military Orphan | F. Carberry | |
| 89 | 1854 | | Fasc. II | | Rāmanārāyaṇa Vidyāratna | Brahmasūtra | Sanskrit, Vedānta | | | |
| 90 | 1854 | XI | Part II, Fasc. X | 96 | E. Röer | Uttaranaiṣadhacarita | Sanskrit, Poetry | Baptist Mission | J. Thomas | 10 annas / 1 shill. 8 d |
| 91 | 1854 | | Fasc. III | | A. Sprenger | Túsys list of Shy'ah books | Arabic | | | |
| 92 | 1854 | | Part I, Fasc. I | 100 | E. Röer, E.B. Cowell | Taittirīya Saṃhitā | Sanskrit, Veda | Baptist Mission | J. Thomas | 10 annas / 1 shill. 8 d |

| | | | | | | | | | | |
|---|---|---|---|---|---|---|---|---|---|---|
| 93 | 1854 | | Part I, Fasc. VI | | A. Sprenger et al. | A Biographical Dictionary of Persons | Arabic | | | |
| 94 | 1855 | XVI | Fasc. I | 164 | A. M. Fitz-Edward Hall | Sāṇkhyapravacanabhāṣya | Sanskrit, Sāṇkhya | Baptist Mission | J. Thomas | 10 annas / 1 shill. 8 d |
| 95 | 1855 | | Fasc. V | | Mohammad Wajih et al. | A Dictionary of Technical Terms | Arabic | Bengal Military Orphan | F. Carberry | |
| 96 | 1855 | | Fasc. III | | W. Nassau Lees | The Conquest of Syria | Arabic | | | |
| 97 | 1855 | XVI | Fasc. II | 96 | A.M. Fitz-Edward Hall | Sāṇkhyapravacanabhāṣya | Sanskrit, Sāṇkhya | Baptist Mission | J. Thomas | 10 annas / 1 shill. 8 d |
| 98 | 1855 | | Fasc. IV | | W. Nassau Lees | The Conquest of Syria | Arabic | | | |
| 99 | 1855 | | Fasc. IX | | M. Sadeedood-deen Khan et al. | Soyúty's Itqán | Arabic | | | |
| 100 | 1855 | | Fasc. VI | | Mohammad Wajih et al. | A Dictionary of Technical Terms | Arabic | Bengal Military Orphan | F. Carberry | |
| 101 | 1855 | I | Fasc. VII | | A. Sprenger et al. | A Biographical Dictionary of Persons | Arabic | | | |
| 102 | 1855 | | Fasc. V | | W. Nassau Lees | The Conquest of Syria | Arabic | | | |
| 103 | 1855 | | Fasc. VI | | W. Nassau Lees | The Conquest of Syria | Arabic | | | |
| 104 | 1855 | | Fasc. X | | M. Sadeedood-deen Khan et al. | Soyúty's Itqán | Arabic | | | |
| 105 | 1855 | | Fasc. II | 96 | A.M. Fitz-Edward Hall, Bāpudeva Śāstrī | Sūryasiddhānta | Sanskrit, Astronomy | Baptist Mission | C.B. Lewis | 10 annas / 1 shill. 8 d |
| 106 | 1855 | | Part I, Fasc. VIII | | A. Sprenger et al. | A Biographical Dictionary of Persons | Arabic | | | |
| 107 | 1855 | | Fasc. IV | | A. Sprenger | Túsys list of Shy'ah books | Arabic | | | |
| 108 | 1855 | | Fasc. VII | | Mohammad Wajih et al. | A Dictionary of Technical Terms | Arabic | Bengal Military Orphan | F. Carberry | |

(continued)

**Table 5.2** (continued)

| No. | Year | Vol. | Part, Fasc. | Pages | Editor/Author | Title | Lang., Genre | Press | Typographer | Price |
|---|---|---|---|---|---|---|---|---|---|---|
| 109 | 1855 | | Fasc. VIII | | Mohammad Wajih et al. | A Dictionary of Technical Terms | Arabic | Bengal Military Orphan | F. Carberry | |
| 110 | 1855 | | Fasc. I | 108 | A. von Kremer | History of Muhammad's campaigns | Arabic | | | |
| 111 | 1855 | I | Fasc. IX | | A. Sprenger et al. | A Biographical Dictionary of Persons | Arabic | | | |
| 112 | 1855 | | Fasc. II | 96 | A. von Kremer | History of Muhammad's campaigns | Arabic | | | |
| 113 | 1855 | | Fasc. III | 96 | A. von Kremer | History of Muhammad's campaigns | Arabic | | | |
| 114 | 1855 | | Fasc. I | 108 | Rev. K.M. Bannerjea | Mārkandeya Purāṇa | Sanskrit, Purāṇa | Bishop's College | T.J. M'Arthur | 10 annas / 1 shill. 8 d |
| 115 | 1855 | | Fasc. III | 96 | A. M. Fitz-Edward Hall, Bāpudeva Śāstrī | Sūryasiddhānta | Sanskrit, Astronomy | Baptist Mission | J. Thomas | 10 annas / 1 shill. 8 d |
| 116 | 1855 | | Fasc. I | 152 | A. M. Fitz-Edward Hall | The Tale of Vāsavadattā | Sanskrit, Poetry | Baptist Mission | J. Thomas | 10 annas / 1 shill. 8 d |
| 117 | 1855 | | Part I, Fasc. II | 96 | E. Röer E.B. Cowell | Taittirīya Saṃhitā | Sanskrit, Veda | Baptist Mission | J. Thomas | 10 annas / 1 shill. 8 d |
| 118 | 1855 | | Fasc. IX | | Mohammad Wajih et al. | A Dictionary of Technical Terms | Arabic | Bengal Military Orphan | F. Carberry | |
| 119 | 1855 | | Part I, Fasc. III | 96 | E. Röer E.B. Cowell | Taittirīya Saṃhitā | Sanskrit, Veda | Baptist Mission | J. Thomas | 10 annas / 1 shill. 8 d |
| 120 | 1855 | XI | Part II Fasc. XI | 96 | E. Röer | Uttaranaiṣadhacarita | Sanskrit, Poetry | Baptist Mission | J. Thomas | 10 annas / 1 shill. 8 d |
| 121 | 1855 | | Fasc. IV | 96 | A. von Kremer | History of Muhammad's campaigns | Arabic | | | |

| | | | | | | | | | | |
|---|---|---|---|---|---|---|---|---|---|---|
| 122 | 1855 | | Part I, Fasc. IV | 96 | E. Röer E.B. Cowell | Taittirīya Saṃhitā | Sanskrit, Veda | Baptist Mission | J. Thomas | 10 annas / 1 shill. 8 d |
| 123 | 1855 | | Part I, Fasc. X | | A. Sprenger et al. | A Biographical Dictionary of Persons | Arabic | | | |
| 124 | 1855 | XI | Part II, Fasc. XII | 72 | E. Röer | Uttaranaiṣadhacarita | Sanskrit, Poetry | Baptist Mission | J. Thomas | 10 annas / 1 shill. 8 d |
| 125 | 1855 | | Part I, Fasc. I | 96 | Rajendralal Mittra | Taittirīya Brāhmaṇa | Sanskrit, Veda | Baptist Mission | J. Thomas | 10 annas / 1 Shill. 8 d. |
| 126 | 1855 | | Part I, Fasc. II | 96 | Rajendralal Mittra | Taittirīya Brāhmaṇa | Sanskrit, Veda | Baptist Mission | J. Thomas | 10 annas / 1 Shill. 8 d. |
| 127 | 1855 | | Fasc. II | 96 | Rev. K.M. Bannerjea | Mārkandeya Purāṇa | Sanskrit, Purāṇa | Bishop's College | T.J. M'Arthur | 10 annas / 1 shill. 8 d |
| 123 | 1855 | | Part I, Fasc. XI | | A. Sprenger et al. | A Biographical Dictionary of Persons | Arabic | | | |
| 129 | 1855 | | Fasc. X | | Mohammad Wajih et al. | A Dictionary of Technical Terms | Arabic | Bengal Military Orphan | F. Carberry | |
| 130 | 1855 | | Fasc. II | 96 | A. M. Fitz-Edward Hall | The Tale of Vāsavadattā | Sanskrit, Poetry | Baptist Mission | J. Thomas | 10 annas / 1 shill. 8 d |
| 131 | 1855 | | Part I, Fasc. V | 96 | E. Röer, E.B. Cowell | Taittirīya Saṃhitā | Sanskrit, Veda | Baptist Mission | J. Thomas | 10 annas / 1 shill. 8 d |
| 132 | 1855 | | Fasc. XI | | Mohammad Wajih et al. | A Dictionary of Technical Terms | Arabic | Bengal Military Orphan | F. Carberry | |
| 133 | 1855 | | Part I, Fasc. VI | 96 | E. Röer E.B. Cowell | Taittirīya Saṃhitā | Sanskrit, Veda | Baptist Mission | J. Thomas | 10 annas / 1 shill. 8 d |
| 134 | 1855 | | Part I, Fasc. VII | 96 | E. Röer E.B. Cowell | Taittirīya Saṃhitā | Sanskrit, Veda | Baptist Mission | J. Thomas | 10 annas / 1 shill. 8 d |
| 135 | 1856 | II | Part III, Fasc. III | 140 | E. Röer | Bṛhadāraṇyaka Up. (tr.) | English, Veda | Baptist Mission | J. Thomas | 10 annas / 1 Shill. 8 d. |

(continued)

**Table 5.2** (continued)

| No. | Year | Vol. | Part, Fasc. | Pages | Editor/Author | Title | Lang., Genre | Press | Typographer | Price |
|---|---|---|---|---|---|---|---|---|---|---|
| 136 | 1856 | | Part I, Fasc. XII | | A. Sprenger et al. | A Biographical Dictionary of Persons | Arabic | | | |
| 137 | 1856 | | Part I, Fasc. VIII | 96 | E. Röer E.B. Cowell | Taittirīya Saṃhitā | Sanskrit, Veda | Baptist Mission | J. Thomas | 10 annas / 1 shill. 8 d |
| 138 | 1856 | | Part II, Fasc. I | | A. Sprenger et al. | A Biographical Dictionary of Persons | Arabic | | | |
| 139 | 1856 | | Fasc. V | 96 | A. von Kremer | History of Muhammad's Campaigns | Arabic | | | |
| 140 | 1856 | | Fasc. III | 96 | Rev. K.M. Bannerjea | Mārkandeya Purāṇa | Sanskrit, Purāṇa | Bishop's College | T.J. M'Arthur | 10 annas / 1 shill. 8 d |
| 141 | 1857 | XVI | Fasc. III | 68 | A. M. Fitz-Edward Hall | Sāṇkhyapravacanabhāṣya (tr.) | English, Sāṇkhya | Baptist Mission | J. Thomas | 10 annas / 1 shill. 8 d |
| 142 | 1859 | | Fasc. II | 96 | Īśvaracandra Vidyāsāgara | Sarvadarśanasaṃgraha | Sanskrit, Doxography | Baptist Mission | C.B. Lewis | 10 annas / 1 Shill. 8 d. |
| 143 | 1859 | V | Fasc. III | 96 | Rajendralal Mitra | Lalitavistara | Sanskrit, Buddhism | Baptist Mission | C.B. Lewis | 10 annas / 1 Shill. 8 d. |
| 144 | 1859 | V | Fasc. IV | 96 | Rajendralal Mitra | Lalitavistara | Sanskrit, Buddhism | Baptist Mission | C.B. Lewis | 10 annas / 1 Shill. 8 d. |
| 145 | 1859 | V | Fasc. V | 96 | Rajendralal Mitra | Lalitavistara | Sanskrit, Buddhism | Baptist Mission | C.B. Lewis | 10 annas / 1 Shill. 8 d. |
| 146 | 1859 | | Fasc. IV | 108 | A. M. Fitz-Edward Hall, Bāpudeva Śāstrī | Sūryasiddhānta | Sanskrit, Astronomy | Baptist Mission | C.B. Lewis | 10 annas / 1 Shill. 8 d. |
| 147 | 1859 | | Part I, Fasc. III | 96 | Rajendralal Mittra | Taittirīya Brāhmaṇa | Sanskrit, Veda | Baptist Mission | C.B. Lewis | 10 annas / 1 Shill. 8 d. |

| | | | | | | | | | | |
|---|---|---|---|---|---|---|---|---|---|---|
| 148 | 1859 | | Fasc. III | 112 | A. M. Fitz-Edward Hall | The Tale of Vāsavadattā | Sanskrit, Poetry | Baptist Mission | C.B. Lewis | 10 annas / 1 Shill. 8 d. |
| 149 | 1859 | | Part I, Fasc. IX | 96 | E. Röer E.B. Cowell | Taittirīya Saṃhitā | Sanskrit, Veda | Baptist Mission | C.Bh. Lewis | 10 annas / 1 shill. 8 d |
| 150 | 1859 | | Part I, Fasc. IV | 96 | Rajendralal Mittra | Taittirīya Brāhmaṇa | Sanskrit, Veda | Baptist Mission | C.B. Lewis | 10 annas / 1 Shill. 8 d. |
| 151 | 1859 | | Part I, Fasc. V | 96 | Rajendralal Mittra | Taittirīya Brāhmaṇa | Sanskrit, Veda | Baptist Mission | C.B. Lewis | 10 annas / 1 Shill. 8 d. |
| 152 | 1859 | | Part I, Fasc. VI | 96 | Rajendralal Mittra | Taittirīya Brāhmaṇa | Sanskrit, Veda | Baptist Mission | C.B. Lewis | 10 annas / 1 Shill. 8 d. |
| 153 | 1859 | | Part I, Fasc. VII | 96 | Rajendralal Mittra | Taittirīya Brāhmaṇa | Sanskrit, Veda | Baptist Mission | C.B. Lewis | 10 annas / 1 Shill. 8 d. |
| 154 | 1859 | | Part I, Fasc. VIII | 96 | Rajendralal Mittra | Taittirīya Brāhmaṇa | Sanskrit, Veda | Baptist Mission | C.B. Lewis | 10 annas / 1 Shill. 8 d. |
| 155 | 1859 | | Part I, Fasc. IX | 96 | Rajendralal Mittra | Taittirīya Brāhmaṇa | Sanskrit, Veda | Baptist Mission | C.B. Lewis | 10 annas / 1 Shill. 8 d. |
| 156 | 1859 | | Fasc. XII | | Mohammad Wajih et al. | A Dictionary of Technical Terms | Arabic | Bengal Military Orphan | F. Carberry | |
| 157 | 1860 | | Part I, Fasc. X | 96 | E. Röer E.B. Cowell | Taittirīya Saṃhitā | Sanskrit, Veda | Baptist Mission | C.B. Lewis | 10 annas / 1 shill. 8 d |
| 158 | 1860 | | Fasc. XIII | | Mohammad Wajih et al. | A Dictionary of Technical Terms | | Bengal Military Orphan | F. Carberry | |
| 159 | 1860 | | Fasc. XIV | | Mohammad Wajih et al. | A Dictionary of Technical Terms | Arabic | Bengal Military Orphan | F. Carberry | |
| 160 | 1860 | | Part I, Fasc. XI | 116 | E. Röer, E.B. Cowell | Taittirīya Saṃhitā | Sanskrit, Veda | Baptist Mission | C.B. Lewis | 10 annas / 1 shill. 8 d |

(continued)

**Table 5.2** (continued)

| No. | Year | Vol. | Part, Fasc. | Pages | Editor/Author | Title | Lang., Genre | Press | Typographer | Price |
|---|---|---|---|---|---|---|---|---|---|---|
| 161 | 1860 | | Part II, Fasc. XII | 104 | E.B. Cowell | Taittirīya Saṃhitā | Sanskrit, Veda | Baptist Mission | C.B. Lewis | 10 annas / 1 Shill. 8 d. |
| 162 | 1860 | | Fasc. XV | | Mohammad Wajih et al. | A Dictionary of Technical Terms | Arabic | Bengal Military Orphan | F. Carberry | |
| 163 | 1860 | | Fasc. IV | 96 | Rev. K.M. Bannerjea | Mārkandeya Purāṇa | Sanskrit, Purāṇa | Bishop's College | | 10 annas / 1 shill. 8 d |
| 164 | 1860 | | Fasc. VII | | W. Nassau Lees | The Conquest of Syria | Arabic | | | |
| 165 | 1860 | | Fasc. XVI | | Mohammad Wajih et al. | A Dictionary of Technical Terms | Arabic | Bengal Military Orphan | F. Carberry | |
| 166 | 1860 | | Part II, Fasc. XIII | 96 | E.B. Cowell | Taittirīya Saṃhitā | Sanskrit, Veda | Baptist Mission | C.B. Lewis | 10 annas / 1 Shill. 8 d. |
| 167 | 1860 | | Fasc. VIII | | W. Nassau Lees | The Conquest of Syria | Arabic | | | |
| 167 | 1860 | | Fasc. XVII | | Mohammad Wajih et al. | A Dictionary of Technical Terms | Arabic | Bengal Military Orphan | F. Carberry | |
| 169 | 1860 | | Fasc. V | 96 | Rev. K.M. Bannerjea | Mārkandeya Purāṇa | Sanskrit, Purāṇa | Bishop's College | | 10 annas / 1 shill. 8 d |
| 170 | 1860 | | Fasc. XVIII | | Mohammad Wajih et al. | A Dictionary of Technical Terms | Arabic | Bengal Military Orphan | F. Carberry | |
| 171 | 1860 | | Part II, Fasc. XIV | 96 | E.B. Cowell | Taittirīya Saṃhitā | Sanskrit, Veda | Baptist Mission | C.B. Lewis | 10 annas / 1 Shill. 8 d. |

| | | | | | | | | | | |
|---|---|---|---|---|---|---|---|---|---|---|
| 172 | | | Fasc. III | | Rāmanārāyaṇa Vidyāratna | Brahmasūtra | Sanskrit, Vedānta | | | |
| 173 | | | Fasc. XIX | | Mohammad Wajih et al. | A Dictionary of Technical Terms | Arabic | Bengal Military Orphan | F. Carberry | |
| 174 | | | Fasc. IV | | Rāmanārāyaṇa Vidyāratna | Brahmasūtra | Sanskrit, Vedānta | | | |
| 175 | | | Part I, Fasc. X | 96 | Rajendralal Mittra | Taittirīya Brāhmaṇa | Sanskrit, Veda | Baptist Mission | C.B. Lewis | 10 annas / 1 shill. 8 d |
| 176 | | | Part I, Fasc. XI | 96 | Rajendralal Mittra | Taittirīya Brāhmaṇa | Sanskrit, Veda | Baptist Mission | C.B. Lewis | 10 annas / 1 shill. 8 d |
| 177 | 1862 | | Fasc. VI | 96 | Rev. K.M. Bannerjea | Mārkandeya Purāṇa | Sanskrit, Purāṇa | Bishop's College | | 10 annas / 1 shill. 8 d |
| 178 | 1862 | | Fasc. V | | Rāmanārāyaṇa Vidyāratna | Brahmasūtra | Sanskrit, Vedānta | | | |
| 179 | 1862 | IV | Fasc. II | 96 | Rajendralal Mittra | Kāmandakīyanītisāraṭīkā | Sanskrit, Dharmaśāstra | | | |
| 180 | 1862 | | Part II, Fasc. XV | 96 | E.B. Cowell | Taittirīya Saṃhitā | Sanskrit, Veda | Baptist Mission | C.B. Lewis | 10 annas / 1 Shill. 8 d. |
| 181 | 1862 | | Fasc. II | 96 | Rajendralal Mittra | Chandogya Up. (tr.) | English, Veda | Baptist Mission | C.B. Lewis | 10 annas / 1 Shill. 8 d. |
| 182 | 1862 | | Fasc. XX | | Mohammad Wajih et al. | A Dictionary of Technical Terms | Arabic | Bengal Military Orphan | F. Carberry | |
| 183 | 1862 | | Fasc. VII | 96 | Rev. K.M. Bannerjea | Mārkandeya Purāṇa | Sanskrit, Purāṇa | Bishop's College | | 10 annas / 1 Shill. 8 d. |
| 184 | 1862 | | Fasc. VI | | Rāmanārāyaṇa Vidyāratna | Brahmasūtra | Sanskrit, Vedānta | | | |
| 185 | 1862 | | Part II, Fasc. XVI | 96 | E.B. Cowell | Taittirīya Saṃhitā | Sanskrit, Veda | Baptist Mission | C.B. Lewis | 10 annas / 1 Shill. 8 d. |

(continued)

**Table 5.2** (continued)

| No. | Year | Vol. | Part, Fasc. | Pages | Editor/Author | Title | Lang., Genre | Press | Typographer | Price |
|---|---|---|---|---|---|---|---|---|---|---|
| 186 | 1862 | | Fasc. VII | | Rāmanārāyaṇa Vidyāratna | Brahmasūtra | Sanskrit, Vedānta | | | |
| 187 | 1862 | | Fasc. IX | | W. Nassau Lees | The Conquest of Syria | Arabic | | | |
| 188 | 1862 | | Part II, Fasc. XII | 96 | Rajendralal Mittra | Taittirīya Brāhmaṇa | Sanskrit, Veda | Baptist Mission | C.B. Lewis | 10 annas / 1 Shill. 8 d. |
| 189 | 1862 | | Part II, Fasc. XIII | 60 | Rajendralal Mittra | Taittirīya Brāhmaṇa | Sanskrit, Veda | Baptist Mission | C.B. Lewis | 10 annas / 1 Shill. 8 d. |
| 190 | | | Part II, Fasc. XIV | | Rajendralal Mittra | Taittirīya Brāhmaṇa | | Baptist Mission | C.B. Lewis | 10 annas / 1 Shill. 8 d. |
| 191 | | | Part II, Fasc. XV | | Rajendralal Mittra | Taittirīya Brāhmaṇa | | Baptist Mission | C.B. Lewis | 10 annas / 1 Shill. 8 d. |
| 192 | | | Part II, Fasc. XVI | | Rajendralal Mittra | Taittirīya Brāhmaṇa | | Baptist Mission | C.B. Lewis | 10 annas / 1 Shill. 8 d. |
| 193 | 1864 | | Part II, Fasc. XVII | 96 | E.B. Cowell | Taittirīya Saṃhitā, second kāṇḍa | 1866 | Baptist Mission | C.B. Lewis | 10 annas / 1 Shill. 8 d. |
| 194 | 1864 | | Fasc. VIII | | Rāmanārāyaṇa Vidyāratna | Brahmasūtra | Sanskrit, Vedānta | | | |
| 195 | 1864 | | Fasc. IX | | Rāmanārāyaṇa Vidyāratna | Brahmasūtra | Sanskrit, Vedānta | | | |
| 196 | 1864 | | Fasc. XVII | | Rajendralal Mittra | Taittirīya Brāhmaṇa | | Baptist Mission | C.B. Lewis | 10 annas / 1 Shill. 8 d. |

| | | | | | | | | | | |
|---|---|---|---|---|---|---|---|---|---|---|
| 197 | 1864 | | Fasc. XVIII | | Rajendralal Mittra | Taittirīya Brāhmaṇa | | Baptist Mission | C.B. Lewis | 10 annas / 1 Shill. 8 d. |
| 198 | 1864 | | Fasc. X | | Rāmanārāyaṇa Vidyāratna | Brahmasūtra | Sanskrit, Vedānta | | | |
| 199 | 1864 | | Fasc. XI | | Rāmanārāyaṇa Vidyāratna | Brahmasūtra | Sanskrit, Vedānta | | | |
| 200 | 1864 | | Fasc. XII | | Rāmanārāyaṇa Vidyāratna | Brahmasūtra | Sanskrit, Vedānta | | | |
| 201 | 1864 | | Fasc. XIII | | Rāmanārāyaṇa Vidyāratna | Brahmasūtra | Sanskrit, Vedānta | | | |
| 202 | 1864 | | Part II, Fasc. XVIII | 96 | E.B. Cowell | Taittirīya Saṃhitā | Sanskrit, Veda | Baptist Mission | C.B. Lewis | 10 annas / 1 Shill. 8 d. |
| 203 | 1865 | | Part II, Fasc. XIX | 96 | E.B. Cowell | Taittirīya Saṃhitā | Sanskrit, Veda | Baptist Mission | C.B. Lewis | 10 annas / 1 Shill. 8 d. |
| 204 | 1865 | | Fasc. XIX | | Rajendralal Mittra | Taittirīya Brāhmaṇa | Sanskrit, Veda | Baptist Mission | C.B. Lewis | 10 annas / 1 Shill. 8 d. |
| 205 | 1865 | | Part II, Fasc. II | | A. Sprenger et al. | A Biographical Dictionary of Persons | Arabic | | | |
| 206 | 1865 | IV | Fasc. III | | Rajendralal Mittra | Kāmandakīyanītisāraṭīkā | Sanskrit, Dharmaśāstra | | | |
| 207 | 1865 | | Part II, Fasc. III | | A. Sprenger et al. | A Biographical Dictionary of Persons | Arabic | | | |
| 208 | 1865 | | Part II, Fasc. IV | | A. Sprenger et al. | A Biographical Dictionary of Persons | Arabic | | | |
| 209 | 1865 | | Part II, Fasc. V | | A. Sprenger et al. | A Biographical Dictionary of Persons | Arabic | | | |

(continued)

**Table 5.2** (continued)

| No. | Year | Vol. | Part, Fasc. | Pages | Editor/Author | Title | Lang., Genre | Press | Typographer | Price |
|---|---|---|---|---|---|---|---|---|---|---|
| 210 | 1865 | | Fasc. XX | | Rajendralal Mittra | Taittirīya Brāhmaṇa | Sanskrit, Veda | Baptist Mission | C.B. Lewis | |
| 211 | 1865 | | Part II, Fasc. VI | | A. Sprenger et al. | A Biographical Dictionary of Persons | Arabic | | | |
| 212 | 1865 | | Fasc. I | 228 | Pramadas Mittra | Sāhityadarpaṇa (tr.) | English, Poetics | Baptist Mission | C.B. Lewis | 10 annas / 1 Shill. 8 d. |
| 213 | 1865 | | Fasc. II | 220 | Pramadas Mittra | Sāhityadarpaṇa (tr.) | English, Poetics | Baptist Mission | C.B. Lewis | 10 annas / 1 Shill. 8 d. |
| 214 | 1865 | | Part II, Fasc. VII | | A. Sprenger et al. | A Biographical Dictionary of Persons | Arabic | | | |

**Table 5.3** Bibliotheca Indica, New Series, 1861–1865 (d=pence, Fasc.=Fasciculus, Rs.=rupees, Shill.=shillings, tr.=translated)

| No. | Year | Fasc. | Editor | Title | Publ. year | Press | Typographer |
|---|---|---|---|---|---|---|---|
| 1 | 1861 | Fasc. I | Bāpudeva Śāstrī | Sūryasiddhānta (tr.) | English, Astronomy | Baptist Mission | C.B. Lewis |
| 2 | 1861 | Fasc. I | Saiyid Ahmad Khan | Tárikh-i Feroz-Sháhí of Ziaa al-Din Barni | Persian | | |
| 3 | 1861 | Fasc. II | Saiyid Ahmad Khan | Tárikh-i Feroz-Sháhí of Ziaa al-Din Barni | Persian | | |
| 4 | 1861 | Fasc. I | Jayanārāyaṇa Tarkapañcānana | Vaiśeṣikadarśana | English, Vaiśeṣika | Baptist Mission | C.B. Lewis |
| 5 | 1861 | Fasc. II | Jayanārāyaṇa Tarkapañcānana | Vaiśeṣikadarśana | English, Vaiśeṣika | Baptist Mission | C.B. Lewis |
| 6 | 1861 | Fasc. III | Jayanārāyaṇa Tarkapañcānana | Vaiśeṣikadarśana | English, Vaiśeṣika | Baptist Mission | C.B. Lewis |
| 7 | 1861 | Fasc. III | Saiyid Ahmad Khan | Tárikh-i Feroz-Sháhí of Ziaa al-Din Barni | Persian | | |
| 8 | 1861 | Fasc. IV | Jayanārāyaṇa Tarkapañcānana | Vaiśeṣikadarśana | English, Vaiśeṣika | Baptist Mission | C.B. Lewis |
| 9 | 1861 | Fasc. IV | Saiyid Ahmad Khan | Tárikh-i Feroz-Sháhí of Ziaa al-Din Barni | Persian | | |
| 10 | 1861 | Fasc. V | Jayanārāyaṇa Tarkapañcānana | Vaiśeṣikadarśana | English, Vaiśeṣika | Baptist Mission | C.B. Lewis |
| 11 | 1861 | Fasc. I | J. R. Ballantyne | Sāṇḍiliya Sūtras | Sanskrit, Arthaśāstra | ? | ? |
| 12 | 1861 | Fasc. I | A. M. Fitz-Edward Hall | Daśarūpa | Sanskrit, Dramaturgy | Baptist Mission | C.B. Lewis |
| 13 | 1861 | Fasc. II | L. Wilkinson, Bāpudeva Śāstrī | Siddhānta Śīromaṇi (tr.) | English, Astronomy | Baptist Mission | C.B. Lewis |
| 14 | 1861 | Fasc. V | Saiyid Ahmad Khan | Tárikh-i Feroz-Sháhí of Ziaa al-Din Barni | Persian | | |
| 15 | 1861 | Fasc. VI | Saiyid Ahmad Khan | Tárikh-i Feroz-Sháhí of Ziaa al-Din Barni | Persian | | |
| 16 | 1861 | Fasc. I | W.H. Morley | Tárikh-i Baihákí | Persian | | |
| 17 | 1861 | Fasc. I | Rev. K. M. Banerjea | Nāradapañcarātra | Sanskrit, Pañcarātra | Bishop's College | |
| 18 | 1861 | Fasc. II | W.H. Morley | Tárikh-i Baihákí | Persian | | |
| 19 | 1861 | Fasc. I | E. B. Cowell | Kauśitaki Brāhmaṇa | Sanskrit, Veda | | |
| 20 | 1861 | Fasc. II | E. B. Cowell | Kauśitaki Brāhmaṇa | Sanskrit, Veda | | |

(continued)

**Table 5.3** (continued)

| No. | Year | Fasc. | Editor | Title | Publ. year | Press | Typographer |
|---|---|---|---|---|---|---|---|
| 21 | 1861 | Fasc. III | W.H. Morley | Tárikh-i Baihákí | Persian | | |
| 22 | 1861 | Fasc. IV | W.H. Morley | Tárikh-i Baihákí | Persian | | |
| 23 | 1861 | Fasc. VII | Saiyid Ahmad Khan | Tárikh-i Feroz-Sháhí of Ziaa al-Din Barni | Persian | | |
| 24 | 1861 | Fasc. II | A. M. Fitz-Edward Hall | Daśarūpa | Sanskrit, Dramaturgy | Baptist Mission | C.B. Lewis |
| 25 | 1862 | Fasc. II | Rev. K. M. Banerjea | Nāradapañcarātra | | Bishop's College | |
| 26 | 1862 | Fasc. V | W.H. Morley | Tárikh-i Baihákí | Persian | | |
| 27 | 1862 | Fasc. VI | W.H. Morley | Tárikh-i Baihákí | Persian | | |
| 28 | 1862 | Fasc. III | L. Wilkinson, Bāpudeva Śāstrī | Siddhānta Śīromaṇi (tr.) | English, Astronomy | | |
| 29 | 1862 | Fasc. VII | W.H. Morley | Tárikh-i Baihákí | Persian | | |
| 30 | 1862 | Fasc. I | Premacandra Tarkavāgīśa | Kāvyadarśa | Sanskrit, Poetics | | |
| 31 | 1862 | Fasc. VIII | W.H. Morley | Tárikh-i Baihákí | Persian | | |
| 32 | 1862 | Fasc. I | J. R. Ballantyne | Sāṇkhyakārikā (tr.) | English, Sāṇkhya | Baptist Mission | C.B. Lewis |
| 33 | 1862 | Fasc. II | Premacandra Tarkavāgīśa | Kāvyadarśa | Sanskrit, Poetics | | |
| 34 | 1862 | Fasc. III | Rev. K. M. Banerjea | Nāradapañcarātra | | Bishop's College | |
| 35 | 1862 | Fasc. I | E. B. Cowell | Maitri Upaniṣad | Sanskrit, Veda | | |
| 36 | 1862 | Fasc. IX | W.H. Morley | Tárikh-i Baihákí | Persian | | |
| 37 | 1862 | | W.N. Lees, Mawlavies Abd-Al-Haqq, Gholam Qadir | The Nokhbat Al-Fikr and Nozhat Al-Nazr | Persian | | |
| 38 | 1862 | Fasc. III | Premacandra Tarkavāgīśa | Kāvyadarśa | Sanskrit, Poetics | | |
| 39 | 1863? | Fasc. IV | Premacandra Tarkavāgīśa | Kāvyadarśa | Sanskrit, Poetics | | |
| 40 | 1863? | Fasc. II | E. B. Cowell | Maitri Upaniṣad | Sanskrit, Veda | | |
| 41 | 1863? | Fasc. V | Premacandra Tarkavāgīśa | Kāvyadarśa | Sanskrit, Poetics | | |
| 42 | 1863? | Fasc. I | W. Nassau Lees, M. Khadim Hosain, 'Abd Al-Hai | Tabaqát-i Náisiri | Persian | | |

| | | | | | | | |
|---|---|---|---|---|---|---|---|
| 43 | 1863? | Fasc. II | W. Nassau Lees, M. Khadim Hosain, 'Abd Al-Hai | Tabaqát-i Náisiri | Persian | | |
| 44 | 1863 | Fasc. I | Maheshachandra Nyayaratna | Mīmāṃsā Darśana | Sanskrit, Mīmāṃsā | Bishop's College | |
| 45 | 1863 | Fasc. III | W. Nassau Lees, M. Khadim Hosain, 'Abd Al-Hai | Tabaqát-i Náisiri | Persian | | |
| 46 | 1864 | Fasc. I | Navadvīpacandra Gosvāmī, Jayanārāyaṇa Tarkapañcānana | The Śaṇkara Vijaya | Sanskrit, Vedānta | Bangala | |
| 47 | 1864 | Fasc. IV | W. Nassau Lees, M. Khadim Hosain, 'Abd Al-Hai | Tabaqát-i Náisiri | Persian | | |
| 48 | 1864 | Fasc. I | W. Nassau Lees, Munshi Ahmed Ali | Wís o Rámín: a Romance of Ancient Persia | English, Persian | | |
| 49 | 1864 | Fasc. II | W. Nassau Lees, Munshi Ahmed Ali | Wís o Rámín: a Romance of Ancient Persia | English, Persian | | |
| 50 | 1864 | Fasc. V | W. Nassau Lees, M. Khadim Hosain, 'Abd Al-Hai | Tabaqát-i Náisiri | Persian | | |
| 51 | 1864 | Fasc. I | H. Kern | Bṛhatsaṃhitā | Sanskrit, Astronomy | | |
| 52 | 1864 | Fasc. III | W. Nassau Lees, Munshi Ahmed Ali | Wís o Rámín: a Romance of Ancient Persia | English, Persian | | |
| 53 | 1864 | Fasc. IV | W. Nassau Lees, Munshi Ahmed Ali | Wís o Rámín: a Romance of Ancient Persia | English, Persian | | |
| 54 | 1864 | Fasc. II | H. Kern | Bṛhatsaṃhitā | Sanskrit, Astronomy | | |
| 55 | 1864 | Fasc. I | Rāmanārāyaṇa Vidyāratna | Śrautasūtra of Aśvalāyana | Sanskrit, Dharmaśāstra | | |
| 56 | 1864 | Fasc. I | Jayanārāyaṇa Tarkapañcānana | Nyāyadarśana | Sanskrit, Nyāya | Baptist Mission | C.B. Lewis |
| 57 | 1864 | Fasc. I | W. Nassau Lees, M. Ahmad 'Ali | Muntakhat al-Tawáríkh | Persian | | |
| 58 | 1864 | Fasc. II | W. Nassau Lees, M. Ahmad 'Ali | Muntakhat al-Tawáríkh | Persian | | |
| 59 | 1864 | Fasc. III | H. Kern | Bṛhatsaṃhitā | Sanskrit, Astronomy | | |
| 60 | 1864 | Fasc. I | Rajendralal Mittra | Taittirīya Āraṇyaka | Sanskrit, Veda | Baptist Mission | C.B. Lewis |

(continued)

**Table 5.3** (continued)

| No. | Year | Fasc. | Editor | Title | Publ. year | Press | Typographer |
|---|---|---|---|---|---|---|---|
| 61 | 1864 | Fasc. II | Rāmanārāyaṇa Vidyāratna | Śrautasūtra of Aśvalāyana | Sanskrit, Dharmaśāstra | | |
| 62 | 1865? | Fasc. III | W. Nassau Lees, M. Ahmad 'Ali | Muntakhat al-Tawáríkh | Persian | | |
| 63 | 1865? | IV | H. Kern | Bṛhatsaṃhitā | Sanskrit, Astronomy | | |
| 64 | 1865? | Fasc. IV | W. Nassau Lees, M. Ahmad 'Ali | Muntakhat al-Tawáríkh | Persian | | |
| 65 | 1865? | Fasc. V | W. Nassau Lees, M. Ahmad 'Ali | Muntakhat al-Tawáríkh | Persian | | |
| 66 | 1865? | Fasc. III | Rāmanārāyaṇa Vidyāratna | Śrautasūtra of Aśvalāyana | Sanskrit, Dharmaśāstra | | |
| 67 | 1865 | Fasc. II | Jayanārāyaṇa Tarkapañcānana | Nyāyadarśana | Sanskrit, Nyāya | Baptist Mission | C.B. Lewis |
| 68 | 1865 | Fasc. V | H. Kern | Bṛhatsaṃhitā | Sanskrit, Astronomy | | |
| 69 | 1865 | Fasc. IV | Rāmanārāyaṇa Vidyāratna | Śrautasūtra of Aśvalāyana | Sanskrit, Dharmaśāstra | | |
| 70 | 1865 | Fasc. III | Jayanārāyaṇa Tarkapañcānana | Nyāyadarśana | Sanskrit, Nyāya | Baptist Mission | C.B. Lewis |
| 137 | 1868 | Fasc. II | Navadvīpacandra Gosvāmī, Jayanārāyaṇa Tarkapañcānana | The Śaṅkara Vijaya | Sanskrit, Vedānta | Baptist Mission | C.B. Lewis |
| 138 | 1868 | Fasc. III | Navadvīpacandra Gosvāmī, Jayanārāyaṇa Tarkapañcānana | The Śaṅkara Vijaya | Sanskrit, Vedānta | Baptist Mission | C.B. Lewis |

## References

Bhattacharya, Durgaprasad, and Rama Deb Roy Malakar. 1977. Agricultural Wages in Bengal and Bihar, 1793–1972. English. *Social Scientist* 6 (1): 65–73. ISSN: 09700293.

Bibliotheca Indica. Catalogue of Works. 1929. *Bibliotheca Indica. A Collection of Oriental Works Published by the Asiatic Society of Bengal. Catalogue of Works relating to Indian Culture.* Calcutta: Asiatic Society.

Bose, Pannalal. 1953. Foreword. In *Our Heritage. Bulletin of the Department of Postgraduate Training and Research, Sanskrit College, Calcutta*, vol. 1, i–ii. July–December.

Broadberry, Stephen, and Bishnupriya Gupta. 2006. The Early Modern Great Divergence: Wages, Prices and Economic Development in Europe and Asia, 1500–1800. In *The Economic History Review*. New Series 59–1, 2–31.

Buchanan, Rev. Claudius. 1805. *The College of Fort William in Bengal*. London: Printed for T. Cadell, W. Davies, Strand, by W. Bulmer, and Co. Cleveland-Row.

Colebrooke, Edward. 1839. Notices of the Life of Henry Thomas Colebrooke, Esq., by His Son. *Journal of the Royal Asiatic Society of Great Britain and Ireland* 5 (2): 1–60.

Drewery, Mary. 1979. *William Carey. A Biography*. Grand Rapids, Michigan: Zondervan.

Emmott, D.H. 1965. Alexander Duff and the Foundation of Modern Education in India. *British Journal of Educational Studies* 13 (2): 160–169.

Gaskell, Philip. 1972. *A New Introduction to Bibliography*. Oxford: Clarendon Press.

Graheli, Alessandro. 2024. The Choice of Devanāgarī. In Felicitation Volume in Honor of Dominik Wujastyk. Ed. by Philipp Maas and Anthony Cerulli. Studia Indologica Universitatis Halensis. Halle: Universitätsverlag Halle-Wittenberg, pp. 257–287.

Halhed, Nathaniel Brassey. 1778. *A Grammar of the Bengal Language*. Hoogly, Bengal: Endorse Press, Bengal Presidency.

Hall, Fitzedward. 1859. *A Contribution Towards an Index to the Bibliography of the Indian Philosophical System*. Calcutta: Published by Order of the Gov't N.-W. P. at the Baptist Mission Press.

Hoerner, Fred. 1995. "A Tiger in a Brake": The Stealth of Reason in the Scholarship of Sir William Jones in India. *Texas Studies in Literature and Language* 37 (2): 215–232.

Jacquemont, Victor. 1844. *Voyage dans l'Inde pendant les années 1828 à 1832*. Vol. 1. Paris: Typographie de Firmin Didot Fréres, Imprimeurs de l'Institut de France.

Jones, Williams, et al. 1798. *Dissertations and Miscellaneous Pieces, Relating to the History and Antiquities, the Arts, Sciences, and Literature of Asia. Volume the fourth: being the whole of the articles contained in the fourth volume of the Asiatic Researches*. Reprint (originally printed in Calcutta). London: Vernor and Hood.

Kesavan, B.S. 1985. *History of Printing and Publishing in India. A Story of Cultural Re-awakening*, vol. 1. New Delhi: National Book Trust, India.

Khan, M. Siddiq. 1961. William Carey and the Serampore Books (1800–1834). *Libri* 11 (3): 197–280.

Lanman, Charles Rockwell. 1920. India and the West with a Plea for Team-Work among Scholars. *Journal of the American Oriental Society* 40: 225–247.

Lewis, Charles Bennett. 1869a. *On the Present Condition of Indian Missions*. Calcutta: Baptist Mission Press.

Lewis, Charles Bennett. 1869b. *To E.B. Underhill, Esq., LL.D. Secretary to the Baptist Missionary Society*. Calcutta: Baptist Mission Press.

Lewis, Charles Bennett. 1871. *Memoirs of the Rev. John Thomas. First Baptist Missionary to Bengal: A Fragment*. Calcutta: Baptist Mission Press.

Losty, J.P. 1990. *Calcutta, City of Palaces. A Survey of the City in the Days of the East India Company, 1690–1858*. London: The British Library, Arnold Publishers.

Master, Alfred. 1946. The Influence of Sir William Jones upon Sanskrit Studies. *Bulletin of the School of Oriental Studies, University of London* 11 (4): 798–806.

Michael, Thomas, ed. 2010. *Standard Catalog of World Coins, 1701–1800*. 5th ed. Iola, WI: Krause Publications.
Mitra, R., A.F.R. Hoernle, and P.N. Bose (1885). Centenary Review of the Asiatic Society of Bengal. From 1784 to 1883. In *Proceedings of the Special Centenary Meeting of the Society*. Printed by Thacker, Spink and Co. Calcutta: Asiatic Society of Bengal.
Mukherjee, S.N. (1964). Sir William Jones and the British Attitudes Towards India. *The Journal of the Royal Asiatic Society of Great Britain and Ireland* 1: 37–47.
Naik, Bapurao S. 1971. *Typography of Devanagari*. Bombay: Directorate of Languages.
Nicholls, George. 2005. *Sketch of the Rise and Progress of the Benares Pathshala*, ed. Harishchandra Mani Tripathi. From a manuscript dated 1848, published in 1906. Varanasi: Sampurnanand Sanskrit University.
O'Conor, J.E. 1886. *Prices and Wages in India*. Calcutta: Government of India, Department of Finance and Commerce.
Ogborn, Miles. 2007. *Indian Ink. Script and Print in the Making of the English East India Company*. Chicago: University of Chicago Press.
Ridding, C.M. 1931. Cowell and His Pupils. *Bulletin of the School of Oriental Studies, University of London* 6 (2): 461–468.
Rocher, Rosane. 1983a. *Orientalism, Poetry, and the Millenium. The Checkered Life of Nathaniel Brassey Halhed (1751–1830)*. Delhi: Motilal Banarsidass.
Rocher, Rosane. 1993b. British Orientalism. In: *Orientalism and the Postcolonial Predicament*, ed. Carol A. Breckenridge and Peter van der Veer, 215–249. Philadelphia: University of Pennsylvania Press.
Rocher, Rosane, and Ludo Rocher. 2013. *Founders of Western Indology. August Wilhelm von Schlegel and Henry Thomas Colebrooke in Correspondence 1820–1837*. Abhandlungen fur Die Kunde des Morgenlandes 84. Wiesbaden: DMG, Harrassowitz Verlag.
Ross, Fiona G.E. 1999. *The Printed Bengali Character and its Evolution*. Richmond, UK: Curzon Press.
Roy, Ananta Kumar. 1921. *Vidyasagar. The Great Indian Educationist and Philanthropist*. Calcutta: Ray & Co.
Said, Edward W. 1979. *Orientalism*. Originally published in 1978. New York: Vintage Books, A Division of Random House.
Shaw, Graham. 1981. *Printing in Calcutta in 1800*. London: The Bibliographical Society.
Sinha, Nandalal, ed. 1911. *The Vaiśeṣika Sūtras of Kaṇāda. With the Commentary of Śaṇkara Miśra and Extracts from the Gloss of Jayanārāyaṇa*. Translation, without Sanskrit text. Allahabad: Pāṇini Office Bhuvaneshvari Ashram Bahadurganji.
*Thacker's Directory*. 1866. *Thacker's Directory. For Bengal, the North-Western Provinces, the Punjab, the Central Provinces, the Rajpootana States, Oude, and British Burmah. For 1866*. Calcutta: Thacker, Spink, and Co.
van Bijlert, Victor A. 1996. Sanskrit and Hindu National Identity in Nineteenth Century Bengal. In *Ideology and Status of Sanskrit*, ed. Jan E.M. Houben, 347–366. Leiden: Brill.
Waterfield, Henry. 1866. *East India (Progress). Statement of the Moral and Material Progress of India, 1864–1865*. Available at the British Library, OIH 330.954. London: India Office.
Wilkins, Charles. 1808. *A Grammar of the Sanskrita Language*. London: Bulmer and Co, Cleveland-Row, St. James's. 50

# Chapter 6
# Editing a Foundational Work on Classical Indian Medicine: The Printed Editions of the *Carakasaṃhitā* in Context

**Karin Preisendanz**

*Dedicated with great appreciation and gratefulness to the memory of Professor G. Jan Meulenbeld (May 28, 1928 – March 26, 2017), Knight of the Order of the Netherlands Lion, Basham medalist and Yavanacharya of Ayurveda.*

**Abstract** This paper presents a first study of the history of the printed editions of the *Carakasaṃhitā*, one of the two foundational early-classical works of traditional medicine in South Asia (Āyurveda) written in Sanskrit, ever since the publication of its *editio princeps* by the Bengali physician–scholar Gangadhar Ray in 1868. First, a general survey of fifty-six editions published until 2011, not counting second and

Work on the present paper was made possible by the generous support of the FWF (Austrian Science Fund) in the context of FWF projects P17300-G03, P19866-G15 and P23330-G15. The many substantial results of collaborative work that were achieved in these projects, published as well as not yet published, together with the various materials accumulated in the combined project archives, form the unique and precious basis of the research conducted for this paper. Furthermore, the paper has greatly benefited from the remarks, suggestions and questions coming from the discussant and the audience after its presentation at the conference 'Shaping the sciences of the ancient world. Text criticism, critical editions and translations of ancient and medieval scholarly texts (18th–20th centuries),' organized by the ERC Advanced Grant research project 'Mathematical sciences in the ancient world' (SAW/ERC Grant Agreement No. 269804) on June 17–21, 2013. I have also taken up a number of valuable points made by other speakers in their presentations and by the audience in the subsequent discussions as well as in the final discussion. My thanks go to all of them. Furthermore, I am indebted to the amazing Karine Chemla and to Christine Proust, Agathe Keller, Florence Bretelle-Establet, Stéphane Schmidt and other members and associates of the SAW project team for the many inspiring discussions from which I have learnt so much. I would also like to thank Oliver Frey for his competent and patient assistance in designing the diagrams presented in this paper, for the technical preparation of the illustrations and tables, and for important observations and suggestions on both the diagrams and the tables. Further thanks go to

K. Preisendanz (✉)
Faculty of Philological and Cultural Studies, Department of South Asian, Tibetan and Buddhist Studies, University of Vienna, Vienna, Austria
e-mail: karin.preisendanz@univie.ac.at


A. Keller, K. Chemla (eds.), *Shaping the Sciences of the Ancient and Medieval World*, Archimedes 69, https://doi.org/10.1007/978-3-031-49617-2_6

further editions, remakes and reprints, is presented and analyzed from various points of view: the supplementation of the text with commentaries and translations, and the involved choice of language; scripts, regions and specific places on the timeline; and persons, such as editors, publishers and printers, and other involved actors. The latter aspect is treated in much detail with a historical focus on the pre-independence period of the present Republic of India, and reveals the relevant scholarly and family networks, with a wealth of prosopographical information gained from the exploration of the rich, hitherto almost completely ignored introductions to the editions of the work written in Sanskrit and Hindi, many of which are presented for the first time in English translation. They also form the substantial textual basis for the subsequent exploration of the multi-faceted motivation and purpose behind the production of the pre-independence editions, commentaries and attached translations of the *Carakasaṃhitā* in the colonial context. They are mainly the preservation of a traditional scientific heritage, the shaping of a national scientific identity, and the accommodation of 'foreign knowledge' in the ancient medical knowledge. Three appendices present (1) a brief tabular survey of the editions of the *Carakasaṃhitā* published until 2011, (2) an analysis of the intricate history of an important early edition of Cakrapāṇidatta's *Āyurvedadīpikā*, the only completely preserved premodern commentary in Sanskrit, and (3) a historical discussion of references to plastic surgery and brain surgery in ancient India made by Ramprasad, author of a Hindi commentary on the *Carakasaṃhitā* attached to an edition published in 1911, in his introduction.

## Abbreviations

| | |
|---|---|
| AS | *Aṣṭāṅgahṛdayasaṃhitā*, edited in Parāḍakara (1939) |
| Ci | Cikitsāsthāna of the *Carakasaṃhitā* |
| CS | *Carakasaṃhitā*, edited in Jādavaji Trikamji Āchārya (1941) |
| CS (CE) | Critical edition of the *Carakasaṃhitā*, Vimānasthāna Chapter 8, in Preisendanz et al. (forthcoming) |
| *DCSMSSU* | see *A Descriptive Catalogue of the Sanskrit Manuscripts* […] (1996) |
| Ind | Indriyasthāna of the *Carakasaṃhitā* |
| Ka | Kalpasthāna of the *Carakasaṃhitā* |

Burkhard Quessel and Pasquale Manzo of the British Library who provided invaluable and unflagging support in gaining access to the treasures of this great library. I am also indebted to Somjit Dasgupta who kindly assisted me in the reading of the introductions to the editions of the *Carakasaṃhitā* by Harinath Sarkar (E7) and Satishchandra Sharma (E16), and discussed some of the questions arising from them with me. Finally, I thank Alexandra Scheuba for her painstaking assistance in giving this paper its final touch and in the proof reading.

Secondary literature that has come to my notice since 2016, the year when this paper was first submitted to the editors of the present volume, could only sporadically be considered in the following. All references to Sharma 1975 have been augmented by the references to the so-called reprint edition of 2016; however, in many cases the text there differs from that of the first edition of 1975 and should be consulted for corrections and especially additions.

| | |
|---|---|
| Ni | Nidānasthāna of the *Carakasaṃhitā* |
| NS | *Nibandhasaṅgraha* by Ḍalhaṇa, edited in Jādavaji Trikamji Āchārya (1938) |
| Śā | Śārīrasthāna of the *Carakasaṃhitā* |
| Si | Siddhisthāna of the *Carakasaṃhitā*, edited in Jādavaji Trikamji Āchārya (1941) |
| SS | *Suśrutasaṃhitā*, edited in Jādavaji Trikamji Āchārya (1938) |
| Sū | Sūtrasthāna of the *Carakasaṃhitā*, edited in Jādavaji Trikamji Āchārya (1941) / Sūtrasthāna of the *Suśrutasaṃhitā*, edited in Jādavaji Trikamji Āchārya (1938) / Sūtrasthāna of the *Aṣṭāṅgahṛdayasaṃhitā*, edited in Parāḍakara (1939) |
| Vi | Vimānasthāna of the *Carakasaṃhitā* |

## 6.1 A Brief Introduction to the *Carakasaṃhitā*

The *Carakasaṃhitā* is one of the two early-classical foundational works on traditional Indian medicine (Āyurveda) and constitutes the most important basis for the further development of the medical science in India, especially in the area of internal medicine which is the main subject of this work written in Sanskrit. According to the *Carakasaṃhitā*'s own testimony, the eternal knowledge (*veda*) about life (*āyus*), that is, the (full) human life span or life force, presented in it was first proclaimed by God Brahmā, who transmitted this body of knowledge to further divine figures, namely, Prajāpati, the two Aśvin-s and Indra.[1] After this mythological genealogy of Āyurveda in general, the narrative of the *Carakasaṃhitā* continues with its own specific legendary genealogy which at some point turns into a historical account. When the sages of yore worried about the obstructive appearance of disease among human beings, they sent sage Bharadvāja as an envoy to approach Indra in his heavenly abode who granted him the knowledge of Āyurveda.[2] Upon his return to earth Bharadvāja, who had grasped it very quickly, imparted his knowledge to the assembled co-sages,[3] among them the sage and physician Punarvasu of the Ātreya clan. Punarvasu Ātreya subsequently taught the 'Life Science' to six students, and Agniveśa, the most outstanding among them, first composed a fundamental systematic scholarly treatise (*tantra*) on the basis of his teacher's instruction.[4] The medical teachings of the *Carakasaṃhitā* are said to be those of Agniveśa's treatise.[5]

---

[1] See CS Sū 1.4–5ab. For an illustration of the sequence Brahmā – Aśvin-s, see below, Fig. 6.21. For the special mythological role of the Aśvin-s as surgeon–healers, see below, p. 273.

[2] See CS Sū 1.6–23. See again Fig. 6.21.

[3] See CS Sū 1.24–29. For a translation of CS Sū 1.24–25 see below, n. 348.

[4] See CS Sū 1.30–32. The other disciples are Bhela, Jatūkarṇa, Parāśara, Hārīta and Kṣārapāṇi. They subsequently composed their own treatises; see CS Sū 1.33ab$_1$.

[5] For a more detailed outline of the origin of Āyurveda and the knowledge contained in the *Carakasaṃhitā* from the emic perspective, see Preisendanz (2018, Section 7.5.1).

However, the work is generally transmitted under the name *Carakasaṃhitā*, which is generally understood to mean 'Caraka's Compendium' because an early substantial redaction of Agniveśa's treatise is said to have been accomplished by a certain medical man named Caraka.[6] Inasmuch as the terms *saṃskartṛ* ('redactor') and *saṃskṛta* ('redacted') used in this context are derived from verbal root √ *kṛ* with the preverb *sam*, meaning 'to put together,' 'to prepare, make up' and 'to decorate, refine,' they provide us with a rough idea of the tasks of a redactor and editor in the traditional Indian context.[7]

Again according to the *Carakasaṃhitā*'s own testimony, still another decisive redaction and even partial restoration of the work occurred at the hands of a certain Dṛḍhabala,[8] who may be dated to the period between 300 and 500 CE.[9] The work thus has a complicated – mythical, legendary and historical – history of original conception and early transmission, composition, redaction and reconstruction. The composition and subsequent first redaction of the work – excluding the portions newly composed on the basis of older material by Dṛḍhabala[10] – probably took place during the first two centuries CE. Some layers of text may even go back to an older medical treatise dating to the last century before the beginning of the Common Era.[11]

The second early-classical medical work, the *Suśrutasaṃhitā*, which is focused on surgery, has a similar mythological and legendary narrative of the origin of the science and the work itself,[12] but does not provide details on its more recent textual history at the hands of human authors and redactors.[13]

[6]The indigenous tradition that the title's first component, the word *caraka* (literally: 'wandering'; 'a wanderer'), relates to a famous physician and important redactor of Agniveśa's treatise called Caraka, appears rather late in medical literature, see Meulenbeld (1999, IA, 106 and 109). On the word *caraka* / Caraka in Vedic literature see Witzel (1985, 114–116), on the possible connection between the medical tradition of the *Carakasaṃhitā* and the Caraka school of the Black Yajurveda see Witzel (1985, 120–121). See also Preisendanz (2018, Section 7.5.3, p. 206).

[7]For more details on Caraka's activity as a redactor of Agniveśa's treatise, see Preisendanz (2018, Section 7.5.1, pp. 201–202).

[8]See CS Si 12.37 cd–41ab. For further details, see Preisendanz (2018, Section 7.5.1, p. 202).

[9]See Meulenbeld (2010). For further details on Dṛḍhabala's activity, see Preisendanz (2018, Section 7.5.1, pp. 202–203). For an illustration showing Caraka and Dṛḍhabala, see below, Fig. 6.22.

[10]See again Preisendanz (2018, Section 7.5.1, p. 203).

[11]For a brief discussion of the chronology of the *Carakasaṃhitā* and the possible roles of Punarvasu Ātreya, Agniveśa, Caraka and Dṛḍhabala in its composition, see Preisendanz (2018, Section 7.5.1, p. 203).

[12]See SS Sū 1.3–6 and 19–20, and the summary in Meulenbeld (1999, IA, 203). For a comparison and interpretation of the two narratives, including further mythological details added in the later medical tradition and other sources, in terms of cultural and religious history, see Filliozat (1964, 2–25). Filliozat denies that any historical memory is preserved in them (11, 25), except for the mention of the figure called Caraka (16–21).

[13]The redactor Nāgārjuna, who reconstructed part of the *Suśrutasaṃhitā* and can be considered a 'structural' counterpart of Caraka, is only mentioned by Ḍalhaṇa, the eleventh-century commentator on the *Suśrutasaṃhitā* (NS 1b, 3–7 on SS Sū 1.1).

## 6.2 The Printed Editions of the *Carakasaṃhitā*

### *6.2.1 General Survey*

The first edition of the *Suśrutasaṃhitā* was published from 1835 to 1836 by Madhusudan Gupta, a pandit associated with the Calcutta Sanskrit College.[14] It thus predates that of the *Carakasaṃhitā* by some thirty years.[15] Even so, probably due to the greater, more general importance of the *Carakasaṃhitā*, a larger number of editions of the Sanskrit text of the latter foundational work have been published to date. To my knowledge, since the publication of its *editio princeps* by Gangadhar Ray in 1868,[16] which may have been incomplete,[17] altogether fifty-six editions of the *Carakasaṃhitā* appeared in print, not counting the second, third or further editions of a given book, numerous reprints and several remakes.[18] Among them, several are

---

[14] See Roth (1872, 441). Further on Madhusudan Gupta, see Sinha (1993, 151) and Bose (1994).

[15] On this, see Pecchia (2022a, 119–120).

[16] E1. See below, p. 211. Sigla such as 'E1' refer to the number of the printed edition in Part IV, Chapter B.2 ('An Annotated Bibliography of the Printed Editions of the *Carakasaṃhitā*') in Preisendanz et al. (forthcoming), produced in the context of the FWF projects mentioned in the acknowledgements above. Brief bibliographical data relating to the printed editions referred to in this paper are provided in the table in Appendix 6.1 below, together with some further basic information relevant to the issues discussed here. Unless otherwise stated in the immediate context, in the following the dates of these editions are given after the sigla wherever this may be relevant for historical orientation. Dates joined by dashes indicate that the book was published over a period of several years. For the treatment of names of editors, etc., in the present paper, see the remarks on column 2 (Editor) of the mentioned table. Different from the treatment of place names in this bibliographical table (column 3 [Place]), the old names of major Indian cities, such as Calcutta, Bombay and Benares, will generally be used below because of the historical focus of this paper on the pre-independence period. All other place names are given in their present-day forms. – Reflections on Gangadhar Ray's philological–editorial practice concerning the *Carakasaṃhitā* in its wider intellectual and cultural context are found in Pecchia (2022b), with a small study of a selected variant in section 6 and a more general typological analysis of variants (called 'innovations') in section 3. The latter analysis is based on Appendix I of Pecchia (2022b), which consists of two comparative lists of variants to two sample passages taken from the eighth chapter of the Vimānasthāna. These lists focus on the textual innovations found in the manuscript of the *Carakasaṃhitā* prepared by Gangadhar Ray himself (see below, n. 120), designated as his first edition by Pecchia (2022a, 115; 2022b, 802), and in the *editio princeps*. Even though this is not clearly stated so in the asterisked initial note and n. 12 of Pecchia (2022b), these data are all taken from the extensive collaborative work to be published in Preisendanz et al. (forthcoming), Part II, Chapter C.1 ('The Critically Edited Text of CS Vimānasthāna Chapter 8') and Part III, Chapter A.2 ('Variants to the Text as Edited in Jadavji Trikamji 1941').

[17] On this possibility see Pecchia (2022a, 113–114; 2022b, 800).

[18] See the annotated bibliography in Preisendanz et al. (forthcoming) (Part IV, Chapter B.2) referred to in n. 16 above. It is also referred to by its mere (earlier) title in Pecchia (2022a, 124, n. 34), but without any further bibliographical details or acknowledgement; the transcriptions of the title pages of printed editions of the *Carakasaṃhitā* together with his *Jalpakalpataru* by Gangadhar Ray in Pecchia (2022a, Appendix I) are taken from there. This bibliography is based on the important information provided by Meulenbeld (1999, IB, 3–6) and information that could be gathered by the present author and members of the FWF project teams from 2001 onwards in various libraries in Europe and India. New editions, further editions, remakes and reprints that have appeared after 2011 are not considered there and in the present paper.

incomplete: their editors presumably had to abandon the project of a complete edition for various reasons[19] or could not complete it in their life time, or they had decided to publish only excerpts or selected books from among the eight books of the *Carakasaṃhitā*.[20] In some cases, publication may still be in progress.[21] Incompleteness can be observed especially with editions that are accompanied by translations.[22] By comparison, there are only forty-two editions of the *Suśrutasaṃhitā* listed by G. Jan Meulenbeld,[23] to which one may add at least another two.[24] However, by way of caution it has to be pointed out that in the world of Sanskrit texts printed in India it is sometimes difficult to ascertain without close scrutiny of the details of the text whether an alleged second or further edition of a particular book presents an edited, i.e., revised and modified, improved or augmented version of the edited text or other parts of the book, or whether it is a basically unchanged new impression or even a downright reprint camouflaged as a new edition which does not offer anything new beyond the modification or replacement of the original title page and the addition or subtraction of preliminaries, such as prefaces, supplements, such as appendices, etc. In the last case, the book may also be presented by the publisher as a new original publication of the text. Among these three types, the proliferation of editions especially of the last two types that can be observed in some historical periods testifies to the importance and relevance of an edited text in the eyes of all involved actors.[25]

An analysis of diverse features of these fifty-six editions of the *Carakasaṃhitā* can be expected to result in a better understanding of the history of the text in the modern period from various points of view, and to bring to light specific historical developments and configurations – political, social, economic, technical, institutional, educational, literary, linguistic and intellectual – which conditioned and shaped the changing environment of traditional Sanskritic learning in nineteenth to

---

[19] For example, Avinashchandra's edition of the *Carakasaṃhitā* together with the *editio princeps* of Cakrapāṇidatta's *Āyurvedadīpikā* published from 1888 onwards (E6) may have been abandoned because of the edition of the *Āyurvedadīpikā* by Harinath Sarkar which was begun in 1892 and probably finished, even though not completed shortly after 1919 (E7). Both editors worked in Calcutta and it is therefore probable that Avinashchandra quit the field after the first fascicles of Harinath Sarkar's edition and the criticism of his own efforts by this scholar had come to his notice. In the introduction to his edition, Harinath Sarkar mentions that Avinashchandra's edition of a few chapters of the Sūtrasthāna of the *Carakasaṃhitā* together with the *Āyurvedadīpikā* contains many errors and misprints (E7, Bhūmikā, 7). Furthermore, it may be that Avinashchandra was also discouraged by Harinath Sarkar's verdict that the (not further characterized) manuscripts of the *Āyurvedadīpikā* he had set out to obtain from various regions were all corrupt (ibid.). At least between 1895 and 1897, Avinashchandra was a sponsor or subscriber of E7; see below, p. 267. On the *Āyurvedadīpikā*, see below, p. 201, for further details concerning E7 p. 233f. and Appendix 6.2.

[20] For excerpts, see E24 (1926), for selected books, see, e.g., E25 (1926), E43 (1979) and E56 (2011). The Cikitsāsthāna with its clear focus and fundamental perspective on treatments seems to have been most popular for editions of a single book only.

[21] See, e.g., E53 (2004).

[22] See, e.g., E4 (1878??).

[23] See Meulenbeld (1999, IB, 311–314).

[24] See the two translations published along with an edition of the Sanskrit text by Priyavrat Sharma (1999–2001) and K. R. Srikantha Murthy (2000–2002) adduced by Wujastyk (2013, 144).

[25] See further the remarks on the timeline of publications below, p. 228ff., and Fig. 6.13.

twentieth-century India, in general and in particular with regard to the indigenous medical science. Given the large number of text editions, not all of them can be addressed in this paper and I will therefore concentrate on approximately the first sixty-five years in my treatment of details. My linguistic limitations when it comes to the diverse regional languages of modern South Asia involved in the making of these editions will furthermore influence the choice of material treated here in detail.

### 6.2.2 Elucidation, Translation and the Choice of Language

In terms of their major content, the Sanskrit text was published only seven times by itself, that is, in roughly 13% of the editions. In the remaining forty-nine editions, the text has been supplemented with commentaries, also in the form of paraphrases and brief notes, and translations (see Fig. 6.1). Among these, thirty-eight contain one or more commentaries or a commentary together with a translation, whereas eleven are accompanied by a translation only (see Fig. 6.2). The editions published only with one or more commentaries number thirty, the ones with a combination of commentary and translation amount to nineteen (see Fig. 6.3).[26] A closer look at the thirty-eight editions of the *Carakasaṃhitā* with commentaries or annotation reveals that eleven of them are supplemented with a pre-modern commentary written in Sanskrit, mostly the eleventh-century *Āyurvedadīpikā* by the Bengali physician–scholar Cakrapāṇidatta which is the only completely preserved pre-modern commentary on the work. In four cases, the editors added a modern commentary to the pre-modern one,[27] and in one case the editors and translators of the text of the *Carakasaṃhitā* also translated or closely paraphrased substantial parts of Cakrapāṇidatta's commentary into English even though this is not announced on the title pages of the volumes.[28] Of the remaining twenty-seven editions, twenty contain a modern commentary only and seven a modern commentary together with a translation (see Fig. 6.4).

The dominance of editions accompanied by translations and modern commentaries, paraphrases, extensive explanations and notes is significant (see Fig. 6.5). The figures

[26] On the occasional difficulty to distinguish between translation and commentary in a modern South Asian vernacular, see below, n. 59.

[27] In one case, E54 (2005), the modern commentary elucidates Cakrapāṇidatta's medieval commentary.

[28] See E42 (1976–2002). On the title pages and at the beginning of their Preface to Volume I (p. XXI), the editors and translators speak of a 'critical exposition based on Cakrapāṇidatta's Āyurveda Dīpikā.' However, this does not match the evidence and the editors' own remarks later on in the Preface (p. XLIVf.). See also the Foreword by Shivsharma (see below, n. 163), who refers to one of the two editors, Bhagwan Dash, as translator both of the basic text and the commentary. Because the focus of the present paper is on the basic text, the partial translation of the *Āyurvedadīpikā* in E42 is not considered separately in the figures relating to the content of the fifty-six editions of the *Carakasaṃhitā*. E52 (2004) contains a translation of the *Āyurvedadīpikā* which is also declared on the title page. Some brief sections of the *Āyurvedadīpikā* have been translated as part of the English commentary on the *Carakasaṃhitā* in E53 (2004), but they are not identified as such. On the sometimes difficult distinction between translation and commentary, see again n. 59 below.

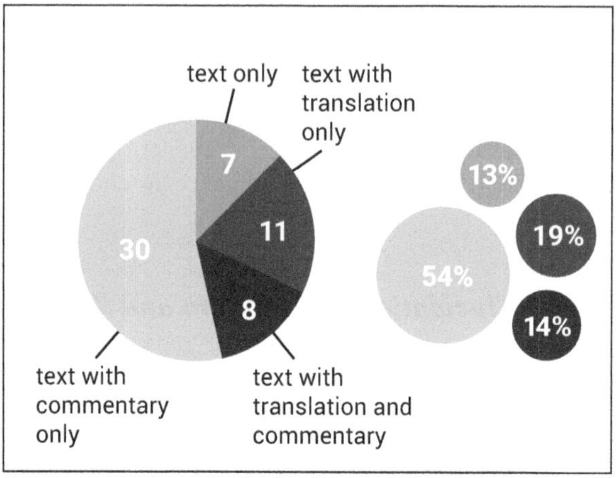


**Fig. 6.1** 56 editions – contents – with/without translation and/or commentary

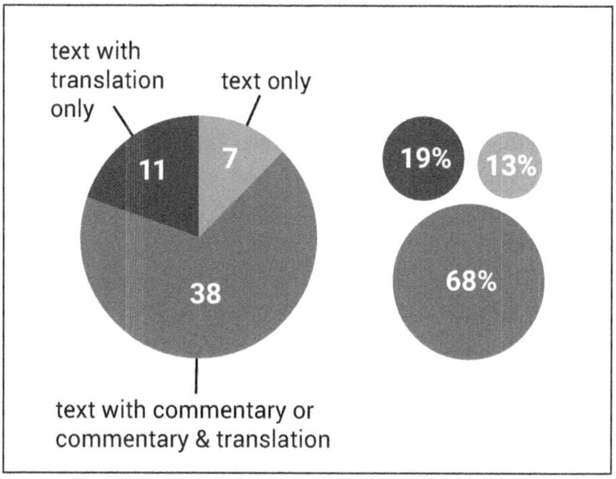


**Fig. 6.2** 56 editions – contents – with/without commentary

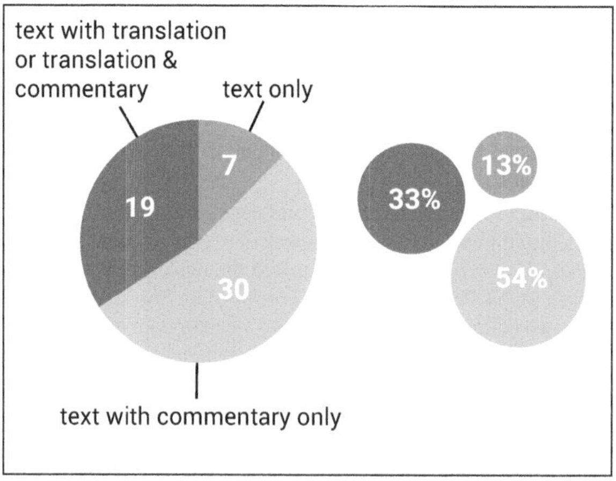


**Fig. 6.3** 56 editions – contents – with/without translation

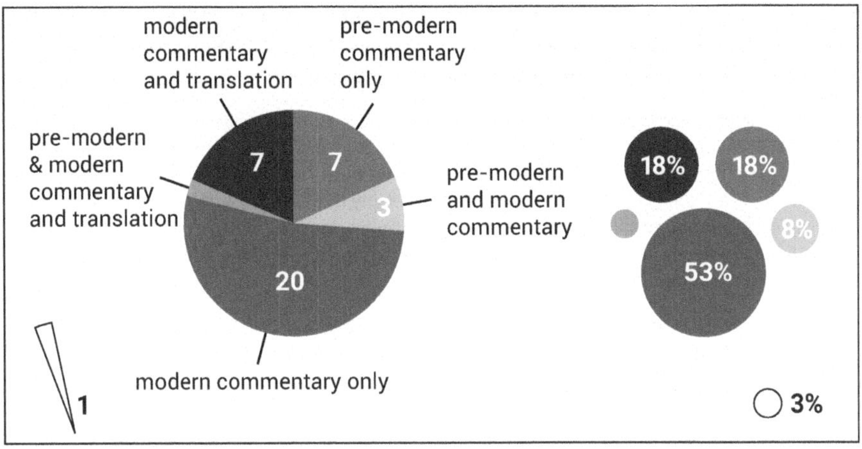


**Fig. 6.4** 38 editions with commentaries – type of commentary, with/without translation

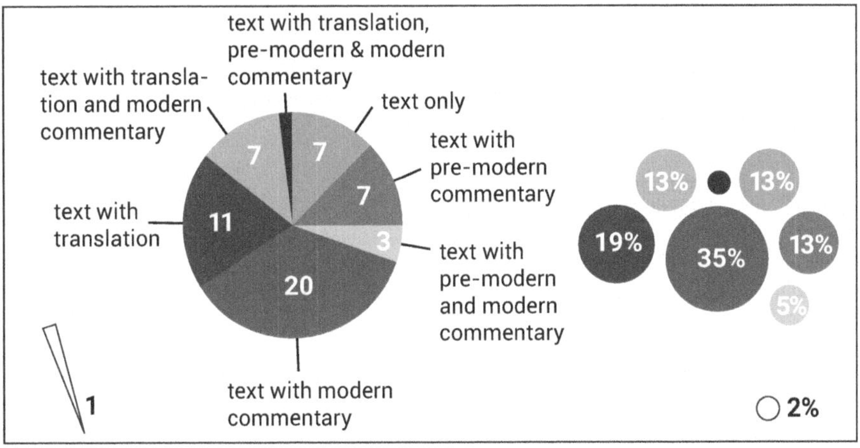


**Fig. 6.5** 56 editions – contents – summary

reflect the general perception of editors and publishers that in their own day the targeted readers of the ancient source of medical knowledge and practice published by them required translation and/or further elucidation of some kind to gain access to it.[29] At the same time, this strategy makes knowledge about the medical teachings of the *Carakasaṃhitā* 'public knowledge'[30] and amounts to a democratization of

[29] This parallels the analysis provided by Ulrike Stark (2009, 195–197) of the publication and mass production of religious works in translation only, or together with commentaries, with or without the original Sanskrit text, starting in the mid-nineteenth century, by the famous Naval Kishore Press in Lucknow (see below, n. 238). See also Stark (2012, 407–409) on the medical works published with translation into Hindi or with Hindi commentary by Naval Kishore.

[30] See Stark (2012, 409).

medical knowledge. For example, in 1911 the Bombay editor, printer and publisher Khemraj Shrikrishnadas[31] observes in the acknowledgement prefixed to his edition with Ramprasad's Hindi commentary called Prasādanī that

> now, in accordance with the time, great necessity of a Hindi commentary on it (i.e., the *Carakasaṃhitā*) has emerged. ... It is a matter of joy that in this commentary the mentioned respected physician has produced such a straightforward commentary even at extremely difficult places that people will understand the meaning of this book without effort. We give many thanks to the respected physician for performing this task which is beneficial for all and hope that he will further produce commentaries in the vernacular on the very best books and take part in making Āyurveda current.[32]

The physician–scholar Ramprasad himself, the author of the Hindi commentary,[33] states in his introduction that by means of the Sanskrit commentaries of Cakrapāṇidatta and Gangadhar Ray[34]

> it is not difficult for scholars to profit from the worth of this book. However, for people who only know the vernacular there is no other means to know the purport of 'Caraka' besides a commentary in the vernacular. Even though a couple of commentaries in the Hindi language have been printed already earlier on, in many places they do not explain the essential nature of the book well and thus were not esteemed by the lovers of Āyurveda. Therefore the illustrious respected merchant Khemraj Shrikrishnadas, who has the proprietary rights to the Śrīveṅkaṭeśvara Steam Press, consigned this book to me in Saṃvat 1966 (i.e., AD 1909) in order to produce a straightforward excellent commentary that follows the basic text in the Hindi language.[35]

---

[31] Further on Khemraj Shrikrishnadas see below, p. 231.

[32] See E17 (1911, next edition 1922), Dhanyăvād: *samayānukūl ab is kī hindī ṭīkā kī baṛī avaśyaktā ho gaī hai. ... ānand kī bāt hai, ki is ṭīkā meṁ ukt vaidyăjī ne atikaṭhin sthaloṁ par bhī aisī saral ṭīkā banāī hai ki log vinā pariśram is granth kā abhiprāy samajh jāyeṁge. is sarvopakārak kāryă karne ke liye ham vaidyăjī ko anek dhanyăvād dete haiṁ aur āśā karte haiṁ ki aur bhī uttam uttam granthoṁ kī bhāṣāṭīkā banā āyurved ke pracār karne meṁ āp bhāg liyā kareṁge*. In general, I am following the original, sometimes antiquated spellings in quotations taken from introductions, etc., written in Hindi. However, nowadays uncommon compounding of words is not reflected in the transcriptions from Devanagari script.

[33] The title page of E17 states that Ramprasad Sharma, son of Dwarakadas, was a resident of Taksal (in modern Himachal Pradesh) which belonged to Patiala, a former princely state that is now part of Punjab, and the royal physician (see Fig. 6.16 below, p. 250). He was born in 1882 (Samvat 1939) and the father of Shivsharma, another prominent physician–scholar of the twentieth century; see Sharma (1975, 223, 563, 597, 624; 2016, 242, 590, 624, 656) and Atridev Vidyālaṅkār (1976, 626, 635), who speaks of Ramprasad's Hindi translation (*anuvād*), not commentary, of the *Carakasaṃhitā* (626), and identifies him as one of the main authors of the Śrīveṅkaṭeśvara and Lakṣmīveṅkaṭeśvara Presses (651). On Shivsharma, who was first active in Lahore and moved to Bombay after partition, see Sharma (1975, 626–627; 2016, 658–659) and Atridev Vidyālaṅkār (1976, 626, 635). He is also the editor of the *Carakasaṃhitā* together with his father's (edited) Hindi commentary published in 1932 by the Lakṣmīveṅkaṭeśvara Steam Press (E29), an edition that seems to be a remake of E17.

[34] See below, p. 211ff.

[35] See E17 (1911), Bhūmikā, 7–8: *... jis se is granth kī yogyătā se vidvān logoṁ ko lābh uṭhānā kaṭhin nahīṁ parantu keval bhāṣāmātră jānnevāloṁ ko "carak kā" bhāv jānne ke liye bhāṣāṭīkā ko choṛ aur bhī koī upāy nahīṁ. yadyapi ek do ṭīkāeṁ hindī bhāṣā meṁ pahile bhī chap cukī haiṁ parantu ve bahutsī jagah granth ke marm ko acchī tarah na samjhāne ke kāraṇ āyurved rasikoṁ*

His aim was thus 'to make the most difficult purports suited to the understanding of common people.'[36]

In 1898, in his Hindi introduction to the first edition published by him with a Hindi commentary,[37] Khemraj Shrikrishnadas even observes that not all contemporary medical men may profit from the *Carakasaṃhitā* because of the difficulty arising from its having been composed by a great sage. He therefore had the outstanding Pandit Mihirchandra[38] compose a Hindi commentary on it and thus produced a book useful for everybody which he printed in his own press and published.[39] This perception may have been coupled with an awareness of the fact that the traditional teacher-to-disciple transmission of the *Carakasaṃhitā* had largely been interrupted, both in terms of the theoretical study of the Sanskrit text and the practice of its teachings.[40] As has been stated by Ulrike Stark for the case of religious literature printed together with commentaries in the vernaculars in nineteenth-century India, '[i]n the transition from practices of oral exposition to private silent reading, the interpretive function came to be vested in the text itself.'[41]

The translations that accompany editions of the Sanskrit text of the *Carakasaṃhitā* and make it linguistically accessible[42] are in five major South Asian vernaculars

*ko ādarāṇīyă na huīṁ isliye yah pustak "śrīveṅkaṭeśvar" sṭīm pres ke svatvādhikārī śrīmān seṭh khemrāj śrīkṛṣṇădāsjī ne saṃvat 1966 meṁ hindībhāṣā meṁ mūlānusār saral uttam ṭīkā banāne ke liye mujhe diyā.* See the similar statements on already existing Hindi translations of the *Bhāvaprakāśa* and on his own, more suitable translation in Kalicharan Sharma's preface to his edition and translation of this medical work in Stark (2012, 408).

[36] See E17, Bhūmikā, 8: ... *kaṭhin se kaṭhin bhāvoṁ ko sarvsādhāraṇ ke samajhne yogyă karne meṁ ....*

[37] See E11 (1898).

[38] As can be gathered from the concluding statement of the book (E11, p. 1480), Mihirchandra Mishra, son of Harisahaya and his wife Ramaraksha, belonged to the Gauḍavaṃśa of Brahmins and lived in the village of Lank (Lāṅkh) southwest of Muzaffarnagar in present Uttar Pradesh (*haridvāradakṣiṇadiksthitamujaḥpharnagaraprāntavartilāṅkhagrāmanivāsigauḍavaṃśāvataṃsamiśropāhvaśrīmaddharisahāyāṅgajaśrīrāmarakṣātanūdbhavapaṇḍitamihiracandrakṛtabhāṣāvivṛttisahitam idaṃ carakatantraṃ samāptam*).

[39] ... *parantu maharṣi praṇīt hone se kaṭhintā ke kāraṇ ādhunik sarv vaidyăjan is se lābh nahīṁ uṭhā sake. isliye paṇḍit pravar śrī mihircandrăjī ke pās is kī bhāṣāṭīkā nirmāṇ karā sarvopayogi yah granth nij "śrīveṅkaṭeśvar" yantrālay meṁ mudrit kar prakāśit kiyā hai* (Prastāvanā [Prologue], opposite the first of a series of illustrations at the beginning of this edition; see below, p. 278ff.). See also the introductions to E13 (1900) and E18 (1911) treated below, p. 262f.

[40] Cf. also Kalicharan Sharma's preface to his edition and translation of the *Bhāvaprakāśa* translated in Stark (2012, 408). Unfortunately, his preface or introduction to E14 (1900) is not available to me.

[41] See Stark (2009, 195), summing up the observations by Philip Lutgendorf on the increasing importance of commentaries and other kinds of explication incorporated into editions of the Awadhi religious poem *Rāmcaritmānas*.

[42] See Stark (2009, 192) for the case of translations of religious classics and medieval texts, and especially for the practice of the Naval Kishore Press, Lucknow (see below, n. 238), to have such texts translated into Hindi and Urdu (195–198). For the increasing popularity of bilingual Sanskrit–Hindi publications in nineteenth-century northern India, Stark refers to the observations made by Christopher King (197).

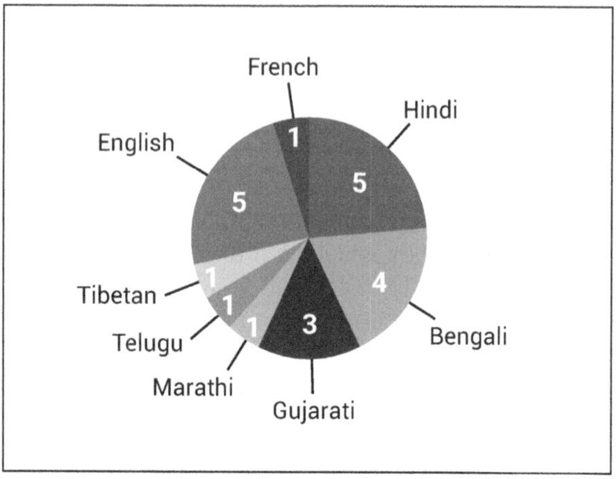


**Fig. 6.6** 19 editions with 21 translations – languages

(*bhāṣā*), characteristically led by Hindi and Bengali, the two languages with most speakers, and in modern Tibetan,[43] a language that has gained some importance in South Asia because of the Tibetan exile community in contemporary India (see Fig. 6.6). In terms of quantity, the five translations into Hindi are on a par with translations into English, the language of the former colonial rulers and nowadays alongside Hindi an official language of the Indian central government. There is a single, incomplete translation into another European language, French, published together with the Sanskrit text. This full edition of two books of the *Carakasaṃhitā* is the only one published so far outside India and at the same time the most recent edition.[44] A monumental and ambitious edition is accompanied by a translation into three languages: Hindi, English and Gujarati, the language of the region of Western India where the Shree Gulabkunverba Ayurvedic Society, which published this multi-volume compilation, was active.[45] As will be seen below,[46] the languages chosen for commentaries and other kinds of elucidation in modern languages present us with a rather different picture: here, Hindi is clearly the predominant language, chosen in eleven cases, followed by (modern) Sanskrit in seven cases.

Concerning the languages of translations accompanying the text of the *Carakasaṃhitā*, Bengal asserts its place as the home of the modern renaissance of classical Āyurveda through the attention devoted to this and other foundational medical works by nineteenth-century scholars and physician–scholars. This is

[43] See E55 (2006).

[44] See E56 (2011). Complete or incomplete translations into South Asian, European and other languages published without the Sanskrit text are outside the scope of this paper and are thus not considered here.

[45] See E37 (1949). The Shree Gulabkunverba Ayurvedic Society also managed the Shree Gulabkunverba Ayurved Mahavidyalaya in Jamnagar which is nowadays a college of the Gujarat Ayurved University, Jamnagar. The multiple translation in E37 accounts for the overall number of twenty-one translations accompanying only nineteen editions.

[46] See p. 210f.

corroborated by the rather early date of the four editions with translations into Bengali, all published in Calcutta. Among these, an incomplete one is by a certain Dharmadas Sen Gupta[47] who can be identified with the physician–scholar Dharmadas, born in 1862 in Burdwan District, Bengal, who studied Āyurveda with his maternal uncle Pareshnath Sen Gupta, a student of Gangadhar Ray, in Benares,[48] and remained there until his death in 1935.[49] He became the first principal of the Ayurvedic College founded there in 1924–1925.[50] According to Prabhakar Chatterjee, Pareshnath Sen Gupta had a deep knowledge of the *Carakasaṃhitā*,[51] and both Priyavrat Sharma and Baldev Upādhyāy mention that Dharmadas achieved special erudition with regard to the *Carakasaṃhitā* and was even called 'Master of Caraka' (*carakācāryă*) for this reason.[52] G. Jan Meulenbeld tentatively assumes that the book was published in 1878,[53] which would be just about a decade after the publication of the *editio princeps* by Gangadhar Ray and make it the first edition of the text together with a translation. However, Dharmadas Sen Gupta would have been much too young to achieve even an incomplete edition and translation, and indeed there is some evidence for a later publication date, namely, 1894.[54] If this is the actual year of publication of Dharmadas Sen Gupta's edition-cum-translation, it would not be the first edition with a Bengali translation. Two others would definitely have preceded it, namely, a little known first attempt by Vamacharan Barat, a professional physician,

---

[47] See E4 (1878??). The copy of Dharmadas Sen Gupta's edition with Bengali translation preserved in the India Office Library of the British Library (VT.982) covers only the first book of the *Carakasaṃhitā*, the Sūtrasthāna, and was accessed on September 19, 1895. The title page identifies the book as 'Part I' which implies that there was the plan to publish at least one more volume. Further volumes could not yet be located physically. However, the old handwritten card catalogue of the National Library, Calcutta, records a two-volume Bengali translation of the *Carakasaṃhitā* by Dharmadas Sen Gupta, published in 1894, with the call number 180.Eb.89.6–7. Even though the book that bears this call number on its spine is actually a copy of Cakrapāṇidatta's *Cikitsāsaṅgraha*, an independent medical work by the above-mentioned commentator on the *Carakasaṃhitā*, this is clear proof for the existence of at least a second volume because it is highly improbable that the translation was published twice, once with the Sanskrit text and afterwards without it.

[48] Through his large circle of students and their students, Pareshnath Sen Gupta became the founder of the 'Benares tradition' of Gangadhar Ray's school and made Benares the second center of Āyurveda next to Calcutta. See Sharma (1975, 222, 550, 552, 594; 2016, 241, 576, 578, 622).

[49] See Atridev Vidyālaṅkār (1976, 616), Gupta (1976, 372: 'Dharmadas Gupta of Benares') and especially Sharma (1975, 550–552, 559; 2016, 576–578, 586) and Baldev Upādhyāy (1983, 667).

[50] See Sharma (1975, 559; 2016, 586) and Baldev Upādhyāy (1983, 907).

[51] See Chatterjee (1959a, 400).

[52] See Sharma (1975, 550, 559, 594; 2016, 576, 586, 622) and Baldev Upādhyāy (1983, 667).

[53] See Meulenbeld (1999, IB, 3).

[54] Evidently, the date of the two-volume publication given on the index card of the National Library, Calcutta, i.e., 1894 (see above, n. 47), does not match the date tentatively assumed by Meulenbeld on unknown grounds, i.e., 1878. However, it comes very close to the date scribbled onto the cover of the India Office Library copy of the book, i.e., September 19, 1895, which may be the date of its accession. Normally, printed books from India reached the India Office Library rather fast. See also Harinath Sarkar's unfamiliarity with Dharmadas Sen Gupta's edition and translation pointed out below, n. 327.

which only covered part of the Sūtrasthāna[55] and the complete one by Avinashchandra published from 1883 to 1884.[56] A complete edition with translation was also published by Yashodanandan Sarkar in 1894; however, a copy of its first edition could not yet be seen.[57] A further edition of the Sanskrit text with Bengali translation was later on published by Satishchandra Sharma in 1904.[58]

Editions with translations into Hindi[59] set in only at the very end of the nineteenth century, with Shrikrishnalal's translation in 1898.[60] According to the title

---

[55] See E7 (1892ff.), Bhūmikā, 6. The project was cut short by Vamacharan Barat's untimely death.

[56] See E5. In this case, the first volume, up to and including the fifth book of the *Carakasaṃhitā*, was first published in 1883. After the completion of the second part, that is, the edition and translation of the remaining three books starting with the Cikitsāsthāna, this second part was published together with the first part as a complete edition-cum-translation in 1884. Thus, one may speak here of a follow-up edition-cum-translation of the first part of the text, augmented to form a first complete edition-cum-translation of the text. In his introduction to E7 (1892ff.), Harinath Sarkar reports that Avinashchandra's translation received great praise, but points out some deficiencies of the text and relates that Avinashchandra – himself not content with this state of affairs – had started a new edition together with Cakrapāṇidatta's *Āyurvedadīpikā* (Bhūmikā, 7). See n. 19 above for this edition (E6, 1888ff.).

[57] See E8. Meulenbeld (1999, IB, 3) gives the date of the first edition of this edition-cum-translation as 1894, referring to the copy of the British Library with the call number 14044.b.2. However, BL 14044.b.2 is a copy of the second edition and dated 1910. The date 1894 for the first edition of E8 is documented by the copy of a catalogue excerpt, attached to this copy of its second edition, which refers to the first edition of 1894. The provenance of the excerpt, probably a contemporary bookseller's catalogue, could not yet be identified. Furthermore, in his list of editions of the *Carakasaṃhitā* Girindranath Mukhopadhyaya refers to this edition and provides the date of publication (1894) as well as the name of the printing press (Baṅgabāsī); see Mukhopadhyaya (1922–1929, III, 614).

[58] See E16.

[59] In the case of Hindi language publications, the distinction between translation (*anuvād*, literally: 'repetition') and commentary is sometimes problematic. The involved actors may consider a close paraphrase of a Sanskrit text in Hindi augmented with explanatory insertions and additions to be a commentary on this text. For example, in E47 (1983, 1988) and E50 (1998?*, reprint 2006) such an explanatory paraphrase in Hindi immediately follows the text of the *Carakasaṃhitā* in Sanskrit. This is followed by further explication under the heading *vaktavyă* ('remarks'). It is unclear whether both sections in Hindi constitute the Hindi commentary, the *Carakacandrikā* and the *Vaidyamanoramā*, respectively, or only the second. For practical purposes, such translation-cum-elucidations in Hindi have not been counted as translations of the *Carakasaṃhitā* in the present survey, unless they are explicitly declared to be translations on the title pages of the books. Another case is E54 (2005) where the Hindi title page speaks of a Hindi commentary on Cakrapāṇidatta's *Āyurvedadīpikā* by Harishchandra Singh Kushwaha called *Āyuṣī*, whereas the English title page calls the supposed Hindi commentator 'Editor–Translator.' In his Prastāvanā, under the heading 'In connection with the Hindi commentary "Āyuṣī",' Harishchandra Singh Kushwaha himself first speaks about his 'transformation' into Hindi of the *Āyurvedadīpikā*; see Prastāvanā (p. 24): ... *maiṁ ne uskā (āyurved-dīpikā) hindī rūpāntaraṇ karne kā niścay kiyā*. Then he continues: 'In this commentary, a straightforward translation of the *Āyurvedadīpikā* has been presented, together with a translation of the basic stanzas of the *Carakasaṃhitā*' (*is vyākhyā meṁ carak-saṃhitā ke mūl ślokoṁ ke anuvād ke sāth hī āyurved-dīpikā kā saral hindī anuvād prastut kiyā gayā hai*). See also the case of E13 (see n. 382 below). For a similar situation, see above, p. 201, on the case of the 'critical exposition [scil. of the *Carakasaṃhitā*] based on Cakrapāṇidatta's Āyurveda Dīpikā' in E42 (1976–2002) which is actually a partial translation of this pre-modern Sanskrit commentary into English.

[60] See E12 (1898).

pages of the two-part book, Shrikrishnalal was a resident of Mathura in modern Uttar Pradesh and also translated the *Suśrutasaṃhitā* and other medical works. The book was published in Mathura by Shyamlal Shrikrishnalal, owner of a printing press there and most probably the son of the translator who himself was the editor of the Sanskrit text.[61] Just one year before, in Bombay, the publication of the single edition with annotated translation into Marathi was started in the form of a quarterly journal by Shankar Daji Pade (1867–1909),[62] who was prominently and enthusiastically engaged in the renaissance of Āyurveda in Maharashtra and Gujarat and established the All India Ayurvedic Congress in 1907.[63] The publication of Sanskrit texts and translations in monthly, bimonthly or quarterly journals was rather common in the nineteenth and early twentieth century and allowed the avid readers faster access to their reading material than if the editor or author would have first completed the entire work. After completion, the fascicles were often united and published in book form, as was done also in the case of Shankar Daji Pade's edition of the *Carakasaṃhitā* in several volumes.[64]

Translations into English together with the Sanskrit text began to appear only in the post-colonial period, the first one being the above-mentioned edition with multiple translation published by the Shree Gulabkunverba Ayurvedic Society.[65] However, a first monumental English translation by a Bengali Sanskrit scholar, Kisari Mohan Ganguli, the famous English translator of the national epic *Mahābhārata*, had been published before by itself, without the text, in sixty-eight fascicles from 1890 to 1925. Until 1905, this translation appeared under the name of its editor, Avinashchandra.[66] From 1906 onwards, after his death, the remaining

---

[61] The prefaces to the two parts of this edition-cum-translation are signed jointly by Gangavishnu Shrikrishnadas, of the Lakṣmīveṅkaṭeśvara Press, Bombay, and Shyamlal, of the Śyāmakāśī Press, Mathura, as commission agents for the book (see the heading *pustak milne kā ṭhikānā* above the two names). The prefaces bear the explicit note that the title pages of this edition were printed by the Lakṣmīveṅkaṭeśvara Press, Bombay, which explains the very similar design of the beautiful title pages of E12 and E11 (1898). Gangavishnu Shrikrishnadas was a close relative of the editor and publisher of E11, Khemraj Shrikrishnadas (see below, p. 231), and himself the editor and publisher of an edition of the *Carakasaṃhitā* published by his Bombay press in 1932 (E30).

[62] See E9 (1897ff.); for the title page of the first issue, see Fig. 6.20 below. This was soon changed to a monthly journal and from some point onwards, the edition was published in the *Āryabhiṣak* book series (see below, n. 372). For instance, the text, translation and commentary of the Siddhisthāna, the last book of the *Carakasaṃhitā*, were published in 1901 as part 34, that of Cikitsāsthāna 1–5 in 1902 as part 41, that of Cikitsāsthāna 20–30 in 1906 as part 49, and that of Śārīrasthāna 7–8 in 1907 as part 58. After the text of the individual instalments of the *Carakasaṃhitā* with translation and commentary in E9, we find, before the appendix (see again n. 372 below) and with a new pagination, instalments of a Marathi treatise on the advantages and disadvantages of medicinal plants (*Auṣadhīguṇadoṣa*).

[63] See Sharma (1975, 550, 562, 621, picture 12 between pp. 620 and 621, 624–625; 2016, 576, 589, 653, picture on p. 651, 656–657), Atridev Vidyālaṅkār (1976, 633–634) and Ashok Vaidya (2010).

[64] See E9 ($^{2}$1902ff., $^{3}$1914ff.).

[65] See E37 (1949); see above, p. 206.

[66] See Jacobi (1908).

fascicles were published by Avinashchandra's son Pareshnath Sharma.[67] Earlier, from 1883 to 1884, Avinashchandra had published the text of the *Carakasaṃhitā* with his own translation into Bengali.[68] The choice of English at this time and also later on was clearly motivated by the desire to reach the largest possible range of general readers unfamiliar with Sanskrit,[69] or not sufficiently familiar with this classical language, inside and outside India,[70] and to present the complete work in an accessible way to Indian practitioners of Āyurveda, to scholars of Āyurveda nationally and internationally,[71] and to a general international audience interested in the globalized phenomenon of Āyurveda, but unfamiliar with Sanskrit.[72] However, at the time of Kisari Mohan Ganguli, translating the *Carakasaṃhitā* into English meant much more than simply making the work accessible to those unfamiliar with Sanskrit: for the revivalist movement of Āyurveda, translations into English of foundational works of Āyurveda provided it with textual authority[73] *vis à vis* the highly negative attitude of the colonial rulers towards traditional Indian medicine who denied Āyurveda the status of a science or system of medicine.[74]

In the case of the thirty-one modern commentaries on the basic text, Hindi, with its much larger distribution than Bengali, is clearly the preferred vernacular language, with eleven commentaries or annotations written in this language or a share of roughly 35% of editions accompanied by modern commentaries (see Fig. 6.7). Commentaries in Hindi were written from 1898 onwards, starting with the

[67] See Part XXXVIII (Library of the Karl-Sudhoff-Institut für Geschichte der Medizin und Naturwissenschaften, Universität Leipzig, call number II 8242). Starting with Part XLIX (1908), Avinashchandra's son calls himself Kavirāj Pareshnath Sharman Kavibhūṣaṇa (call number II 8243). The small extract of a letter written in 1913 by A. F. R. Hoernle to a certain Pareshnath Kavibhūṣaṇa, which is printed on the inner front cover of the last fascicle of E7 fully documented by me (1919), is most probably addressed to him. He has to be distinguished from Gangadhar Ray's student Pareshnath Sen Gupta; see above, p. 207. The editors of E50 (1998?, reprint 2006) list this translation as an incomplete one (Bhūmikā, [20]); see also E52 (2004), Introduction, XIII, where K. R. Srikantha Murthy states that it was published in 1891.

[68] See above, p. 208.

[69] See, e.g., Shivsharma's remark on the partial translation of Cakrapāṇidatta's commentary in E42 (1972): 'This book will probably be the first to offer a standard English translation of Cakrapāṇi's important work to those having no direct access to the Sanskrit language' (Foreword, v).

[70] See especially the case of E37 (1949).

[71] See Priyavrat Sharma in his introduction to E44 (1981): 'Looking to the above facts and the gradually increasing curiosity to know about Indian Medicine all over the world, it was felt necessary to bring forth an edition of the Caraka Saṃhitā with a faithful and literal English translation after carefully editing the text' (Introduction, xxxii). He also thanks the 'publishers, who are devoted sincerely to the noble cause of propagating the mission of Caraka all over the world, for bringing out this novel edition of the Caraka Saṃhitā' (ibid.).

[72] See E42 (1976–2002), E44 (1981–1994) and E 52 (2004) by well-known Indian exponents of Āyurveda.

[73] See Arnold (2000, 178).

[74] For a highly critical review of seven fascicles of Kisari Mohan Ganguli's translation, see Roth (1894b). See also Müller's review (1930a) written after the publication of the last fascicle. Müller notes that the translator attempts to adjust the medical teachings of ancient India to the modern development of medicine in Europe.

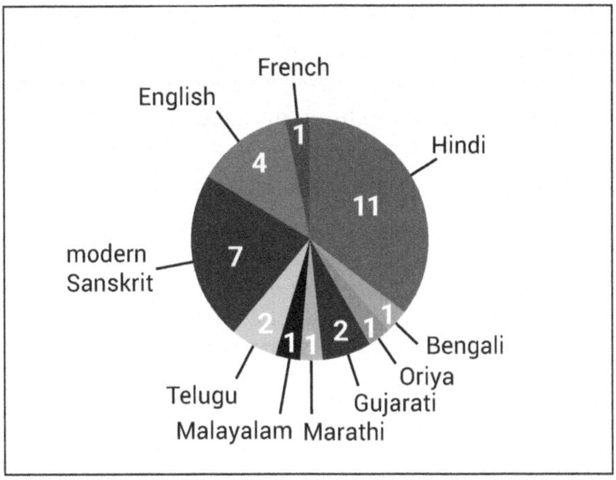


**Fig. 6.7** 31 modern commentaries – languages

commentary by Pandit Mihirchandra which was published in Bombay[75] and followed closely on the heels of Shankar Daji Pade's edition with Marathi translation and notes, published at the same place.[76] In Bengal, that is Calcutta and its surroundings, on the other hand, where we find the already mentioned four early editions with translations of the *Carakasaṃhitā* into the vernacular[77] two important modern commentaries on the Sanskrit text were composed in Sanskrit. The first one, entitled the 'Wishing Tree of Scholarly Controversy' (*Jalpakalpataru*), by Gangadhar Ray already accompanied the *editio princeps* by the same physician–scholar.[78] Born in 1798[79] in a village in what is now Bangladesh, Gangadhar Ray was a true polymath and authored about ninety works, original treatises and commentaries in different areas of Sanskritic learning as well as Sanskrit poems, plays and religious hymns of praise.[80] In the traditional environment of Sanskrit scholars and especially

[75] See E11 (1898) and above, p. 205.

[76] See E9 (1897) and above, p. 209.

[77] See above, p. 207f., also for a discussion of the dates of these editions.

[78] See E1 (1868). On the title of this commentary, see below, p. 236. For an annotated list of known manuscripts of the *Jalpakalpataru*, see Pecchia (2022a, 133); in n. 64 it should have been stated more clearly that this list is taken from Part II, Chapter A.2 ('Inventory and Concordance of Known Manuscripts of the CS and its Commentaries') of Preisendanz et al. (forthcoming).

[79] Sharma (1975, 221; 2016, 240) gives the year 1799 as his birth year, whereas Gupta (1976, 371) has 1789 which may be a typographical error.

[80] For literature on Gangadhar Ray, who passed away in 1884 or 1885, see Preisendanz (2013, 66, n. 4) and Pecchia (2022a, 114, with n. 14). To this may be added the information in Sharma (1975, 221–222, 549–550; 2016, 240–241, 575–576) and the life sketch in Atridev Vidyālaṅkār (1976, 612–614). Both sources give a list of the most important medical works by Gangadhar Ray, numbering ten and eleven, respectively, because Sharma does not include the *Jalpakalpataru* here, and genealogical trees showing his students and their students, which supplement the tree given in Gupta (1976, 372). There is also a series of articles devoted to this reputed physician–scholar by Prabhakar Chatterjee, himself a grand-disciple of Gangadhar Ray through Haranchandra Chakraborty (Gupta, ibid.), who is identical with Harayanchandra Chakraborty (see Atridev

physician–scholars, he must have considered Sanskrit the most suitable linguistic means to communicate his interpretation of the ancient scientific text and share his profuse thoughts on it with his colleagues, in Bengal as well as elsewhere in India.[81] However, the great erudition of its author has led to a mixed reception of this first modern commentary on the *Carakasaṃhitā*. Rudolf Roth's early assessment of the *Jalpakalpataru* is worthwhile quoting here:

> He (i.e., Gangadhar Ray) supplements the book (i.e., *editio princeps* of the *Carakasaṃhitā*) with a commentary he calls Jalpakalpataru, which is unnecessarily extensive and adduces so many irrelevant matters that the first fascicle of 384 pages (Samvat 1925 = 1869) does not contain more than a single chapter of the text and a small part of the beginning of the second, which is about the seventieth or eightieth part of Caraka.[82]

This is not only the isolated opinion of a European scholar. For example, Dattaram Narayan Pathak, the presumed editor of an edition accompanied by his own Hindi commentary,[83] also offers polite criticism of this commentary and reports the

---

Vidyālaṅkar 1976, 615) and was one of the chief students of Gangadhar Ray (see Sharma 1975, 223; 2016, 242). They were published in seven issues of the journal *Nagarjun* from 1958 to 1959. I am greatly indebted to the late Professor G. Jan Meulenbeld for sharing rare copies of instalments IV (February 1959) to VII (June 1959) with me (see Chatterjee 1959a, b, c, d). Chatterjee (1959d, 816–817) provides a list of writings by Gangadhar Ray with 76 items – unfortunately riddled with typos and other terrible mistakes which greatly disfigure the titles – among which the *Jalpakalpataru* on the *Carakasaṃhitā* ranks first. According to Chattopadhyay (1995, 5), he also wrote a short commentary on part of his own *Jalpakalpataru*. In spite of all their flaws and the ensuing uncertainties, it is clear that the combined lists of Chatterjee and Chattopadhyay amount to a much larger number of works than the listed 76 and 74, respectively. The first eleven items of Chatterjee's list are identical with the eleven items of Atridev Vidyālaṅkār's list of medical works. Meulenbeld (1999, IA, 186) lists two further medical works, one of which is also found in Chattopadhyay's list. Much research remains to be done on these mostly unpublished writings many of which are deposited in manuscript form in the library of the Calcutta Sanskrit College. See also below, n. 119.

[81] On the possible relationship between the *Jalpakalpataru* and Gangadhar Ray's editorial work, see Pecchia (2022a, 115–116; 2022b, 816–817).

[82] See Roth (1872, 441–442): 'Er (scil. Gangadhar Ray) begleitet das Buch mit einem Commentar, den er Ǵalpakalpataru heisst und welcher so unnütz breit ist und so viel gleichgiltiges herbeizieht, dass in dem ersten Heft von 384 Seiten (Samvat 1925 = 1869) nicht mehr als e i n Kapitel des Textes und ein kleiner Anfang des zweiten, etwa der siebenzigste oder achtzigste Theil des Caraka enthalten ist.' Roth refers here to the first fascicle of E1 (1868) with a slightly divergent calculation of its Gregorian publication date. Conversely, precisely the extensive information provided in the *Jalpakalpataru* on the teachings of Sāṃkhya, Nyāya and other philosophical traditions is highly appreciated by the compiler of the *Records in the Bengal Library*, Vol. 1879–1881, as manifest in his remark on the publication of an 1879 instalment of Gangadhar's *editio princeps* quoted in Pecchia (2022b, 817).

[83] Dattaram Narayan Pathak, who appears as Pandit Dattaram Narayan Chaube on the title page of E13, was a resident of Mathura and translated many books on Āyurveda into Hindi which were published by the Śrīveṅkaṭeśvara Press in Bombay (Sharma 1975, 223; 2016, 242; Atridev Vidyālaṅkār 1976, 649, 651). At the end of the Siddhisthāna of E13, he provides a genealogy of his family of Pathak Brahmins from the oldest known ancestor by name of Gopinath up to the fifth generation after him which is that of Dattaram's father Narayan. Further on Dattaram Narayan Pathak, see pp. 233, 262f., and 275f. below.

critical response of others in his presentation of the Sanskrit commentaries on the *Carakasaṃhitā*:

> Then there remains the commentary (called *Jalpakalpataru*) of the well-known Pandit Gangadhar Kaviratna. It is recent and extremely difficult to be grasped (hard). In comparison with the other books, the commentary on the book on treatment is excellent and also rather straightforward. It is like a mirror for modern instructors (students). However, many doctors are of the opinion that out of egoism Gangadhar has done injury to the meaning at some places. For this very reason, even though his language and knowledge are excellent, his opinion may not enjoy undoubted acceptance at all places. Furthermore, great carelessness has occurred in its printing.[84]

Yogindranath Sen, a grand-disciple of Gangadhar Ray who was very familiar with the *Jalpakalpataru*,[85] speaks of it with high praise and much reverence, but acknowledges its unsuitability for beginners as well as its size and prolixity:

> Next we come to Acharya Gangadhara. His admirable commentary, the Jalpakalpataru, though an unmistakable indication of his vast erudition and versatile genius, is, we are afraid, not quite suited to the wants of the beginner in Charaka. Jalpakalpataru deals not with Ayurveda alone, but with Philosophy, the Upanishadas, Rhetoric and other similar subjects. In fact, it is a masterly treatise discussing the most abstruse questions of Philosophy of Ayurveda, and as such, though of a very great interest to the advanced student of Medical Science, its scope and development, are [sic] too difficult for an average beginner in Ayurveda. Moreover, dealing with so many subjects, its bulk is too heavy and trying to the patience of most reader.[86]

Similar criticism has been repeated frequently,[87] even though in general this commentary by a founding father of modern Āyurveda is highly praised.

---

[84] See E13 (1900), Bhūmikā, 3: *rahī ab prasiddh paṇḍit gaṅgādhar kaviratn kī (jalpakalpataru nāmak) ṭīkā. so navīn aur atyant kliṣṭ (kaṭhor) hai. anyă sthān kī apekṣā cikitsāsthān kī ṭīkā uttam hai aur kuch saral bhī hai. yah ādhunik śikṣak (vidyārthiyoṁ) ko darpaṇ ke samān hai. parantu anek vaidyoṁ kā mat hai ki gaṅgādhar ahaṅkārvaś ho kaī jagah arth kā anarth kar ḍālā hai. is se us kī bhāṣā aur jñān uttam hone par bhī us kā matāmat sab jagah niḥsandeh grahaṇ nahīṁ karā jāy. tathā us ke chapne meṁ adhik asāvdhānī ho gaī hai.* See also E8 ($^2$1910), Vijñāpan, 3. Except for the last two paragraphs (p. 4, line 5ff.) the Announcement of E8 ($^2$1910), written in Bengali, is identical with the Introduction of E13 (1900), written in Hindi, p. 1, line 19ff. Until a copy of the first edition of E8 (1894) will have been located and examined, it has to remain open which of the two is the source and which the copy. However, it is much more probable that Yashodanandan Sarkar, editor and translator of E8, relied on Dattaram Narayan Pathak's introduction when the second edition of his own work was published. Thus, reference will only be made to E13 in the following.

[85] See below.

[86] E21 (1920), Critical Notice, iii. For the discussion of 'abstruse philosophical questions' Yogindranath refers advanced students to 'the admirable and unrivalled treatise of Acharya Gangadhara' because he intentionally omitted such discussion in his own commentary *Carakopaskāra*. See Critical Notice, iii–iv.

[87] More recently, Atridev Vidyālaṅkār (1976, 600) politely appreciates the author's great erudition which is sparkling on every page, but considers the commentary too difficult to be penetrated by the mind of a common student (*gaṅgādharjī kā pāṇḍityă pratyek pṛṣṭh par jhalaktā hai, parantu vah itnā kaṭhin hai ki sāmānyă śiṣyă kī buddhi us meṁ nahīṁ ghus pātī*). Furthermore, he briefly addresses the opinion prevalent in Bengal that Āyurveda cannot be acquired without knowledge of

Sanskrit is also the natural idiom for Yogindranath Sen (1871–1931),[88] as author of the *Carakopaskāra* ('Ornament of Caraka'), a valuable Sanskrit commentary on the *Carakasaṃhitā* from Calcutta published in three volumes between 1920 and 1930.[89] Yogindranath Sen was a grand-disciple of Gangadhar Ray through his father Dwarkanath Sen (1843–1906)[90] and had been involved in the publication of the second edition of the *editio princeps* of the *Carakasaṃhitā* in 1908, together with Gangadhar Ray's *Jalpakalpataru*. According to the title page of one of its first fascicles, he 'thoroughly revised and corrected' this edition supervised by his father and a certain Rajendra Narayan Sen, who was a student of his father,[91] and published by Gangadhar Ray's grandson Tryambakeshwar Ray.[92]

Short explanatory notes in Sanskrit on the *Carakasaṃhitā* were published in 1926 and from 1937 to 1938 in two other centers of Sanskrit learning in colonial India, Lahore and Benares. The first one, the *Auṣadhavivṛti* ('Exposition of Medicine') by Sadanand Ghildiyal, is a commentary on four excerpts from the Sūtrasthāna and the Vimānasthāna, the first and third book of the *Carakasaṃhitā*, respectively.[93] Sadanand Ghildiyal was a student of (Kavirāj)[94] Narendranath Mitra of Lahore.[95] The latter (born 1874) may be identical with the editor of an edition of the *Carakasaṃhitā* together with Cakrapāṇīdatta's *Āyurvedadīpikā* published in 1929 in Lahore, (Āyurvedacakravartī Kavirāj) Narendranath (Śāstrī),[96] and further with (Kavirāj) Narendranath to whom his student Jaydev, another editor and a Hindi commentator of the *Carakasaṃhitā* working in Lahore, expresses his great gratitude: thanks to his teacher Narendranath's fitting recommendations and blessing,

philosophy (ibid.). See also E44 (1981–1994), Introduction to Volume III (1985), vii. – Others see this differently. For instance, Vrajvallabh Sharma acknowledges in his introduction to E18 (1911) that in a situation where well-preserved and complete ancient commentaries are unavailable this modern commentary of the well-known respected pandit assists in understanding the meaning of the *Carakasaṃhitā* (Bhūmikā, 5: *aisī daśā meṁ prasiddh paṇḍit Gaṅgādhar śāstrījī kī "Jalpakalpataru" nāmak ādhunik ṭīkā hī is granth kā arth samajhne meṁ sahkārīkāraṇ hai*). For much praise and a defense of the *Jalpakalpataru*, see Chatterjee (1959a, 406–407).

[88] See Meulenbeld (1999, IA, 200). Meulenbeld mentions that the year of Yogindranath Sen's death is often erroneously given as 1918 (1999, IB, 308, n. 619). This is the case, e.g., in Sharma (1975, 223; 2016, 242) and Atridev Vidyālaṅkār (1976, 616), where the date is specified as (July 1,) 1918.

[89] See E21 (1920, 1922, 1930). The last volume ends with the commentary on Ci 13. More on the *Carakopaskāra* see below, p. 223ff. On this clear and straightforward commentary, see Atridev Vidyālaṅkār (1976, 601) and E44 (1981–1994), Introduction to Volume III (1985), vii–viii.

[90] See Atridev Vidyālaṅkār (1976, 615 and the genealogical tree on p. 613) and Gupta (1976, 372). See also E47 (1983, 1988), Bhūmikā (Introduction), 21. According to Sharma (1975, 557; 2016, 583), he was Dwarkanath Sen's oldest son. The date of Dwarkanath Sen's death is given as February 11, 1909 in Sharma (1986, 1).

[91] See again the genealogical tree in Atridev Vidyālaṅkār (1976, 613).

[92] On the *Carakopaskāra* see also below, p. 223ff., and p. 247, with Fig. 6.15.

[93] See E24 (1926).

[94] On this title, see below, p. 242ff.

[95] See Sharma (1975, 464; 2016, 486) and Atridev Vidyālaṅkār (1976, 625).

[96] See E27.

his commentary, the *Tantrārthadīpikā*, had become complete.[97] On the title page of his edition, Narendranath is identified as the son of a certain Mihirchandraratna, resident of Datarpur, a former princely state in the foothills of the Western Himalaya. The editor of the *Auṣadhavivṛti*, Atmanand Sharma Jassar, may have been a relative of its author, Sadanand Sharma Ghildiyal.

The other modern commentary in Sanskrit referred to above, the *Bhāgīrathī Ṭippaṇī* ('Ganges Gloss') by Taradatt Pant, was published in two volumes from 1937 to 1938 in Benares.[98] Pandit Taradatt Pant is presented as a master (*ācārya*) of literature and Āyurveda[99] on the Sanskrit and English title pages of this edition. He may therefore be identified with Pandit Taradatt Pant, M.A. and master of literature, who was a disciple of Giridhar Sharma Caturvedi and became chief instructor at the Royal Sanskrit College in Jodhpur.[100] A further, incomplete commentary which only covers the Sūtrasthāna of the *Carakasaṃhitā* was also published there in 1937.[101] It is the *Carakapradīpikā* ('Small Lamp Illuminating Caraka') by Jyotishchandra Sarasvati,[102] who was a grand-disciple of Gangadhar Ray through Haranchandra

---

[97] See E32 (1934; *[2]1944; *[3]194?; *[4]1948 and numerous further, post-independence editions and reprints), Preface, VIII: *parantu śraddheyă guru kavirāj narendranāthjī mitră ke ucit parāmarśoṁ aur āśīrvād se yah vyākhyā pūrṇ huī hai.* On Jaydev, see Atridev Vidyālaṅkār (1976, 625–626, 665–666). According to Sharma (1975, 222–223; 2016, 241–242), Jaydev stands in the 'Gurukul Kangri tradition' (of Haridwar) of Gangadhar Ray's school, in a student–teacher lineage starting from Pareshnath Sen Gupta's student Shyamadas (see the chart in Gupta 1976, 372). On Pareshnath Sen Gupta, see above, p. 207, on Shyamadas see Sharma (1975, 551, 558; 2016, 577, 584–585; 1986, 3–4), Atridev Vidyālaṅkār (1976, 616) and Gupta (1976, 376). Sharma claims (1975, 551; 2016, 577) that Shyamadas, who was born in 1876, was a younger brother of Dharmadas Sen Gupta (see above, p. 207), which would be chronologically consistent. However, this is problematic as Sharma himself states that Dharmadas Sen Gupta's father was the physician Kashiprasann Sen (1975, 559; 2016, 586; see also Baldev Upādhyāyă, 1983, 667) and Shyamadas's father the renowned physician Annadaprasad (1975, 558) corrected to Jagannathprasad in Sharma (2016, 584). The information in Sharma (1975, 558) agrees with the information given in Atridev Vidyālaṅkār (1976, 616); however, Shyamadas's lifetime is given there as 1864 to 1934 (BS 1271 to 1341). The many reprints made Jaydev's commentary very popular; see also the remarks by Sharma (1975, 223; 2016, 242). On Narendranath Mitra (born 1874) see Sharma (1975, 597; 2016, 624, with Mishra instead of Mitra) and Atridev Vidyālaṅkār (1976, 625, 653).

[98] See E34.

[99] On the title *ācārya*, see below, p. 245f.

[100] See Upādhyāy (1983, 616).

[101] See E33.

[102] The context in which this commentary is listed by Chatterjee (1959a, 407) implies that it was composed in Sanskrit, which is not clear from the mention of this edition by Meulenbeld (1999, IB, 5). Even though this is an incomplete commentary, it is often mentioned together with other Sanskrit commentaries, pre-modern and modern; see E47 (1983, 1988), Bhūmikā (p. 22), E50 (1998?; reprint 2006), Bhūmikā (p. 19), E52 (2004), Introduction (p. 13), and E54 (2005), Prastāvanā (p. 24). See also Sharma (1975, 223; 2016, 242 with the typo Sarasvami; 1986, 4).

(Harayanchandra) Chakraborty.[103] Presumably, Jyotishchandra Sarasvati was also the editor of the basic text.[104]

Thus, Sanskrit as a language of modern elucidation takes the second place after Hindi.[105] This may have been due not only to the attitude of the authors and editors of the commentaries towards Sanskrit as a medium of expression, as assumed above, but also to their expectation that in the eyes of the readers and students of the *Carakasaṃhitā* a Sanskrit commentary or gloss participates in the prestige and authenticity of pre-modern Sanskrit commentaries on the basic text and thus enhances the value and attractiveness of an edition accompanied by it. Taken together, the number of editions with pre-modern and modern commentaries written in Sanskrit would even surpass that of editions published with an elucidation in Hindi[106] and attests to a continuing great regard for this language in circles of traditional medical learning.

Three editions with commentaries written in English are from the post-colonial period, published in 1976[107] and 2004.[108] To these one may add the two volumes of 'Critical Notes' on the *Carakasaṃhitā*, which contain paraphrases of comments by earlier commentators as well as his own comments, published by Priyavrat Sharma as a supplement to his edition of the text with translation.[109] In their consideration of the pre-modern Sanskrit commentaries and general intention, they clearly represent pieces of modern scholarship by prominent exponents of the academic study of Āyurveda conducted mainly in Benares/Varanasi.

The language of the prefaces, introductions, etc., to the edited text may be modern Sanskrit, the regional vernacular, or English.[110]

---

[103] See the genealogical tree in Gupta (1976, 372). According to Atridev Vidyālaṅkar (1976, 615), Harayanchandra Chakraborty died in 1935; he practiced surgery following the *Suśrutasaṃhitā* (Sharma 1986, 2). Jyotishchandra Sarasvati's name is missing in Sharma's genealogical chart (1975, 222; 2016, 241). Meulenbeld (1999, IA, 194) reports that in this work 'the views of Gaṇanāthasena are repeatedly referred to.' Sharma (1975, 223; 2016, 242) specifies that Jyotishchandra Sarasvati did not agree with Gananath Sen's too modern reflections and thus refuted them when an occasion arose. Further on Gananath Sen see Sharma (1975, 488–489, 558; 2016, 510–511, 585; 1986, 3) and Atridev Vidyālaṅkar (1976, 617).

[104] The publisher, S. K. Sarasvati, must have been a relative, probably a son.

[105] The mentioned five commentaries have been printed in seven editions of the basic text.

[106] There are altogether seven editions with three pre-modern Sanskrit commentaries which raises the overall number of editions with elucidation in the Sanskrit language to fourteen. One further edition combines a pre-modern with a modern commentary in Sanskrit, the *Jalpakalpataru*; see E26 (1927–1933).

[107] See E42 (1976–2002) by Ramkaran Sharma and Bhagwandash. On the nature of their elucidation of the text, see above, n. 28.

[108] See E52 by K. R. Srikantha Murthy and E53 by R. Udupa et al.

[109] See Volumes III (1985) and IV (1995) of E44.

[110] Sanskrit prefaces, introductions, etc., to the edited text are found in the *editio princeps* (E1, Calcutta 1868; this is actually Gangadhar Ray's poetic and flowery story of how and why he undertook the task of composing the modern-Sanskrit commentary *Jalpakalpataru*), E2 (Bombay 1876ff.), E6 (Calcutta 1888ff.), E10 (Calcutta 1897), E15 (Bombay 1903?), E22 (Bombay 1922, $^{2}$1935, $^{3}$1941), E23 (Bombay $^{2}$1922, $^{3}$1933), E26 (Calcutta 1927–1933), E27 (Lahore 1929), E30

### 6.2.3 Scripts, Regions and Places on the Timeline

As is well known, Sanskrit texts can be written and therefore printed practically in all regional scripts of South Asia, in some cases with the addition of some special, traditionally modified characters. The poem 'The Pageant of the Seasons' (*Ṛtusaṃhāra*) by Kālidāsa, said to be the first Sanskrit work ever printed, was set with Bengali type.[111] It was edited by Sir William Jones, the founder of the Asiatic Society of Bengal, and printed in Calcutta at the Honorable Company's Press by Edmund Morris, superintendent of the press from 1787 onwards, in 1792. The set of Bengali letters used for this edition was presumably the famous typeface prepared by Chrles Wilkins, the first superintendent of the Honorable Company's Press.[112] A survey of the scripts in which the *Carakasaṃhitā* was edited shows that in 78% the modern Devanagari script was chosen, i.e., for 44 out of 56 editions (see Fig. 6.8). If one counts second and further editions, revised editions, remakes and reprints, the figure rises to 86%, for 95 out of a total of 110 known editions[113] (see Fig. 6.9). In the modern period, Devanagari, which is in use in large parts of Northern and Central India and in Maharashtra and the modern script especially for Hindi and its dialects as well as for Marathi, dominated the production of printed Sanskrit books in South Asia and outside the subcontinent in academic publishing from early on. The first Devanagari font successfully cast in India was that of the Chronicle

---

(Bombay 1932), E31 (Bombay $^{2}$1935, $^{3}$1941), E34 (Benares 1937–1938), E36 (Lahore $^{2}$1940–1941), E41 (Varanasi 1969–1970, $^{2}$1983), E45 (Jaipur 1982) and E46 (Varanasi 1983). Bengali is the language of the preliminaries in E7 (Calcutta 1892ff.), E8 (Calcutta $^{2}$1910) and E16 (Calcutta 1904), Hindi in E11 (Bombay 1898), E12 (Mathura 1898, $^{2}$1903), E13 (Bombay 1900), E14 (Lucknow 1900), E17 (Bombay 1911, $^{2}$1922) and E18 (Bombay 1911), E32 (Lahore 1934, etc.), E47 (Varanasi 1983, 1988), E50 (Delhi 1998?, $^{R}$2006) and E54 (Varanasi 2005), Marathi in E9 (Poona 1897ff., $^{2}$1902ff., $^{3}$1914ff., $^{R}$1926), Telugu in E25 (Bezwada 1926, $^{2}$1935–1941) and E28 (Bezwada 1930), modern Tibetan in E55 (Varanasi 2006), French in E56 (Paris 2011), and English in E6 (Calcutta 1888ff.), E21 (Calcutta 1920–1930), E37 (Jamnagar 1949), E42 (Varanasi 1976–2002), E44 (Varanasi 1981–1994), E52 (Varanasi 2004), E53 (Manipal 2004) and E55 (Varanasi 2006). Due to the fact that in some cases the surrogates of editions available to me unfortunately do not contain the preliminaries, the above rough survey is incomplete even with regard to those editions that were accessible at the time of finalizing this paper. Conjectured information has also not been included here. See the bracketed language abbreviations and empty cells in Column 8 (P/I) of the bibliographical table in Appendix 6.1.

[111] See Shaw (1981, 149, item 202).

[112] See Shaw (1981, 69–71). See also Ogborn (2007, 226, 239, 241). Wilkins was also responsible for the first successful development of a Bengali script font; see Ogborn (2007, 239–250).

[113] The figure of 110 editions also includes re-editions that can be inferred without doubt from the ordinal numbers of available editions and other book-internal and external indicators. Of course, some reprints, especially more recent ones, and regional editions with a low number of copies and limited distribution may have been missed. Furthermore, in the case of early editions published in fascicles, it is sometimes uncertain whether the available dated fascicles are part of complete editions, part of editions that were never completed or selected partial reprint runs. Thus, the figure of 110 editions has to be understood as an approximation. See also n. 18 above.

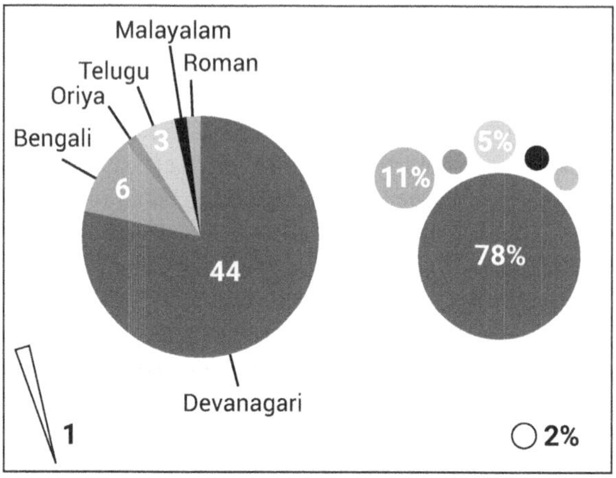


**Fig. 6.8** 56 first editions – scripts

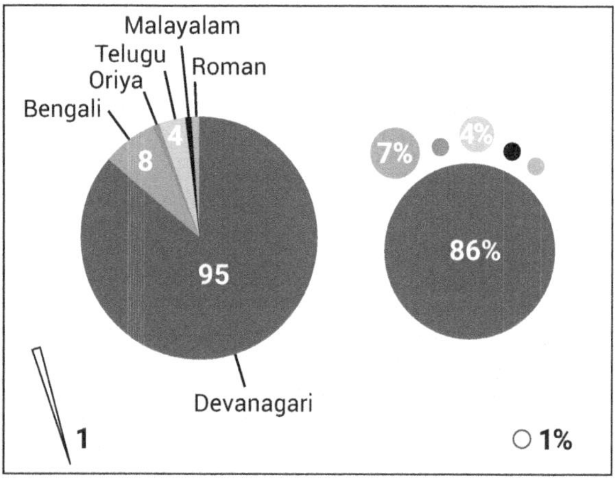


**Fig. 6.9** All 110 editions – scripts

Press in Calcutta, used for some Sanskrit verses printed in a book published in 1789.[114] This general development, which has led to the widespread erroneous perception that Devanagari is the 'Sanskrit script,' was at the expense of the regional scripts and presumably driven by the intention of the modern editors, series editors and publishers to make the texts available to a wider readership, first of all in the large geographical area where Devanagari script was used for the vernaculars. The tendency can already be observed in the first half of the nineteenth century when the printing of Sanskrit texts steadily increased in volume. The many Sanskrit texts that were first published by the Asiatic Society of Bengal in its *Bibliotheca Indica* series from 1849 onwards were set in the non-

[114] See Shaw (1981, 32). On other early Devanagari fonts cast in Calcutta before Charles Wilkins created his celebrated type in 1808 after he had returned to England, see Shaw (1981, 33–34).

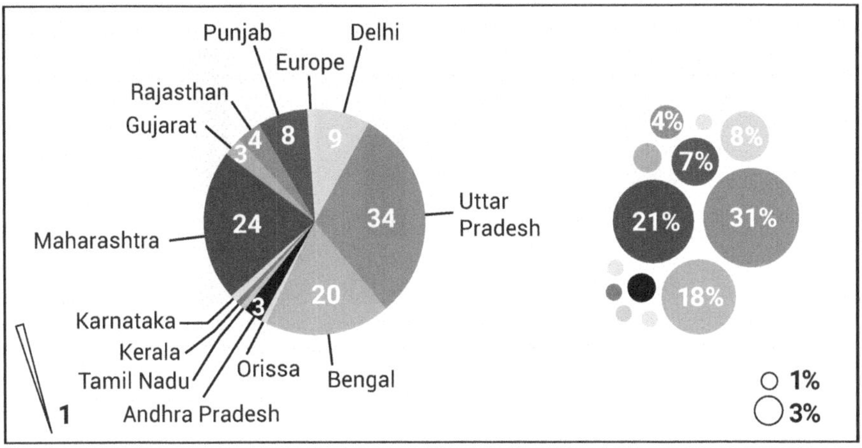


**Fig. 6.10** All 110 editions – regions

local Devanagari font.[115] The later, equally important series of Sanskrit text editions from South India, where a variety of South Indian scripts are used, were also regularly printed in Devanagari.[116] Indigenous Sanskrit scholars who turned to the modern medium of print vs. manuscripts had therefore to acquire familiarity with this script, beyond their regional script, from at least the late nineteenth century onwards. After independence, with the politically motivated propagation of Hindi as a second language for all Indians and the resultant much wider spread of knowledge of Devanagari on the subcontinent, access to Sanskrit texts predominantly printed in Devanagari became easier also for educated readers outside the scholarly or academic circles.[117]

In the case of the printed editions of the *Carakasaṃhitā* we can clearly see the Devanagari trend described above, especially when we look again at the seminal region of Bengal which led the production of *Carakasaṃhitā* texts until the end of the nineteenth century and continued to be very productive in the pre-independence period. Out of the 110 first and further editions, revised editions, etc., of the text, twenty, or roughly 18%, were published in Bengal (see Figs. 6.10 and 6.11), twelve of them being first editions (see Figs. 6.11 and 6.12) However, only eight

[115] Further on this series, see Alessandro Graheli's contribution to this volume, Chap. 5.

[116] See, e.g., the *Government Oriental Library Series*, *Bibliotheca Sanskrita* (Mysore) (from 1893 onwards), R. S. Vadhyar and Sons (Kalpathy, Palakkad) (from 1894 onwards) and the *Madras Sanskrit Series* (from 1898 onwards), mentioned by Patil (2013, 114), to which one may add the *Trivandrum Sanskrit Series* (from 1905 onwards) and the still later *Madras Government Oriental Series*.

[117] On the usage of local scripts in the early history of printing in South Asia, beginning with the efforts of Christian missionaries in the early eighteenth century, their standardization and role in emerging identity politics in the nineteenth century, as well as the dominance of the Devanagari script – in which European scholars of Sanskrit were originally and significantly involved – and the subsequent Hindu nationalistic glorification of this script, see Brandt (2020, 4–9).

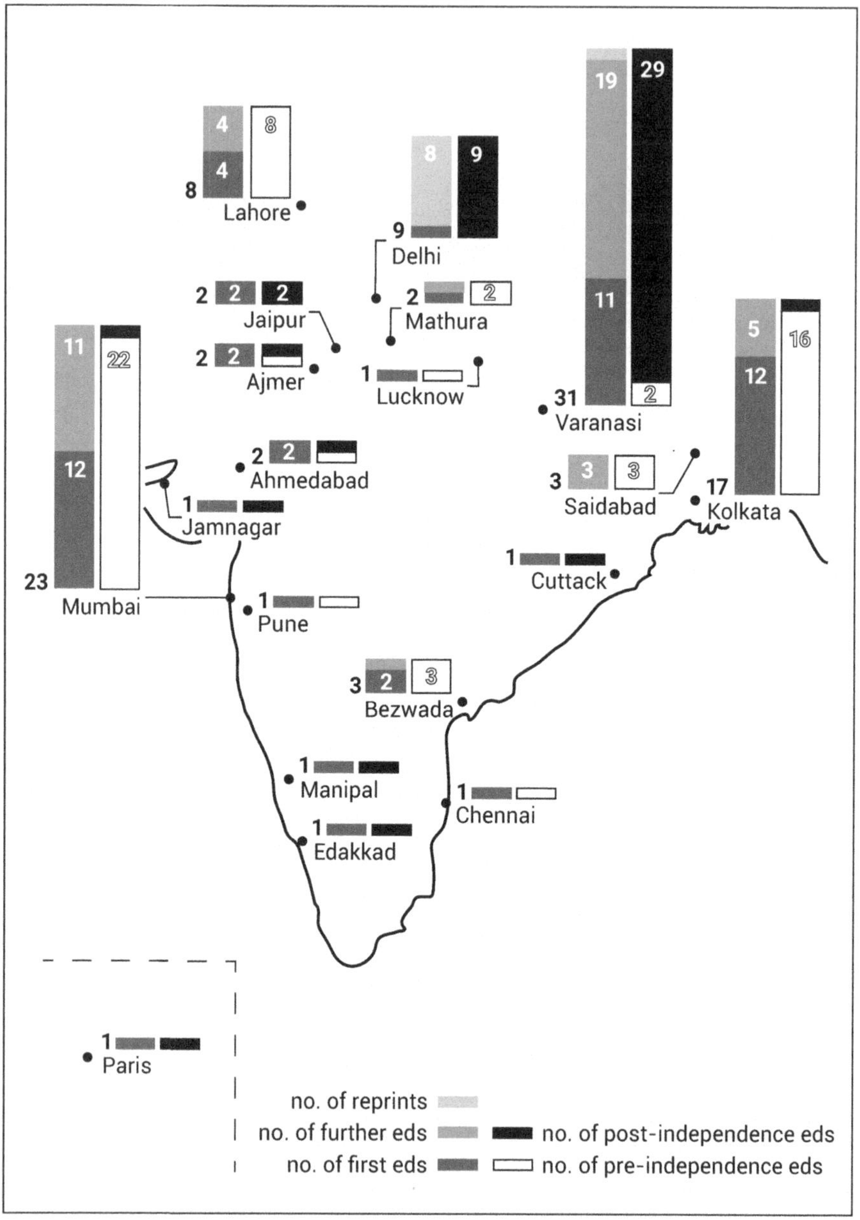


**Fig. 6.11** All 110 editions – places – types – period

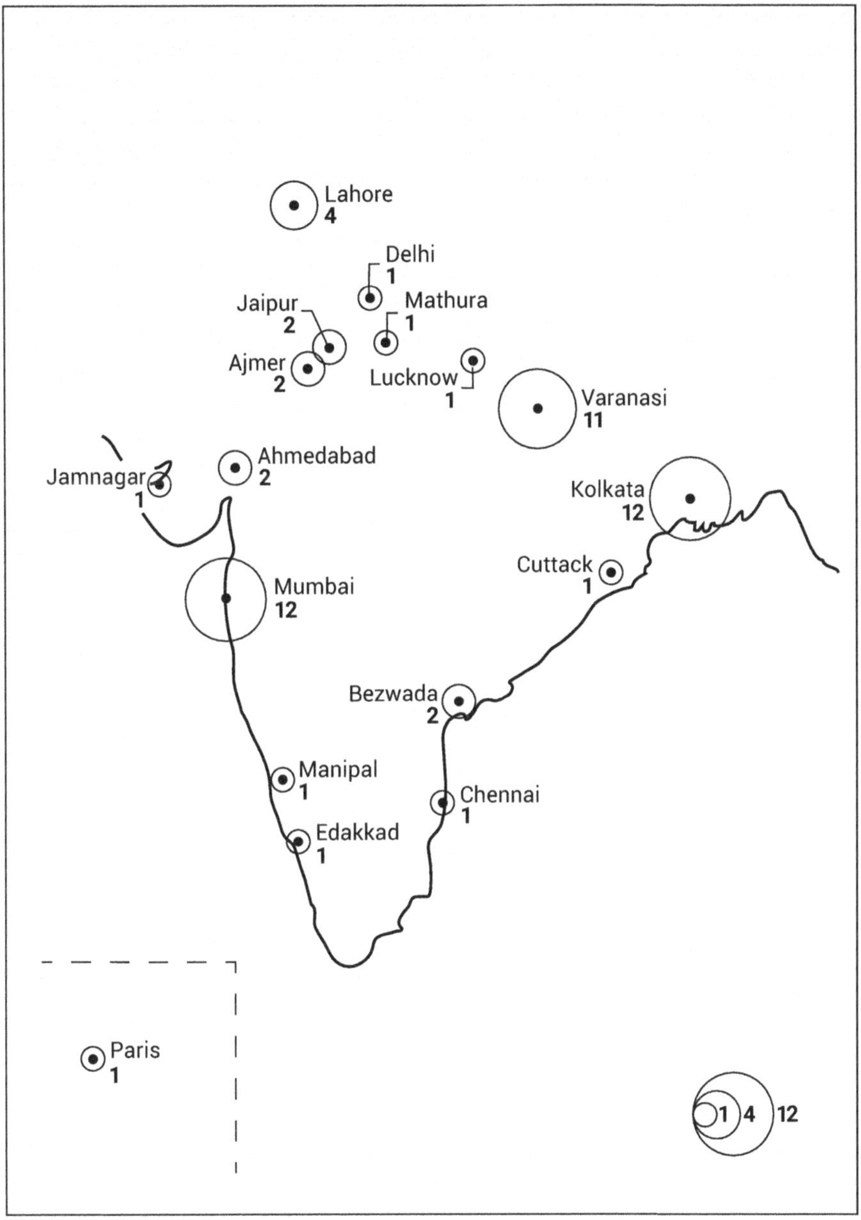


**Fig. 6.12** 56 first editions – places

of these books were printed in Bengali characters (see again Fig. 6.9). The first fascicle of the *editio princeps* by Gangadhar Ray appeared in Bengali script in 1868,[118] but was – at least partially – duplicated in Devanagari script in the same year. The unnumbered follow-up edition of 1878 was in Bengali script, but duplicated in 1884 in Devanagari script. However, the manuscripts of Gangadhar Ray's own works in Sanskrit I could see are written in Bengali script,[119] as is one of the manuscripts on which he based his edition of the *Carakasaṃhitā* and which was most probably written by himself.[120] The second edition of the *editio princeps* that was posthumously published by his grandson Tryambakeshvar Ray in 1908 is also a Devanagari edition.[121] Characteristically, the subsequent editions from Calcutta, the hub of book production in Bengal, were all printed in Devanagari script, except those with accompanying full Bengali translation or explanatory notes in Bengali.[122] In the latter case, the choice of script for the Sanskrit text is obviously guided by the wish to address the non-specialist readers in the linguistic region and encourage them to turn to the Sanskrit original, which is more accessible in this way, for reference, whereas the specialists who can study the Sanskrit text directly and more or less independently, possibly with the help of a pre-modern or

[118] See E1.

[119] They include manuscripts of various parts of the *Jalpakalpataru* (handlist of the Vaidyaka section, ms. nos. 150, dated 1840 [BS 1247], and 151), a manuscript of the *Mṛtyuñjayasaṃhitā* (ibid., no. 153), which is a reorganized version of the *Carakasaṃhitā*, rather than a succinct explanation of its contents (Pecchia 2022b, 816–817), and a manuscript of the *Vaiśeṣikasūtra* (handlist of the Nyāya-Vaiśeṣika section, no. 1401), all preserved in the library of the Calcutta Sanskrit College. It can be assumed rather safely that Gangadhar Ray also penned his other works in his native Bengali script.

[120] According to the colophons at the end of the Nidānasthāna and Vimānasthāna in ms. no. 107465 of the Saraswati Bhawan Library, Sampurnanand Sanskrit Vishwavidyalaya, Varanasi (acc. no. 107465, *DCSMSSU* no. 108824, p. 98), this *Carakasaṃhitā* manuscript was written in 1838 by a certain Gaṅgādhara, who calls himself 'a good healer' (*subhiṣak*). See also Pecchia (2022a, 115). For further details, see the detailed information on ms. *V2*$^{b}$ in Part II, Chapter A.1 ('Descriptions of the Manuscripts Containing the Text of *Carakasaṃhitā* Vimānasthāna Chapter 8'), Section 2.1, in Preisendanz et al. (forthcoming). See also the description in Pecchia (2022b, Appendix II), which to a large extent relies on the collaboratively achieved description there, however without adequate acknowledgment of this fact in the initial asterisked note of Pecchia's paper.

[121] It seems that the mentioned editions in Devanagari script had not come to the notice of Vaman Kesheo Datar when he wrote his prefatory remarks on E22 (1922), because he characterizes the *Jalpakalpataru* edited alongside the *Carakasaṃhitā* as having been printed in Bengali characters and – presumably for this reason – as being little current and difficult to obtain; see Prāsthāvikam, 1: *vaidyavaragaṅgādharakṛtā ṭīkā vaṅgākṣarair mudritā, sāpi viralapracārā durlabhā ca.* This is unexpected because Vaman Kesheo Datar probably was personally acquainted with Yogindranath Sen who was directly involved in the publication of the second edition of the *Carakasaṃhitā* together with his grand-teacher's commentary in 1908 (see above, p. 214). Yogindranath Sen obviously kept Vaman Kesheo Datar informed on the progress of the publication of his own commentary *Carakopaskāra* (see below) and provided him with a manuscript of the Cikitsāsthāna of Cakrapāṇidatta's *Āyurvedadīpikā* when he visited Calcutta in search of manuscripts of this text. See E22 (1922), Prāstāvikam, 2. Vaman Kesheo Datar was also not informed about the post-1905 progress of the edition of the *Carakasaṃhitā* along with the *Āyurvedadīpikā* by Harinath Sarkar; see below, Appendix 6.2.

[122] See E4 (1878??), E5 (1883–1884), E8 (1894, $^{2}$1910, *1930), E16 (1904) and E48 (1984–1989).

modern commentary in Sanskrit supplied with it, are assumed to be able to deal easily with Devanagari, a non-regional script. A similar motivation can be presupposed for the choice of the regional script for the editions in Oriya,[123] Telugu,[124] and Malayalam[125] script, all with translations or modern commentary in the regional language.

The case of the Sanskrit commentary *Carakopaskāra* by the Calcutta-based physician–scholar Yogindranath Sen referred to above[126] deserves special mention in the context of the issue of scripts. Bhagwansahay Trivedi, the secretary of the Shrisvamilakshmiram Trust based in Jaipur, Rajasthan, which published the *Carakopaskāra* on *Carakasaṃhitā*, Cikitsāsthāna 1–20 together with the basic text, repeatedly refers to the issue of script in his Publisher's Preface (Prakāśanīyaṃ Prākkathanam) written in Sanskrit when he recounts the story of the publication of this commentary. It is worthwhile to paraphrase this interesting narrative in some detail because it also nicely presents a segment of the pan-Indian network of traditional physician–scholars and the special relationships that obtained between them and finally led to an edition of the Cikitsāsthāna of the *Carakasaṃhitā* with seven further chapters of the *Carakopaskāra* that had not been available before.[127] Lakshmiram, the later founder of the Jaipur trust named after him, first studied in Jaipur. After he had obtained his Master's (*ācārya*) degree,[128] he moved to Calcutta where he became a student of Dwarkanath Sen and befriended his son Yogindranath Sen.[129] The best student of Lakshmiram, Narayanprasad Sharma, also studied in Jaipur for his *ācārya* degree and then, at the order of Lakshmiram, set out for Calcutta to serve Yogindranath Sen and continue his medical studies with him. He also learnt the Bengali language and script well. Thus, Yogindranath Sen had him convert his commentary, which was originally written in Bengali script, from Bengali into Devanagari characters 'due to an idea useful for all' and published it to the extent that it had been converted, that is, up to and including the comments on the thirteenth chapter of the Cikitsāsthāna.[130] Immediately after his return to Jaipur,

---

[123] See E51 (2000), with an Oriya commentary.

[124] See E19 (1914, 1918–1920), with Cakrapāṇidatta's *Āyurvedadīpikā* and Telugu explanation, E25 (1926, $^{2}$1935–1941), with a Telugu translation, and E28 (1930), with a Telugu commentary.

[125] E43 (1979), with a Malayalam commentary.

[126] See above, p. 214.

[127] See above, n. 89.

[128] See below, p. 245f.

[129] See again above, p. 214. On Lakshmiram (born 1873 [Samvat 1930]), see the chart in Gupta (1976, 372), Sharma (1975, 222, 557, 561–562; 2016, 241, 583, 588), Atridev Vidyālaṅkār (1976, 624) and his concise biography by Maṅgaldās Svāmī (1940). In this biography, Lakshmiram's study period in Calcutta is not mentioned. Instead, the author describes a visit of Lakshmiram to Pareshnath Sen Gupta in Benares (40). As teacher of Jayramdas and grand-teacher of Ramprakash, Laksmiram became the founder of the 'Jaipur tradition' of Gangadhar Ray's school.

[130] See E45 (1982), Prakāśanīyaṃ Prākkathanam, 1f.: *mūlato baṅgākṣarair likhitasyāpi tadetasya carakopaskāragranthasya sarvopayogibhāvanayā nāgarākṣareṣu saṃkalanaṃ kārayitvā cikitsāsthānacaturdaśādhyāyaparyantaṃ prakāśanaṃ svayaṃ śrīyogīndranāthasenamahodayair niṣpāditam.... śrīprasādamahodayāḥ śrīyogīndranāthasenamahodayānāṃ sevāsu kalikātānagaraṃ pra-*

it was not possible for Narayanprasad Sharma to continue the conversion of the remaining parts of the *Carakopaskāra* from Bengali to Devanagari letters, nor could he continue with its publication.[131] Then Yogindranath Sen, who was very concerned about the full publication of the commentary in precisely this way, passed away. Afterwards, his cousin Jnanendranath Sen[132] safeguarded the manuscript of the *Carakopaskāra* which was available up to and including the Kalpasthāna of the *Carakasaṃhitā* and wrote himself a commentary on the final book, the Siddhisthāna. He was in correspondence with Jayramprasad, Lakshmiram's successor at the Government Ayurveda University and a student of his, about the publication of the commentary, but passed away before any agreement could be reached. Subsequently, the complete manuscript was with a certain I. K. Sen, most probably a relative, who worked at the office of the newspaper *National Herald*.[133] Jayramprasad wrote him many times, but never received a reply and was very much worried that the manuscript of this work by a fellow-student of his own teacher might get lost. Then I. K. Sen had a dream in which Jnanendranath Sen appeared to him and told him to swiftly send the complete manuscript to Jayramdas. That night I. K. Sen took the train to Jaipur and presented the manuscript to Jayramdas. The latter thought that it should be published by the Shrisvamilakshmiram Trust and instructed his student Ramprakash to edit it, but there remained the problem of its conversion from Bengali to Devanagari characters.[134] Fortunately, although he was already in his eighties, Narayanprasad Sharma who was then overseeing the temple of Lakshman Dwara in Jaipur tackled the problem and with the zeal appropriate for a young man magically rejuvenated the body of the *Carakopaskāra* through its conversion from Bengali to

*sthāpitaḥ. tatraiva baṅgadeśīyāṃ bhāṣāṃ lipiṃ ca samyag abhyasya śrīsenamahodayānāṃ tattvāvadhāne taiś carakopaskāro 'yaṃ yāvān saṃkalito baṅgākṣarebhyo nāgarākṣareṣu tāvān eva prakāśitaḥ.*

[131] See E45 (1982), Prakāśanīyaṃ Prākkathanam, 2: *atha śrīprasādamahodayānāṃ kalikātāto jayapurapratyāgamanānantaraṃ naiva sambhavam abhūd avaśiṣṭacarakopaskārasya nāgarākṣareṣu saṃkalanaṃ nāpi ca prakāśanam.*

[132] In E45 (1982), Jnanendranath Sen is identified as Yogindranath Sen's younger brother: see *teṣāṃ svargagamanānantaraṃ carakopaskārasāhityaṃ surakṣitam avatasthe tadīyānujānāṃ śrījñānendranāthasenamahodayānāṃ sakāśe* (Prakāśanīyaṃ Prākkathanam, 2). However, this is a misunderstanding probably based on the ambiguous wording in Atridev Vidyālaṅkār (1976, 614–615). Jnanendranath Sen was a nephew of Yogindranath Sen's father Dwarkanath Sen who also studied under him and thus was a fellow-student of Yogindranath Sen; see Sharma (1975, 222, 561; 2016, 241, 587). He was first active in Patna, but then moved to Haridwar where he founded the 'Haridwar tradition' of Gangadhar Ray's school at the Rishikul Ayurvedic College; see Sharma (1975, 222, 544, 561; 2016, 241, 570, 587–588) and Atridev Vidyālaṅkār (1976, 615). In his introduction to E21 (1920) (Critical Notice, iv), Yogindranath Sen thanks his cousin Kavirāj Jnanendranath Sen Kaviratna, B.A., for assisting him with the Table of Contents. The latter also acted as publisher and printer, at the Vidyodaya Press, of E21 (1920, $^{2}$1922).

[133] The *National Herald* was founded in 1938 in Lucknow; from 1968 onwards, it was published in New Delhi.

[134] See E45 (1982), Prakāśanīyaṃ Prākkathanam, 3: ... *tāvat samupasthitā tadetasya nāgarākṣareṣu samākalanasya samasyā.*

Devanagari characters.[135] Printing started during the lifetime of Jayramdas, but he did not live to see the publication of the first volume with the commentary on Cikitsāsthāna 1–20 edited by Ramprakash who had become his successor.

Even though there were many other obstacles to the editing and publication of the *Carakopaskāra*, a major issue was the necessity to get the work published in Devanagari characters which was clearly felt by the main actors in this saga, starting with the author of the commentary himself. Furthermore, even among the engaged physician–scholars of twentieth-century Jaipur who strove so much to put the work into print missing familiarity with the Bengali script was evidently the rule and the initiative to acquire it the exception.

The clear dominance of the Devanagari script used in editions of the *Carakasaṃhitā* deserves some further analysis and interpretation. From the point of view of the places of publication, it is the combined result of two phenomena. One is the intensive publication activity from 1897 onwards that occurred in Maharashtra, that is, in Bombay (see Fig. 6.13), with text editions that are frequently accompanied by translations into Hindi and Marathi and by commentaries and explanations in these languages. Before that, a first, probably incomplete edition of the *Carakasaṃhitā* in Devanagari script was started in fascicles in Bombay by Anna (Ananta) Moreshvar Kunte as early as 1876.[136] In the Punjab, Lahore was a traditional place of Sanskrit learning and became a lively center of publishing in the decades before independence and partition. Altogether eight Devanagari editions of the *Carakasaṃhitā* were produced there within a timespan of roughly twenty years after 1926 (see Fig. 6.13), four of them being first editions (see Figs. 6.11 and 6.12).[137] Benares, although located in the United Provinces (of Agra and Oudh) (approximately present-day Uttar Pradesh and Uttarakhand) and the heart of the Hindi speaking region, and an important place for Sanskrit publishing in general, comes into the picture for *Carakasaṃhitā* editions rather late, with only two Devanagari editions published in the thirties, one with the modern Sanskrit commentary *Carakapradīpikā*,[138] and one with the explanatory notes called *Bhagīrathī Ṭippaṇī*.[139] The bulk of *Carakasaṃhitā* editions in Devanagari published in Benares (present-day Varanasi), which makes it the place with the largest number of editions, came out after independence, namely, thirty-one, many of them further edi-

[135] See E45 (1982), Prakāśanīyaṃ Prākkathanam, 3: *aśītikalpavṛddhāvasthāyām api navayuvakocitotsāhena tair niravacchinnapariśrameṇa sampāditaś ca carakopaskārakalevarasya baṅgākṣarebhyo nāgarākṣareṣu kāyākalpaḥ*. See also p. 2: *granthasyāsya baṅgākṣarebhyaḥ parivartya nāgarākṣareṣu samākalanasya śreyas tāvat samāgacchati jayapurasthalakṣmaṇadvārāmandirasya vartamānādhyakṣān vayovṛddhān vidvadvarān ācāryaśrīnārāyaṇaprasādaśarmamahodayān prati*.

[136] See E2.

[137] E24 (1926) (excerpts), with explanatory notes in Sanskrit, E27 (1929), with the *Āyurvedadīpikā* commentary, E32 (1934; *²1944; *³194?; *⁴1948; numerous further, post-independence editions and reprints published elsewhere), with the Hindi commentary *Tantrārthadīpikā*, and E36 (*date unknown, ²1940–1941) with *Āyurvedadīpikā* and Jejjaṭa's *Nirantarapadavyākhyā*.

[138] See E33 (1937), discussed above, p. 215f.

[139] See E34 (1937–1938), addressed above, p. 215.

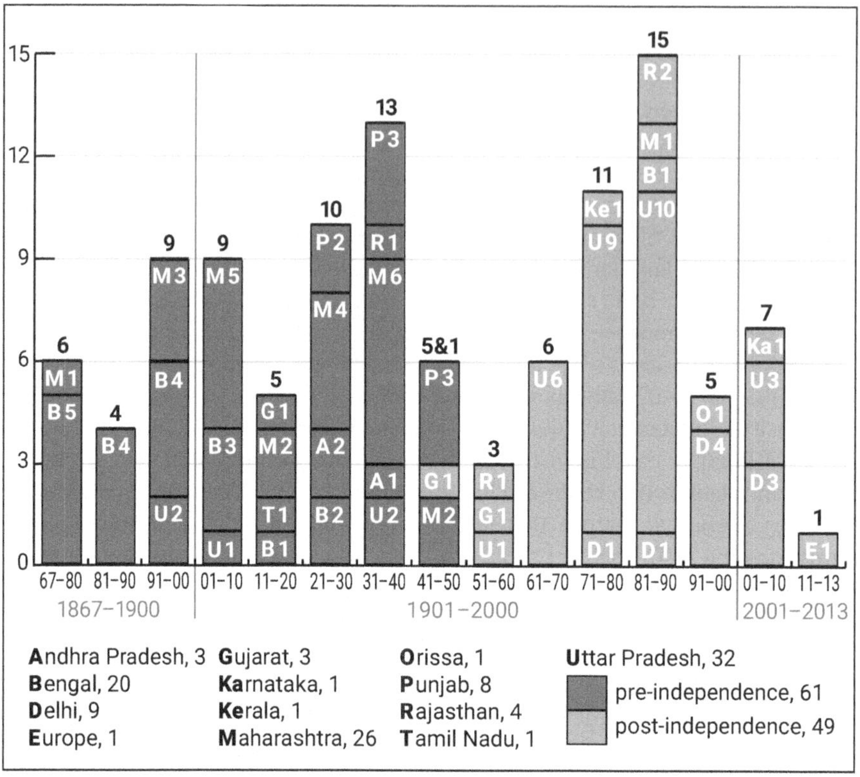


**Fig. 6.13** All 110 editions – decades – regions

tions, reprints and reprints presented as further editions, also of editions first published elsewhere in India in Devanagari script (see Fig. 6.11). Even so, among places with first editions of the text Benares/Varanasi still occupies the second rank after Bombay/Mumbai and Calcutta/Kolkata with eleven editions (see Fig. 6.12).[140] The state-funded growth of academic institutions there and the widening scope of their research activities increased the modern academic study of Āyurveda. This development was accompanied by the founding of new publishing houses in the area of Sanskrit studies in general. There were also some Sanskrit publishing houses that after partition moved to Benares from Lahore which had become part of Islamic West Pakistan. Motilal Banarsidas(s), the publisher of seven out of eight *Carakasaṃhitā* editions that came out in Lahore, was one of them.[141]

[140] In the rough survey of Ayurvedic works in general published in nineteenth-century India presented by Singh and Joshi, Calcutta ranks first, Bombay second and Benares third; see Singh and Joshi (1982, 315).

[141] See E27 (1929), E32 (1934; *²1944; *³194?; *⁴1948; further editions and reprints published in Benares/Varanasi and later on in Delhi) and E36 (*date unknown, ²1940–1941). See also the general remark on this publishing house in Sharma (1975, 586; 2016, 614) and Atridev Vidyālaṅkār (1976, 652–654).

The vibrant editing and especially re-editing and reprinting activity in Benares/Varanasi, which accounts for the dominance of modern Uttar Pradesh in terms of publication of editions of any kind of the *Carakasaṃhitā* (see Fig. 6.10), seems to have entailed that after independence not a single edition of the *Carakasaṃhitā* was published in the Bombay Presidency, which became a part of post-independence Maharashtra, or more precisely, from Bombay/Mumbai. In the present context, this city practically stands for what is now Maharashtra because, as can be seen from a comparison between Figs. 6.10 and 6.11, only one edition from Maharashtra was not published in Bombay, but in Poona (modern Pune).[142] The publication of the second-earliest first edition of the work and the beginning of a steady flow of publication of first and further editions in Bombay from 1897 onwards has already been addressed above.[143] This development made Bombay/Mumbai a place with the same number of first editions as Calcutta/Kolkata (12) (see Fig. 6.12). In terms of editions in general, it holds the second place after Benares/Varanasi with twenty-three against thirty-one editions and therefore accounts for the prominence of Maharashtra as a region where editions of the *Carakasaṃhitā* were published (see Fig. 6.10). However, only one of them was published after independence (see Figs. 6.11 and 6.13). Similarly, after 1947 only one edition, a first edition with Bengali explanations, was published in Bengali script in Bengal, more precisely in Calcutta/Kolkata[144] (see Fig. 6.13), which had been so dominant before, with altogether seventeen editions, twelve of them first editions (see Figs. 6.11 and 6.12), making Bengal the third-ranking region in terms of editions of any kind (see Fig. 6.10).[145]

Institutions of learning devoted to the study of Āyurveda in Rajasthan contributed further editions in Devanagari, the local script.[146] And even the three editions from Gujarat[147] present the Sanskrit text in Devanagari script,[148] presumably because it is rather similar to the Gujarati script which was used for the accompanying translation, commentary and explanatory notes in Gujarati language. Thus, a multiplicity of factors contributed to the overall Devanagari dominance in *Carakasaṃhitā* editions.

When we take a general look at the timeline of the publication of the text of the *Carakasaṃhitā*, including second and further editions and reprints, irrespective of the addition of translations and commentaries to the text and of the issue of script,

---

[142] That is, the 1920 reprint of E9 discussed above, p. 209.

[143] See p. 225.

[144] See E48 (1984–1989).

[145] The difference of three editions between Calcutta and Bengal (17 vs. 20) is explained by the fact that two follow-up editions and the second edition of the *editio princeps* (E1) were published in Saidabad, Berhampore (Murshidabad District), north of Calcutta, to where Gangadhar Ray had moved in 1829 (see below, p. 231). See above, p. 222.

[146] See E35 (1937), with Hindi translation; E38 (1954, 1957, 1962), with Hindi translation; E45 (1982), with Yogindranath Sen's modern Sanskrit commentary *Carakopaskāra*; E49 (1990), with Śivadāsa Sena's fragmentary early-modern Sanskrit commentary *Tattvapradīpikā*.

[147] E20 (1916, 1931), with Gujarati translation and commentary; E37 (1949), with Hindi, Gujarati and English translation; E39 (1958–1963), with Gujarati translation and commentary.

[148] However, in the second volume of E20 the Sanskrit text is not printed alongside the Gujarati translation and commentary.

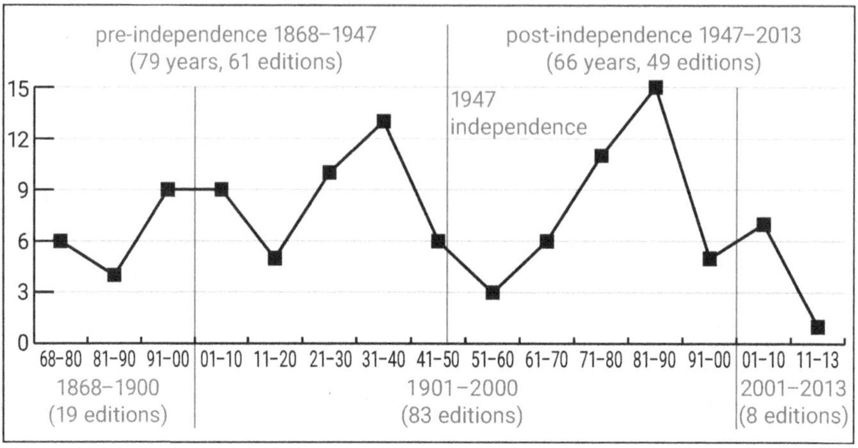


**Fig. 6.14** All 110 editions – decades

we can observe a rather steady flow of publications ever since the *editio princeps* appeared in 1868 (see Fig. 6.14).[149] There are three clear periods that show a remarkable increase of publication activity. One is from the 1890s until the first decade of the twentieth century, the second during the 1920s and 1930s, the period before World Word II, and the last began in the seventies and reached a high point during the 1980s. During the first period, publication activity was especially dense in the five years from 1896 to 1900, with seven editions published in consecutive years. During the second period, there were many years, also consecutive ones, from 1922 onwards until 1938 in which two, three or even five editions of the text appeared in print. Even during World Word II, six editions came out between 1939 and 1944. After independence, the eleven years between 1979 and 1989 show the highest density of publications in the third period, with twenty editions, including many further editions and reprints. Over the seventy-nine years from the date of the *editio princeps* until independence, altogether sixty-one editions came out, during the sixty-six years after independence and until the endpoint of this survey, 2013, the text of the *Carakasaṃhitā* was edited forty-nine times. This would amount to an identical average of roughly 1.3 editions per year for each of the two large periods in terms of Indian history and thus testifies to an unflagging interest of editors, publishers and readers in editions of this foundational ancient text. The authenticity and prestige as well as the canonical status gained by the *Carakasaṃhitā* through its first

[149] On the data at the basis of this timeline, see n. 113 above. Editions that appeared in fascicles are only taken into account for the year of publication of the first fascicle because the publication history of individual subsequent fascicles is rather difficult to trace. Furthermore, some dates of first and follow-up editions, or of individual volumes, are unknown, uncertain or doubtful. Thus, the placement of some editions on the timeline is conjectural.

appearance in print[150] are thus being continuously maintained and multiplied. At the same time, this results in a continuous increase of the 'textual community' that uses the *Carakasaṃhitā* as an actual or rhetorical point of reference for its professional activities and involvement.[151]

Several reasons for the increase of publication activity related to the text of the *Carakasaṃhitā* during the three clearly discernible periods have already been suggested above in the discussion on scripts and regions on the timeline of editions. Bengal, with the colonial capital Calcutta and its many modern and traditional institutions of learning, was the intellectual hub of British India and center of book publishing during the nineteenth century, which can be confirmed for the special case of the editions of the *Carakasaṃhitā* (see again Figs. 6.11 and 6.13). The first period of significant increase in this case can be accounted for by the fact that a printing industry had also developed in Maharashtra (then Bombay Presidency), especially in Bombay, and publishers of Sanskrit books, among them the owners of well-known and influential presses such as the Nirṇayasāgara Press (founded in the 1860s)[152] and the Śrīveṅkaṭeśvara Steam Press (established in 1871),[153] became interested in publishing important scientific texts and thus joined their colleagues in Calcutta in publishing editions of the *Carakasaṃhitā* and other classical medical texts. The second period of increase in editions of the *Carakasaṃhitā* may be seen under several aspects. These are the modernization of printing and further spread of the printing industry in all parts of the subcontinent, and the rise of nationalism and national pride in indigenous achievements in the context of the intensification of the independence movement.[154] The publication activity from Maharashtra, that is, Bombay, with books in Devanagari accessible to a wide readership remained strong, but now also other regions, such as Andhra Pradesh, with Bezwada (modern Vijayawada), in the South (then part of the Madras Presidency),[155] Uttar Pradesh, with Benares, in the North, Rajasthan, with Ajmer (then located in the small British province of Ajmer–Merwara), in the West,[156] and especially the Punjab, with Lahore, come into the picture and contributed to the general output. Bengal, however, has only two editions to its credit

[150] See Weiss (2015, 655) for the case of works belonging to the genre of religious literature, specifically the South Indian Śaiva canon in the making, with reference to Stuart Blackburn's observations on the publication of Tamil classics in the nineteenth century.

[151] See Weiss (2015, 657), relating to a South Indian religious community and with reference to Brian Stock's research on the implications of literacy for heretical groups in early medieval Europe.

[152] See below, pp. 238 and 261f. On the history of the Nirṇayasāgara Press, see P. B. Kulkarni's Marathi-language book *Nirṇayasāgar cī akṣar sādhanā* published by the Nirṇayasāgara Press in 1967 which is linguistically inaccessible to me.

[153] See below, p. 239.

[154] See further Sect. 6.2.5.2 below.

[155] One edition was published in Madras (modern Chennai) in Tamil Nadu (then also in Madras Presidency). However, it was only printed there, in Telugu characters and not in the regional Tamil (Grantha) script, together with a Telugu commentary. Thus it can be considered a further regional edition belonging to Andhra Pradesh seen as a cultural–linguistic area. See E19 (1914, 1918–1920).

[156] See E35 (1937).

in the first decade of the discussed period, which will be the last until the 1980s. This turning point in the history of a crucial place for the publication of editions of the *Carakasaṃhitā* may be explained by the fact that in 1931 New Delhi was inaugurated and became functional as the new capital of British India. The post-independence rise of Benares in modern Uttar Pradesh as an important center also of modern, as opposed to traditional institutions of learning and as a place of publication of Sanskrit books in Devanagari in all areas, with many new publishing houses, has already been referred to above and is reflected by the large numbers of editions of the *Carakasaṃhitā* that began to appear there in the 1960s. Many of these were further editions and reprints. However, there were also important and influential original publications with editions of the basic text together with new translations and commentaries produced by Benares/Varanasi-based physician–scholars. This trend as a whole made the city the dominant place of publication of *Carakasaṃhitā* editions in the post-independence period. Furthermore, we can see how further regions of modern India – Kerala, Karnataka and Orissa – now come out with their own editions, in the regional scripts and with translations or commentaries by regional scholars.[157] Rajasthan stands out in this respect, with two out of three editions published by the Shrisvamilakshmiram Trust in Jaipur.[158] Finally, Delhi and New Delhi in the National Capitol Territory of Delhi come into the picture as places where mainly reprints of editions of the *Carakasaṃhitā* are published.

### 6.2.4 Persons: Editors, Publishers and Printers, and Other Involved Actors

The main actors involved in publishing the editions of the *Carakasaṃhitā* are its editors, publishers and printers, to whom we can add sponsors, commissioners and other persons who initiated an edition. Up to the 1920s, editors and owners of printing presses (*yantra*) or printing houses (*mudraṇālaya*),[159] some of which already produced a large range of publications by the end of the nineteenth century, were often identical with the publishers.[160] Among the editions of the *Carakasaṃhitā*, the above-

[157] See E43 (1979), E51 (2000) and E53 (2004).

[158] E45 (1982) and E49 (*1990). On E45, see above, p. 223ff.

[159] In the modern vernaculars, the Sanskrit word *yantra* ('instrument, device, machine, implement') and the modern Sanskrit word *mudraṇālaya* ('abode/house of sealing/stamping/printing') are most commonly used as loanwords to refer to print shops and print works. To simplify matters, I will only use the Sanskrit forms of these two words and of names of presses in the following and disregard the pronunciation of the loanwords and their modified transliteration in the various vernaculars.

[160] See Ghosh (2006, 125) on this merging of roles from the mid-nineteenth century onwards in Calcutta in connection with the 1867 Copyright Act (Act XXV). If the printing firms acted as publishers and held the copyright, as opposed to the authors or editors, this ensured a cheaper

mentioned first edition with a Hindi commentary, Mihirchandra's *Vivṛti*,[161] was published and printed by the above-mentioned Khemraj Shrikrishnadas, co-founder and owner of the famous Śrīveṅkaṭeśvara Steam Press in Bombay, and most probably also edited by him.[162] Similarly, he may not only have published and printed the *Carakasaṃhitā* together with Ramprasad's Hindi commentary called *Prasādanī*, but also have edited it for publication.[163] Gangavishnu Shrikrishnadas, a brother of Khemraj Shrikrishnadas, together with whom he founded the Śrīveṅkaṭeśvara Steam Press in 1871,[164] may also have taken over all three functions in the publication of the *Carakasaṃhitā* by the Lakṣmīveṅkaṭeśvara Steam Press in Bombay.[165] Already in the case of the *editio princeps* by the physician–scholar Gangadhar Ray,[166] the next edition of this book was published and printed by his son Dharanidhar Ray[167] at his Pramādabhañjana Press ('The Press that Dispels Mistaken Judgements') in Saidabad (Berhampore, Bengal)[168] where Gangadhar Ray had moved to in 1829.[169] The wording of the title page suggests that the printing press was owned by the publisher.[170] Especially in the early days of printing in Calcutta, ownership of a printing press was also a means by which an author could ensure the correctness of printing of his publications.[171]

In the case of the second edition of the book, Gangadhar Ray's grandson Tryambakeshvar Ray figures as (presumed) editor and publisher.[172] Many more examples for this dual role could be adduced. The editors and translators Dharmadas

---

production of books. The 1867 Copyright Act introduced ownership rights over printed products and required that all publications be registered; see Ghosh (2006, 123). – At least in the early history of printing in Calcutta, correctness of printing was safeguarded when printers also acted as publishers and thus had a personal interest in the publication. See the quote from Haji Mustapha (1789) in Ogborn (2007, 257).

[161] See above, p. 205.

[162] See E11 (1898).

[163] See E17 (1911, next edition 1922). See above, p. 204f.

[164] See Galewicz (2020, 174, n. 416; 243).

[165] See E30 (1932). According to Galewicz (2020, 244), the Lakṣmīveṅkaṭeśvara Steam Press was established by Gangavishnu Shrikrishnadas after the two brothers had gone their own ways. See, however, n. 236 below.

[166] See above, p. 222.

[167] See Chatterjee (1959d, 815) and Chattopadhyay (1995, 12). See also Pecchia (2022a, 114, 117; 2022b, 800).

[168] See E1 (next editions 1878 and 1884).

[169] See Chattopadhyay (1995, 2). Gupta mentions the year 1819 as the year of Gangadhar Ray's earlier move to Calcutta (1976, 371). Thus, he stayed there for only 10 years.

[170] It is, however, doubtful whether Gangadhar Ray himself already owned and operated a printing press, as stated by Chatterjee (1959a, 406) who speaks of the edition of the *Carakasaṃhitā* 'published by Gangadhar Ray himself in his Saidabad Press.' See also Pecchia (2022a, 117).

[171] See again the quote from Haji Mustapha (1789) in Ogborn (2007, 257).

[172] See E1 ($^{2}$1908ff.). On this, see also Pecchia (2022a, 114, 117; 2022b, 800).

Sen Gupta,[173] Avinashchandra[174] and Satishchandra Sharma[175] have already been mentioned. The cases of Dattaram Narayan Pathak, presumed editor and Hindi commentator of the *Carakasaṃhitā*, as well as owner of an Ayurvedic bookstore and clinic, of Harinath Sarkar,[176] editor and owner of an Ayurvedic bookstore and pharmacy, of the editor and bookstore owner Vrajvallabh Sharma, and of Jethalal Devshankar Dave, presumed editor, Gujarati translator and commentator of the *Carakasaṃhitā* and owner of an Ayurvedic pharmacy, will be addressed below.[177] It can be assumed that the editors who acted as publishers were also involved in the dissemination of the books through their businesses.

As can be seen in the case of the editions of the *Carakasaṃhitā*, from the end of the nineteenth century onwards owners of bookstores (*pustakālaya*, *granthālaya*)[178] were also active as publishers. From the pre-independence period, one may mention the bookseller and publisher Yajneshvar Gopal Dikshit in Pune,[179] the Das Brothers, booksellers and publishers in Anarkali, Lahore,[180] and Sundarlal Jain who owned the Punjab Saṃskṛta Pustakālaya 'Motilal Banarsidas'[181] founded by his uncle Motilal Jain in Lahore and ran the store's own Bombay/Mumbai Sanskrit Press.[182] In Calcutta, the Giriśa Pustakālaya ('The Bookstore of the One who Reclines on the Mountain [i.e., God Śiva]') in Cornwallis Street (now Bidhan Sarani)[183] and in Bezwada (Vijayawada in modern Andhra Pradesh) the Āndhra Granthālaya[184] were places where editions of the *Carakasaṃhitā* were published. The latter bookstore evidently had its own printing press. In the case of the edition with Ravidatt's Hindi translation, its presumed editor, the scholar Vrajvallabh Sharma, master of a

---

[173] See E4 (1878??); on Dharmadas Sen Gupta see above, p. 207.

[174] See E6 (1888ff.), the partial first edition of the *Carakasaṃhitā* with Cakrapāṇidatta's *Āyurvedadīpikā* (see above, n. 19); see also below, p. 234. On Avinashchandra see also above, pp. 208 and 210.

[175] See E16 (1904); on Satishchandra Sharma see above, p. 208.

[176] For Harinath's surname Sarkar, see his signature at the end of his introduction (E7 [1892ff.], Bhūmikā, 8). Vaman Kesheo Datar refers to him as 'Harinath Mishra' in the prefatory remarks on E22 (1922) (Prāstāvikam, 1–2); see below, n. 421.

[177] Further editor-cum-publishers are Anna Moreshvar Kunte (E2, 1876ff.), Jivanand Bhattacharya (E3, 1877, $^{2}$1896), Debendranath Sen Gupta and Upendranath Sen Gupta (E10, 1897) and Narendranath Sen Gupta and Balaichandra Sen Gupta (E26, 1927, 1928, 1933). On these two pairs of editors see also below, p. 240ff.

[178] Literally: 'abode/house of books / literary compositions.' Again, I will only use the Sanskrit forms of these two Sanskrit loanwords in the following.

[179] E9 ($^{2}$1902ff., $^{3}$1914ff., reprint 1926).

[180] E24 (1926).

[181] The Sanskrit or Sanskritic names of bookstores and pharmacies are spelled here according to their Sanskrit pronunciation, irrespective of their pronunciation and specific transliteration in the regional languages.

[182] See n. 141 above.

[183] E7 (1892ff.). The precise address was 21 Cornwallis Street. The bookstore was owned by Taraprasanna Mukhopadhyaya.

[184] E25 ($^{2}$1935–1941).

bookstore in Mumbai ('The Old Bookstore'),[185] acted as publisher, but the book was printed at the Nirṇayasāgara Press.[186] In Benares, a first edition of the text was published by Jaykrishnadas Haridas Gupta of the Chowkhamba Sanskrit Series Office, a bookstore and publishing house, in 1937–1938.[187] Other editions were distributed from Ayurvedic bookstores, such as the Anātha Pustakālaya ('The Bookstore for Those without Protector [i.e., Orphans / the Poor]') in Mathura,[188] which was owned by Dattaram Narayan Pathak, publisher and Hindi commentator, and presumably also editor, of the *Carakasamḥitā*,[189] and attached to the Anātha Cikitsālaya ('The Hospital [literally: the abode/house of medical treatment] for Those without Protector'), most probably Dattaram Narayan Pathak's own clinic.

This brings us to Ayurvedic pharmacies as publishing institutions. Several fascicles of the second, much more advanced but still incomplete edition of the *Carakasaṃhitā* together with the *Āyurvedadīpikā* by Kavirāj[190] Harinath Viśārada[191] were published out of the Viśārada Auṣadhālaya ('The Expert's Pharmacy [lit.: 'the abode/house of medicinal plants']') in Calcutta; earlier fascicles were published from the Giriśa Pustakālaya mentioned above. The name of the pharmacy reminds of the degree 'expert' (Sanskrit *viśārada*)[192] held by the editor and publisher of this edition. It is therefore very probable that this physician–scholar owned the Viśārada Auṣadhālaya, which first was located in Cornwallis Street,[193] like the Giriśa Pustakālaya, and then moved to Madan Mitra Lane, a small lane running parallel to Cornwallis Street.[194] Further fascicles were published out of the Āyurvedaratnākara Auṣadhālaya ('Jewel Mine of Āyurveda Pharmacy') at still another address in

[185] The wording on the title page of E18 (1911) indicates that the bookstore was actually owned by Hariprasad, son of Bhagirath, who was also a scholar. In his introduction to this edition, Vrajvallabh Sharma refers to Pandit Hariprasad Bhagirath as owner and book publisher. See below, p. 270.

[186] See E18 (1911). On this press, see above, p. 229.

[187] E34.

[188] E13 (1900). The book was printed by the Nirṇayasāgara Press in Bombay, but distributed from Mathura.

[189] See above, p. 212f., and below, n. 234. See also below, pp. 262f. and 275f.

[190] On this title of Bengali physician–scholars, see below, p. 242ff.

[191] See E7 (1892ff.). On Avinashchandra's first incomplete edition with this commentary see above, n. 19.

[192] In the hierarchy of traditional degrees, *viśārada* is located at the lowest level. See Michaels (2001, 8, Fig. 2). We encounter this title, in the specialized form *āyurvedaviśārada* ('expert on Āyurveda') in the Vijñāpanam (p. 2) of E10 (1897), also with Prasannakumar Bandyopadhyaya, a teacher at the Calcutta Ayurvedic College who furthermore bore the general title *vidyāratna*. See below, nn. 264 and 283.

[193] The precise address was 61 Cornwallis Street.

[194] The move to 25 Madan Mitra Lane in the Shimla ward of Calcutta must have taken place after 1897, the year when fascicle 21 (unnumbered) was published, and before 1902, when fascicle 22 came out. 25 Madan Mitra Lane is also Harinath Sarkar's address that appears – together with his signature after the information on the rules for subscription and advance payment – printed on the back covers of fascicles 23–26 (1902) and 27&28 (1905).

Cornwallis Street, an important thoroughfare in North Calcutta.[195] As the address 192/2 Cornwallis Street is the same as Harinath Sarkar's address given for inquiries and orders on the inner back cover of fascicle 35 (Śaka 1835, i.e., 1913) of the edition, it can be assumed that he had again moved his pharmacy between 1905 and 1907 and furthermore changed its name.[196] This address is also found as Harinath Sarkar's address in an extract from the famous newspaper *Amrita Bazar Patrika* (January 23, 1913) printed on the back cover of the second fascicle of the Kalpasthāna (1913) of this edition, in which an appeal is made for financial support to allow the financially depleted editor-cum-publisher to continue his valuable and highly appreciated work. Furthermore, it is possible that the incomplete first edition of the *Carakasaṃhitā* with Cakrapāṇidatta's *Āyurvedadīpikā* edited and published by Avinashchandra[197] was also published from his own pharmacy (and clinic) just a few houses away at 200 Cornwallis Street because this address provided on the title pages of this edition reappears after an announcement printed in its sixth fascicle.[198] This announcement provides information on the progress and prices of this edition, of Avinashchandra's two editions of the *Carakasaṃhitā* with translations into Bengali and Hindi, of his edition as well as Bengali and Hindi translations of the *Suśrutasaṃhitā*, and of his medical journal *Cikitsāsammilanī*. It is signed by 'Pyarimohan Sen Kavirāj, Manager.' The surname and title of Pyarimohan very much suggest that he was the manager of a pharmacy attached to Avinashchandra's clinic, whereas the nature of the announcement indicates that this pharmacy functioned as Avinashchandra's publication house.

---

[195] The last fascicle of the Siddhisthāna in the copy of E7 owned by the British Library (14043. dd.15) was published out of 'the pharmacy,' without any further specification, in 192/2 Cornwallis Street. However, because this is the same address as that of the Āyurvedaratnākara Auṣadhālaya which still appears on the title page of the penultimate fascicle of the British Library copy published in 1917 (Śaka 1839), it can be assumed that this pharmacy continued to function as the 'publishing house' for E7 until the end of Harinath Sarkar's editorial activity. E7 is a fascinating example of an early edition of the *Carakasaṃhitā* published in fascicles out of various 'publishing houses' and printed by different presses and printers in late nineteenth and early twentieth-century Calcutta. For this reason, and because copies of this edition are rather rare and difficult to access, a synopsis of the precious copy 14043.dd.15 of the British Library is provided in Appendix 6.2 below.

[196] Until 1910, the superintendent of the Āyurvedaratnākara Auṣadhālaya was Kavirāj Gopalchandra Ray. He was a good friend of Harinath Sarkar who had actually encouraged him to edit the *Carakasaṃhitā* with Cakrapāṇidatta's commentary and fully supported him in this endeavor. See E7, Bhūmikā, 8. From 1897 onwards (unnumbered fascicle 21), Gopalchandra Ray had already functioned as superintendent of the Viśārada Auṣadhālaya, first in Cornwallis Street and then in Madan Mitra Lane. In 1909 (fascicle 33), he was joined by a second superintendent and later on manager of the pharmacy, Kavirāj Purnachandra Kaviratna (for this title, see below, p. 244). For the last two fascicles of E7 fully documented by me, fascicles 2 and 3 of the Siddhisthāna (1917 and 1919), we find Kavirāj Surendranath Kaviratna in this dual position.

[197] See E6 (1888ff.). Contrary to Singh and Joshi (1982, 314), the *Āyurvedadīpikā* was not edited together with the *Jalpakalpataru* in the *editio princeps* by Gangadhar Ray.

[198] See below, n. 406.

'The Karvetnagar Mansion of Āyurveda' (Karvetnagar Āyurveda Mandira) in Karvetnagar, Madras, where the first edition of the *Carakasaṃhitā* in Telugu script was published, may also have been a pharmacy.[199] Finally, Jethalal Devshankar Dave, the translator of the *Carakasaṃhitā* into Gujarati and author of a Gujarati commentary on it, who presumably also edited the basic text, calls himself the publisher of this book and further mentions that he is the proprietor of The Dave Chemical Works and Dave Ayurvedic Pharmacy in Ahmedabad (Gujarat). The pharmacy can thus be considered the publishing house.[200] In all of the above cases,[201] we can assume that the pharmacies were places from where the editors and publishers disseminated the books to other physician–scholars and general interested readers.

Furthermore, institutions devoted to the study and practice of Āyurveda, such as the already mentioned Shree Gulabkunverba Ayurvedic Society in Jamnagar (Gujarat), acted as publishing houses.[202] For the more recent period, the Shrisvamilakshmiram Trust in Jaipur has to be mentioned again.[203] Publishing houses in the modern sense dominate the scene after independence,[204] and some editions now appear in well-known general Sanskrit book series, such as the *Kashi Sanskrit Series*,[205] the *Chowkhamba Sanskrit Studies*[206] or the *Haridas Sanskrit Series*,[207] or in special series devoted to Sanskrit medical works, such as the *Vidyabhavan Ayurved Series*,[208] the *Chaukhamba Āyurvijñān Granthamālā*,[209] the *Kashi Ayurveda Series*,[210] and the *Śrīsvāmīlakṣmīrāmanidhigranthamālā*.[211] The *Purātanavaidyakagranthasaṃgraha*, the 'Collection of Books on Ancient Medicine,' of which at least nine numbers were published with an edition of the

199 See E19 (1914, 1918–1920). The two known volumes cover the text of Sūtrasthāna only.

200 See E20 (1916, 1931).

201 See also the case of the pharmaceutical firm C. K. Sen and Company which acted as the publishing house for E26 (1927–1933); see below, p. 240f.

202 See E37 (1949) and above, p. 206.

203 See above, p. 223ff., and n. 158. See also the general mention in Sharma (1975, 586–587; 2016, 614). Other regional publishing institutions after independence are the Sastum Sahitya Vardhak Karyalaya in Ahmedabad (E39, 1958–1963) and the Dharma Grantha Store in Cuttack (E51, 2000).

204 See E38 (Arya Sahitya Mandal Ltd., Ajmer), E40 (1961–1962 and numerous further editions) (Chowkhamba Vidyabhavan, Varanasi), E41 (1969–1970, [2]1983) and E42 (1976–2002) (Chowkhamba Sanskrit Series Office, Varanasi), E26 ([2]1984–1985, reprint 1991), E44 (1981–1994) and E46 (1983) (Chowkhamba Orientalia, Vanarasi), E47 (1983, 1988; Chaukhamba Surabharati Prakashan, Varanasi), E48 (1984–1989) (Nabapatra Prakashan, Kolkata) and E56 (2011) (Les Belles Lettres, Paris).

205 See E41 (1969–1970, [2]1983).

206 See E42 (1976–2002).

207 See E34 (1937–1938).

208 See E40 (1961–1962 and numerous further editions).

209 See E47 (1983, 1988).

210 See E26 (reprint 1991).

211 See E45 (1982) and E49 (*1990).

Sūtrasthāna of the *Carakasaṃhitā* and a partial translation of the *Suśrutasaṃhitā*, may also have been intended as a series.[212]

Until independence a large number of printing presses in Calcutta, Bombay, Pune and other places were involved in the printing of complete *Carakasaṃhitā* editions or individual fascicles. Especially in the early period, many of them were small presses with picturesque and programmatic names. For instance, in Calcutta, where in the middle of the nineteenth century a veritable explosion of book production and the establishment of new indigenous printing presses occurred,[213] we find 'The Jewel Mine of Knowledge Gained through Colloquies Press' (Saṃvādajñānaratnākara Press) which printed the *editio princeps* of the *Carakasaṃhitā* in 1868.[214] The distinctive name of the press refers to the technical name of a traditional type of scholarly debate in ancient India called 'colloquy' (*saṃvāda*), a friendly and honest discussion with the aim to jointly discover the truth and protect its adequate knowledge. This is reminiscent of the unusual title of Gangadhar Ray's Sanskrit commentary on the *Carakasaṃhitā*, namely, 'The Wishing Tree of Scholarly Controversy' (*Jalpakalpataru*), which is here printed along with the text of the *Carakasaṃhitā*. In this title, Gangadhar Ray refers to a further type of traditional debate called *jalpa*, which is a verbal contest in the form of contentious scholarly debate where two opponents take a position and counter-position and strive to prove their own position and disprove that of the opponent.[215] The Saṃvādajñānaratnākara Press was owned by Bhubanchandra Basak who also acted as publisher.[216] The back cover of a fascicle printed in 1871, which is pre-

[212] See E2 (1876ff.). The call number and related information on a copy of E2 held by the (American) Surgeon General's Office Library, no. 91641 (Med. Hist., WZ, 290, C6965, nos. 1–9, 1877) suggests that there were at least nine numbers published. This could be confirmed in the online catalogue of this library published by the U.S. National Library of Medicine. I am grateful to Kenneth Zysk, Copenhagen, for providing me with a photocopy of his own photocopy of this rare book.

[213] Darnton speaks about a turning point in the production of another segment of the Indian book market, namely, books in the Bengali language, in 1852, with fifty new books per year. In 1857, the year of James Long's survey of printed Bengali books, forty-six presses were working in Calcutta, producing 322 new books in just six months; see Darnton (2002, 241). Ghosh (2006, 121) reports that according to the 1881 *Census of the Town of Calcutta and Its Suburbs*, over three thousand printers and more than a thousand bookbinders were operating in Calcutta and its suburbs in this year. According to Ogborn (2007, 268), in the Battala area of North Calcutta, where cheap and popular literature in Bengali was printed, 'hundreds of tiny printshops lined the narrow lanes' in the middle of the eighteenth century.

[214] See E1 and above, p. 222.

[215] For a discussion of the character of these two types of debate in the basic context of the *Nyāyasūtra*, a foundational philosophical work with which Gangadhar Ray was very familiar, see, e.g., Preisendanz (2000, 233–237).

[216] According to Ghosh, referring to Graham Shaw's detailed study of printing in the second half of the nineteenth century in Dacca, a significant number of Basaks, members of a weaver subcaste (see Datta 1981, 145), became involved in the new technology and book trade in this city because the import of machine-made goods deprived them of their traditional livelihood; see Ghosh (2006, 123). In eighteenth-century Calcutta, there was also an extended Basak family of influential land-

served in the copy of the *editio princeps* in the library of the Societé Asiatique, Paris,[217] confirms this. There and on the flyleaf of this copy printed in 1870, we also see the precise address of Bhubanchandra Basak's press, namely, 8 Nimtala Ghat Street, which is in Ahiritola, Jorabagan.[218] From the slightly unclear wording on the title page of the Devanagari script version of the *editio princeps* it seems that the editor had some personal relationship with the publisher and printer who may have requested him to prepare the *Carakasaṃhitā* together with his commentary for printing.[219] However, details could not yet be asserted and thus the question remains open whether Gangadhar Ray was somehow involved in the founding or naming of the press.[220]

Other similarly named presses are 'The Jewel of Knowledge Press' (Vidyāratna Press)[221] and 'The Rise of Knowledge Press' (Vidyodaya Press)[222] in Calcutta, 'The Press that Dispels Mistaken Judgements' (Pramādabhañjana Press)[223] in Saidabad, 'The Pleasure of Knowledge Press' (Vidyāvilāsa Press)[224] in Benares, and in

---

owners and traders (see Banerjee 1989, 19 and Robb 2014, 197). Further research is required to determine whether Bhubanchandra Basak may have been a member of this family. On Bhubanchandra Basak's role as printer–publisher of books in Sanskrit belonging to various genres, including modern commentaries written in Sanskrit, see Pecchia (2022a, 116–117). For a list of books published by his press, which started to function in 1861 (Pecchia 2022a, 134, n. 63, with reference to Stark's Chapakhana website), during the period from 1868 to 1893, see Pecchia (2022a, Appendix III)..

[217] See no. 788 of the collection of Jean Filliozat.

[218] Both the name of Bhubanchandra Basak, as founder of the press, and this address are also seen on the last page of this fascicle above a list of other books in Sanskrit and Hindi published by this press in Devanagari letters.

[219] See the title pages of the copies 279.6.H.14 (India Office Library) of the British Library and no. 788 of the collection of Jean Filliozat at the Societé Asiatique which cover Sū 1.1 to 18.18ab: … *kālikātāyām / śrīyuta bhuvanacandravasākamahodayasya prārthanayā / taddvāraiva saṃvāda-jñānaratnākarākhyayantre / 1925 saṃvatsare / mudritārabdham /*. See also Pecchia (2022a, 119) and her more general remarks on the preparation of Sanskrit medical books at the request of publishers, and on a dual function of their publishing activity (ibid.).

[220] It is striking that Anna Moreshvar Kunte, the editor of E2 (1876ff.), refers first to Jivanand Bhattacharya whose first edition of the *Carakasaṃhitā* appeared in 1877 (E3) and then only to Bhubanchandra, and not to Gangadhar Ray, as the first among 'the eminences residing in Bengal devoted to the good of their country who are currently making an effort to print the compendium of Āyurveda (i.e., the *Carakasaṃhitā*)'; see Vijñapti before fascicles 1–5 (covering Sū 1–30) printed by the Ganpat Krishnaji Press in Bombay in 1877 (Library of Congress call number SanA1198 and private collection of Kenneth Zysk, Copenhagen, see above, n. 212): *svadeśahitaratānāṃ vaṃgadeśasthānāṃ śrīmajjīvānandabhuvanacandrādimahodayānām āyurvedasaṃhitāmudrāṃkane prayatno vartate.*

[221] This printing press was located in Chitpur Road, the 'Pater Noster Row' of Calcutta in the middle of the nineteenth century; see Darnton (2002, 244). On Chitpur Road, now Rabindra Sarani, an ancient arterial road used by pilgrims on their way from the north to the Kalighat temple in the south, see also Dutta (2003, 47–52).

[222] See E1 (1868ff.), E5 (1883–1884) and E21 (1920–1930).

[223] See E1 (1878, 1884) and above, p. 231.

[224] See E34 (1937–1938).

Bombay 'The Friend of Knowledge Press' (Jñānamitra Press).[225] Another printing press bears the name of Sarasvatī, the Goddess of Learning,[226] which was owned by the editor and publisher himself, Jivanand Bhattacharya,[227] who printed a large number of important Sanskrit texts with it, many of them first editions.[228] These names reflect the great expectations held by the founders and owners of the presses in terms of an increased wide spread of knowledge through the dissemination of printed books, as opposed to the more restricted traditional transmission of knowledge by means of oral instruction and manuscripts. Other designations show that some presses were obviously devoted to printing medical books,[229] such as the Dhanvantari Machine Press and the Dhanvantari Steam Machine Press, later on the Dhanvantari Electric Machine Press, in Calcutta.[230] The names refer to Dhanvantari, the divine physician of the gods, who according to tradition assumed the human form of Divodāsa, King of Kāśī (Benares), and promulgated the medical knowledge transmitted in the *Suśrutasaṃhitā*.[231] The Dhanvantari Press was located in the small Kalutola ward of Calcutta, close to the Calcutta Medical College (then Medical College, Bengal) and its hospital, and to several other hospitals; the area – prior to the 1880s a large slum – was also known for its pharmaceutical production.[232] Another example for a typical name of a printing press that indicates its specialization is 'The Medicine Steam Machine Press' (Bhaiṣajya Steam Machine Press) which was located in Raja Naba Krishna Street in North Calcutta.[233] In Bombay, we find the names of famous general Sanskrit presses that printed and mostly also published the *Carakasaṃhitā*, such as the Nirṇayasāgara Press, which has six different editions to its credit, three of them with further editions,[234] the

[225] See E2 (1876ff.).

[226] See E3 (1877).

[227] See below, p. 250ff.

[228] See Roth (1894a, 138). The second edition of E3 (1896) was printed by the Nārāyaṇa Press where also fascicles 27 to 29 (1905) of E7 were printed by Benimadhab Ghosh, who may have been the owner or the printer entrusted with the task.

[229] Pecchia (2022a, 117) refers to an obviously specialized Ayurved Press in Kolkata that was active already in 1878.

[230] See E10 (1897) and E26 (1927–1933). See also the Dhanvantari Printing Press in Etakkad, Kerala (E43; 1979).

[231] See SS Sū 1. For the myth of Dhanvantari's birth, see below, p. 257, for a late-nineteenth century drawing of Dhanvantari in the form of a human physician, see below, p. 278.

[232] See Banerjee (2016, 48) and Gupta (1976, 374).

[233] See E16 (1904). The presumed editor and translator into Bengali, Satishchandra Sharma, lived in the same street (no. 4), just a few houses away from the printing press (no. 25); he gives his address in an advertisement for a medical preparation called *someśvar rasāyan* placed on the outer back cover of the book.

[234] See E13 (1900), actually published by its (presumed) editor Dattaram Narayan Pathak already mentioned above (p. 212), E15 (1903?), E18 (1911), published by the editor Vrajvallabh Sharma, E22 (1922, $^2$1935, $^3$1941), E23 (*date unknown, $^2$1922, $^3$1933) and E31 (*1933, $^2$1935, $^3$1941). See also the general reference to this publishing house and press in Sharma (1975, 587; 2016, 614) and Atridev Vidyālaṅkār (1976, 654), and below, p. 261f.

Śrīveṅkaṭeśvara Steam Press[235] and the related Lakṣmīveṅkaṭeśvara Steam Press.[236] It is significant that different from the many religious works published by the Śrīveṅkaṭeśvara Press, the scientific work *Carakasaṃhitā* was not published in the traditional *pothī* format, a narrow oblong format reminiscent of the format of palm-leaf manuscripts which is also the usual format of paper manuscripts,[237] but in modern European-style book format. Finally, the edition of the *Carakasaṃhitā*, accompanied by a Hindi commentary, by Kalicharan Sharma and Kshamapati Sharma was published by the well-known Naval Kishore Press in Lucknow.[238] Pandit Kalicharan Sharma was a teacher at the Kenning College, Lucknow, and also translated the sixteenth-century Sanskrit medical work *Bhāvaprakāśa* and an eighteenth-century popular health manual called *Amṛtsāgar* written in Marwari dialect into Hindi for the Naval Kishore Press.[239] He may possibly be identified with the early twentieth-century medical author Kalicharan, son of Ramnidhi, who wrote a treatise on eye diseases published, together with a Hindi translation, in 1904 in Lucknow.[240]

Throughout the publication history of the *Carakasaṃhitā*, the editors were for the most part physician–scholars, not generally versatile traditional Sanskrit scholars or pandits,[241] that is, they were traditionally learned practitioners of Āyurveda.

---

[235] See E11 (1898) and E17 (1911, next edition 1922). See also above, p. 229, on these two printing presses and publishing houses.

[236] See E29 (*1932, *reprint 1989) and E30 (1932). Gangavishnu Shrikrishnadas was a brother of Khemraj Shrikrishnadas. For his untitled Sanskrit preface to E30, he could thus freely borrow from Ramprasad's Hindi introduction to E17, the edition with Ramprasad's Hindi commentary published by Khemraj Shrikrishnadas. Sharma (1975, 586; 2016, 614) refers to their important publishing activity in the field of Āyurveda, and Atridev Vidyālaṅkār (1976, 651) states that both presses were actually owned by Khemraj Shrikrishnadas who published some 300 books on Āyurveda.

[237] See Stark (2009, 190) on this traditional format preferred in the early period of printing religious texts in India and the cultural authority the novel printed books may have derived from it.

[238] See E14 (1900). For a book-length study on this influential North Indian press and publishing house named after its owner Naval Kishore, see Stark (2012). See also Stark (2009) on the printer-cum-publisher Naval Kishore as a modern 'commercial' patron of religious literature. On the medical tracts and manuals in Sanskrit, Hindi and Urdu published by Naval Kishore, see Stark (2012, 406–410).

[239] See Stark (2012, 211, 408–409). In his introduction to the second edition (1904) of the translation of the *Bhāvaprakāśa* (first edition 1894) translated in Stark (2012, 408), Kalicharan Sharma acknowledges the assistance of Vajapeyi Kshamapati Sharma. As can be gathered from the lists provided in the appendices in Stark (2012), his translation activity for the Naval Kishore Press extended far beyond medical literature to the epic *Mahābhārata* (1889) (478) and Kālidāsa's famous poem *Kumārasambhava* (1890) (479). Kalicharan also translated an abridged version of the Urdu *Dāstān-e Amīr Hamzā* (or *Hamzānāma*) into 'easy colloquial Hindi language' (1879) (470, see also 442) and a textbook on geography from Arabic (1865) (461). Furthermore, he authored a textbook on geography (1869) (462) and two textbooks on letter writing (1868) (461).

[240] See Meulenbeld (2000, IIA, 390 and IIB, 407).

[241] The designation 'pandit' (Sanskrit *paṇḍita*), which we also find with some physician–scholars related to editions of the *Carakasaṃhitā*, is not an official title; see Michaels (2001, 11). On the pandit as a traditional scholar and teacher, see Michaels (2001, 3–4, 12–13, with further references) and Aklujkar (2001).

In Bengal, we find many Sen Gupta-s and Sen-s among the editors and other relevant actors involved in editions of the *Carakasaṃhitā*, who belong to the Baidya (Sanskrit *vaidya* 'physician, doctor') community which traditionally practiced medicine. Dharmadas Sen Gupta[242] and Yogindranath Sen[243] are two prominent names. Two pairs of editors who have not yet been mentioned by name, Debendranath Sen Gupta and Upendranath Sen Gupta,[244] and Narendranath Sen Gupta and Balaichandra Sen Gupta,[245] also belonged to this community of closely and remotely related physician–scholars. Prabhakar Chatterjee refers to Debendranath and Upendranath Sen Gupta as sons of the influential physician, scholar and entrepreneur Chandrakishor Sen Gupta from Burdwan (modern Bardhaman) District in Bengal, a younger contemporary of Gangadhar Ray, who had a large practice in Calcutta and ran a famous company (Messrs. C. K. Sen and Co.) that produced Ayurvedic medicines.[246] Having become a very rich man, Chandrakishor Sen Gupta also sponsored the publication of books on Āyurveda, among them the reprint of Gangadhar Ray's *Jalpakalpataru* on the *Carakasaṃhitā*.[247] With 'reprint,' Chatterjee must refer to the edition of the *Carakasaṃhitā* by Narendranath Sen Gupta and Balaichandra Sen Gupta,[248] which was the first edition that contained the *Jalpakalpataru* after the *editio princeps* of the *Carakasaṃhitā* had appeared in 1868ff., with some follow-up editions and a second edition of 1908ff.[249] Narendranath and Balaichandra Sen Gupta mention the great expense involved in the publication of the book twice in their editorial announcement (*vijñāpana*),[250] but they do not refer to a munificent sponsor. However, on the title page of the third volume of this edition, we find mention of C. K. Sen and Company, Ltd., of 29, Kalutola Street, Calcutta, at a place where one would expect information about the publishing house.[251] It is therefore probable that

---

[242] Translator and most probably editor of E4 (see above, p. 207).

[243] Commentator and editor of E21 (see above, p. 214).

[244] See E10 (1897).

[245] See E26 (1927, 1928, 1933).

[246] Atridev Vidyālaṅkār (1976, 678) mentions it among the most well-known Ayurvedic pharmacies in India. The company still exists today, located on Chittaranjan Avenue in the Kalutola ward of Calcutta (see above, p. 238).

[247] See Chatterjee (1959d, 812–814).

[248] E26 (1927, 1928, 1933).

[249] E1 (1868ff., 1878, 1884, ²1908ff.).

[250] See E26, Vijñāpanam, 1: *saty apy atra nitarāṃ vyayabāhulye … vayam atra dhanaiḥ prāṇaiś ca prayatāmahe*. Such reference to financial commitment on the part of the publisher is also frequently found on the title pages of the translations of religious texts published by the Naval Kishore Press, Lucknow; see Stark (2009, 196).

[251] See the information on the title page of the copy owned by the library of the Bhandarkar Oriental Research Institute, Pune (call number 9890): si, ke, sen eṇḍ kompānī liḥ / 29 naṃ kaluṭolā ṣṭrīṭ, kalikātā. The book appeared on the full moon day of the month of Jyeṣṭha in Śaka 1855, i.e., in June 1933. See also the short book notice on the completion of the edition by Müller (1934), where C. K. Sen & Co. Ld. is mentioned as the publishing house. At the same time, Müller identifies the two editors as publishers ('… von den beiden Herausgebern und Verlegern, *Narendra Nath Sen* und *Balai Chandra Sen* …').

also the first two volumes were financed by C. K. Sen.[252] According to Brahmanand Gupta,[253] Chandrakishor Sen Gupta opened a dispensary for affordable Ayurvedic medicines in 1878 which is said to have been the first in Calcutta.[254] Large-scale production began in 1898 after he had moved the successful dispensary to Kalutola.[255] At the time when Narendranath and Balaichandra Sen Gupta wrote their editorial announcement, he could still have been alive, but would have been at least around seventy years of age.

Furthermore, the two editor-cum-publishers also express their great joy about the outstanding effort their brothers made towards this edition.[256] This may be a reference to Debendranath and Upendranath Sen Gupta, the editors of an edition of the *Carakasaṃhitā* published thirty years earlier[257] and the sons of Chandrakishor Sen Gupta mentioned by Chatterjee. Furthermore, it is striking that both editions were printed by the same press, the Dhanvantari (Steam/Electric) Machine Press in Calcutta,[258] a coincidence that may be considered another link between the two pairs of editors. Furthermore, both pairs of editors express their gratitude to Pandit Chandrashekhar: Debendranath and Upendranath Sen Gupta refer to him as an excellent good friend to whom they are devoted, a teacher of poetry (*kāvyatīrtha*)[259] and skilled in poetry (*kāvyacuñcu*),[260] and a 'jewel of a physician–scholar' (*kaviratna*).[261] Narendranath and Balaichandra Sen Gupta, for their part, express their highest respect for him when they acknowledge the great effort and labor he invested in the editing and revising of the *Carakasaṃhitā*. They identify him as their teacher at the Ayurvedic College, call him 'a lion roaming the forest of the science

[252] In his short book notice on the first two volumes of this edition, Müller (1930b, 229) supplies the information about C. K. Sen and Co. Ld. in brackets after the title.

[253] See Gupta (1976, 374).

[254] See Arnold (2000, 178).

[255] On this Calcutta ward, see above, p. 238.

[256] See E26, Vijñāpanam, 2: *saṃskaraṇe 'sminn asmadbhrātāro 'py atīva pariśramaṃ kṛtavantaḥ. tena vayam ānandasandohaṃ labhāmahe.*

[257] See E10 (1897).

[258] See above, p. 238.

[259] See below, p. 246, for the title *tīrtha*. In the following, all titles and degrees will be given in their Sanskrit form, with diacritics, except for the case of the very common title *kavirāja* for which the Bengali form Kavirāj is used.

[260] Sinha (1993, 145) refers to *cuñcu* as a traditional title, as in *nyāyacuñcu* ('One Skilled in Logic'). See also the pandit Nandkumar Nyāyacuñcu, mentioned by Hatcher (2005, 703). Dual competence in poetry and medicine is also displayed by the titles of the deceased Kavirāj Kunjbihari Kāvyatīrtha Dhanvantari (see above, p. 238) to whom Debendranath and Upendranath Sen Gupta refer with gratitude in their Vijñāpanam (p. 2).

[261] See below, p. 244, for this title. See E10, Vijñāpanam, 2: *suhṛdvarabhaktibhājanaśrīyuktapaṇḍitacandraśekharakāvyatīrthakāvyacuñcukaviratnena ... atra mahāñ chramaḥ kṛtaḥ. atas teṣām api sakāśe cirakṛtajñāḥ smaḥ.*

of Āyurveda'[262] and add the titles *kavirāja* and *kaviratna* to his name.[263] The stylistic parallels in the ornate Sanskrit prose of the two editorial announcements are also striking.[264] However, such duplication would be admissible if the later editors were reproducing a text written by close relatives. It can therefore be conjectured that Narendranath and Balaichandra Sen Gupta were related to Debendranath and Upendranath Sen Gupta, the sons of Chandrakishor Sen Gupta, who may have been their elder brothers and in charge of the company of their aged father.[265] Debendranath and Upendranath would have been of roughly the same age as their friend and colleague Chandrashekhar who became the teacher of their younger brothers.

The often multiple traditional titles and honorifics that adorn the names of a large number of editors as well as translators, commentators and publishers clearly identify them as physician–scholars. These epithets, which are sometimes combined with modern titles, provide the editions, translations, commentaries and books as a whole with prestige and authority. The title *kavirāja* (literally: 'King of/among the Wise'),[266] which may be prefixed or affixed to the personal name, is special to Bengal and was awarded to excellent physicians after an examination.[267] It is borne by the editor of the *editio princeps*, Gangadhar Ray, who appears as Gangadhar Kavirāj Kaviratna and Vaidya Gangadhar Ray Kaviratna Kavirāj, by his son and grandson who were his early editors and publishers, Dharanidhar Ray Kavirāj and Kavirāj Tryambakeshvar Ray, and by others who were involved in its second edition, namely, Mahāmahopādhyāya Kavirāj Dwarkanath Sen Kaviratna, Kavirāj Rajendra Narayan

[262] This may refer to the traditional title 'Lion' (in a particular area of knowledge) (see below, n. 317) or simply be a literary trope.

[263] See E26, Vijñāpanam, 1–2: *upasaṃhārakāle vaktavyam – asyāś carakasaṃhitāyāḥ saṃskaraṇe saṃśodhane cāsmadāyurvedavidyālayādhyāpakenāyurvedaśāstrāṭavīsaṃcaraṇapañcānanena paramapūjyapādaśrīyuktakavirājacandraśekharakaviratnenāsādhāraṇo yatnaḥ śramaś ca kṛtaḥ. ato vayaṃ tatsamīpe cirakṛtajñatāpāśena baddhāḥ smaḥ.*

[264] The first part of the acknowledgements in the Vijñāpanam of E10, which relates to Prasannakumar Bandyopadhyaya who was the editors' teacher at the Ayurvedic College, has clearly been copied by the editors of E26 with some minor changes and adjusted to refer to their own teacher (see the previous note). Cf. E10, Vijñāpanam, 2: *upasaṃhārasamaye vaktavyam – asyāś carakasaṃhitāyāḥ saṃskaraṇe viśodhane cāsmadāyurvedavidyālayādhyāpakenāyurvedaviśāradena vidyāratnopādhikaparamapūjyapādaśrīyuktaprasannakumāravandyopādhyāyenāsādhāraṇo yatnaḥ śramaś ca kṛtaḥ. ato vayaṃ tatsamīpe cirakṛtajñatāpāśena baddhāḥ smaḥ.* The acknowledgement of the editors' indebtedness to Chandrashekhar in the Vijñāpanam of E10 (see above, n. 261) was also copied in the Vijñāpanam of E26 and adjusted to refer to the two physician–scholars Kunjbihari and Kartikchandra Gupta. See E26, Vijñāpanam, 2: *āyurvedaśailīkuśalena kavirājaśrīyuktakuñjavihārikāvyatīrthadhanvantariṇā tathā kavirājavariṣṭhaśrīkārtikacandraguptenātra mahāñ chramaḥ kṛtaḥ. atas tayor api sakāśe cirakṛtajñāḥ smaḥ.*

[265] Chatterjee (1959d, 813) mentions that Chandrakishor Sen's sons Upendranath and Debendranath greatly assisted him in the propagation of the medicines produced by his company. They were thus involved in the family business already at a time when their father was still in charge.

[266] The dominant meaning of the word *kavi* in Sanskrit is 'poet,' but in the present context *kavi* clearly refers to a wise man or thinker which is its primary referent. On the traditional *kavirāja*-s of Bengal, their way of practice and training, see Gupta (1976, 368–369).

[267] See Liétard (1903, 91) and especially Cordier (1899, 556–557 = Roşu 1989, 428–429). On the place of the *Carakasaṃhitā* in the curriculum of the Kavirāj-s see below, n. 476.

Sen Kaviratna and Kavirāj Yogindranath Sen Vidyābhūṣaṇa,[268] later on called Vaidyaratna Kavirāj Pandit Yogindranath Sen Vidyābhūṣaṇa, M. A.[269] Many of the editors, translators, commentators and publishers referred to above were *kavirāja*-s: Dharmadas Sen Gupta Kavirāj,[270] Avinashchandra Kaviratna Kavirāj or Kavirāj Avinashchandra Kaviratna,[271] Kavirāj Harinath Viśārada,[272] Kavirāj Yashodanandan Sarkar,[273] Kavirāj Debendranath Sen Gupta and Kavirāj Upendranath Sen Gupta,[274] Kavirāj Satishchandra Sharma Kavibhūṣaṇa,[275] Kavirāj Narendranath Sen Gupta and Kavirāj Balaichandra Sen Gupta,[276] and Āyurvedacakravartī Kavirāj Narendranath Śāstrī.[277] We also encounter the title with other involved actors, such

---

[268] See E1 and its further editions.

[269] See the Devanagari title page of E21 and the post-positioned chapter headings of the *Carakopaskāra* in E45. On the title *vidyābhūṣaṇa*, see n. 285 below.

[270] E 4, see above, p. 207.

[271] E5 and E6, see above, p. 208f.

[272] E7, see above, p. 233.

[273] E8, see above, p. 208.

[274] E10, see above, p. 241.

[275] E16, see above, pp. 208 and 232.

[276] E26, see above, p. 240.

[277] E27, see above, p. 214. Others are Kavirāj Atridev Gupta Vidyālaṅkāra Bhiṣagratna, editor of E35 (1937) and Hindi translator, and Kavirāj Brajendrachandra Nag, editor of E48 (1984–1989) and Bengali commentator. E35 is listed by Meulenbeld (1999, IB, 5, siglum *aa) as containing Sūtrasthāna, Nidānasthāna and Vimānasthāna, but was not seen by him. Only the third part of the book in the collection of the India Office Library, British Library (call number IO San.B.1542), could be seen by me. It covers Cikitsāsthāna 16 to 30, Kalpasthāna and Siddhisthāna and is dated 1937 (Samvat 1994), According to the card catalogue of the British Library, the date is 1938, which is actually when the book was received, and there is no mention of the incompleteness of the edition. As there is no introduction to the third part, further details on the identity of the editor and translator could not be gathered. Atridev Gupta is mentioned several times by Priyavrat Sharma as a medical author of the period between 1935 and 1945 (1975, 554; 2016, 580) and as the author of several books published by the Hindi Samiti in Lucknow (1975, 442; 2016, 462: *Bhaiṣajyasaṃhitā*) and the Hindi Sahitya Sammelan in Allahabad (1975, 444; 2016, 464: *Bhaiṣajyakalpanā*, 1975, 471; 2016, 493: *Rasaśāstra*). He also refers to Atridev Vidyālaṅkar as a Hindi commentator on the *Carakasaṃhitā*. Atridev Vidyālaṅkār, the author of the *Āyurved kā bṛhat itihās* ([2]1976), when speaking about himself in this history of Āyurveda, mentions that he wrote about thirty books, *inter alia* on the preparation of medicines (*bhaiṣajyakalpanā*) and including Hindi translations of the *Carakasaṃhitā* and other classical foundational compendia (1976, 666–667). Furthermore, he gives his own name, after Jaydev Vidyālaṅkār (see above, p. 214), as a student of Dharanidhar, student of Shyamadas (see above, n. 97), in the genealogical tree of Gangadhar Ray's school (613). Meulenbeld, for his part (2000, 791), lists the book *Rasaśāstra* (Hindi Sahitya Sammelan, Prayāg [= Allahabad] 1961) and a history of Āyurveda (*Āyurved kā itihās*) (1978, by the same publisher) as publications of Atridev Vidyālaṅkār, next to his *Āyurved kā bṛhat itihās* which was first published in 1960. It can therefore be safely assumed that the two authors are the same and that Atridev sometimes published under his surname Gupta, which in Bengal is actually the name of a community belonging to the Baidya (*vaidya*) community and related to the Sen Gupta-s (see above, p. 240), and sometimes with the title *vidyālaṅkāra* added to it or merely to his personal name. On the frequent blurring of the distinction between Hindi translation and commentary, which may account for the conflicting information about Atridev's authorship of a commentary on the *Carakasaṃhitā* on the one hand, and of a translation of the *Carakasaṃhitā* on the other, see above, n. 59.

as Kavirāj Gopalchandra Ray, Kavirāj Purnachandra Kaviratna (Bhiṣagratna Sāṃkhyabhūṣaṇa Śāstrī) and Kavirāj Surendranath Kaviratna, the managers and superintendents of the Āyurvedaratnākara Auṣadhālaya in Calcutta where many fascicles of Kavirāj Harinath Viśārada's edition of the *Carakasaṃhitā* with the *Āyurvedadīpikā* were published.[278]

As can be seen from the above examples, additional titles that identify the editors, etc., as physician–scholars are *kaviratna* and *vaidyaratna*, 'Jewel of a Physician / among Physicians.' The latter title also adorns the name of Ramprakash, the above-mentioned editor of previously unpublished parts of Yogindranath Sen's *Carakopaskāra* (Vaidyaratna Ramprakash Svami, M. A., Bhiṣagācārya).[279] This title was awarded by the government in recognition of the public service rendered by well-known physicians. According to Sharma, Yogindranath Sen's father Dwarkanath Sen[280] was the first physician who was honored in this way.[281] The practically synonymous designation *bhiṣagratna* ('Jewel of a Healer / among Healers') can be noted with Kavirāj Atridev Gupta Vidyālaṅkāra Bhiṣagratna[282] and much later with the name of Pandit Brahmashankar Mishra Bhiṣagratna.[283] Vaman Kesheo Datar, who produced an edition of the *Carakasaṃhitā* with the complete text of Cakrapāṇidatta's *Āyurvedadīpikā* which is still very popular in Maharashtra,[284] was awarded the title 'Ornament of Physicians' (*vaidyabhūṣaṇa*),[285] and the similar title 'Ornament of Āyurveda' (*āyurvedabhūṣaṇa*) to Śrīpāda Krishnamurti Śāstrī, who edited the Cikitsāsthāna of the *Carakasaṃhitā* together with his Telugu

[278] E7, see above, p. 233, and Appendix 6.2 below.

[279] E45; see above, p. 224.

[280] See above, p. 214.

[281] See Sharma (1975, 620; 2016, 650).

[282] Translator and presumably editor of E35 (1937); see above, n. 277.

[283] He was a member of the editorial board of E40 (1961, 1962). The specialized titles *kaviratna*, *vaidyaratna* and *bhiṣagratna* are paralleled by similar titles in other fields of knowledge, such as *nyāyaratna* ('Jewel of Logic'), *tarkaratna* ('Jewel of Reasoning') or *smṛtiratna* ('Jewel of Traditional Law'), and were awarded after years of specialized study; see, e.g., Sinha (1993, 17, 50, 155). Most professors at the Calcutta Sanskrit College were 'jewels' in their respective field of specialization (118–119). There was also the general title *vidyāratna* ('Jewel of Knowledge') given to outstanding scholars of many disciplines (76). In the nineteenth century, the title was awarded by the Calcutta Sanskrit College to students who were successful in all their examinations in various fields of learning (143).

[284] See E22 (1922, $^{2}$195, $^{3}$1941).

[285] This title is awarded as a prestigious degree in Āyurveda by some contemporary Indian institutions of learning. However, the significance of the title *vaidyabhūṣaṇa* in Vaman Kesheo Datar's time and place still needs to be ascertained.

translation.[286] The title *mahāmahopādhyāya* held by Dwarkanath Sen was the highest title awarded by the government to exceptional Sanskrit scholars in general.[287]

With the institutionalization of the teaching of Āyurveda at Sanskrit colleges and universities and at other institutions of higher Sanskrit learning, we also find the title 'Master of Āyurveda' (*āyurvedācārya*). The specialized degree of *ācārya*, which roughly corresponds to an M. A. degree, is awarded by Sanskrit universities and other institutions of Sanskrit education with post-graduate courses.[288] At the same time, the word is used, with or without specification, for the general title of 'professor' at a Sanskrit college, university or other institution of higher Sanskrit education.[289] Examples of *āyurvedācārya*-s among the actors involved in editions of the *Carakasaṃhitā* are Āyurvedācārya Pandit Shivsharma,[290] Āyurvedācārya Jaydev, who bore the additional title *vidyālaṃkāra* ('Adornment of Knowledge'),[291]

---

[286] See E25. On Śrīpāda Krishnamurti (1894–1953) see Sharma Ganti and A. V. K. Lakshmi Prasanna (2019) who state, without providing further publication details, that this (edition with) translation constituted the third and last of three volumes, with volume one covering the Sūtrasthāna and volume two comprising the translation of Nidānasthāna to Indriyasthāna, and Kalpasthāna to Siddhisthāna. The full name, Āyurvedabhūṣaṇa Bhiṣagvara Śrīpāda Krishnamurti Śāstrī, also contains another title, *bhiṣagvara* ('Excellent Healer'). Both titles were awarded to Śrīpāda Krishnamurti even before graduation from Madras Ayurveda College, after the respective examinations conducted by the Nikhila Bhāratīya Āyurveda Pariṣad. See further Bhiṣagvara Haripad Sen Gupta Śāstrī who is given thanks for his great effort in the Vijñāpanam (p. 2) of E10 (1897). According to Sinha (1993, 131), in the nineteenth century the more general title *vidyābhūṣaṇa* ('Ornament of Knowledge') was awarded by the Calcutta Sanskrit College to outstanding students who gained the first place in the scholarship examinations. Yogindranath Sen may have been one of them; see above, p. 243. The title was also awarded by learned societies; see Sinha (1993, 31) on the case of Mahāmahopādhyāya Satishchandra Ācārya Vidyābhūṣaṇa.

[287] See, e.g., Sinha (1993, 23) on this distinction awarded by the Lieutenant–Governor of Bengal to Rajkrishna Tarkapañcānana. This normally happened at an advanced stage in the career of a scholar; see the cases mentioned in Sinha (1993, 50, 51, 52) and, e.g., the case of Mahāmahopādhyāya Gopinath Kavirāj (Mukhopadhyaya 1990, 2). Sinha (1993, 32) points out the recipient's unusually young age of thirty-six in the case of Mahāmahopādhyāya Satishchandra Ācārya Vidyābhūṣaṇa, and Jha (1992, 51) the age of forty years in the case of Mahāmahopādhyāya Ganganath Jha. Umesh Mishra was forty-eight when he was honored with this title; see Important Dates in the Life of Mahāmahopādhyāya Dr. Umesha Mishra, 2, and Saksena 1968. According to Atridev Vidyālaṅkār (1976, 614), Dwarkanath Sen was in the last year of his life when he received the award at sixty-three. Sharma (1986, 1) states that the title was awarded to him in 1906; however, according to Sharma this would have been three years before his death. See above, n. 90. He was the first physician–scholar who was thus honored by the government (Sharma 1986, 1).

[288] See Michaels (2001, 9 and Figs. 1 and 2) and, e.g., Jha (1992, 57). On the Sanskrit word *ācārya* ('one to be approached'), that is, a teacher, see Scharfe (2002, 90–91). On the traditional teacher in ancient India, see also Mookerji (1951, 200–201).

[289] See Michaels (2001, 7). The word *ācārya* may also simply refer to a teacher at a traditional Sanskrit school (ibid.).

[290] Editor, E29; see above, n. 33.

[291] Editor(?) and commentator, E32; see above, p. 214 and n. 97. Specialized forms of this general title, such as *nyāyālaṅkāra* ('Adornment of Logic') and *tarkālaṅkāra* ('Adornment of Reasoning'), are also frequently seen. Sinha (1993, 215) reports that, e.g., Pandit Chandrakanta Tarkālaṅkāra received this title after he had completed his studies in a number of traditional subjects at various traditional schools (*ṭol*-s) in Nabadwip (Bengal).

Sāhityāyurvedācārya Pandit Taradatt Pant[292] who evidently also held an *ācārya* degree in literature (*sāhitya*), and furthermore a *tīrtha* degree in grammar[293] and a *śāstrī* degree in philosophy,[294] and Āyurvedācārya Pandit Haridatt (Sharma) Śāstrī, the editor of the *Cakarasaṃhitā* with the fragmentary early Sanskrit commentary by Jejjaṭa who was active in Lahore and after partition, from 1955 onwards, in Maharashtra and then Delhi.[295] Vaidyaratna Ramprakash, M. A., Bhiṣagācārya bears the related title *bhiṣagācārya* ('healer–master').[296] The title 'Great Instructor in Āyurveda' (*āyurvedamahopādhyāya*) identifies Shankar Daji Śāstrī Pade / Shankar Śāstrī Daji Pade as a physician.[297] Here too, the general title *mahopādhyāya* ('Great Instructor') is preceded by the name of the special field of study of its bearer. The degree of *mahopādhyāya* roughly corresponds to an M. A. degree.[298] 'Universal Sovereign of Āyurveda' (*āyurvedacakravartin*), as seen with Āyurvedacakravartī Kavirāj Narendranath Śāstrī,[299] on the other hand, is a specialized form of the title *cakravārtin* ('Sovereign') and belongs to the highest level of degrees awarded by traditional institutions of Sanskrit learning.[300] Especially in more recent times, teaching and researching practitioners involved in editing the *Carakasaṃhitā* also attach English titles, such as 'M. A.,' 'Ph. D.' and 'Dr.,' to their names and indicate membership in professional

---

[292] Editor and commentator, E34; see above, p. 215.

[293] Vyākaraṇatīrtha. On the traditional degree of *tīrtha* (lit. 'passage, ford,' metaphorically referring to a preceptor or teacher), which is slightly higher than the *śāstrī* degree, see Michaels (2001, 8, 10 and Fig. 2).

[294] Darśanaśāstrī. The *śāstrī* ('one versed in the sciences') degree is usually the first degree obtained at modern institutions of Sanskrit learning and precedes the *ācārya* degree. It roughly corresponds to the B. A. degree. See Michaels (2001, 9 and Figs. 1 and 2).

[295] Editor, E36; see Sharma (1975, 563–564; 2016, 590) and Atridev Vidyālaṅkār (1976, 622, 626). Haridatt Sharma (for the surname Sharma see the signature at the end of the introduction [*upodghāta*] to E36 [$^{2}$1940–1941]) may be identical with Pandit Haridatt Sharma Śāstrī who greatly assisted Ramprasad in the correction of his Hindi commentary on the *Carakasaṃhitā*; see E17 (1911, $^{2}$1922), Bhūmikā, 8. In this case, he would still have been a young man when he rendered this service to Ramprasad and at least in his sixties when came to Maharashtra from Lahore. On Ramprasad, see above, p. 204f. Priyavrat Sharma himself, the most outstanding Indian exponent of Āyurveda in the second half of the twentieth century and a multiple editor, translator and commentator of the *Carakasaṃhitā* (E44, 46 and 49), is also an Āyurvedācārya; on Priyavrat Sharma, see Baldev Upādhyāyă (1983, 672) and Narasimha Murthy (2012). See further the double qualification of Balram Swami, M.A., Nyāyāyurvedācārya, co-editor, E45.

[296] Editor, E45; see above, p. 224f.

[297] Editor, translator and commentator, E9 ($^{2}$1902ff., $^{3}$1914ff., reprint 1926), and editor, E15; see above, p. 209.

[298] See Michaels (2001, 8, Fig. 2). In the orthodox–Brahminical culture of ancient India, an instructor or preceptor (*upādhyāya*) was a narrow specialist who teaches particular subjects without possessing the spiritual authority of an *ācārya*; see Scharfe (2002, 137–138). On their difference in rank and function, see also Mookerji (1951, 200–201, 219, 238) and Chatterjee (1999, 101). On the reverse situation and different functions of the *upādhyāya*/*upajjhāya* and *ācārya* in the Buddhist culture of India, see Mookerji (1951, 402–403, 541–542) and Scharfe (2002, 133–138, 156).

[299] Editor, E27; see above, p. 214.

[300] See Michaels (2001, 8, Fig. 2).

associations, academic functions, positions in the medical system, etc.,[301] which provides the editions with authority in a modern scientific and academic setting. Already Anna Moreshvar Kunte,[302] who does not refer to any traditional or modern titles he may have borne, identified himself as anatomical instructor at Grant College,[303] and on the English title page of his edition of the *Carakasaṃhitā* with his commentary *Carakopaskāra*,[304] Yogindranath Sen not only combines his traditional titles with the European title 'M. A.,' but also presents some of his most important professional functions and positions to the reader (see Fig. 6.15). This suggests that he lived in two or even more worlds, the traditional world of physician–scholars and Sanskrit pandits on the one hand and the modern worlds of professional Ayurvedic associations and institutions, and of academic and educational institutions in general on the other.

Other involved actors added the designation of their profession to their name, that is, *vaidya* ('physician') or *cikitsaka* ('doctor'), presumably to stress that practical clinical experience was part of their qualification as editors, translators or commentators of the *Carakasaṃhitā*. Shankar Daji Shastri Pade (Shankar Shastri Daji Pade)[305] adds the professional designation *cikitsaka* to his name, except for some fascicles of the second edition of his book. From the second edition onwards, he also mentions his title *āyurvedamahopādhyāya*.[306] The designation *vaidya* was used by the famous physician–scholar Yadav Sharma, son of Trivikram (Trikamji) and editor of two editions of the *Carakasaṃhitā*.[307] On the title pages of the first[308] and on the Devanagari title page of the second in its third edition,[309] he only refers to his

[301] See, e.g., Pandit Rajeshvardatt Śāstrī, Āyurvedaśāstrācārya (i.e., 'Master in the Science of Āyurveda'), D. Sc. (Āyurveda), Professor and Head, Dept. of Āyurveda, College of Medical Sciences, Chief Physician, Sir Sunderlal Hospital, Benares Hindu University, co-editor, E40. On this student of Dharmadas Sen Gupta (see above, p. 207f.) and highly respected scholar and physician in the 'Benares tradition' of Gangadhar Ray's school see Sharma (1975, 222, 552, 554, 559–561, 594; 2016, 241, 578, 580, 586–587, 622), Atridev Vidyālaṅkār (1976, 671) and Baldev Upādhyāyă (1983, 667). Further examples are Dr. (Pandit) Gangasahay Pandey, A.M.S., professor and physician, Ayurveda Department, Benares Hindu University, editor, E40 and E41 (see Atridev Vidyālaṅkār 1976, 647 and Baldev Upādhyāyă 1983, 672), and Dr. Brahmanand Tripathi, Sāhityācārya, Āyurvedācārya, M. A., Ph. D., D. Sc. A., editor and commentator, E47.

[302] Editor, E2 (1876ff.); see above, p. 225.

[303] That is, Grant Medical College in Bombay, founded in 1845 to replace the earlier Native Medical Institution; see Arnold (2000, 63). The title page of Bhiṣagācārya Harishastri Paradkar Vaidya's edition of the *Aṣṭāṅgahṛdayasaṃhitā* (see Parāḍakara Vaidya 1939 in the References below) reveals that Anna Moreshvar Kunte was a doctor and had earned the European titles B. A. and M. D. He is said to have been the first M. D. graduate of the College in 1876. Already in the mid-1880s, about one third of the medical students at Grant Medical College were Hindus; see Arnold (2000, 64).

[304] See E21 (1920–1930).

[305] Editor, translator and commentator (E9) and editor (E15).

[306] E9 ($^2$1902ff.), etc.; see above, p. 246.

[307] See E23 and E31.

[308] See E23 ($^2$1922, $^3$1933).

[309] See E31 ($^3$1941).

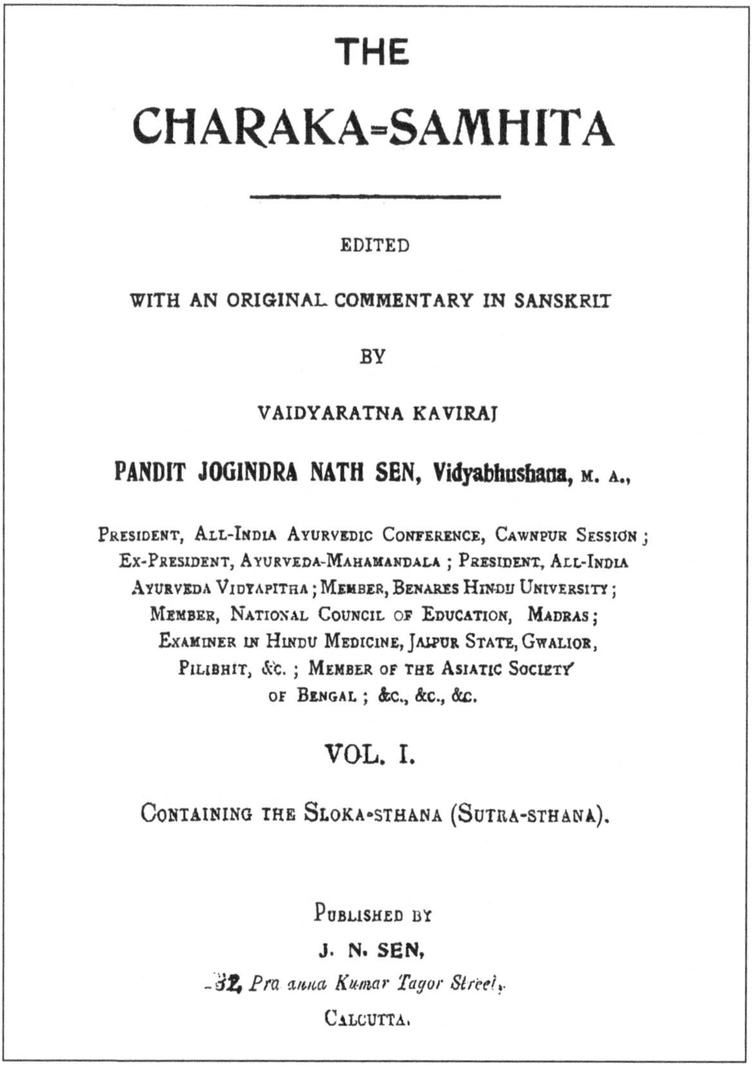
THE

CHARAKA-SAMHITA

EDITED

WITH AN ORIGINAL COMMENTARY IN SANSKRIT

BY

VAIDYARATNA KAVIRAJ

PANDIT JOGINDRA NATH SEN, Vidyabhushana, M. A.,

PRESIDENT, ALL-INDIA AYURVEDIC CONFERENCE, CAWNPUR SESSION; EX-PRESIDENT, AYURVEDA-MAHAMANDALA; PRESIDENT, ALL-INDIA AYURVEDA VIDYAPITHA; MEMBER, BENARES HINDU UNIVERSITY; MEMBER, NATIONAL COUNCIL OF EDUCATION, MADRAS; EXAMINER IN HINDU MEDICINE, JAIPUR STATE, GWALIOR, PILIBHIT, &c.; MEMBER OF THE ASIATIC SOCIETY OF BENGAL; &c., &c., &c.

VOL. I.

CONTAINING THE SLOKA-STHANA (SUTRA-STHANA).

PUBLISHED BY

J. N. SEN,

32, Pra anna Kumar Tagor Street,

CALCUTTA.

**Fig. 6.15** English title page of E21 (1920–1930)

unspecified *ācārya* degree.[310] However, on the Roman title pages in English of the second and third editions of the latter edition,[311] he adds the professional designation *vaidya* to his name. He was probably aware of his fame as a physician–scholar in Western India and thus added this clarification for the sake of non-Indian readers only

[310] See above, p. 245.

[311] See E31 ([2]1935, [3]1941); the first edition of 1933 could not yet be seen by me.

whom he addressed with the English title page.[312] More recently, for instance, Bhagwandash, editor, translator and commentator of the *Carakasaṃhitā*,[313] is presented as 'Vaidya Bhagwan Dash, Deputy Adviser in Ayurveda, Ministry of Health & Family Planning, New Delhi,' and M. Narayanan[314] as 'M. Narayanan Vaidyar, "A" Class Registered Medical Practitioner.' Highest traditional prestige was communicated with the title 'royal physician' (*rājavaidya*) held by K. S. Śāstrī, editor of part of the *Carakasaṃhitā* and author of a Telugu commentary on it,[315] and by Pandit Ramprasad of Patiala, the above-mentioned author of the Hindi commentary *Prasādanī*.[316] On the ornate Sanskrit title page of Kshemraj Shrikrishnadas's edition of the *Carakasaṃhitā* with this commentary (see Fig. 6.16), Ramprasad is also called a savior or restorer of Āyurveda (or 'one who uplifts Āyurveda') (*āyurvedoddhāraka*) and a 'Lion among Physicians' (Vaidyapañcānana),[317] a 'Jewel of a Physician / among Physicians' (Vaidyaratna)[318] and 'Physician–Instructor' (Vaidyopādhyāya).[319]

From the above brief survey of involved actors, it becomes clear that there was a strong professional interest in editing the *Carakasaṃhitā* which may account for the many editions in various shapes. For the early period of publication of the text, when the involved physician–scholars were also publishers of their editions, we can assume that the books were distributed by them at their practices or from the pharmacies, possibly even small bookstores, attached to them.[320] Even though the professional identity of some editors could not yet be identified or identified with certainty,[321] it is evident that 'mere' Sanskrit

---

[312] On Yadav Sharma (born in 1881), a.k.a. Jadavji Trikamji, see Sharma (1975, 405–406, 563; 2016, 425–426, 589–590) and Atridev Vidyālaṅkār (1976, 629–630). See also Vaidya Jethalal Devshankar Dave, editor, translator and commentator of E20 (see above, p. 235)

[313] See E42.

[314] Editor and commentator, E43.

[315] See E19.

[316] See E17 and above, p. 204 with n. 33.

[317] This is a traditional title rarely found with physician–scholars. The specialized title 'Lion' is more often seen with scholars in other traditional areas of knowledge, such as logic (*nyāya*) (Nyāyapañcānana) and reasoning (*tarka*) (Tarkapañcānana). Cf., however, Narendranath and Balaichandra Sen Gupta's designation of Pandit Chandrashekhar as 'a lion roaming the forest of the science of Āyurveda' (above, p. 241f.).

[318] See above, p. 244.

[319] On the title 'Instructor' see above, n. 298.

[320] See above, p. 233ff., for evidence that could be gathered from the title pages of some editions or prefaces.

[321] See, e.g., Pandit Atmanand Sharma Jassar, editor of E24 (see above, p. 215); Śāstrī Girijashankar Mayashankar, editor and Gujarati translator and commentator (E39); Dr. Satyadev Dube, co-editor of E49, and Pandit Purushottam Kar Sharma, editor and Oriya commentator on the first two books (Sūtrasthāna and Nidānasthāna) of the *Carakasaṃhitā* (E51). Pandit Kalicharan Sharma, who edited the *Carakasaṃhitā* with a Hindi commentary together with Pandit Kshamapati Sharma (E14), was a teacher at the Kenning College, Lucknow. He may possibly be identified with the twentieth-century medical author Kalicharan, son of Ramnidhi, who wrote a treatise on eye diseases published in 1904 together with a Hindi translation in Lucknow; see Meulenbeld (2000, IIA, 390 and IIB, 407). Pandit Jaydev Sharma, editor of E38 published in Ajmer (Rajasthan) from 1954

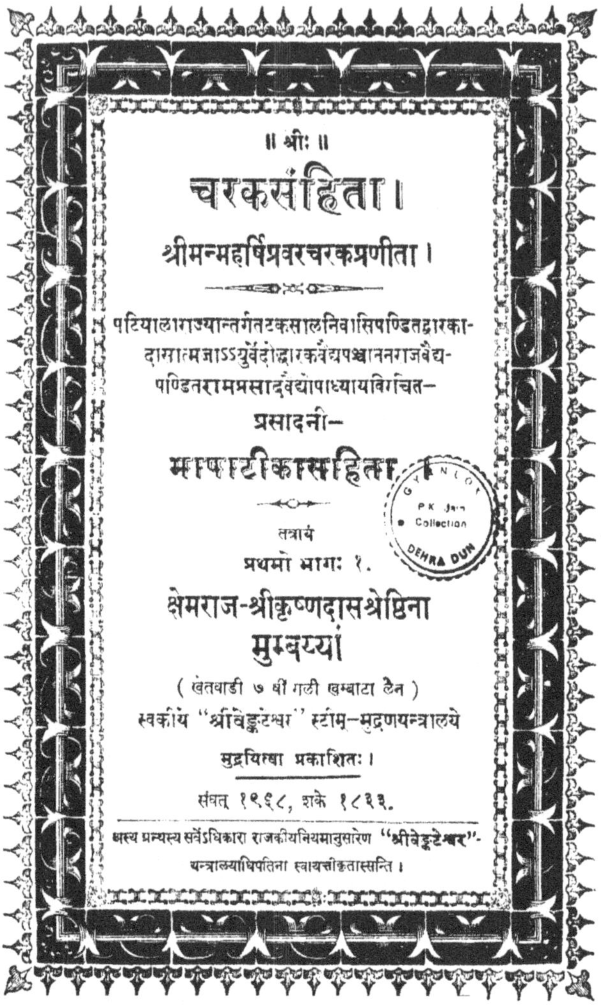
॥ श्रीः ॥

चरकसंहिता।

श्रीमन्महर्षिप्रवरचरकप्रणीता।

पटियालाराज्यान्तर्गतटकसालनिवासिपण्डितद्वारका-
दासात्मजाऽऽयुर्वेदोद्धारकवैद्यपञ्चाननराजवैद्य-
पण्डितरामप्रसादवैद्योपाध्यायविरचित-

प्रसादनी-

भाषाटीकासहिता।

तत्रायं

प्रथमो भागः १.

क्षेमराज-श्रीकृष्णदासश्रेष्ठिना

मुम्बय्यां

( खेतवाडी ७ वीं गली खम्बाटा लैन )

स्वकीयं "श्रीवेङ्कटेश्वर" स्टीम्-मुद्रणयन्त्रालये

मुद्रयित्वा प्रकाशितः।

संवत् १९६८, शके १८३३.

अस्य ग्रन्थस्य सर्वेऽधिकारा राजकीयनियमानुसारेण "श्रीवेङ्कटेश्वर"-
यन्त्रालयाधिपतिना स्वायत्तीकृतास्सन्ति।



**Fig. 6.16** Title page of E17 (1911)

scholars, that is, traditional or modern textual scholars who are not also practitioners of Āyurveda, are a small minority among the editors of the *Carakasaṃhitā*. A very prominent example is Jivanand Bhattacharya, better known as Jivanand Vidyāsāgara Bhattacharya (see Fig. 6.17), with the very prestigious title *vidyāsāgara*

to 1962, may possibly be identified with the physician–scholar Jaydev Sharma Śāstrī, author of the *Siddhabhaiṣajyamañjūṣā* (completed in 1929–1930 and published in 1932), who hailed from Ramgarh in Marwar (*marupradeśa*) (the former princely state of Jodhpur) in Rajasthan. See Meulenbeld (2000, IIA, 401–403).

**Fig. 6.17** Jivanand Bhattacharya, frontispiece of E3 ($^{2}$1896)

('Ocean of Knowledge') added to his personal name,[322] who first edited the *Carakasaṃhitā* on the basis of a manuscript belonging to the Kashmiri recension of the text (K).[323] Jivanand Bhattacharya was superintendent (*kulapati*) of

[322] This title was awarded by the Calcutta Sanskrit College. Another Calcutta-based *vidyāsāgara* was Ishvarchandra Vidyāsāgara Bandyopadhyaya, the famous polymath, social reformer and principal of the Calcutta Sanskrit College; see, e.g., Sinha (1993, especially 152–154), Hatcher (1996) and Hatcher (2005, 703ff.), and, most comprehensively, Bhattacharya (n.d.). He received this title at the early age of 19 after having passed his exams in the area of *smṛti*, i.e., the science of traditional Hindu law, before the Law Committee then headed by Henry Thoby Prinsep, superintendent and remembrancer of legal affairs; see Bhattacharya (n.d., 35).

[323] See E3 (1877). On the two recensions of the *Carakasaṃhitā* (K and E), see Maas (2010a, *passim*, especially 76–79 and the stemma on p. 65; 2010b, 10–14).

the Free Sanskrit College, Calcutta, and a true polymath who edited, published and printed almost two hundred texts in various areas of Sanskrit literature, many of them first editions, and wrote many commentaries in Sanskrit on them.[324] It is interesting to see that Harinath Sarkar, who briefly discusses this edition in the introduction to his own edition of the text together with the *Āyurvedadīpikā*,[325] is highly critical of Jivanand Bhattacharya's edition of the complete *Carakasaṃhitā*. He refers to many errors and left-out parts, a judgment that may be owed to the fact that Jivanand Bhattacharya very closely followed his far from perfect manuscript[326] and thus also reproduced some obviously unacceptable readings. Besides, Harinath Sarkar was most probably not aware of the existence of another recension of the text different from the one familiar to him and other Bengali physician–scholars at the time and seen in the two editions that came out in Calcutta after the one by Jivanand Bhattacharya.[327] Thus, he attributes the deficiency of the latter edition to the fact that its editor is not a scholar–physician, but a Sanskrit scholar (even though he may be the superintendent of a Sanskrit college) and demands that books relating to Āyurveda should be printed by trained and experienced physician–scholars only. This rather harsh criticism may have induced Jivanand Bhattacharya to follow the widely accepted text of the Eastern recension of the *Carakasaṃhitā* (E) in his second edition published in 1896.[328]

From a prosopographical point of view, the survey has also shown numerous family relationships and relationships through traditional teacher–student lineages among editors and other involved actors. Sometimes these relationships are congruent

[324] He was the owner of the Saraswati Press; see above, p. 238. The most recent example for an editor who is not also a physician is Michel Angot, editor, translator and commentator of the Sūtrasthāna and Śārīrasthāna of the *Carakasaṃhitā* (E56).

[325] See E7 (1892ff.), Bhūmikā.

[326] This manuscript could not be identified with any of the manuscripts belonging to the Kashmir recension that were utilized in Preisendanz et al. (forthcoming) for the critical edition of *Carakasaṃhitā* Vimānasthāna Chapter 8. Although there is no preface to this edition, the title pages of both the first and second edition state that Jivanand Bhattacharya edited and published the text 'after he had inspected many manuscripts.' See E3 (1877): … *vi, e, upādhidhāriṇā / Śrī-Jīvānanda Vidyāsāgara Bhaṭṭācāryyeṇa / bahūni ādarśapustakāni samālocya / saṃskṛtā prakāśitā ca /* …. However, especially in the case of the first edition there is no evidence of this and we may thus be dealing with a phrase here that had been taken over from other editions produced by him.

[327] That is, E5 (1883, 1884) and E6 (1888ff.). Both editions, printed alongside a Bengali translation and Cakrapāṇi's commentary, respectively, are briefly discussed by Harinath Sarkar (E7 [1892ff.], Bhūmikā, 6). Interestingly, he extensively discusses Gangadhar Ray's *Jalpakalpataru* (5), but does not mention that Gangadhar Ray had furnished his commentary with a complete edition of the *Carakasaṃhitā* (E1, 1868ff., next editions 1878 and 1884). The fact that the well-informed Harinath Sarkar is silent on Dharmadas Sen Gupta's edition with Bengali translation (E4, 1878??) confirms that its date has to be revised. See the discussion above, p. 207.

[328] It could not be determined whether for this second edition he relied on one or more manuscripts belonging to the Eastern recension or based the constitution of the text on one or more of the other editions that had come out until 1896.

because students were often taught by their fathers and other close relatives inasmuch as medical knowledge was traditionally handed down in the family. Sons and grandsons, disciples and grand-disciples edited, supervised, revised and published further editions of first editions of the *Carakasaṃhitā* by their father and grandfather, teacher and grand-teacher, edited and published their translations of the *Carakasaṃhitā* or their commentaries on it, and produced their own translations and commentaries. The *editio princeps* by Gangadhar Ray is an especially rich example of this and may be taken as the starting point for sketching out further strands of networking among physician–scholars. After a follow-up edition had been published by Gangadhar Ray's son Dharanidhar Ray in 1878,[329] the second edition of 1908[330] was 'thoroughly revised and corrected' by Yogindranath Sen, a grand-disciple of Gangadhar Ray,[331] supervised by Gangadhar Ray's disciples Dwarkanath Sen, Yogindranath Sen's father and his teacher,[332] and Rajendra Narayan Sen,[333] and edited and published by Gangadhar Ray's grandson Tryambakeshvar Ray (see Fig. 6.18).[334] Yogindranath Sen's own edition of the *Carakasaṃhitā* and Sanskrit commentary on it, the *Carakopaskāra*,[335] was first converted from Bengali characters to Devanagari by Narayanprasad Sharma, a student of his who was a student of his friend and fellow-student (under his father Dwarkanath Sen) Lakshmiram. The first two parts of this edition[336] were published and printed by Yogindranath Sen's cousin Jnanendranath Sen.[337] After Yogindranath Sen's death, the manuscript of unpublished parts of this edition with commentary was obtained by Jayramprasad, student and successor of Lakshmiram, from I. K. Sen, probably a relative of Jnanendranath Sen who was in possession of the manuscript after the death of his cousin. Further parts of the *Carakopaskāra* were converted to Devanagari by Narayanprasad Sharma and edited together with Ramprakash, a student of Jayramprasad.[338] Some other, less elaborate family and teacher–student relationships among actors involved in the history of publication of the *Carakasaṃhitā*

---

[329] See above, pp. 222 and 231.

[330] E1 (²1908ff.).

[331] See above, p. 214.

[332] See again above, p. 214.

[333] Different from the information given on the title page of E1 (²1908ff.), the genealogical chart in Gupta (1976, 372) shows Rajendra Narayan Sen as a student of Dwarkanath Sen, not as his fellow-student. Rajendra Narayan Sen is missing in the genealogical trees of Gangadhar Ray's school given by Sharma (1975, 222; 2016, 241) and Atridev Vidyālaṅkār (1976, 613). They do not mention him elsewhere. Meulenbeld (2000, IIA, 466) refers to a certain Rajendra, author of a work called *Aṣṭāṅgaśuddhi*, published in 1949, but could not provide any further details. It is thus not possible to connect the two persons.

[334] See also the brief mention in Pecchia (2022a, 114; 2022b, 800), where, however, Yogindranath Sen is erroneously identified as a student of Gangadhar Ray.

[335] E21 (1920, 1922, 1930), see above, p. 214.

[336] Unfortunately, the only copy of the very rare third part seen by me lacks a title page.

[337] See above, n. 132.

[338] For the full story from the point of view of the involved scripts, see above, p. 223ff.

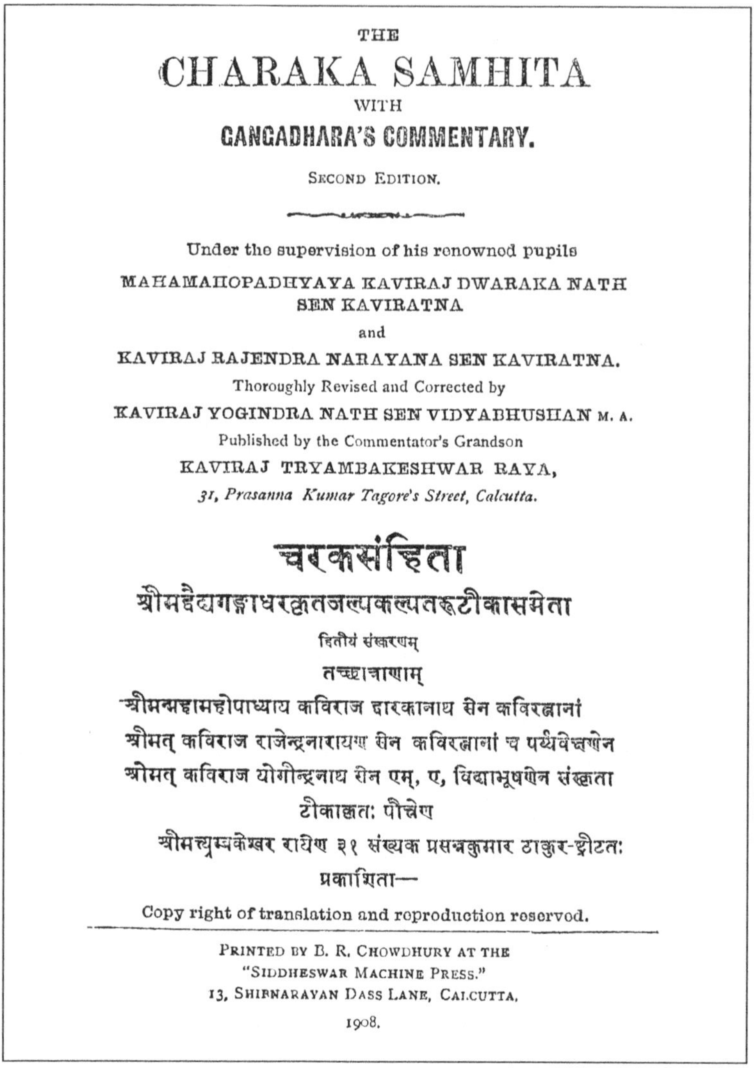
THE

CHARAKA SAMHITA

WITH

GANGADHARA'S COMMENTARY.

SECOND EDITION.

Under the supervision of his renowned pupils

MAHAMAHOPADHYAYA KAVIRAJ DWARAKA NATH SEN KAVIRATNA

and

KAVIRAJ RAJENDRA NARAYANA SEN KAVIRATNA.

Thoroughly Revised and Corrected by

KAVIRAJ YOGINDRA NATH SEN VIDYABHUSHAN M. A.

Published by the Commentator's Grandson

KAVIRAJ TRYAMBAKESHWAR RAYA,

*31, Prasanna Kumar Tagore's Street, Calcutta.*

चरकसंहिता

श्रीमद्वैद्यगङ्गाधरकृतजल्पकल्पतरुटीकासमेता

द्वितीयं संस्करणम्

तच्छात्राणाम्

श्रीमन्महामहोपाध्याय कविराज द्वारकानाथ सेन कविरत्नानां

श्रीमत् कविराज राजेन्द्रनारायण सेन कविरत्नानां च पर्य्यवेक्षणेन

श्रीमत् कविराज योगीन्द्रनाथ सेन एम्, ए, विद्याभूषणेन संस्कृता

टीकाकृत: पौत्रेण

श्रीमत्त्र्यम्बकेश्वर रायेण २१ संख्यक प्रसन्नकुमार ठाकुर-ष्ट्रीटत:

प्रकाशिता—

Copy right of translation and reproduction reserved.

PRINTED BY B. R. CHOWDHURY AT THE

"SIDDHESWAR MACHINE PRESS."

13, SHIBNARAYAN DASS LANE, CALCUTTA.

1908.

**Fig. 6.18** Title page of E1 (²1908ff.)

have been pointed out above.[339] Deeply entrenched in their traditional networks, these actors must have been motivated by filial piety and reverence for their professional teachers and lineages.[340] However, they were also clearly driven by further motives.

[339] See pp. 207, 209, 223f. and 241f.

[340] For reasons of space, the professional and academic network of scholars involved in the production of editions of the *Carakasaṃhitā* cannot be fully treated here in a systematic way. Even in the nineteenth and early twentieth century, this network extended far beyond their places of residence and geographical–cultural regions in South Asia, reaching in some cases all the way to Europe. The

## 6.2.5 *Motivation and Purpose in the Colonial Context*

### 6.2.5.1 Fighting Decline, Loss and Neglect of a Great Treasure

In their hardly studied introductions, the early editors, publishers or printers often deplore the 'worn-down' state or even loss of the ancient medical works by great physician–sages, such as Ātreya Punarvasu and his successors,[341] which had motivated them to edit and publish the *Carakasaṃhitā*. In this context, and also in the context of the praise of Āyurveda and the *Carakasaṃhitā*, they frequently refer to traditional notions and values of the Brahminical cultural tradition. An early example for this kind of discourse is the editorial announcement by the editor-cum-publisher of the second edition of the *Carakasaṃhitā*, Anna Moreshvar Kunte.[342] Kunte sees bad fate and the force of time as causes of this deplorable condition of the traditional medical treatises, but also blames the spread of European allopathic medicine. In spite of this, because of its excellence for both the healthy and the sick the *Carakasaṃhitā* has to be saved from oblivion and made available again. After all, the four traditional ends of man in Brahminical culture – religious merit, material well-being, sensual gratification and salvation – can only be obtained by means of an unimpaired healthy body, and it was precisely for this final purpose that the physician–sages composed their treatises:

> In this world region of Bharata,[343] in former times many compendia of the science of medical treatment[344] were composed by excellent healers, such as Punarvasu, Parāśara, Bhṛgu, Bhela, Agniveśa and Hārīta,[345] which convey the content matter of the knowledge about life,[346] are means for the achievement of religious merit, material well-being, sensual grati-

relevant information can be gathered especially from the acknowledgements and other references in the prefaces, introductions, etc. They mainly relate to assistance and support rendered in the production of the editions which included putting manuscripts of the *Carakasaṃhitā* or its commentaries, in the original or as copies, at the disposal of the editors.

341 See above, p. 197.

342 See above, pp. 225 and 247.

343 That is, in South Asia or geographical India. On the world regions of the continent called 'Black Plum Island' (Jambūdvīpa) in epic–puranic cosmography, see, e.g., Kintaert (2012, 94–101, Table 1 and Figs. 2–4).

344 That is, the traditional body of expert medical knowledge.

345 According to the narrative of the origin of Āyurveda in the *Carakasaṃhitā*, Parāśara, Bhela, Agniveśa and Hārīta are four of Ātreya Punarvasu's six disciples (CS Sū 1.29–30). See above, p. 197.

346 See above, p. 197, on the precise meaning of the compound word *āyurveda*.

> fication and salvation,[347] and are the best resort for the healthy and the sick.[348] Because of the negative development of fate, the spread of the Greek science of physicians[349] and the greatness of time, these compendia are now in a faded condition, like the stars at dawn.[350]

With variations and changing emphases, these points and sentiments emerge again and again in the preliminaries to editions of the *Carakasaṃhitā*. For example, Debendranath and Upendranath Sen Gupta[351] state that 'for us who pursue life[352] it (i.e., the *Carakasaṃhitā*) is exceedingly useful for the pacification of all diseases which stand in the way of religious merit and material well-being.'[353] In this context,

---

[347] The achievement of the four ends of man as the purpose of Āyurveda is already implied in the narrative of the origin of Āyurveda in the *Carakasaṃhitā*; see CS Sū 1.15cd–16ab: 'The most important fundamental cause of religious merit, material well-being, sensual gratification and salvation is health (lit.: 'freedom from disease'). Diseases take it away, and also a better life' (*dharmārthakāmamokṣāṇām ārogyaṃ mūlam uttamam / rogās tasyāpahartāraḥ śreyaso jīvitasya ca //*). On 'life,' that is, the full life span or life force of man (*āyus*), as a means to achieve the 'group of three' (*trivarga*), see AS Sū 1.2: 'A person who desires life which is a means to achieve religious merit, material well-being and pleasure should pay highest attention to the teachings of the knowledge about life' (*āyuḥ kāmayamānena dharmārthasukhasādhanam / āyurvedopadeśeṣu vidheyaḥ paramādaraḥ //*). From the historical point of view, the 'group of three' is at the basis of the four ends of man (*puruṣārtha*), also called the 'group of four' (*caturvarga*). On the 'group of three' in Āyurveda, see Roşu (1978).

[348] For this attribute, see the characterization of Āyurveda in CS Sū 1.24–25: 'Soon he (i.e., Bharadvāja), possessing a great intellect, properly and completely understood the limitless triple-branched knowledge about the full human life span inasmuch as he, the sage, had fixed his mind on it, the knowledge about causes, symptoms and medicines, the best resort for the healthy and the sick, triple-stranded, eternal and meritorious, which the grandfather (i.e., God Brahmā) had already understood' (*hetuliṅgauṣadhajñānaṃ svasthāturaparāyaṇam / trisūtraṃ śāsvataṃ puṇyaṃ bubudhe yaṃ pitāmahaḥ // so 'nantapāraṃ triskandham āyurvedaṃ mahāmatiḥ / yathāvad acirāt sarvaṃ bubudhe tanmanāḥ muniḥ //*).

[349] That is, European medical science. On the generalized usage of the Sanskrit word *yavana* 'Ionian,' i.e., Greek, for foreigners, see Halbfass (1988, 176–177 and 185–186).

[350] See E2 (1876ff.), Vijñaptiḥ (Announcement): *bharatavarṣe 'smin purā punarvasuparāśarabhṛgubheḍāgniveśahārītaprabhṛtibhir bhiṣagvarair āyurvedārthabodhakā dharmārthakāmamokṣasādhanāḥ svasthāturaparāyaṇā bahvyaś cikitsāśāstrasaṃhitā viracitāḥ. daivadurvipākād yāvanavaidyaśāstrapracārāt kālamāhātmyāc ca tā jīrṇā vibhātatārakā iva tiṣṭhanti.*

[351] See above, p. 240ff.

[352] This is a reference to the first part of CS Sū Chapter 11 where the triad of human pursuits is treated, i.e., the pursuit of life (*prāṇaiṣaṇā*), the pursuit of wealth and the pursuit of the 'other world,' that is, purposeful activity in view of a renewed existence in another setting. See Preisendanz (2013, 99–100).

[353] It should be noted that the two editor-cum-publishers mention only the first two of the four ends of man here (or the first two members of the 'group of three'; see above, n. 347). Variations on this topos are found in a number of introductions to other editions of the *Carakasaṃhitā*. See, for instance, E11 (1898), Prastāvanā by Khemraj Shrikrishnadas, editor and publisher, and proprietor of the Śrīveṅkaṭeśvara Press (see above, p. 231): *sūkṣmā buddhi se vicar kar ke dekhiye to caturvarg* (*dharm arth kām mokṣ*) *kā śarīr se bhinn anyă sādhak nahīṁ hai. parantu vah ādhi vyādhi rahit hone se kriyāoṁ meṁ yathāvat pravṛtt hotā hai. isliye sarv prakār se tiskī rakṣā karnī ucit hai. isī kāraṇ ... bhagavān dhanvantarijī ne bhī śarīr ke rakṣārth hī avatār dhāraṇ kar āyurved prakaṭ kiyā hai*. Cf. also Dattaram Narayan Pathak's (see above, p. 212 and n. 234) introduction to E13 (1900) where he refers to CS Sū 1.15cd–16ab and quotes the second half of AS Sū 1.2 (see above, n. 347):

they also praise the *Carakasaṃhitā* as the brightest among the jewels of Āyurveda[354] that emerged at the time of the churning of the ocean which is a mine of jewels,[355] referring to the epic–puranic myth of the churning of the ocean by gods and demons to obtain ambrosia which first yielded a number of other precious items, animate and inanimate.[356] Even though Āyurveda is not explicitly mentioned as one of these items, in many versions of the myth Dhanvantari, the physician of the gods and promulgator of Āyurveda,[357] is one of the divine beings who emerges from the ocean in the process of its churning, holding a jar or bowl full of ambrosia in his hand.[358] Implicitly, the *Carakasaṃhitā* as a foundational work on Āyurveda thus belongs to the mythical 'jewels.'

---

*yah āyurved–śāstrā ābālvṛddh ko sadaiva aur sab se pratham jānnā atyāvaśyak hai ki 'dharmārthakāmamokṣoṁ kī* ***jaṛ*** *uttam ārogyā hī hai, us ārogyā ke tathā kalyāṇ aur jīvan ke haraṇkartā rog hai'. is vāste maharṣiyoṁ ne likhā hai ki "āyurvedopadeśeṣu vidheyaḥ paramādaraḥ" arthāt āyurved hī dharmārthakām kā mūl–rūp hone se prāṇimātrā ko āyurvedopadeś (āyurvedoktavidhi) meṁ paramādar karnā cāhiye* (Bhūmikā, 1). See further E17 (1911), Bhūmikā, 1 (*āyurved ke upādeśoṁ ko param ādar se dhāraṇ karnā cāhiye. yah kyoṁ? isliye ki, yah dharm, arth, kām aur mokṣ in cār puruṣārthoṁ kī ādhārbhūt manuṣyā kī ārogyātā kī prāpti aur āyu kī rakṣā ke liye hai*), E18 (1911), Bhūmikā, 1 (*ab yah khol kar kahne kī koī āvaśyākatā nahīṁ hai ki dharm, arth, kām aur mokṣrūp cār prakār kā puruṣārth bhī unhīṁ pravṛttiyoṁ ke bhītar samāyā huā hai. yah bāt bhī āpā mār prasiddh hai ki pravṛttimātrā kā pradhān sādhan keval śarīr hī hai. yadi vah* (*śarīr*) *rugṇ ho kar aśakt athavā gatāyu ho jāy to koī kisī prakār kī pravṛtti nahīṁ kar saktā*) and E30 (1932), untitled preface, 1 (*viditacaram evedaṃ sarveṣām anena śarīreṇaiva sarve puruṣārthāḥ sidhyanti. ata eva vadanti cāptā mahāntaḥ "śarīram ādyaṃ khalu dharmasādhanam" iti. dharmārthakāmamo kṣākhyacaturvidhasādhanabhūte 'smiñ charīre …*).

[354] See also E18 (1911), Bhūmikā, 2: 'The crest jewel of these jewels of books is this extremely ancient book going back to the sages by name of *Carakasaṃhitā*' (*un hī granthratnoṁ kā maulimaṇi yah "caraksaṃhitā" nāmak ati prācīn ārṣ granth hai*), and E27 (1929), Prākkathanam, 1, with reference to Dṛḍhabala (see above, p. 198): 'After he had seen that this invaluable jewel had practically perished …' (*tad etad anarghyaṃ ratnaṃ śīrṇaprāyaṃ vilokya …*).

[355] See E10 (1897), Vijñāpanam, 1: *iha khalu carakasaṃhitā ratnākaramanthanotthitāyurvedaratneṣūjjvalatamaṃ ratnam. prāṇaiṣiṇām asmākaṃ dharmārthaparipanthisarvarogopaśamāyeyam ekāntenaivopayoginī.*

[356] See Bedekar (1967) and Rüping (1970). In the versions of the *Skandapurāṇa*, these items are called 'jewels'; see 1.1.9.19, 57 and 59 and 1.1.11.54 of the edition published by the Śrīveṅkaṭeśvara Press and quoted by Bedekar (14, 19, 44). On this edition, see Adriaensen et al. (1998, 3–4). Bedekar (1967, 59) also refers to a related popular stanza recited during the marriage ceremony which lists fourteen 'jewels' emerging from the ocean, adding two additional items to the standard puranic number of twelve items. For a translation of the version of the myth in the epic *Mahābhārata*, see van Buitenen (1973, 72–76) and O'Flaherty (1975, 273–280); for a translation of the version in the *Viṣṇupurāṇa*, see Wilson (1980, 106–115, especially n. 8), abbreviated and without the notes reproduced by Renou (1962, 168–170), and for the *Skandapurāṇa* versions Tagare (1992, 64–73, 86–90).

[357] See above, p. 238.

[358] The jar or bowl of ambrosia is counted as the next item that emerges from the ocean. See the synoptic table in Bedekar (1967, 46–47). For a further reference to the myth in the *Vāyupurāṇa* in the context of the story of Dhanvantari's birth, which is not included in this table, see Bedekar (45). Additional sources are treated in Rüping (1970).

Especially the aspect of the decline or loss of the *Carakasaṃhitā* addressed by Anna Moreshvar Kunte recurs frequently in introductions to editions of the *Carakasaṃhitā* and represents an important motive for their authors to edit and publish the text. Thus, Debendranath and Upendranath Sen Gupta exclaim: 'Alas! such a great jewel has practically been lost due to the nature of time,'[359] and the two publishers Shyamlal Shrikrishnalal and Gangavishnu Shrikrishnadas refer to a versified Sanskrit saying when they state more generally that

> like the midday sun, the excellence of Āyurveda is evident not in India alone, but in the whole world. However, 'the course of time is crooked.' Even the sun happens to be eclipsed and clouds may also encroach upon it. In the same way, also our Āyurveda was covered by darkness and ceased to provide light.[360]

In Maharashtra, a region that played a major part in the emergence of the national movement in the late nineteenth century, Shankar Daji Pade, the prominent editor, translator and commentator of the *Carakasaṃhitā* (see Fig. 6.19),[361] emphasizes the role of foreigners in this deplorable state of affairs. In the introduction to his edition of the text of the *Carakasaṃhitā* by itself, he claims in eloquent and poetic Sanskrit that Indian books in general came to great grief at their hands, even though he also blames his fellow Indians:

> Once translations, explanations, etc., of Indian[362] works had come to their knowledge, foreigners became great lovers of Indian lore, came to our country under various pretexts, collected a great number of our books and took them to their respective countries. Even today, they are found for the most part in the particular country of Germany located in Europe. The first printed book of the *Ṛgveda* together with its commentary, so they say, which Max Müller had caused to be printed, came here from Europe. ... Books of this kind, absolutely singular even though they are many, are now obtained there which are not even known by name here. In the said manner, our large collection of old books disappeared.

---

[359] See E10 (Vijñāpanam, 1): *aho kāladharmeṇaitādṛśaṃ mahāratnaṃ viluptaprāyam*! On the Bengali physician–scholars Debendranath and Upendranath Sen Gupta, see above, p. 240ff.

[360] See E12 (1898), Nivedan ('communication'): *āyurved kī utkṛṣṭatā ek bhārat kyā sampūrṇ bhūmaṇḍal par madhyāhn ke sūryă ke samān prakāśit hai parantu "kālasya kuṭilā gatiḥ" sūryă ko bhī grahaṇ lagtā hai aur us par bādal bhī ā jāte haiṁ, isī tarah se hamārā āyurved bhī tamasācchādit ho kar prakāś karne se baiṭh rahā thā.* The sentence 'the course of time is crooked' is taken from the stanza 'Stones swim on water and men kill the demons (and not the other way around), monkeys do a (wonderful) job – the course of time is crooked!' See Böhtlingk (1873, 587, no. 7451) and Hara (1997, 529). The stanza points out the unpredictable course of almighty Time and specifically refers to a famous episode in the epic *Rāmāyaṇa* where forest monkeys assist king Rāma and miraculously build a bridge across the sea to the island of Laṅkā where the demon king Rāvaṇa holds his queen Sītā imprisoned. On Shyamlal Shrikrishnalal, see above, p. 209, on Gangavishnu Shrikrishnadas p. 231. Similarly, in the context of his extensive description and analysis of the much deplored decline of the study, teaching and understanding of Āyurveda, which cannot be addressed in detail here, Ramprasad (see above, p. 204) repeatedly refers to the strange and unpredictable, inescapable workings of Lord Time (*samay bhagvān, kāl bhagvān*) and his winding course. See E17 (1911), Bhūmikā, 6.

[361] See above, pp. 209 and 211.

[362] Literally: 'belonging to [the world region of] Bharata'; see above, n. 343.

**Fig. 6.19** Shankar Daji Pade

Immediately afterwards, however, innumerable books of our gods[363] were even openly set on fire by some Mogul rulers, foremost among them Aurangzeb,[364] who fought against Indian lore and religion. Also for this reason a great loss occurred. Subsequently, the rich land of Bharata[365] was occupied by the present English rulers. After they had established government medical colleges, they spread just their own accepted European science of medical treatment. After the general maxim 'What is agreeable to the king, is agreeable to the subjects,' residents of the world region of Bharata who, for their part, were ardently longing for a career as government physicians and thus disposed to disregard the extremely ancient and extraordinary[366] science of medical treatment coming from the gods which had reached them in regular succession in their own families,[367] became attached only to the study of the English science of medical treatment. It seems that for this reason the custom of studying and teaching our Indian science of medical treatment has now practically been lost.[368]

---

[363] That is, of our Sanskrit books.

[364] The Mogul emperor Aurangzeb ruled from 1658 until his death in 1707. See also the reference to Aurangzeb and the destruction of books, specifically on Āyurveda, during his period in E17 (1911), Bhūmikā, 6: *ab bhī bahutere yogyă puruṣ paropakārī sadvaidyă aur āyurved kī mahimā ko jānnevāle haiṁ jin kī kṛpā se auraṅjebī jamāne ke mahāāghāt se bace hue granth is unnatśīl śrībhārat sarkār ke śubh rājyă meṁ baṛī āsānī se chap chap kar prāpt hone lage haiṁ.*

[365] See again above, n. 343.

[366] The word *alaukika* (literally: 'unworldly, not of this world') may also allude to the non-human or superhuman origin of the medical science (see above, p. 197). At the same time, the negated adjective 'worldly, secular' (*laukika*) may suggest that the medical science is Vedic (*vaidika*), i.e., part of or derived from the Vedic revelation of the Aryans (see below, n. 372).

[367] This refers to the traditional setting of the transmission of medical knowledge; see above, p. 252ff.

[368] See E15 (1903?), Bhūmikā, 1–2: *bhāratīyanibandhānuvādatātparyādīn avagamya viṣayāntaravāsino janā bhāratīyavidyākāmukatarā bhūtvāsmajjanapade nānāvyājair āgatavanto vividhopāyaiḥ saṃghaṭīkṛtyāsmadgranthanicayān nijanijadeśaṃ prāpayām āsuḥ. idānīm api te bāhulyena yūropāntargatajarmanīdeśaviśeṣe vartante. ... etādṛśā bahavo 'py apūrvatarā granthās tatredānīm upalabhyante ye khalv atra nāmnāpy aprasiddhāḥ santi. uktarītyāsmākaṃ mahatī prācīnagrantharāśir vilayam upagatā. tato 'nantaram auraṅjibpramukhaiḥ katipayamugalamahārājair bhāratīyavidyādharmapratibhaṭair asmadgīrvāṇagranthās tv aparisaṅkhyeyāḥ sākṣāt prajvālitā eva. ato 'pi mahatī hāniḥ saṃvṛttā. tad anv idānīntanāṅglaprabhubhir*

One may not agree with Shankar Daji Pade's nationalistic attitude from the historical point of view: the repression of Hindus and Hindu cultural practices and destruction of Hindu religious places and objects during Aurangzeb's reign in the second half of the seventeenth century, which are often exaggerated in Hindu nationalistic discourse, certainly precede the modern European interest in Indian Sanskrit literature and the collection of Sanskrit 'books,' i.e., manuscripts, by European scholars which commenced at the beginning of the nineteenth century. The systematic survey and collection of Sanskrit manuscripts did not become an official agenda of the colonial government until the 1860s and the manuscripts were deposited in existing or newly established repositories in India itself.[369] In general, the number of Sanskrit manuscripts that ended up in European and other extra-Indian repositories is negligible when compared to the huge number of manuscripts still preserved in Indian repositories and private collections, in spite of the much deplorable, continuing loss due to the lack of sufficient means as well as outright neglect and ignorance.[370] Even so, Shankar Daji Pade is right in pointing out the loss of many traditional medical works and in stressing, like Anna Moreshvar Kunte,[371] the spread and attractiveness of European allopathic medicine which he sees as a major cause of the decline of traditional study of Āyurveda and the lack of interest in the medical works that have survived. He continues:

> Now then, after immeasurable jewels of literary compositions have disappeared because of the bad fate due to that much, this makes one contemplate also on what is happening inasmuch as it may be the cause of the disappearance of all of them. Even today, some extraordinary noble[372]

---

*iyaṃ bhāratavasundharā sanāthīkṛtā. tair hi rājakīyavaidyakapāṭhaśālāsaṃsthāpanapuraḥsaraṃ nijābhimatayūropīyacikitsāśāstrasyaiva samprasāraḥ kṛtaḥ. yad rājñaḥ priyaṃ tad eva rāṣṭrasyeti sāmānyanyāyād rājakīyabhiṣakpadavīlolupā api bhāratavarṣīyā nijakulaparamparāgatam atiprācīnam alaukikaṃ gīrvāṇacikitsāśāstram avahelanaśīlā āṅglacikitsāśāstrādhyayana eva saṃlagnā babhūvuḥ. ato hetor asmadbhāratīyacikitsāśāstrasya paṭhanapāṭhanapracāro 'pīdānīṃ luptaprāya eva saṃdṛśyate.*

[369] See the early documentation of this process by Gough (1878).

[370] For a rough overview of the still overwhelmingly large treasure of Sanskrit manuscripts preserved in India and the world, see the *New Catalogus Catalogorum* (Raghavan et al. 1949–2015) and the database of the National Mission for Manuscripts, New Delhi.

[371] See above, p. 255f.

[372] The recurrent use of the adjective 'noble' (*ārya*) in Shankar Daji Pade's introduction is terminological and extremely pregnant. It reflects his understanding of Āyurveda as a medical tradition that goes back to the Vedic people of ancient South Asia, a group of immigrant tribes who called themselves *ārya*-s ('Aryans'), and has been transmitted by their descendants until the present. Similarly, on the title page of E4 (1878??) the *Carakasaṃhitā* is introduced as 'The Medical Science of the Ancient Aryans,' which reminds of the title of an important revivalist treatise published in 1895 by Bhagvat Sinhji, the Maharaja of Gondal, namely, *Aryan Medical Science*; see Arnold (2000, 176–177). On the notion of *ārya* in the context of traditional Indian xenology, see Halbfass (1988, 175–187 and 331–332). In the context of the Hindu nationalistic discourse and of the discourse of the Āyurveda revivalists in which Shankar Daji Pade participated, however, the 'Aryans' and their culture were emphatically claimed to be indigenous to South Asia; see, e.g., Bergunder (2002, 165–172) and Bryant (2002). Among other medical journals, Shankar Daji Pade published a monthly journal, or rather book series (*pustakāvali*), in Marathi called *Āryabhiṣak* ('The Noble Healer') from 1889 onwards, which was continued for about two years after his death; he also ran a Gujarati journal by the same name. See Sharma (1975, 585 and 625; 2016, 612 and

medical books composed by Caraka, Suśruta, Hārīta,[373] Vāgbhaṭa[374] and others are shining solely because of the exceeding greatness of their merits. Amongst the ones that are still available this book called *Carakasaṃhitā*, the most ancient and most venerable one, the one most esteemed by the good, a mine of all medical topics, shines high above all. One should know that this compendium is not only something to be esteemed just by us Indians; English Crest Jewels of Physicians[375] living in Europe, America and other continents also show it their highest respect. This is evident for all of us. ... Now, while savants living on other continents, being wise men, exceedingly celebrate the greatness of our science of medical treatment, many persons who have not as much as smelled at the medical books even though they have been born in a noble lineage and who reportedly fancy themselves pandits prattle in the manner of a frog in a well: 'This noble medicine is useless; exceedingly unsubstantial and devoid of an understanding of the true nature of things it does not even deserve to be called a science.' Alas! it is a great misfortune for the world region of Bharata that without any reason these newly learned great men attribute ignorance to the ancient noble great men who possess extraordinary insight and despise the extraordinary books composed by them.[376]

In the light of this situation, he had engaged himself in publishing the *Carakasaṃhitā* already before, with translation and commentary, and now continues to do so with the support of the owner and superintendent of the Nirṇayasāgara Press,[377] Tukaram Javaji:

657). For information by the manager on *Āryabhiṣak* and a list of issues detailing the individual instalments, with prices, of the books published in this series from 1889 to 1892, see the advertisement at the end of the first fascicle of E9 (1897ff.), pp. (7)–(8) of the appendix, after a list of these books and other medical books offered for sale at the Āryabhiṣak Office and other places, with opinions of well-known physicians and scholars. On the designation of Hindi as the 'noble vernacular' (*āryabhāṣā*) as an expression of Hindu revivalist sentiment, see Stark (2009, 196).

[373] See above, n. 345.

[374] Vāgbhaṭa is the name of two authors of two early-medieval foundational works on Āyurveda, the *Aṣṭāṅgahṛdayasaṃhitā* and the *Aṣṭāṅgasaṃgraha*.

[375] 'Crest Jewel' (*śiromaṇi*) is another traditional title which nowadays roughly corresponds to an M. A., M. Phil. or M. Litt. degree. See Michaels (2001, 8, Fig. 2). It may be supplemented with the specification of the area of expertise of its bearer, but was also used by itself as a distinctive title expressing highest proficiency. For instance, the great medieval logician Raghunātha Śiromaṇi is often just called Śiromaṇi in the works of other logicians. Shankar Daji Pade probably uses the word in this traditional sense here.

[376] See E15 (1903?), Bhūmikā, 2: *etāvatā durdaivavaśād vilayam upāgatavatsv aparimitaprabandharatneṣu samupasthitāv api sakalasaṃhārahetau hanta samādhānajanakaṃ caitad. adyāpi caraka–suśruta–hārīta–vāgbhaṭādipraṇītāḥ katipayalokottarāryavaidyakagranthāḥ kevalaṃ tu nijaguṇamāhātmyātiśayenaiva virājante. upalabdheṣu madhye carakasaṃhitābhidhāno 'yaṃ granthaḥ prācīnatamaḥ pūjyatamaḥ paramasanmānanīyo yāvadvaidyakaviṣayāṇām ākaraḥ sarvopari virājate. na khalv ayaṃ saṃhitāsmākaṃ bhāratīyānām eva kevalaṃ mānyā, api tu yūropa–amerikādidvīpāntarīyāṅglabhiṣakśiromaṇibhir api tasyāḥ paramādaraḥ kriyata iti naḥ pratyakṣaṃ sarveṣām. ... samprati yadā hi dvīpāntarīyakṛtadhiyo vipaścito 'smadīyacikitsāśāstramāhātmyam atitarām upagāyanti tadāryavaṃśasamudbhavā apy anāghrātāryavaidyakagranthā bhūyāṃsaḥ kila paṇḍitammanyā kūpamaṇḍūkanyāyena 'aprayojakam idam āryavaidyakam asārataraṃ tattvāvabodhavihīnaṃ śāstrasaṃjñāyā apy anarham iti' pralapanti. hanta mahaddaurbhāgyam idaṃ bhāratavarṣasya yad ete navyaśikṣitamahāpuruṣā niṣkāraṇam evālaukikaprajñeṣu prācīnāryapuruṣeṣv ajñatvāropaṃ kṛtvā tatpraṇītālaukikagranthāṃs tiraskurvanti.*

[377] See above, pp. 229 and 238.

> Because of the unsurpassed usefulness of this book, some years go I started the printing of its translation into the language of Maharashtra[378] especially as a service to all people who are not familiar with the language of the gods (i.e., Sanskrit). Complying with the request of doctors and other lovers of treatises on medical treatment who desired just the basic text because the book together with the translation could not be obtained at a low price, I communicated to my best friend, the owner of the Nirṇayasāgara Press, that we should publish this book. And he, who had already attained fame through the ceaseless publication of difficult to get and very much unknown books, emulating his exceedingly highly munificent father,[379] accepted my suggestion and printed this extremely venerable great book with very pleasing iron letters at his own printing press which is renowned all over India. In this matter, the favours of the eminent illustrious Tukaram Javaji, who bears the nickname 'J. P.,'[380] the supervisor of the mentioned printing press, are sparkling outstandingly. I pray to God that in this very way his excellent fame through the ceaseless publication of ancient and not yet available books may sparkle to the highest degree.[381]

For Dattaram Narayan Pathak[382] the basic problem lies elsewhere. He highlights the extreme difficulty of the text of the *Carakasaṃhitā* which he considers a defect: even with excellent things this or that defect necessarily remains although such defects are minor, unessential ones, just like a spot on the moon that disappears among its rays. Due to its difficulty, in his day the *Carakasaṃhitā* did not remain in use in studying and teaching. This state of affairs is also at the basis of the completely unjustified defamation of the work by fools who pretend to be physicians and are enimies of good sense. However, Dattaram Narayan Pathak feels extremely lucky that its translation into the vernacular[383] has come to be so much charming and attractive that upon acquiring even a little bit of general knowledge of Sanskrit

---

[378] That is, into Marathi. See E9 (1897ff.) and above, pp. 209 and 211.

[379] This must refer to the founder of the Nirṇayasāgara Press, Javaji Dadaji, who also developed his own types; see Galewicz (2020, 218). After Tukaram Javaji, his son Pandurang Javaji and grandson Satyabhamabai Pandurang continued to print editions and further editions of the *Carakasaṃhitā*. See the editions listed above in n. 234.

[380] Alternatively, one could understand 'holding the title "J. P.".' I fail to understand to what the English abbreviation may refer in either alternative.

[381] See E15 (1903?), Bhūmikā, 4: *niratiśayopayogitvād asya granthasya viśeṣato gīrvāṇabhāṣānabhijñalokopakārārtham asmābhir mahārāṣṭrabhāṣāyāṃ gatakatipayavarṣebhyo 'syānuvādamudraṇaṃ prārabdham. sānuvādapustakasyālpamūlyālabhyatayā mūlamātram abhikāṅkṣamāṇānāṃ cikitsakānām itareṣāṃ ca cikitsāśāstreṣv anurāgavatām anunayam urarīkṛtyāsmatsuhṛttamebhyo nirṇa<ya>sāgaramudraṇālayādhipatibhyo niveditam asmābhiḥ prakāśanīyo 'yaṃ grantha iti. anavaratadurlabhāprasiddhataragranthaprakāśanena labdhakīrtayaś ca te 'nukṛtya nijapitaraṃ nitarāṃ vadānyataram aṅgīkṛtyāsmatsūcanāṃ svakīye bhārataviśrutamudraṇālaye suramyair āyasākṣarair imaṃ paramamānyamahāgranthaṃ mudrayāmbabhūvuḥ. asmin viṣaye nirdiṣṭamudraṇālayādhyakṣāṇāṃ nāma je. pī. ityupādhidhāriśrīyutatukārāmajāvajīmahodayānām atīvopakṛtayaḥ samullasanti. evam eva teṣām anavarataprācīnānupalabdhagranthaprakāśanena suyaśaḥ samullasatutarām ity abhyarthayāmaḥ parameśam.*

[382] See above, pp. 212f. and 233.

[383] Even though on the title pages of E13 the *Carakacandrodayā* is said to be a Hindi commentary (*hindībhāṣāṭīkā*) on the *Carakasaṃhitā*, Dattaram Narayan Pathak calls it a translation (*bhāṣānuvād, bhāṣāntar*) here and later on at the end of his Introduction (see Bhūmikā, 6). See above, n. 59, on the problem of the distinction between Hindi commentary and translation.

literature the book can be comprehended easily from seeing its translation.[384] Similarly, Vrajvallabh Sharma, presumably the editor of Ravidatt's Hindi translation of the *Carakasaṃhitā*,[385] after having briefly described the unavailability of well-preserved and complete ancient commentaries on the *Carakasaṃhitā*, states that

> just as commentaries on it are difficult to get hold of, people who may be able to offer instruction on this difficult-to-understand book are indeed rare on the face of the earth. This is so because the book is not such that a scholar specialized in a single area of knowledge may know its secret; just a broadly learned scholar versed in the six philosophical traditions may do so. For this very reason the majority of contemporary physicians are not familiar with the core of such a most excellent book on the subject matter of medical treatment. With the purpose of making all common physicians understand the secret of this book, the book publisher and master, the illustrious Pandit Hariprasad Bhagirath, because of his resolve to make good for this lack concerning Ayurvedic medicine, had it adorned by Pandit Ravidatt, resident of Beri,[386] with a translation into the vernacular. I for my part have had it printed in the very well-known printing house Nirṇayasāgara.[387]

In general, the editors demand that attention be directed again at Āyurveda as taught in the *Carakasaṃhitā* so that this precious knowledge may flourish.[388] The explicitly targeted audience and beneficiaries of the editions range from physicians and those interested in the medical science to learned persons in general[389]; an edition

[384] See E13 (1900), Bhūmikā, 1: *suhṛdgaṇ! yāvanmātr uttam vastu haiṁ un meṁ ek na ek avăguṇ avaśyă rahtā haiṁ "nimajjatīndau kiraṇeṣv ivāṅkaḥ" arthāt is granth kī atikliṣṭātā hī is kā avăguṇ hai. arthāt iske kaṭhin hone se hī ājkal paṭhanpāṭhan meṁ yah granth pracalit nahīṁ rahā. is vāste bahut se vaiśākhnandan vaidyăbruv kah uṭhte haiṁ ki carak meṁ tau (nām* **baṛe** *aur darśan* **thoṛe**) *kuch bhī nahīṁ haiṁ. parantu un akal ke duśmanoṁ kī nindā karne se kyā yah granth nindănīyă ho saktā hai? kadāpi nahīṁ. ... parantu ham logoṁ kā param saubhāgyă hai ki carakpraṇīt carakasaṃhitā kā bhāṣānuvād aisā sulalit aur hṛdayaṅgam huā hai ki saṃskṛtsāhityă meṁ* **thoṛā** *bhī sāmānyă jñān hone par is kā bhāṣāntar dekhne se granth sahaj hī bodhgamyă ho saktā hai.*

[385] See above, p. 232f.

[386] Without further details, it is difficult to determine which of the several villages and towns by name of Beri in the large Hindi speaking area was Ravidatt's place of residence. Among them, Beri in Madhya Pradesh would be closest to Mumbai where the edition with translation was published. However, Beri in present-day Haryana, a village south of Rohtak, is the best candidate because Pandit Ravidatt is most probably identical with Pandit Ravidatt, a medical practitioner from Rohtak District, who was commissioned by Naval Kishore with Hindi translations of several important Ayurvedic works which were published by him in the early 1890s. See Stark (2012, 481).

[387] See E18 (1911), Bhūmikā, 5–6: *jis prakār is kī ṭīkā durlabh haiṁ usī prakār is durbodh granth ke adhyāpak bhī is jagatītal meṁ viral hī haiṁ. kyoṁki, yah granth aisā nahīṁ hai ki ṣaḍdarśanniṣṇāt vyāpak paṇḍit ko choṛ, ekdeśī paṇḍit is kā rahasyă jān sake. yahī kāraṇ hai ki ājkal ke adhikāṃś vaidyă cikitsāviṣayak aise sarvottam granth ke marm se aparicit haiṁ. sarv sādhāraṇ vaidyoṁ ko is granth kā rahasyă samjhāne ke uddeś se pustakprakāśak sva. śrīyut paṃ. hariprasād bhagīrathjī vaidyaksambandhī is abhāv ko pūrṇ karne ke saṅkalp se verīnivāsī paṃ. ravidatt se bhāṣānuvād se samalaṅkṛt karāyā. so ham ne suprasiddh nirṇaysāgar chāpkhāne meṁ chapvāyā hai.*

[388] See also the wish expressed at the end of the publisher's dedication in E9 ($^{3}$1914ff., $^{R}$1926).

[389] See the Prologue (Prastāvanā) to E11 (1898) where Khemraj Shrikrishnadas addresses the learned people (*vidvajjan*) in his signature phrase (*vidvajjan kṛpābhilāṣī – khemrāj śrīkṛṣṇădās, "śrīveṅkaṭeśvar" mudraṇālayădhyakṣ – mumbaī*).

may even be dedicated to the 'whole world.'[390] Printing and disseminating the text of the *Carakasaṃhitā* is the chosen strategy in achieving this goal, which is coincident with the goal of saving Āyurveda from decline and disappearance. As can be gathered from the above quotation from Shankar Daji Pade's introduction to his reduced edition of the *Carakasaṃhitā*, that is, of the text by itself without translation and notes,[391] an affordable price, that is, financial accessibility, is an important desideratum in this endeavor[392] and requires sufficient means on the part of the involved actors. In the case of Shankar Daji Pade's text-only edition, the munificence of the printer-cum-publisher, Tukaram Javaji of the Nirṇayasāgara Press, made the moderate price of two rupees for this complete edition possible. Thus,

> this suchlike book, manifest / brought out like the sun among medical works, whose application is fruitful and which gives fast results, inasmuch as – relative to its greatness – it can be obtained at a low price, should be collected by doctors and other lovers of ancient noble[393] books after they have ceased to be stingy. This is my extremely important announcement.[394]

In the case of Shankar Daji Pade's earlier edition with Marathi translation and commentary, which was first published in a quarterly journal,[395] already part of its first edition was financially supported by Maharaja Sayaji Rao Gaekwad III, ruler of the princely state of Baroda in Western India from 1875 to 1939 and a great

---

[390] See again the dedicatory page of E9 ($^3$1914).

[391] See above, p. 258f.

[392] See also Stark (2009, 194–195) on the case of religious books in nineteenth-century northern India. For the diametrically opposed route taken by the followers of the nineteenth-century South Indian Śaiva mystic Ramalinga Adigal in order to maximize the prestige of his work, rather than to ensure wide readership, see Weiss (2015, 665 and 677).

[393] See above, n. 372.

[394] See the conclusion of the Introduction to E15 (1903?) (Bhūmikā, 4): *etādṛśo vaidyakanibandheṣu bhāskara iva prakāśito 'yaṃ granthaḥ saphalayogaḥ śīghraphalado nijamahattvāpekṣayā-lpamūlyalabhyaś cikitsakair itaraiś ca prācīnāryagranthānurāgibhir dravyakṛpaṇatāṃ tyaktvā saṅgrahītavya iti garīyasī no vijñaptiḥ*. Similarly, Gangavishnu Shrikrishnadas, at the end of his preface to E30 (1932) (p. 2), says: 'This suchlike compendium here has been printed at my "Lakṣmīveṅkaṭeśvara" press and is now for sale. May you therefore experience delight of the mind's eye after you have made my effort bear fruit and taken it at the price as indicated' (*etādṛśīyaṃ saṃhitāsmadīya"lakṣmīveṅkaṭeśvara"mudraṇālaye saṃmudrya vikrayyāsti. ato matpariśramaṃ saphalīkṛtya yathoddiṣṭamūlyenemāṃ gṛhītvā ca manonayanānandam anubhavantu bhavantaḥ*). Unfortunately, the original price of this edition could not be ascertained. The original maximum list price was written in the center of the slightly faded oval stamp of Khemraj Shrikrishnadas (Khetwadi?) placed on the beautiful title page of the copy in the private library of Kenneth Zysk, Copenhagen. However, it has been crudely overwritten, probably by the Indian bookseller who offered the old book for sale. As can be seen from the information printed at the very end of the book, after the last stanzas of the Siddhisthāna, the book could be obtained either from Gangavishnu Shrikrishnadas at the Veṅkaṭeśvara Steam Press in Kalyan or from Khemraj Shrikrishnadas at the Śrīveṅkaṭeśvara Steam Press in Khetwadi, both in Bombay. On Gangavishnu and Khemraj Shrikrishnadas see above, p. 231, and n. 236.

[395] E9 (1897ff.); see above, p. 209.

philanthropist.[396] He also sponsored its third edition and the reprint in a multi-volume book.[397] The revenue from the sale of other books on medicine and medicinal plants from the Āryabhiṣak office advertised at the end of some individual issues of the earlier edition[398] must have also contributed to cover the expenses of publication. However, the cost for the whole set was much higher, with twenty rupees for all fourteen parts of the reprint.[399]

Chandrakishor Sen Gupta's generous financing of the reprint of Gangadhar Ray's *Jalpakalpataru* on the *Carakasaṃhitā* and of the publication of many other books on Āyurveda has already been addressed above.[400] Less wealthy physician–scholars who were not talented entrepreneurs and acted as editors and publishers in the early history of publication of the *Carakasaṃhitā* must have done this at great personal expense. Even in the case of editions that were published by subscription and other kinds of advance payment by interested readers,[401] additional sponsors had to be found to alleviate the financial struggle of the editor-cum-publishers. Harinath Sarkar's edition of the *Carakasaṃhitā* with Cakrapāṇidatta's *Āyurvedadīpikā*[402] is a good example for this. Already in the introduction to the first fascicle, Harinath Sarkar points out the great expense involved in the publication of his edition, with four or maybe even eight print sheets planned for monthly

---

[396] Reference to the Maharaja's generous support figures prominently on the cover of the fascicle with the text, translation and commentary of the Siddhisthāna of the first edition of E9, published in 1901 as part 34 of the *Āryabhiṣak* series. Shankar Daji Shastri Pade's name appears only in small letters in a band in the lower part of the physician–scholar's crest (with his entwined ornamental initials SDP instead of a shield) printed beneath this acknowledgement. The Maharaja's financial support is also highlighted on the very elaborately illustrated dedication page before the table of contents of the Siddhisthāna. See also the dedication pages of the fascicle with the text of Cikitsāsthāna 1–5, published in 1902 as *Āryabhiṣak* 41, the last fascicle (no. 4) of the Cikitsāsthāna (20–30), published in 1906 as *Āryabhiṣak* 49, and the last fascicle (no. 2) of the Śārīrasthāna (7–8), published in 1907 as *Āryabhiṣak* 58.

[397] See the slightly differently decorated dedication pages of E9 ($^{3}$1914ff.) and E9 (reprint 1926) published by the Pune bookseller Yajneshvar Gopal Dikshit. It could not be ascertained whether the second printing, which was also published by Yajneshvar Gopal Dikshit, had similarly been sponsored by this ruler.

[398] See above, n. 372, for book advertisements. Medicinal plants sold by the Āryabhiṣak Office itself are advertised at the end of book 1, issues 4–5 (January and February 1898) of E9. At the end of issue 6 (September 1898), we find a three-page advertisement by the Āryamahauṣadhālaya, a pharmacy owned by the physician Krishnashastri Bilvalkar located in Anjarle in Dapoli District southwest of Pune. External advertisement, probably as a source of revenue for the publishers, is furthermore found in E8 ($^{2}$1910) where Ayurvedic medicines produced by B. Basu and Company, 79 Harrison Road, Calcutta (now Mahatma Gandhi Street), are extensively advertised.

[399] For example, the price of the fifth part (Vimānasthāna) of the second edition (dated 1913) was one rupee, that of the first part (Sūtrasthāna 1 to 11) of the third edition (dated 1914) two rupees.

[400] See above, p. 240f.

[401] For an especially interesting case of publication by subscription out of financial necessity in eighteenth-century India, see Ogborn (2007, 255–256).

[402] E7 (1892ff.); see above, p. 233f.

publication.[403] There was an option to either annually subscribe to the publication, whose fascicles were indeed initially published monthly, or to pre-order a complete or partial set, both options with advance payment. The price was twenty rupees for the complete set, like for the reprint of Shankar Daji Pade's edition with translation and commentary, which was reduced to twelve rupees 'for the poor'[404] and was 'payable in advance.'[405] From the beginning in 1892, the price of the annual subscription was six rupees, reduced to three rupees for needy persons, such as students, as becomes clear from the bilingual Hindi–Bengali announcements as well as the single announcement in Hindi on progress and pricing.[406] The income from advance payments for complete sets and annual subscriptions obviously did not suffice to cover the expenses because the outer back cover of the last fascicle of the text of the Sūtrasthāna, the first book of the *Carakasaṃhitā*, in this edition not only lists the

[403] See E7, Bhūmikā, 8. One sheet must have had eight pages. E6 (1888ff.) was also printed with a rhythm of four sheets per month (each having eight pages); see the announcement (*vijñāpan*) in part 6 referred to in n. 406 below.

[404] Cf. E6 (1888ff), Preface (pp. 1–2): 'Although there is already a Bengali translation of the work, I intend also to publish the same in parts, for the convenience of the poorer classes of the native community who cannot afford to purchase this voluminous work all at once' and Bhūmikā (p. 2): *saty api carakasaṃhitāyā vaṅgānuvāde na hi durgatair yugapat tādṛśādhikamūlyena kretuṃ śakyate. ato 'smābhis taiḥ sugamāya punar api khaṇḍaśaḥ sa vaṅgānuvādaḥ pracārayituṃ saṃkalpitaḥ.* In the case of religious texts, the affordability of printed editions even for very poor people was considered even more important; see the pronounced statement of Pandit Madanmohan Malviya quoted in Stark (2009, 201).

[405] See the 'rules in connection with the Carakasaṃhitā' once provided in English (fascicles 23&24), otherwise in Bengali, and the anouncements of progress and on pricing printed on the inner front and inner or outer back covers of most fascicles of E7, as indicated in Appendix 6.2 below. These prices did not change over the years as long as the subscription system was in place, i.e., until 1905 (fascicles 29&30). From 1907 onwards (fascicle 31), only a one-time advance payment of twenty-five rupees, reduced to fifteen rupees for needy persons, was solicited for the whole book, a price which included standard postage in Bengal until 1917 (Siddhisthāna fascicle 2).

[406] Similarly, E6 (1888ff.), the immediately preceding edition of the *Carakasaṃhitā* with the *Āyurvedadīpikā* by Avinashchandra (see above, especially n. 19), was published in instalments of four print sheets each and available by subscription with advance payment of three rupees for a whole year (twelve fascicles). The total price was given as twenty rupees. See the Announcement (Vijñāpan) signed by Pyarimohan Sen on the inner front cover of the last fascicle of this edition seen by me, no. 6 (covering Sū 8.9–11.14), which is undated. However, the date 3–94, i.e., March 1894, is written on the cover of part 6 of the copy of the British Library (call number 14043.dd.1) and the stamp of the British Museum affixed on its last page gives the date of January 12, 97. Fascicle no. 5 was published in 1891 according to the list of printed editions in E37 (1949) (see item 8), which makes 1892 or 1893 a possible date for no. 6. The list also lists a seventh fascicle, but without a date. The text of the copy in the library of the Bhandarkar Oriental Institute, Poona (call number 1761), formerly owned by the High School Library, Poona, extends to Sū 14.8, but the title pages of the individual fascicles are not preserved and the title page only gives the date 1295 of the Bengali era (= 1888 CE), i.e., the date of publication of the first fascicle. On this title page, the price per fascicle is given as one rupee. – The Āyurved-Granthmālā series of editions with Hindi translations of fundamental Ayurvedic works, published by the Ārya-Sāhitya Maṇḍala, was also offered by subscription to merchants. E35 (1937) was its first volume, complete in three parts. Subscribers had to pay a deposit of one rupee and subsequently got all individually published parts of books at a 25% discount. A copy of E35 would thus have cost a subscriber nine rupees.

names and addresses of those individuals who paid various due amounts which helped to ensure the continued publication of the edition, but also gives information on some general sponsors.[407]

Harinath Sarkar's sponsors and subscribers were mostly from Calcutta and its vicinity and other parts of Bengal; the well-known physician–scholars Yogindranath Sen[408] and Avinashchandra[409] were among them.[410] However, there were also subscribers and other advance payers from Patna and other places in Bihar, and a few from Benares and even Jaipur; among the subscribers from the latter city we find *inter alia* Lakshmiram's[411] name.[412] Furthermore, the Mysore Government is mentioned as a paying customer.[413] The outer back cover of the 'three and a half parts after part 17,' i.e., of the unnumbered and undated jointly published fascicles that cover the text of the Nidānasthāna, the second book of the *Carakasaṃhitā*,[414] acknowledges the financial support of Maharaja Manindrachandranandi of Kasim

[407] See the outer back cover of fascicles 15–17 which were published jointly in 1895. The outer back covers of the subsequent undated fascicles 18 to 20, which contain more than half of the text of the Vimānasthāna, the third book of the *Carakasaṃhitā*, merely list individual persons who paid their dues, with addresses and amounts. See again Appendix 6.2. The dual strategy of soliciting subscriptions with advance payment and inviting munificent sponsors is also referred to by Stark (2009, 193).

[408] See above, p. 214 and passim.

[409] See above, especially pp. 208ff.

[410] See the list on the back cover of fascicle 18.

[411] See above, p. 223.

[412] See again the list on the back cover of fascicle 18.

[413] See the lists on the back covers of fascicles 19 and 20.

[414] From the fascicle numbering it is obvious that the text of the Nidānasthāna was not published immediately after that of the Sūtrasthāna which ends with fascicle 17. As can be seen in Appendix 6.2, beginning with fascicle 18 the text of the Vimānasthāna sets in. The printing press was no longer the Siddheśvara Press, but the Jyotiṣaprakāśa Press. This press printed all fascicles of the Vimānasthāna except for the unnumbered penultimate fascicle 21 of 1897 which was set at the Purāṇa Press where also the 'three and a half parts after part 17' with the text of the Nidānasthāna were printed. The described irregularity in the sequence of fascicles and Sthāna-s is explained by Harinath Sarkar in his announcement on the inner back cover of the jointly printed 'three and a half parts after part 17': Even though it would have been appropriate to publish the Nidānasthāna after the Sūtrasthāna, at the request of several venerable persons the publication of the Vimānasthāna was undertaken after the Sūtrasthāna. Then, because of the extreme eagerness of the subscribers, even before the Vimānasthāna had been completedly published the Nidānasthāna was published in one go, but was not assigned the regular sequential number 21 because it comes after the Sūtrasthāna. It can thus be concluded that the text of the Nidānasthāna was printed after the undated fascicle 20 (text of Vimānasthāna 5.37–8.6/7, pp. 65–96) and before the unnumbered fascicle 21 (text of Vimānasthāna 8.6/7–14/2, pp. 97–128) of 1897, and that the Purāṇa Press which was entrusted with printing the Nidānasthāna also obtained the order to print the subsequent fascicle 21. This change of printer is coincidental with the change from the Giriśa Pustakālaya to the Viśārada Auṣadhālaya as publishing house. Afterwards, Harinath Sarkar returned to the Jyotiṣaprakāśa Press which printed the last fascicle of the Vimānasthāna (22) and the first four fascicles of the Śārīrasthāna, the fourth book of the *Carakasaṃhitā* (23&24, 25, 26), in 1902. Later, in 1910, 1913 and 1915, this press also printed the two fascicles of the Kalpasthāna and the first fascicle of the Siddhisthāna, the seventh and eighth book, respectively.

Bazar just north of Berhampore, another great philanthropist and patron of learning who also sponsored and maintained some hospitals in the area. This comes right after Harinath Sarkar's edition of the *Carakasaṃhitā* was no longer published out of the bookstore Giriśa Pustakālaya, but from his own pharmacy, the Viśārada Auṣadhālaya.[415] Furthermore, A. F. Rudolf Hoernle (1841–1918), a British–German scholar who made important contributions *inter alia* in the field of ancient Indian medicine,[416] may also have provided financial support to Harinath Sarkar's publication project in its final stage, as can be understood from the above-mentioned extract from the Calcutta journal *Amrita Bazar Patrika* (January 23, 1913) printed on the back cover of the second fascicle of the Kalpasthāna (1913) of this edition[417]:

> Harinath's edition is almost complete but at this stage his labours are much hampered owing to the want of funds. Harinath has exhausted his slender purse for this noble work and, as he is not a 'pushing man,' his countrymen should, as the good Dr. Hornle, come forward with some pecuniary help to enable him to complete the work and preserve the great monument of the ancient Indian glory.

The advertisements for the Viśārada Auṣadhālaya,[418] later on for the Āyurvedaratnākara Auṣadhālaya,[419] and for the Giriśa Pustakālaya[420] printed on the covers of many of the fascicles were obviously meant to attract new customers among the readers and thus to eventually facilitate the financing of the publication whose price was considered high even in 1922.[421]

A further sponsor of an edition of the *Carakasaṃhitā* may possibly have been Raja Vinaykrishna Dev (Benoy Krishna Deb) of Shobhabazar in Calcutta whose photograph is shown in a vignette after the title page of the above-mentioned edition with Bengali translation by Satishchandra Sharma.[422] The presumed editor, translator and publisher of the book solemnly dedicates it to him on a special dedication page. However, contrary to what one would expect, neither there nor in his subsequent brief announcement (*vijñāpan*) does he explicitly mention any financial support received from this prominent member of an influential Calcutta family,[423] but merely refers to the great expense involved for him in publishing the *Carakasaṃhitā*

[415] See above, p. 232f, and here, n. 414.

[416] On August Friedrich Rudolf Hoernle see Stache–Rosen (1990, 95–96).

[417] See above. p. 234.

[418] See above, p. 233.

[419] See again above, p. 233.

[420] See above, p. 232. After the function of publishing house had passed from the Giriśa Pustakālaya to the Viśārada Auṣadhālaya following the publication of fascicle 20, the former still placed two further advertisements in fascicles 21 and 22.

[421] See E22 (1922), Prāstāvikam, 1–2: *iyam api ṭīkā prathamam atītakatipayābdair vaṅgadeśasthaiḥ śrīkavirājaharināthamiśraiḥ khaṇḍaśaḥ prakāśayitum ārabdhā nādyāpy avasānabhūmiṃ gatā ca bahumūlyatvān na sarvasulabhā.*

[422] E16 (1904). See above, p. 208.

[423] The two mansions of the Shobhabazar Rajas were actually located in the same street where Satishchandra Sharma lived; see above, n. 233. On Vinaykrishna Dev, see, e.g., Arp (2000, 74 and n. 251).

together with its translation. The book with more than 1500 pages was sold at the price of six rupees. Similarly, Avinashchandra exuberantly dedicates his partial edition of the *Carakasaṃhitā* with Cakrapāṇidatta's *Āyurvedadīpikā*[424] to Chamarajendra Wadiyar X, Maharaja of Mysore from 1868 (with his actual rule starting only in 1881) to 1894 ('H. H. Maharaja Chamajendra Woodayaz Bahadur G. C. S. I. Maharajah of Mysore'), mentioning *inter alia* his 'love and unbounded liberality towards the cultivation of Sanskrit science and literature,' but does not acknowledge any support by the Maharaja in his English Preface and its Sanskrit counterpart, the Bhūmikā.[425]

By way of their financial contribution, editor-cum-publishers, printer-cum-publishers, publishers and sponsors, but also subscribers, advance payers and other customer can be seen as patrons of the publication of the *Carakasaṃhitā* who thus contributed to its rescue from decline and loss.[426] However, the activity of editing

---

[424] See E6 (188ff.)

[425] See the copy of E6 (reaching up to Sū 14.8) in the library of the Bhandarkar Oriental Research Institute, Poona, mentioned in n. 406 above, and the copy with the call number VT 1390 of the British Library (reaching up to Sū 4.14). In the Bhūmikā, however, in a sentence not found in the Preface, Avinashchandra states: 'Now, should the excellent well-wishers for the country who are devoted to knowledge, the kings and great kings, compassionately throw a favourable glance on the matter under discussion, then – having seen its completion without obstacles – my hope will be fulfilled' (*atha vidyānurāgiṇāṃ deśahitaiṣivarāṇāṃ rājñāṃ mahārājānāṃ cātra prastutaviṣaye sakṛpaṃ kaṭākṣapātaḥ syāt, tarhi nirvighnenāsya parisamāptim ālokya phalavaty asmākam āśā sampatsyate*). This sounds like an indirect call for royal sponsoring of his publication plans in the area of medicine. Incidentally, it is striking that the Mysore government was among the buyers of Harinath Sarkar's edition of the *Carakasaṃhitā* with the *Āyurvedadīpikā* (E7) between 1895 and 1897; see above, p. 265f. The Maharaja died from diphtheria at the very end of 1894 on a visit to Calcutta (see Ikegame 2013: 101; see also p. 65 on his sudden death), which may possibly have also been the year when the title and dedication pages for the fascicles united in the above-mentioned copies of E6 were printed. However, further research is required to determine whether Avinashchandra's dedication indicates some specific concrete relationship to Chamarajendra Wadiyar X, and whether the latter had a special interest in Āyurveda and the *Carakasaṃhitā* (and possibly the *Āyurvedadīpikā*) and may therefore have had an order placed for Harinath Sarkar's edition in 1894. The Maharaja was also somehow connected to Calcutta as one of the financial supporters of Vivekananda's famous trip to Chicago in 1893 where the latter attended the Parliament of the World's Religions from September 11 to 27. See Vivekananda's letter to the Maharaja of Mysore from Chicago dated June 23, 1894 in Pavitrananda (1948, no. 64, 108–111, at pp. 108 and 110). In a letter to another supporter, Alasinga Perumal, from New York dated March 6, 1895, Vivekananda refers to the Maharaja's death; see Pavitrananda (1948, no. 117, 210). Further on Vivekananda's encounter with the Maharaja in Bangalore, see *The Life of Swami Vivekananda* (1974, 241–244) and Dhar (1975, 359–364, where the two letters are also referred to), and specifically on the issue of Chamarajendra's financial support of Vivekananda's trip to America *The Life of Swami Vivekananda* (1974, 243) and, more extensively, Dhar (1975, 361 and 363). For the prices of E6, see above, n. 406.

[426] More recently, the inclusion of obligatory readings from the *Carakasaṃhitā* in the modern Ayurvedic curriculum and official examination schedule of Ayurvedic colleges and universities also contributes to this aim. At the same time it promotes the ambitions of contemporary editors and ensures the continuing interest of publishers in bringing out further and novel editions of the *Carakasaṃhitā*. It remains to be seen whether recent attempts to spread knowledge of the *Carakasaṃhitā* among Ayurvedic physicians through diagnostic and therapeutical software

the *Carakasaṃhitā* was pivotal in this recovery of the ancient scientific heritage which turns out to be some kind of social practice. Moreover, editing the *Carakasaṃhitā* may be seen as a tool in the creation and shaping of a national scientific identity, or even of a broader cultural identity encompassing past scientific achievements as part of a newly created, or re-created, cultural memory, as will be outlined in the following.

#### 6.2.5.2 Shaping a National Scientific Identity

One can note a further prominent sentiment on the part of the editors that appears related to the issue of salvaging the *Carakasaṃhitā* and Āyurveda from decay and loss, and becomes manifest in their exuberant praise of Āyurveda. It is great pride in this ancient indigenous scientific tradition coupled with the conviction of its superiority *vis à vis* the European medical science which had been introduced by the colonial rulers and presented by them as undoubtedly much more advanced and efficient.[427] Thus, Vrajvallabh Sharma,[428] referring to the verse 'What exists here (i.e., in the treatise composed by Agniveśa and redacted by Caraka) exists elsewhere, what does not exist here exists nowhere' that is transmitted at the end of the *Carakasaṃhitā* in part of the textual tradition,[429] writes on the *Carakasaṃhitā* as follows:

running on databases built upon information extracted from the classical foundations of Āyurveda, including, of course, the *Carakasaṃhitā*, will have concrete and permanent results. See Gadgil (2010) on the *Triskandhakośa*, which is also available in print in more than 14,000 pages. In general, the repeated reprints of some editions of the *Carakasaṃhitā* seem to indicate that there is indeed a stable market for the text of the *Carakasaṃhitā*.

[427] On the critical, negative or even hostile attitude of the British towards Āyurveda, see, e.g., Arnold (2000, 66–69, 182–185).

[428] See above, p. 233.

[429] See CS Si 12.54ab: *yad ihāsti tad anyatra yan nehāsti na tat kvacit*. This verse is borrowed from the epic *Mahābhārata* where it appears in stanza 1.56.33 and, towards the end of the text, in stanza 18.5.38: *dharme cārthe ca kāme ca mokṣe ca bharatarṣabha / yad ihāsti tad anyatra yan nehāsti na tat kvacit //*. 'Concerning religious merit as well as material well-being, sensual gratification as well as salvation, Bull among the descendants of Bharata, what exists here (i.e., in the epic *Mahābharata*) exists elsewhere, what does not exist here exists nowhere.' That is, the many issues relating to the four ends of man (see above, p. 255f.) treated in the *Mahābhārata* are not exclusively found there, but also elsewhere; at the same time, there are no such issues as may not be found there but elsewhere. Together with a variation of the verse CS Si 12.53cd ('Medical treatment, oh Agniveśa, is directed towards the good of the healthy and the sick' [*cikitsitaṃ vahniveśa svasthāturahitaṃ prati*]), the verse appears as a motto on the title pages of E8 ($^{2}$1910), E9 (1897ff.) (see Fig. 6.20) (see also the quotation at the end of its introduction, p. 6) and E13 (1900) (Ni, Vi, Śā, Ka and Si) and is quoted to emphasize the comprehensiveness of the *Carakasaṃhitā*, for instance, in E17 (1911) (Bhūmikā, 4) and E30 (1932) (untitled preface, 2). It has also been adapted in several other Sanskrit treatises to express their claim to an all-encompassing scope.

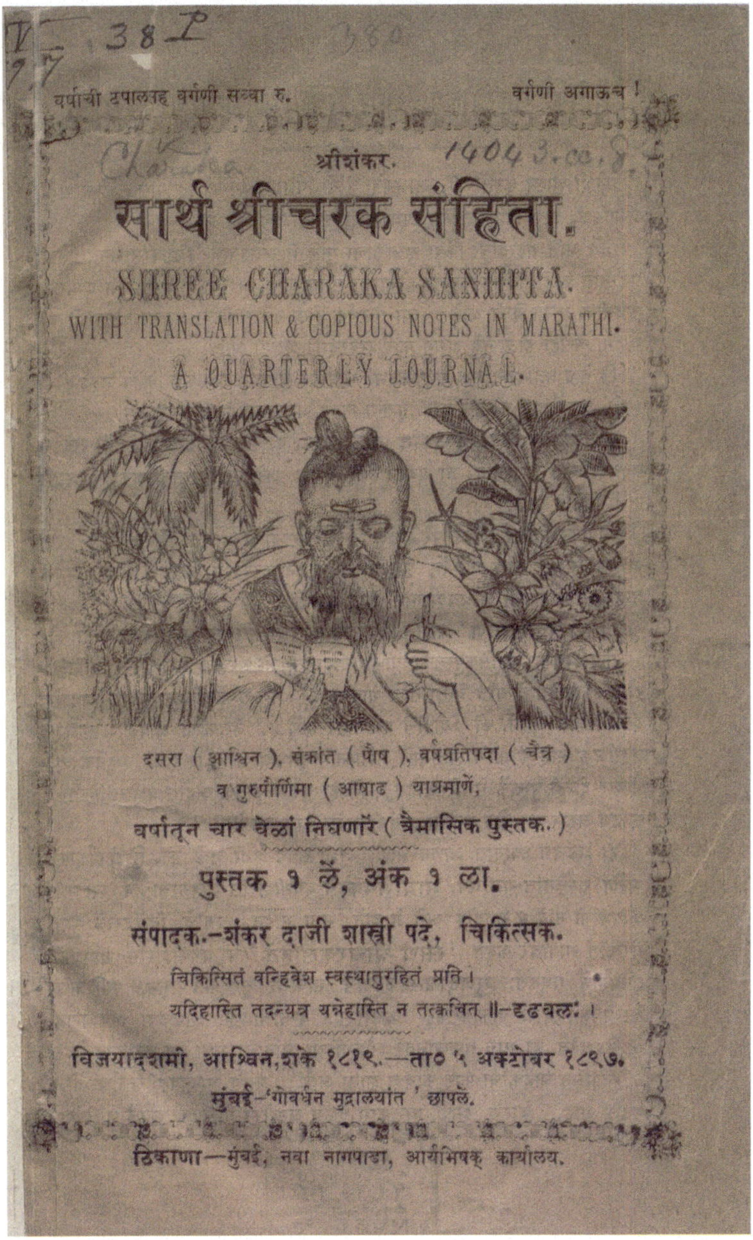

वर्षाची टपालसह वर्गणी सव्वा रु. वर्गणी अगाऊच !

श्रीशंकर.

सार्थ श्रीचरक संहिता.

SHREE CHARAKA SANHITA.

WITH TRANSLATION & COPIOUS NOTES IN MARATHI.

A QUARTERLY JOURNAL.

दसरा ( आश्विन ), संक्रांत ( पौष ), वर्षप्रतिपदा ( चैत्र )

व गुरुपौर्णिमा ( आषाढ ) याप्रमाणें,

वर्षांतून चार वेळां निघणारें ( त्रैमासिक पुस्तक. )

पुस्तक १ लें, अंक १ ला.

संपादक.–शंकर दाजी शास्त्री पदे, चिकित्सक.

चिकित्सितं वन्हिवेश स्वस्थातुरहितं प्रति।

यदिहास्ति तदन्यत्र यन्नेहास्ति न तत्कचित् ॥–दृढबलः ।

विजयादशमी, आश्विन, शके १८१९.—ता० ५ अक्टोबर १८९७.

मुंबई–'गोवर्धन मुद्रालयांत' छापलें.

ठिकाणा—मुंबई, नवा नागपाडा, आर्यभिषक् कार्यालय.

**Fig. 6.20** Title page of E9 (1897ff.)

In summary, it is enough to say on the matter of its superiority that in its own subject matter it is unique of its kind because the medical treatments, such as the application of enema and so on, that have been written down in it are evidently not found in another book even of our Ayurvedic medicine, let alone in those of European and Unani medicine. For this very rea-

> son, at the conclusion of the book, together with other things, the following statement which has the nature of its eulogy[430] has been written with great pride: 'What exists here exists elsewhere, what does not exist here exists nowhere.' To explain, what has been said in this book that indeed is found elsewhere, but what is not found in it that is not found anywhere. Enough, from just that much it is not difficult to impress on one's mind that absolutely all matters concerning medicine have been written down in it. Therefore, concerning its subject matter this is a book that is beautiful in all limbs[431] and perfectly endowed with all matters. Even foreigners have become covetous of it on account of its excellence and have translated it into their languages. At the beginning of the eighteenth century of the Christian era, first its translation into Arabic took place; subsequently, the one into Latin[432] has taken place.[433]

Similarly, Ramprasad, the above-mentioned author of a Hindi commentary on the *Carakasaṃhitā*,[434] emphasizes the completeness of Āyurveda, in spite of its delapidated state, and attributes the contemporary progress of European medicine in India to the imitation of numerous Ayurvedic medical treatments.[435] In this context, he

[430] In religious and other traditional Sanskrit literature, one often finds a versified eulogy (*praśasti*) of the text placed at its very end or even at the end of individual chapters.

[431] This epithet refers to the eight 'limbs' or branches that make up the 'body' of the medical science. They are listed in CS Sū 30.28 as follows: medical treatment of the body (i.e., especially internal medicine, but also dermatology and other areas of medical treatment not mentioned in the following), fine surgery (i.e., eye surgery and other kinds of surgery done with fine instruments), removal of foreign objects from the body (i.e., general surgery), pacification of poisoning through antidotes (i.e., toxicology), knowledge about demons (i.e., treatment of mental disorders), care of young children (i.e., obstetrics and pediatrics), the elixir of life (i.e., life-prolonging and rejuvenation therapy) and strengthening (i.e., potency therapy) (*tasyāyurvedasyāṅgāny aṣṭau. tadyathā-kāyacikitsā śālākyaṃ śalyāpahartṛkaṃ viṣagaravairodhikapraśamanaṃ bhūtavidyā kaumārabhṛtyakaṃ rasāyanaṃ vājīkaraṇam iti*).

[432] I am not aware of any translation of the *Carakasaṃhitā* into Latin. Possibly, some confusion has occurred here with the Latin translation of the *Suśrutasaṃhitā* by Franz Hessler which was published from Erlangen in 1844, 1847 and 1850 (*pace* Pecchia 2022a, 120: 'between 1844 and 1848'). The Arabic translation of the *Carakasaṃhitā* by 'Abdullāh ibn 'Alī, from an earlier Persian translation, was probably made around the year 800. See Meulenbeld (1999, IA, 116) and, for the approximate date of the Arabic translation, the references provided by him in note 8 on Part I, Chapter 11 (1999, IB, 203). For details, see Kahl (2015, 18–19).

[433] See E18 (1911), Bhūmikā, 5: *samaṣṭirūp se is kī uttamtā ke viṣay meṁ itnā hī kah denā bahut hai ki apne viṣay meṁ yah apne ḍhaṅg kā ek hī hai. kyoṁki is meṁ basti ādik jo cikitsā likhī gaī hai vah ḍākṭarī aur yūnānī kī to bāt hī kyā ? sākṣāt hamāre āyurvaidyak ke bhī anyă granth meṁ nahīṁ hai. isī se is granth kī samāpti meṁ is kī praśastisvarūp aur bātoṁ ke sāth yah vākyă baṛe abhimān se likhā gayā hai ki "yad ihāsti tad anyatra yan nehāsti na tat kvacit" arthāt jo is granth meṁ kahā gayā hai vahī sarvatră hai kintu jo is meṁ nahīṁ hai vah kahīṁ nahīṁ hai. bas, itne hī se yah dhyānsthă kar lenā kaṭhin nahīṁ hai ki vaidyaksambandhī sabhī viṣay is meṁ likhe gaye haiṁ. is se apne viṣay kā yah sarvāṅgsundar aur sarv viṣaysampann granth hai. is kī uttamtā se lubdh ho kar videśiyoṁ ne bhī apnī bhāṣāoṁ meṁ ise anuvādit kiyā hai. īsā kī āṭhvīṁ śatābdī ke ārambh meṁ pratham is kā anuvād arabī meṁ huā phir lāṭin meṁ huā hai.*

[434] See above, pp. 204f. and 231.

[435] See E17 (1911), Bhūmikā, 5–6, especially at p. 6, where he states ironically: '… Having imitated all these medical practices of Āyurveda, European medicine has today been advancing (i.e., progressing) in this lucky advanced kingdom' (… *yah sab āyurved ke cikitsā kā anukaraṇ kar ke hī āj unnatśīl śubh rājyă meṁ ḍākṭarī vidyā kī unnati ho rahī hai*). On the completeness of the *Carakasaṃhitā*, *Suśrutasaṃhitā*, etc., see also p. 7: *carak, suśrut ādi granthoṁ se aisā kaun viṣay bacā hai jo sthūl vā sūkṣmă rūp se in ke bhītar na bharā ho.*

also refers to traditional practices in the area of Ayurvedic fine surgery and general surgery[436] which even now contemporary advanced physicians will not even have dreamt of. For illustration, he adduces some wonderful 'surgical' practices described in Sanskrit literature, such as the refitting of Dakṣa's cut-off head by the Aśvin-s, the attaching of Brahmā's head, and the splitting of Bhoja's head and the extraction of souls from inside the skull.[437]

Enhancing the notion of the superiority of their own medical tradition still further, some editors claim that the medical sciences current in all countries eventually originate from Āyurveda or even from the *Carakasaṃhitā* itself. For example, Debendranath and Upendranath Sen[438] state that

> of all compendia the *Carakasaṃhitā* is the most excellent, like the faculty of thinking is the most excellent of the sense faculties[439] and consciousness the most excellent of the

[436] See above, n. 431.

[437] See E17 (1911), Bhūmikā, 6: *āyurved ke śalyăśālākyă jānnevāloṁ ne jo kāryă kiye haiṁ un ko abhī unnatśīl cikitsakoṁ ne svapnă meṁ bhī nahīṁ dekhā hogā. jaise aśvinīkumāroṁ kā dakṣ kā kaṭā huā śir lagā denā, brahmā kā mastak joṛnā, bhoj kā mastak cīr kar kapāl ke bhītar se jīvoṁ kā nikālnā ādi anek prakār kī kriyāyeṁ kaisī vicitră thīṁ.* For a discussion of these three 'surgical' operations, see Appendix 6.3 below.

[438] See above, p. 240ff.

[439] In the pre-philosophical religious literature, the mind or the faculty of thinking (*manas*) is often mentioned together with various other essential human faculties. Its precise position *vis à vis* these other faculties and especially the sense faculties (*indriya*) has been much discussed in the history of Indian philosophy. In the *Carakasaṃhitā*, we find more than one point of view on this issue. Debendranath and Upendranath Sen clearly adopt the position that *manas* is the sixth sense faculty next to the five faculties of smell, taste, vision, touch and hearing which is implied in CS Sū 26.43(1) (*tatra madhuro rasaḥ … ṣaḍindriyaprasādanaḥ …*). This doctrine of six sense faculties is explicitly mentioned as a counter-position to the doctrine of five sense faculties in CS (CE) Vi 8.33 (= CS Vi 8.37) (*… pratitantrasiddhānto nāma tasmiṃs tasmiṃs tantre tat tat prasiddhaṃ. yathā … pañcendriyāṇi yathānyatra, anyatra ṣaḍ indriyāṇi …*). However, references to the former are scarce, whereas references to the latter are abounding in the *Carakasaṃhitā* in a variety of contexts, for instance in the description of the human metabolism (CS Sū 28.4: *puṣyanti tv āhārarasāt … pañcendriyadravyāṇi dhātuprasādasaṃjñakāni …*). Next to this, we find the Sāṃkhya position of ten faculties, namely, five cognitive faculties (i.e., sense faculties) and five faculties of action, in the Śārīrasthāna. There, too, *manas* figures separately from these faculties (*indriya*); see, e.g., CS Śā 1.17 (*punaś ca dhātubhedena caturviṃśatikaḥ smṛtaḥ / mano daśendriyāṇy arthāḥ prakṛtiś cāṣṭadhātukī //*), 1.22ab (*indriyeṇendriyārtho hi samanaskena gṛhyate*), 1.56ab (*karaṇāni mano buddhir buddhikarmendriyāṇi ca*) and 1.64 (*buddhīndriyāṇi pañcaiva pañca karmendriyāṇi ca / samanaskāś ca pañcārthā vikārā iti saṃjñitāḥ //*). Even when the number of sense faculties is not specified, *manas* is mentioned separately from them and clearly conceived as something categorically different. This can be seen especially in epistemological contexts, such as the characterization of sense perception in CS Sū 11.20 (*ātmendriyamanorthānāṃ sannikarṣāt pravartate / vyaktā tadātve yā buddhiḥ pratyakṣaṃ sā nirucyate //*) and Vi 4.4 (*pratyakṣaṃ tu khalu tad yat svayam indriyair manasā copalabhyate*) or in the explanation of the sensation of pleasure and pain in CS Śā 1.138 (*ātmendriyamanorthānāṃ sannikarṣāt pravartate / sukhaduḥkham …*). Furthermore, the *manas* and the sense faculties are often listed as separate items alongside the body (e.g., CS Ni 1.35: *jvaras tu khalu … dehendriyamanastāpakaraḥ …*) and other factors relating to the production of cognition, such as the objects of cognition, cognition itself and the Self (e.g., CS Sū 11.33: *evaṃ pramāṇaiś caturbhir upadiṣṭe punarbhave dharmadvāreṣv avadhīyeta. tadyathā – dehendriyamanorthabuddhyātmaparīkṣāyām … iti*). The two also appear side by side, but separately in the context of the theory of the three humours; see, e.g., CS Sū 12.8 (*vāyuḥ … niyantā praṇetā ca manasaḥ, sarvendriyāṇām*

> elements.[440] Furthermore, scholars of antiquity agree unanimously that it is the root of the sciences of medical treatment prevalent in all countries.[441]

Similarly, Shankar Daji Pade[442] claims that

> the method of counteracting disease of the Arab, Greek[443] and other countries is indeed one that has found its root in the Indian science of medical treatment. On this matter there is certainly no controversy at all on the part of leading contemporary researchers, even if they may be European savants.[444]

Now, if Āyurveda is the root of all other medical sciences, it follows that it is the oldest among them and thus enjoys the high prestige of historical precedence. This is pointed out by several editors, among them Yadav Sharma[445] who states that

> here in this world, certainly ahead of the other countries, in the world region of Bharata[446] the extremely compassionate and most trustworthy great sages, for the cessation of disease for the sake of the people tormented by the four kinds of disease[447] and for the purpose of the continuance of health, divided the knowledge about life (*āyurveda*) into eight branches,

---

*udyojakaḥ ... bhavaty akupitaḥ. kupitas tu khalu ... mano vyāharṣayati, sarvendriyāṇy upahanti ...*), Sū 16.6ab (*buddhīndriyamanaḥśuddhir mārutasyānulomatā*) and CS (CE) Vi 8.74–75 (= CS Vi 8.89–90) (*kāryaṃ dhātusāmyam. tasya lakṣaṇaṃ vikāropaśamaḥ. parīkṣā tv asya ... sarvākārair manobuddhīndriyāṇāṃ cāvyāpattir iti. kāryaphalaṃ sukhāvāptiḥ. tasya lakṣaṇaṃ manobuddhīndriyaśarīratuṣṭiḥ*).

440 This is a reference to CS Śā 1.16ab where man is said to consist of six elements (*dhātu*): space, wind, fire, water and earth, and consciousness (*khādayaś cetanāṣaṣṭā dhātavaḥ puruṣaḥ smṛtaḥ*). See further CS Śā 4.6, in the context of embryology: 'Furthermore, one should realize that the embryo is a transformation of space, wind, fire, water and earth and embodies the seat of consciousness. Thus, by this reasoning the embryo is a conglomeration of the five great elements and embodies the seat of consciousness. For it (i.e., consciousness) is called its sixth element' (*garbhas tu khalv antarikṣavāyvagnitoyabhūmivikāraś cetanādhiṣṭhānabhūtaḥ. evam anayā yuktyā pañcamahābhūtavikārasamudāyātmako garbhaś cetanādhiṣṭhānabhūtaḥ. sa hy asya ṣaṣṭho dhātur uktaḥ*). See also CS Śā 6.4 (*tatra śarīraṃ nāma cetanādhiṣṭhānabhūtaṃ pañcamahābhūtavikārasamudāyātmakaṃ samayogavāhi ...*). On the five elements and the terminology applied to them (*bhūta, mahābhūta, dhātu*), see Preisendanz (2010).

441 See E10 (1897), Vijñāpanam, 1: *indriyāṇāṃ mana iva, bhūtānāṃ cetaneva saṃhitānāṃ carakasaṃhitā śreṣṭhatamā, parantu sakaladeśapracalitānāṃ cikitsāśāstrāṇāṃ mūlam iti pratnatattvavidbhir ekavākyena svīkṛtam*. This statement follows immediately upon the initial sentences of the Vijñāpanam quoted above, n. 355.

442 See above, p. 209.

443 See above, n. 349.

444 E15 (1903?), Bhūmikā, 1: *bhāratīyacikitsāśāstramūlībhūtaiva araba-yūnānaprabhṛtideśānāṃ rukpratikriyārītir ity atrādhūnikaśodhakāgraṇīnāṃ nāma yūropīyaviduṣām api nāsti kāścid vipratipattiḥ*. This attitude may be compared to the notion found in eighteenth-century China that all European science goes back to Chinese science mentioned in Han Qi's contribution to the present volume, Chap. 4. Han Qi also refers to the Chinese claim that Western scholars agree with this point of view.

445 See above, p. 247.

446 See above, n. 343.

447 Yadav Sharma probably refers here to the fourfold classification of disease taught in CS Sū 20.3: disease caused by external factors and disease caused by one of the three humours wind, bile and phlegm, that is, internal factors (*catvāro rogā bhavanti āgantuvātapittaśleṣmanimittāḥ ...*).

called medical treatment of the body, fine surgery, surgery, care of young children, the fundamental systematic scholarly treatise on antidotes, knowledge about demons, the elixir of life and strengthening.[448] Many fundamental systematic scholarly treatises were then composed in this and that branch. ... And among the books of the ancient science of medical treatment that are nowadays seen in the world this fundamental systematic scholarly treatise (i.e., the *Carakasaṃhitā*) is the most ancient of all. This is the opinion of historians.[449]

Dattaram Narayan Pathak,[450] for his part, agrees with 'certain far-sighted physicians' that before studying Āyurveda a certain amount of general familiarity with European medical science (*ḍākṭarīśāstră*) is necessary. In this connection, he refers to English books on botany and *materia medica*, mentioning the names of Rauksvarā, i.e., the British surgeon William Roxburgh, the 'Father of Indian Botany' (1751–1815), and Ḍiṃmak, most probably Henry Peers Dimmock (1857–1921) of the Bombay Medical Service who served as Principal of Grant Medical College in Bombay.[451] These books have a practical value inasmuch as they assist in the determination of the different names of medicinal plants in Latin, Arabic, Farsi, Hindi, Bengali, Marathi, Gujarati, and so on. Dattaram Narayan Pathak also acknowledges the help of English books in the determination of diseases. Even though the specific characterization of diseases is superior in Āyurveda, detailed description of diseases is evident in the English books.[452] However, with much confidence the editor makes it sufficiently clear that even though there may be some dependence (in current practice?), knowledge of Āyurveda does not everywhere rely on European medical knowledge and points out several of the numerous issues and areas that demonstrate

Another fourfold classification is found in CS Ni 1.12(4), namely, into curable, incurable, mild and violent diseases (... *vyādhayaḥ ... caturvidhāḥ sādhyāsādhyamṛdudāruṇabhedena*).

[448] See above, n. 431.

[449] E23 ($^{2}$1922), Nivedanam: *iha khalu jagatītaradeśebhyaḥ pūrvam eva bhāratavarṣe caturvidhavyādhiprapīḍitānāṃ janānāṃ vyādhinivṛttaye svāsthyānuvartanārthaṃ ca paramakāruṇikair āptatamair maharṣibhiḥ kāyacikitsāśalyaśālākyakaumārabhṛtyāgadatantrabhūtavidyārasāyanavājīkaraṇākhyeṣv aṣṭeṣv aṅgeṣv āyurvedaṃ saṃvibhajya teṣu teṣv aṅgeṣu bahūni tantrāṇi viracitāny āsan ... tantraṃ caitaj jagati sampraty upalabhyamāneṣu prācīnacikitsāśāstragrantheṣu sarvebhyaḥ prācīnataram itītihāsavidāṃ matam.* See also E23 ($^{3}$1933), Nivedanam, 1.

[450] See above, p. 212.

[451] On the pragmatic British interest in Āyurvedic drugs, see Arnold (2000, 68 and 181–182).

[452] E13 (1900), Bhūmikā, 2: *koī koī dūrdarśī vaidyoṁ kī sammati hai ki āyurved paṛhne ke pūrv ḍākṭarīśāstră kā sāmānyă abhyās thoṛābahut vaidyă ko avaśyă kar lenā cāhiye, kyauṁki udbhidvidyā (auṣadh-parīkṣaṇādi viṣay sab) pūrṇtayā iṅgrejī meṁ hī miltī hai. ata eva udbhid-auṣadh kā svarūp jānnā hoy tau rauksvarā ḍiṃmak ādi ḍākṭaroṁ ke granth paṛhne se hī jñān hoygā. kyauṁki ek udbhid (auṣadh) ke anyă anyă nām jaise lāṭin, aravī, phārsī, hindī, baṅgālī, mareṭhī, gujrātī ityādi iṅgrejīgranth dekhne se hī niścay hoyge. tathā rognirṇay ke viṣay meṁ bhī iṅgrejīgranthoṁ kī sahāyătā lenī ucit hai, kyauṁki rogoṁ ke viśeṣlakṣaṇ āyurved meṁ uttam hone par bhī rogoṁ kā vistārpūrvak varṇan iṅgrejīgranthoṁ meṁ hī dekhne meṁ ātā hai.* See also E17 (1911), Bhūmikā, 7: 'If an appropriate step would take place towards Āyurveda related study, then Āyurveda can again reach this very advanced state. For advancement (i.e., progress), there is no necessity to take anything from outside' (*yadi āyurvedīyă śikṣā kā yathocit prabandh ho jāy to phir bhī āyurved usī unnat avasthā meṁ pahuñc sakta hai. unnati ke liye kuch bāhar se lāne kī āvaśyăkatā nahīṁ*).

the uniqueness and superiority of Āyurveda and make reliance on European medical science, even though its physiology may be leading in various matters, completely unnecessary inasmuch as it can offer neither solutions nor assistance.[453] The same is true of Hakim (i.e., Unani or Arabic) and homeopathic medicine.[454] He even refers to 'a well-known Englishman residing in America who now writes that through the medical treatment of Caraka being prevalent there would not be so much untimely death on earth as is presently happening.'[455] A similar 'foreign' acknowledgement of the superiority of the treatments described in the *Carakasaṃhitā* is referred to by Vaman Kesheo Datar[456] after he has praised the *Carakasaṃhitā* as the most excellent systematic scholarly treatise on Āyurveda and stated that in spite of its antiquity even until today it appears as if new in terms of the subject matter of medical treatment – nothing deficient is found in it[457]: 'A certain American scholar has claimed that if the medical treatment taught by Caraka would be adopted immediately in the science of medical treatment in all countries, other systematic scholarly treatises that remove disease and protect health would not be necessary.'[458] Generally speaking, as Avinashchandra puts it in the Preface to his edition of the *Carakasaṃhitā* with Cakrapāṇidatta's *Āyurvedadīpikā*,

> European scientific men do no longer dare to pronounce the works of Rishi Agnibes and Susruta as the outcome of the quackery of a savage and ignorant age. Western scientific men who have been all along most intolerant of Eastern science have, though rather late, begun to realize the truth that the East is the Source of Light.[459]

---

[453] E13 (1900), Bhūmikā, 2: ... *kintu ḍākṭarī-śārīrak aur aur bātoṁ meṁ agrăgāmī ho kar bhī pūrvokt bātoṁ ke liye hamāre āyurved ke mukh kī tarah dekhega. ... in sab bātoṁ kā varṇan yadyapi ḍākṭarī meṁ vidyămān hai parantu us se hamārā samădhān aur sahāyătā kadāpi nahīṁ ho saktī. ... ham pratijñāpūrvak sāhas ke sāth kahte haiṁ ki ḍākṭarī śārīr-tattvă adyāpi us śreṇī ko nahīṁ pahuṁcā ki ukt sthaloṁ kī vyākhyā meṁ hamārī sahāy karai.*

[454] E13 (1900), Bhūmikā, 2: *hakīmī ḍākṭarī tathā homiyopethik śāstră kā jñān is jagah ham ko kuch bhī sahāyătā nahīṁ kartā. ... is jagah tau sab hī ne hamāre ṛṣivākyă ko prāmāṇik mānā hai.*

[455] E13 (1900), Bhūmikā, 2: *ek vikhyāt amerikāvāsī iṅgrej samprati likhtā hai ki carak kī cikitsā pracalit hone se pṛthvī meṁ itnī akālmṛtyu nahīṁ hoy ki jitnī ab ho rahī hai.* This statement is repeated in slightly different words by Vrajvallabh Sharma in E18 (1911), Bhūmikā, 5: *ek amerikāvāsī prasiddh aṅgrejmahāśay kī sammati hai ki yadi "carak" kī cikitsā sarvatră kām meṁ lāyī jāy to jagat meṁ is samay ke samān akālmṛtyuyeṁ na hoṁ.*

[456] See above, p. 244.

[457] See E22 (1922), Prāstāvikam, 1: *prācīnatve saty apy asya tantrasyādyāvadhi cikitsāviṣaye navīnatvam iva pratibhāti, na kim api parihīnam atra vidyate.*

[458] See E22 (1922), Prāstāvikam, 1: *kaścanāmerikan paṇḍitas tu cikitsāśāstre sarveṣu deśeṣu yadi carakoktā cikitsā sapadi gṛhītā syāt tarhy anārogyadhvaṃsakānāṃ ca svāsthyasaṃrakṣakāṇām anyeṣāṃ tantrāṇāṃ nāvaśyakateti pratijñātavān.*

[459] See E6 (1888ff.), Preface (p. 1). For the Sanskrit version see the immediately following Bhūmikā (p. 1): *na hīdānīṃ pāścātyadarśanavido 'pi suśrutādīnāṃ prācīnatantrāṇām asāratvaṃ pravañcanāmūlatvaṃ ca pratipādayitum utsahante. yāvad etāvantaṃ kālaṃ pāścātyadarśanavidaḥ prācyadeśīyeṣv āyurvedādiśāstreṣūdāsīnā evāsan. gacchatā kālena te 'pi punaḥ prācyadeśa eva sarveṣāṃ jñānālokānāṃ nidānam iti pratipādayanti.*

The editing and publishing of the *Carakasaṃhitā* can thus be seen as an act of self-assertion and reassurance, especially during the pre-independence period, and as 'part of the contestation of the hegemonic claims of Western medicine by indigenous medical practitioners and scholars.'[460] Even though to a lesser degree, this situation continues into the post-colonial period. In the eyes of its editors, editing and publishing the ancient foundational text now becomes part of some international competition in building a special national reputation when the antiquity of sciences is at stake.

#### 6.2.5.3 Accommodating the 'Foreign Knowledge' in the Ancient Medical Knowledge

It will have to be examined separately and in detail to which extent contemporary works on European medicine were accessible and familiar to the above-mentioned editors who explicitly refer to it, and how precisely they anchored the modern 'foreign knowledge' in their own ancient knowledge embodied by the *Carakasaṃhitā* and consequently in their editions. This accommodation may be expressed in their shaping of the edited text itself; in their accompanying translations or paraphrases of it or in their commentaries; in the addition of para-textual elements of different kinds, such as the title pages, introductions and prefaces, etc., already studied under certain aspects in the present paper, chapter and section headings, and glossaries; or in the illustrations with which the editors and publishers supplemented the edited text.

In a preliminary way, I would like to briefly address these illustrations which are a special feature of some early editions of the *Carakasaṃhitā*.[461] To my knowledge, in this context the first anatomical drawings appear in the commentary on the Sūtrasthāna in Shankar Daji Pade's edition with Marathi translation and commentary.[462] They are clearly adapted from drawings in books on European medicine, and understandably so as there is hardly any evidence for a tradition of anatomical illustrations in South Asian medical manuscripts.[463] The drawings show cross sections of the male and female reproductive organs together with the bladder which are supplied with numbers for the identification of their individual parts.[464] The most important among the latter are identified by Shankar Daji Pade using Sanskrit and Marathi as well as English and Latin terms (in Devanagari and Roman script, respectively) in a synthesis of traditional and modern knowledge which is also continued in his subsequent explanations in Marathi. Furthermore, thirty-two anatomi-

---

[460] See Stark (2012, 406). As has been observed by Arnold (2000, 176), medicine was the 'most practical, and therefore most contentious, area of engagement between Hindu and Western science.'

[461] For a rather recent edition of the Sūtrasthāna of the *Carakasaṃhitā* supplied with tables, diagrams and illustrations see E53 (2004).

[462] E9 (1897ff.).

[463] See Wujastyk (2008, 201–204 and 2009, 206–207).

[464] See E9, part 1 of Sū, pp. 34 and 36.

cal drawings are found at the end of the text, translation and commentary of the Śārīrasthāna, that is, the book of the *Carakasaṃhitā* 'that relates to the body,' with subsequent extensive explanation, again including some Latin and English terms in Roman script.[465] This is followed by further illustrations related to obstetrics, that is, of the female reproductive organs, conception, pregnancy and birth, again with Marathi commentary using Latin and English terms, some of them with Devanagari transcription.[466] Finally, the correlation between traditional indigenous knowledge and modern 'foreign' knowledge is expressed in a schematic and systematic way in the form of an English–Marathi glossary of medical terms, with Devanagari transcriptions of the English and Latin terms in Roman script, arranged according to the Latin alphabet.[467] This structure of the glossary suggests that Shankar Daji Pade considered the European medical terminology and knowledge as the point of departure which implies a certain element of primacy of this terminology. This and his entire approach as described above, most probably aimed at guiding Ayurvedic practitioners towards integrating aspects of modern European medicine into their traditional indigenous medical knowledge and practice, shows that Shankar Daji Pade belonged to those physician–scholars whom Charles Leslie called 'syncretists' and 'integrationists.'[468]

From the above perspective, the illustrations in Khemraj Shrikrishnadas's edition together with Mihirchandra's Hindi commentary[469] are especially interesting. These illustrations start with two drawings showing the veneration of God Śiva in his Ardhanārīśvara (half man, half woman) form, together with his mount, the bull Nandin, under the heading 'Half-bodied Form of Śiva,'[470] and God 'Dhanvantari, the King of Physicians,'[471] in human form in his clinic surrounded by patients and

[465] See E9, part 2 of Śā (1907), after p. 60. The explanation of the thirty-two drawings extends over 128 separately numbered pages. In the second edition of this fascicle (1914), at least in the copy San.D.376 of the British Library, these drawings are missing, except for the first drawing which shows the human skeleton.

[466] See E9, part 2 of Śā (1907), 129–148 (second pagination). This series of drawings is completely missing in the copy of the second edition of this fascicle (1914) owned by the British Library (San.D.376).

[467] See E9, part 2 of Śā (1907), 149–158 (second pagination, with some misprints affecting the page numbers). The glossary is also found at the end of the second edition of this fascicle (1914), with a separate pagination (1–8).

[468] See Arnold (2000, 177). This is also suggested by Shankar Daji Pade's use of the word 'Aryan' in connection with Āyurveda and its foundational works; see above, p. 260 and n. 372.

[469] See E11 (1898) and above, p. 205.

[470] *Ardhāṅgī śiv kī mūrti*. I fail to see the significance of this prominent first illustration, unless the editor and publisher of this and other editions of the *Carakasaṃhitā* venerated this form of Śiva as his personal deity, in spite of the Vaiṣṇava name of his father (Shrikrishnadas) and the Vaiṣṇava name of his press, the Śrīveṅkaṭeśvara Press.

[471] *Dhanvantari vaidyă rāj kī mūrti*. On Dhanvantari, see above, p. 238. This drawing does not really fit here because Dhanvantari does not figure in the traditional narrative on the origin of the *Carakasaṃhitā*, but rather in that on the origin of the *Suśrutasaṃhitā*. See also below, n. 473. In E12 (1898, ²1903), however, where we find a very similar set of illustrations, a scene with Caraka among his patients and assistants has been added to the one with Dhanvantari; see below, n. 475.

busy assistants. The following plates depict the mythological and legendary origin of Āyurveda and the *Carakasaṃhitā*[472] in a unique way,[473] showing, for instance, the mythological stages of transmission from God Prajāpati mounted on his vehicle, the goose, to the twin-gods called Aśvins ('those characterized by horses'), who are depicted as centaurs, and from God Indra mounted on his elephant Airāvata to sage Bharadvāja (Fig. 6.21). They also show Caraka the redactor and Dṛḍhabala the restorer of the *Carakasaṃhitā* in learned conversation (Fig. 6.22). Next comes a curious set of nine line drawings illustrating different types of fever in the form of gruesome fever demons, one of them the 'triple-headed fever' (Fig. 6.23). Quite unexpectedly, these plates with an evidently mythological content are followed by a plate showing some contemporary medical instruments, labeled in English written in Devanagari script, and several plates with anatomical drawings. The anatomical plates show the brain, spine and main internal organs (1), the alimentary canal (2) (Fig. 6.24),[474] the lungs and the skeleton (3), the eye ball (4), the brain (5), the bones (6), the nerves (7) and the vessels of the trunk (8). Within the plates, each bodily part is labeled or explained mainly with Sanskrit-derived Hindi words and English terminology, partially written in Devanagari script for those readers who may not be very familiar with English Latin-derived medical terminology especially when written in Roman script. These plates were thus drawn and labeled with an integrationistic intention to accommodate certain elements of modern European medicine, such as specific knowledge about the brain and the nerves whose functions were not

---

[472] See above, p. 197f.

[473] Here, too, we see a special place assigned to Dhanvantari (see above, n. 471) which gives him precedence over God Brahmā in the chain of transmission of medical knowledge inasmuch as in the third panel, Dhanvantari is shown in divine form in the top register under the heading 'Own Form of Dhanvantari, Proud of Āyurveda' (*āyurvedābhimānī dhanvantarisvarūpam*). Facing him, a monkey is depicted holding a jar in his hand under the heading 'Own Form of Āyurveda' (*āyurvedasvarūpam*). This matches the iconography of Āyurveda according to the Hemādrivrata Khaṇḍa of the *Viṣnudharmottarapurāna* as quoted in Taranatha Tarkavacaspati (1881, s. v. *devatāpratimā*, p. 367b). The monkey must be Hanumān holding the jar of ambrosia (see above, p. 257, and the explicit labeling of the same kind of jar at the feet of the two Aśvin-s as *amṛtaghaṭaḥ* in Fig. 6.21) and an allusion to two episodes in the epic *Rāmāyaṇa* which describe Hanumān's miraculous flight from the island of Laṅkā to the Himalayas and back to bring urgently needed medicinal herbs for the treatment of Rāma, his brother Lakṣmaṇa and the many monkeys who had been wounded in Rāma's battle against the demon king Rāvaṇa, ruler of Laṅkā. See *Rāmāyaṇa* Yuddhakāṇḍa 61 and 89, translated in Goldman et al. (2010, 326–332 and 416–419). For a third version where Hanumān's feat is cut short because of the appearance of the bird Garuḍa Vainateya, God Viṣṇu's mount, who heals Rāma and Lakṣmaṇa by merely touching them, see Yuddhakāṇḍa 40.1–40. Brockington and Brockington consider only the core of the episode in Yuddhakāṇḍa 89 – where the mortally injured Lakṣmaṇa is restored to life – as part of the oldest version of the *Rāmāyaṇa* reconstructed by them (2006, 339). On the three episodes see also Goldman et al. (2010, Introduction, 65–67). On the herbs fetched by Hanumān, see below, Appendix 6.3. The described upper register of the panel has been used as the cover illustration of Cerulli (2012). In the lower register of the third panel, the artist shows God Brahmā instructing Prajāpati.

[474] For technical reasons, this figure is taken from E18 (1911), and not from E11 (1898), as can be seen from the different frame and style of heading. However, the differences, e.g., in the handwriting, are minimal and difficult to spot unless one looks for them. See below, n. 475, for parallel versions of the plates.

**Fig. 6.21** The transmission of Āyurveda from Prajāpati to the Aśvins (top register), and from Indra to sage Bharadvāja (bottom register), E11 (1898), panel 4

**Fig. 6.22** Caraka and Dṛḍhabala in conversation, E11 (1898), panel 6 (bottom register)

**Fig. 6.23** The 'triple-headed fever' demon, E11 (1898), panel 9

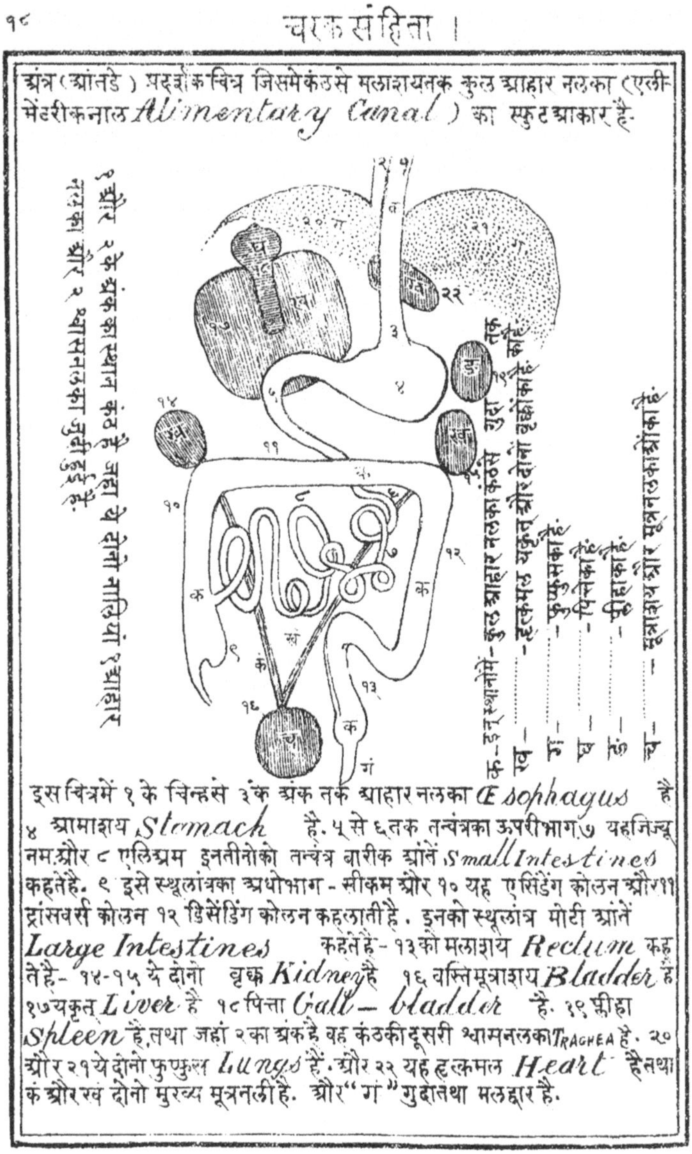


**Fig. 6.24** Alimentary canal, E18 (1911), panel 18

known in Āyurveda, in the ancient anatomical knowledge of Āyurveda, and to correlate certain other, common elements. At the same time, one could also understand that the anatomical plates are meant to anchor European medical knowledge in Āyurveda and suggest the origin of the former in the latter, making the former dependent on the latter and thus eventually less prestigious.[475] From another point of view, modernization and traditionalism meet in these illustrations, and the tension between the two discourses is minimalized. The entire choice of the compiled illustrations, their sequence and particular contents also reflect the editors' ideas about the readers they intend to address with their editions. In the present cases, they are most probably traditional practitioners who are not sufficiently versed in Sanskrit to follow the text without commentary or translation[476] and non-professionals interested in Āyurveda, both cherishing a belief in the divine origin of Āyurveda and in fever demons and at the same time being to some degree already exposed to modern European medicine and ready for its integration into their traditional knowledge.

[475] The same set of illustrations as in E11, redrawn and with some minor corrections of the text, is also found in another, later edition of the *Carakasaṃhitā* published by Khemraj Shrikrishnadas at the Śrīveṅkaṭeśvara Press, namely, the one with Ramprasad's Hindi commentary; see E17 (1911, next edition 1922) and above, p. 204f. E12 (1898, $^{2}$1903), the edition with Shrikrishnalal's Hindi translation (see above, p. 208f.), which was printed in the same year as E11 by another press, the Mumbaimitra Press in Mathura, and is very similar to E11 in terms of its title page and layout, also has a number of plates after the editorial communication (*nivedan*), a long history of Āyurveda in Hindi written by the translator, the advertisement of three other books published by the publisher–printer Shyamlal Shrikrishnalal, and the table of contents. The first set of plates depicts the origin of Āyurveda in a similar style as in E11, but in a less naïve way in the mythological part. This is followed by two plates that show Dhanvantari and Caraka treating patients in their respective clinics, the first of which is very similar to the matching plate in E11 described above. The following set of nine plates with fever demons is also very closely related to those in E11. The anatomical plates are missing in E12. As the prefaces to the two parts of this edition-cum-translation are also signed by Gangavishnu Shrikrishnadas of the Lakṣmīveṅkaṭeśvara Press, Bombay, who functioned as commission agent for the book and printed the two beautiful title pages (see above, n. 61), and was furthermore a brother of Khemraj Shrikrishnadas (see above, n. 236), it is obvious that the two editions E11 and E12 were somehow coordinated and that the design of some of the plates was shared in this context. This was most probably not the case with the full set of illustrations very similar to those of E11 that follows the foreword of E14 (1900), the edition with the Hindi commentary by Kalicharan Sharma and Kshamapati Sharma published in Lucknow (see above, p. 239). Furthermore, in E18 (1911), the edition with Ravidatt's Hindi translation published by the Nirṇayasāgara Press in Bombay (see above, p. 232f.) and in the same year as Khemraj Shrikrishnadas's first edition with Ramprasad's Hindi commentary (E17), we find a full set of very similar, but not identical and in some respect slightly cruder illustrations which were clearly drawn based on the model of the illustrations in E11 and E17.

[476] At least for the Kavirāj-s of Bengal, Cordier reports in a letter from India written in 1898 that they 'basically' or 'in principle' (*à fond*) studied the *Carakasaṃhitā*; see Roşu (1989, LXXXIV). Elsewhere Cordier mentions that the word-by-word study of the *Carakasaṃhitā* was part of the Kavirāj curriculum (Cordier 1899, 556 = Roşu 1989, 428). In 1903, Liétard states more explicitly that the study of the *Carakasaṃhitā* by the contemporary Kavirāj-s takes place only by way of 'meagre manuals' with extracts from the ancient foundational works, while at the same time the latter are constantly reedited and translated into modern Indian languages (Liétard 1903, 91 = Roşu 1989, 272).

## 6.3 Conclusion

For reasons of space, the paratextual elements of the editions of the *Carakasaṃhitā*, such as the preliminaries (title pages, dedication pages, announcements, prefaces, introductions, etc.), the textual supplements (e.g., bilingual glossaries) and the graphic supplements to the text in the form of illustrations and diagrams could not be explored further in the present paper. Various aspects of further paratextual elements, such as running titles and headings, footnotes, tables of contents and indices, and illustrations and ornaments that are not directly related to the content of the text and may appear at various places in the books[477] also deserve examination and analysis. Moreover, a fuller treatment of the issue of the pricing of the editions is required. Other important aspects of the editions, such as the concrete shape of the printed text, its layout, formal structuring and content-segmentation in relation to the text in traditional South Asian manuscripts,[478] could not be addressed at all here. Furthermore, it will have to be explored how the editors of the *Carakasaṃhitā* saw their role and chief goal from a philological or text-critical point of view, which methodology they adopted and possibly described, and which were their actual editorial practices as well as their linguistic practices with regard to their editorial activity as such and its details. The general linguistic practices of editors and other involved actors, such as publishers, that are directly related to the books also deserve attention and interpretation, such as the Sanskrit or Sanskrit-derived terminology chosen to designate prefaces, introductions, etc., and tables of contents, appendices, etc., inasmuch as these practices further reflect on how they understood their activities and roles as editors, etc. Continuities and discontinuities with traditional philological practices and terminology will have to be considered in this. Another important issue closely connected to that of the editors' perception of themselves as editors and of their methodology are the sources of the editions, that is, the textual or other witnesses utilized by the editors for their editions of the text: what were the editors' attitudes towards them, how did they evaluate and utilize these witnesses, and how did they choose and obtain them in the first place, especially in the case of manuscripts? In this context, further facets and aspects of the networks among physician–scholars in colonial India which have already been touched upon in the present paper and their extension to places outside South Asia will also become manifest.[479] Finally, for more comprehensive results it would be necessary to fully

---

[477] See, e.g., the only indirectly related drawing of a physician–scholar, maybe even a physician–sage, on the cover of E9 (1897ff.) in Fig. 6.20 above. On this figure, see also n. 486 below.

[478] For a first attempt to come to grips with the content-related segmentation of the *Carakasaṃhitā*, the numbering of segments and naming of chapters in the history of its transmission, with reference to some manuscripts and printed editions, see Preisendanz (2018), Sections 7.4.5–6, 7.4.8 and 7.5.

[479] The systematic study of acknowledgements found in prefaces, introductions, etc., will also further contribute to the prosopography of physician–scholars involved in editions of the

consider also the editions published after independence which have been treated with less detail here, to examine the editions of other medical works produced by the editors of the *Carakasaṃhitā*, and to conduct a similar study on the editions of the second foundational work of the early-classical period, the *Suśrutasaṃhitā*.

In spite of the need to do further research as outlined above, a multi-faceted picture of the editors and other actors involved in the publication of editions of the *Carakasaṃhitā* from its *editio princeps* until the eve of independence has emerged from the present study. Furthermore, first steps were taken towards the determination of the place of the editions of the *Carakasaṃhitā* in the cultural history of colonial India from the second half of the nineteenth century onwards. Some aspects of this case history exemplify general phenomena in the history of the printed book in colonial India and of printed editions of Sanskrit texts in general, both of which are not yet studied sufficiently, while others specifically relate to the practically unexplored history of editions of medical texts as opposed to texts belonging to other genres of Sanskrit literature. As is highlighted by the case of the editions of the *Carakasaṃhitā*, the key figures are in their majority practicing physicians with an excellent specialized knowledge of their professional scientific tradition and its texts, and for the most part not pandits, that is, traditional textual scholars and teachers with broad and profound expertise in several bodies of traditional knowledge, such as Sanskrit grammar and poetic theory, the various philosophical traditions and socio-religious norms (law), or in the belles lettres. Different from the pandits, these physician–scholars were not sought out early by the British orientalists as informants, resource persons or consultants, employed as educators in modern institutions of Sanskrit learning and commissioned to edit Sanskrit texts, or employed in various Company or government offices. They were not seen as 'authoritative guardians and interpreters of Sanskritic knowledge'[480] in the same way as the pandits in general and thus played a different cultural and social role. Furthermore, they were not patronized in the same way by the British with the aim to 'naturalise' the Company's or Crown's presence and ruling authority, as Dodson has formulated it.[481] Because Āyurveda, their area of expertise, was considered by far inferior to European medical science or not even recognized as a science,[482] it was not turned into an 'instrument of colonial rule,' different from the field of Indian languages and other areas of Sanskritic knowledge.[483] Thus, knowledge production in the area of Āyurveda largely remained in the hands of physician–scholars, even after this situ-

*Carakasaṃhitā*. In the present paper, this kind of information has been considered only from some perspectives.

[480] Dodson (2007, 44).

[481] See Dodson (2007, 45).

[482] For instance, courses on Āyurveda were initially offered next to courses in European medicine at the Sanskrit College in Calcutta, but then abolished in 1834; see Mukhopadhyaya (1922–1929, II, Introduction, 15–16). According to Singh and Joshi (1982, 318) and Arnold (2000, 62), the *Carakasaṃhitā* was studied in these courses. In 1902, Cordier reports that courses on Āyurveda were officially suppressed at Sanskrit Colleges; see Cordier (1902, 491 = Roşu 1989, 511).

[483] Cf. Dodson (2007, 53).

ation had changed with the establishment of Ayurvedic Colleges and other modern institutions of Ayurvedic learning, in the beginning mostly through their own initiative and that of sympathizing non-governmental sponsors, in order to counteract the perilous effect the simultaneous introduction and spread of European medicine had on the practice of Āyurveda and the continuation of its teaching and study in general.[484]

In their endeavor to revive indigenous interest in Āyurveda and to encourage and strengthen its teaching, study and practice 'for the good of all,' a corollary of the rise of Hindu nationalism,[485] the physician–scholars self-confidently took advantage of modernity with its new medium of print[486] introduced by the colonial rulers and utilized it for their own agenda which included the editing and dissemination of neglected foundational works of their traditional science such as the *Carakasaṃhitā*.[487] Sometimes on their own financial risk, but sometimes also supported by publishers, subscribers and munificent sponsors, they strove to make the ancient *Carakasaṃhitā* generally accessible to fellow physician–scholars and practitioners, particularly to those for whom it had so far played no role in their professional training and daily practice, and to non-specialists. Towards this end, they printed the text also in the regional scripts and at affordable prices, and added translations and commentaries in Sanskrit and the regional languages. They proved to be successful in their endeavor because in this way the *Carakasaṃhitā* became part of a re-created cultural memory and instrumental in the shaping of a new national scientific identity through the assertion of its comprehensiveness and great scientific value, and thus of its validity *vis à vis* the rival European medical science. As Arnold has suggested, foundational works such as the *Carakasaṃhitā* were even central to establishing claims for the validity and rationality of traditional Indian science as such.[488] Various integrative strategies aimed at the reconciliation and alignment of particular elements of the two medical knowledge systems were part of this validation of the ancient work and are another context where modernity and tradition meet in editing the *Carakasaṃhitā*.

---

[484] For a vivid and engaged account of the decline of Ayurvedic practice and education and of its reestablishment and institutionalisation in colonial India, with a regional focus on Bengal, see Mukhopadhyaya (1922–1929, Introduction).

[485] See Arnold (2000, 176).

[486] See also the remarks in Pecchia (2022a, 118, 120) on the embrace of modernity, manifested by the new technology of printing, on the part of the Vaidya community, and, beyond that, the specialist printed book as a sign of their group identity within a colonial middle class. Pecchia (2022a, 118) points out that the sage-like figure depicted on the title page of E9 (1897ff.) in Fig. 6.20 above holds a European-style book in his hand, not a manuscript in the wide-spread traditional pothī format.

[487] See the similar role ascribed to the early publishers of Sanskrit medical works in Pecchia (2022a, 119).

[488] See Arnold (2000, 176).

# Appendices

## *Appendix 6.1: A Survey of the Editions of the* Carakasaṃhitā

The following survey contains basic information on all editions (including second and other re-editions) explicitly referred to in the present paper. Empty slots indicate that information is not available, whereas '–' refers to the confirmed absence of an item. '?' means that a detail is uncertain, '??' that it is doubtful. Conjectured information is placed between square brackets.

As in the present paper, for practical reasons the names of editors, translators, publishers, etc., provided in Sanskrit and vernacular languages have been anglicized. In this, the regional pronunciation is considered unless the title page, etc., gives some version of the name in Roman script. Already anglicized names of persons have been unified and slightly adjusted, wherever necessary, and all names have been made uniform in terms of the word divisions, for the sake of a clearer distinction between personal names and surnames of various kinds. Titles and honorifics attached to the names of persons have been omitted, except in those cases where personal names and surnames appear only as initials.

Sanskritic names of printing presses and publishing institutions are consistently spelled with diacritics and according to their Sanskrit pronunciation.

Column 1 (No.):
The sigla refer to the number of the printed edition in Part IV, Chapter B.2 ('An Annotated Bibliography of the Printed Editions of the *Carakasaṃhitā*') in Preisendanz et al. (forthcoming). A prefixed asterisk indicates an edition listed by Meulenbeld (1999, IB, 3–6) but not been seen by him. If I could see such an edition or at least obtain further information on it, this asterisk has been removed. If an edition not listed by Meulenbeld could be found, an asterisk is appended to the siglum, even if the edition itself could not be seen by me. Naturally, this is not done in the case of editions that appeared after Meulenbeld had published his list. A superscripted number following the siglum refers to the number of the edition of the book. Superscripted 'N' refers to a next, but unnumbered edition ('follow-up' edition), superscripted 'D' to a duplicate edition, and superscripted 'R' to a reprint. For the fuzziness involved in the distinction between first and further editions and reprints, see above, p. 200. The information provided here generally follows the evidence of the respective book itself, that is, it is mainly based on the title page, imprint page, preface, introduction, etc.

Column 2 (Editor):
If an editor also functions as translator or commentator, this is indicated by the abbreviations 'tr.' and 'c.,' respectively. Names in round brackets indicate explicitly mentioned secondary editors, e.g., of further editions or reprints. If a name appears in square brackets, it is conjectured that the translator or commentator also edited the text of the *Carakasaṃhitā* included in the publication, even though the title page or preface, introduction, etc., does not say so explicitly. Sometimes, in the case of unseen editions for which only very rudimentary bibliographical data could be gathered, title pages are not available. In such cases, if only the name of a translator or commentator is known it is similarly conjectured that the same person also edited the text of the *Carakasaṃhitā* printed alongside the translation or commentary. It should be noted that in several cases the surnames of editors do not appear on the title pages and have been obtained from the preface, introduction, etc., or from other sources.

Column 3 (Place):
For the sake of uniformity, the modern English names of places are used here, regardless of the form and language of the place name as found in the edition itself.

Column 4 (Publisher/Press):
Especially in the early period of the publication of editions of the *Carakasaṃhitā*, publishers and printers, editors and publishers, and publishing houses or other publishing institutions and printing presses were often identical. See above, Sect. 6.2.4. If next to the name of the publisher the name of a printer is mentioned in a particular edition or if the publisher is expressly stated to have also been the printer of the book, this name is included in round brackets after the name of the printing press.

Column 5 (Year):
All years are given in CE years. Wherever a CE year had to be calculated from a year of an Indian era, I have followed Renou and Filliozat (1985, 736–737) for the approximate calculation: subtraction of 57 years from Samvat (i.e., Vikramasamvat) years (assuming expired years), addition of 78 years to Śaka years (assuming expired years), and addition of 593 years to Bengali era years (BS) (assuming current years). The original dates are given in round brackets. Depending on the month of publication, which is normally unknown, a book may not have appeared until the first months of the subsequent year of the Common Era. If a Common Era year has been supplied alongside the year of the Indian era, the former date was adopted as authoritative. In those cases where an edition appeared in several volumes published in different, not necessarily consecutive years, only the first and last years are given, joined by a dash. Sometimes, especially when editions were published in fascicles, the year of the completion of the book is unknown; here, 'ff.' follows after the initial year.

Column 6 (C/E):
This column provides information on the language of commentaries, elucidations or annotations that accompany the Sanskrit text of the *Carakasaṃhitā*. The following abbreviations are used: B = Bengali, E = English, F = French, G = Gujarati, H = Hindi, M = Marathi, Ma = Malayalam, O = Oriya, S = Sanskrit, T = Telugu.

Column 7 (T):
This column gives information on the language of translations of the *Carakasaṃhitā* accompanying its Sanskrit text. The following abbreviations are used: B = Bengali, E = English, F = French, G = Gujarati, H = Hindi, M = Marathi, O = Oriya, T = Telugu, Ti = modern Tibetan.

Column 8 (P/I):
Here, the language of prefaces, introductions, etc., is recorded. For the abbreviations, see under Columns 6 and 7.

Column 9 (S):
This column indicates the script in which the Sanskrit text of the *Carakasaṃhitā* is printed in the respective edition. The following abbreviations are used: D = Devanagari, B = Bengali script, G = Gujarati script, Ma = Malayalam script, O = Oriya script, R = Roman (Latin) script, and T = Telugu script. It is self-evident that the commentaries, translations and prefaces, introductions, etc., are printed in the scripts regularly used for the chosen language. Sanskrit commentaries, etc., are printed in Devanagari script.

Column 10 (E):
Here, the extent of the text of the *Carakasaṃhitā* covered in the respective edition is given. The abbreviation 'C' refers to an edition that could be confirmed to have been completed. If an edition is explicitly stated to cover only one or more particular books (*sthāna*) of the *Carakasaṃhitā*, the abbreviations for these books used elsewhere in the present paper are used. Of course, it cannot be excluded that further volumes with further books were or will be published in the case of more recent editions. Similarly, the extent of incomplete editions in general, as seen, is indicated in this column, with their own segment numbers. If information on further fascicles of incomplete editions could be obtained, but without indication of the portion of the text they cover, the field is not filled. 'ex.' stands for 'excerpts.'

| No. | Editor | Place | Published/Press | Year | C/E | T | P/I | S | E |
|---|---|---|---|---|---|---|---|---|---|
| E1 | Gangadhar Ray (c.) | Kolkata | Bhubanchandra Basak; Saṃvādajñāna-ratnākara Press (Bhubanchandra Basak) | 1868ff. (Samvat 1925ff.) | S | – | S | B | |
| E1$^{0}$ | Idem | Ibidem | Idem | Idem | S | – | S | D | |
| E1$^{N}$ | Idem | Saidabad | Dharanidhar Ray; Pramādabhañjana Press | 1878 (Samvat 1935) | S | – | | B | C |
| E1$^{N}$ | Idem | Ibidem | Idem | 1884 (BS 1291) | S | – | | D | |
| E1$^{2*}$ | [Dwarkanath Sen, Rajendra Narayan Sen] | Saidabad & Kolkata | Tryambakeshvar Ray; Kaṇikā Press & Siddheśvara Machine Press (B.R. Chowdhury) | 1908ff. | S | – | | D | |
| E2* | Ananta (Anna) Moreshvar Kunte | Mumbai | Ananta (Anna) Moreshvar Kunte; Jñānamitra Press (N.W. Ghumre, Vasudev Narayan) & Ganpat Krishnaji's Press (Atmaram Kanhoba) | 1876ff. | – | – | S | D | Sū 1–30 |
| E3 | Jivanand Bhattacharya | Kolkata | Jivanand Bhattacharya; Sarasvatī Press | 1877 | – | – | – | D | C |
| E3$^{2}$ | Idem | Ibidem | Jivanand Vidyasagar Bhattacharya; Nārāyaṇa Press | 1896 | – | – | – | D | C |
| E4 | [Dharmadas Sen Gupta] (tr.) | Kolkata | Dharmadas Sen; Vālmīki Press (Bishvanath Nandi) | 1878?? | – | B | | B | Sū |
| E5 | Avinashchandra (tr.) | Kolkata | Lalbihari De; Vidyāratna Press (Arunoday Ghosh) | 1883–1884 (Samvat 1940–1941) | – | B | | B | C |
| E6 | Avinashchandra | Kolkata | Avinashchandra; Jyotiṣaprakāśa Press (Gopalchandra Ghoshal) | 1888ff. (BS 1295ff.) | S | – | E, S | D | parts of Sū |
| E7 | Harinath Sarkar | Kolkata | Taraprasanna Mukhopadhyaya, Giriśa Pustakālaya, and Harinath Sarkar, Viśārada Auṣadhālaya & Āyurvedaratnākara Auṣadhālaya; various presses and printers in Kolkata, see Appendix 6.2 | 1892–1919 (Śaka 1814–1841) | S | – | B | D | Sū – Ci 4.12. Ka – Si 7.34 |

(continued)

| No. | Editor | Place | Published/Press | Year | C/E | T | P/I | S | E |
|---|---|---|---|---|---|---|---|---|---|
| E8 | [Yashodanandan Sarkar] (tr.) | Kolkata | Baṅgabāsī Press | 1894 | – | B | [B] | [B] | C |
| E8$^{2}$ | Idem | Ibidem | Natbar Chakrabarti; Baṅgabāsī Electro Machine Press | 1910 (BS 1317) | – | B | B | B | C |
| E8$^{3*}$ | Idem | [Ibidem] | | 1930 | [–] | B | [B] | [B] | C |
| E9 | Shankar Daji Pade (tr., c.) | Mumbai | Shankar Daji Pade, Āryabhiṣak Kāryālaya; Govardhana Press | 1897ff. | M | M | M | D | C |
| E9$^{2}$ | Idem | Pune | Yajneshvar Gopal Dikshit, Bookseller; Yaśvant Press (Acyut Cintaman Bhat) | 1902ff. | M | M | M | D | C |
| E9$^{3}$ | Idem | Ibidem | Idem | 1914ff. | M | M | M | D | C |
| E9$^{R}$ | Idem (Krishnashastri Karve) | Pune | Yajneshvar Gopal Dikshit, Bookseller; Hanumān Press (Lakshman Bhaurao Kokate) | 1926 | M | M | M | D | C |
| E10 | Debendranath Sen Gupta, Upendranath Sen Gupta | Kolkata | Debendranath Sen Gupta, Upendranath Sen Gupta; Dhanvantari Machine Press (Haridas Bandhyopadhyaya) | 1897 (Śaka 1819) | – | – | S | D | C |
| E11 | [Khemraj Shrikrishnadas] | Mumbai | Khemraj Shrikrishnadas; Śrīveṅkaṭeśvara Press (Khemraj Shrikrishnadas) | 1898 | H | – | H | D | C |
| E12 | [Shyamlal Shrikrishnalal] | Mathura | Shyamlal Shrikrishnalal; Mumbaimitra Press, Lakṣmīnārāyaṇa Press (L.N. Press); title pages by Lakṣmīveṅkaṭeśvara Press, Mumbai | 1898 | – | H | H | D | C |
| E12$^{2}$ | Idem | Ibidem | Shyamlal Shrikrishnalal; Mumbai Mitra Press, V.K. Press | 1903 | – | H | H | D | C |
| E13 | [Dattaram Narayan Pathak] (c.) | Mumbai (Mathura) | Dattaram Narayan Pathak, Anātha Cikitsālaya & Pustakālaya; Nirṇayasāgara Press | 1900 (Samvat 1957) | H | – | H | D | C |

| | | | | | | | | | |
|---|---|---|---|---|---|---|---|---|---|
| E14 | [Kalicharan Sharma, Kshamapati Sharma] (c.) | Lucknow | Naval Kishore Press | 1900 | H | – | H | D | |
| E15 | Shankar Daji Pade | Mumbai | Nirṇayasāgara Press | 1903? | – | – | S | D | C |
| E16 | [Satishchandra Sharma] (tr.) | Kolkata | Satishchandra Sharma; Bhaiṣajya Steam Machine Press (Upendranath Mandal) | 1904 (BS 1311) | – | B | B | B | C |
| E17* | [Khemraj Shrikrishnadas] | Mumbai | Khemraj Shrikrishnadas, Śrīveṅkaṭeśvara Steam Press [Khemraj Shrikrishnadas] | 1911 (Samvat 1968, Śaka 1833) | H | – | H | D | C |
| E17$^{N}$* | Idem | Mumbai | Khemraj Shrikrishnadas; Śrīveṅkaṭeśvara Steam Press [Khemraj Shrikrishnadas] | 1922 (Samvat 1979, Śaka 1844) | H | – | H | D | C |
| E18 | [Vrajvallabh Sharma] | Mumbai | Vrajvallabh Sharma; Nirṇayasāgara Press | 1911 | – | H | H | D | C |
| E19 | [K.S. Shastri] (c.) | Karvetnagar, Chennai | Karvetnagar Ayurved Mandir; Kapālī Press | 1914, 1918–1920 | S, T | | | T | Sū |
| E20 | [Jethalal Devshankar Dave] (tr., c.) | Ahmedabad | Jethalal Devshankar Dave, Dave Chemical Works & Dave Ayurvedic Pharmacy; Satyanārāyaṇa Printing Press | 1916, 1931 | G | G | | D, G | C |
| E21 | Yogindranath Sen (c.) | Kolkata | J.N. Sen; Vidyodaya Press (J.N. Sen) | 1920–1930 | S | – | E | D | Sū – Ci 14.180 |
| E22 | Vaman Kesheo Datar | Mumbai | Pandurang Javaji; Nirṇayasāgara Press | 1922 | S | – | S | D | C |
| E22$^{2}$ | Idem | Ibidem | Idem | 1935 | S | – | S | D | C |
| E22$^{3}$ | Idem | Ibidem | Satyabhamabai Pandurang; Nirṇayasāgara Press | 1941 | S | – | S | D | C |
| *E23 | Yadav Sharma | Mumbai | [Pandurang Javaji; Nirṇayasāgara Press] | | – | – | [S] | D | C |
| E23$^{2}$ | Idem | Ibidem | Pandurang Javaji; Nirṇayasāgara Press (Ramchandra Yesu Shedge) | 1922 | – | – | S | D | C |
| E23$^{3}$ | Idem | Ibidem | Idem | 1933 | – | – | S | D | C |

(continued)

| No. | Editor | Place | Published/Press | Year | C/E | T | P/I | S | E |
|---|---|---|---|---|---|---|---|---|---|
| E24 | Atmanand Sharma Jassar | Lahore | Das Brothers | 1926 | S | – | – | D | ex. |
| E25* | [Krishnamurti] (tr.) | Bezwada | Vāṇī Press | 1926 | – | T | T | T | Ci |
| E25$^{2}$* | Idem | Ibidem | Āndhra Granthālaya Press; Vavilla Press (Chennai) (V. Venkateswara of V. Ramaswamy & Sons) | 1935–1941 | – | T | T | T | Ci |
| E26 | Narendranath Sen Gupta, Balaichandra Sen Gupta | Kolkata | Narendranath Sen Gupta, Balaichandra Sen Gupta; Dhanvantari Steam Machine Press (Dinnath Deb) / Dhanvantari Electric Machine Press | 1927–1933 (Śaka 1849–1855) | S | – | S | D | C |
| E26$^{2}$ | Idem | Varanasi | Chaukhamba Orientalia | 1984–1985 | S | – | S | D | C |
| E26$^{R}$ | Narendranath Sen Gupta, Balaichandra Sen Gupta | Varanasi | Chaukhamba Orientalia | 1991 | S | – | S | D | C |
| E26$^{R2}$ | Idem | Ibidem | Idem; Globe Offset Press, New Delhi | 2002 | S | – | S | D | C |
| E26$^{R}$ | Idem | New Delhi | Rashtriya Sanskrit Sansthan | 2011 | S | – | S | D | C |
| E27 | Narendranath | Lahore | [Motilal Banarsidas], Punjab Saṃskṛta Pustakālaya; The Bombay Sanskrit Press | 1929 | S | – | S | D | C |
| E28* | [Venkatachalpatiprasad] (c.) | Bezwada | Shri Krishna Devaraya, Āndhra Bhāṣhā Nilayamu; Saubhāgya and Bhavānī Press (Katur Ramamurti) | 1930 | – | – | T | T | Ind |
| *E29 | Shivsharma | Mumbai | Lakṣmīveṅkaṭeśvara Steam Press | 1932 | H | – | [H] | D | [C] |
| E29$^{R}$ | Idem | Mumbai | Khemraj Shrikrishnadas Bombay Praka-shana | 1989 | H | – | [H] | D | [C] |
| E30* | [Gangavishnu Shrikrish-nadas] | Mumbai | Gangavishnu Shrikrishnadas; Lakṣmī-veṅkaṭeśvara Steam Press | 1932 | – | – | S | D | C |
| *E31 | Yadavji Trikamji | Mumbai | [Pandurang Javaji]; Nirṇayasāgara Press | 1933 | – | – | [S] | D | C |
| E31$^{2}$ | Idem | Ibidem | [Pandurang Javaji]; Nirṇayasāgara Press | 1935 | – | – | S | D | C |
| E31$^{3}$ | Idem | Ibidem | Satyabhamabai Pandurang; Nirṇayasā-gara Press | 1941 | – | – | S | D | C |
| E31$^{R}$ | Idem | Delhi | Munshiram Manoharlal; Mehra Offset Press | 1981 | – | – | S | D | C |

| | | | | | | | | | |
|---|---|---|---|---|---|---|---|---|---|
| E32 | [Jaydev] (c.) | Lahore | Motilal Banarsidas, Punjab Saṃskṛta Pustakālaya; Bombay Sanskrit Press | 1934 (Samvat 1991) | H | – | H | D | C |
| *E32$^2$ | Idem | Ibidem | Idem | 1944 | H | – | H | D | C |
| *E32$^3$ | Idem | Ibidem | Idem | 194? | H | | H | D | C |
| *E32$^4$ | [Jaydev] (c.) | | [Motilal Banarsidas] | 1948 | H | | H | D | C |
| E33 | [Jyotishchandra Sarasvati] (c.) | Varanasi | S.K. Saraswati; Indian Press | 1937 | S | – | | D | Sū |
| E34 | Taradatt Pant (c.) | Varanasi | Jaykrishnadas Haridas Gupta, The Chowkhamba Sanskrit Series Office; Vidyāvilāsa Press | 1937–1938 | S | – | S | D | C |
| E35 | [Atridev Gupta] (tr.) | Ajmer | Ārya Sāhitya Maṇḍal Ltd.; The Fine Art Printing Press, Ajmer (Mathuraprasad Shivhare) | 1935–1937 | – | H | | D | C |
| *E36 | Haridatt Sharma | Lahore | Motilal Banarsidas, Punjab Saṃskṛta Pustakālaya; Bombay Sanskrit Press | | S | – | | D | C |
| E36$^2$ | Idem | Lahore | Motilal Banarsidas, Punjab Saṃskṛta Pustakālaya; Bombay Sanskrit Press (Shantilal Jain) | 1940–1941 | S | – | S | D | C |
| E37 | Shree Gulabkunverba Ayurvedic Society | Jamnagar | Shree Gulabkunverba Ayurvedic Society; Āyurvedamudraṇālaya | 1949 | – | H, E, G | E | D | C |
| E38 | [Jaydev Sharma] | Ajmer | Ārya Sāhitya Maṇḍal Ltd. | 1954–1962 | | H | | D | C |
| E39* | [Girijashankar Maya-shankar] (tr., c.) | Ahmedabad | Sastum Sāhitya Vardhak Kāryālaya | 1958–1963 | S | G | | D | [C] |
| E40 | Rajeshvardatt et al. | Varanasi | The Chowkhamba Vidyabhavan | 1961–1962 | H | – | S, H | D | C |
| E41 | Gangasahay Pandey | Varanasi | Chowkhamba Sanskrit Series Office | 1969–1970 | H | – | S | D | C |
| E41$^2$ | Idem | Ibidem | Idem | 1983 | H | – | S | D | C |
| E42 | Ramkaran Sharma, Bhagwandash (tr., c.) | Varanasi | Chowkhamba Sanskrit Series Office | 1976–2002 | E | E | E | D | C |

(continued)

| No. | Editor | Place | Published/Press | Year | C/E | T | P/I | S | E |
|---|---|---|---|---|---|---|---|---|---|
| E43* | [M. Narayanan] (c.) | Kuttikkakam, Etakkad | Dhanvantari Press | 1979 | Ma | – | | Ma | Ci |
| E44 | Priyavrat Sharma (tr., c.) | Varanasi | Chaukhamba Orientalia; Srigokul Press | 1981–1994 | E | E | E | D | C |
| E45* | Ramprakash, Narayan-prasad; Balram, Mohanlal Gothecha | Jaipur | Shrisvamilakshmiram Trust; Shankar Art Printers & Oriental Printers and Publishers | 1982 (Samvat 2039) | S | – | S | D | Ci 1–20 |
| E46* | Priyavrat Sharma | Varanasi | Chaukhamba Orientalia | 1983 | – | – | S | D | C |
| E47 | Brahmanand Tripathi (c.) | Varanasi | Chaukhamba Surabharati Prakashan; Shriji Press | 1983, 1988 | H | – | H | D | C |
| E48* | Brajendrachandra Nag (c.) | Kolkata | Nabapatra Prakashan | 1984–1989 | B | – | | B | C |
| E49 | Priyavrat Sharma, Satyadev Dube | Jaipur | Shrisvamilakshmiram Trust | 1990 | S | – | | D | Sū |
| E50* | Vidyadhar Shukla, Ravidatt Tripathi (c.) | Delhi | Chaukhamba Sanskrit Pratishthan | 1998? | H | – | H | D | C |
| E50[R] | Idem | Ibidem | Idem; A.K. Lithographers | 2006 | H | – | H | D | C |
| E51 | [Purushottam Kar Sharma] (c.) | Cuttack | Dharma Grantha Store | 2000 | O | O | | O | Sū, Ni |
| E52 | K.R. Srikantha Murthy | Varanasi | Chaukhambha Orientalia; Charu Printers | 2004 | S | E | E | D | Sū, Ni |
| E53 | R. Udupa et al. (tr., c.) | Manipal | Dr. U. Krishna Muniyal Memorial Trust; Navayuga Press (Udupi) | 2004 | E | E | E | D | Sū |
| E54 | Harishchandra Singh Kushwaha (c.) | Varanasi | Chaukhambha Orientalia; Charu Printers | 2005 | H | – | H | D | Sū – Ind |
| E55 | Lobsang Tenzin, Lobsang Norbu (tr.) | Varanasi | Central Institute of Higher Tibetan Studies; Sattnam Printing Press | 2006 | – | Ti | Ti, E | D | Sū 1–15 |
| E56 | Michel Angot (tr., c.) | Paris | Les Belles Lettres | 2011 | F | F | F | R | Sū, Śā |

## *Appendix 6.2: A Synopsis of the Fascicles of Harinath Sarkar's Edition of the* Carakasaṃhitā *with Cakrapāṇidatta's* Āyurvedadīpikā *Begun in 1892 (E7) (British Library 14043. dd.15)*

Vaman Kesheo Datar, who published the first complete edition of the *Āyurvedadīpikā* in 1922 (E22),[489] assumed that Harinath Sarkar's not yet completed edition had reached only the end of the Indriyasthāna, which was actually published in 1905 in fascicle 29 of E7, and briefly addressed the problem of the scarcity and fragmentariness of manuscripts of the *Āyurvedadīpikā* he had to face when editing the Cikitsāsthāna and Siddhisthāna.[490] It seems that information on the further fascicles of E7 published until 1919 and in the following year(s) had not yet reached his place of residence, Nashik in modern-day Maharashtra. According to Müller (1930b), Harinath Sarkar's edition remained incomplete due to the editor's death. He refers to 1892–1905 as the years of publication of this edition. Thus it is evident that the fascicles of E7 published from 1907 onwards from the Āyurvedaratnākara Auṣadhālaya – fascicles 31–35 (Cikitsāsthāna 1.1–4.12), fascicles 1–2 of the Kalpasthāna, and fascicles 1–3 and subsequent fascicles of the Siddhisthāna – had also not yet come to Müller's notice when he wrote his short book notice on the first two volumes of E26 (1927, 1928) in 1930. The case of Vaman Kesheo Datar addressed above shows that these fascicles must have been difficult to obtain even in India. Even though this is hard to understand, even Yogindranath Sen[491] does not seem to have been aware of the progress and full extent of E7 when he wrote the introductory Critical Notice to the first volume of his edition of the *Carakasaṃhitā* with his own Sanskrit commentary *Carakopaskāra* (1920). There he states, with reference to the *Āyurvedadīpikā*: 'A portion of this commentary has been published in Calcutta, and though more than two decades ago, the editor held out a promise of publishing the remaining portions in the near future, those expectations have, unfortunately, not yet been realized.'[492] At least some time between 1895 and 1897 Yogindranath Sen was a subscriber of Harinath Sarkar's edition,[493] and one may assume that even after he might have discontinued the payment of his subscription fee he would have stayed informed about the progress of this edition which evidently covered much more than just 'a portion' of the text in 1920. Yadavji Trikamji[494] for his part makes it a point to specifically thank his friend Vaidyabhūṣaṇa Pandit Govardhansharma Changani in Nagpur for providing him with copies of Cikitsāsthāna chapters 1–3, the Kalpasthāna and Siddhisthāna chapters 1–10 of E7 for his own work on E31 ($^3$1941), an edition of the *Carakasaṃhitā* together with the

---

489 See above, p. 244.

490 See E22, Prāstāvikam, 1–2.

491 See above, p. 214.

492 See E21 (1920), Critical Notice, iii.

493 See above, p. 267.

494 See above, p. 247.

*Āyurvedadīpikā*.[495] Much less informed is the erudite Girindranath Mukhopadhyaya who in the late 1920s merely refers to 1892–1893 as the publication date of a not further specified Calcutta edition of the *Carakasaṃhitā* with the *Āyurvedadīpikā*.[496] It is striking that he does not care to mention the name of the editor here, different from the cases of the other eight editions he lists. Harinath Sarkar's name as editor appears only in a footnote to Mukhopadhyaya's list of commentaries on the *Carakasaṃhitā*, where the publication date of Harinath Sarkar's edition is given as 1895.[497] More recently, for instance, Singh and Joshi (1982, 321) state that only the text of the Sūtrasthāna and Vimānasthāna has been published in E7.

The last fascicle of E7 in the copy of the British Library, London (call number 14043.dd.15), is from 1919 (Śaka 1841) and ends with 7.34 of the Siddhisthāna (i.e., CS Si 7.63), the eighth and final book of the *Carakasaṃhitā*. This leaves the rest of the text of this chapter and that of the remaining chapters 8 to 12 unaccounted for. However, a copy owned by the National Library, Calcutta (call number 180.Eb.91.9), which was unfortunately not fully documented by me *in situ* in 2015, contains six fascicles of the Siddhisthāna and thus must have twice as much text as the three fascicles of the Siddhisthāna of the copy in the British Library. Yadavji Trikamji's abovementioned reference to the text of Siddhisthāna chapters 1–10 published in E7 also shows that Harinath Sarkar's edition must have progressed beyond the first three fascicles of the Siddhisthāna seen in the copy of the British Library.

The edition of the sixth book, the Cikitsāsthāna, was most probably never completed. Fascicles 31 to 35, published in 1907 (Śaka 1829, two fascicles), 1909 (Śaka 1831), 1910 (Śaka 1832) and 1913 (Śaka 1835), cover only the beginning of this book, that is, chapters 1 to 3 and the beginning of chapter 4, out of altogether 30 chapters. This state of affairs is also noted by Yadav Sharma[498] in his communication (*nivedanam*) at the beginning of E23 ($^2$1922) and explicitly stated by the manager and superintendent of the Āyurvedaratnākara Auṣadhālaya, Purnachandra, on the covers of two fascicles of E7 published from 1913 onwards (fascicle 35; fascicle 2 of the Kalpasthāna) and by his successor, Surendranath, on the covers of the last two fascicles of the copy of the British Library dated 1917 and 1919 (fascicles 2 and 3 of the Siddhisthāna).

The following synopsis provides an overview of the most essential details of E7 as seen in the copy of the British Library, London (call number 14043.dd.15).

---

[495] See his introduction (*carakasaṃhitāyās tṛtīyāvṛtter upodghātaḥ*), p. 17.

[496] See Mukhopadhyaya (1922–1929, III, 614). The Preface to the relevant third volume of Mukhopadhyaya's *History of Indian Medicine* published in 1929 is dated November 1, 1929. In it, Mukhopadhyaya states that '[t]he material here [i.e., in the third volume] presented has been collected several years ago,' but also expresses his anxiety to make the publication 'as complete as possible' in terms of the 'new information' that 'is being published in India, in Europe and in America' (p. ix). His reference to Vaman Kesheo Datar's edition of the *Carakasaṃhitā* with the *Āyurvedadīpikā* (E22) of 1922 and to the first volume of E21, Yogindranath's edition with his *Carakopaskāra*, published in 1920 (III, 613–616), shows that he indeed attempted to update the information on the editions of the *Carakasaṃhitā* and its commentaries before the third volume of his *History* was finally printed.

[497] See Mukhopadhyaya (1922–1929, III, 615, n. 3)

[498] See above, p. 247.

| Part | Publishing house | Year | Text | Pages | Cover 1b | Cover 2a | Cover 2b | Press | Printer |
|---|---|---|---|---|---|---|---|---|---|
| Vol. 1 | | | | | | | | | |
| 1 | Giriśa Pustakālaya, 21 Cornwallis St, Calcutta | Śaka 1814 (1892) | Introduction (Bhūmikā), signed by Kavirāj Harinath Sarkar; Sū 1.1–1.36 | 1–24 | Expert opinion on the CS by Kamakhyanath Tarkavāgīśa, Sanskrit College, Calcutta & Kavirāj Kaliprasanna Sen, Kambali Tola, Calcutta | Prices and subscription rules, signed by Kavirāj Harinath Viśārada, 61 Cornwallis St, Calcutta | Advertisements for Viśārada Auṣadhālaya, signed by Ramchandra Sarkar (superintendent) & Giriśa Pustakālaya, signed by Taraprasanna Mukhopadhyaya | Gupta Press, 221 Cornwallis St, Calcutta | Prahladchandra Das |
| 2 | Idem | Idem | Sū 1.37–2.25 | 25–56 | Idem | Idem | Idem | Idem | Idem |
| 3 | Idem | Idem | Sū 2.26–5.20 | 57–88 | Prices and subscription rules, signed by Kavirāj Harinath Viśārada, 61 Cornwallis St, Calcutta | | Idem | Siddheśvara Press, 38 Shibnarayan Das Lane, Calcutta | Siddheshvar Pan |
| 4 | Idem | Idem | Sū 5.21–7.39 | 89–120 | Idem | Expert opinion on the CS by Kedarnath Ghosh, Hindu School, Calcutta | Idem | Idem | Idem |
| 5 | Idem | Idem | Sū 7.39–9 | 121–152 | Idem | | Idem | Idem | Idem |
| 6&7 | Idem | Idem | Sū 10.1–14.6 | 153–216 | Book notice from *New India*, January 15, 1894 (English); prices and subscription rules, signed by Kavirāj Harinath Viśārada, 61 Cornwallis St, Calcutta | Corrigenda (Bhramasaṃśodhanam) | Idem | Idem | Idem |

(continued)

| Part | Publishing house | Year | Text | Pages | Cover 1b | Cover 2a | Cover 2b | Press | Printer |
|---|---|---|---|---|---|---|---|---|---|
| 8 | Idem | Idem | Sū 14.6–16 | 217–248 | Idem | Book notice on the CS from *The Statesman*, May 3, 1893 (English) | Idem | Idem | Idem |
| 9 to 12 | Idem | Idem | Sū 17.1–26.28 | 249–376 | Prices and subscription rules, signed by Kavirāj Harinath Viśārada, 61 Cornwallis St, Calcutta | Book notices on the CS from *New India*, January 15, 1894 and *The Statesman*, May 3, 1893 (English) | Idem | Idem | Idem |
| 13&14* | Idem | Idem | Sū 26.29–27.157 | 377–440 | Idem | Idem | Idem | Idem | Idem |
| 15–17 | Idem | Śaka 1817 (1895) | Sū 27.158–30; Table of Contents (Sūcīpatra) of Sū | 441–533 | Prices and subscription rules, unsigned | Idem | List of sponsors and persons who paid the fee, with addresses and amounts | Idem | Idem |
| 18 | Idem | n.d. | Vi 1.1–3.12 | 1–32 | Prices and subscription rules and special notice on pricing, signed by Kavirāj Harinath Viśārada, 61 Cornwallis St, Calcutta | Advertisements for Viśārada Auṣadhālaya, signed by Ramchandra Sarkar (superintendent) & Giriśa Pustakālaya, signed by Taraprasanna Mukhopadhyaya | List of persons who paid the fee, with addresses and amounts | Jyotiṣaprakāśa Press, 5 Shimla St, Calcutta | Aghornath Chaudhuri |

* Added by hand on the cover.

| | | | | | | | | | |
|---|---|---|---|---|---|---|---|---|---|
| 19 | Idem | n.d. | Vi 3.13–5.36 | 33–64 | Prices and subscription rules, signed by Kavirāj Harinath Viśārada, 61 Cornwallis St, Calcutta | Advertisements for Viśārada Auṣadhālaya, signed by Sharacchandra Mitra (superintendent) & Giriśa Pustakālaya, signed by Taraprasanna Mukhopadhyaya | Idem | Idem | Idem |
| 20 | Idem | n.d | Vi 5.37–8.6/7 | 65–96 | Idem | Idem | Idem | Idem | Idem |
| [21] | Viśārada Auṣadhālaya, 61 Cornwallis St, Calcutta | Śaka 1819 (1897) | Vi 8.6/7–14/2 | 97–128 | | Prices and subscription rules, unsigned | Advertisements for Viśārada Auṣadhālaya, signed by Kavirāj Gopalchandra Ray (superintendent) & Giriśa Pustakālaya, signed by Taraprasanna Mukhopadhyaya | Purāṇa Press, 21 Balram Ghosh St, Calcutta | Gopalchandra Lahiri |
| 22 | Viśārada Auṣadhālaya, 25 Madan Mitra Lane, Calcutta | Śaka 1824 (1902) | Vi 8.14/2–26 (end) | 129–164 | | Idem | Idem | Jyotiṣaprakāśa Press, 5 Shimla St, Calcutta | Kumudbandhu Ray |
| 23&24 | Idem | Idem | Śā 1.1–2.20/4 | 1–64 | | Prices and subscription rules, signed by Kavirāj Harinath Viśārada, 25 Madan Mitra Lane, Calcutta | Prices and subscription rules (English), unsigned | Idem | Krishnachandra Chattopadhyaya |

(continued)

| Part | Publishing house | Year | Text | Pages | Cover 1b | Cover 2a | Cover 2b | Press | Printer |
|---|---|---|---|---|---|---|---|---|---|
| 25 | Idem | Idem | Śā 2.20/5–4.7/5 | 65–96 | | | Prices and subscription rules, signed by Kavirāj Harinath Viśārada, 25 Madan Mitra Lane, Calcutta | Idem | Harimohan Das |
| 26 | Idem | Idem | Śā 4.7/4/1–5 | 97–128 | | | Idem | Idem | Idem |
| 27&28 | Idem | Śaka 1827 (1905) | Śā 6–8 | 129-194 | | | Idem | Nārāyaṇa Press, 36 Mechua Bazar St, Calcutta | Benimadhab Ghosh |
| "3 1/2 parts after 17" | Viśārada Auṣadhālaya, 61 Cornwallis St, Calcutta | n.d. | Table of Contents (Sūcīpatra) of Ni; Ni 1.1–8 | 1–112 | | Announcement (Vijñāpan) on progress of publication, signed by Kavirāj Harinath Viśārada (publisher) | Note on sponsor Maharaja Manindra Chandranandi Bahadur of Kasim Bazar, signed by Kavirāj Harinath Viśārada (publisher), 61 Cornwallis St, Calcutta | Purāṇa Press, 21 Balram Ghosh St, Calcutta | Gopalchandra Lahiri |
| 29&30 | Viśārada Auṣadhālaya, 25 Madan Mitra Lane, Calcutta | Śaka 1827 (1905) | Ind 1–12 | 1–72 | | Prices and subscription rules, signed by Kavirāj Harinath Viśārada [lower half of page torn off] | [not available] | Nārāyaṇa Press, 36 Mechua Bazar St, Calcutta | Benimadhab Ghosh |

| | | | | | | | | |
|---|---|---|---|---|---|---|---|---|
| Vol. 2 | | | | | | | | |
| 31 | Āyurvedaratnākara Auṣadhālaya, 192/2 Cornwallis St, Calcutta | Śaka 1829 (1907) | Ci 1.1.1–3.19 | 1–32 | Prices and rules for advance payment, unsigned | Corrigendum (Śuddhi-patram) | Advertisement for Āyurvedaratnākara Auṣadhālaya, signed by Kavirāj Gopal-chandra Ray (super-intendent) | Ghosh Press | Manmathnath Ghosh, 38 Shib-narayan Das Lane, Calcutta |
| 32 | Idem | Idem | Ci 1.3.19–2.4.4 | 33–64 | Prices and rules for advance payment, signed by Kavirāj Gopalchandra Ray (superintendent) | Idem | Idem | Idem | Idem |
| 33 | Idem | Śaka 1831 (1909) | Ci 2.4.4–3.54 | 65–96 | | Prices and rules for advance payment, signed by Kavirāj Purnachandra (super-intendent) | Idem | Idem | Idem |
| 34 | Idem | Śaka 1832 (1910) | Ci 3.54–139 | 97–128 | Corrigendum (Śuddhi-patra) for Ni and Ci | Prices and rules for advance payment, signed by Kavirāj Purnachandra (super-intendent) | Idem | Idem | Idem |

(continued)

| Part | Publishing house | Year | Text | Pages | Cover 1b | Cover 2a | Cover 2b | Press | Printer |
|---|---|---|---|---|---|---|---|---|---|
| 35 | Idem | Śaka 1835 (1913) | Ci 3.140–4.12 | 129–160 | | | Note on progress and pricing, signed by Kavirāj Harinath Viśārada, 192/2 Cornwallis St, Calcutta & Kavirāj Purnachandra Bhiṣag-ratna Sāṃkhyabhū-ṣaṇa, Śāstrī (manager and superintendent) (Hindi) | Siddheśvara Press; according to cover 1a: Kaviratna Press | Avinash-chandra Mandal, 13 Shib-narayan Das Lane, Calcutta |
| Kalpa 1 | Idem | Śaka 1832 (1910) | Ka 1–6.11 | 1–32 | | Advertisement for Āyurvedaratnākara Auṣadhālaya, signed by Kavirāj Gopalchandra Ray (superintendent) | Prices and rules for advance payment, signed by Kavirāj Purnachandra (superintendent) | Jyotiṣaprakāśa Press, 5 Shimla St, Calcutta | Surendranath Mukhopadhyaya |
| Kalpa 2 | Idem | Śaka 1835 (1913) | Ka 6.12–12 | 33–79 | Note on progress and pricing, signed by Kavirāj Purnachandra Bhiṣagratna Sāṃkhya-bhūṣaṇa, Śāstrī (manager & superintendent) | Extract from *Amrita Bazar Patrika*, January 23, 1913 (English) | Extract cont. from 2a; extract from *The Journal* (Sunday edition of *The Englishman*), March 9, 1913 (English) | Jyotiṣaprakāśa Press, 9 Ram-tanu Basu Lane, Calcutta | Idem |

| | | | | | | | | | |
|---|---|---|---|---|---|---|---|---|---|
| Siddhi 1 | Idem | Śaka 1837 (1915) | Si 1.1–2.14 | 1–32 | Extract from *Bangabasi Patrika*, Agraharan 22, 1903 (expert's letter) | Extract cont. from 1b | Extract cont. from 2a; small list of corrections for Ci | Idem | Idem |
| Siddhi 2 | Idem | Śaka 1839 (1917) | Si 2.15–4.33 | 33–64 | Extracts from correspondence of A.F.R. Hoernle with Pareshnath (May 4, 1912) & Harinath (October 13, 1913) (English and Bengali tr.) | Extract from correspondence of A.F.R. Hoernle with Harinath (March 23, 1914) | Announcement of progress and on pricing, signed by Kavirāj Surendranath Kaviratna (manager and superintendent) (Hindi and Bengali) | Siddheśvara Machine Press, 1 Shibnarayan Das Lane, Calcutta | Avinashchandra Mandal |
| Siddhi 3 | Auṣadhālaya, 192/2 Cornwallis St, Calcutta | Śaka 1841 (1919) | Si 4.33–7.34 | 65–96 | Idem | Idem | Idem | Siddheśvara Machine Press, 11 Yadunath Sen Lane, Calcutta | Idem |

## *Appendix 6.3: Ramprasad's References to Plastic Surgery and Brain Surgery in Ancient India*

Reference to plastic surgery in the form of the surgical reattachment of cut-off heads and exchange of heads as well as to advanced brain surgery in ancient India is frequently found in secondary literature especially by Indian scholars in order to demonstrate the far advanced state of Indian medicine in hoary times.[499] The issue appears again and again in contemporary nationalistic discourse.[500] Thus, it is worthwhile to briefly comment on the three specific examples adduced by Ramprasad, the Hindi commentator on the *Carakasaṃhitā* edited in 1911 by Khemraj Shrikrishnadas,[501] in his introduction[502] and on their literary background.[503]

The first surgical feat mentioned by Ramprasad refers to a group of myths in which God Śiva beheads Dakṣa, the central one among them being the story about Dakṣa's sacrifice.[504] However, in this puranic myth Śiva himself, God Brahmā or the gods, without specification, replace the removed head with a ram's head. Thus Ramprasad may refer to a version of the myth I could not identify in which the twin gods called Aśvin-s are involved as surgeon–healers, like in the old Vedic myth of the cut-off head of the personified and anthropomorphized sacrifice which was healed by the Aśvin-s. He may also have simply conflated the two myths.[505] In Vedic religion, Dakṣa is the deified notion of ritual dexterity (*dakṣa*) and thus may also

---

[499] See, e.g., Bhagwan Dash (1984, xv–xvii), Bhagwan Dash (1989, 7–8), Sharma (2005c, 781) and Sharma (2005a, 150–151).

[500] On this, see, e.g., Khair (2015, 399). See also the numerous online media reports on Narendra Modi's claim, in a speech in October 2014, that ancient India knew plastic surgery, such as attaching an elephant's head to a human body, and had genetic science in the form of stem cell research and reproductive genetics.

[501] E17 (1911, next edition 1922).

[502] See above, p. 272f.

[503] For an illustrated overview and very balanced treatment of the difficult issue of actual surgical practices in pre-modern India, with references, see Wujastyk (1995, 24–29). See also Wujastyk (2003, 65–72) and, most recently, on a type of pre-modern plastic surgery famous in European medicine ever since the publication of Hessler's Latin translation of the *Suśrutasaṃhita* (see above, n. 432), Wujastyk et al. (2023).

[504] See the discussion in O'Flaherty (1973, 128–129, with references), Kramrisch (1981, 317–330, especially 322–325 relating to the version in the *Śivapurāṇa*) and Vettam Mani (1975, 194a), with reference to the version of the *Devībhāgavatapurāṇa* where Vīrabhadra and Bhadrakālī, terrible emanations of the furious Śiva, tear off Dakṣa's head.

[505] On the Aśvin-s, who figure in the traditional account of the transmission of Āyurveda (see above, p. 197), as divine healers and physicians per excellence, see Preisendanz (2015, 117). On the myth of the cut-off head of sacrifice in its wider cultural–historical context, see Preisendanz (2015, 125–129); for interpretations of the myth and the Vedic symbolism of the severed head of sacrifice from the point of view of history of religion see Heesterman (1985, Chapter 3) and Witzel (1987, 411–413). For a survey of the miraculous medical practices ascribed to the Aśvin-s mentioned in Vedic literature, see Sharma (2005b), with a list of surgical practices on pp. 32–33 that include the rejoining of the severed head of sacrifice and the head transplantations performed on Dadhīci/Dadhyañc. It is not entirely clear whether Sharma considers the references to these practices as purely mythological or as mythological but nevertheless indicative of actual surgical practices.

stand for the sacrifice as such.[506] There is also a myth about the cutting off of the head of Dadhīci/Dadhyañc, its replacement by the head of a horse and the final refitting of the original head all achieved by the Aśvin-s in Ṛgvedic mythology.[507]

Secondly, Ramprasad refers to the various puranic myths of the beheading of God Brahmā, or his mythological 'predecessor' Prajāpati, by God Śiva himself or his terrible forms Rudra and Bhairava.[508] Depending on the individual myth and its concrete version, Brahmā loses one head which is replaced with four heads by Śiva himself, or loses one of his five heads which then sticks to the hand of Śiva in the form of a skull he has to use as his begging bowl during his wanderings in expiation for this brahminicide. Alternatively, Śiva keeps the fifth head in this hand to prevent damage to the earth, or he or his wife Pārvatī madly dance with it in their hands, without returning it to its owner. Furthermore, in an epic story, Śiva presents Brahmā's head as the powerful magic weapon called the Pāśupata weapon to Arjuna, one of the five Pāṇḍava brothers of the *Mahābhārata*. In iconography, Brahmā is regularly shown with four heads and he is called 'four-faced' (*caturānana*). In view of all this, reattachment of Brahmā's cut-off head to his body and – if Ramprasad's qualification 'by the Aśvin-s' is taken to relate to all three feats – any involvement of the Aśvin-s in such a feat remain unaccounted for and even seem to be out of place in these myths. However, again it may be that Ramprasad refers to some otherwise unknown version of one of the myths relating to Śiva's beheading of Brahmā that was current among physician–scholars at his time.

Finally, Ramprasad mentions the splitting of Bhoja's head and the extraction of souls or living creatures (*jīv*). This is most probably a reference to an episode in the *Bhojaprabandha*, a very popular literary composition in prose and verse of the sixteenth century attributed to the poet Śrīballāla of Benares which relates various stories and anecdotes centering on the figure of King Bhoja of Malwa (ca. eleventh century).[509] In the twelfth and last chapter of this fictional work, we find the rather strange story about King Bhoja cleansing his skull with the water of some pond on which occasion a tiny juvenile fish called *śaphara* (Cyprinus sophore, i.e., Puntius sophore [Hamilton 1822], pool barb) entered the skull. After one year, the resulting terrible headache of the king was finally healed by the divine Aśvin-s themselves who administered a powder to make him lose consciousness (*mohacūrṇa*), removed his head, caught the multitude of small fish that had settled there in the meantime, reattached the head with the help of some medicinal herb called 'conjoiner' (*sandhānakaraṇī*) and revived it with the herb 'reviver' (*saṃjīvanī*).[510] Both herbs have pregnant names and are also among the herbs brought by Hanumān from the

[506] For literature on Dakṣa, see Oberlies (1998, 187, n. 183).

[507] See, e.g., Bhattacharji (1970, 240) and Oberlies (1998, 394; 2023, 178).

[508] See O'Flaherty (1973, 123–127), Vettam Mani (1975, 148) and Kramrisch (1981, 250–265).

[509] See Winternitz (1920, 352; English translation in Winternitz 1985, 387–388).

[510] See Saradaprosad Vidyabhusan (1926, story 71, 222–227) and Gray (1950, 93–95). See also the slightly modified summary of this 'operation' in Bhagwan Dash (1984, xvii) who like Ramprasad speaks of the opening of the skull, not of the removal of the head, and refers to physicians, not to the divine Aśvin-s, as agents.

Himalayas to the island of Laṅkā to heal Rāma, Lakṣmaṇa and others who were wounded in the battle with the forces of the demon king Rāvaṇa.[511] In the version of this popular story in *Rāmāyaṇa* Yuddhakāṇḍa 61, four herbs are brought, among them the 'reviver of the dead' (*mṛtasaṃjīvanī*) and the 'conjoiner' (*sandhānī*) (61.32–33). The names of these two herbs in the version of Yuddhakāṇḍa 89 are precisely the ones used by Śrīballāla here in the *Bhojaprabandha*: *saṃjīvanī* and *sandhānakaraṇī* (89.16). In the episode in Yuddhakāṇḍa 40, only two herbs are mentioned that should have been brought by Hanumān. One of them is the 'reviver' (*saṃjīvakaraṇī*).[512] It is thus clear that Śrīballāla is borrowing here from the version in Yuddhakāṇḍa 89. Furthermore, concerning the 'conjoiner' Goldman et al.[513] refer to the commentators on the Rāmāyaṇa who explain the function of this herb as 'conjoining the vessels even of a cut-off head and a trunk as they were before'[514] and 'connecting cut-off heads, etc., as they were before'[515] which obviously relates to the context of battle. One commentator explains that after the application of the 'reviver of the dead' (*mṛtasaṃjīvanī*) and another herb that assists in extracting arrows from the bodies, the 'conjoiner' is applied in order to heal the injured skin.[516] The first-mentioned function of the 'conjoiner' fits the context in Śrīballāla's story well. It goes without saying that this story is 'science fiction'[517] and that the author does not intend to refer to any surgical operation performed on King Bhoja, even though Bhagwan Dash regrets that 'details of these two types of drugs[518] are not available now.'[519]

## References

*A Descriptive Catalogue of the Sanskrit Manuscripts Acquired for and Deposited in the Sampurnanand Sanskrit University (Sarasvatī–Bhavana) Library, Varanasi During the Years 1951–1981*. Vol. 12,2: *Jaina-bhakti-sampradāya-āyurveda-kāmaśāstra-śilpakalā-saṅgīta-nīti-dhanurveda-pañjī-praśasti-citra-deśībhāṣā-Manuscripts*, 1996. Varanasi: Sampurnanand Sanskrit University.

Adriaensen, R., H.T. Bakker, and H. Isaacson, eds. 1998. *The Skandapurāṇa. Volume I. Adhyāyas 1–25*. Groningen: Egbert Forsten. (Supplement to Groningen Oriental Studies).

---

511 See above, n. 473.

512 Some manuscripts also add a reference to the 'conjoiner' here (*sandhānakaraṇī, sandhinī*), paired with the 'reviver' / 'reviver of the dead' (*jīvanī, mṛtasaṃjīvanī*).

513 See Goldman et al. (2010, notes on Yuddhakāṇḍa 60.32–33, p. 1092).

514 *chinnaśiraḥkabandhayor api yathāpūrvaṃ nāḍīsandhānakaraṇīm*. See the *Amṛtakataka* by Mādhava Yogīndra and the *Tilaka* by Nāgeśa Bhaṭṭa quoted by Goldman et al.

515 *chinnaśiraḥprabhṛtīnāṃ yathāpūrvaṃ saṃyojikām*. See Vaṃśīdhara's *Śiromaṇi* quoted by Goldman et al.

516 See *viśalye kṛte tvacaḥ sandhānaṃ karoti*, quoted by Goldman et al.

517 Wujastyk (1995, 24).

518 That is, the powder that made the king lose consciousness (called an 'anaesthetic powder' by Bhagwan Dash) and the *saṃjīvanī* (also understood as referring to a powder by Bhagwan Dash).

519 See Bhagwan Dash (1984, xvii).

Aklujkar, Ashok. 2001. *Paṇḍita* and Pandits in History. In *The Pandit. Traditional Scholarship in India* (= Festschrift Parameswara Aithal), ed. Axel Michaels, 17–38. New Delhi: Manohar. (South Asia Institute, New Delhi Branch, Heidelberg University, South Asian Studies 38).

Arnold, David. 2000. *Science, Technology and Medicine in Colonial India*. Cambridge: Cambridge University Press. (The New Cambridge History of India III, 5).

Arp, Susmita. 2000. *Kālāpāni: Zum Streit über die Zulässigkeit von Seereisen im kolonialzeitlichen Indien*. Stuttgart: Franz Steiner Verlag. (Alt- und Neu-Indische Studien 52).

Ashok Vaidya. 2010. Shastri Shankar Daji Pade. *Journal of Ayurveda Integrative Medicine* 1 (2): 132–135.

Atridev Vidyālaṅkār. 1976. *Āyurved kā bṛhat itihās*. 2nd ed. Lucknow: Hindī Samiti, Uttar Pradeś Śāsan. (Hindī Samiti Granthmālā 33).

Baldev Upādhyāyă. 1983. *Kāśī kī pāṇḍityă paramparā (kāśīsthă saṃskṛt vidvānoṁ ke jīvan-carit evaṃ sāhityik avadānoṁ kā prāmāṇik vivaraṇ) [1200 ī. – 1880 ī.]*. Vārāṇasī: Viśvăvidyālay Prakāśan.

Banerjee, Sumanta. 1989. *The Parlour and the Streets. Elite and Popular Culture in Nineteenth Century Calcutta*. Calcutta: Seagull Books.

———. 2016. *Memoirs of Roads. Calcutta from Colonial Urbanization to Global Modernization*. New Delhi: Oxford University Press.

Bedekar, V.M. 1967. The legend of the churning of the ocean in the epics and the Purāṇas: A comparative study. *Purāṇa* 9 (1): 7–61.

Bergunder, Michael. 2002. Umkämpfte Vergangenheit. Anti-brahmanische und hindu-nationalistische Rekonstruktionen der frühen indischen Religionsgeschichte. In *"Arier" und "Draviden"*, ed. Michael Bergunder and Rahul Peter Das, 135–180. Halle: Verlag der Franckeschen Stiftungen zu Halle. (Neue Hallesche Berichte 2).

Bhagwan Dash. 1984. Preface. In A.F. Rudolf Hoernle, *Studies in the Medicine of Ancient India. Part I: Osteology or the Bones of the Human Body*, v–xx. New Delhi: Concept Publishing Company. Rpt. 1994.

———. 1989. *Fundamentals of Ayurvedic Medicine*. 7th revised ed. Delhi: Konark Publishers. Rpt. 1999.

Bhattacharji, Sukumari. 1970. *The Indian Theogony. A Comparative Study of Indian Mythology from the Vedas to the Purāṇas*. London: Cambridge University Press. Rpt. Delhi, etc.: Motilal Banarsidass, 1988.

Bhattacharya, France. n.d. *Paṇḍita Īśvaracandra Vidyāsāgara (Iswarchandra Vidyasagar) (1820–1891): la tradition au service d'un humanisme moderne*. Available for download at http://ceias.ehess.fr/index.php?1192.

Böhtlingk, Otto. 1873. *Indische Sprüche. Sanskrit und Deutsch*. Zweite vermehrte und verbesserte Auflage. Theil III: *ma – ha nebst Nachträgen*. St. Petersburg: Kaiserliche Akademie der Wissenschaften.

Bose, Debasis. 1994. Madhusudan Gupta. *Indian Journal of History of Science* 29 (1): 31–40.

Brandt, Carmen. 2020. The Roman script for South Asian languages. In *Wege durchs Labyrinth. Festschrift zu Ehren von Rahul Peter Das*, ed. Carmen Brandt and Hans Harder, 1–36. Heidelberg/Berlin: CrossAsia e-books.

Brockington, John, and Mary Brockington. 2006. *Rāma the Steadfast*. London: Penguin Books. (Penguin Classics).

Bryant, Edwin. 2002. Disput um die Vergangenheit. Indoarische Ursprünge und moderner nationalistischer Diskurs. In *"Arier" und "Draviden"*, ed. Michael Bergunder and Rahul Peter Das, 206–231. Halle: Verlag der Franckeschen Stiftungen zu Halle. (Neue Hallesche Berichte 2).

Cerulli, Anthony. 2012. *Somatic Lessons. Narrating Parenthood and Illness in Indian Medical Literature*. Albany: State University of New York Press. (SUNY Series in Hindu Studies).

Chatterjee, Mitali. 1999. *Education in Ancient India*. New Delhi: D. K. Printworld. (Reconstructing Indian History & Culture 17).

Chatterjee, P. 1959a. Kaviraj Gangadhar Roy Kaviratna – IV. *Nagarjun* (February 1959): 399–407.

———. 1959b. Kaviraj Gangadhar Roy Kaviratna – V. *Nagarjun* (March 1959): 501–506.
———. 1959c. Kaviraj Gangadhar Roy Kaviratna – VI. *Nagarjun* (May 1959): 685–693.
———. 1959d. Kaviraj Gangadhar Roy Kaviratna – VII. *Nagarjun* (June 1959): 811–818.
Chattopadhyay, Rita. 1995. *Gaṅgādhara Kavirāja: A 19th Century Indian Polymath in Oblivion*. Calcutta: Sanskrit Pustak Bhandar. Rpt. in *Contributions of the Traditional Paṇḍits of Bengal Towards Growth, Nourishment and Development of Sanskrit Studies*, Part I, ed. Karunasindhu Das, 266–281. Kolkata: Sanskrit Sahitya Parishat, 2012.
Cordier, Palmyr. 1899. Médecins et médecine au Bengale. *Annales d'hygiène et de médecine coloniales* 2: 555–562.
———. 1902. L'enseignement médical dans l'Inde ancienne. Temps védico-brahmaniques. *Bulletin de la Société française d'histoire de la médecine* 1: 177–191.
Darnton, Robert. 2002. Book production in British India, 1850–1900. *Book History* 5: 239–262.
Datta, Bhabataran. 1981. *A Linguistic Study of Personal Names and Surnames in Bengali*. Calcutta: University of Calcutta.
Dhar, Sailendra Nath. 1975. *A Comprehensive Biography of Swami Vivekananda*. Madras: Vivekananda Prakashan Kendra.
Dodson, Michael S. 2007. *Orientalism, Empire, and National Culture. India, 1770–1880*. New York: Palgrave Macmillan. (Cambridge Imperial and Post-Colonial Studies).
Dutta, Krishna. 2003. *Calcutta. A Cultural and Literary History*. New Delhi: Roli Books. (Lotus Collection).
Filliozat, J. 1964. *The Classical Doctrine of Indian Medicine. Its Origins and its Greek Parallels*. Trans. Dev Raj Chanana. Delhi: Munshiram Manoharlal. (Originally published as *La doctrine classique de la médecine indienne. Ses origines et ses parallèles grecs*. Paris: Imprimerie nationale, 1949)
Gadgil, Dilip. 2010. Understanding Ayurveda. *Journal of Ayurveda and Integrative Medicine* 1 (1): 77–80.
Galewicz, Cezary. 2020. *Kingdoms of Memory, Empires of Ink. The Veda and the Regional Print Cultures of Colonial India*. Kraków: Jagiellonian University Press.
Ghosh, Anindita. 2006. *Power in Print. Popular Publishing and the Politics of Language and Culture in a Colonial Society, 1778–1905*. New Delhi: Oxford University Press.
Goldman, Robert P., Sally J. Sutherland Goldman and Barend A. van Nooten, trans. 2010. *The Rāmāyaṇa of Vālmīki. An Epic of Ancient India*. Vol. VI: Yuddhakāṇḍa. 1st Indian edition. Delhi: Motilal Banarsidass Publishers.
Gough, Archibald Edward. 1878. *Papers Relating to the Collection and Preservation of the Records of Ancient Sanskrit Literature in India*. Calcutta: Office of the Superintendent of Government Printing.
Gray, Louis H., trans. 1950. *The Narrative of Bhoja (Bhojaprabandha) by Ballāla of Benares*. New Haven: American Oriental Society. (American Oriental Series 34).
Gupta, Brahmananda. 1976. Indigenous medicine in nineteenth- and twentieth-century Bengal. In *Asian Medical Systems. A Comparative Study*, ed. Charles Leslie, 368–378. Berkeley/Los Angeles/London: University of California Press.
Halbfass, Wilhelm. 1988. *India and Europe. An Essay in Understanding*. Albany: State University of New York Press.
Hara, Minoru. 1997. A note on *Dharmasya Sūkṣmā Gatiḥ*. In *Beyond Orientalism. The Work of Wilhelm Halbfass and Its Impact on Indian and Cross-Cultural Studies*, ed. Eli Franco and Karin Preisendanz, 515–532. Amsterdam/Atlanta: Rodopi. (Poznań Studies in the Philosophy of the Sciences and the Humanities 59).
Hatcher, Brian A. 1996. *Idioms of Improvement. Vidyāsāgar and Cultural Encounter in Bengal*. Calcutta: Oxford University Press.
———. 2005. What's become of the Pandit? Rethinking the history of Sanskrit scholars in colonial Bengal. *Modern Asian Studies* 39 (3): 683–723.
Heesterman, J.C. 1985. *The Inner Conflict of Tradition. Essays in Indian Ritual, Kingship, and Society*. Chicago/London: The University of Chicago Press.

Ikegame, Aya. 2013. *Princely India Re-imagined. A Historical Anthropology of Mysore From 1799 to the Present*. London: Routledge.

Important dates in the life of Mahāmahopādhyāya Dr. Umesha Mishra. 1968. *The Journal of the Ganganatha Jha Research Institute Allahabad* 24 (1–4) (= *Umeshamishra Commemoration Volume*): 1–3.

Jacobi, Hermann. 1908. Pandit Kisari Mohan Ganguli. *Zeitschrift der Deutschen Morgenländischen Gesellschaft* 62: 132–133.

Jādavaji Trikamji Āchārya, ed. 1938. *Suśrutasaṃhitā of Suśruta with the Nibandhasaṅgraha Commentary of Śrī Ḍalhaṇāchārya and the Nyāyacandrikā Pañjikā of Śrī Gayadāsāchārya on Nidānasthāna*. 3rd ed. Bombay: Nirṇaya Sāgar Press.

———, ed. 1941. *The Charakasaṃhitā of Agniveśa. Revised by Charaka and Dṛidhabala. With the Āyurveda-Dīpikā Commentary of Chakrapāṇidatta*. 3rd ed. Bombay: Nirṇaya Sāgar Press.

Jha, Hetukar. 1992. *Ganganath Jha*. New Delhi: Sahitya Akademi. (Makers of Indian Literature).

Kahl, Oliver. 2015. *The Sanskrit, Syriac and Persian Sources in the Comprehensive Book of Rhazes*. Leiden: Brill. (Islamic Philosophy, Theology and Science. Texts and Studies 93).

Khair, Tabish. 2015. India 2015: Magic, Modi and Arundhati Roy. *The Journal of Commonwealth Literature* 50 (3): 398–406.

Kintaert, Thomas. 2012. On the role of the lotus leaf in South Asian cosmography. *Wiener Zeitschrift für die Kunde Südasiens/Vienna Journal of South Asian Studies* 54 (2011–2012): 85–120.

Kramrisch, Stella. 1981. *The Presence of Śiva*. Princeton: Princeton University Press. Rpt. Delhi: Motilal Banarsidass, 1988.

Liétard, Gustave. 1903. Review of J. Jolly, *Medicin*, Straßburg 1901. *Janus* 8: 90–94.

Maas, Philipp A. 2010a. Computer aided stemmatics – The case of the fifty-two text versions of Carakasaṃhitā Vimānasthāna 8.67–157. *Wiener Zeitschrift für die Kunde Südasiens/Vienna Journal of South Asian Studies* 52–53 (2009–2010): 63–119.

———. 2010b. On what became of the *Carakasaṃhitā* After Dṛḍhabala's revision. *eJournal of Indian Medicine* 3: 1–22.

Maṅgaldās Svāmī. 1940 [Samvat 1997]. *Āyurvedmārtaṇḍ vaidyăratn prāṇācāryă śrī lakṣmīrām jī svāmī bhiṣagācāryă kā saṅkṣipt jīvan-caritr*. Jaypur: Śrī Dādū Mahāvidyālay.

Meulenbeld, G. Jan. 1999. *A History of Indian Medical Literature*. Vol. IA: *Text*, Vol. IB: *Annotation*. Groningen: Egbert Forsten. (Groningen Oriental Studies 15).

———. 2000. *A History of Indian Medical Literature*. Vol. IIA: *Text*, Vol. IIB: *Annotation*. Groningen: Egbert Forsten. (Groningen Oriental Studies 15).

———. 2010. Reflections on some Oddities of the Carakasaṃhitā. In *From Turfan to Ajanta. Festschrift for Dieter Schlingloff on the Occasion of his Eightieth Birthday*, 2 vols., ed. Eli Franco and Monika Zin, 715–724. Lumbini: Lumbini International Research Institute.

Michaels, Axel. 2001. Traditional Sanskrit learning in Contemporary India. In *The Pandit. Traditional Scholarship in India* (= Festschrift Parameswara Aithal), ed. Axel Michaels, 3–16. New Delhi: Manohar. (South Asia Institute, New Delhi Branch, Heidelberg University, South Asian Studies 38).

Mookerji, Radha Kumud. 1951. *Ancient Indian Education. Brahmanical and Buddhist*. 2nd ed. London: Macmillan.

Mukhopadhyaya, Girindranath. 1922–1929. *History of Indian Medicine. Containing Notices, Biographical and Bibliographical, of the Ayurvedic Physicians and Their Works on Medicine. From the Earliest Age to the Present Time*, 3 vols. Calcutta: University of Calcutta. Rpt. New Delhi: Munshiram Manoharlal Publishers.

Mukhopadhyaya, Govinda Gopal. 1990. *A Great Savant. Mahamahopadhyaya Gopinath Kaviraj*. Calcutta: Jagadiswar Pal.

Müller, Reinhold F.G. 1930a. Book notice on Kaviratna, Avinash Chandra. *English Translation of Charaka-Samhita*. Calcutta. 200, Cornwallis Str. [1891–1925] [...]. *Mitteilungen zur Geschichte der Medizin und der Naturwissenschaften* 29: 31–32.

———. 1930b. Book notice on *Caraka-Saṃhitā* ... Kavirāja Śrīnarendranātha Senaguptena Kavirāja Śrībalāicandra Senaguptena ca sampāditā ... Kalikātā. Calcutta (C. K. Sen and Co.

Ld.) Khaṇḍa I, ś. 1849 [1927] [...] Khaṇḍa II, ś. 1850 [1928] [...]. *Mitteilungen zur Geschichte der Medizin, der Naturwissenschaften und der Technik* 29: 229.

———. 1934. Book notice on *Caraka-Saṃhitā – cikitsita-kalpa-siddhināmakasthāna – tṛtīyaḥ khaṇḍaḥ*. Calcutta: C. K. Sen & Co. Ltd. 1933 [...]. *Mitteilungen zur Geschichte der Medizin, der Naturwissenschaften und der Technik* 33 (1) (Nr. 155): 207.

Narasimha Murthy. 2012. Acharya Priyavrata Sharma – A pioneer in creation of new authentic ayurvedic literature in 20th century. *Annals of Ayurvedic Medicine* 1 (3) (July–September 2012): 113–116.

National Mission for Manuscripts, New Delhi. Database of manuscripts. https://www.pandulipipatala.nic.in

Oberlies, Thomas. 1998. *Die Religion des Ṛgveda. Erster Teil – Das religiöse System des Ṛgveda*. Wien: Sammlung de Nobili. (Publications of the De Nobili Research Library 26).

———. 2023. *The Religion of the Ṛgveda*. Oxford: Oxford University Press.

O'Flaherty, Wendy Doniger. 1973. *Asceticism and Eroticism in the Mythology of Śiva*. Delhi: Oxford University Press.

———. 1975. *Hindu Myths. A Source Book Translated From the Sanskrit*. Harmondsworth: Penguin Books. (The Penguin Classics).

Ogborn, Miles. 2007. *Indian Ink. Script and Print in the Making of the English East India Company*. Chicago/London: The University of Chicago Press.

Parāḍakara Vaidya, Bhiṣagāchārya Hariśāstrī, ed. 1939. *Aṣṭāṅgahṛdayam (the Core of Octopartite Āyurveda) Composed by Vāgbhaṭa With the Commentaries (Sarvāṅgasundarā) of Aruṇadatta and (Āyurvedarasāyana) of Hemādri*. Collated by Late Dr. Aṇṇā Moreśwara Kuṇṭe, B. A., M. D., Demonstrator of Anatomy, Grant Medical College, and Kṛṣṇa Rāmchandra Śāstrī Navare. 6th ed. Bombay: Nirṇaya Sāgar Press.

Patil, Parimal. 2013. The historical rhythms of the Nyāya–Vaiśeṣika knowledge system. In *Periodization and Historiography of Indian Philosophy*, ed. Eli Franco, 91–125. Wien: Sammlung de Nobili. (Publications of the De Nobili Research Library 37).

Pavitrananda, Swami, ed. 1948. *Letters of Swami Vivekananda*. 4th ed. Mayavati, Almore: Advaita Ashrama.

Pecchia, Cristina. 2022a. Ayurveda, philology and print. On the first printed edition of the Carakasaṃhitā and its context. *South Asian History and Culture* 13 (1): 112–134.

———. 2022b. With the eye of a scholar and the insight of a physician: Gangadhar Ray Kaviraj and the *Carakasaṃhitā*. In *'Verità e bellezza'. Essays in Honour of Raffaele Torella*, ed. Francesco Sferra and Vincenzo Vergiani, 797–839. Napoli: UniorPress.

Preisendanz, Karin. 2000. Debate and independent reasoning vs. tradition: On the precarious position of early Nyāya. In *Harānandalaharī. Volume in Honour of Professor Minoru Hara on His Seventieth Birthday*, ed. Ryutaro Tsuchida and Albrecht Wezler, 221–251. Reinbek: Dr. Inge Wezler Verlag für Orientalistische Fachpublikationen.

———. 2010. Mahābhūtas. In *Brill's Encyclopedia of Hinduism*, ed. Knut A. Jacobsen et al., Vol. 2, 806–817. Leiden: Brill. (Handbook of Oriental Studies/Handbuch der Orientalistik, Section 2: South Asia, 22.2).

———. 2013. Logic, debate and epistemology in ancient Indian medical science – An investigation into the history and historiography of Indian philosophy Part I. In *Medical Texts and Manuscripts in Indian Cultural History*, ed. Dominik Wujastyk, Anthony Cerulli and Karin Preisendanz, 63–139. New Delhi: Manohar.

———. 2015. Between affirmation and rejection. Attitudes towards the body in Ancient South Asia from the *Ṛgveda* to early classical medicine. In *Thinking the Body as a Basis, Provocation and Burden of Life. Studies in Intercultural and Historical Contexts*, ed. Gert Melville and Carlos Ruta, 113–144. Berlin/Boston: de Gruyter Oldenbourg. (Challenges of Life. Essays on Philosophical and Cultural Anthropology 2).

———. 2018. Text segmentation, chapter naming and the transmission of embedded texts in South Asia, with special reference to the medical and philosophical traditions as exemplified by the *Carakasaṃhitā* and the *Nyāyasūtra*. In *Pieces and Parts in Scientific Texts*, ed. Florence Bretelle-Establet and Stéphane Schmitt, 159–220. Cham: Springer. (Why the Sciences of the Ancient World Matter 1).

Preisendanz, Karin, Philipp A. Maas, and Cristina Pecchia. (forthcoming). *A Critical Edition of Carakasaṃhitā Vimānasthāna Chapter 8.*

Raghavan, V., K. Kunjunni Raja, N. Veezhinathan, E.R. Rambai and Siniruddha Dash, eds. 1949–2015. *New Catalogus Catalogorum. An Alphabetical Register of Sanskrit and Allied Works and Authors.* Chennai: University of Madras. (Madras University Sanskrit Series 18, 26, 28–35, 37–65).

*Rāmāyaṇa.* P.L. Vaidya, ed. 1971. *The Yuddhakāṇḍa. The Sixth Book of the Vālmīki-Rāmāyaṇa. The National Epic of India.* Baroda: Oriental Institute.

Renou, Louis, ed. 1962. *Hinduism.* New York: George Braziller. (Great Religions of Modern Man).

Renou, Louis, and Jean Filliozat. 1985. *L'Inde classique. Tome II, avec le concours de Paul Demiéville, Olivier Lacombe and Pierre Meile.* Paris: École Française d'Extrême-Orient. (Bibliothèque de l'École Française d'Extrême-Orient 3).

Robb, Peter. 2014. *Useful Friendship. Europeans and Indians in Early Calcutta.* New Delhi: Oxford University Press.

Roşu, Arion. 1978. Études āyurvediques I. Le *trivarga* dans l'Āyurveda. *Indologica Taurinensia* 6 (= *Proceedings of the Third World Sanskrit Conference* [*Paris, 20–25 June 1977*]): 255–260.

———, ed. 1989. *Gustave Liétard et Palmyr Cordier. Travaux sur l'histoire de la médecine Indienne,* Documents réunis et présenté par Arion Roşu. Paris: de Boccard. (Publications de l'Institut de Civilisation Indienne, Série in-8°, Fascicule 56).

Roth, Rudolf. 1872. Indische Medicin. Caraka. *Zeitschrift der Deutschen Morgenländischen Gesellschaft* 26: 441–452. Rpt. in *Rudolf von Roth. Kleine Schriften,* ed. Konrad Meisig, 467–478. Stuttgart: Steiner Verlag, 1994. (Glasenapp-Stiftung 36).

———. 1894a. Review of *Nibandhasangraha a Commentary on the Sushrutasanhita by Dallana Mishra.* Edited and published by Pandit Jibananda Vidyasagara. Third Edition. Calcutta, Saraswati Press 1891. *Zeitschrift der Deutschen Morgenländischen Gesellschaft* 48: 138–140.

———. 1894b. Review of *Charaka-Samhitâ translated into English. Published by Avinash Chandra Kaviratna etc.* Calcutta, printed by D. C. Dass & Co. *Zeitschrift der Deutschen Morgenländischen Gesellschaft* 48: 140–142.

Rüping, Klaus. 1970. *Amṛtamanthana und Kūrma-Avatāra.* Wiesbaden: Otto Harrassowitz. (Schriftenreihe des Südasien-Instituts der Universität Heidelberg).

Saksena, B.R. 1968. Preface. *The Journal of the Ganganatha Jha Research Institute Allahabad* 24 (1–3) (= *Umeshamishra Commemoration Volume*).

Saradaprosad Vidyabhusan. 1926. *The Bhoj-Prabandha of Sree Ballal (With English Translation).* 3rd ed. Calcutta: S.C. Auddy&Co., Booksellers and Publishers.

Scharfe, Hartmut. 2002. *Education in Ancient India.* Leiden/Boston/Köln: Brill. (Handbook of Oriental Studies/Handbuch der Orientalistik, Section 2: India/Indien, 16).

Sharma, Priyăvrat. 1975. *Āyurved kā vaijñānik itihās.* Vārāṇasī: Caukhambhā Oriyanṭāliyā. (Jaykṛṣṇadās Āyurved Granthmālā 1).

———. 1986. Teaching and practice of Āyurveda in Calcutta. *Āyurveda Newsletter* 4: 1–4.

———. 2005a. *Six Decades of Ayurveda (1941–2000). A Collection of the Selected Articles and Lectures of Prof. Priya Vrat Sharma.* Eds. Satya Deo Dubey and Anugrah Narain Singh. Delhi: Chaukhamba Sanskrit Pratishthan. (The Vrajajiwan Ayurvijnan Granthamala 37).

———. 2005b. Ashwins and Their Miracles. In Sharma 2005a, 29–38. Originally published in 1964.

———. 2005c. Light on Ayurveda. In Sharma 2005a, 774–794. Originally published in 1953.

———. 2016. *Āyurved kā vaijñānik itihās.* Corrected and enlarged edition. Vārāṇasī: Caukhambhā Oriyanṭāliyā. (Jaykṛṣṇadās Āyurved Granthmālā 1).

Sharma Ganti, K.S. Dutta, and A.V.K. Lakshmi Prasanna. 2019. Life profile and contribution of Vaidya Sripada Krishnamurthy Sastry to Ayurveda literature. *Āryavaidyan. A Quarterly Journal on Ayurveda and Allied Sciences* 32 (4): 54–60.

Shaw, Graham. 1981. *Printing in Calcutta to 1800. A Description and Checklist of Printing in Late 18th-Century Calcutta.* London: The Bibliographical Society.

Singh, Gurdip, and P.D. Joshi. 1982. Impact of European science and technology on the development of modern Āyurveda during 19th century. *Indian Journal of History of Science* 17 (2): 313–325.

Sinha, Samita. 1993. *Pandits in a Changing Environment. Centres of Sanskrit Learning in Nineteenth Century Bengal*. Calcutta: Sarat Book House.
Stache–Rosen, Valentina. 1990. *German Indologists. Biographies of Scholars in Indian Studies Writing in German. With a Summary of Indology in German Speaking Countries*, 2nd revised ed. by Agnes Stache–Weiske. New Delhi: Max Mueller Bhavan.
Stark, Ulrike. 2009. Publishers as patrons and the commodification of Hindu religious texts in nineteenth-century North India. In *Patronage and Popularisation, Pilgrimage and Procession. Channels of Transcultural Translation and Transmission in Early Modern South Asia. Papers in Honour of Monika Horstmann*, ed. Heidi Rika Maria Pauwels, 189–203. Wiesbaden: Harrassowitz Verlag. (Studies in Oriental Religions 58).
———. 2012. *An Empire of Books. The Naval Kishore Press and the Diffusion of the Printed Word in Colonial India*. 2nd ed. Ranikhet: Permanent Black.
Tagare, G. V., trans. 1992. *The Skanda-Purāṇa. Part I*. Delhi: Motilal Banarsidass Publishers. (Ancient Indian Tradition and Mythology Series [Purāṇas in Translation] 4).
Taranatha Tarakavachaspati. 1881. *Vachaspatya. A Comprehensive Sanskrit Dictionary*. Vol. 16. Calcutta: The Saraswati Press.
*The Life of Swami Vivekananda by His Eastern and Western Disciples*. 1974. 8th ed. Calcutta: Advaita Ashrama.
van Buitenen, J.A.B. 1973. *The Mahābhārata. I. The Book of the Beginning*. Chicago/London: The University of Chicago Press.
Vettam Mani. 1975. *Purāṇic Encyclopedia. A Comprehensive Dictionary with Special Reference to the Epic and Purāṇic Literature*. Motilal Banarsidass. Rpt. 1989: Delhi.
Weiss, Richard S. 2015. Print, religion, and canon in colonial India: The publication of Ramalinga Adigal's Tiruvarutpa. *Modern Asian Studies* 49 (3): 650–677.
Wilson, H. H., trans. 1980. *The Viṣṇu Purāṇa. A System of Hindu Mythology and Tradition*. Translated from the original Sanskrit and illustrated by notes chiefly derived from the other Purāṇas. Enlarged and arranged by Nag Sharan Singh. New Delhi: Nag Publishers. (Translation and notes originally published London: John Murray, 1840).
Winternitz, Moriz. 1920. *Geschichte der indischen Litteratur*. Dritter Band: *Die Kunstdichtung – die wissenschaftliche Litteratur – neuindische Litteratur – Nachträge zu allen drei Bänden*. Leipzig: C. F. Amelangs Verlag.
———. 1985. *History of Indian Literature*. Volume III. Part One: *Classical Sanskrit Literature*. Part Two: *Scientific Literature*. Trans. Subhadra Jha. Delhi: Motilal Banarsidass.
Witzel, Michael. 1985. Materialien zu den vedischen Schulen. I. Über die Caraka-Śākhā. *Studien zur Indologie und Iranistik* 7: 109–132.
———. 1987. The Case of the Scattered Head. *Studien zur Indologie und Iranistik* 13–14: 363–415 (= *Festschrift Wilhelm Rau zur Vollendung des 65. Lebensjahres dargebracht von Schülern, Freunden und Kollegen*, eds. Heidrun Brückner, Dieter George †, Claus Vogel, and Albrecht Wezler).
Wujastyk, Dominik. 1995. Medicine in India. In *Oriental Medicine. An Illustrated Guide to the Asian Arts of Healing*, ed. Jan Van Alphen and Anthony Aris, 19–37. London: Serindia Publications.
———. 2003. *The Roots of Ayurveda. Selections from Sanskrit Medical Writings*. Revised ed. London: Penguin Books.
———. 2008. A body of knowledge: The Wellcome Ayurvedic anatomical man and his Sanskrit context. *Asian Medicine* 4: 201–248.
———. 2009. Interpreting the image of the human body in premodern India. *International Journal of Hindu Studies* 13 (2): 189–228.
———. 2013. New manuscript evidence for the textual and cultural history of early classical Indian medicine. In *Medical Texts and Manuscripts in Indian Cultural History*, ed. Dominik Wujastyk, Anthony Cerulli, and Karin Preisendanz, 141–157. New Delhi: Manohar.
Wujastyk, Dominik, Jason Birch, Andrey Klebanov, Madhu K. Parameswaran, Madhusudan Rimal, Deepro Chakraborty, Harshal Bhatt, Vandana Lele, and Paras Mehta. 2023. *On the Plastic Surgery of the Ears and Nose. The Nepalese Version of the Suśrutasaṃhitā*. Heidelberg: Heidelberg Asian Studies Publishing.

# Part III
# Shaping Specific Features of Scientific Texts

# Chapter 7
# Representing Numbers and Quantities in Editions of Mathematical Cuneiform Texts

**Christine Proust**

**Abstract** Numbers and quantities are omnipresent in administrative, economic and mathematical texts produced in the Ancient Near East. However, the philological work ignored numbers and quantities. The conventions adopted by philologists for representing the cuneiform signs attached to numbers and quantities have varied greatly according to the date and genre of the sources, the editors' choices, the constraints imposed by publishers and other parameters. There is little consensus on conventions for the representation of quantitative data and this is particularly visible in recent editions of mathematical cuneiform texts. At the same time, a wider consensus has emerged regarding the representation of sexagesimal place-value notation, a numerical system used essentially in mathematical texts. This contribution explores how historians of cuneiform mathematics have approached the problems raised by the representation of numbers and quantities in modern editions. The conclusion opens a debate on current issues at stake in editing mathematical cuneiform texts.

## Abbreviations

BDTNS Database of Neo-Sumerian Texts (http://bdts.filol.csic.es/, accessed 14 June 2017).

CDLI Cuneiform Digital Library Initiative (http://cdli.ucla.edu/, accessed 14 June 2017).

ETCSL Electronic Text Corpus of Sumerian Literature (http://etcsl.orinst.ox.ac.uk/, accessed 14 June 2017).

The research leading to these results has received funding from the European Research Council under the European Union's Seventh Framework Programme (FP7/2007–2013) / ERC Grant Agreement No. 269804.

C. Proust (✉)
CNRS and Université Paris Cité – Laboratoire SPHERE UMR 7219 – case 7093, Paris Cedex 13, France





A. Keller, K. Chemla (eds.), *Shaping the Sciences of the Ancient and Medieval World*, Archimedes 69, https://doi.org/10.1007/978-3-031-49617-2_7

| | |
|---|---|
| MCT | Neugebauer and Sachs (1945) |
| Obv | Obverse |
| ORACC | Open Richly Annotated Cuneiform Corpus (http://oracc.museum.upenn.edu/, accessed 14 June 2017). |
| Rev. | Reverse |
| SPVN | sexagesimal place-value notation |

## 7.1 Introduction

Numbers and quantities are omnipresent in administrative, economic and mathematical texts produced in the Ancient Near East. In a way, writing began with numbers and quantities (Nissen et al. 1993). The conventions adopted by philologists for representing the cuneiform signs attached to numbers and quantities have varied greatly according to the date and genre of the sources, the editors' choices, the constraints imposed by publishers and other parameters. There is little consensus on conventions for the representation of quantitative data and this is particularly visible in recent editions of mathematical cuneiform texts. At the same time, a wider consensus has emerged regarding the representation of sexagesimal place-value notation, a numerical system widely used in mathematical cuneiform texts. This contribution explores how historians of cuneiform mathematics have approached the problems raised by the representation of numbers and quantities in modern editions. The conclusion opens a debate on current issues at stake in editing mathematical cuneiform texts.

The representation of numbers and quantities in editions of cuneiform texts was one of Neugebauer's first concerns when he entered the field of ancient mathematics: '… an inapt transcription of the numbers could render them incomprehensible and completely distort the meaning of the calculation.' (Neugebauer 1933–1934: 221). This chapter is an invitation to continue and expand Neugebauer's reflection.

What I mean by 'numbers' encompasses two types of numbers: first, the numbers written in non-positional systems, which were used in cuneiform texts for expressing quantities such as measurement values (examples are examined in Sect. 7.3 below) or counting discrete items such as things, animals or individuals (see Sect. 7.4); second, the numbers written in sexagesimal place-value notation (SPVN hereafter) which were used in cuneiform texts for calculation (see Sect. 7.5).[1]

Elaborating relevant conventions for the representation of the cuneiform signs in modern script was not an easy task for pioneers of cuneiform studies. Not only the writing systems but also the concepts are different in ancient and modern contexts. For example, from the point of view of ancient scribes, which concepts correspond to what we call a 'number', a 'quantity' or a 'measurement unit'? What is the

[1] For further details and arguments on this distinction, see (Proust 2008).

difference between a metrological and a numerical system in the framework of ancient practices? The fact that modern definitions do not adhere exactly to ancient concepts is one of the main difficulties, and the risk is that our modern notations erase the specificities of ancient categories.[2] However, inventing sophisticated symbolic systems for representing ancient categories may make the editions extremely difficult to read. As Høyrup pointed out in relation to the translation of technical terminology used in cuneiform problem texts, 'considerations of intelligibility put some constraints on the principle of conformity' (Høyrup 1990: 62). For his part, Friberg developed a system of transliteration which may be the most faithful to the original, but which was not adopted widely, probably because of its sophistication.[3]

Does the diversity of conventions adopted by historians for representing numbers and quantities reflect the diversity of practices by ancient scribes, the diversity of interpretations by historians or the diversity of academic habits? Is it desirable to search for a consensus among historians in order to standardize conventions? Or would such consensual conventions produce a uniform system of representation which would conceal the diversity of practices of ancient scribes? How did the development of digital databases and, more generally, of large international projects change the way in which this problem was approached? Any scholar who works on mathematical cuneiform texts is faced with these questions. Revisiting the discussions and practices developed during the short history of editing cuneiform mathematics may help to raise awareness of the meanings conveyed by the conventions used today and perhaps improve the readability of textual editions beyond a narrow circle of specialists.

The edition of cuneiform texts includes several stages, each of which raises specific issues: photographs, hand copies, transliteration, translation and interpretation. It must be noted that the notion of 'edition' in Assyriology is rather vague and that the practices which fall into this category are variable. Some 'editions', particularly the earliest ones, are limited to hand copies, while others include the complete process, from photographs to interpretation. In many cases, only part of the process is offered as an 'edition'.[4] Today, the system of representation of a cuneiform text using the modern Latin alphabet is almost always a 'transliteration', that is, a system which represents the cuneiform signs. However, other systems were adopted in the history of these editions, such as transcription or conform transliteration,[5] and sometimes mixtures of the different systems. To sum up, the process can be described by the diagram in Table 7.1.

[2] This subtle 'algebra of comprehension' was perceived in 1926 by Landsberger as follows: 'Most often we have to content ourselves with partial equations of the type: Part of the Babylonian concept *x* corresponds with part of our concept *y*.' (Landsberger 1976: 6).

[3] Friberg explains in detail and justifies his conventions in (Friberg 1993: 385–386).

[4] Hilprecht (1906), the earliest book devoted to editing mathematical texts, provides only a detailed catalogue and hand copies. By contrast, Neugebauer's editions of mathematical cuneiform texts include all the stages: photographs, hand copies, transliterations, translations, and mathematical interpretation (Neugebauer 1935–1937). Note that the process of edition has been heavily impacted by new technologies. For example, in many recent publications, digital photographs replace traditional hand copies.

**Table 7.1** From clay tablet to interpretation

| Cuneiform text on clay tablet | |
|---|---|
| Photograph | |
| Hand copy | |
| Representation in modern Latin alphabet | Conform transliteration (Friberg) |
| | Transliteration |
| | Transcription |
| Translation | |
| Interpretation | |

In this chapter, I shall focus mainly on transliteration and translation, even if the other stages also may equally require a discussion. In Sect. 7.2, I evoke the history of the distinction between transcription and transliteration. In Sect. 7.3, through the example of the notation of measurements of surface and capacity, I propose an analysis of the way in which the transliteration of numbers and quantities convey interpretations. In Sect. 7.4, I discuss the issue of both transliteration and translation of non-positional numbers through considerations on the version of the Sumerian King List noted on the Old Babylonian prism Ashm. 1923-444. In relation to the transliteration and translation of numbers written in sexagesimal place-value notation (Sect. 7.5), I shall introduce the discussion with an extract from a relevant article published by Neugebauer in 1933, *Zur Terminologie der Mathematischen Keilschrifttexte*.[6]

## 7.2 Representing a Cuneiform Text in the Modern Latin Alphabet

What is the reason for the great diversity of the systems of representation of cuneiform texts adopted in various editions? Some of the causes are inherent to the ancient writing systems adopted in the Ancient Near East, which differ dramatically from modern alphabetic script and arithmetical symbols. The distance between the original texts on clay tablets, written with cuneiform signs in both phonetic and logographic scripts, and the representations of them on paper (or digital) media, written in modern alphabetic scripts, is necessarily considerable. Moreover, the relationships between written and spoken languages took different

[5] Friberg added an intermediate phase between copy and transliteration that he termed as 'conformal transliteration, in which the *individual cuneiform signs* of the original texts have been replaced by their Akkadian or Sumerian transliterated counterparts. […] An effort is thereby made to place the transliteration of the cuneiform signs *in their correct positions* within outlines of the obverses and reverses of the clay tablets.' (Friberg 1993: 385; emphasis by the author).

[6] At the time the article was published, the University of Göttingen was enduring dramatic upheavals. In April 1933, Neugebauer refused to swear an oath to the Nazis and left Germany.

forms in ancient practices with texts. These relationships are not well elucidated by present-day historians since 'we do not have at our disposal informants speaking the original languages' (Gelb 1948: 2). In addition, several languages were recorded in cuneiform script, namely Sumerian and Akkadian in the case of mathematical texts.

As pointed out in the introduction, different systems were adopted to represent cuneiform texts using the modern Latin alphabet: transliteration, transcription and various hybrids of these two systems. It is nowadays considered that, broadly speaking, transliteration encodes graphemes and phonemes while transcription encodes words, but this distinction was not always so clear.[7] The distinction between the two systems, borrowed from linguistic studies, became clear in the field of Assyriology from the mid-twentieth century, when the development of large international projects, such as the Chicago Akkadian Dictionary, revealed the necessity for a stable system of representation of the texts that would be widely accepted by the community of Assyriologists. The International Commission on Transliteration of Cuneiform was created at the Twenty-first Congress of Orientalists in Paris in 1948 for this purpose. Ignace J. Gelb elaborated a 'Memorandum'[8] to 'provide Assyriologists with a number of questions and proposals which were to serve as a basis for consideration of an international system of transliteration and transcription of cuneiform writing and language' (Gelb 1948: 1). His concern was also to 'put a stop to the unbridled expansion of the existing transliteration system' (Gelb 1948: 1). Gelb's Memorandum clarified numerous questions concerning the representation of cuneiform signs. In particular, it clarifies the distinction between transliteration and transcription in these terms:

> Transliteration is a form of graphic transfer wherein one sign (or a combination of alphabetic signs and artificial symbols) stands for each character of the writing we are recording, while 'transcription' is a form of graphic transfer wherein one sign (or a combination of alphabetic signs and artificial symbols) stands for each phoneme of the language we are recording (Gelb 1948: 2; Gelb 1951: Sect. 7.4).

However, it should be noted that Gelb's Memorandum does not consider the cuneiform signs used to represent quantitative data. As for transliteration of numbers and measurement values, only recently did the staff of the Cuneiform Digital Library Initiative (CDLI) begin acting to 'put a stop to the unbridled expansion of the existing transliteration system', to use Gelb's words.

How did the distinctions clarified by Gelb apply to mathematical cuneiform texts? In mathematical texts dating back to the Old Babylonian period onward, the words are written either phonetically with signs denoting the syllables of the Akkadian language or with Sumerograms (signs which represent Sumerian words,

[7] The cdli:wiki encyclopedia defines transliteration and translation as follows: "Whereas transliteration attempts to mimic the written form of the cuneiform sign in question, often misrepresenting the linguistic or grammatical structure to some degree, transcription focuses on the grammatical or linguistic structure at the expense of the written orthography".

[8] (Gelb 1948), with complementary considerations in (Gelb 1951).

sometimes completed with grammatical elements noted on a syllabic principle).[9] It is widely accepted that the Sumerograms were spoken in Akkadian. However, some mathematical texts are written only in Sumerograms. How these 'Sumerian' texts were spoken is not always clear, and remains the subject of discussion.

How can the Sumerograms which are supposed to have been spoken in Akkadian be represented in publications? Thureau-Dangin, like many Assyriologists of his generation, opted for representing Sumerograms, when used in Akkadian contexts, with their Akkadian phonetic counterpart. For example, the Sumerogram transliterated a-ša$_3$, which means 'field' in Sumerian, was transcribed by Thureau-Dangin as '*eqlum*', ('field' in Akkadian; see Table 7.2). Neugebauer, 'against the *opinio communis*', objected strongly to this practice and argued that the 'Akkadization' destroys essential clues.

> While the way of transcribing numerals is merely a matter of convention, the question of the transcription of ideograms of mathematical terms falls within a purely philological field, in which a solid tradition has already been established. I am aware that with the following I am going against the *opinio communis* [...]. The matter touched upon here is not a mere triviality, but a fundamentally important point for mathematical-astronomical texts. Firstly: In certain circumstances, from the transcription one might no longer be able to determine

**Table 7.2** YBC 4710 #1 according to Neugebauer and Thureau-Dangin[a]

| Neugebauer 1935–1937: I, 402 | Thureau-Dangin 1938: 148 | Translation (mine) |
|---|---|---|
| [a-š]à 1 (eše) gán | *[eql]um ebel ikîm* | The surface is 1 *eše GAN.* |
| [uš] *ša a-na* ba-zi nu<-zu> | *ina [siddim] mala assuḫu <ul(a) îde>* | What I subtracted from the length (I don't know). |
| [íb-]tag$_4$ uš *ù šá* zi | *[ša]pilti šiddim ù šá assuḫu* | The remaining length, with what I subtracted, |
| ì-kú-ma sag íb-si$_8$ | *uštakil-ma <ana> pûtim ašši* | I crossed then to the width I raised. |
| *ù* 1 (eše) gán a-šà | *ù ebel ikîm eqlam* | Otherwise, 1 *eše GAN,* the surface, |
| KI *šá* uš ba-zi kú | *itti šá <ina> šiddim assuḫu uštakil* | with what I subtracted from the length, I crossed. |
| ì-kú (?) DUP (?) SAG (?) daḫ 3,12 | ì-k ú d u b-s a g *ûṣib 3.12* | To what I crossed before, I added, it is 3:12. |
| *šá* uš-ta ba-zi | *sá ina šiddim assuḫu* | What I subtracted from the length |
| u-gù íb-tag$_4$ uš 6 dirig | *eli šapilti šiddim* 6 *îter* | exceeds the remaining length of 6. |

[a]The modern conventions in transliterations are as follows: italics denote Akkadian words, while plain fonts denote Sumerian logograms (usually an expanded font is used for Sumerian, as in the case of Thureau-Dangin, but Neugebauer simply used plain font). Signs whose meaning or pronunciation remains unidentified are noted in capitals. Brackets indicate damaged (e.g., [a-š]à) or broken (e.g., [uš]) signs. Signs inside triangular brackets are missing on the tablet and have been restored by the editor (e.g., <zu>)

[9]Modern scholars generally call such signs 'ideograms', or occasionally 'logograms'. I prefer to follow Gonzalo Rubio and use the neutral term 'Sumerogram' (Rubio 2007: 44), which does not prejudge the function or the reading of the sign.

which character was contained in the text [...] Furthermore: how should one proceed in cases in which entire Sumerian phrases occur in Akkadian texts? Why should one Akkadize them? The same principle would lead one to eradicate the Latin and French in writings of scholars of the [seventeenth] century who essentially wrote in German. And finally: with the process of complete Akkadization one destroys the fundamental factual role of ideograms: namely that they function entirely like mathematical symbols... (Neugebauer 1932–1933: 222, translated by Sandra Hoehn modified).

Original text: Während die Transkriptionsweise der Zahlen eine reine Angelegenheit der Konvention ist, greift die Frage nach der Transkription von Ideogrammen mathematischer Termini in ein rein philologisches Gebiet ein, in dem sich schon eine feste Tradition gebildet hat. Ich bin mir bewusst, dass ich mit dem Folgenden gegen die opinio communis verstoße, [...] Es ist keine bloße Nebensächlichkeit, die hier gestreift wird, sondern ein für mathematisch-astronomische Texte ganz wesentlicher Punkt. Zunächst: man kann unter Umständen aus der Transkription nicht mehr erkennen, welches Zeichen im Text stand [...]. Dann: wie hat man sich in Fällen zu verhalten, wo ganze sumerische Phrasen in akkadischen Texten stehen? Warum muss man sie akkadisieren? Mit demselben Recht müsste man das Lateinische oder Französische aus im Prinzip deutschen Schriften von Gelehrten des 17. Jahrhunderts ausmerzen. Und schließlich: Man zerstört durch das absolute Akkadisieren eine fundamentale sachliche Rolle der Ideogramme: dass sie nämlich vollkommen wie mathematische Symbole wirken,...

As a consequence, the representation of the same cuneiform text may appear very different, depending on the edition. Table 7.2 shows the representations by Neugebauer and Thureau-Dangin of the first problem of tablet YBC 4710, an Old Babylonian mathematical document kept in the Yale Babylonian Collection, written almost entirely in Sumerograms. Neugebauer used the term 'transcription' to describe his system of representation of cuneiform texts, even if it is actually a 'transliteration' according to Gelb's definitions. Thureau-Dangin used a hybrid system: he transliterated the Akkadian words but generally transcribed the Sumerograms with their presumed Akkadian pronunciations; sometimes, however, when he could not identify the Akkadian pronunciation, he transliterated the Sumerograms.

Subsequently, after the general principles stated in Gelb's Memorandum were adopted by Assyriologists, Neugebauer's point of view came to prevail over that of Thureau-Dangin. It is interesting to note that a controversy which aroused among specialists of cuneiform mathematics turned out to present significant challenges for the whole field of Assyriology. Neugebauer and Gelb defended more or less the same principles for the representation of cuneiform texts. However, their arguments were not exactly the same. Neugebauer perceived the Sumerograms (which he called 'ideograms') as types of mathematical symbols in some contexts and his arguments were specific to mathematical texts.

To sum up, the diversity of transliterations of the Sumerograms has manifold causes. Some are inherent to ancient practices, and derive from the complexity of the systems of writing, mixing phoneticism and logography, the relations between the written and spoken languages, and the fact that ancient scribes attributed different functions to various classes of graphemes. Other causes of divergences are the interpretations of these relations and functions by modern historians, sometimes in a rather controversial mode. The ultimate causes result from personal choices in conventions adopted for representing cuneiform signs in modern alphabet and languages. We shall see in Sect. 7.3 of this chapter how these different phenomena impacted the transliteration of measurement values in some editions of texts.

Translations are even more sensitive to interpretation than transliteration and raise other kinds of problems. I shall not deal here with the discussion on the translation of technical terms related to mathematical terminology.[10] I shall confine my focus to the translation of numbers and quantities only (see Sects. 7.4 and 7.5).

For the sake of clarity, transliteration and translation must be considered separately. In editions of texts, copies, transliteration, translations and commentaries appear clearly separated. However, it must be underlined that these different stages are closely interdependent and that editors constantly navigate from one stage to another, for example, in correcting a copy or transliteration on the basis of interpretation. The transliteration and translation are thus continuously impacted by the interpretation and vice versa. The conventions adopted for transliteration and translation reflect this process of navigating back and forth through the flow of understanding.

## 7.3 How Transliterations Reflect Interpretations

The above-mentioned discussion about the diversity of conventions adopted by historians for representing numbers and quantities can be illustrated with an example. It concerns the transliteration of cuneiform signs used in mathematical texts dated from the Old Babylonian period (c.a. 2000–1600 BCE) for quantifying the surface of a field. In the first line of the text quoted above (YBC 4710 #1; see Table 7.2), the two cuneiform signs  represent the measurement of the surface of a rectangular field of about 2 ha. These two signs, which are also found in many similar mathematical contexts, were transliterated in the following ways by different authors.

| Cuneiform signs |  |
|---|---|
| Neugebauer 1935–1937: I, 402 (YBC 4710, obv. l. 1) | 1 (eše) gán |
| Thureau-Dangin 1938: 148 (YBC 4710, l. 1) | *ebel ikîm* |
| Neugebauer and Sachs 1945: 111 (A 24194, rev. iii, l. 2) | 1(ešè)$^{iku}$ |
| Høyrup 2002: 204 (YBC 4713, obv. i, l. 1) | 1(ešè)$^{iku}$ |
| Robson 2004: 34 (Ashm. 1931.137, rev. ix) | 0.1.0 $^{ganá}$ |
| Friberg 2007: 117 (MS 2735) | $1_e$ aša$_5$ |
| CDLI[11] | 1(eše$_3$) GAN$_2$ |

[10] For an approach to this discussion focused on procedures for solving quadratic problems, see Høyrup (2002).

[11] The CDLI (http://cdli.ucla.edu/, accessed 14 June 2017), is the database of cuneiform documents developed by the University of California (Los Angeles), the Max Planck Institute for the History of Science, Berlin, and Oxford University, under the supervision of Robert K. Englund, Jürgen

As we can see, the same signs are transliterated very differently according to each different publication.[12] Different kinds of reasons, discussed previously in Sect. 7.2 and which I analyze in more detail below, explain this diversity. This example encapsulates most of the problems confronting the editor of a mathematical cuneiform text. Note that, by a curious irony of history, the conventions recently adopted by the CDLI revive those adopted by Neugebauer at the beginning of his career.

Some of the discrepancies observed in the transliterations above reflect minor evolutions in the habits of Assyriologists, such as the use of diacritics (acute and grave accents) to distinguish homophone signs (different signs with the same pronunciation). For example, 'gán' has been replaced with '$gan_2$' and 'ešè' with '$eše_3$'. This evolution, stimulated by the increasing use of digital databases, is quite consensual and does not raise serious problems.[13]

Other differences reflect uncertainties concerning the pronunciation of Sumerian words and the function of the cuneiform signs. For example, the sign  is supposed to correspond to the Akkadian word '*ikîm*' according to Thureau-Dangin, to a surface unit pronounced 'gan' for Neugebauer in 1935 and to the Sumerian word 'aša' (field) for Friberg. The transliteration $GAN_2$, in uppercase letters, adopted by the CDLI, indicates that for the editors of this database, the pronunciation of the sign  in the context of surface measurements is unknown.

The function of the graphemes is not analyzed in the same manner by the authors quoted above. Neugebauer and Sachs, Høyrup and Robson consider the sign  as a determinative (i.e, a graphical semantic indicator with no counterpart in the spoken language); for this reason, this sign is represented, in their transliterations, as superscripts. We have just mentioned other divergent analyses of the function of the sign : Neugebauer in 1935 thought it was a surface unit and Friberg saw it as a lexical word meaning 'field'. Note that Neugebauer's analysis of the function of this grapheme changed between 1935 and 1945.

The function of the sign  is also unclear. Does it represent a number? A measurement value? A measurement unit? The transliteration adopted by Neugebauer in 1935 suggests that  represents the number (6) of surface units 'gán' (about 3600 $m^2$). By contrast, no measurement unit appears in transliterations offered by

Renn and Jacob Dahl. Among about one million inscribed artifacts that have probably been excavated to date in the Ancient Near East, mainly in Iraq, Syria, Turkey and Iran, more than half are registered in the CDLI, many already documented with photos, copies and transliteration.

[12] In my own transliterations, I follow CDLI conventions.

[13] As early as 1948, Gelb advocated the use of numbers instead of diacritics (Gelb 1948: 13, Sect. 56).

Neugebauer and Sachs, Høyrup, Robson and Friberg. Indeed, for them, the sign incorporates both numerical and metrological information: this sign alone represents a complete measurement value (number + measurement unit). Pursuing this logic, Robson (2004: 34) imported the positional notation adopted by many Assyriologists for capacities. She therefore represented the sign by 0.1.0; however, she seems to have abandoned this idea in her more recent publications.

In the example just examined, we have seen that the same signs, found in similar contexts, were interpreted and thus transliterated differently by historians of mathematics. Sometimes, the diversity of transliterations reflects the diverse habits of academic milieus, and different sectors of Assyriological studies do not follow the same conventions. For example, the editors of administrative texts represented the measurements of capacity with a positional system, while the same cuneiform signs are represented differently by historians of mathematics, who generally follow CDLI conventions (see example in Table 7.3 and further details in Michel in Chap. 8 of the present volume).

The analysis of the cuneiform sequences examined above is beyond the scope of the present chapter.[14] These examples merely aim at launching a discussion. We can observe that the diversity of conventions for transliteration does not reflect any diversity of practices by ancient scribes, but only modern historians' diversity of interpretations or habits. This shows that these transliterations are not neutral and rely greatly on interpretation. We are far from the ideal system which 'faithfully [records] the original writing, avoiding as much as possible the elements of interpretation' (Gelb 1948: 3, Sect. 12), and which is probably unattainable.

The diversity of transliterations in modern editions contrasts with the strong coherence and stability of the ancient metrological notations. Indeed, the same system of notations of measurement values for surfaces, lengths, weights and, to a lesser extent, for capacity, is attested in a variety of cuneiform sources, whether economic or mathematical, at least from the middle of the third millennium to the

**Table 7.3** Transliteration of the Ur III administrative text BM 19027, obverse, column *i*, section 6, line 8

| Signs (copy in King 1898, plate 27) | Transliteration in BDTNS (No. 000147) | Transliteration in CDLI[a] (No. 108470) |
|---|---|---|
| 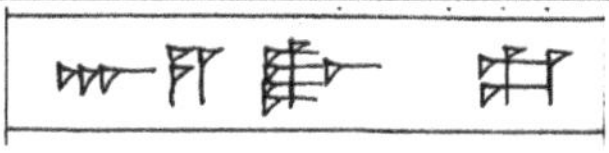 | 3.3.4 gur | 3(aš) 3(barig) 4(ban$_2$) gur |

[a]CDLI applies the following basic rule: in cuneiform texts, the quantities are generally expressed by repeating the same sign several times. CDLI's transliteration indicates the number of repeated signs, followed immediately in parentheses (without a space) by the name of the repeated signs. For example, the sign is transliterated 3(aš) because the horizontal wedge represents a unit 'one' (for counting the capacity units 'gur'), 'aš' in Sumerian, and we have three such signs.

[14] For further discussion about the function of the sign GAN$_2$ in quantification of surfaces, see the bibliography in (Proust 2020).

middle of the second millennium, with minor (but meaningful) variations. Does this rather extensive standardization of metrology in ancient texts require uniform principles for representing numbers and quantities in modern editions? How can we improve our modern conventions of transliteration by a better understanding of the ancient concepts related to the science of writing? I have suggested elsewhere (Proust 2009) that in this regard, we would benefit greatly from considering the way in which the ancient scribes themselves developed in-depth reflection on the classification of the cuneiform signs used in notations of numbers and quantities, notably in metrological lists and tables.

The examples discussed above have suggested that what is a number is not always clear, because certain signs may incorporate both numerical and metrological information. For example, a unique sign such as , transliterated '4(ban$_2$)' according to CDLI conventions, designates a capacity value. The sign simultaneously incorporates the specification of the capacity unit, the *ban*,[15] the value of which is about 10 L, and the number of such capacity units, 4, to be counted in order to express this measurement of capacity. The sign  can be considered neither as designating a number nor a measurement unit. However, such an ambiguous situation is the exception rather than the rule in cuneiform documentation, at least in the texts dated after the middle of the third millennium. Quantities are usually expressed by a number followed by a measurement unit. For example the measurement of capacity noted  and transliterated '4(diš) sila$_3$' according to CDLI conventions, is composed of two signs, a four-wedges sign used for counting, and a measurement unit, the *sila*, with a value of about 1 L. In this case, the four-wedges sign  can be considered as a number. This kind of number is omnipresent in cuneiform texts, whether mathematical, economic, lexical, historical or other. These numbers are used in cuneiform texts not only for counting measurement units (*sila* and many others) but also for counting discrete elements (things, animals or individuals).

## 7.4 Numbers for Counting

As an example, let us look more closely at the problems raised by the transliteration and translation of numbers used for counting years. A good case study is provided by the so-called 'Sumerian King Lists', a list of kings' names with the duration of

[15] The pronunciation *ban* for the capacity unit equivalent to 10 *sila* (about 10 L) is evidenced in lexical lists. The word *ban* is also the name of the bushel used for measuring the corresponding capacity of grain. The name of the standard vessel containing 1 *ban* of grain is noted phonetically '$^{\text{giš}}$ba-an' ('giš' is the determinative of wood objects) in economic, lexical and mathematical texts (see, for example, the mathematical text YBC 4669, obv. I, line 9, published in (Neugebauer 1935–1937: I, 514–516)).

their corresponding reigns, attested in documents dating from the Old Babylonian period.[16] The first sections of the Sumerian King List enumerate the durations of the reigns of the legendary Antediluvian Kings, which are attributed with extravagant lengths. As an example, Table 7.4 shows the transliteration and translation by Langdon (1923) of the first 17 lines as they appear on face 1 of the Weld–Blundell prism (Ashm. 1923-444).[17] We shall not enter into discussion here of the historical significance of such a text, but only focus on the way in which the numbers of years are transliterated and translated in Langdon's edition, and on some historiographical consequences of these numerical representations.

In Langdon's transliteration and translation, the numbers of years are converted into our modern decimal system, with nothing to indicate the way in which these numbers are written in cuneiform sources. Table 7.5 provides a representation of the cuneiform notations of the numbers, the corresponding transliteration and translation by Langdon, and the transliteration according to the CDLI. In the last column, I have added a modern representation of the quantities conveyed by the cuneiform numbers; this modern representation uses decimal notation but clearly shows the arithmetical structure of the original numbers.

If we look at the cuneiform signs, we see that the numbers of years are expressed by the sign  (šar$_2$), which represents 3600 (that is, 60 × 60), and the sign  (šar'u), which represents 36,000, that is, ten times more than one 'šar$_2$' sign. Note that the relationship between the numbers represented by the signs 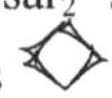 (šar$_2$) and  (šar'u) is graphically indicated by the insertion of the sign u (), which represents 10, inside the former to obtain the latter. The total number of years of the first two reigns, which appears in line 7, is the sum of the numbers of years found in lines 4 and 5. This total can be obtained by simply pasting together the signs 'šar$_2$' and 'šar'u' collected in these lines. The total number of years of the following three reigns, which appears in line 17, is the sum of the numbers of years found in lines 12, 14 and 15. Collecting the 'šar$_2$' and 'šar'u' signs found in these lines provides two 'šar'u' signs and ten 'šar$_2$' signs, which are converted into three 'šar'u' signs in line 17. Then, the numbers of years of legendary reigns and dynasties seem to be simple arrangements of the signs 'šar$_2$' and 'šar'u'.

What is the meaning of these long durations of reigns of legendary kings? Some authors have developed esoteric speculations or arithmetical considerations to

[16] The Sumerian King List is not a mathematical text but has been often examined with interest for its mathematical content. For example, extracts of the Sumerian King List were published for the first time by Hermann Hilprecht along with mathematical texts (Hilprecht 1906), and the text was remarked upon by historians of mathematics (see, for example, Young 1988, Steiner 1988, and Friberg 2007: Chap. 9).

[17] Langdon 1923; see also the photograph, copy and transliteration listed on CDLI under the number P384786. The edition provided online by ETCSL under the number 2.1.1 is a composite text based not only on the prism Ashm 1923–444, but also on other sources. ETCSL adopts the modern numbering of homophones, instead of diacritics (example: $e_3$ in ETCSL, è in Langdon edition). The transliteration and translation of the numbers in ETCSL are the same as in Langdon.

**Table 7.4** Copy, transliteration and translation by Langdon of the Sumerian King List on the prism Ashm. 1923-444, face 1, lines 1–17, with original notations (Langdon 1923: 8)

| | | | |
|---|---|---|---|
| 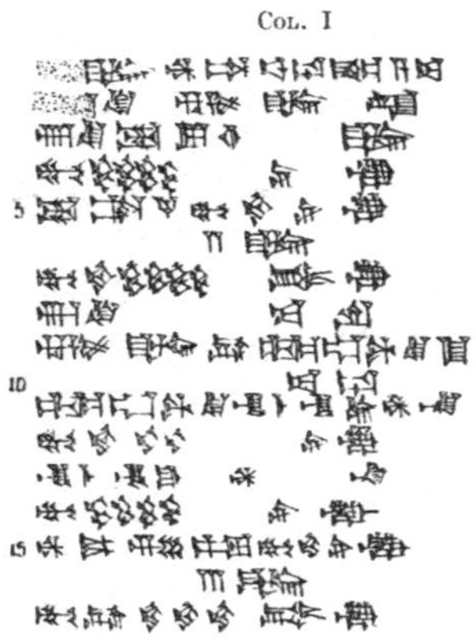 | 1. | [nam]-lugal an-ta è-dé-a-ba | Rulership which from heaven descended. |
| | 2. | [Eri]du[ki] nam-lugal-la | At Eridu rulership (began). |
| | 3. | Eridu[ki] $a_2$-lu-lim lugal | At Eridu Alulim was king. |
| | 4. | mu 28,800 ni-ag | He ruled for 28,800 years. |
| | 5. | á-lal-gar mu 36,000 ni-ag | Alagar ruled 36,000 years. |
| | 6. | 2 lugal | Two kings. |
| | 7. | mu 64,800 íb-ag | 64,800 years they ruled. |
| | 8. | Eridu- (ki) ba-šub | Eridu was overthrown. |
| | 9. | nam-lugal-bi Bad-tibira-(ki)-šú | The kingship to Badtibira |
| | 10. | ba-gín | passed. |
| | 11. | Bad-tibira-(ki) En-me-en-lù-an-na | At Badtibira, Enmenluana |
| | 12. | mu 43,200 $i_3$-ak | ruled 43,200 years. |
| | 13. | En-me-en-gal-an-na | Enmengalanna |
| | 14. | mu 28,800 ni-ag | ruled 28,800 years. |
| | 15. | [d]Dumu-zi-sid mu 36,000 ni-ag | Dumuzid, ruled 36,000 years. |
| | 16. | 3 lugal | Three kings. |
| | 17. | Mu-bi 108,000 ni-ag | They ruled 108,000 years. |

**Table 7.5** Numbers of years on Ashm. 1923-444, face 1, lines 1-17[a]

| | Cuneiform signs | Langdon | Transliteration by the CDLI | Arithmetical structure of the cuneiform notation |
|---|---|---|---|---|
| Line 4 (Reign 1) | 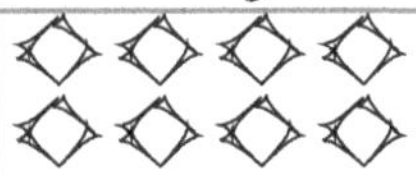 | 28,800 | 8(šar$_2$) | 8 × 3600 |
| Line 5 (Reign 2) |  | 36,000 | 1(šar'u) | 36,000 |
| Line 7 (Reigns 1 + 2) | 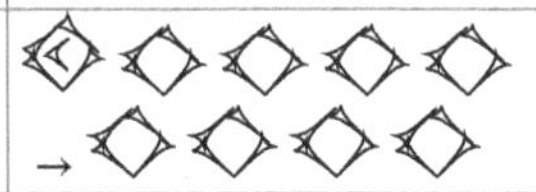 | 64,800 | 1(šar'u) 8(šar$_2$) | 36,000 + 8 × 3600 |
| Line 12 (Reign 3) | 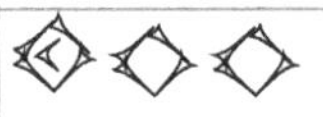 | 43,200 | 1(šar'u) 2(šar$_2$) | 36,000 + 2 × 3600 |
| Line 14 (Reign 4) | 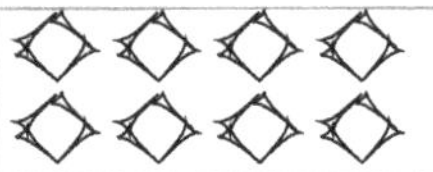 | 28,800 | 8(šar$_2$) | 8 × 3600 |
| Line 15 (Reign 5) |  | 36,000 | 1(šar'u) | 36,000 |
| Line 17 (Reigns 3 + 4 + 5) |  | 108,000 | 3(šar'u) | 3 × 36,000 |

[a]As mentioned above, the translation principle used by CDLI clearly shows the number of times that a given sign is repeated, adding the name of the relevant sign in parenthesis. The transliteration '8(šar$_2$)' means that the sign 'šar$_2$' (the value of which is 3600) is repeated eight times. Thus the arithmetical structure is 8 × 3600 (as shown in the last column)

explain these figures. For example, Dwight Young found connections between the Sumerian King List and certain mathematical texts by analyzing some arithmetical properties of the numbers of years in the Sumerian King Lists (Young 1988).

> Aside from virtually astronomical figures for the antediluvians, among the postdiluvian rulers there are a good many reigns of incredible duration. It is not merely because of their largeness that some of these numbers appear artificial. Etana's 1,560 years, to cite the longest, is but the sum of the two preceding reigns: 840 + 720. Certain spans (e.g., 120, 180, 360, 420) seem to have arisen simply as multiples of 60. Other large numbers may be recognized as squares: 900, the square of 30; 625, the square of 25; 400, the square of 20; 324, the square of 18; 100, the square of 10. Even among smaller figures, the square of six appears more frequently than one might expect; in particular, the sequence 36 6 36 at the end of Uruk I arouses suspicion. (Young 1988: 123)

These considerations are based on the modern decimal representation of the numbers of years as provided by Langdon. The fact that the reigns are multiples of 60 or $60^2$ results from the numerical system used, which has a sexagesimal structure. Then, perhaps, no conclusion can be drawn about the mathematical significance of these features.[18]

This example opens just a small window on the way in which the interpretation may depend on conventions of representation. The complete decimalization of numbers in Langdon's transliteration and translation induced certain kinds of arithmetical considerations (e.g., those of Young). Unlike Langdon's, the transliteration by the CDLI represents the cuneiform signs (see the fourth column of Table 7.5) and enables reconstruction of the original cuneiform text from the transliteration. Assuredly, this should guide the interpretation toward other directions. A translation showing the arithmetical structure of the cuneiform notation, similar to that which I suggest in the last column of Table 7.5, may avert the temptation to interpret numbers uniquely on the basis of modern decimal representation. Moreover, a representation which takes into consideration the sexagesimal structure of numbers may facilitate the restitution of the operations on numbers carried out by ancient scribes.

Other kinds of problems are raised by transliteration and translation of the sexagesimal place-value notations found essentially in mathematical texts.

## 7.5 Sexagesimal Place-Value Notation

The article written by Otto Neugebauer (1932–1933) on 'transcription of mathematical and astronomical cuneiform texts'[19] is a good starting point for discussing practices of transliteration and translation of the sexagesimal place-value notation (thereafter SPVN) from the early twentieth century until the present day. This short publication includes two parts: 'Numbers' and 'Technical Terms'. The section on 'Technical Terms' criticizes Thureau-Dangin's transcription of the Sumerograms into Akkadian (see Sect. 7.2). In the part on 'Numbers', Neugebauer is actually interested only in SPVN and bypasses the other numerical systems and metrological notations. This focus on SPVN reflects a general attitude held by Neugebauer and many historians after him: from a mathematical point of view, there is nothing of interest in metrological lists and tables and, more generally, in metrology, notations of measurement values and non-positional numbers. For example, in his editions of 1935–1937, Neugebauer decided not to publish metrological texts partly because 'for the history of mathematics in the narrow sense of the term, [metrological texts] do not bring any kind of result' (Neugebauer 1935–1937: I, 5). Pursuing the same idea, in the introduction to their 1945 edition (MCT), Neugebauer and Sachs considered that metrological lists and tables only had a practical interest, in contrast to problem-texts, which were truly mathematical.[20] Associated with the lack of interest in metrological texts expressed in the introduction to MCT, the only numbers that merit the authors' attention are SPVN. Nothing is said about non-positional

[18] See also (Steiner 1988) and the criticism of Young and Steiner in (Friberg 2007: 242).

[19] (Neugebauer 1933–1934) (Original title: Zur Transkription mathematischer und astronomischer Keilschrifttexte). The English translation of the extracts quoted in this chapter is due to Sandra Hoehn, whom I thank warmly, and slightly edited by the polisher.

numbers, even though they are omnipresent in mathematical texts as parts of the expression of measurement values.[21]

In the section titled 'Numbers', Neugebauer offers an extremely rigorous discussion of the conventions to be adopted for representing SPVN. First, he discusses the representation of digits, advocating and subsequently rejecting the idea of transcribing the digits with Roman numerals. Second, he discusses the punctuation mark to be used as the 'delimiter for sexagesimal digits' and, after suggesting several possibilities, he opts for the comma. For example, a cuneiform notation such as  is transcribed, following Neugebauer's conventions, as 44,26,40. Neugebauer was widely followed by other scholars, although the symbol chosen as a 'delimiter for sexagesimal digits' in transliteration is often a dot rather than a comma. For example, Thureau-Dangin, as well as many other authors, transcribes the example above as 44.26.40. The disadvantage of Neugebauer's system (comma-separator) and Thureau-Dangin's system (dot-separator) is that in writings in English and other modern languages, they do not allow the reader to distinguish immediately between decimal and sexagesimal numbers. Some authors (Friberg and Robson, for example) use a blank space as a separator between sexagesimal digits (e.g., 44 26 40). This raises another problem since, in this case, the separator between digits is the same as the separator between numbers. Perhaps the best choice for representing a separator between sexagesimal digits in transliterations should be the one we use nowadays for counting time in digital documents and video, that is, a colon (e.g., 44:26:40). But the choice of the 'delimiter for sexagesimal digits' is actually a minor detail.

Much more important is the problem of orders of magnitude. As clearly stated by Neugebauer (1932–1933: 221), 'By way of principle, cuneiform numbers are not assigned any order of magnitude, and therefore any mention of an order of magnitude must by all means be avoided in transcriptions (original text: Da Keilschriftzahlen irgendwelche Stellenwertbezeichnung grundsätzlich nicht zukommt, so ist sie auch in der Transkription eines Textes absolut zu vermeiden.)'. This recommendation by Neugebauer reflects his constant concern to make the transliteration a mirror of the cuneiform signs; he was widely followed on this point. However, he adds immediately (*ibid*: 221):

> To me, only in the translation of a text does indication of order of magnitude seem justified. [...] To indicate the order of magnitude, [...] one needs only a single convention: after the fundamental unit, one ought to put a semicolon instead of a comma (which can be omitted if the last digit means "one"). Hence, 1;40 = 1 + 2/3, 1,40; = 1,40 = 100. However, if none of the extant digits is a fundamental unit, one has to make use of zeros, for example 0;40 = 2/3, 4,0 = 240, 1,0,40 = 3.640.

[20] 'The practical needs satisfied by the metrological texts are evident' while 'Problem-texts, however, sometimes exhibit only very remote connections, if any, with practical questions' (Neugebauer and Sachs 1945: 1).

[21] 'The number system used in the mathematical texts has two characteristic properties: (a) it is based on sixty (hence the term "sexagesimal") and (b) it uses a place-value notation' (Neugebauer and Sachs 1945: 2).

> Original text: Erst in der Übersetzung eines Textes scheinen mir positionelle Angaben gerechtfertigt. […] Zum Ausdruck eines Stellenwertes bedarf es […] bloß der einzigen Übereinkunft: hinter der Grundeinheit setze man ein Semikolon statt eines Kommas (das wegbleiben kann, wenn die letzte Ziffer Einer bedeutet). Dann ist 1;40 = 1 + 2/3, 1,40; = 1,40 = 100. Ist aber keine der vorhandenen Ziffern den Grundeinheiten entnommen, so hat man entsprechend Nullen einzuschalten, z. B. 0;40 = 2/3, 4,0 = 240, 1,0,40 = 3640.

There are two strong assumptions here. First, that it is justified to restore the orders of magnitude of SPVN in translations, although this is not indicated in the cuneiform texts. Second, that there is a 'fundamental unit' in a number noted in SPVN. Neugebauer regarded the specification of a 'fundamental unit' in a number in SPVN as necessary because he wanted to make the numbers in SPVN 'equal' to quantities, for example, in expressions containing fractions ('1;40 = 1 + 2/3') or decimal integers ('4,0 = 240').

Some scholars disagreed with Neugebauer on the way in which the 'order of magnitude' of SPVN should be represented in translations. While Neugebauer used semicolons and zeros, Thureau-Dangin, followed by Høyrup, used an extended system of degree-minute-second (for example, 53ʿ31°6′40″ in Thureau-Dangin (1938: 175) or 14′ 48″ 53‴ 20⁗ in Høyrup (2002: 195)). Some authors, such as Bruins or Friberg, were reluctant to restore the orders of magnitude in translations and preferred to indicate them only in interpretations and commentaries. Whatever the case, Neugebauer's successors followed him unquestioningly in relation to the essential points stated in the quotation above: there is a 'fundamental unit' in a number noted in SPVN, and this 'fundamental unit', which is not indicated in cuneiform texts, must be restored at some point, either in translations or only in interpretation. Following Neugebauer, most historians work on SPVN with an order of magnitude, while the cuneiform texts exhibit work on SPVN without any order of magnitude. In other words, the SPVN generally handled by historians are quantities. This practice therefore eluded a fundamental question: Did ancient scribes handle SPVN as quantities?[22]

These considerations shed light on Neugebauer's major influence on the way in which the numbers in SPVN are perceived by modern historians. Thus, it would be interesting to understand why Neugebauer took for granted the necessity 'to indicate the order of magnitude' of SPVN in translations and interpretations despite his rigorous approach to the philology of numbers. Perhaps the answer is given by the title of his paper from 1932–1933: 'Concerning the transcription of mathematical and astronomical cuneiform texts (*Zur Transkription mathematischer und astronomischer Keilschrifttexte*),' and, echoing his title, by the last section of the part on numbers, which deals with a 'proposition concerning numbers in astronomical texts'. He summarizes his suggestions for transliteration of SPVN as follows (Neugebauer 1932–1933: 222): 'As a result, a truly consistent and also typographically economic transcription system of all sexagesimal numbers would be achieved'. (Original text: Damit wäre eine wirklich einheitliche und auch drucktechnisch sparsame

[22] I addressed this issue in depth in (Proust 2008) and (Proust 2013), and Carlos Gonçalves discussed it in (Gonçalves 2022). My interpretation is that, in most cases, the SPVN are handled as floating numbers and do not represent any quantity, while Gonçalves offers a more balanced answer.

Umschriftweise aller Sexagesimalzahlen erreicht.) For Neugebauer, the problems raised by the transliteration and translation of SPVN are the same for mathematical texts as for astronomical texts. In other words, he considers that both genres of texts reflect the same practices with SPVN. The basic reason why the necessity 'to indicate the order of magnitude' is so consensual among historians lies, perhaps, in the fact that Neugebauer suggested a uniform vision of ancient practices with SPVN in mathematics and astronomy, and that this vision was widely accepted.

## 7.6 Conclusion

The different examples analyzed in the previous sections show contrasting practices in editions of cuneiform texts. On the one hand, we have seen the variety of conventions adopted by historians for representing measurement values noted in cuneiform texts. On the other hand, we can observe a relative uniformity of conventions for representing the SPVN. Different phenomena explain this situation rather chaotic. The diversity of the conventions adopted to transliterate measurement values sometimes results from the diversity of the interpretations (as in the case of the measurement of surface) and sometimes from different cultures in academic milieus, for example those of Assyriologists and those of historians of mathematics (as in the case of the measurement of capacity). In both cases, the diversity of conventions does not reflect a diversity of practices by ancient actors, but divergences between historians. It leads us to ask whether the Babel of notations erected by modern historians is not a serious obstacle to the dissemination, transmission and popularization of cuneiform mathematics. As for the SPVN, the situation is reversed. There is considerable consensus among historians about accepting, following Neugebauer, that the transliteration must be the same for the SPVN found in Old Babylonian mathematics and for the SPVN found in mathematical astronomy. It raises the question, however, whether this uniformity of representation concealed the diversity of practices with SPVN, which may have varied in these different contexts.

To what extent must the transliteration reflect the interpretations and preferences of individual editors? On the one hand, transliteration should be no more than a system of representation of cuneiform signs. A consensus among specialists on common conventions can be considered necessary, first for making the transliterations intelligible and, second, for enabling the use of large databases. On the other hand, transliterations cannot be completely neutral and necessarily reflect a basic level of interpretation. I have shown above how interpretation impacts the transliteration in the case of the representation of a measurement value. Many other examples can be found. For example, the editors of cuneiform texts indicate, in the transliteration of each sign, whether this sign is interpreted as an Akkadian syllable or a Sumerogram. In the same way, it seems to me essential that the transliteration of a number should indicate whether this number is noted in the source in sexagesimal place-value notation or in another non-positional system.[23] The right balance between these two requirements—(impossible) neutrality on the one hand and a (minimal) basic level of interpretation on the other hand—is not always easy to determine. Establishing a consensual set of conventions for the transliteration of 'mathematical notations' (i.e., for all kinds of numbers), was one of the goals of recent discussion in the CDLI project.[24] In any case, such decisions should not be individual but should result from the collective work of a community on the philology of numbers and quantities.

## Primary Sources

| Museum number | Primary publication | CDLI n° |
|---|---|---|
| A 24194 | Neugebauer and Sachs 1945: text T | P254383 |
| Ashm. 1923–444 | Langdon 1923 | P384786 |
| Ashm. 1931–137 | Robson 2004: n°19 | P368260 |
| BM 19027 | King 1898 | P108470 |
| MS 2735 | Friberg 2007: 127 | P251759 |
| YBC 4710 | Neugebauer 1935: 402 | P255017 |
| YBC 4713 | Neugebauer 1935: 421 | P255020 |

## References

Cuneiform Digital Library Initiative. 2008. *Transliteration and the diacritics*. http://cdli.ox.ac.uk/wiki/doku.php?id=sumerian:transliteration_and_the_diacritics. Accessed 14 June 2017.

Friberg, Jöran. 1993. On the structure of cuneiform metrological table texts from the first millennium. In *Die Rolle der Astronomie in den Kulturen Mesopotamiens*, ed. H.D. Galter, 383–405. Graz: GrazKult.

———. 2007. A remarkable collection of Babylonian mathematical texts. In *Manuscripts in the Schøyen collection: Cuneiform texts*, vol. I. New York: Springer.

Gelb, Ignace J. 1948. *Memorandum on transliteration and transcription of cuneiform*, Submitted to the 21st international congress of orientalists, Paris. Chicago: Oriental Institute, University of Chicago.

[23] See more details in (Proust 2009).

[24] Conventions for the transliteration of numbers used in measurement values and discrete counting were established at the beginning of the CDLI Project by Robert K. Englund and Peter Damerow. The conventions for transliterating sexagesimal place-value notation, or "math protocol", were established after the technical meeting of the CDLI held in Berlin in May 2008 and subsequently posted online by Steve Tinney. This "math protocol" follows Thureau-Dangin system: sexagesimal digits are separated by dots. For example, the number  is transliterated as 44.26.40; see more on this point in the Sect. 7.3 of this chapter. The last (July 2014) of the conventions for "Mathematical Notation", adopted in the databases belonging to the ORACC Projects (http://oracc.museum.upenn.edu/) are available at http://oracc.museum.upenn.edu/doc/help/editinginatf/maths/index.html

———. 1951. *Second memorandum on transliteration and transcription of Cuneiform*, Submitted to the 161st meeting of the American Oriental Society, Philadelphia. Chicago: Oriental Institute, University of Chicago.

Gonçalves, Carlos. 2022. Quantification and computation in the mathematical texts of old Babylonian Diyala. In *Cultures of computation and quantification in the ancient world. Numbers, measurements, and operations in documents from Mesopotamia, China and South Asia*, ed. Karine Chemla, Agathe Keller, and Christine Proust, 691–730. Cham: Springer.

Hilprecht, Herman Vollrat. 1906. *Mathematical, metrological and chronological tablets from the Temple library of Nippur*, Babylonian expedition 20/1. Philadelphia: University of Pennsylvania.

Høyrup, Jens. 1990. Algebra and naive geometry. An investigation of some basic aspects of old Babylonian mathematical thought. *Altorientalische Forschungen* 17: 27–69, 262–354.

———. 2002. *Lengths, widths, surfaces. A portrait of old Babylonian Algebra and its Kin*. Berlin/ London: Springer.

King, Leonard W. 1898. *Cuneiform texts from Babylonian tablets in the British Museum*. Vol. III. London: British Museum.

Landsberger, Benno. 1926. Die Eigenbegrifflichkeit der babylonischen Welt. *Islamica* 2: 355–372.

———. 1976. The conceptual autonomy of the Babylonian world. In *Sources and monographs, monographs on the ancient near east*. Trans. T. Jacobsen, B.R. Foster, and H. von Siebenthal. Malibu: Undena Publications.

Langdon, Stephen. 1923. *The Weld–Blundell collection: Historical inscriptions, containing principally the chronological Prism W-B 444*, Oxford editions of cuneiform texts. Vol. II. London: Oxford University Press.

Neugebauer, Otto. 1932–1933. Zur Transkription mathematischer und astronomischer Keilschrifttexte. *Archiv für Orientforschung* 8: 221–223.

———. 1933–1934. Zur Terminologie der mathematischen Keilschrifttexte. *Archiv für Orientforschung* 9: 199–204.

———. 1935–1937. *Mathematische Keilschrifttexte I–III. Quellen und Studien zur Geschichte der Mathematik, Astronomie und Physik*. Berlin: Springer.

Neugebauer, Otto, and Abraham J. Sachs. 1945. *Mathematical cuneiform texts*. New Haven: American Oriental Series and American Schools of Oriental Research.

Nissen, Hans J., Peter Damerow, and Robert Englund. 1993. Archaic bookkeeping. In *Writing and techniques of economic Administration in the ancient near East*. Chicago: University of Chicago Press.

Proust, Christine. 2008. Quantifier et calculer: usages des nombres à Nippur. *Revue d'Histoire des Mathématiques* 14 (2): 143–209.

———. 2009. Numerical and metrological graphemes: From cuneiform to transliteration. *Cuneiform Digital Library Journal* 2009: 1.

———. 2013. Du calcul flottant en Mésopotamie. *La Gazette des Mathématiciens* 138: 23–48.

———. 2020. Early-dynastic tables from Southern Mesopotamia, or the multiple facets of the quantification of surfaces. In *Mathematics, administrative and economic activities in the ancient worlds*, Why the sciences of the ancient world matter 5, ed. K. Chemla and C. Michel, 345–395. Dordrecht: Springer.

Robson, Eleanor. 2004. Mathematical cuneiform tablets in the Ashmolean Museum, Oxford. *SCIAMVS* 5: 3–66.

Rubio, Gonzalo. 2007. Writing in another tongue: Alloglottography and scribal antiquarianism in the ancient near East. In *Margins of writing, origins of cultures: Unofficial writing in the ancient near East and beyond*, ed. S.L. Sanders, 33–70. Chicago: The Oriental Institute.

Steiner, Gerd. 1988. Der 'reale Kern' in den 'legendären' Zahlen von Regierungsjahren der ältesten Herrscher Mesopotamiens. *Acta Sumerologica* 10: 129–152.

Thureau-Dangin, François. 1938. *Textes Mathématiques Babyloniens*. Leiden: Ex Oriente Lux.

Young, Dwight W. 1988. A mathematical approach to certain dynastic spans in the Sumerian King list. *Journal of Near Eastern Studies* 47 (2): 123–130.

# Chapter 8
# Numbers and Quantities in Editions of Economic and Administrative Cuneiform Texts From the Beginning of the Second Millennium BCE: The Case of the Capacity System

**Cécile Michel**

**Abstract** Numbers and quantities are omnipresent in cuneiform texts, and have been transliterated and translated by Assyriologists since the nineteenth century in many diverse ways. This study focuses on editions of administrative texts dated to the first half of the second millennium BCE, and on the measuring system for capacity, whose interpretation by modern editors has resulted in a multitude of different transliterations and translations. It shows how transliteration and translation conventions have an impact on the understanding of ancient numerical and metrological systems, and suggests that editions of cuneiform texts should reflect, as far as possible, the original ways of expressing numbers and quantities.

## Abbreviations

| | |
|---|---|
| AbB 1 | see Kraus (1964) |
| AKT 5 | see Veenhof (2010) |
| ARMT 7 | see Bottéro (1957) |
| ARMT 21 | see Durand (1983) |
| ATHE | see Kienast (1960) |
| BE 6/2 | see Poebel (1909) |

The research leading to these results has received funding from the European Research Council under the European Union's Seventh Framework Programme (FP7/2007–2013)/ERC Grant Agreement No. 269804.

C. Michel (✉)
CNRS, ArScAn-HAROC (UMR 7041), Maison des Sciences de l'Homme Mondes,
21 allée de l'Université, Nanterre Cedex, France
e-mail: cecile.michel@cnrs.fr



A. Keller, K. Chemla (eds.), *Shaping the Sciences of the Ancient and Medieval World*, Archimedes 69, https://doi.org/10.1007/978-3-031-49617-2_8

| | |
|---|---|
| BIN 4 | see Clay (1927) |
| CTMMA 1 | see Spar (1988) |
| ICK 1 | see Hrozný (1952) |
| ICK 2 | see Matouš (1962) |
| Kt 00/k | inventory number of Kültepe texts excavated in 2000 |
| KTS 1 | see Lewy (1926) |
| Prag I | see Hecker et al. (1998) |
| UET 5 | see Figulla (1953) |
| TC 2 | see Thureau-Dangin (1928) |
| TCL 1 | see Thureau-Dangin (1910) |
| TCS 1 | see Sollberger (1966) |
| YOS | see Grice (1919) |

Numbers and quantities are omnipresent in the cuneiform texts of the Ancient Near East, especially in administrative and economic tablets. They were also widely used in royal inscriptions for counting captured cities, prisoners and booty, often with some exaggeration (De Odorico 1995). Kings played with numbers. Esarhaddon (280–669 BCE) decided to restore Babylon, which had been destroyed by his father, before the expiry of the seventy-year period imposed by the god Marduk. For this, he inverted the two signs used to write 70 (𒁹 𒌋, thus 60 + 10) and obtained 11 (𒌋 𒁹, thus 10 + 1) years. This allowed him to start the work right away. On another level, in private correspondence, numbers could be used in figurative expressions to refer to very small or very large quantities, as in this example from an Old Assyrian letter (from the nineteenth century BCE): 'Why did you not send me your greetings on a tablet (even) two fingers (long)?'[1] (Hrozný 1952: ICK 1, 17:4–12).

For many years, when editing cuneiform texts, most Assyriologists focused on the vocabulary and the grammar, and thus on the meaning of the texts inscribed on clay tablets. Numbers and quantities were considered as marginal. However, some of the first scholars tried to understand the numerical and metrological systems used in daily accounts. This study focuses on editions of administrative and economical texts dated to the first centuries of the second millennium BCE—a period considered as the classical Babylonian age—in order to analyze how Assyriologists became familiarized with these elements and how they treated them from the end of the nineteenth century to the present day. Most of these texts are written in the two main Akkadian dialects: Babylonian and Assyrian, using a cuneiform script, with a great majority of syllabic signs. The cuneiform signs were borrowed from the Sumerians, who used mainly logograms to write down their language.

Numbers are rarely written in sexagesimal place-value notation in these texts[2]; more often, they are written in an additional system for counting discrete inanimate or animate items or for expressing quantities. This chapter aims to explain how

---

[1] Hrozný 1952, ICK 1, 17:4–12, *mi*$_3$*-nu-um*: *a-ni-tum*, *ša ṭup-pa*$_2$*-am* 2 *u*$_2$*-ba-an*, *šu-lu-um-ku-nu*, *la*$_2$ *tu*$_3$*-še*$_2$*-ba-la*$_2$*-ni-ni*.

[2] See Proust in this volume for the definition of sexagesimal place-value notation (SPVN).

some editorial notations adopted by the Assyriologists tend to hide the various practices of the ancient scribes. We shall investigate several questions: How do some of the chosen transliteration conventions make some writings identical when, in fact, they are different in the cuneiform texts? How can notation be a source of errors? How can transliteration and translations norms delete evidence, in particular about computation practices? How did Biblical, Classical and even modern studies influence Assyriologists in their translations and rendering of measurement units?

This chapter starts with a brief overview of the use of numbers in the first half of the second millennium BCE, both in the Babylonian and Assyrian economical texts, and in the various ways the modern editors transliterated and translated them. The measuring system for capacity has been chosen as the object of this study because it was the most complicated measuring system used by the ancient Mesopotamians; its interpretation by modern editors resulted in a multitude of different transliterations and translations.

## 8.1 Editing Numbers in Administrative and Economic Texts

In this first part, we will explore the numerical systems used in cuneiform texts dated to the beginning of the second millennium BCE and the transliterations and translations of numbers used by the editors of these texts. Our goal is to analyze how some editorial conventions make similar ancient notations that were different, or conceal notation practices linked to computation methods.

### *8.1.1 The Editing of Cuneiform Texts*

The rediscovery of the ancient Near Eastern civilizations is a recent phenomenon: the decipherment of Akkadian written in cuneiform script dates back to somewhat 160 years. Indeed, contrary to Greek and Latin, there is no tradition. Everything had to be rediscovered (Lion and Michel 2009a). Since they cannot follow on from the Mesopotamian scribes, Assyriologists endeavor to make the ancient writings accessible to the scientific community; they are still in the process of deciphering cuneiform clay tablets.

Until the end of the twentieth century, hand copies of the tablets in two dimensions were privileged. At the end of the nineteenth century, copies were sometimes even "normalized", with very nice regular signs obtained using lead cuneiform signs made, for example, by the Imprimerie Nationale. Indeed, scholars were more interested in the content of the text than its materiality and the context in which it had been written. There was no problem in using the lead signs to represent royal inscriptions that had been composed with regular signs engraved on stone, but it was much more difficult to use them for producing copies of clay tablets, where there is more ambiguity in reading the signs because writing changed according to periods, areas, scribes' schools, etc.[3] Some researchers who have specialized in a specific group of tablets are even able to recognize the different scribes according to their script. For around the past twenty years, scholars have preferred to use digital photographs for this first step in the study of cuneiform tablets. However, since cuneiform clay tablets are three-dimensional objects with no flat surface, at least six photographs are needed to cover the whole inscribed object. New three-dimensional technologies are now being developed to portray the tablets in their entirety.

The second step is the transliteration: every cuneiform sign is associated with its equivalent in the Latin alphabet. In Akkadian, the majority of the written signs correspond to a syllable (noted in italics in the transliterations and transcriptions hereafter), but some signs are Sumerian logograms that were pronounced in the Akkadian language. The present contribution follows the Cuneiform Digital Library Initiative (CDLI) transliteration system for numbers and quantities.[4] The final steps are the translation of the text and its interpretation.[5]

## 8.1.2 *Numerical Systems Used for Numbers and Quantities*

Compared with the writing system using logograms and syllables, signs for numbers seem very simple: there were basically very few signs, used in a mixed decimal and sexagesimal system, to transcribe numbers and quantities in the economic and administrative texts. The main signs were  for 1 (diš) and 60 (geš$_2$),  (u) for 10 and  (aš) for 1, in cases when the unit was placed before specific measurement units. Their relationships are presented below in Table 8.1, with their transcription in Sumerian (System S). Some higher multiples of sixty could be represented by specific signs.

**Table 8.1** System S

|  | ×10 ← |  | ×6 ← |  | ×10 ← |  | ×6 ← |  | ×10 ← |  or  |
|---|---|---|---|---|---|---|---|---|---|---|
| šar'u 36,000 | | šar$_2$ 3600 | | geš'u 600 | | geš$_2$ 60 | | u 10 | | aš or diš 1 |

[3] Lead cuneiform fonts were used to publish tablets; see, for example, R. F. Harper (1892–1914) in the series *Assyrian and Babylonian Letters* (Chicago).

[4] The Cuneiform Digital Library Initiative (CDLI) uses a translation system for numbers and quantities that respects the ancient texts and transliterates every cuneiform sign: http://cdli.ucla.edu/ (accessed 6 April 2015). My transliterations given in the tables of this chapter follow the CDLI notation system, except when numerical signs appear before a numerical word written in Akkadian, for example: 1 *me-at*. The transcriptions and transliterations of cited authors follow their own publications.

[5] See Table 7.1, p. 320, in Christine Proust's Chap. 7 in this volume.

However, signs for these high numbers are rarely found in administrative and economic texts (see, for example, the use of the sign geš'u for 600 below).

### *8.1.3 Transliterations of Numbers in the Texts from the Beginning of the Second Millennium BCE*

The Assyrian merchants who were settled in central Anatolia, at Kaneš, during the nineteenth century BCE most often used an additive decimal system, while the Mari administrative scribes (in the eighteenth century BCE) used both System S and a decimal notation for tens (up to fifty), hundreds and thousands. How do the transliterations of these two text corpuses highlight these different practices?

#### 8.1.3.1 Examples from Kaneš (Central Anatolia, Nineteenth Century BCE)

The Assyrian merchants used a decimal additive system, writing 6(u)—that is, six times the sign u—for 60 instead of 1(geš$_2$), and then writing 70, 80 and 90 as 7(u), 8(u) or 9(u). The numbers 100 and 1000 were expressed in terms of their Akkadian names, *meat* and *lim* respectively. Thus, in the following example (shown in Table 8.2), the number 162 is written as '1 hundred 62 textiles'.

However, in some rare instances, a scribe used the sign 𒁹 (geš$_2$) for 60 when noting a value that also used the words *lim* and *meat* for a thousand and a hundred, thus mixing the decimal and sexagesimal notations. In the following text (Table 8.3), the number 2670 is written '2 thousands 6 hundreds + 1 × 60 + 10'. In this example, the latter two signs obviously have to be read separately and not as a single sign: *geš'u* (600), since they follow the notation '6 hundreds.'

But the reading geš$_2$ (60) for the vertical wedge (𒁹) remains exceptional in the Old Assyrian texts and is not always obvious. In the following text (Table 8.4), the same sign could be read '60' and we would thus translate it as '1 hundred breads …, 1 × 60 + 50 breads…, 1 × 60 + 50 breads…, 1 × 60 + 50 breads'. However, it is rather used as an abbreviation of 1 *meat*, thus of 100, and we should read it as:

**Table 8.2** Mixed writing with words and numbers

| Copy by Clay (1927: BIN 4, 185:1) | Transliteration using CDLI norms |
|---|---|
|  | 1 *me-at* 6(u) 2(diš) tug$_2$$^{hi-a}$ |

**Table 8.3** Attestation of the use of geš$_2$ for 60 in Old Assyrian

| Copy by Lewy (1926: KTS 1, 58a:10) | Transliteration using CDLI norms |
|---|---|
| 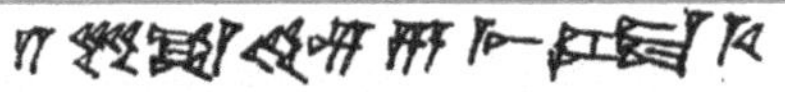 | 2 *li-im* 6 *me-at* 1(geš$_2$) 1(u) |

**Table 8.4** Two possible transliterations for ICK 1, 134:2–5

| Copy by Hrozný (1952: ICK 1, 134:2–5) | 𒁹 read geš$_2$ = 60? | 𒁹 read diš = 1 (*me-at*) = 100? |
|---|---|---|
| 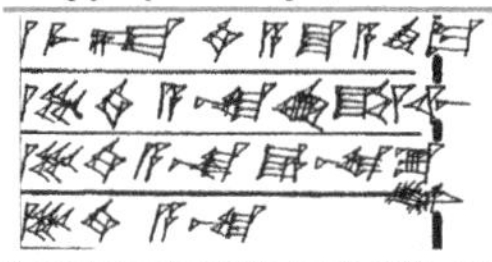 | 1 *me-at* ninda … | 1 *me-at* ninda … |
| | 1(geš$_2$) + 50 ninda … | 1(*me-at*) 50 ninda … |
| | 1(geš$_2$) + 50 ninda … | 1(*me-at*) 50 ninda … |
| | 1(geš$_2$) + 50 ninda … | 1(*me-at*) 50 ninda … |

**Table 8.5** The number 60 written with 6(u)

| Copy by Matouš (1962: ICK 2, 341:1–3) | Transliteration using CDLI norms |
|---|---|
| 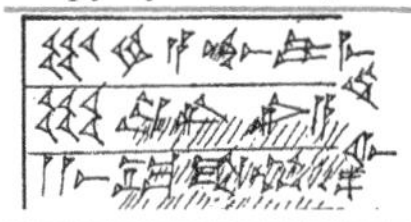 | 5(u) ninda *A-šur$_3$-i-mi$_3$-/ti$_2$* |
| | 6(u) ninda *I$_3$-li$_2$-a-/lim* |
| | 1 *me-at Ša-x-ki* |

'1 hundred breads…, 1 (hundred) 50 breads…, 1 (hundred) 50 breads…, 1(hundred) 50 breads'.

The hypothesis for the reading 1 of the vertical wedge (diš) is confirmed by the use, in similar lists of bread distributions, of six times the sign u (10) for 60, as in the following example (Table 8.5): '50 breads: Aššur-imittī, 60 breads: Ilī-ālim, 1 hundred (breads): Ša-x-ki'.

#### 8.1.3.2 Examples from Mari (Middle Euphrates, Eighteenth Century BCE)

During the same period, the numerical system (System S) was often used in the Old Babylonian and Syrian texts. In the following administrative text from Mari (Syria), which gives an account of personnel, it is used to write the numbers of men and women (Table 8.6). The transcription by Jean Bottéro (1957) of the first line shows the use of the sign 𒁹 (geš$_2$) for 60; there are '1 × 60 + 16 men (and) 53 women'. By transcribing the sign 1(geš$_2$) as 60, Jean Bottéro intends to help us to understand the sexagesimal system better. The total number of men and women stated in the last line of the extract introduces a decimal additive system with the Sumerian logogram ME (which is read in Akkadian as *meat*) corresponding to 'one hundred'. The total amounts to: '1 hundred 29 men and women'. Thus the scribe used two different systems for writing numbers in the same document.[6]

In his transcription of numbers, Jean Bottéro gathers the hundred and the sixties on one side, and the tens and units on the other, thus structuring the number, separating the decimal component from the non-decimal. Such structuring does not exist in the cuneiform text as it is reflected in the transliteration following the CDLI norms. In his introduction to ARMT 7 (1957: iv–v), Jean Bottéro explains that he uses the

[6] Note that the transcription by Jean Bottéro (1957) uses the Akkadian translations (in italics), whereas my transliteration indicates the Sumerian logograms used (in plain Roman).

**Table 8.6** (Column 1) Text ARMT 7, 157:1–5, (Column 2) Transcription by J. Bottéro, and (Column 3) Transliteration following the CDLI norms

| Bottéro 1957 (ARMT 7, 157:1–5) | | Transliteration using CDLI norms |
|---|---|---|
| 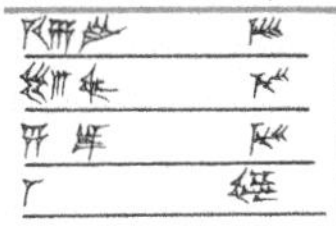 | 60(+)16 *awîlû*(meš)<br>53 *sinnišâtum*(meš)<br>(…) | 1(géš) 1(u) 6(diš) lu$_2$-meš<br>5(u) 3(diš) munus-meš |
|  | *naphar* 1 ME 29 *awîlû ù sinnišâtum*(m[eš]) | šunigin 1 me 2(u) 9(diš) lu$_2$ u$_3$ munus-m[eš] |

**Table 8.7** Text read by M. Birot and transliteration following the CDLI norms

| Birot (1969, no. 28:2, 28) | Transcription by Birot (1969) | Transliteration using CDLI norms |
|---|---|---|
|  | *80* I *40* I $^{d}$*Adad-šar-rum* | 1(geš$_2$) 2(u) I 4(u) I $^{d}$im-*šar-rum* |
|  | *1223 šûmu*$^{sar}$ *ša qû*$^{?}$*-tù*$^{?}$*-pu* | 2(geš'u) 2(u) 3(diš) sum$^{sar}$ *ša qú-tù-pu* |

**Table 8.8** Text read by J.-M. Durand and transliteration using the CDLI norms

| Durand (1983: ARMT 21, 63:6) | | Transliteration using CDLI norms |
|---|---|---|
| 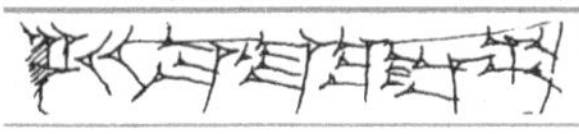 | 260 *ma-la-ku ša* GU$_4$ | 4(geš$_2$) 2(u) *ma-la-ku ša* gu$_4$ |

conventions of his predecessors, who published ARMT Vol. 1–6. However, in these volumes, there is no discussion of the transliteration of numbers.[7] After Jean Bottéro's editions of the Mari texts, publications no longer showed the numerical system used by the ancient scribes.

In some instances, the sign 𒁹 (geš$_2$ = 60) could be followed by the word *šūši*, literally meaning 'sixty' in Akkadian, as in the following document (Table 8.7), an account for onions, which is extremely interesting because the scribe wrote the numbers in columns marked by a vertical line. Unfortunately, we do not know the meaning of these columns and why two different amounts of onions are listed for each person. The edition of this text, which was completed by Maurice Birot in 1969, respects its alignment in columns indicated by ' I '. The total amount of harvested onions is expressed by System S, using, for example, the sign 𒐞 (geš'u = 600) as follows, in the second line of the text quoted, '2 × 600 + 23 onions that have been plucked'. Maurice Birot's transcription (1969) goes one step further than that of Jean Bottéro: he gives the result directly in our modern decimal system, thus concealing the ancient numerical notation.[8]

Subsequent editors of Mari texts followed the same tradition, rendering numbers directly in our decimal numerical system, as in the next example (Table 8.8). This

[7] However, on Bottéro (1957: 347–363), there are detailed comments about the different metrological systems and how they are rendered in the transliterations.

[8] There is no indication in the commentaries about the transliteration of numbers in this volume.

text records a distribution of meat. The total number of beef portions distributed to the cooks is written '$4 \times 60 + 20$' according to the sexagesimal system (System S). But the editor of the text writes '260' in his transliteration.

These examples reveal the main concern of the Assyriologists, who are more interested in the quantity or value of things than in how these quantities or values are written. Their transcriptions or transliterations conceal the various scribes' practices, since numbers and quantities are transformed directly into our numerical system. If we ignore the differences between these systems, the conversion of quantities from the various cuneiform notations into our system can be a source of errors.

### 8.1.4 Wedges in Rows of Three

Moreover, the reading of some numbers may be wrong. The wedges forming a number are usually grouped in rows of three for an immediate visual reading. Thus, in the text edited by Jean Bottéro (1957: ARMT 7, 157) presented Table 8.6, the '9' in the number 129 (last line), is made of nine wedges organized by three rows of three. In texts originating from Mari, on the Middle Euphrates, for example, signs representing numbers were not always grouped by three; the same observation can be made for the Old Assyrian texts. In the following example (Table 8.9), the number '8' in the sentence 'in 2 hundred 38 textiles' is written with two rows of four wedges each:

There is no mention of the disposition of wedges in numbers in the transliterations. But it is important to note that the arrangement of wedges in two rows has regularly been a source of mistakes in textual editions. Indeed, there are often confusions between '7' and '8', or '8' and '9'.

### 8.1.5 Translations of Fractions

In mathematical texts, scribes usually converted fractions of measurement units into integers according to the sexagesimal place-value notation (SPVN) system so that they could use their numerical tables.[9] However, fractions were widely used in economical texts and SPVN rarely appears in this corpus. During the first half of the second millennium BCE, scribes used a limited number of fractions.

There were some specific signs for:

1/3  1/2  2/3  5/6 

In the Old Babylonian system, inverses ($1/n$) were expressed by 'igi-n-gal$_2$':

Example: 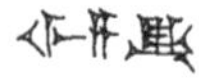 igi-4-gal$_2$ = 1/4

[9] See Proust in this volume (Chap. 7).

**Table 8.9** Old Assyrian example in which the number eight is written with two rows of wedges

| Copy Thureau-Dangin (1928: TC 2, 52:1) | Transliteration |
|---|---|
| 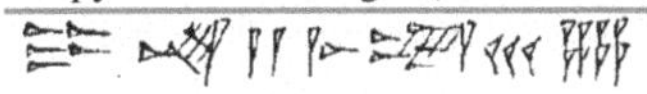 | *i-na* 2 *me-at* 3(u) 8(diš) $tug_2^{hi\text{-}a}$ |

However, the Old Assyrian merchants did not use this formula and created specific signs for 1/4 and 1/6:

1/4 1/6

They wrote 3/4 using the subtractive notation 1 $la_2$ 1/4, which we can translate as '1 minus 1/4'. They had no sign for 1/5, which was a source of errors in their computations (Michel 1992).

The Old Assyrian merchants always preferred to express quantities in fractions of the upper unit rather than integers of lower units (Michel, forthcoming). This is a very important point in trying to reconstruct the way they were computing. Some scholars, very respectful of the ancient practices of writing numbers and quantities in their transliterations, choose to give their own interpretation of the text in their translations, highlighting the point which, according to them, is the most interesting. For example, they change fractions into integers, concealing the existence of these fractions.

In the following examples, Klaas Veenhof (2010) translates 2/3 *mana* as 40 *gin* (shekels), 1/2 *mana* as 30 *gin*, and 1/3 *mana* as 20 *gin* (shekels), because of a specific interest in the silver value of the transactions:

| | |
|---|---|
| Veenhof (2010: AKT 5, 76:3) | |
| 2/3 *ma-na* 4 gín kù-babbar | 44 shekels of silver |
| Veenhof (2010: AKT 5, 54:1–3) | |
| 2 *e-ma-re-e a-na* 1/2 *ma-na*: 3 gín kù-babbar 2 *e-ma-re a-na* 1/3 *ma-na* 5 1/2 gín kù-babbar | 2 donkeys for 33 shekels of silver, 2 donkeys for 25 1/2 shekels of silver |

In the following example, the choice made for the translation is inconsistent because one fraction is changed into an integer (1/3 *mana* as 20 *gin*) but not the following one (1/2 *mana*):

| | |
|---|---|
| (Veenhof 2010: AKT 5, 61:13–14)<br>šu-nigin 1/3 *ma-na* kù-[gi]<br>*ù* 1/2 *ma-<na>* kù-babbar | In all 20 shekels of gold<br>and 1/2 minas of silver |

Presumably with the same idea in mind, some authors change a subtractive notation into an additive one in their translations, as in the following examples (Table 8.10).

**Table 8.10** Transliteration and translation of extracts of texts: examples of changes of notation from subtractive to additive in the translation

| Larsen and Møller (1991, no. 4:33) | Günbattı (2004) Kt 00/k 10:iii 9' | Dercksen (2008, 90) Prag I 512:20–21 | Dercksen (2008, 90) ATHE 12:2–3 |
|---|---|---|---|
| 1 lá 1/4 gín *a-mu-tum* | 7 lá 1/3 gín-ta kù-babbar | 20 lá 1 *na-ru-uq, še-am* | *ša* 8 1/3 *ma-na* 2 lá 1/4 gín, kù-babbar |
| 3/4 shekels of iron | 6 2/3 shekels of silver per piece | 19 sacks of barley | that 8 minas 21 3/4 shekels of silver |

In the last example (on the right of the table), the author converts 1/3 *mana* into 20 *gin* (shekels) and adds the sum to the result of the subtraction 2 minus 1/4 *gin*. Even though the transliteration is very respectful of the original notation, its total transformation in the translation conceals the computational logic by means of which the merchants obtained the result; it makes these texts unusable for historians of mathematics unfamiliar with cuneiform and Akkadian documentation.

## 8.2 Mesopotamian Capacity System Used in Editions of Administrative and Economic Texts

### *8.2.1 The Capacity System*

Among the metrological systems used by ancient Mesopotamians, the capacity system is perhaps the most complicated one because of the following characteristics (Powell 1990: 492–508; Huehnergard 1997: 584):

- The capacity system has many local variants.
- The ratio between the different measurement units is never the same.
- This metrological system uses specific numerical signs which integrate both the numerical value and the measurement unit.

With some simplifications, the capacity system used at the beginning of the second millennium BCE consisted of the following terms (see Table 8.11).

This system uses specific numerical signs for the different measurement units:

- The measurement unit gur (*gur*) uses a variant of system S in which the unit is represented with the horizontal wedge 𒀸 (aš).
- The measurement units sila$_3$ (*sila*), gin$_2$ (*gin*) and še (*še*) use the standard system S in which the unit is represented with the vertical wedge 𒁹 (diš).
- The measurement unit bariga (*bariga*), whose name is rarely written ( sign 'PI'), ranges from 1 to 4 as follows:

| | | | | | | | |
|---|---|---|---|---|---|---|---|
|  | 1 bariga |  | 2 bariga |  | 3 bariga |  | 4 bariga |

**Table 8.11** The capacity system used at the beginning of the second millennium BCE

|  | | 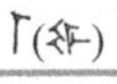 | |  | |  | |  | |  |
|---|---|---|---|---|---|---|---|---|---|---|
| gur | ←5← | bariga | ←6← | ban$_2$ | ←10← | sila$_3$<br>c. | ←60← | gin$_2$ | ←180← | še |
| 300<br>*sila* | | 60 *sila* | | 10<br>*sila* | | 1<br>litre<br>1 *sila* | | 1/60<br>*sila* | | 1/10800<br>*sila* |

– The measurement unit ban$_2$ (*ban*) ranges from 1 to 5 as follows (graphic variants existed for 4 and 5 ban$_2$):

| | | | | | | | | | |
|---|---|---|---|---|---|---|---|---|---|
|  | 1 ban$_2$ |  | 2 ban$_2$ |  | 3 ban$_2$ |  | 4 ban$_2$ |  | 5 ban$_2$ |

– The measurement unit sila$_3$ uses the cuneiform sign QA , whose Akkadian name is *qû*(*m*). All three writings (sila$_3$, QA, *qû*) have been used since the decipherment of cuneiform writing began.

Editors have rendered the complexity of the Mesopotamian metrological system for capacity in many different ways.

### 8.2.2 *In the Frame of the Debate Concerning the Existence of Sumerian*

In order to analyze how Assyriologists understood this metrological system, transliterated these quantities and translated them, we need to go back to the pioneering works.[10] Several factors hampered the understanding of Mesopotamian numerical and metrological systems at the end of the nineteenth century. The existence of the Sumerian language was strongly denied by some scholars who wanted to prove that writing was invented by Semitic peoples.[11] Among them, Joseph Halévy tried to show that Sumerian was not a language, but a writing system invented by Assyrian scribes to note their language logographically (Halévy 1874).

Numbers and quantities played a key part in this debate. Indeed, some defenders of the existence of Sumerian as a language argued that "Mesopotamian" expression of numbers and quantities made use of a sexagesimal system while Semitic

[10] This section relies mainly on Old Babylonian texts. In fact, most of the ancient publications of the Old Assyrian texts dating to the first half of the nineteenth century appeared only in cuneiform copies. There were only a few systematic editions of some of these texts (e.g., Eisser and Lewy 1930–1933). And the products traded by the merchants were quantified with the measuring units for weight, rarely for capacity.

[11] For a detailed overview of this debate, see (Cooper 1991).

languages used a decimal system. Among the scholars who believed in the existence of Sumerian was the French Assyriologist François Lenormant (1837–1883). From early involvement in Greek archaeology and numismatics, he came to the field of Near Eastern studies to find help to understand Greek monetary weights (1868).

### 8.2.3 *Trying to Understand Mesopotamian Metrology as a Whole System: The Pioneers*

An initial important work in Mesopotamian metrology was published by Carl-Friedrich Lehmann-Haupt (1861–1938) in 1889, when he was only 27 years old. He found a link between the 'foot' and the 'brick', and gave the value of a cubit (*kuš*) as 49.5 cm. He made a connection between the capacity measurement unit QA (*sila*) and the length measurements, and he tried to determine the correct value of the *mana*, a unit of weight.

Jules Oppert (1825–1905) was another of the very first Assyriologists who tried to understand Mesopotamian metrology. He had contributed to the decipherment of Akkadian written in cuneiform. When he participated in an expedition to Mesopotamia in the 1850s (Lion and Michel 2009b), he wrote letters on Babylonian measurement and authored some thirty articles dealing with Mesopotamian metrology, trying to show that it constituted a coherent system, which was based on the exact value of the cubit. In 1894, he presented a comprehensive survey of his reconstructions (Oppert 1894). After establishing the value of the length measurement units at the time of the Assyrian king Sargon II (721–705 BCE) based on the constructions on the site of Khorsabad (the ancient city of Dūr-Šarrukēn) and the sizes of bricks excavated there, he developed some hypotheses on the capacity system. He compared the *gur* of the cuneiform texts with the '*cor*' of the Bible, which corresponds to 300 l. He declared the *sila* to be the basic capacity measure. However, he privileged the *bariga* and worked out the ratio between the *bariga* and the *ban*, which he expressed as one-sixth of a *bariga*. This implied a rather complicated notation, introducing fractions with sixths of a *bariga*, for which he used the notation '*pi*' (Table 8.12).

**Table 8.12** (First column) Notation of capacities used by J. Oppert; (Second column): the corresponding reading in *bariga* according to J. Oppert; (Third column) Transliteration following the CDLI norms

| F. Thureau-Dangin (1894) | | Transliteration using CDLI norms |
|---|---|---|
| Le sixième ou l'hecte d'un *pi* s'exprime par  | 1/6 *bariga* (10 *sila*) | 1(ban$_2$) |
| 2 sixièmes par  | 2/6 *bariga* (20 *sila*) | 2(ban$_2$) |
| 3 —  | 3/6 *bariga* (30 *sila*) | 3(ban$_2$) |
| 4 —  ou  | 4/6 *bariga* (40 *sila*) | 4(ban$_2$) |
| 5 —  ou  | 5/6 *bariga* (50 *sila*) | 5(ban$_2$) |

His explanations went as follows: 'Cinq *pi* forment un *guru* ou *cor*, le κόρος des Grecs.' (Five *pi* make one *guru* or *cor*, the κόρος of the Greeks.) This sentence reveals that the translations of Mesopotamian units of measure were contaminated by editors of ancient classical texts, mainly in Biblical Hebrew and Greek. Indeed, Jules Oppert translates the Babylonian measurements as follows: 'cor' for *gur*, 'médimne' for *bariga* and 'hecte' for *ban*, and leaves *qa* untranslated. The first is borrowed from the Bible and the following two from ancient Greek words. In the writings of the scholars of the nineteenth and twentieth centuries CE, the translation choices for Mesopotamian capacity measurements have never been consistent.

At the time when they started to edit texts systematically, scholars tried to explain the system they adopted for the transliteration and translation of the measurement units. For example, Bruno Meissner (1868–1947), when publishing Old Babylonian private contracts in 1893, used the work of Jules Oppert and presented the following table which converted everything into ḲA (*sila*) še (barley) (Meissner 1893: 98–101, Table page 100):

**Bruno Meissner's table of capacity measurement units. (Meissner 1893, 100)**

As we can see, Bruno Meissner did not distinguish between the different measurement units, namely the *ban* and the *bariga*; such an approach has the drawback of hiding the structure of the capacity system and the ratio between two units of measurement.

Vincent Scheil (1858–1940) was known for his masterly publication of the Hammurabi Code in 1902. In the same year he published some Old Babylonian texts found at Sippar: in his transliterations and translations of the texts, he left capacity quantities in cuneiform without translating them (Scheil 1902: 110).

**Scheil's edition of an Old Babylonian text. (Scheil 1902)**

He was not sure how to understand the value and the measurement unit coded in the sign. He cited a lexical text listing containers made of wood (Scheil 1902, text No. 53). The text gives the measurements for 30, 20, 10, 6, 2, 1, 1/2, and 1/3 *sila*. He noted that the name of the measurement unit QA (*sila*) only follows numbers under 10. For numbers over 10, it is omitted; he left the text in cuneiform as follows:

**Cuneiform signs in Scheil's edition of text no. 53. [$^{giš}$3(ban$_2$), $^{giš}$2(ban$_2$), $^{giš}$1(ban$_2$)]**

This observation suggests a change of measurement units upon reaching 10, a fact already demonstrated by Jules Oppert. Somewhat later, Vincent Scheil explained that, depending on the period under consideration, the ratio between the *ban* and the QA (*sila*) was not always equal to ten, since he had found a *ban* of 5 QA (*sila*) (Scheil 1918: 85).

### 8.2.4 A Value for the QA (sila) Measure

François Thureau-Dangin (1872–1944), another pioneer in Sumerian studies and Mesopotamian mathematics, tried to determine the exact capacity of the measurement unit QA (*sila*) and considered it as the main unit of the capacity system. In an article published in 1909 (Thureau-Dangin 1909: 90), he demonstrated that all the measurement units used by Sumero-Akkadians formed a closed system in which measurement units of length, capacity and weight were linked. For the capacity, he made a first computation based on the silver vase of Enmetena (king of Lagaš, c.a. 2420 BCE), assuming that the royal *gur* was equal to the cubic cubit (*kuš*). Later on, in 1912, he published a short note about an inscribed *alabastrum* from Susa giving a QA (*sila*) value of about 0.81 L. In 1921, in another article, Thureau-Dangin adjusted his computations once again (Thureau-Dangin 1921). According to him, during the Neo-Babylonian period (sixth century BCE), the *ban* was equivalent to 6 *sila* and presumably weighed 10 *mana* of water (Thureau-Dangin 1912: 24–25); thus 1 *sila* would correspond to 84 1/6 cl (Thureau-Dangin 1921: 131, 136). In his table of the capacity system, he distinguished the royal *gur* in which 1 *ban* was equivalent to 10 *sila*, and another *gur* in which 1 *ban* was equivalent to 5 or 6 *sila*.

**Table 8.13** Edition of an Ur III tablet by F. Thureau-Dagin and transliteration using CDLI norms

| F. Thureau-Dangin (1894) | | Transliteration using CDLI norms |
|---|---|---|
|  | 204 QA de boisson fermentée commune, | 3(bariga) 2(ban$_2$) 4 sila$_3$ kaš ús |
| | 7500 QA de boisson fermentée inférieure, | 2(u) 5(aš) (gur) kaš ú-sa ús |
| | 224 QA de grains rôtis, | 3(bariga) 4(ban$_2$) 4 sila$_3$ ninda |

## 8.2.5 *How to Transliterate Quantities in the Capacity System: Conversion into sila or gur?*

Even though Thureau-Dangin was correct in his reconstruction of the structure of the capacity system, in his edition of the ancient texts, he was unsure about the way to render quantities expressed in the capacity system. In 1894, he translated an Ur III list of food distribution to employees and converted all the quantities in QA (*sila*) as shown above (Table 8.13; Thureau-Dangin 1894: 134–135).

He did not offer any transliteration. In his translation, the reason why he converted all the values into QA (*sila*) was presumably because of its proximity to the liter, perhaps in order to facilitate the reading of the text for those interested in economic history. The translation of these lines should be read as follows: '3 × 60 + 24 *sila* of ordinary beer, 25 *gur* of ordinary wort, 3 *bariga* 4 *ban* 4 *sila* of bread'.

A few years later, in 1904, when publishing contracts from the city of Šuruppak dated to the third millennium BCE, Thureau-Dangin (1904: 151) used Jules Oppert's notation, and wrote, for example: "1 2/6 še" for "80 sila$_3$".

In 1911, in his publication of Old Babylonian tablets preserved in the Louvre, Thureau-Dangin opted to keep in his transliteration the cuneiform signs corresponding to the capacity measures that for us are expressed in *bariga* and *ban*, translating them as a fraction of *gur* (1 *bariga* and 4 *ban* correspond to one third of a *gur*)[12]:

| | |
|---|---|
| 3 *še-ḫar-ra* ..... | 3 [gur] de grain .... |
| *ṣibat* 1 *gur*  [*u*]*ṣ*(-*ṣap* ?) | — comme intérêt d'un gur, 1/3 de gur il ajoutera — |

**Thureau-Dangin's edition of an Old Babylonian tablet with capacity measures. (Thureau-Dangin 1911: 76)**

Other editors of ancient texts used slightly different systems to transliterate measurement values expressed in the capacity system. Arno Poebel (1881–1910), for example, published a book in 1909 containing his edition of various Old Babylonian documents. Instead of converting everything into *sila* as Thureau-Dangin had done in 1894, he converted all the quantities into *gur*, as Thureau-Dangin had done in 1911 in the text just mentioned, but, contrary to the latter, he did not add "gur" in

[12] See also Thureau-Dangin (1909: 79–80), where he gives equivalences between the three measures and notes that the number of QA (*sila*) in a *gur* could vary according to the period considered.

**Table 8.14** Edition of an Old Babylonian tablet by A. Poebel and transliteration using CDLI norms

| Poebel (1909: BE 6/2, 55) copy and transliteration | | Transliteration using CDLI norms |
|---|---|---|
|  | $\frac{3}{30}$ *she shu-ti-a*<br>*lù-$^{\text{šu}}$usán-sur*<br>$\frac{1}{5}$ + $\frac{4}{30}$ *she sham su-e-sir-xun*<br>$\frac{1}{5}$ *she ki-dish(?)*<br>$\frac{1}{5}$ + $\frac{1}{30}$ *she i-na gish-ba-an*<br>*lù-gash-tin-na* | 3(ban$_2$) še …<br>…<br>1(bariga) 4(ban$_2$) še …<br>1(bariga) še …<br>1(bariga) 1(ban$_2$) še …<br>… |

**Table 8.15** Edition of an Old Babylonian text by E. Ebeling

| Thureau-Dangin (1910: TCL 1, 49:5–7) | Transliteration Ebeling (1913, 152) | Transliteration of quantities using CDLI norms |
|---|---|---|
| 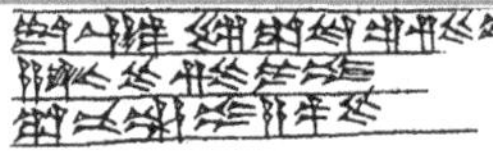 | i-na 1 7/10 $^{\text{še}}$gur ša te-e-zi-ba 4/5+1/30 še zi-i-tum ša-bi-il-tum 4/5 + 2/30 še | …1(aš) 3(bariga) 3(ban$_2$) še gur…<br>4(bariga) 5(ban$_2$) še …<br>… 4(bariga) 2(ban$_2$) še |

**Table 8.16** Transliteration by R. Jestin and transliteration using CDLI norms

| Copy Jestin (1952, Plate III, Text No. 23) | Transliteration Jestin (1952, 190) | Transliteration using CDLI norms |
|---|---|---|
| 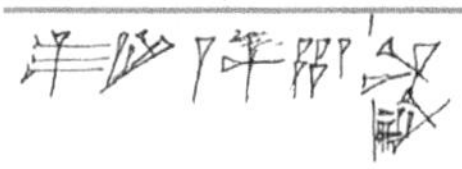 | *30* NINDA *75* SILÀ<br>/KAŠ | 3(ban$_2$) ninda 1(bariga) 1(ban$_2$) 5 sila$_3$<br>/kaš |

his transcription. Thus, 1 *bariga* became 1/5 (of a *gur*) and 1 *ban*, 1/30 (of a *gur*) (Table 8.14).

Erich Ebeling used the same system in 1913 but made a mistake in the second line below (Table 8.15).

Line 5: 1 7/10 *gur* of Erich Ebeling's transliteration correspond to 1 *gur* 3 *bariga* 3 *ban* expressed in *gur*, thus 1 + 3/5 + 3/30 *gur*.
Line 6: 4/5 + 1/30 has to be corrected to 4/5 + 5/30. Indeed, the copy must be read as: 4 *bariga* 5 *ban.*
Line 7: 4/5 + 2/30 correspond to 4 *bariga* 2 *ban* expressed in *gur*.

In the following years, Assyriologists used to transliterate quantities expressed in the capacity measurement units directly into *sila*, as Raymond Jestin did in his publication and edition of Mari texts (Middle Euphrates, eighteenth century BCE, Table 8.16).

During the first half of the twentieth century CE, scholars chose either to transliterate capacities into *gur*, the highest measurement unit of the Mesopotamian capacity system, or into *sila*, a unit whose value is often considered as corresponding more or less to a liter, but could change according to place and date. Such transliterations, which tend to present capacities as expressed in a uniform way, conceal

**Table 8.17** Copy and transliteration by W. Leemans of an Old Babylonian tablet

| Leemans 1954, 62, Text YOS 5, No.169 | Transliteration using CDLI norms |
|---|---|
| Line 3: 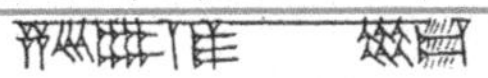<br>278 (gur) 1 (bariga) 3 (bán) še-gur, | 4(geš$_2$) 3(u) 8(aš) 1(bariga) 3(ban$_2$) še gur |

the various units of the metrological system for capacity and the different ratios linking each measurement unit together.

### 8.2.6 *Fidelity to the Original Text*

A new step was taken in the 1950s when several scholars decided to reflect the cuneiform text more closely while transliterating these quantities. This was the case with Wilhelmus Leemans, who published an edition of Old Babylonian texts from Larsa in 1954 (Table 8.17).

Wilhelmus Leemans transliterated Line 3 as follows: 278 (gur) 1 (bariga) 3 (ban$_2$) še-gur. If we use the CDLI transliteration norms, we obtain: 4(geš$_2$) 3(u) 8(aš) 1(bariga) 3(ban$_2$) še gur, which can be understood as: (4 × 60 + 38) *gur* 1 *bariga* 3 *ban*. The system of transliterating quantities adopted by Leemans had the advantage of showing the complexity of the system for measuring capacity.[13]

### 8.2.7 *A Completely New Positional Notation*

A dramatic change was introduced in the first half of the 1960s by Fritz Kraus and Edmond Sollberger, who imposed a completely new system in the transliteration of the values for capacity measurement. The great majority of Assyriologists still use this system today. It converts the measurement units for capacity into a positional notation in which the absence of a unit is marked by a 'zero'. This system was initially proposed in 1964 by Fritz Kraus when he published the first volume of a series dedicated to a comprehensive edition of the Old Babylonian letters preserved in various museums:

> Bemerkt sei noch, dass die Zahlenangaben bei Hohlmassen nicht, wie es oft geschieht, in silà-Einheiten umgerechnet, sondern im Anschluss an das babylonische System äusserlich ähnlich unseren Dezimalzahlen umschrieben sind; 1 'PI' erscheint als 0,1; 1 bán als 0,0.1; 3,3.3 gur bedeutet also 3 gur 3 'PI' 3 bán. (Kraus 1964: XI)
>
> It should also be noted that the figures for capacities are not converted into silà units, as is often the case, but are, after the Babylonian system, externally similar to our decimal numbers; 1 'PI' appears as 0,1; 1 ban as 0,0.1; 3,3.3 gur means 3 gur 3 'PI' 3 bán. (Kraus 1964: XI)

[13] This transliteration of capacity measures was also adopted by Otto Neugebauer (see C. Proust's Chap. 7 in the present volume).

He therefore proposed the following equivalences:

| | | |
|---|---|---|
| 1 *gur* | is transliterated | 1 |
| 1 *bariga* | is transliterated | 0,1 |
| 1 *ban* | is transliterated | 0,0.1 |

This positional system, with a minor variation (no comma but a point after the number of *gur*), was also used by Edmond Sollberger in 1965 when he published three Ur III tablets:

| | |
|---|---|
| Sollberger (1965: 27, no. 3:1): | *14.3.0. še-giz-ì gur* |
| translation: | 14 *gur* 3 PI sesame oil rations of the troops |
| CDLI transliteration style: | 1(u) 4(aš) 3(bariga) še-giš-$i_3$ gur |

Edmond Sollberger explained this system of transliteration of capacity measurements—with the introduction of a fourth position for the *sila*—in the introduction to his volume, published a year later (Sollberger 1966: 7):

> Quantities expressed in gur and its subdivisions (nigida and bán) are not transliterated as integers and fractions but as a set of three numbers: thus 1 gur is 1.0.0, 1 nigida is 0.1.0, 1 bán is 0.0.1. When sìla are mentioned, a fourth number is added followed by the word sìla. This, with a slight variation in the punctuation is also the system adopted by Kraus in recent publications. (Sollberger 1966: 7)

However, these transliteration conventions are only used for the capacity system. Indeed, Edmond Sollberger specifies: 'For areas, rather inconsistently, I add bùr or ešè in brackets after the numbers, thus: 2(bùr) 1(ešè) iku.' (Sollberger 1966: 7)

Such a positional notation for transliterating quantities noted according to the capacity system, does not reflect the construction of the capacity system in ancient times. It was, however, adopted by the editors of the Mari corpus. Jean-Marie Durand, in 1983, used it in Volume 21 of the *Archives Royales de Mari*: he followed Fritz Kraus's system with a comma after the quantity of *gur* (Durand 1983: ARMT 21).[14] Note that in the Mari corpus, 1 *gur* equals 120 *sila*, and not 300 *sila* (as mentioned above, in Sect. 8.2.1).

This notation has been adopted by many contemporary Assyriologists. One example is Dominique Charpin's transliteration of long Old Babylonian accounts from Ur (Table 8.18).

The changing ratios between measurement units of the capacity system, depending on the period under consideration, added to the complexity of the readings. Marvin Powell summarized them in 1990 as shown in (Table 8.19).

Besides the *gur*, which has a specific size at Mari, two standard sizes are known for the *ban*, 6 *sila* or 10 *sila*, the latter being the most common one. In his long

[14] This system is also used by Charpin and Durand (1983). This new notation was employed earlier by Dominique Charpin for Old Babylonian texts. See, for example, Charpin (1982). It is also in use for the edition of Ur III tablets (twenty-first century BCE); see Sigrist (1984: 33–35).

**Table 8.18** Copy of an Old Babylonian tablet by H. Figulla and transliteration by D. Charpin

| Copy by Figulla (1953: UET 5, 507) | | Transliteration by Charpin (1986: 307–309) |
|---|---|---|
| 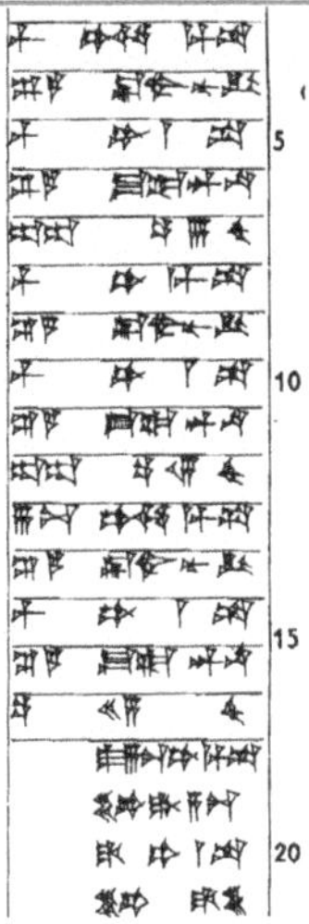 | | 0,0.1 kaš sig$_5$ 1 1/2-ta |
| | 4 | si-gar é-kiš-nu-gál |
| | | 0,0.1 kaš 1-ta |
| | 6 | si-gar kisal-sag-an-na |
| | | èš-èš u$_4$-8-kam |
| | 8 | 0,0.1 kaš 1-ta |
| | | si-gar é-kiš-nu-gál |
| | 10 | 0,0.1 kaš 1-ta |
| | | si-gar kisal-sag-an-na |
| | 12 | èš-èš u$_4$-15-kam |
| | | 8 sila$_3$ kaš sig$_5$ 1 1/2-ta |
| | 14 | si-gar é-kiš-nu-gál |
| | | 0,0.1 kaš 1-ta |
| | 16 | si-gar kisal-sag-an-na |
| | | u$_4$-25-kam |
| | 18 | 0,0.3.8 sila$_3$ kaš 1 1/2-ta |
| | | še-bi 0,0.5.4 sila$_3$ |
| | 20 | 0,0.4 kaš 1-ta |
| | | še-bi 0,0.4 še |

**Table 8.19** Summary of ratios between measurement units in the basic capacity systems (Powell 1990: 494)

| System | Ratios of units | | Size of guru in sila |
|---|---|---|---|
| Fāra lidga / gur maḫ | sila —10→ ban —6→ bariga —4→ lidga —2→ gur maḫ | —2400→ guru | 5,20,0,0 |
| Standard gur saĝĝal | sila —10→ ban —6→ bariga —4→ gur | —3600→ guru | 4, 0,0,0 |
| Lagaš gur saĝĝal | sila —6→ ban —6→ bariga —4→ gur | —3600→ guru | 2,24,0,0 |
| Standard Akkad-OB | sila —10→ ban —6→ bariga —5→ gur | —3600→ guru | 5, 0,0,0 |
| Standard NB-LB | sila —6→ ban —6→ bariga —5→ gur | | |

article, however, Marvin Powell (1990) does not discuss the modern transliterations of quantities measured with the capacity system.

Finally, in 1988, Marcel Sigrist suggested to slightly modify the notation of capacity measurement values in modern transliterations by inserting a distinct sign between the *gur* and the other units, thus changing Fritz Kraus's comma into a semi-colon, as we can read in the foreword of (Spar 1988: LII):

> Complex capacity measures are expressed through this system in the transliterations, while in the translations the whole capacity is expressed in *sìla*.

The 'system' in question is presented a few lines earlier, beginning with a 'Table of Correspondences for Capacity Measures' Spar (1988: LII):

| *gur* | | PI | | *bán* | | *sìla* |
|---|---|---|---|---|---|---|
| | | | | | | 1 |
| | | | | 1 | = | 10 |
| | | 1 | | 6 | = | 60 |
| 1 | = | 5 | = | 30 | = | 300 |

> The system is expressed in the following manner:
>
> gur PI *bán* *sìla*
>
> 0; 0. 0. 0
>
> (…) Thus, 1;3.0.0. in the transliteration stands for 1 *gur*, 3 PI (0 *bán*, and 0 *sìla*) or 480 *sìla* altogether, as it would be indicated in the translation. (Spar 1988, *apud* Sigrist: LII)

The decision was taken not to apply this notation when a quantity was expressed in one unit (*gur* or *sila*):

> When not found in a list with other measures, simple measures in *gur* or *sìla* will be transliterated by the number followed by the capacity measure, such as 3 *gur* or 3 *sìla* (Sigrist 1988: LII).

## 8.2.8 Impact of the New Notation

This type of notation, is now widely used by Assyriologists in their editions of ancient texts. However, it is really problematic, since it modifies the system used by the ancient Mesopotamians and conceals data which are essential clues for the reconstruction of their computational methods. It introduces **a positional notation** which did not exist in the capacity metrological system and makes use of the 'zero', which was invented much later (Proust 2009). It also hides measurement units that may have different sizes according to place and date, as for the *gur* and the *ban*.

Some scholars in the mid-1950s, including Wilhelmus Leemans, Otto Neugebauer and several others, translated quantities expressed with units of capacity as closely as possible to the cuneiform text. They thus used systems of notation which permitted them to avoid the introduction of notions that had not been used in these contexts, such as a positional notation, and also to avoid creating anachronisms. These scholars inspired the CDLI editorial board when it constructed its charter for the transliteration of cuneiform texts.

## 8.3 Conclusions

At the beginning of the second millennium BCE, large institutions, as well as private entrepreneurs, produced a considerable amount of administrative and economic tablets containing numbers and quantities. The writing of numbers followed different systems, sexagesimal and/or decimal, according to the area and context under consideration. The capacity system, the most complex part of the Mesopotamian metrological system, also had different local variations. The pioneers of Assyriology were unsure how to render these variations; even today, there is still no consensus on the transliteration and the translation of capacities. When deciphering a new script such as cuneiform, which was used for more than three millennia in a very large area and by a dozen languages belonging to different linguistic families, one can easily understand that numbers and quantities are not the main concern of scholars, except for those who have a special interest in mathematics. Moreover, in the brief history of Assyriology, numbers and quantities are often the elements which modern editors have tended to neglect: their reading is regularly wrong.

Nowadays, in most cases, the transliterations and translations of numbers occurring in Old Babylonian and Old Assyrian texts fail to highlight variations in numerical systems, since they directly use the modern one. Respecting the original way of writing numbers, at least in transliterations, allows historians to understand how the ancient administrators and merchants perceived numbers and quantities and computed with them. Ideally, even the paleography of a number should be taken into account. For example, in the Old Assyrian society, it seems that merchants who had received a scribal education used to arrange the wedges forming a number in rows of three, while others who had no formal learning did not use more than two rows of wedges. This is one of the clues for distinguishing the hands of different scribes.

Text editors have rendered the complexity of the Mesopotamian capacity metrological system in a great variety of ways for the past 160 years. We can distinguish different phases in the perception of this issue. At the time of the decipherment of cuneiform, the sexagesimal system was used as an argument for the existence of the Sumerian language (as the origin of cuneiform) as distinguished from the Semitic one, which used a decimal system. In the second half of the nineteenth century, scholars tried to understand the functioning of the numerical and metrological system, to discover the ratio between the different measurement units, to discern links between the different metrological systems, and to give the modern equivalent of ancient measurement units. The proposed translations for the measurement units were greatly influenced by Greek and Biblical Hebrew, and today there is still no consensus about the translation of various measuring units. During the first three decades of the twentieth century, text editors chose either to convert all quantities expressed in the capacity system into *sila*, the smallest measurement unit occurring most frequently in the texts, or to convert them into *gur*, the biggest measurement unit for capacity. During the subsequent twenty years, they preferred transliterating them into *sila*. In the 1950s, some scholars decided to respect the original notation, not for the numbers but for quantities, giving the value for each measurement unit without conversion, which clearly revealed the complexity of the system. But in the 1960s, the transliteration system proposed by Fritz Kraus and Edmond Sollberger, which consisted of the invention of a new positional notation, was widely adopted by Assyriologists. It is still in use today. This notation, which has no links to the ancient one, creates a new level of complexity in the understanding of the Mesopotamian capacity system; it also introduces the zero, which was invented much later. For anyone outside the world of Assyriology and not familiar with the works of Fritz Kraus and Edmond Sollberger, it is unintelligible.

We hope to have shown in this article that editions of cuneiform texts should reflect, as far as possible, the original way of expressing numbers and quantities. The CDLI has proposed a system which, even if it seems a little "heavy" at first glance, has the great advantage of respecting the cuneiform notation; it can be followed by Assyriologists in transliterations, at least for complicated metrological systems such as the capacity system.

Numbers and quantities in cuneiform texts deserve increased attention because they can tell us much about the knowledge possessed by the authors of the texts, who were either administrators or merchants, including such questions as the numerical and metrological systems in use, methods of computation, the level of literacy and so forth.

## References

### *Primary Sources*

Birot, M. 1969. *Tablettes Économiques et Administratives d'Époque Babylonienne Ancienne Conservées au Musée d'Art et d'Histoire de Genève*. Paris: P. Geuthner.

Bottéro, J. 1957. *Textes Administratifs de la Salle 110*, Archives royales de Mari VII. Paris: P. Geuthner.

Clay, A.T. 1927. *Letters and transactions from Cappadocia*, Babylonian inscriptions in the collection of J. B. Nies, Part 4. New Haven: Yale University Press.

Durand, J.-M. 1983. *Textes Administratifs des Salles 134 et 160 du Palais de Mari*, Archives Royales de Mari XXI. Paris: P. Geuthner.

Figulla, H.H. 1953. *Letters and documents of the old Babylonian period Ur excavation texts*, 5. Philadelphia: The Trustees of the Two Museums.

Grice, E.M. 1919. *Records from Ur and Larsa dated in the Larsa Dynasty*, Yale Oriental Series. Babylonian Texts Vol. V. New Haven: Yale University Press.

Harper, R.F. 1892–1914. *Assyrian and Babylonian letters belonging to the K collection of the British Museum*. Vol. I–XIV. Chicago: The University of Chicago Press.

Hecker, K., G. Kryszat, and L. Matouš. 1998. *Kappadokische Keilschrifttafeln aus den Sammlungen der Karlsuniversität Prag*. Prag: Univerzita Karlova.

Hrozný, B. 1952. *Inscriptions Cunéiformes du Kultépé, Vol. I (ICK 1)*, Monografie Archivu Orientálního Vol. XIV. Prague: Státní Pedagogické Nakladatelství.

Kienast, B. 1960. *Die altassyrischen Texte des orientalischen Seminars der Universität Heidelberg und der Sammlung Erlenmeyer*. Berlin: De Gruyter.

Kraus, F.R. 1964. *Briefe aus dem British Museum (CT 43 und 44)*, Altbabylonische Briefe 1. Leiden: Brill.
Larsen, M.T., and E. Møller. 1991. Five old Assyrian texts. In *Marchands, Diplomates et Empereurs. Études sur la Civilisation Mésopotamienne Offertes à Paul Garelli*, eds. D. Charpin and F. Joannès, 227–252. Paris: Editions Recherches sur les Civilisations.
Lewy, J. 1926. *Die altassyrischen Texte von Kültepe*. Istanbul: Keilschrifttexte in den Antiken Museen zu Stambul.
Matouš, L. 1962. *Inscriptions Cunéiformes de Kültepe*. Vol. 3. Prague: Editions de l'Académie tchécoslovaque des sciences.
Poebel, A. 1909. *Legal and business documents from the time of the First Dynasty of Babylon*, The Babylonian expedition of the University of Pennsylvania 6/2. Philadelphia: University of Pennsylvania.
Sollberger, E. 1966. *The business and administrative correspondence under the Kings of Ur*, Texts from Cuneiform Sources 1. New York: J. J. Augustin Publisher.
Thureau-Dangin, F. 1910. *Lettres et Contrats de la Première Dynastie Babylonienne*, Textes cunéiformes du Louvre I. Vol. 13. Paris: Librairie Orientaliste Paul Geuthner.
———. 1928. *Tablettes Cappadociennes. Deuxième série*, Textes Cunéiformes XIV. Paris: Librairie Orientaliste Paul Geuthner.
Veenhof, K.R. 2010. *The archive of Kuliya, son of Ali-abum (Kt. 92/k 188–263), (Ankara)*, Kültepe Tabletleri V. Ankara: Türk Tarih Kurumu.

## Secondary Literature

Charpin, D. 1982. Marchands du palais et marchands du temple à la fin de la I° dynastie de Babylone. *Journal Asiatique* 207: 25–65.
———. 1986. *Le clergé d'Ur au siècle d'Hammurabi (XIX°–XVIII° siècles av. J.-C.)*, Hautes Etudes Orientales 22. Genève: Droz.
Charpin, D., and J.-M. Durand. 1983. Relectures d'ARMT VII. In *Mari Annales de Recherches Interdisciplinaires*, 2, eds. J.-M. Durand and J. Margueron, 75–115. Paris: Editions Recherches sur les Civilisations.
Cooper, J. 1991. Posing the Sumerian question: Race and scholarship in the early history of Assyriology. *Aula Orientalis* 9: 47–66.
De Odorico, M. 1995. *The use of numbers and quantifications in the Assyrian Royal Inscriptions*. Helsinki: Helsinki University Press.
Dercksen, J.G. 2008. Observations on land use and agriculture in Kaneš. In *Old Assyrian studies in memory of Paul Garelli*, Old Assyrian Archives Studies 3. Publications de l'Institut historique-archéologique néerlandais de Stamboul 112, ed. C. Michel, 139–157. Leiden: Nederlands Instituut voor het Nabije Oosten.
Ebeling, E. 1913. Altbabylonische Briefe (Suite). *Revue d'Assyriologie et d'Archéologie Orientale* 10: 105–155.
Eisser, G., and J. Lewy. 1933. *Die altassyrischen Rechtsurkunden vom Kültepe*, Mitteilungen der Vorderasiatischen Gesellschaft 33 and 35 (1930–1935). Leipzig: J. C. Hinrich.
Figulla, H. H. 1953. *Letters and documents of the old Babylonian period Ur excavation texts* 5. London/Philadelphia: The Trustees of the Two Museums.
Günbattı, C. 2004. Two treaty texts found at Kültepe. In *Assyria and beyond. Studies presented to Mogens Trolle Larsen*, Publications de l'Institut historique-archéologique néerlandais de Stamboul 100, ed. J.G. Dercksen, 249–268. Leiden: Nederlands Instituut voor het Nabije Oosten.
Halévy, J. 1874. Observations critiques sur les prétendus Touraniens de la Babylonie. *Journal Asiatique* juin 1874: 461–536.

Huehnergard, J. 1997. *A grammar of Akkadian*, Harvard Semitic Museum Studies 45. Atlanta: Scholars Press.

Jestin, R. 1952. Textes économiques de Mari (IIIe dynastie d'Ur). *Revue d'Assyriologie et d'Archéologie Orientale* 46: 185–202.

Larsen, M. T. and Møller, E. 1991. Five old Assyrian texts. In *Marchands, Diplomates et Empereurs. Études sur la Civilisation Mésopotamienne Offertes à Paul Garelli*, eds. D. Charpin and F. Joannès. Paris: Editions Recherches sur les Civilisations: 227–252.

Leemans, W.F. 1954. é-DUB-ba dans les textes économiques de Larsa. *Revue d'Assyriologie et d'Archéologie Orientale* 48: 57–66.

Lehmann-Haupt, C.-F. 1889. *Über altbabylonisches Maass und Gewicht und deren Wanderung*, Verhandlungen der Berliner Gesellschaft für Anthropologie, Ethnologie und Urgeschichte. Berlin: Berliner anthropologische Gesellschaft.

Lenormant, F. 1868. *Essai sur un Document Mathématique Chaldéen, et à cette Occasion sur le Système des Poids et Mesures de Babylone*. Paris: A. Lévy.

Lion, B., and C. Michel, eds. 2009a. *Histoires de Déchiffrements. Les Écritures du Proche-Orient à l'Égée*, Les Hespérides. Paris: Editions Errance.

Lion, B., and Michel, C. 2009b. Jules Oppert et le syllabaire akkadien. In *Lion and Michel 2009a*: 81–94. http://halshs.archives-ouvertes.fr/halshs-00781407.

Meissner, B. 1893. *Beiträge zum Altabylonischen Privat Urkunden*. Leipzig: J.C. Hinrichs.

Michel, C. 1992. Les fractions dans les tablettes économiques du début du second millénaire en Assyrie et en Babylonie. In *Histoire de fractions, fraction d'histoire*, Science Networks: Historical Studies X, eds. P. Benoit, K. Chemla, and J. Ritter, 87–101. Basel: Springer. http://halshs.archives-ouvertes.fr/halshs-00821242.

———. forthcoming. Weighing units and weights in the context of trade from Upper Mesopotamia to Anatolia (nineteenth and eighteenth centuries BCE). In *Cultures of quantification and computation*, eds. K. Chemla, A. Keller, and C. Proust. Cham: Springer.

Oppert, J. 1894. Les mesures de Khorsabad. *Revue d'Assyriologie et d'Archéologie Orientale* 3: 89–104.

Powell, M.A. 1990. Maße und Gewichte. In *Reallexikon der Assyriologie*, ed. D.O. Edzard, 457–530. Berlin: Walter de Gruyter.

Proust, C. 2009. Numerical and metrological graphemes: From cuneiform to transliteration. *Cuneiform Digital Library Initiative Journal* 2009: 1. http://.cdli.ucla.edu/pubs/cdlj/2009/cdlj2009_001.html.

Scheil, V. 1902. *Une Saison de Fouilles à Sippar*. Cairo: Imprimerie de l'Institut français d'archéologie orientale.

———. 1918. Notules XLIX La mesure *(giš) BA-AN*. *Revue d'Assyriologie et d'Archéologie Orientale* 15: 85–86.

Sigrist, M. 1984. *The Sattukku in Esumesa during the Isin and Larsa period, Bibliotheca Mesopotamica* 11. Malibu: Undena Publications.

Sollberger, E. 1965. Three Ur-dynasty documents. *Journal of Cuneiform Studies* 19: 26–30.

Spar, I. 1988. Notation system for capacity measures. In *Cuneiform texts in the Metropolitan Museum of Art*, ed. I. Spar, vol. 1. New York: The Metropolitan Museum of Art.

Thureau-Dangin, F. 1894. La comptabilité agricole en Chaldée au IIIe millénaire. *Revue d'Assyriologie et d'Archéologie Orientale* 3: 118–146.

———. 1904. Contrats archaïques provenant de Šuruppak. *Revue d'Assyriologie et d'Archéologie Orientale* 6: 141–154.

———. 1909. L'U, le Qa et la Mine, leur mesure et leur rapport. *Journal Asiatique* 1909: 79–90.

———. 1911. Ašduni-erim roi de Kiš. *Revue d'Assyriologie et d'Archéologie Orientale* 8: 65–79.

———. 1912. Notes assyriologiques, XVII. La mesure du qa. *Revue d'Assyriologie et d'Archéologie Orientale* 9: 24–25.

———. 1921. Numération et métrologie sumériennes. *Revue d'Assyriologie et d'Archéologie Orientale* 18: 123–142.

# Chapter 9
# Reduction of Absurdity: Notes on the Editorial Transformations of Greek Diagrams

**Reviel Netz**

**Abstract** This chapter studies a subsection of Greek mathematical diagrams: the diagrams used for proofs by contradiction. In some of those proofs, the configuration required for the assumption of the proof is visually impossible. It is shown that manuscript diagrams tend to emphasize the impossibility of such configurations. This is different from many modern editions, where the diagram tries to represent the case as close to plausibility as is possible. In general, it is noted that such diagram traditions tend to be stable, in both manuscript and print transmissions. The major rupture in the tradition occurred during the Renaissance, in both print and manuscript, and seems to represent a specifically Renaissance attitude of disdain towards the medieval transmission.

How did modern editors transform early Greek diagrams? We start from the evidence. In the extant manuscripts—almost entirely from the Middle Ages and the Renaissance[1]—we find diagrams accompanying ancient Greek mathematical texts. Our questions are straightforward: are these diagrams the same as those in modern editions? If not, how were they transformed?

These questions are straightforward but even 20 years ago, we could barely discuss them. Until recently, hardly any scholarly attention was paid to the ancient diagram and so we had to put together the very basic pieces of information. Since then, several scholars have theorized and collected a significant amount of data. Though this project is still far from completion, a general approach may now be

The research leading to these results has received funding from the European Research Council under the European Union's Seventh Framework Program (FP7/2007–2013)/ERC Grant Agreement No. 269804.

[1] The handful of exceptions are the Euclid papyri—not all yet published—in which a little over fifteen diagrams are attested in fragmentary, mediated form (see Note 4).

R. Netz (✉)
Department of Classics, Stanford University, Stanford, CA, USA



A. Keller, K. Chemla (eds.), *Shaping the Sciences of the Ancient and Medieval World*, Archimedes 69, https://doi.org/10.1007/978-3-031-49617-2_9

characterized as the consensus view among scholars of ancient diagrams—it is perhaps, however, not sufficiently well known beyond this very narrow circle. It will be useful, therefore, to present again this consensus view.

This chapter, then, does two things. First, it discusses the general practices of the ancient diagram and its transformations. Second, it tries to account for those transformations by offering a specific case study.

The case study chosen here is an extreme, if rare, case. These are the diagrams of the proof by contradiction or 'reduction to absurdity', where, based on a hypothesis, an absurd result is obtained, thus refuting the hypothesis. In this group of manuscript diagrams, we see a tendency to maximize the absurdity of the resulting situation by depicting a manifestly impossible configuration. Conversely, in modern editions, we see a tendency to reduce such absurdities, to make the drawing appear less manifestly impossible. This speaks to a new meaning of the mathematical diagram.

The structure of the paper is as follows. In Sect. 9.1, I present the ancient practice while postponing (for reasons which will become clear below) the question of its reconstruction—that is, ignoring manuscript variations, and presenting my reconstruction of source diagrams in dogmatic fashion. In Sect. 9.2, I begin by describing the transformations effected by modern editors. After I produce this evidence, I finally bring in the variation among the extant Renaissance manuscripts. A key observation is that the variation among the Renaissance manuscripts is similar (though not identical) to the transformation found in modern printed editions and so we need to understand the nature and historical meaning of this continuity between late manuscripts and print.

## 9.1 Early Diagrams of Proof by Contradiction: Stating a Practice

### *9.1.1 The Overall Case of Diagrams, Ancient and Edited*

As noted above, I am about to present something of a consensus opinion: it is shared by a number of scholars who have looked, over the last 20 years, into the tradition, in several languages, of the Greek mathematical diagram. This means primarily that this is the view of the series of studies pursued and aided by Ken Saito.[2]

The starting observation is that the extant diagrams in various copies of the same work are usually very similar to each other. We can be therefore absolutely certain that the great majority of manuscript diagrams were copied from a shared archetype

[2]The essential texts are Saito (2006) and Saito et al. (2014), best accessed through http://www.greekmath.org/diagrams/diagrams_index.html, accessed 18 August 2017. References to "Saito", without date attached, are to this website. Also important are Saito (2012), Saito and Sidoli (2012), De Young (2005), van Leeuwen (2012). My own contribution is Netz (1999, 2004, 2012).

rather than drawn without reference to a visual source. In practice, then, a shared archetype can be reconstructed (with caveats that extend even beyond the normal ones of text criticism).

What do we find in this shared archetype? Saito and Sidoli (2012) list three central areas:

1. Overspecification. What is meant by this is that the diagrams are more particular (more 'specified', as a consequence) than the text. Thus, when the text mentions a 'triangle' without further specification, the diagram will typically depict an isosceles triangle. The tendency indeed is towards more symmetrical and simpler objects.
2. Indifference to visual accuracy. This means primarily that the diagram gets some measured property 'wrong'. In the more trivial cases, lines that are specified to be equal (or unequal) to each other are drawn as unequal (or equal). More interesting cases involve 'wrong' angles or even wrong curvatures (a straight line represented by a curved line or vice versa).
3. Diagrams in solid geometry. Here the observation is very simple: most solid configurations are drawn in very schematic planar versions that do not support the direct perceptual reconstruction of any solid image.

The various observations cohere around a clearly defined practice: diagrams are not pictures but are instead schematic representations. All of this should be understood as a non-controversial description of the evidence found in the extant sources.

What is the historical significance of such observations? First of all, how far back do they go? Many scholars tend to think we have evidence essentially for an early practice. Let me state the case for this now in my own voice (not claiming to speak for a consensus: such questions have not been discussed as thoroughly over the last two decades).

For several authors, we have more than a single copy from the central Byzantine centuries (from the tenth to the thirteenth centuries). Advanced mathematical texts are not copied incessantly; furthermore, following stasis in the second half of the sixth century, the Byzantine world spun into catastrophic collapse throughout the early seventh century from which it has never fully recovered.[3] It seems that manuscript-making, especially of sophisticated non-Christian works, severely plummeted from about the mid-sixth century to mid-ninth century. Our sources begin soon thereafter and their shared roots likely extend beyond these troubled times.[4] Thus the likelihood is that the shared archetype of our diagrams is Early Byzantine or indeed Late Ancient.

Now bear in mind, once again, the strong similarity we find between the extant diagrams. This allows our reconstruction in the first place. Another consequence is

[3] For a brief overview, see Wickham (2005: 30–32).

[4] The one exception is the important case of Euclid's *Elements*, which might have been absorbed into educational practice, explaining how its copying did go on through the 'Dark Ages': for, in fact, we have an extant fragment of Euclid's *Elements* (in Palimpsest), the British Library manuscript BL add. Ms. 17,211, from the seventh or eighth century.

that we learn that creativity in the copying of diagrams was the exception and not the norm. As we will see below, the same conservative tendency to preserve diagrams is evident also in the great majority of diagrams in print (which are, once again, we find, typically taken over from previous printed versions). The overall conservative tendency in the transmission of diagrams is easy to account for in practical, cognitive terms: it is easier to copy a model, when available, than to draw a fresh one. Conversely, it is hard to unsee a diagram present to the eye and to ignore it while making one's own version. Either way, we find that it is rare for diagrams to be changed; hence the default assumption ought to be that diagrams were preserved, not radically transformed.

Further, we should note that a handful of ancient diagrams survive on papyrus. These are only from Euclid and even then typically from low-grade classroom productions. Still, it is remarkable that this evidence—small and problematic as it is—does suggest overall that the features we see in reconstructed archetypes are continuous with those of ancient diagrams.[5]

There is more we can say. The schematic character of manuscript diagrams is consistent with ancient scribal practices as a whole (in Greek antiquity, the text, too, was produced with 'schematic', minimalistic elaboration, with only a single 'upper case' set of letters, no variation of color and hardly any of size, and famously no punctuation or even word division[6]). Further, while the practice is strange from the perspective of a modern mathematician, it is a plausible diagrammatic strategy: if diagrams are to be employed as part of the logic of the text, then they ought to be used for their scope of validity, which is, roughly speaking, 'topological'; hence one should rely on diagrams to the extent that they are schematic.[7] Greek diagrams fulfill, in this sense, at least one function that Greek mathematicians needed from their diagrams.

Add all this together and my verdict would be that the evidence we have from medieval sources can be used to reconstruct the ancient practice of diagrams. This puts into sharper relief another basic observation: the diagrams in modern editions are nothing like that. Until quite recently indeed, modern publications (whether in critical editions or elsewhere) typically did not include a critical edition of the

[5] Euclid papyri with diagrams are: P. Berol. inv. 17,469 (I.8–10), P. Fay. 9 (I.40), P. Oxy. I.29 (II.5). Especially important are two unedited papyri, one with a trace of a diagram from Book XIII, the other with a set of diagrams for I.18–25 (for the latter, see Cairncross (2012)). For a survey of the evidence, see Sidoli (2015) and Netz (2015: 299–301). Eunsoo Lee is currently producing a critical edition of, among other things, all the Euclidean diagrams for which papyri evidence is available.

[6] For ancient scribal textual practices as a whole, see Johnson (2004). For the question of how visual illustrations fitted within such simple, minimalistic scribal practices, the answer is that they did not. My argument would be that ancient texts were almost entirely unillustrated, the case of mathematical diagrams forming, then, an exception to the rule. See (Netz 2014: 232–241).

[7] This is a separately emerging consensus concerning the logical function of Greek diagrams, originating from Manders (2008) and Netz (1999) and going back to Poincaré (1912 [1963]). For a recent statement of this position—that schematic diagrams can be made logically valid—see Mumma (2010).

diagrams. Instead, they produced their own diagrams, without comment or reference to ancient manuscript authority (and, of course, without collating and noting manuscript variations). These new diagrams typically are quite removed from the medieval manuscript evidence, though, as noted above, they are conservative in their own way: they are usually similar to other diagrams *in print*. This opposition is the starting point for our investigation.

### 9.1.2 The Diagrams of Proof by Contradiction: A Typology

I am about to zoom into a special, rare practice. The proofs by contradiction themselves are, of course, not rare in Greek mathematics; but for reasons which will become clear below, the issues presented here apply only in a limited range of cases.

In general terms, in a proof by contradiction, we seek to prove P by assuming ~P. So—to take two examples—a Greek mathematical author might assume, for the sake of such an argument, that 'if possible' (as the formulaic expression typically goes) a line is a tangent to a spiral at two points; or that, 'if possible', the spiral area is less (or greater) than one-third of the area of the circumscribing circle. Such are the starting point for Archimedes' *Spiral Lines* 13 and 24, respectively (the aim, of course, is to prove that a line can be tangent to a spiral at one point at most or that the spiral area is exactly one-third of the circle). The proofs take off from these assumptions and their *geometrical* operation ends once a contradiction has been achieved. The diagram required for a proof by contradiction is the one in which the false assumption holds.

This creates a curious situation, since the diagram is *required* to be wrong. If indeed diagrams are expected to be correct pictures of the discussed geometrical objects, we face a genuine conundrum since, in fact, a correct picture of the discussed geometrical object may, in some cases, be literally impossible. This will be true, in particular, whenever the false assumption involves a strictly 'topological' element of the configuration. Let me explain with (appropriately) a counter-example. Take once again *Spiral Lines* 24: the false assumption here is that the spiral area is greater (or smaller) than one-third of the circumscribing circle. But in reality, we never seek to draw true spirals (spirals in the diagram tradition are drawn as a sequence of arcs of circles of widening radii).[8] Furthermore, diagrams do not necessarily seek to display correctly metrical relations, not even *equalities* between *line segments*, and most definitely not *inequalities* between *areas*. Even more: since the point of *Spiral Lines* 24 is precisely that the difference between the spiral area and one-third of the circumscribing circle can be made smaller than any assigned magnitude; it is, in fact, in principle impossible to represent this inequality even assum-

[8]This is what we find in the Medieval tradition of Archimedes' *Spiral Lines*, a practice continuous with that of the representation of conic sections as combinations of arcs of circle (see below). It is interesting that the spiral built into the back of the ancient Antikythera Mechanism (first century BCE?) was also constructed as a series of such circle-arcs (Anastasiou et al. 2014).

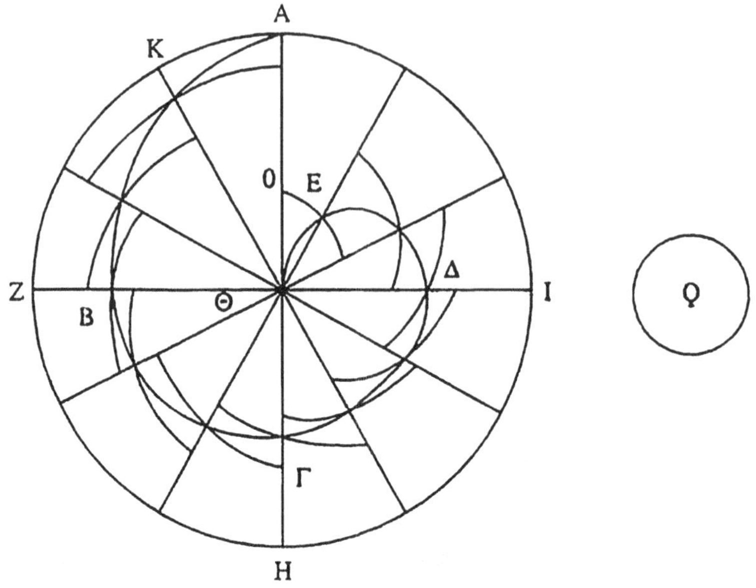


**Fig. 9.1** *Spiral Lines* 24, reconstruction of ancient diagram

ing a perfectly measurable diagram. In short, the false assumption of the diagram has to be, in this case, visually indistinguishable from the correct depiction of reality. Here is the diagram to *Spiral Lines* 24, reconstructed from the manuscript evidence (Fig. 9.1).

Does the Spiral Line area 'look' equal to the circle on the side? Greater? Smaller? This is, in fact, completely immaterial to the proof.

Now contrast this with *Spiral Lines* 13. The manuscript diagrams, we pointed out above, are schematic and 'topological'. And it is precisely at this level that the false assumption of this proposition fails. A line touching a spiral line only once is visually distinct, even in the schematic terms of the manuscript diagram, from a line touching it twice. The diagram can show the difference: but the entire point of the proposition is that the false case cannot, in fact, be produced unless the diagram is wrong. It is visually possible but geometrically impossible to draw *Spiral Lines* 13 as a manuscript diagram.

As usual, it may be useful to consider the contrast between manuscript and print diagrams so as to present the various choices available. For Archimedes' *Spiral Lines*, we have two independent medieval traditions, that of A (lost, but copied frequently, especially in the Renaissance) and C (the well-known Palimpsest). While there is some variety in the copies of A, to which I shall return below, sufficiently many of them—and precisely those that appear overall to be the less creative, visually—cohere around the same configuration, which may therefore be safely reconstructed. This is exactly the same as that of Codex C. We therefore can reconstruct an Early Byzantine/Late Ancient source as below (Fig. 9.2):

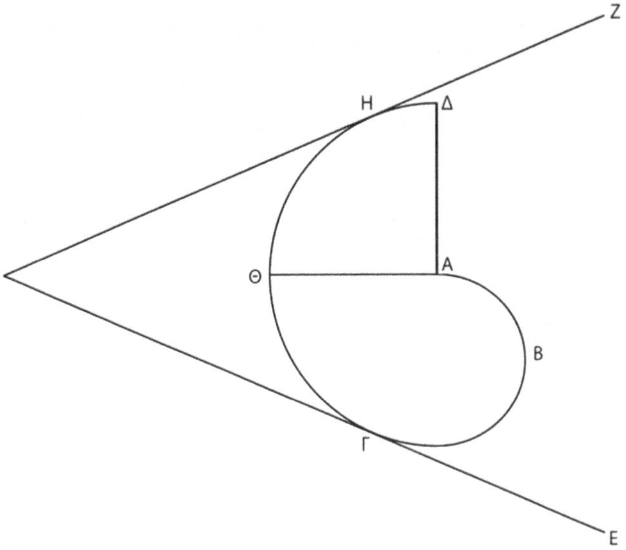


**Fig. 9.2** *Spiral Lines* 13, reconstruction of ancient diagram

**Fig. 9.3** *Spiral Lines* 13, Heiberg's (1913) diagram

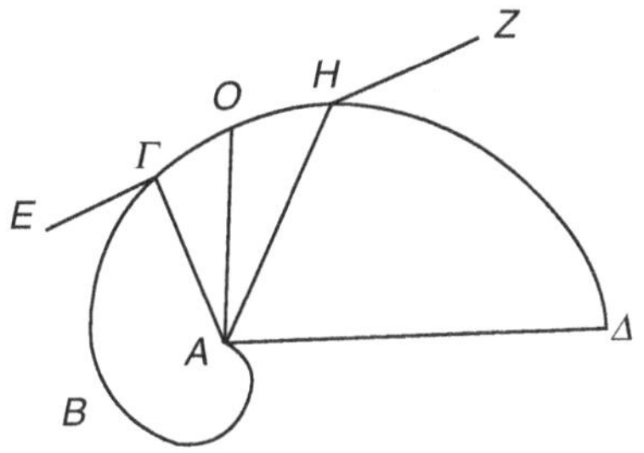


This may be contrasted, for instance, with Heiberg's 1913 drawing (which is similar to that of most diagrams in print)[9] (Fig. 9.3).

This contrast is what I refer to in my title as 'reduction of absurdity'.

The manuscript diagram dramatizes the impossibility of the configuration, drawing an intended tangent as a broken line. It is, indeed, broken with a rather sharp angle, well beyond what the conditions of the configuration demand. The print diagram, conversely, draws the same tangent in the least problematic manner possible, so that the line of the tangent is indeed a line. This diagram effaces the difficulty of the situation so effectively indeed that it becomes untrue to the text in an important topological sense, in that instead of touching the spiral at precisely two points, the tangent instead seems to merge with the spiral for a certain length. Visual reasonableness is here valued above topological (hence, logical) correctness.

[9] In this particular case, Heiberg's diagram seems to have also been mislabeled (O instead of a Θ), but this is immaterial to our argument.

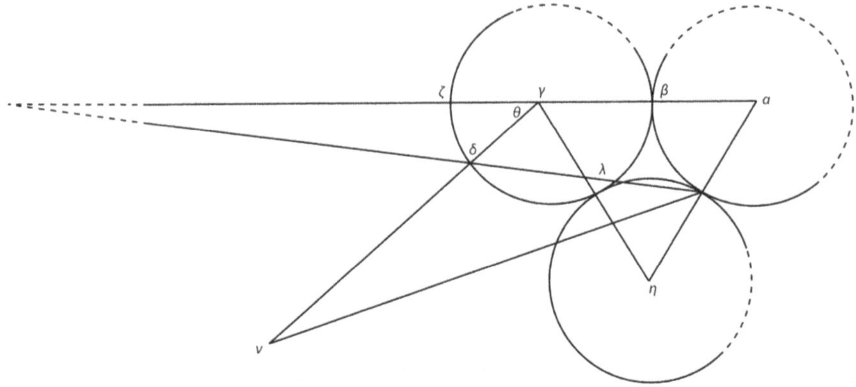


**Fig. 9.4** Pappus IV.13, transcription of manuscript diagram

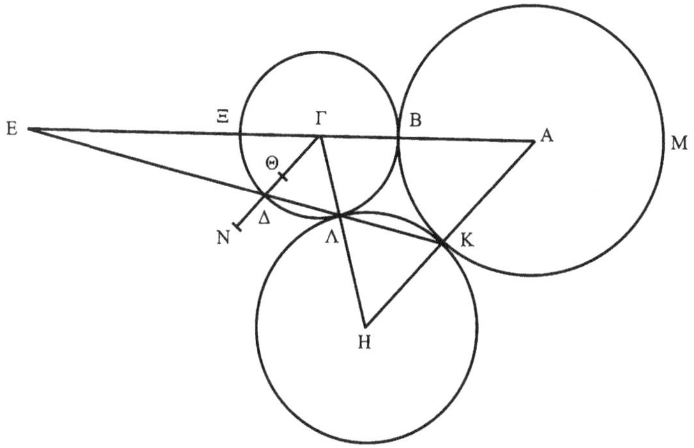


**Fig. 9.5** Pappus IV.13, Hultsch's (1876) diagram

Consider next the following example, from Pappus' *Collection*, IV.13. Here is the manuscript diagram (dotted lines are faded lines that would have been fully visible in the original state of the manuscript) (Fig. 9.4).

And here is the same diagram in Hultsch's critical edition (1876 Vol. I, 211) (Fig. 9.5).

The text of Pappus relies entirely on a single independent manuscript, Vat. Gr. 218. This, however, is a relatively early Byzantine manuscript (tenth century) and both subject matter (demanding) as well as author (obscure) suggest the text was not frequently copied; hence we are even more likely than usual to be proximate to the Late Ancient appearance of the diagram. Since, in this case, the text itself is Late Ancient, the likelihood that we have a diagram close to that of Pappus himself is very significant.

This is good to know, since the difference, in this case, is very subtle indeed. Pappus proves that with three mutually touching circles, as in the figure, with ΓΔ parallel to AK, when AE:EΓ::AB:BΓ, then the line KE passes through Δ. The proof by contradiction assumes that the line KE cuts the line ΓΔ, instead at some other point Θ. (The argument works by showing that if KE cuts ΓΔ at Θ, we have—ΓΔ being parallel to AK—AK:ΓΘ::AE:EΓ so together with the given proportion, we have AB:BΓ::AK:ΓΘ, but AB=AK while BΓ>ΓΘ). The burden of the proof by contradiction is on the position of the point Θ and, indeed, as it gets closer to the point Δ, the ratio becomes nearly correct as BΘ becomes nearly equal to BΓ. The argument of this proposition is, in a sense, quantitative and the 'falsehood' of the position of Θ is, in a sense, quantitative as well. Topologically, the position of Θ is identical in the two diagrams, the medieval (ancient?) and the modern.

Now, while the manuscript diagram is schematic, it is not, strictly speaking, topological (to begin with, the general tendency is to represent straight lines by straight lines; right angles by right angles; etc.) It is therefore not impossible that the position of Θ—very near Δ or very far from Δ—could be meaningful. If so, we have once again the same contrast: a more absurd diagram in the original (one where the failure of the proportion is manifest), a less absurd diagram in the modern version (one where the failure of the proportion is imperceptible). To be clear: I am not sure that the manuscript's position should be read as meaningful (though I find this possible, indeed likely in this case). But it is quite evident that the printed edition does carry a meaning, reducing the absurdity of the position.

Such cases of a purely quantitative but possibly meaningful difference are fairly common in the diagrams of proofs by contradiction. Strictly topological ones are less common. I add a couple more topological examples:

The following is the figure for Apollonius' *Conics* I.31 from Vat. Gr. 206 (according to Heiberg, our only independent manuscript of Apollonius, on which all other manuscripts depend; see Heiberg (1893 Vol. II: lvi)) (Fig. 9.6).

This twelfth-century manuscript, as noted above, is effectively our only source for the Greek text of the *Conics*. This is something of a concern, as we are no longer in early Byzantium, and since Apollonius was not merely a Hellenistic author—thus far removed in time from Byzantium—but also a substantially re-studied one, the subject of an extensive (and apparently intrusive) commentary and edition by the Late Ancient author Eutocius.[10] In this case, then, all we can say is that we likely see the diagram of Eutocius' sixth-century edition.

What we find then is that Eutocius likely drew a diagram (whether independently or based on his sources) with two distinct lines, ΓΔ and ΔE.

This now should be contrasted with Heiberg's (1893) version (Fig. 9.7).

Here, the two line segments become a single line. Now, this proposition involves once again a proof by contradiction. Γ is on the diameter of the hyperbola AB such that AΓ<ΓB; it is to be proved that if a line such as ΓΔ meets the hyperbola at a

[10] On the question of Eutocius and Apollonius' diagrams, see Decorps-Foulquier (1999); for a general statement of the authorial difficulties concerning Apollonius, see Sidoli (2011).

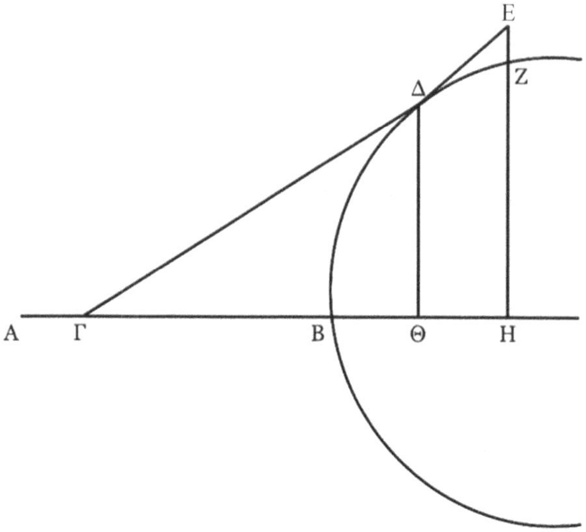


**Fig. 9.6** *Conics* I.31, manuscript's diagram

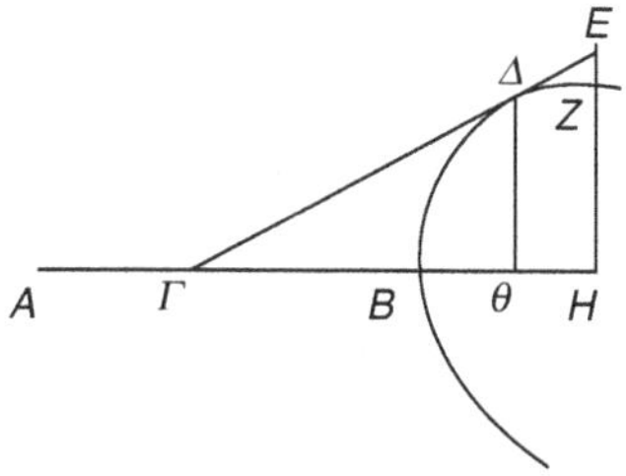


**Fig. 9.7** *Conics* I.31, Heiberg's diagram

point Δ, then it will cut the figure (and not be a tangent at the point Δ). The proof then assumes that ΓΔE is such a line falling outside the hyperbola (but touching it at Δ) and draws a contradiction. The precise arrangement of the points Γ, Δ and E is, in this case, immaterial to the proof (which relies instead on the proportions of the hyperbola with the ordinates ΔΘ, ZH). It is thus immaterial how we draw the sequence of point Γ, Δ and E, as long as E is drawn outside the hyperbola.

There is no difficulty in drawing a 'false' diagram such as the proposition requires. In fact, as noted above, conic sections are drawn in medieval manuscripts uniformly as one or more arcs of circles (in this case, a single one) and all that is called for, then, is to draw ΓΔ as a tangent to a given arc of circle, which, of course, can be done regardless of our choice of position for Γ. It is thus very interesting that Eutocius' diagram produces an *unforced* dramatization of the impossibility of the configuration. It is feasible to draw a visually 'correct' picture satisfying the condition of the refutation but the manuscript diagram creates an exaggerated absurdity by drawing a single line as broken. Heiberg (1893), we note again, removed this and so reduced the absurdity.

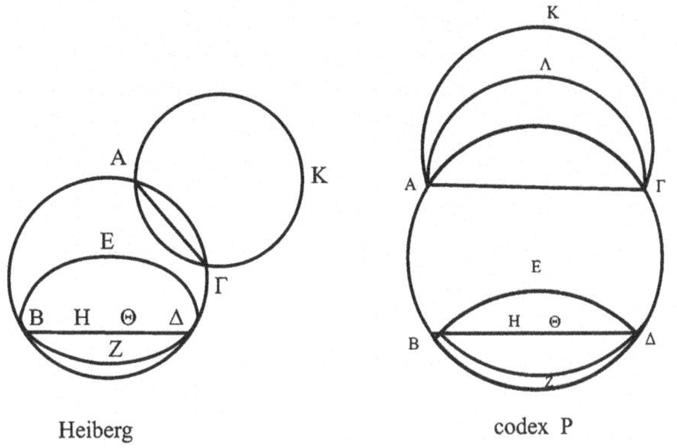


**Fig. 9.8** *Elements* III.13, Saito's comparison

The most striking case of reduction of absurdity was noted by Saito. This involves *Elements* III.13. The following diagram is taken from Saito (2012) and I adapt his analysis (Fig. 9.8).

Note that the diagram of Codex P, chosen by Saito for his analysis, is essentially the same as that of the remaining manuscript diagrams.

What we have here is a proof that a circle does not touch another at exactly two points, which is shown through a proof by contradiction. Specifically, assume the two circles do so externally, so that circle ΑΒΓΔ is touched externally by circle ΑΛΓΚ (once again, Heiberg (1893) does not even print Λ; this is not material to our argument). Now, a line joining two points on the circumference of a circle is internal to that circle so that the line ΑΓ must be internal to both circles, ΑΒΓΔ, ΑΛΓΚ, against the (case) assumption that the touching is external.

The contrast between the manuscript diagram and the print one is easy to understand within the terms of the reduction of absurdity. The manuscript diagram, in depicting an impossible arrangement of the external circle, made it into an impossible circle—in fact, a crescent. The print diagram re-circled the circle but, in the process, messed the argument up completely which, in fact, relied on the external circle being depicted as entirely external. For if the external circle ΑΛΓΚ is allowed to creep within the main circle ΑΒΓΔ, then we also allow the line ΑΓ to become internal to both circles so that the argument is simply false.

This case is interesting (and highlighted by Saito (2012)) because it is so obvious that, in this case, mathematical reason is with the manuscript diagram—making it easier for us to trust the manuscript against its modern revision. This is a very rare state of affairs. Generally speaking, arguments in Greek mathematics do not rely so transparently on visual evidence based on the particularities of the diagram, and it is almost unheard-of for modern editors to transform diagrams against the grain of the argument itself.

We may sum up our survey so far with a typology of diagrams for the proofs by contradiction:

1. The conviction of impossibility relies on the visual evidence of an impossible configuration (our example was *Elements* III.13). In such a case, modern editors who reduce the absurdity are quite simply *wrong*.
2. The *reductio* hypothesis is impossible in a way that absolutely forces a non-pictorial representation but for which the source of evidence is independent of the diagram (our example was *Spiral Lines* 13). Generally speaking, the early diagrams emphasize the absurdity; modern editors reduce it.
3. The *reductio* hypothesis is qualitative and involves the configuration; it is, however, not absolutely necessary to represent it falsely, as it is representable by the choice of somewhat 'wrong' quantitative parameters (our example was *Conics* I.31). What we find is that early diagrams do not avail themselves of the possibility of drawing such diagrams in a plausible, pictorial manner, but modern diagrams do.

Cases 1–3 are what interest us most and they are absolutely rare.

4. The *reductio* hypothesis is quantitative and does not involve the configuration; its quantitative difference from the 'correct' case may, however, be implied by the diagram (our example was Pappus IV. 13). We found that modern editors tend, in some cases, to reduce the absurdity of such diagrams as well. Such cases are not rare in the extant evidence; their significance, however, is not clear, as I will note below.
5. The *reductio* hypothesis is quantitative and is, in principle, not distinguishable by a diagram (our example was *Spiral Lines* 24). This is once again a common case in extant Greek proofs by contradiction and it is one that does not interest us here.

The typology in place, I now move on to discuss the meaning and relative frequency of ancient and modern practices.

### 9.1.3 The Diagrams of Proof by Contradiction: Preliminary Notes

What are the exceptions to our rule? Even in our small subset of Cases 1–3, reduction of absurdity is not universal. Consider *Elements* I.27, where a weaker but provable version of the parallels postulate is proved: if a line meets two other lines so that the alternate angles are equal, the two lines will be parallel. This falls under Case 2 above. The result concerns the topology of incidence and it is naturally shown through a proof by contradiction, but the argument does not rely directly on visual evidence. The impossibility is, however, represented in the manuscripts graphically, as usual, through broken lines. In this sense, the diagram is of a piece with the evidence seen so far. However, the printed editions maintain the ancient practice: absurdity unreduced. I put side by side Paris 2466 and Heiberg (1883) (Fig. 9.9):

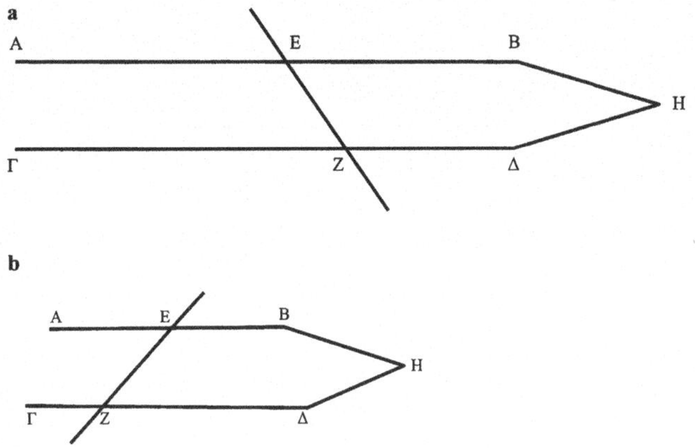


**Fig. 9.9** *Elements* I.27, (**a**) manuscript versus (**b**) Heiberg's (1883) diagram

One could have tilted the parallel lines slightly; or one could have had their point of contact set at a distance, signifying a meeting beyond the boundaries of the shown diagram; or the lines could have been slightly curved. Each of these options would have made the impossibility less present in visual terms. The ancient diagram did not avail itself of such absurdity-mitigating options. Why the moderns do not deviate from the ancients, in this case, is more difficult to say. Perhaps the very iconicity of this representation (having to do with no less than the Parallels Postulate) preserved it from modern transformation.

Case 4 is hard to interpret based on our evidence. There are many cases where the diagram of the proof by impossibility is indeed somewhat more absurd, in quantitative terms, than required. Thus, for instance, *Elements* XI.3. The proposition argues, by impossibility, that the intersection of two planes is a straight line. Otherwise, we could draw two other straight lines from point Δ to point B, each in its own plane, and they would have enclosed an area, hence the absurdity. It is evident in this case that the enclosed area is more 'puffed up' in the manuscripts than it is in Heiberg (1885) who has the non-lines, as it were, somewhat embarrassed to admit that they are different from each other, huddling together instead as close as they can (Fig. 9.10).

Once again, the tendency of ancient diagrams to emphasize such cases is perhaps real, but modern reductions are not universal. Consider now Apollonius' *Conics* II.36: that the line joining the middle points of opposite parallel lines in the opposite sections (two 'branches' of a hyperbola, from our perspective) is a diameter. The proof is that if such is not the case, then another line is (Fig. 9.11).

This would be the line HK. The absurdity resides in that we now need EK=KZ, which can be shown to be false. The diagram makes this quite evident, indeed unfairly so—could it not take a point K nearer the middle so that the inequality would not appear evident to the eye? But compare this to Heiberg (1891) (Fig. 9.12).

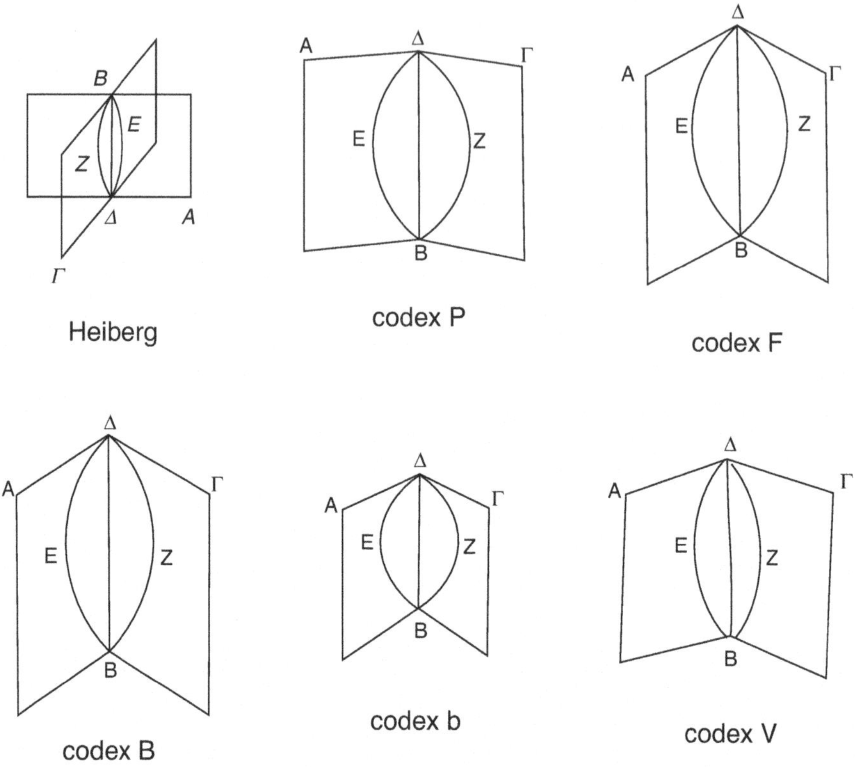


**Fig. 9.10** *Elements* XI.3, diagrams from (**a**) Heiberg (1885) and (**b**) manuscripts, following Saito (2012)

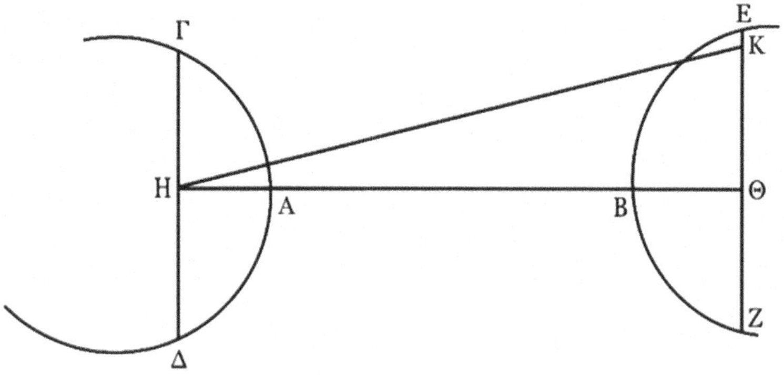


**Fig. 9.11** *Conics* II.36, manuscript's diagram

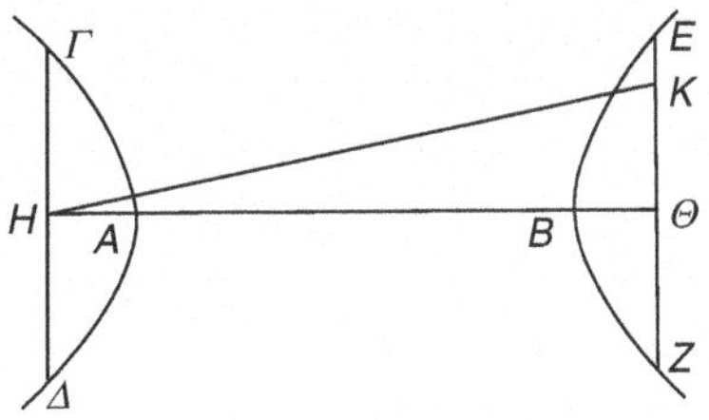


**Fig. 9.12** *Conics* II.36, Heiberg's (1891) diagram

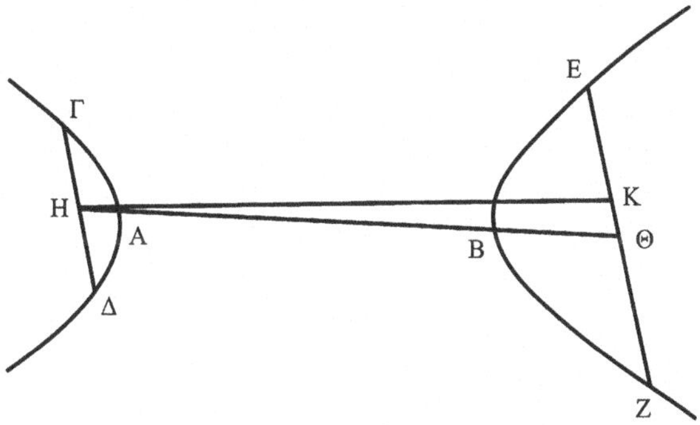


**Fig. 9.13** *Conics* II.36, Halley's diagram

Now, the modern editor did not transform the basic character of the ancient diagram and if Heiberg's K is somewhat nearer the center than the manuscript's, this is because his labels are bigger and so he requires a lower resolution. We see a tendency of Heiberg, in his 1891 edition of Apollonius, to remain (generally speaking) fairly close to the manuscript evidence (without, however, providing critical notes on the diagrams). It is interesting therefore to compare this to the diagram of Halley's edition from 1710 (Fig. 9.13).

Here, we see a clear case of the quantitative reduction of absurdity.

The most prolific setting for genuine topological absurdities is in Apollonius' *Conics* Book IV. This book is related in subject matter to theorems such as Euclid's *Elements* III.13. If Euclid studied the meetings of circles, Apollonius extends this to study the meetings of conic sections as well (this observation was made indeed already in Eutocius' commentary on *Conics* IV, who even pointed out that Apollonius' reliance on proof by impossibility continues Euclid's own practice).[11] Indeed, a couple of propositions (35, 37) come quite close to the reliance of *Elements* III.13 on visual evidence as the grounds of the proof itself. *Conics* IV.35 is a direct extension of Euclid's earlier result: 'a section of a cone or a circumference of a circle will not intersect a section of a cone or a circumference of a circle, not having its convexity in the same direction, at more than two points'.[12] The argument is really not much more than an inspection of the diagram, in the manuscript (Fig. 9.14):

[11] Heiberg (1893: Vol. II, 354, 8–11).

[12] Fried (2002: 37).

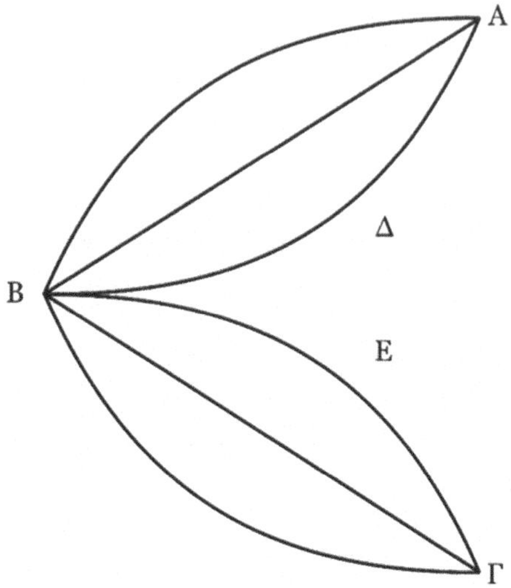


**Fig. 9.14** *Conics* IV.35, manuscript's diagram

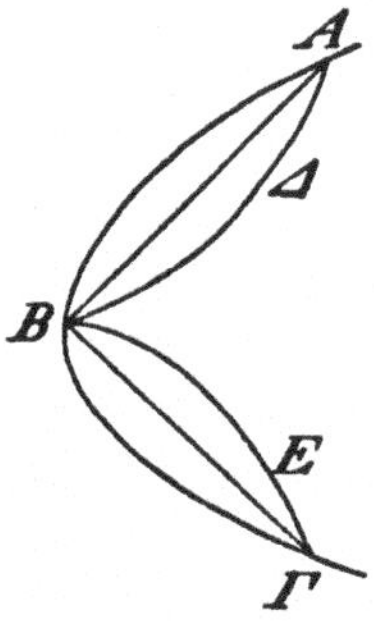


**Fig. 9.15** *Conics* IV.35, Heiberg's diagram

This is very close to the crescent-shape circle in *Elements* III.13 but, in this case, it is only slightly 'normalized' by Heiberg (1893) (Fig. 9.15).

Though once again, we should note Halley's (1710) rather surprising figure (Fig. 9.16).

What sets Halley's figure apart from that of the manuscripts (or of Heiberg) is that his 'false' conic section is everywhere differentiable. Conversely, we recognize that it is the rupture within the curve that dramatizes the absurdity in the manuscripts diagram. Bear this in mind as we consider *Conics* IV.26: sections (or circles) that touch at one point meet each other at no more than two other points. In the manuscript diagram below, A is the point of touch and—the false hypothesis—the two sections cut each other further at B, Γ and Δ (Fig. 9.17).

One of the sections is represented by the nearly complete circle passing through A, B, Γ, Δ. The other 'false' section is represented by the straight line ΔP continued by the arc ABP. Here, even Heiberg 1893 flinches and produces instead a pair of two continuous curves (Fig. 9.18).

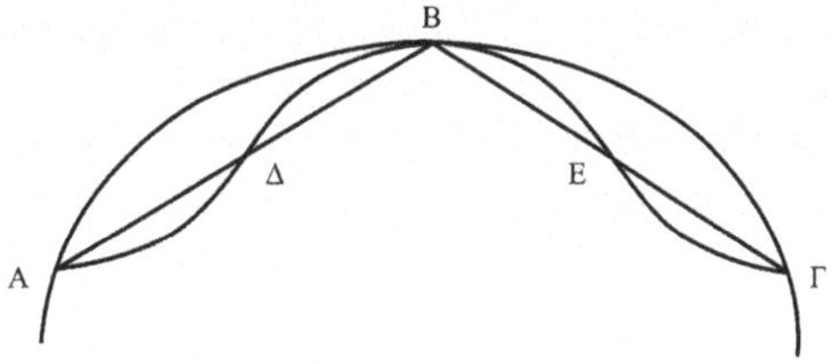


**Fig. 9.16** *Conics* IV.35, Halley's diagram

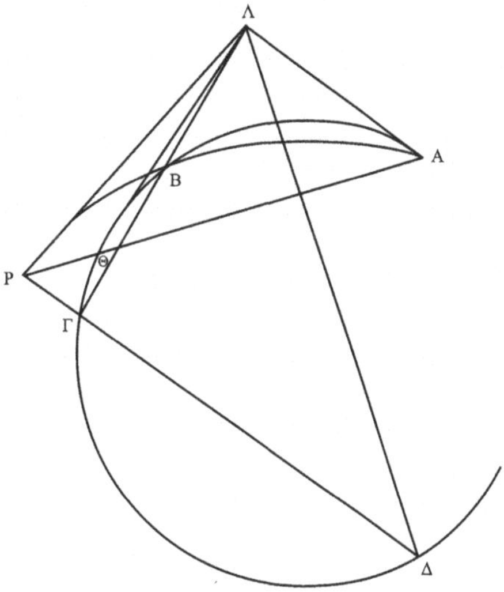


**Fig. 9.17** *Conics* IV.26, manuscript's diagram

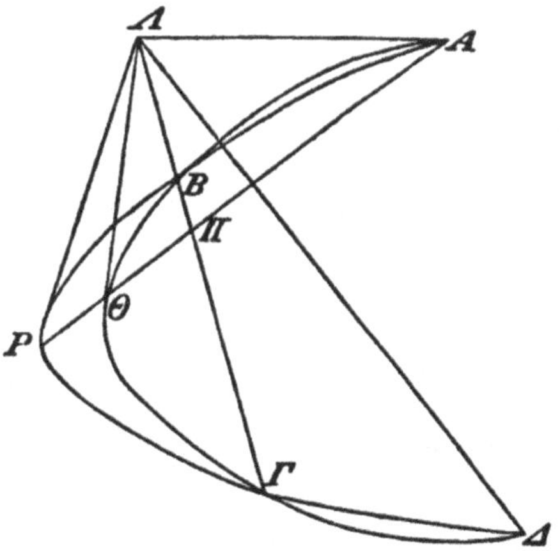


**Fig. 9.18** *Conics* IV.26, Heiberg's 1893 diagram

Heiberg's (1893) originality here may not represent primarily an interest in reducing absurdity. Rather, in general, Heiberg (as the moderns do almost invariably) did draw conic sections as continuous curves and not as arcs of circle.

The variety of ways in which conic sections were represented in the manuscript tradition—as one or more circular arcs—produced certain possibilities for the diagram-maker and I suggest that it was used, once again, in a particular set of proofs to emphasize the absurdity of a configuration.

The opposite sections are a perfectly symmetrical figure and are typically represented as such, for instance *Conics* I.29 (not a proof by contradiction), where we can see the symmetrical arrangement of two equal circles (Fig. 9.19).

Now, many propositions near the end of Book IV deal with the impossibility of the intersecting and touching between opposite sections. A typical figure is that of one of the cases of IV.50, 'if a hyperbola touches one of two opposite sections at one point, the opposite section of the hyperbola will not intersect the other opposite section at more points than two' (Fig. 9.20).

The skeleton of the arrangement is two 'normal' opposite sections, represented by two circular arcs. A hyperbola touches the left of these 'normal' opposite sections, represented by two arcs of circle that are very slightly 'off' at the point A, so that the impression is once again of a circular arc (indeed, on the left side of this diagram, it is not immediately visually obvious which of the two bottom circle arcs is supposed to align with which of the two top circle arcs). On the right-hand side, however, the opposite section to that hyperbola is produced by two very unrelated arcs of circle, indeed not even centered on the axis or symmetrical with each other.

**Fig. 9.19** *Conics* I.29, manuscript's diagram

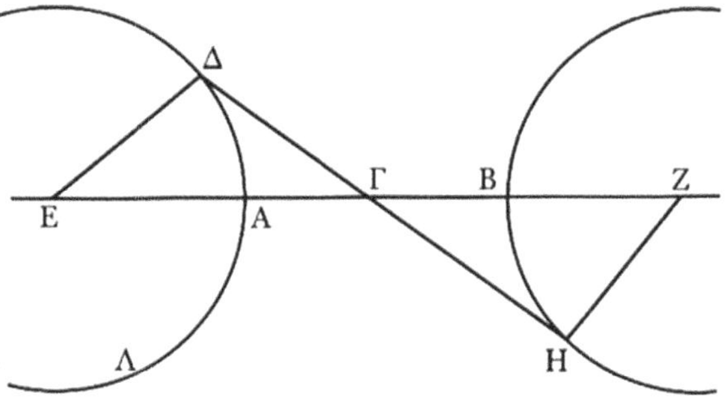


**Fig. 9.20** *Conics* IV.50, manuscript's diagram

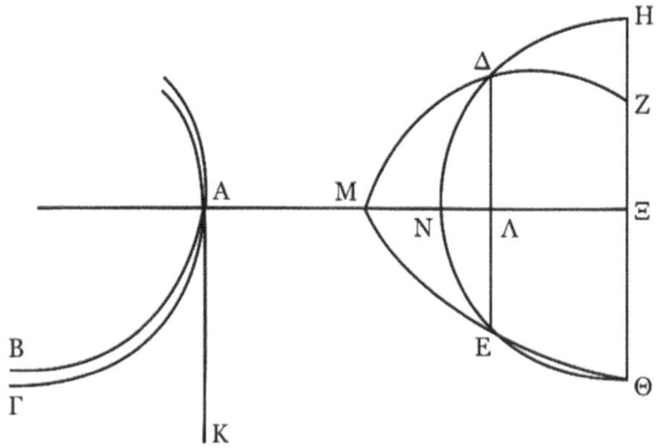

**Fig. 9.21** *Conics* IV.50, Heiberg's (1893) diagram

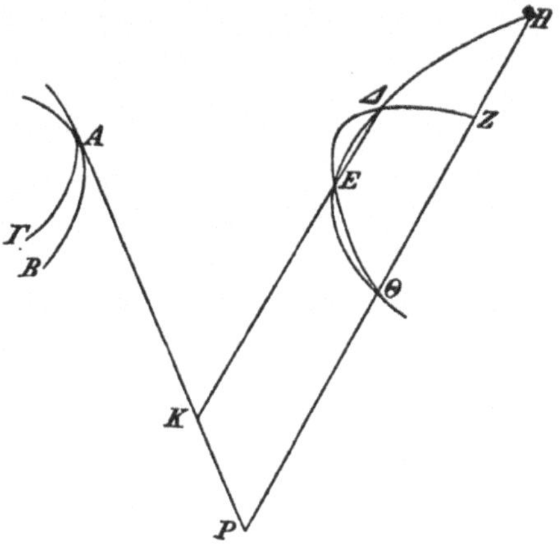


This is, of course, produced so as to provide the required points of intersection, but it would have been very simple for the producer of the diagram to have the left-hand hyperbola, of which this was the opposite, equally made of two highly distinct arcs. Indeed, it is useful to compare Heiberg (1893) at this point (Fig. 9.21).

Heiberg's hyperbolas are more difficult to disentangle, precisely because they are more free-form, but once we recognize which is the opposite of which, we see that with a slight misalignment of the sections Heiberg managed to produce, in fact, two pairs of *symmetrical* opposite sections. The same clearly was possible for the early diagram: once again, then, the modern diagram reduces an absurdity, in this case the absurdity of manifestly asymmetrical opposite sections.

It has been emphasized throughout that the pattern discussed here—manifestly absurd diagrams in the ancient diagrams of proofs by contradiction and its reduction in modern editions—is rare. Let me move on briefly to characterize its incidence. I have surveyed all the geometrical proofs by contradiction in Archimedes, Euclid's *Elements* and Apollonius' *Conics* I-IV. This entire set has about one hundred specimens. Of these, several dozen (especially in Archimedes) belong to Case 5 and do not interest us at all. Of the remaining, the majority belong to Case 4 (e.g., the impossibility will be displayed by having a point closer or further away than its 'real' position, as in the example above from Pappus IV). In such cases, ancient diagrams chose a position 'further away' or more absurd about half of the time. Thus, in Euclid's *Elements*, I find thirty relevant proofs, in fourteen of which the ancient position is markedly absurd, in sixteen it is not. In modern editions, the absurdity of the position is typically (but not always) reduced.

Put differently, for Case 4, in modern diagrams, the position tends to be the least absurd possible, but in the ancient diagrams, there is no systematic tendency to look for a more or less absurd position. This may be interpreted in either of two ways: as a willingness of the ancient diagram to present a more absurd diagram or, alternatively and quite possibly, as evidence that such diagrams should be read non-metrically and thus the more absurd position in the ancient diagram should not be seen as significant. We see here a genuine modern tendency to reduce absurdity, but perhaps no clear, systematic ancient practice.

We therefore need to concentrate, finally, on the cases of more topological absurdities. Indeed, these are absolutely rare. There are, by my count, exactly two in Euclid's *Elements* (I.27 and III.13, both discussed above) and exactly two in Archimedes (*Spiral Lines* 13, as well as one more example to be discussed below, *Floating Bodies* I.8); there are rather more cases in Apollonius' *Conics* I-IV. This pattern may be explained. Greeks discuss questions of a 'topological' character such as the incidence or tangency of lines in a more 'elementary' context. There is relatively little to say, however, about such basic questions with truly simple lines such as straight lines or circles. More complicated curves bring out a more complicated set of relations. Thus, Archimedes' *Spiral Lines* 13 arises in an 'elementary' stage of the treatise, where Archimedes sets up the tools required for the measurement associated with a new curve, the spiral, while Apollonius develops many results of tangency and intersection of the more complicated conic sections. Book IV, as noted above, is especially prolific (from Proposition 24 onwards, almost all propositions are proofs by impossibility involving 'topological' considerations[13]), but there are a handful of such propositions in the first three books, of which I.31 is only an example. The absurdity in such cases is typically not very dramatic (it is not very dramatic in I.31, either) and could often be interpreted alternatively as an example of the pictorial indifference of ancient diagrams. For instance, in *Conics* II.1 the line ΓH is assumed to be, impossibly, a tangent. It is drawn as follows (Fig. 9.22).

Here, the line ΓH does not strike one as a tangent. Indeed, it does not appear as a line at all. The line that is supposed to be drawn from Γ to the curve hits the conic curve 'too early' so that H appears to be free-floating and unrelated to its line. It would have been very easy for the scribe to draw down the line HΘ from the actual point of meeting of the line ΓΔ with the curve. Indeed, since the curve is a circular

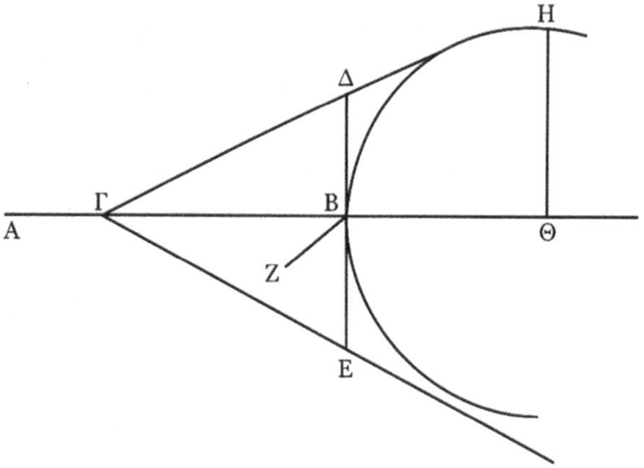


**Fig. 9.22** *Conics* II.1, manuscript's diagram

[13] In the first eighteen propositions, the impossibility is typically of a more quantitative nature. Interestingly, these are all marred in the manuscript by decisive errors: in this case, the early diagrams are clearly wrong.

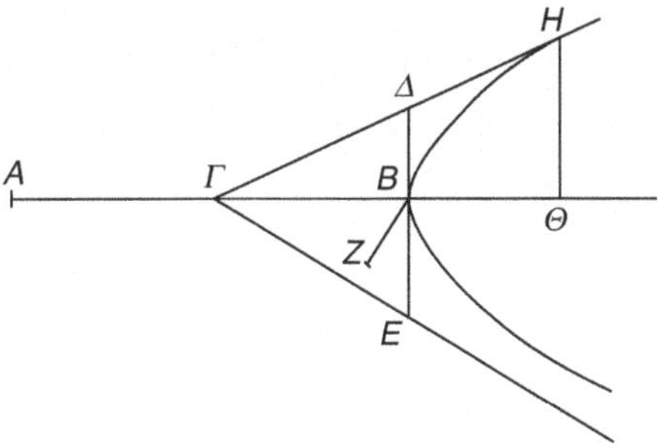


**Fig. 9.23** *Conics* II.1, Heiberg's (1891) diagram

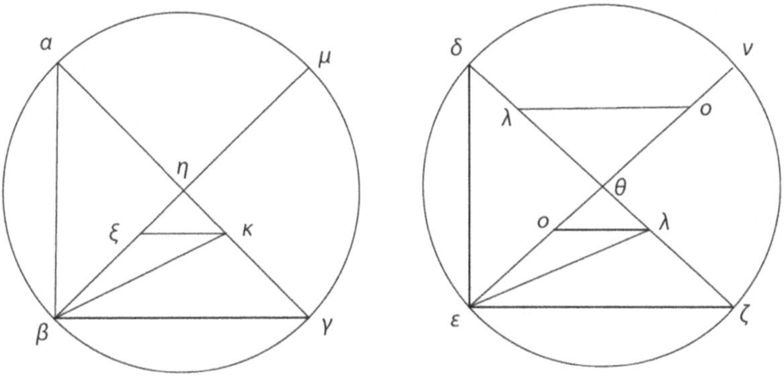


**Fig. 9.24** Pappus VII.216, transcription of manuscript's diagram

arc, it is easy to draw a tangent to it from a given point and, in fact, the line shown as a tangent *is* a tangent. I therefore tend to think that the manner in which the point H is divorced from its proper line is intentional and can be seen as yet another example of an intensified absurdity, though, of course, this is not a settled case. In this case, at any rate, Heiberg (1891) certainly reduced his absurdity (Fig. 9.23).

It should be emphasized that while my complete survey is limited to the three major authors in Greek, I did look haphazardly through Pappus and stopped when I found two examples, one quantitative and the other qualitative. I have shown the quantitative example above; qualitative examples, we now notice, are rare, each with its own character. We should therefore take a look at Pappus VII.216. The following is a transcription of the manuscript diagram (Fig. 9.24).

This is to be contrasted with Hultsch's version in his modern edition (1877: Vol. II, 980) (Fig. 9.25).

The argument at hand involves two right-angled triangles, ABΓ, ΔEZ (represented in the manuscript diagram, as we would expect, by *isosceles* right-angled triangles). The lines BH, EΘ are drawn in the two triangles respectively based on certain quantitative relations and, at the first instance, we need to argue that both lines fall on the same side of the center of the circumscribing circles K, Λ: that is, if H is above the center, so is Θ. We assume the opposite of that to draw an impossibility. This then is a proof of a topological claim, based on impossibility, and the

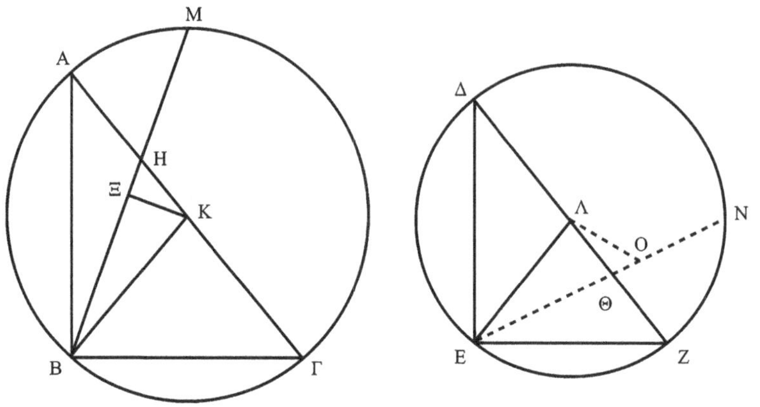


**Fig. 9.25** Pappus VII.216, Hultsch's diagram

manuscript diagram dramatizes this in a truly disorienting way: the two key points H, Θ are drawn at the centers of their respective circles, and it is the centers that are positioned differently in the proof by contradiction stage, one above—Λ the other below—K (the 'correct', lower Λ is in the manuscript diagram for the sake of a following positive argument). This does not involve the more obviously topological rupture of a line broken, as in *Spiral Lines* 13, but the visual appearance of a center of a circle pushed away from its proper position is indeed very evident and so serves, I believe, intentionally to emphasize the absurdity. Surely, the modern editor found it so disturbing (as I do as well) that he removed this effect altogether.

A similar effect, finally, is found in the example with which I wish to conclude. This is from Archimedes' *Floating Bodies* I.8. Archimedes shows, for the first time in this treatise, that a body is unstable under certain conditions, and to show this, he uses, remarkably, the language of the proof of contradiction, allowing that, *if possible*, let the object be tilted in a given manner—remarkable, since the object in fact may tilt!

This is reflected, we see, not only in the language but also in the diagram. In this transcription of the only Greek manuscript diagram (corrected for mislabeled points), the surface of the liquid is ΑΒΓΔ, the immersed object is ΕΖΗ (a spherical segment) (Fig. 9.26).

K is the center of the sphere from which the segment has been taken; Λ is the center of the earth. I propose that this very marked eccentric earth is, in fact, one more case of a proof by contradiction marking itself through an exaggerated visual absurdity, once again, of course, reduced in the modern version, such as Heiberg's (1913) (Fig. 9.27).

But where does Heiberg get his figure? And, in general, how and why did such absurdities become reduced? I move on, then, to a discussion of the transformations of Greek diagrams.

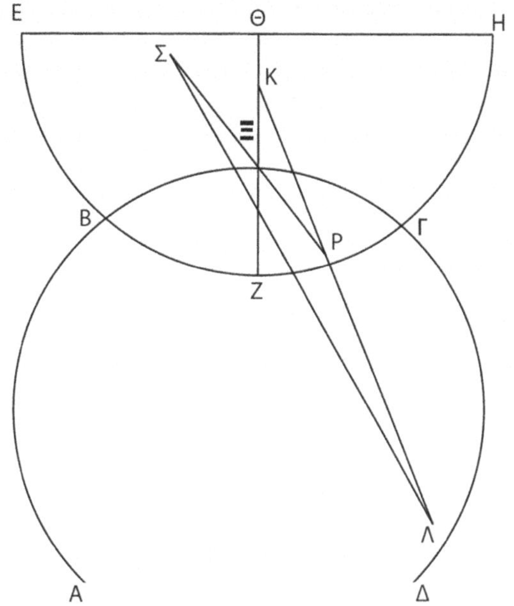


**Fig. 9.26** *Floating Bodies* I.8, reconstruction of ancient diagram based on the Palimpsest

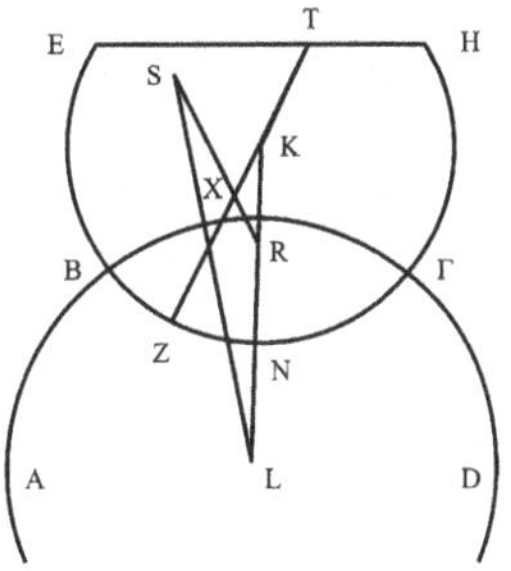


**Fig. 9.27** *Floating Bodies* I.8, Heiberg's (1913) diagram

## 9.2 Early and Modern Diagrams of Proof by Contradiction: Reduction of Absurdity

### *9.2.1 The Print Tradition: Preliminary Notes*

Heiberg looms gigantic over the edition of Greek mathematics and I have appropriately used him above as the emblem of a 'modern edition'. His role, however, in the transformation of early diagrams was minimal—which is to his credit. Having avoided, largely speaking, the task of producing a critical edition of the diagrams, it appears that Heiberg tended to do, as far as the diagrams were concerned, as little as possible, thus avoiding any authorial pretense. He produced little that was new and his small intervention tended in a conservative direction, generally speaking, restoring something of the manuscripts' appearance as *against* modifications in previous printed editions.

In the case of Euclid's *Elements*, Heiberg's intervention was at its minimum, relying on what was closest at hand. As pointed out by Saito and Sidoli (2012), the figures in Heiberg's edition were taken over almost directly from August (1826–1829), a publication produced for use in German gymnasia. This edition—translated into German and bringing in something of the mathematical idiom familiar to school pupils—was yet faithful to the original Greek, not attempting to transform the original text into modern mathematical forms. A dual commitment true to its time and place: technical education alongside philology. This would have been the version most familiar to Heiberg's readers and I imagine that it was the version with which Heiberg himself took his first steps in Greek mathematics.

For Archimedes and Apollonius, the position is more complicated. Nizze (1824) was the functional equivalent, for Archimedes, of August (1826–1829), a school version of the works of Archimedes translated into German, largely a faithful translation rather than a re-casting of the Archimedean material, produced for the same Prussian gymnasia, infused by the same Humboldtian impulse. In this case, however, Heiberg cannot be said to be directly dependent on this set of diagrams and so we need to look more closely at Heiberg's practice (obviously, going beyond the evidence of proofs by contradiction alone). We can point for instance such contrasts: in *Spiral Lines* 18 (the first case) the manuscript figure has two lines approaching each other in such a way that eventually they will meet each other. Nizze (1824) chose to represent them as merely approaching each other, without showing their intersection. Heiberg, however, shows them intersecting (see Fig. 9.29 below). Heiberg follows the manuscript diagram but the manuscripts have them intersecting 'nearby' (at a fairly large angle); Heiberg has them intersecting 'far' (at a fairly small angle.) Thus, Heiberg keeps the manuscript practice but verges towards Nizze's version.

Or consider *Planes in Equilibrium* I.12: both Heiberg as well as Nizze (1824) have two similar triangles but Heiberg's figure, unlike that of Nizze, has the base of the triangle parallel to the base of the page. This is a feature Heiberg takes from the manuscript diagrams but those had the triangles (as one would have expected) also as perfectly isosceles; Heiberg makes them slightly unequal. Such cases can easily be multiplied and what we find is that Heiberg's figures are, overall, similar enough to those of Nizze, and yet it is clear that they were not copied from those of Nizze and indeed there seems to be a certain nod, in Heiberg's diagrams, towards the evidence of the manuscripts (unacknowledged, generally speaking, in the textual apparatus). The same, as noted above, is even more evidently the case with Apollonius, where the figures are indeed very often near-identical to those of the manuscript. Indeed, there was no equivalent to August or Nizze, no gymnasium version of Apollonius: so perhaps, less of an immediate source for Heiberg, hence he felt more at ease making his own diagrams, which, consequently, were based more directly on his manuscript source.

I did not find other alternative sources for Heiberg's figures of Archimedes or Apollonius in earlier printed editions. His figures there are apparently a composite created by Heiberg himself and a compromise between a print tradition and the evidence of manuscripts.

The emphasis is on the notion of a print *tradition*. Just as manuscript diagrams were copied from previous manuscript diagrams, so were print diagrams. Heiberg, we noted, relied on August for his diagrams of Euclid. In fact both August as well as Nizze produced diagrams that are closely related (but are not direct replicas) to those of Peyrard. This scholar—famous for his rediscovery of the important Euclidean manuscript Vat. 190 (now often referred to simply as Codex Peyrard)—was the librarian at the École Polytechnique, infused by the ideology of his time, his place (which anticipated, of course, Prussia's own Humboldtian initiatives of education). Peyrard has produced two remarkable editions of Archimedes (1807) and of Euclid (1814–1818). Those are elegant in appearance and emphasize functionality, with remarkable production values (thus, diagrams are reproduced for each page-opening, so that there is no need to turn the page to look up a figure half-remembered). Now we may put side by side Nizze's (1824) figure for *Spiral Lines* 18 (first case), mentioned above, together with Peyrard (1807) (Fig. 9.28).

This should be compared with Nizze (1824), who has the following diagram (Fig. 9.29).

We find the very same idea of convergence instead of meeting.

Saito and Sidoli (2012: 136–137) pointed out the interesting example of *Elements* XI.12, where two lines are set up as emerging upwards in space. In the manuscripts, a plane of reference is merely understood so that the picture is simply of two straight lines. August (1826–1829) inserted, and Heiberg (1885) copied from him, a plane of reference, clarifying the three-dimensional structure (the dependency can be shown—which is why Saito and Sidoli chose this as their example—as, in this case, the direction of the labels was copied by Heiberg from August as well—and included

**Fig. 9.28** *Spiral Lines* 18, Peyrard's diagram

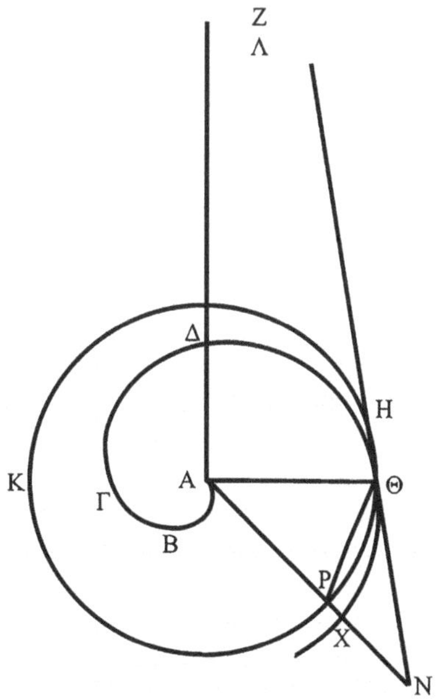

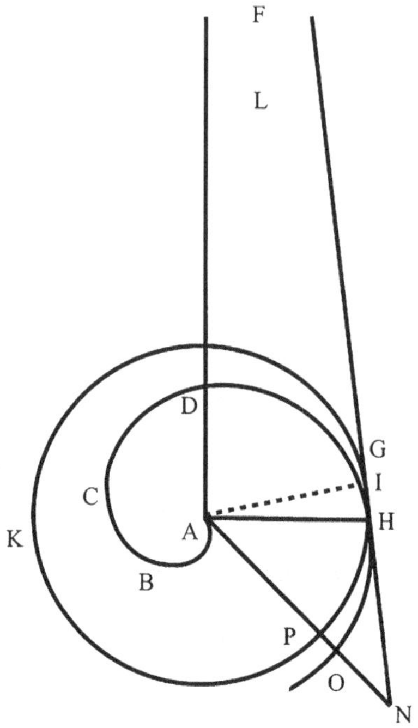


**Fig. 9.29** *Spiral Lines* 18, Nizze's (1824) diagram

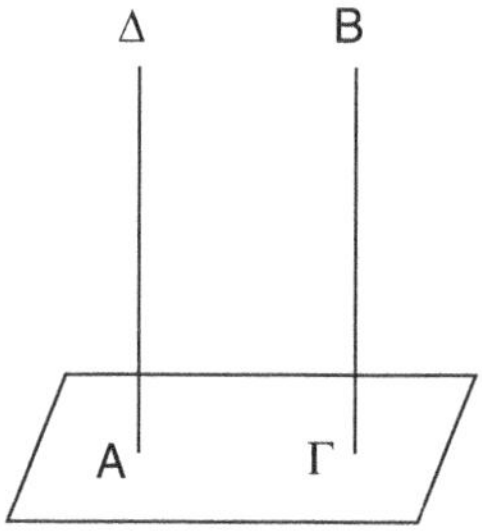


**Fig. 9.30** *Elements* XI.12, Peyrard's diagram

an error introduced by August). It is interesting to note, then, that this plane of reference is found, once again, in Peyrard (1807) as well (Fig. 9.30).

At this point, perhaps one might be tempted to tell a Foucauldian story of a rupture taking place around the year 1800, with a new diagrammatic language emerging to replace that of antiquity. To cut a long story short, this would be wrong. The not-yet-converging lines of *Spiral Lines* 18 are found already in the first printed edition of Archimedes (Basel 1544), and the plane of reference of Euclid's *Elements* XI.12 is found at least as early as Commandino (1572). These are sixteenth-century original departures (in this case, without a basis in the manuscript tradition); both of them then extended through the print tradition in multiple versions. A systematic survey of the print tradition as a whole is beyond the scope of this chapter, but let me add two more examples to the two above.

1. There is a broad practice in the manuscript tradition of Archimedes' *Conoids and Spheroids*, to represent all solid cuts, whether right or oblique, as if they were right. That is: all the solid cuts are represented by combinations of circular arcs resting on a base parallel to the side of the page. I show here Proposition 20, from the Florence manuscript (Laur. 28. 4 54v) (Fig. 9.31).

The print tradition, up to and including Heiberg (1910), represents the oblique cuts obliquely; that is, with the line of the cut represented (more pictorially) as slanted to the base (the figure is also rotated so that it points up). This is Heiberg's version, the same as the print tradition starting from 1544 (Fig. 9.32).

2. Several propositions towards the end of *Elements* VI deal with parallelograms. These are represented in the manuscripts uniformly as rectangles, as can be seen in Saito's transcription of five of the major manuscripts, as well as Heiberg (who represents a slanted parallelogram), of *Elements* VI.24 (Fig. 9.33).

The transformation into slanted parallelograms goes back in this case at least to Faber's 1516 edition and is a constant of the print tradition.

What we find then is that the print tradition is overall very stable. We also find a certain moment of rupture early on in the print tradition of mathematics, that is in the sixteenth century. Most mathematicians were not printed in fifteenth century; Euclid's *Editio Princeps*, from 1482, is produced from the Latin and is largely continuous with the practices of manuscript.

With this at hand, let us now look more closely at the modern reduction of absurdity.

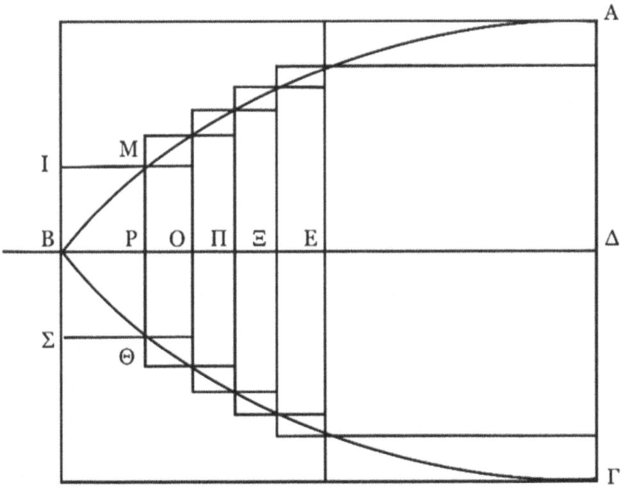


**Fig. 9.31** *Conoids and Spheroids* 20, diagram from Codex D

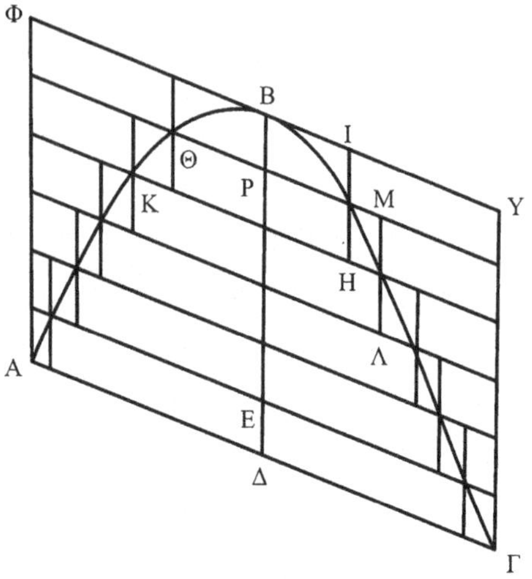


**Fig. 9.32** *Conoids and Spheroids* 20, Heiberg's diagram (1910)

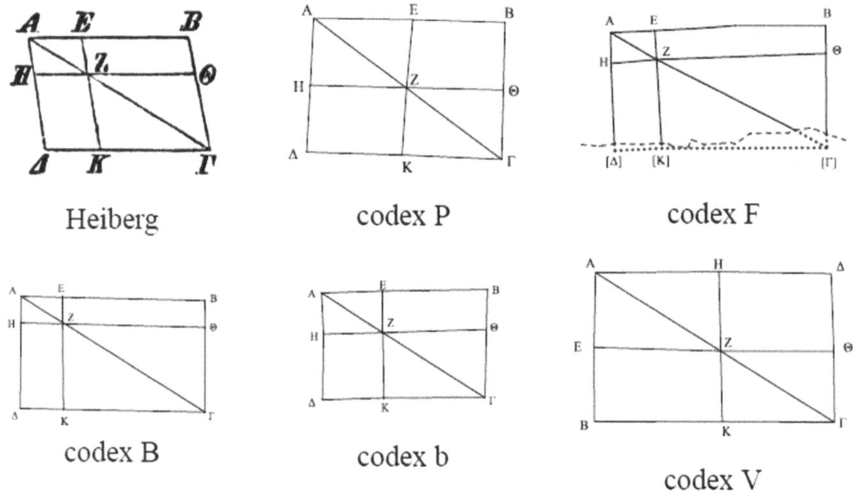


**Fig. 9.33** *Elements* VI.23, diagrams from (**a**) Heiberg (1884) and (**b**) manuscripts, following Saito

### *9.2.2 Absurdity into Print*

We have focused on three categories: (1) where the false configuration cannot be drawn, and this visual fact forms the ground for the argument (*Elements* III.13, *Conics* IV.35, 37); (2) where the false configuration cannot be drawn, but the argument is not directly based on the visual impression of the diagram (*Spiral Lines* 13, *Elements* I.27) and (3) where the false configuration can, with some creativity, be shown and yet is, in principle, configurationally impossible or in more general terms 'wrong' (*Floating Bodies* I.8, Pappus VII.216, *Conics* I.31, II.1, IV.26).

In the first case, the reduction of absurdity is essentially a mathematical error and so I put it aside for the moment. I look first, then, at the seven examples of Cases 2–3. *Elements* I.27 may be skipped, as, in fact (as noted above), the absurdity is never reduced in the print tradition, a rare case of the survival of the ancient practice. Here are the remaining six examples (Fig. 9.34).

This is Commandino's 1566 version of *Conics* I.31 (compare Figs. 9.6 and 9.7). He clearly understood the 'reflection' of the line at point D as a significant element of the diagram and preserved it. Consider next his version of II.1 (compare Figs. 9.22, 9.23, and 9.35).

While this is less clear, my impression is that a straight line is drawn extended from CD onwards, and then an inflecting branch of that line is drawn to touch the curve at point G. The impression of absurdity is subtle, but clearly Commandino felt a need to preserve some element of visual absurdity. Finally, here is Commandino's (1566) IV.50 (compare Figs. 9.20, 9.21, and 9.36).

The absurdity was reduced in this case. This is related to Commandino's general practice in this edition: to produce conic sections in the form of continuous non-circular curves (rather as the combination of circular arcs). He must have felt that, in this case, the absurdity he saw in the manuscripts' diagram was not a mathematical necessity but a reflection of a limited visual technique (that of representing conic curves by circular arcs). This is continuous with Commandino's practice overall. While his diagrams are, generally speaking, produced with sympathy to the manuscript tradition, he invests them with visual technique which is far superior, in sheer

**Fig. 9.34** *Conics* I.31, Commandino's (1566) diagram

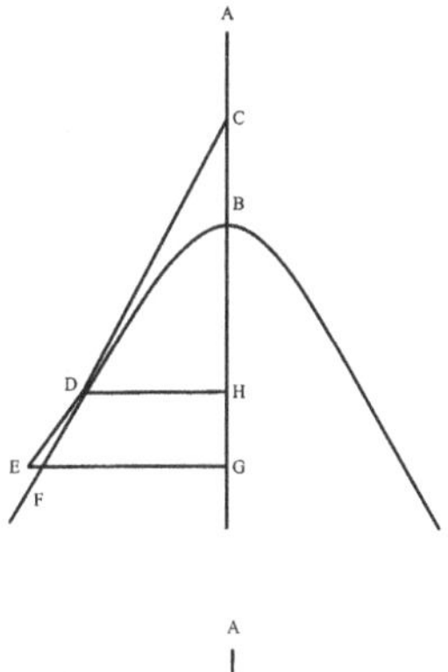


**Fig. 9.35** *Conics* II.1, Commandino's (1566) diagram

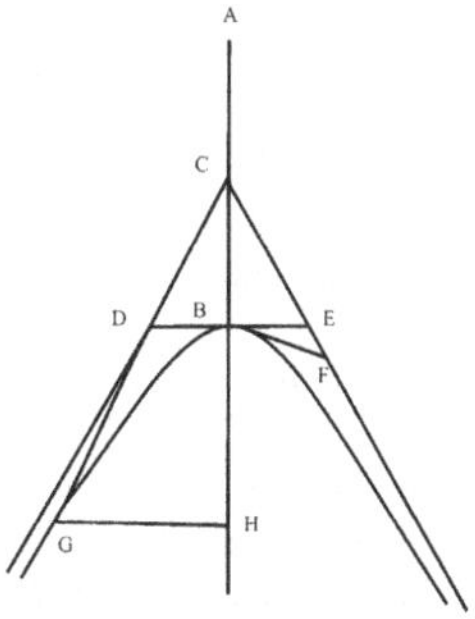

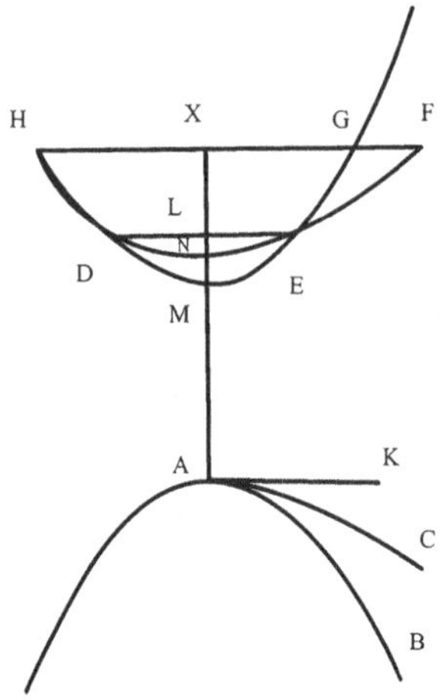


**Fig. 9.36** *Conics* IV.50, Commandino's (1566) diagram

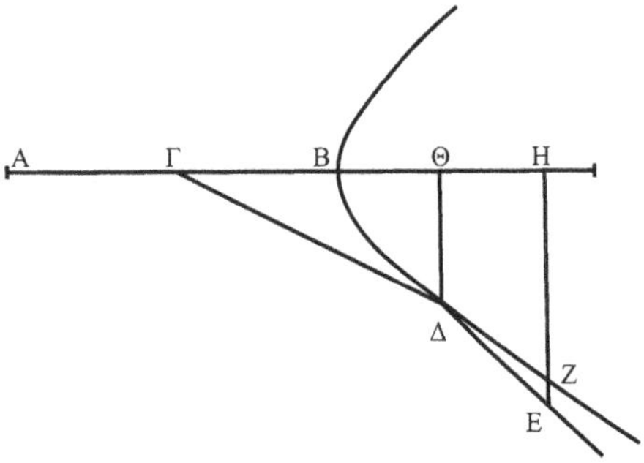


**Fig. 9.37** *Conics* I.31, Halley's (1710) diagram

graphic terms, to that of the ancient manuscripts so that, for instance, a three-dimensional pictorial representation is often introduced, with genuine foreshortening (thus, when solid figures stand on parallelograms, the more 'remote' side of the parallelogram would appear shorter: such perspectival technique is mostly dropped from the later print tradition and is in fact somewhat distracting for the geometrical argument).

Here is Halley (1710) for *Conics* I.31, if anything emphasizing the absurdity (compare Figs. 9.6, 9.7, 9.33, and 9.37)!

While his version of II.1 appears to be influenced by Commandino's, and once again is, if anything, even more flagrantly impossible (compare Figs. 9.22, 9.23, 9.34, and 9.38).

However, IV.26 is once again drawn without a trace of absurdity.

Where did Heiberg get his figures of I.31 and II.1, then? He mentions three modern authors in his introduction to the Conics which, besides the two sources discussed already, also includes Memus, author of a Latin translation of *Conics* I-IV published in Venice in 1537. Here is Memus' figure of I.31 (compare Figs. 9.6, 9.7, 9.33, 9.36, and 9.39).

And here is the figure of II.1 (compare Figs. 9.22, 9.23, 9.34, 9.37, and 9.40).

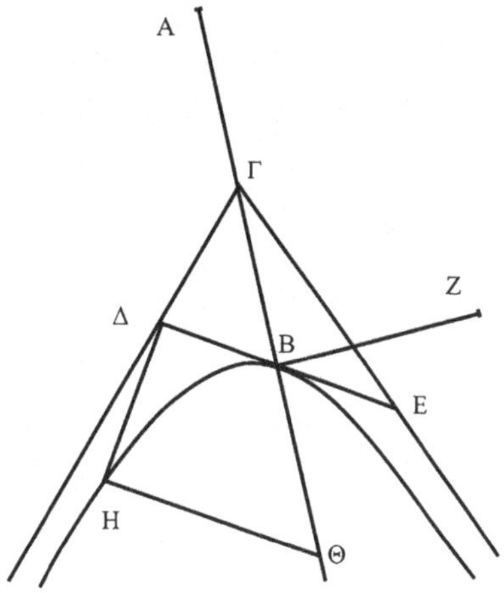


**Fig. 9.38** *Conics* II.1, Halley's (1710) diagram

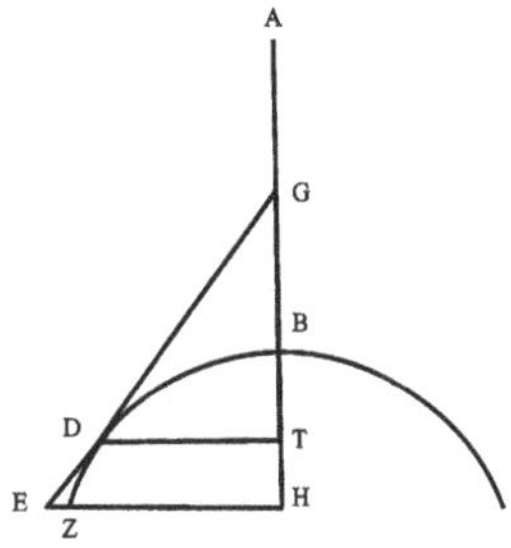


**Fig. 9.39** *Conics* I.31, Memus' (1537) diagram

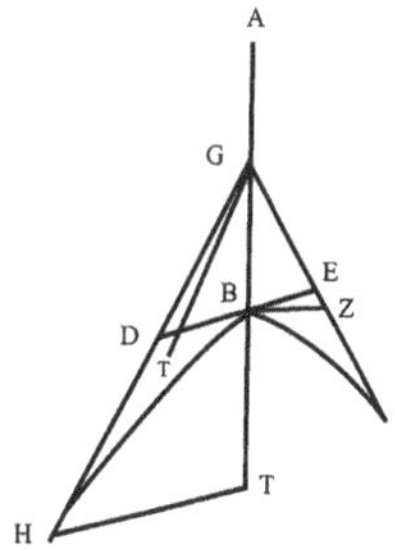


**Fig. 9.40** *Conics* II.1, Memus' (1537) diagram

There seems to be a tiny inflection in I.31 and also a certain doubling of the lines in II.1, but all in all, Memus' figures are much closer to Heiberg's. Either Heiberg followed Memus or he made a decision, on his own, to depart from Commandino and Halley in these two figures. (It will be noted that Memus' representation of conic sections is heterogeneous, sometimes drawing them as curves, sometimes as circular arc combinations; in IV.26, sure enough, he produces curves and thus reduces the absurdity there).

We note the following two conclusions concerning the print tradition of Apollonius. First, even though Heiberg comes across as a reducer of absurdity

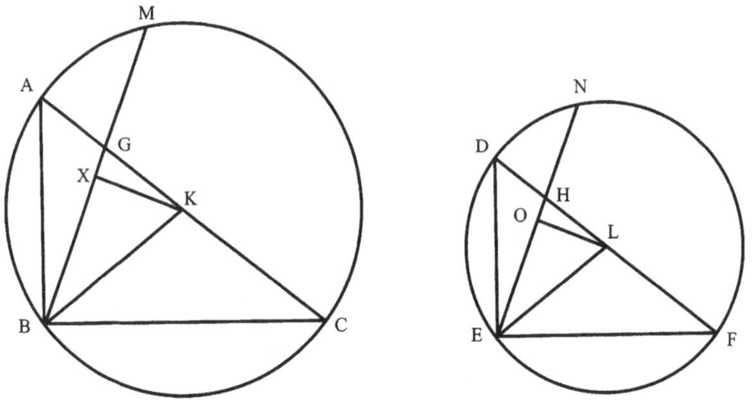


**Fig. 9.41** Pappus VII.216, Commandino's (1588) diagram

relative to his immediate print predecessor Halley, it is also obvious that his diagrams overall are much closer to the manuscript than those of any of his print predecessors. This is evidenced primarily in Heiberg's basic layout (there is a clear tendency in print, before Heiberg, to orient conic sections so that their diameter is upright, whereas in manuscripts, the diameter is typically horizontal: in his Apollonius edition—though not in his Archimedes' edition of *Conoids and Spheroids*—Heiberg went back to the manuscripts' orientation).

Second, we find that the most significant moment of the reduction of absurdity, for Apollonius, occurs in the *Editio Princeps* of 1537.

The following is Commandino's (1588) version of Pappus VII: 216 (compare Figs. 9.24, 9.25, and 9.41).

Absurdity, in this case, has been reduced to the avoidance of depicting the impossible case at all, the two lines now appearing on the *same* side of the centers. Hultsch, then, if anything, restored something of what I take to be the original intention of the diagrams, without however letting in the absurdity.

We move next to the two diagrams from Archimedes. Here is Commandino's version from 1565 of *Floating Bodies* I.8 (I use Clagett's (1976–1978) transcription; compare Figs. 9.26, 9.27, and 9.42).

The center of the earth has been returned to its rightful place; this will be kept in all subsequent printed editions.

The same is true for *Spiral Lines* 13, which already in the Basel printed edition from 1544 has been transformed to a version closely approximating Heiberg's and will remain such, its absurdity reduced, throughout Archimedes' print tradition (note that this printing does not include *Floating Bodies*, then unknown in the Greek).

We may now finally take *Elements* III.13. As noted above, reducing the absurdity of such a figure would be a mathematical mistake and it is in fact preserved by Commandino and other sixteenth-century editions (I did not attempt an exhaustive search). Even a late printing, for example, such as Simson/Ingram (1799) still maintains the character of the manuscript diagram, if with an expanded crescent, which therefore approximates the circle as far it may: this is a quantitative reduction of the

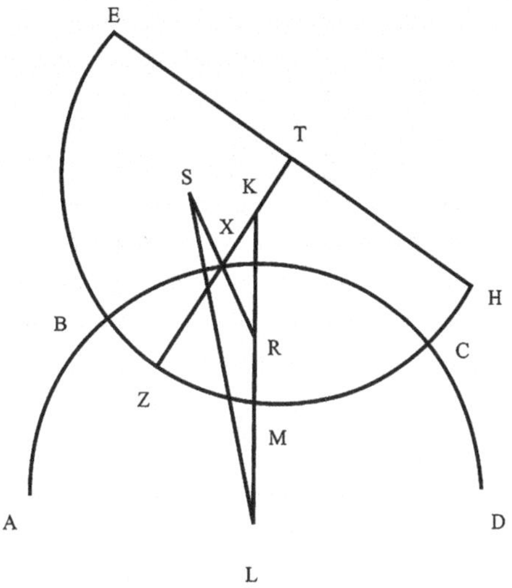


**Fig. 9.42** *Floating Bodies* I.8, transcription of Commandino's (1565) diagram

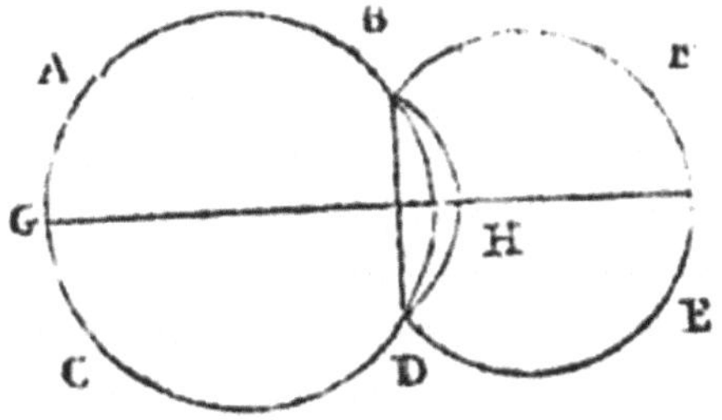


**Fig. 9.43** *Elements* III.13, Simson's/Ingram's (1799) diagram

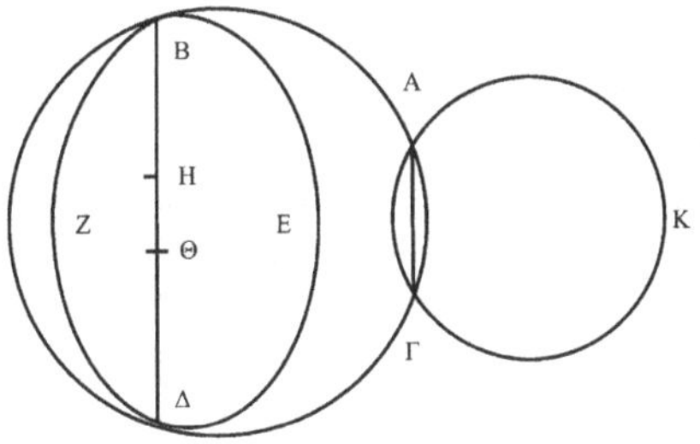


**Fig. 9.44** *Elements* III.13, Peyrard's (1814–1818) diagram

absurdity and yet still maintains the qualitative absurdity of the original (compare Figs. 9.8 and 9.43).

I believe it may be only with Gregory 1703 that the break with the manuscript tradition occurs; this version, for some reason, then got picked up by Peyrard (1814–1818) (Fig. 9.44).

I am at a loss to account for this transformation. At any rate, once made, this transition becomes entrenched: August (1826–1829) did not deviate much from Peyrard (1814–1818), and Heiberg (1881–1885) followed August.

To sum up, then, we find that the only general rule concerning the destinies of diagrams from the sixteenth to the nineteenth century is their stability. The degree of absurdity in the diagrams of proof by impossibility evolves in a non-linear fashion, sometimes becoming more marked, sometimes less so. Much more important than any trend through the centuries from the sixteenth to the nineteenth is the phenomenon of the sixteenth century itself. It is there—right at the beginning of the print mathematical tradition—that we find the main thrust of reduction of absurdity. All of this is consistent with a pattern seen across the evidence for diagrams in manuscript and print: it is a tale of a sharp rupture at the sixteenth century, not one of gradual evolution towards more and more 'modern' forms.

### 9.2.3 Manuscript Variation and the Rejection of Absurdity

So far, I have spoken of the 'manuscript' diagram, referring effectively to the reconstructed archetype. That such an archetype may in general be reconstructed is not controversial, but still it is useful to note the manuscript variation in our sources. This interest narrows our field to Archimedes, as for Euclid relatively little variation is found among the major—medieval—manuscripts (especially on the issues that concern us), while the editions of both Apollonius as well as Pappus are based on a single medieval manuscript. It will remain useful to compare Euclid, Pappus and Apollonius with their several Renaissance manuscript copies, which I hope others will do in the future; I did not undertake this study for this publication.

And so it turns out that as soon as we consider the detail of Archimedes's manuscript tradition, we are struck by a considerable variation. We must first of all address and understand this variation.

The case of *Floating Bodies* is unique and its full explication must be left for another opportunity (I will discuss this in detail in my translation with commentary of *Floating Bodies*). The outline is as follows. Other than the Palimpsest, our sole manuscript authority is the Latin translation by Moerbeke from 1269 (Ottob. Lat. 1850). This, however, does differ from the Palimpsest for *Floating Bodies* I.8, as follows (compare Figs. 9.26 above and 9.45).

In Moerbeke's Latin translation we see three diagrams, not one. What is going on here? Well, we do find the added two diagrams seen in Moerbeke in the Palimpsest as well, but there, they are attached to the *following* Proposition I.9 (Fig. 9.46).

To repeat, the details of the argument are difficult, but what explains those added diagrams, in Moerbeke as well as in the Palimpsest, are passages in the proof of I.8 that are most likely scholiastic interventions that added two more cases for the discussion. The likely account then is that Archimedes had a diagram with an eccentric earth; a scholiast added two cases (and did not reproduce the eccentric earth for his two added cases, as this was irrelevant for his purposes); and Moerbeke's figures erroneously followed the arrangement of the scholiast. This textual history is interesting in and of itself, but its results, as far as we are concerned, are mostly negative. We are reminded again of the labor—and conjecture—hidden under the facile

**Fig. 9.45** *Floating Bodies* I.8, Moerbeke's diagram

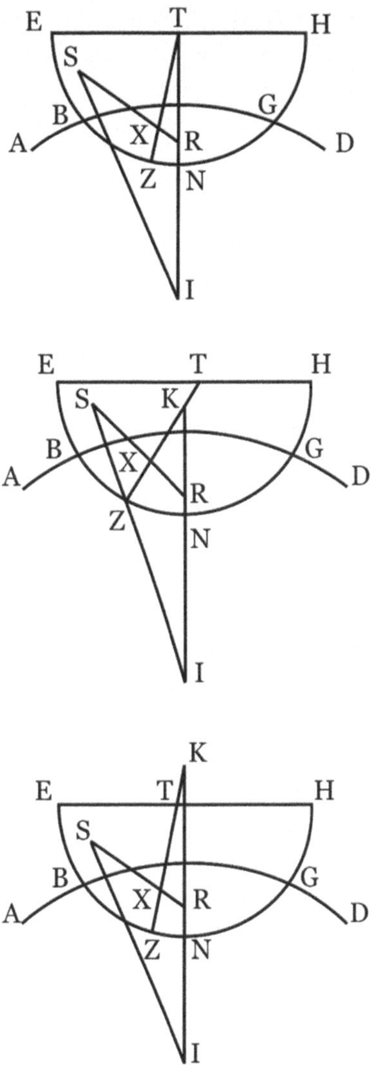


reference to a 'reconstructed archetype'. We are hopefully reassured that this reconstruction is in fact correct. And we notice that in this case, we probably do not see a genuine intentional manuscript variation but, instead, what amounts to a mere scribal error.

For genuine scribal variation, then—for a genuine reduction of absurdity—we need to turn to the case of *Spiral Lines* 13 (for the following observations, compare throughout with Figs. 9.2 and 9.3). Here, we find, however, quite a bit to look at. The diagram shown above as the reconstruction of the manuscripts archetype is essentially the same as Codex C, which is still extant. The only other Byzantine source for this diagram was Codex A, now lost, of which six direct copies are extant,

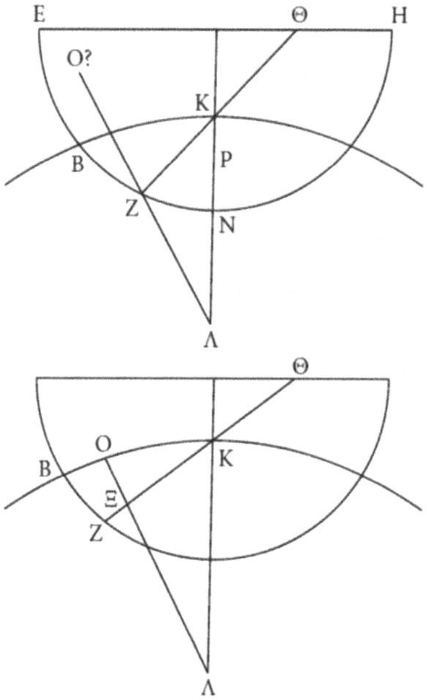


**Fig. 9.46** *Floating Bodies* I.9, transcription of Palimpsest's diagram, detail

including Moerbeke's translation or Heiberg's B, as well as five Greek manuscripts, two in Paris (Heiberg's G and H) and one each in Venice (E), Florence (D) and Rome (Heiberg's 4. Note also that in this case, Moerbeke is a copy of A and not an independent witness).

In three of these manuscripts (E, H and 4), we see once again essentially the same diagram as above. In E, however, a second hand inserted a visual scholion, an alternative diagram labelled in Latin (Fig. 9.47).

While the straight line is still inflected, it is no longer broken and its absurdity reduced.

The fact that this is a second hand means it carries little authority for the reconstruction of the archetype; it does show that some Renaissance reader felt an urge to reduce this absurdity.

More relevant for the reconstruction of the archetype are diagrams drawn by the first hand. Once again, there is considerable variety there as well. Here is Moerbeke (in Clagett's (1976–1978) transcription) (Fig. 9.48).

This is a significant reduction of absurdity, indeed identical to that of Heiberg, down to the precise orientation: the print tradition as a whole follows Moerbeke's lead, though typically aligning the line EZ as parallel to the base of the page.

In Codex D we find (Fig. 9.49).

This is a slightly different reduction of absurdity and perhaps a more rational one: the tangent is drawn as a straight line and not allowed to coalesce with the curve of the spiral line. Notice that the orientation is different from Moerbeke's (though it is the same as that of the second hand in E; was E's second hand perhaps based on D?)

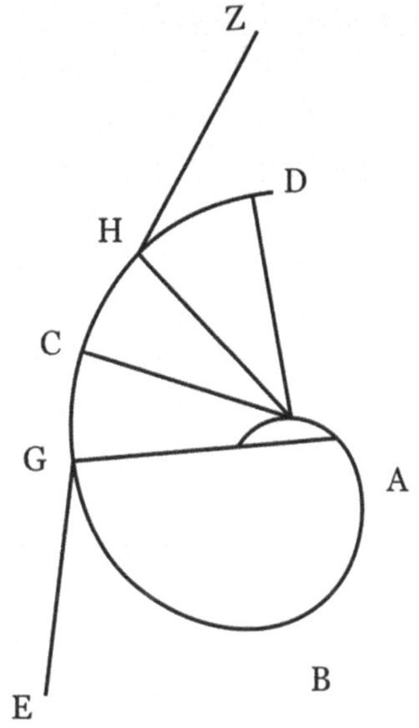


**Fig. 9.47** *Spiral Lines* 13, Codex E, detail of second hand

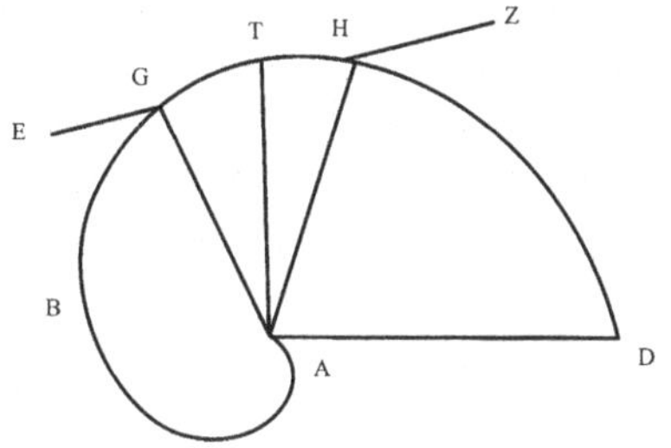


**Fig. 9.48** *Spiral Lines* 13, transcription of Moerbeke's diagram

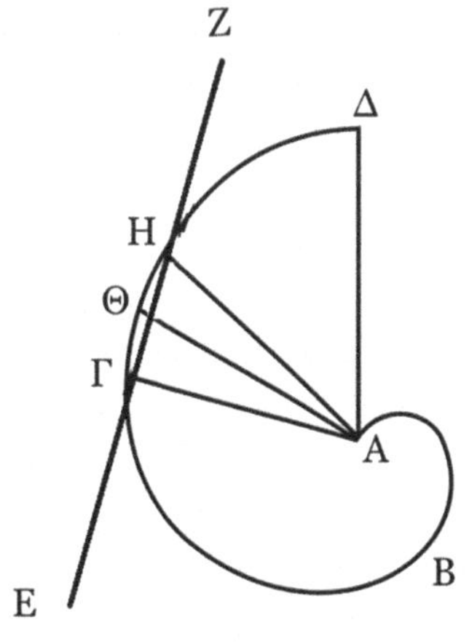


**Fig. 9.49** *Spiral Lines* 13, diagram of Codex D

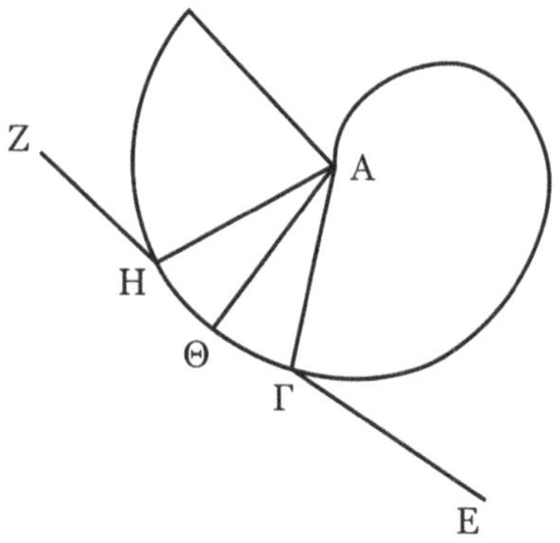


**Fig. 9.50** *Spiral Lines* 13, diagram of Codex G

Finally, yet another reduction of absurdity is found in G (Fig. 9.50).

This is the same diagram idea as that of Moerbeke (or of the print tradition), though with an inverted orientation.

Such then is the significant degree of variation. We have three different topological solutions for the arrangement of the tangent (a single line that never coincides with the curve in D, an inflected line that does coincide with the curve in the second hand of E and a single line coinciding with the curve in B and G). We have three different orientations for the tangent (near-horizontal, pointing right in B; near vertical, somewhat right in D and in the second hand of E; slanted left in G). While the various manuscripts show a shared origin in their basic structure and labeling of the spiral itself (which however can be put down to the textual directions themselves), their divergence otherwise is of a level seldom seen in the diagrammatic tradition and one has to conclude that these diagrams were essentially independent (with the possible exception of an influence of the diagram of D on that of E: it should be borne in mind that we know that D was copied from A while both A and E were in Venice, so that the manuscripts did spend some time not far apart from each other[14]). Put this variation together with the unanimity of E, H and 4 as witnesses for A, as well as with the agreement of that reconstruction with the evidence of C, and the result is as clear as it could be. The archetype shared by A and C, probably an Early Byzantine or a Late Ancient one, has shown a very absurd, very broken tangent. This was copied faithfully into the Byzantine codices but in no less than four occasions, scholars made the choice to insert a distinct, less absurd diagram, alongside or instead of the original diagram.

The emphasis is on 'scholars'. Moerbeke is, of course, a translator, and while his grasp of the mathematical material is shaky from time to time, he clearly understands this proposition and redraws its figure based on his understanding of it. Indeed, in general, his diagrams are distinct from those of other copies of Codex A (in those treatises copied by him from this archetype), which might reflect also his original layout. That is: in other copies, diagrams are overwhelmingly folded into

[14] Heiberg (1915: xi–xii).

the columns of text (probably reflecting the original arrangement on archetype manuscripts and ultimately on papyrus) but Moerbeke often reassigns his figures to the margins, manipulating their dimensions and orientations. Moerbeke also writes in a very compact Gothic script, so that he crams more propositions, hence diagrams, per page, constraining the space further; it is therefore probably relevant that this reduction of absurdity is also a massive reduction of scale. The figure above is from within the column, but this is because it is really tiny.

Codex G is remarkable for its extensive scribal Greek colophon, stating in detail the provenance of the manuscript and reproducing what amounts to a brief glossary of the abbreviations in use. The scribe notes in particular that the manuscript contained many mistakes in the figures (by which he means, as he explains, the textual errors of those figures, with letters substituted for letters such as Λ for A and vice versa; such errors are indeed very common in the manuscript tradition[15]). We find an active scholarly interest and this is, no doubt, the work of a humanist (the manuscript is taken to be from the early sixteenth century).

We have some documentation for the making of Codex D, which appears to have been copied for the Medici library. The Florentine scout, discovering A in Venice, was no less than Poliziano, and the copy was made around 1491–1492.[16] The humanist connections then are obvious and in fact, while we have no colophon matching the scholarly aspirations of Codex G, we have something much more remarkable instead: the written manuscript. The hand of Codex D is very hard to classify, as it is unlike anything normally seen. It takes the letter forms of early Byzantine miniscule and then produces them anachronistically, the letters flowing less rigorously than one would have expected in the tenth century; strangely for such a late Renaissance manuscript, the accents are missing. In a word, it appears that the Renaissance scribe of D made a conscious decision—and a very rare one—not only to copy his text but also to imitate the source's hand to the best of his ability and to remove, into the bargain, all the accents. This choice had important consequences, as Bandini, the librarian and first cataloguer of the Laurenzian Library in the eighteenth century, classified this manuscript—based on the superficial impression of its script—as a very early one (splitting the difference between the conflicting chronological indications, between paper and handwriting, Bandini assigned it to the thirteenth century). Following that indication by Bandini, Heiberg in his dissertation became convinced that the Florence manuscript was the original from which all other manuscripts were made (1879: 126), a conviction carried into Heiberg's first edition (1880).

At one level, the script of Codex D is a mere idiosyncrasy. But it does fit a wider pattern. The lack of accents, a curiosity in and of itself, is of a piece with Poliziano's practices elsewhere: remarkably, in Poliziano's 1489 printing of Callimachus, a conscious choice was made to avoid accents. Poliziano was explicit about his goal,

---

[15] The colophon is G 120v, transcribed in Heiberg (1915) x–xi.

[16] Heiberg (1915) xii.

to print this work '*veteri more sine ullis accentiunculis*'.[17] Codex D emerges from a milieu engaged with the intentionally antiquarian production of Greek, seeking to bypass the manuscript evidence in favor of what was understood, on scholarly grounds, to be the original, authentic form.

This milieu gains a clearer meaning when we consider it not in the context of Greek, but Latin paleography. As is well known, a Carolingian script emerged at about the same time as did Byzantine minuscule; this was gradually replaced by a Gothic script, easier to produce (in its simpler forms) but more difficult to tell apart (all in all, this moves in parallel to the evolution of Greek script in Byzantium which, with its growing reliance on ligature and compact letter forms, does become less and less transparent). Early humanists were attracted by the discovery of Carolingian script, which they mistakenly took to be ancient. Petrarch explicitly raged against what he called '*litterae scholasticae*' (by which he meant Gothic script) and early in the fifteenth century, scholars close to his circle explicitly set out to copy ancient Latin in what they took to be an ancient form of writing, and what we would identify essentially as Carolingian. Such was the power of humanistic ideology that a new form of writing was consciously introduced and consciously adopted, within a generation and across all Europe: humanistic script became established early enough to form the standard backdrop to Latin in print, hence to our own form of writing. All based on a humanistic, contrived, scholarly impulse.[18] What we see then is that at some point in the late fifteenth century, a similar humanistic impulse infused a few scholars of the first generation of true Greek learning in Italy seeking to bypass their models so as to recover a lost, ancient past, in the very scribal act of copying ancient Greek works. Unlike the case of Latin, such efforts were abortive: perhaps at this late stage, no single location could exercise the kind of influence on humanism that Florence had in the early fifteenth century.

At any rate, what we see in the particular cases of Codices D and G (and probably also in the second hand of Codex E) is the humanistic background to this particular reduction of absurdity, which perhaps may be considered as the rejection of absurdity. Clearly, the scholars responsible for those manuscript copies shared a deeply suspicious attitude towards the '*litterae scholasticae*' of the Middle Ages and yearned to recover a lost and mutilated ancient past. They flaunted their ability to see beyond such fragments into the original edifice (hence the philological colophon of Codex G, boasting its ability to decipher medieval abbreviations and to see through medieval errors). It is probably in such a context that we need to understand the scribes' attitude towards a diagram that seemed to them, at first glance, wrong: better to cast it away, then, and re-make it.

This is one particular case, but a similar attitude is, in fact, visible elsewhere in the manuscript tradition of diagrams. To add just a couple of examples from the Archimedean tradition: first, one of the most striking features of the Archimedean diagrams as a whole is the non-pictorial use of arcs for drawing polygons, in the first

[17] Lee (2005: 270).

[18] For all of this, see Bischoff 1990, chapter 10.

book of *Sphere and Cylinder*.[19] Those are seen in both A (reconstructed) and C, and probably go back to Late Antiquity and beyond (it is interesting that in Pappus' Book V, we see a related representation of functionally similar polygons produced as series of zigzag lines). This non-pictorial representation is removed by the very same copies of Archimedes B, D and G, the polygons drawn instead 'correctly' with straight lines. Or consider another small vignette: the figure of *Floating Bodies* I.4 shows two bodies immersed in the spherical ocean. Considered as a schematic arrangement, it is natural to draw the immersed objects merely as an abstract 'magnitude', represented by a rectangle parallel to the base of the page, as in the following edition of Codex C (Fig. 9.51).

Moerbeke in this case copied the same diagram, with the same abstract figures. However, we know that Moerbeke's text of *Floating Bodies* I.1–8 was systematically studied and revised by a Renaissance scholar, Andreas Coner, so that Moerbeke's original diagram is in this case only readable under an ultraviolet light.[20] The diagram in Latin now visible to the naked eye is the following (Fig. 9.52).

The only real difference is that the schematic, abstract rectangles are slightly tilted so that they appear like objects verging towards the center of the earth. A schematic figure—rejected to be made more pictorial.

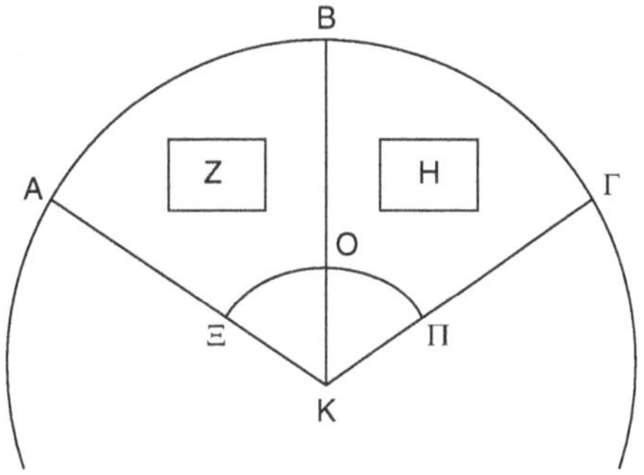


**Fig. 9.51** *Floating Bodies* I.4, transcription of Palimpsest's diagram

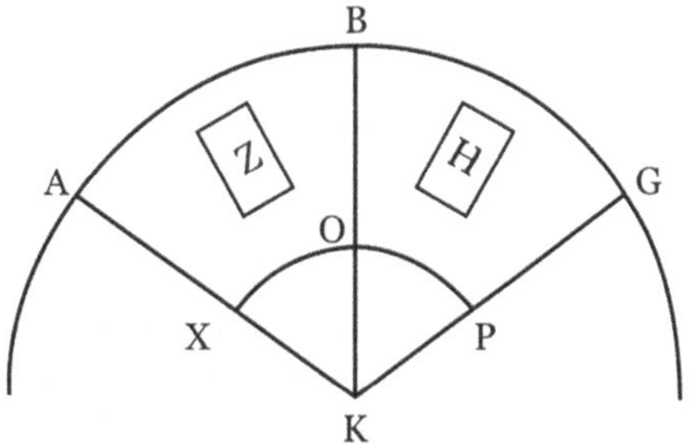


**Fig. 9.52** *Floating Bodies* I.4, current state of diagram in Moerbeke's manuscript

[19] See discussion in Netz (2012: 172–175) and Saito and Sidoli (2012: 145–146); it should be noted that here, again, Moerbeke is a copy of A, not an independent witness.

[20] For Coner's engagement with Moerbeke's manuscript, see Clagett (1976–1978: 525–538).

Finally, we may note the typical Renaissance manuscript project which is Jacob of Cremona's retranslation of Archimedes into Latin. This mid-fifteenth-century translation, directly or indirectly based on Codex A[21] systematically rejects the diagrams of Codex A (this originality is carefully documented by d'Alessandro and Napolitani (2012b)), of course producing straight-line polygons and, generally speaking, drawing, quite simply, its own original figures.

Overall then, in Codices D and G, as well as in Coner's and Jacob's contributions, we see a more interventionist attitude towards manuscript diagrams. This is consonant with a humanism that seeks to leapfrog the perceived opacity and distortions of, literally, a Middle Stage along the transmission of antiquity—that of the Middle Ages.

## 9.3 Conclusion

We operated through this chapter with this notion of a basic opposition between a manuscript tradition and a print tradition. But at this point, we may notice that our evidence for both leads in a similar direction. On the side of print, we noticed a general stable, conservative tradition, with the major exception of the very early stage of mathematical printing in the sixteenth century, which does tend to depart from the preceding manuscript tradition. It would be easy to ascribe this to a fundamental gap separating manuscript from print, perhaps a matter of different media giving rise to different visual forms. But, in fact, we notice a parallel result on the manuscript side as well: as far as we can tell, manuscripts, too, seem to be generally speaking quite conservative in their transmission of diagrams, the major exception having to do with a subset of manuscript copies produced by humanist scholars, i.e. of exactly the same age as that of early print. Early print and humanist manuscript-making are, after all, often continuous with each other: distinct media expressing similar cultural forces.

Books have their fate and generalizations concerning book culture are rash. Let me not create the impression as if I argue for a single story accounting for the entire trajectory of Greek mathematical diagrams: it is sufficient to note that Moerbeke's translation is, in some ways, comparable to my evidence for early print and for humanistic manuscript making—and yet is, of course, a scholastic work par excellence! Another theme touched upon throughout this paper has been that of educational function, especially those coming to the fore in the nineteenth century and so consistent with a more 'pictorial' representation. We noted Heiberg's sensitivity to the variety of sources at his disposal, so that he did go back on occasion to some

[21] It was suggested by d'Alessandro and Napolitani (2012a) that this translation could have been based on a Greek source distinct from any of the known manuscripts. For reasons I briefly laid out in Netz (2017 10 n. 4), I find this suggestion quite unpersuasive. Readers should be warned, however, that according to d'Alessandro and Napolitani, the divergences in the diagrams of Jacob are due not to his originality but to his dependence upon a source other than Codex A.

features of the manuscript tradition, always depending on the nature of the sources, in print and manuscript, he had available. The story of the transmission of diagrams is anything but *simple*.

And yet the case we have focused on in this chapter carries a wider moral. It was the theme of a certain reception of antiquity—that of the humanism of the Renaissance—to perceive a certain solecism, a certain absurdity, in the medieval transmission, and to seek to reduce such absurdity. The aim was to restore antiquity, even going beyond or even going in the face of the evidence. Such beginnings now strike us as crude and naïve, but they were, to say the least, critical. Out of such beginnings, there arose the textual criticism of a more scientific philology, obtaining its current form (arguably, its apex) in the nineteenth century. However, this scientific evolution of philology did not extend, in the nineteenth century itself, to the treatment of diagrams. We thus see in the main editions of Greek mathematics a mix of practices—a sophisticated philology on display in the text and a hyper-critical humanistic tradition, on display (somewhat unwittingly) in the figs. A colossus, half-Madvig, half-Poliziano: Heiberg presents to us this strange, remarkable hybrid. Dwarfed as we are by his example, our task is now to produce a new, scientific corpus of the diagrams, on a par with Heiberg's text.

## Appendix Chapter 9: Visual Sources in Manuscript and in Print

### *A. Visual Sources in Manuscript*

#### Archimedes Manuscripts

Archimedes Palimpsest = Codex C, Private Collection
Moerbeke = B = Ottob. Lat. 1850
D = Laur.28.4
E = Marc. 305
G = Par. Gr. 2360
H = Par. Gr. 2361
4 = Vat. Gr. Pii II.16

#### Euclid Manuscripts

B = Bodl. D'Orville 301
b = Bibl. Comm. Bonon. 18–19
F = Laur. 28.3
V = Vindobon. Gr. 103
P = Vat. Gr. 190
p = Par. Gr. 2466

### Others

Apollonius. Vat. Gr. 206
Pappus. Vat. Gr. 219

## *B. Visual Sources in Print*

### Archimedes

Basel = Venatorius, T. 1544. *Archimedes / Opera, Quae Quidem Extant, Omnia*. Basel.
Commandino, F. 1565. *Archimedes / De Iis Quae Vehuntur in Aqua, Libri Duo*. Bologna.
Peyrard, F. ed. 1807. *Oeuvres d'Archimède*. Paris.
Nizze, E. ed. 1824. *Archimedes. Vorhandene Werke*. Stralsund.
Heiberg, J.L. 1913–5. *Archimedes / Opera Omnia* (2nd ed.). Leipzig.

### Euclid

Commandino, F. 1572. *Euclidis Elementorum Libri XV*. Pesaro.
Peyrard, F. 1814–1818. *Les œuvres d'Euclide*. Paris.
August, E.F. 1826–1829. *Euclidis Elementa*. Berlin.
Heiberg, J.L. 1883–1888. *Euclidis Elementa*. Leipzig.

### Apollonius

Memus, J.B. 1537. *Apollonii Pergaei. Opera*. Venice.
Commandino, F. 1566. *Apollonii Pergaei Conicorum Libri quattuor*. Bologna.
Halley, E. 1710. *Apollonii Pergæi Conicorum Libri octo*. Oxford.
Heiberg, J.L. 1891–1893. *Apollonius / Conica*. Leipzig.

### Pappus

Commandino, F. 1588. *Pappi Alexandrini Mathematicae Collectiones*. Pesaro.
Hultsch, F. 1876–1878. *Pappi Alexandrini Collectionis quae Supersunt*. Berlin.

## References

Anastasiou, Magdalini, John Seiradakis, Christián Carman, and Kyriakos Efstathiou. 2014. The Antikythera mechanism: The construction of the Metonic pointer and the back plate spirals. *Journal for the History of Astronomy* 45: 418–441.

Bischoff, Bernhard. 1990. *Latin paleography*. Cambridge: Cambridge University Press.

Cairncross, Andrew. 2012. Edition with commentary of a school text representing Euclid, Elementa I 18–25. M.A. Classics dissertation, King's College, London.

Clagett, Marshall. 1976–1978. *Archimedes in the middle ages Vol. II: The translations from the Greek by William of Moerbeke*. Philadelphia: American Philosophical Society.

d'Alessandro, Paolo, and Pier Daniele Napolitani. 2012a. *Archimede Latino: Iacopo da San Cassiano e il Corpus archimedeo alla metà del Quattrocento*. Paris: Les Belles Lettres.

———. 2012b. Diagrammi e figure tradotti dal greco in latino — l'Archimede di Iacopo da San Cassiano. *The Humanist* 7: 109–129.

de Young, Gregg. 2005. Diagrams in the Arabic Euclidean tradition: A preliminary assessment. *Historia Mathematica* 32: 129–179.

Decorps-Foulquier, Micheline. 1999. Sur les figures du traité des coniques d'Apollonios de Pergé édité par Eutocius d'Ascalon. *Revue d'histoire des mathématiques* 5: 55–76.

Fried, Michael. 2002. *Apollonius of Perga*, Conics book IV. Santa Fe: Green Lion Press.

Heiberg, Johan Ludvig. 1879. *Quaestiones Archimedeae*. Copenhagen: R. Klein.

———. 1880. *Archimedes. Opera Omnia*. 1st ed. Leipzig: Teubner.

———. 1881–1885. *Euclides. Elementa*. Leipzig: Teubner.

———. 1891–1893. *Apollonius. Conica*. Leipzig: Teubner.

———. 1913–1915. Archimedes. Opera Omnia. 2. Leipzig: Teubner.

Johnson, William. 2004. *Bookrolls and scribes in Oxyrhynchus*. Toronto: University of Toronto Press.

Lee, John. 2005. Dimitrios Doukas and the accentuation of the new testament text of the complutensian polyglot. *Novum Testamentum* 47: 250–290.

Manders, Kenneth. 2008. The Euclidean diagram. In *Philosophy of mathematical practice*, ed. P. Mancosu, 112–183. Oxford: Oxford University Press.

Mumma, John. 2010. Proofs, pictures, and Euclid. *Synthese* 175: 255–287.

Netz, Reviel. 1999. *The shaping of deduction in Greek mathematics*. Cambridge: Cambridge University Press.

———. 2004. The works of Archimedes translated into English. In *The two books on the sphere and the cylinder*, vol. I. Cambridge: Cambridge University Press.

———. 2012. The texture of Archimedes' writings: Through Heiberg's veil. In *The history of mathematical proof in ancient traditions*, ed. K. Chemla, 163–205. Cambridge: Cambridge University Press.

———. 2014. Authorial presence in the ancient exact sciences. In *Writing science: Medical and mathematical authorship in ancient Greece*, ed. M. Asper, 217–253. Berlin: De Gruyter.

———. 2015. Were there Epicurean mathematicians? *Oxford Studies in Ancient Philosophy* 49: 283–319.

———. 2017. *The works of Archimedes translated into English, volume II: On spirals*. Cambridge: Cambridge University Press.

Poincaré, Henri. 1912 [1963]. *Mathematics and science: Last essays*. New York: Dover.

Saito, Ken. 2006. A preliminary study in the critical assessment of diagrams in Greek mathematical works. *SCIAMVS* 7: 81–144.

———. 2012. Traditions of the diagram, tradition of the text: A case study. *Synthese* 186: 7–20.

Saito, Ken, and Nathan Sidoli. 2012. Diagrams and arguments in ancient Greek mathematics: Lessons drawn from comparisons of the manuscript diagrams with those in modern critical editions. In *The history of mathematical proof in ancient traditions*, ed. K. Chemla, 135–162. Cambridge: Cambridge University Press.

Saito, Ken et al. 2014. *Reproduced diagrams from Greek and Arabic manuscripts*. http://www.greekmath.org/diagrams/diagrams_index.html. Accessed 18 Aug 2017.

Sidoli, Nathan. 2011. Apollonius's *conics*: The Greek and Arabic traditions. *Isis* 102: 537–542.

———. 2015. Mathematics education. In *A companion to ancient education*, ed. W.M. Bloomer, 387–400. Oxford: Oxford University Press.

van Leeuwen, Joyce. 2012. *The tradition of the Aristotelian mechanics: Text and diagrams*. Ph.D. dissertation Humboldt University, Berlin.

Wickham, Chris. 2005. *Framing the early middle ages*. Oxford: Oxford University Press.

# Chapter 10
# Editing the Sumerians, How and Why?

**Jerrold S. Cooper**

**Abstract** The science of pedagogy in Babylonia in the early second millennium BCE was based on a two-part curriculum: students began their study of Sumerian by copying and memorizing lists of signs and words, and then progressed to copying connected texts. All surviving manuscripts (in the form of cuneiform tablets) of Sumerian literary compositions from that time are school copies produced by students following the second part of the curriculum. This contribution examines how the editing of these compositions and the representation of individual text witnesses and their variants have evolved in the eight decades since the first scientific edition of a Sumerian literary composition in 1937, and defines what a reconstructed and edited Sumerian composition actually represents.

Writing, the graphic representation of human language, was devised first in the ancient Near East, in the centuries just prior to 3000 BCE, by the Sumerians in the far south of Mesopotamia, in what is now Iraq, and by the Egyptians in the northeastern corner of the African continent.[1] The Sumerian language[2] was at home in southern Babylonia, Babylonia itself being the region between the neck of the Tigris and Euphrates rivers at about the latitude of Baghdad, and the Persian Gulf.

The original version of the chapter has been revised. A correction to this chapter can be found at https://doi.org/10.1007/978-3-031-49617-2_16

The research leading to these results has received funding from the European Research Council under the European Union's Seventh Framework Programme (FP7/2007–2013)/ERC Grant Agreement No. 269804.

[1] For the origins of writing in Mesopotamia and Egypt, see the essays in Houston (2004) and Woods (2010). For the history of the Mesopotamia and neighboring regions, see Van de Mieroop (2015) and Liverani (2014).

[2] See the description of Sumerian in Michalowski (2008).

J. S. Cooper (✉)
The Johns Hopkins University (emeritus), Berkeley, CA, USA
e-mail: anzu@jhu.edu



A. Keller, K. Chemla (eds.), *Shaping the Sciences of the Ancient and Medieval World*, Archimedes 69, https://doi.org/10.1007/978-3-031-49617-2_10

The manuscripts—clay tablets inscribed with cuneiform signs—that I will be discussing are predominantly from Nippur, the early religious capital of Babylonia, with a smaller yet significant number from Ur (of Biblical fame), and much smaller groups of manuscripts from other sites or of unknown provenance.[3] The reason why a discussion of literary manuscripts belongs in a volume on editing ancient *scientific* texts will be put forward after a brief introduction.

Sumerian literature[4] appears, with some minor exceptions, at three moments in Mesopotamian history: the earliest literature is found on several hundred manuscripts from around 2600–2500 BC, very difficult to understand, and frequently written in a cryptographic writing system that has been only partially deciphered. The next large group of Sumerian literary manuscripts, the group that will concern us here, can be dated to a one-hundred-year period in the early second millennium, from approximately 1825–1725 BCE. And then, from the first millennium, there are a large number of Sumerian incantations and liturgical texts, many descended from compositions known from the early second millennium, and most with interlinear Akkadian translations.[5] The earliest written Sumerian—though not literary—appears before 3000 BCE, and the latest Sumerian manuscripts, a few with transliterations into Greek, are dated to near the beginning of our own era.[6]

Sumerian, like Egyptian or Chinese, has an extraordinary record of continuous use spanning over three millennia. But unlike Chinese and Egyptian, after a bit more than the first 1000 years of documentation, Sumerian was a dead language, a learned language, certainly what Ong (1982: 115) has called a male language, taught by men to boys. The teaching was done in schools, which, in the period that interests us here (the early second millennium BCE), were located in private houses and probably never had more than a dozen students, most often less.[7] All of what is considered Sumerian literature in the narrower sense (i.e. excluding magical and liturgical texts) comes from these schools, including mythological texts such as *Inana's Descent to the Netherworld*, the Sumerian tales that were a major source for the Akkadian *Gilgamesh Epic*, hymns to gods and praise songs to kings. There were also minor genres, including scribal dialogues; debates between, say, *Ewe and Grain*, or *Silver and Copper*; and proverbs.[8]

[3] For the particularities of clay tablets as objects and manuscripts, see Taylor (2011).

[4] Rubio (2009), Michalowski (2008: 6–8).

[5] Akkadian is the semitic language that by the early second millennium BCE had replaced Sumerian as the vernacular in Babylonia. Its two main dialects, Babylonian and Assyrian (at home to the north of Babylonia), persisted well into the first millennium BCE, eventually being replaced as the vernacular by Aramaic.

[6] The case for a possible attempt to write prose in the late fourth millennium BCE is made by Civil (2013). For the late tablets with Greek transliterations, see Houston et al. (2003: 454–456) and Oelsner (2014).

[7] For the Babylonian school of the early second millennium BCE, see, e.g., Volk (2000), Robson (2001), Veldhuis (2004: 58–66; 2014: 202–212), George (2005), and Waetzoldt and Cavigneaux (2009: §11).

[8] Rubio (2009), Tinney (2011), Vanstiphout (2003b), Veldhuis (2003, 2004) and Black et al. (2004); this last book is a volume of translations from the Sumerian.

Students began by learning how to shape clay tablets and inscribe wedges, and then learned lists of signs with their values, and a compendium of personal names. There followed thematic word lists, advanced sign lists, metrology and mathematics. Only then did they begin to produce continuous Sumerian texts, starting with model contracts and proverbs, and then a group of four relatively simple literary compositions. All of these were learned by copying a teacher's model, written on a tablet's left-hand column, and copied by the student on the right-hand column. The student's work was erased (smoothed over) and repeated until he got it right. Excerpts of just a few lines of these texts were sometimes written down on small round tablets. Only after mastering all of this material did the student begin to study Sumerian literature in earnest.[9]

Sumerian literary manuscripts, found—when find spots are known—in the same contexts as the practice tablets just described, are primarily of two kinds. They are either large, multi-column tablets containing an entire composition or half of a larger composition, or they are single-column excerpt tablets with from about thirty to one hundred or more lines excerpted from a composition.[10] We imagine that students first learned a composition by writing a series of excerpts, and then, to demonstrate their mastery, wrote out the complete composition on multi-column tablets. For a given composition, the ratio of multi-column tablet manuscripts to single column excerpts is between 1:4 and 1:6.[11] Nothing from this period resembles a library copy or a teacher's master copy; all our manuscripts are the detritus of scribal education. These four thousand or so manuscripts were found on the floors of abandoned school rooms or used as building material in rebuilt school houses. Otherwise, they would have been recycled, soaked in a tub of water until soft and formed into new tablets for student exercises. These manuscripts, then, are the primary evidence for Babylonian pedagogy in the early second millennium BCE, and if pedagogy be a science, then Sumerian literary manuscripts are primary textual evidence for that science.

There are altogether about 6000 manuscripts of Sumerian literature in the larger sense from the first half of the second millennium, and about 4000 represent literary compositions in the narrower sense that I am speaking of today. Over 80% of these manuscripts were excavated at Nippur, and another 10% were found at Ur. About 90% of the manuscripts found at Nippur and Ur fall into the group that has been called curricular, representing about 106 separate compositions,[12] each of which can have anywhere from a few to over 100 manuscripts. Nearly half of these have over twenty manuscripts each; a given line of a composition can have fifteen or more witnesses. Because the manuscripts' origins are so restricted geographically and chronologically, the texts of the Sumerian literary compositions are remarkably stable.

---

[9] See the description of the scribal curriculum in Veldhuis (2014: 204–212).

[10] Delnero (2010) discusses the function of these manuscript types in the study and copying of Sumerian literature.

[11] Based on the table of Delnero (2010: 63).

[12] Delnero (2015); Tinney (2011) puts the number at 'c.a. 175'.

Most variants are either orthographic, or errors typical of manuscript copying. Yet significant variants do occur, and when there are only a few witnesses for a line, even a choice between insignificant variants is not always simple. Clearly, the prime responsibility of an editor is to fully represent the contents of each manuscript, so that the evidence for the editor's choices of readings is available to the reader.

Before presenting a short history of the editing of Sumerian texts, with emphasis on how the representation of multiple manuscripts with their variants has developed, let me clarify what an editor's goal must be, that is, what the resulting text edition represents. The goal is certainly not to reconstruct the composition as it emerged through an author's stylus. We know nothing of the authorship of Sumerian literary compositions, nor, with some exceptions, do we know when or where they were composed. We think that the majority was composed between 100 and 350 years before our manuscripts were copied. Fragments dating to around 2000 BCE of a few of the compositions have turned up at Nippur, and a couple of compositions are found on other fragments dating to c.a. 2500. But most of the tiny corpus of literary manuscripts from 2000, and the vast majority of the larger corpus dating to 2500, consist of compositions that do not appear in the Sumerian literary corpus studied and transmitted in the Babylonian schools of the eighteenth century BCE. We can imagine the context for certain compositions and categories of compositions: royal panegyrics may have been part of court ceremonial under the dynasties of Ur and Isin (twenty-first to twentieth centuries BCE)[13]; the so-called 'epics' of the legendary kings of Uruk—Enmerkar, Lugalbanda and Gilgamesh—may have been written under the patronage of the kings of the Third Dynasty of Ur, who felt a special kinship to those legendary rulers[14]; debates between Silver and Copper and the like may have been court entertainment; hymns to deities could have been used in the cult. But for many compositions, such as the great mythological tales of the gods Inana, Enki, Enlil and Ninurta, scholarly guesses of original context have been few and far between.

What Assyriologists of today aim to produce in a scholarly edition of a Sumerian literary composition is a text that a teacher at Nippur or Ur would recognize as correct, that is, to approximate the as yet undiscovered 'master copy' that the teacher would have taught from, or, if he taught from memory, a text he would recognize as an accurate representation of the composition.[15] The editing process begins with careful study of the individual manuscripts of the composition upon which the reconstruction of the entire composition will be based. Unlike manuscripts in Biblical or Classical corpora, our manuscripts are not the work of scribes intending to produce copies for reference or posterity (Delnero 2012b: 10). Nor are any of the copied Sumerian compositions sacred scripture where absolute fidelity to the text

[13] In a praise hymn to king Shulgi of Ur, Shulgi boasts that he founded academies at Nippur and Ur where scholars would write songs praising him.

[14] Wilcke (2012) has proposed that one of those 'epics' was a drama written to celebrate a specific military victory of Ur in the twenty-first century BCE.

[15] Delnero (2012a) discusses the role of memorization in the transmission of Sumerian literary texts, certainly of great importance, but not as exclusively as he suggests.

would be essential. Because our Sumerian manuscripts are school exercises coming mainly from one place over a short span of time, we almost never can reconstruct manuscript families; manuscripts don't sort into groups that regularly incorporate the same set of variants, even when manuscripts from Ur are thrown into the mix. That is, a given variant can occur in some manuscripts from both Nippur and Ur, but not in others, and usually, there are no Ur manuscripts that regularly pair with individual Nippur manuscripts.

Nippur was first excavated by the University of Pennsylvania at the end of the nineteenth century CE; it was the earliest American archeological expedition to Mesopotamia.[16] Excavations there were resumed in 1948 under the auspices of the American Schools of Oriental Research (now the American Society of Overseas Research), the University of Pennsylvania and the Oriental Institute of the University of Chicago (now the Institute for the Study of Ancient Cultures). These continued until 1990, and the later campaigns were undertaken solely by the Oriental Institute. Of approximately 5000 Sumerian literary manuscripts found at Nippur, 4000 came from the late nineteenth-century excavations, and were divided between the University of Pennsylvania's University Museum (now the Penn Museum) and the Archeological Museum in Istanbul, with several hundred ending up in Jena. But although a considerable corpus of Sumerian manuscripts was available at the beginning of the twentieth century, the first scientific Sumerian text edition didn't appear until 1937. The reasons for this, having to do with museum accession policy and procedures, the fragmentary state of most of the manuscripts, as well as the difficulties of the Sumerian language itself, will not detain us here.

The first scientific editions of Sumerian literary texts were produced by S.N. Kramer, but the story of Sumerian literature has its own Rosalind Franklin or Alice Kober figure (albeit male).[17] In 1923, the Assyriologist Edward Chiera went to Istanbul to copy Sumerian tablets from Nippur, and then spent 3 years copying and studying Nippur tablets in Philadelphia. In 1927, he was called to the University of Chicago's Oriental Institute to work on the Assyrian Dictionary project,[18] and Chicago agreed to publish the several hundred copies he made in Philadelphia. When Chiera died prematurely in 1933, he had not yet prepared the introductions to those volumes, and the task was given to Kramer, then a post-doc at the

[16] For the decidedly non-imperialist origins of American archeology in Iraq, see Cooper (1992). The archeology of Nippur is summarized by Gibson et al. (2001).

[17] Rosalind Franklin's X-ray crystallography was a crucial element leading to the development of Crick and Watson's double-helix model of DNA. Alice Kober's work on Linear B was essential to Michael Ventris's decipherment of that script. Franklin died before the double-helix model was proven and won Crick and Watson their Nobel Prizes; Kober died a few years before Ventris's publication of his decipherment.

[18] The dictionary's history, though mainly its post-war development, is recounted in Reiner (2002), where Chiera's role is mentioned on Page xv. The final volume of the dictionary was published in 2010. It is a dictionary of the Akkadian language (see fn. 5), which early on was called Assyrian, because the earliest European excavations in Mesopotamia were at the great Assyrian capitals of the ninth to seventh centuries BCE. For the same reason, the study of ancient Mesopotamia was called--and still is--Assyriology.

Oriental Institute. Kramer, in his own words, 'got hooked' on Sumerian when preparing those introductions.[19] Working through Chiera's copies, and using the list of unpublished duplicates in Philadelphia and Istanbul compiled by Chiera, Kramer was able to reconstruct several entire Sumerian compositions. The first to be published was *Inana's Descent to the Netherworld*, which appeared in 1937.

Despite Chiera's work, Kramer had only eight manuscripts for his initial edition, just 20% of the forty or so manuscripts known today (many of these would be identified by Kramer himself over the following four decades). He was able to list his manuscripts in a short footnote (see Fig. 10.1). The alphabetical manuscript sigla are assigned according to the first preserved line of the manuscript, and where several manuscripts begin with Line 1, the longest manuscripts are given priority. Variants are given in footnotes, but, as can be seen in Note 2 on his Page 99, although you are told that 'B omits a', you are not told what B has (that is, how much of the line is actually preserved in Manuscript B). This is a shortcoming in the editing of Sumerian manuscripts that would persist until the mid-1970s. But Note 4 reveals a strength in Kramer's method (though 'method' is not what he'd have called it): Even though Manuscript A is a multi-column tablet that originally contained the entire composition, he is not following that manuscript and giving variants from other

98 S. N. KRAMER

TRANSLITERATION AND TRANSLATION [1]

1. [an-gal]-ta [ki-gal-šè] geštúg-ga-ni na-an-g[ub]
From [the great heaven] t[o the *kig*]*allu*, he (?) ga[ve] ear [2] to him (her ?)
2. an a[n-g]al-ta ki-gal-[šè] geštúg-ga-ni na-an-g[ub]
Anu (?) from the great heaven to the *kigallu*, ga[ve] ear [2] to him (her ?)
3. $^{d}$inanna [an-gal-ta ki-gal-šè] geštúg-ga-ni na-an-g[ub]
Inanna [from the great heaven to the *kigallu*], ga[ve] ear [2] to him (her ?)
4. nin-mu an mu-un-šub ki mu-un-šub kur-ra ba-e-a-e$_{11}$ [3]
My Lady, the heaven she forsook, the earth she forsook, to the nether world she descended
5. $^{d}$inanna an mu-un-šub ki mu-un-šub kur-ra ba-e-a-e$_{11}$
Inanna, the heaven she forsook, the earth she forsook, to the nether world she descended
6. nam-en mu-un-šub nam[n]in mu-un-šub kur-ra ba-e-a-e$_{11}$
The *lord*-ship she forsook, the [*la*]*dy*-ship she forsook, to the nether world she descended

1. Reconstructed from the following texts :
Lines 1-207 = A = Ni. 368 (= BE. XXXI, No. 33 ; republished CSRT, No. 33) and CBS 9800 (cf. accompanying photographic reproductions)
1-49 = B = SEM, No. 49
1-48 = C = SEM, No. 50
3-26 = D = BE, XXXI, No. 34.
73-95 = E = SEM, No. 48
116-141 = F = HGT, No. 23
129-137 = G = HGT, No. 24
207-212 = Cf. comment on p. 55
x(212+)1-x58 = H = HGT, No. 22
x58-x63 = Cf. comment on p. 61

For the sake of consistency and because no thorough investigation of the problem involved has as yet been made, all final consonants (except those of words like u$_4$, " day ", šà, " heart ", for which the shorter transcription is more or less accepted by Assyriologists) are reproduced in the transliteration of the Sumerian, despite the numerous indications that even in the more or less artificial Sumerian of the early post-Sumerian period, by no means all the final consonants were pronounced. Similarly the signs BI, GI, NI, etc. are transliterated as bi, gi, ni, etc., despite the fact that we now have proof that they actually represent the syllables be, ge, ne, etc. ; cf. the writer's monograph, " The Sumerian Prefix BE- and BI- in the Time of the Early Princes of Lagaš ".
2. Literally, " raised the ear ".
3. D omits -a- in this and the following lines.

INANNA'S DESCENT TO THE NETHER WORLD 99

7. unug$^{ki}$-ga é-an-na mu-un-šub kur-ra ba-e-a-e$_{11}$ [1]
In Uruk, Eanna she forsook, to the nether world she descended
8. ZA.ḪSLAN-UNU$^{ki}$ [2] -a gi-gu$_{14}$-$^{ki}$na (!) mu-un-šub kur-ra ba-e-a-e$_{11}$
In ZA.USLAN-UNU$^{ki}$, Giguna she forsook, to the nether world she descended
9. [uri$^{ki}$-ma é-dilmun-]na mu-un-šub kur-ra ba-e-a-e$_{11}$ [3]
[In Ur, Edilmun]na she forsook, to the nether world she descended
10. [adaba$^{ki}$] é-šar-ra mu-un-šub kur-ra ba-e-a-e$_{11}$
[In Adab], Ešarra she forsook, to the nether world she descended
11. nibru$^{ki}$-[a dur-an-ki] mu-un-šub kur-ra ba-e-a-e$_{11}$
[In] Nippur, [Duranki] she forsook, to the nether world she descended
12. kiši$^{ki}$-a ḫur-sag-kalam-ma mu-un-šub kur-ra ba-e-a-e$_{11}$
In Kiš, Ḫursagkalamma she forsook, to the nether world she descended
13. a-ga-dè$^{ki}$-a [4] é-UL-maš (!) mu-un-šub kur-ra ba-e-a-e$_{11}$
In Agade, EULmaš she forsook, to the nether world she descended
14. me-imin-bi zag-mu-ni-in-kešd
The seven *rites* she bound to her side
15. me mu-un-kin-kin šu-ni-šè mu-un- [5] -gál
Again and again she sought out the *rites* (and) placed them in her hand
16. me-DU [6] GIR-gub-ba i-im-DU [7]
The... *rites*. .................. .....
17. $^{tug}$šu-gur-ra men-edin-na sag-gá-na [8] mu-un-gál
The *šugurra* garment, the crown of the plain, she set upon her head
18. ḫi-LI sag-ki-na šu-ba-ni-in-ti
Radiance (!) she placed upon her countenance
19. gi-ninda-ḫun-gá-za-gin šu-mi-ni-in-du$_8$ [9]
The... measuring rod of lazuli she gripped in her hand

1. A omits kur-ra ba-e-a-e$_{11}$ in this and the following six lines.
2. B omits -a.
3. Lines 9-13 are omitted in C and D. B where the last five cities are arranged in a different word order lists bàd-tibi[ra]$^{ki}$-a é-MÙ[Š-kalam-ma] instead of Ur and its temple Edilmunna.
4. -a omitted in A.
5. B perhaps U$_5$ for UN ; or is it a bad copy for sign UN ?
6. D, if the copy is correct, has du$_{11}$.
7. D seems to read : um-mi-in-[DU].
8. B seems to omit the sign NA ; more likely, however, the last sign on the line, which looks like a bad MU, may have been NA on the original.
9. This line is not found in the extant text of B and D. In C, line 20 of the obverse corresponds to this line ; the sign GÁN at the beginning of this broken line is a miscopy for GÁ. Instead of šu-mi-ni-in-du$_8$, D seems to have šu-na-[ba]-an-du$_8$.

Fig. 10.1 Kramer 1937: 98–99

[19] See Kramer's account in Kramer (1986: 46–48).

manuscripts in his notes. Rather, he is reconstructing an eclectic best text, preferring other manuscripts to A when he believes they are more correct. When scholars have edited Sumerian texts following the readings of a single best preserved manuscript, they have sometimes produced texts that enshrine erroneous, defective or atypical readings.[20] An unfortunate feature of Kramer's edition is the interlinear English translation, a practice Kramer would soon abandon. Future editions of Sumerian compositions, by Kramer and other others, would put entire translations following the Sumerian text, or, better, on facing pages. Without a connected translation, a Sumerian composition appears to be a series of puzzles requiring line-by-line decipherment; a connected translation was essential for the composition to be received as a work of literature.

Kramer was a pioneer who continued throughout his long career to identify and publish Sumerian manuscripts, as well as to make the Sumerians known to the larger public beyond the academy. But Kramer was not the most careful philologist. His editions could be sloppy, and his lexicography and interpretation were superficial, as he was the first to admit. The standards for fastidious Sumerian text editing were set by Adam Falkenstein at Heidelberg in the 1950s and 1960s. But, as Fig. 10.2 shows, it remained impossible in Falkenstein's editions to ascertain the state of individual manuscripts, since only the variant signs were indicated in the apparatus. The Falkenstein tradition reached its acme at the end of the 1960s in the work of his student, C. Wilcke, whose edition of the Sumerian epic now known as *Lugalbanda II* or *Lugalbanda and the Anzud Bird* (Black et al. 2004) or *The Return of Lugalbanda* (Vanstiphout 2003a) had the added convenience of providing the translation on facing pages (Fig. 10.3). Note, too, that in his Note 1 to line 53, he tells the reader that three manuscripts—AA, H and J—have the sign in question, whereas EE omits it. From this information, one can deduce that the other witnesses to the line—A, FFF and I—must be broken at the line's beginning. The composition Wilcke edits here is one of the few Sumerian literary (in the narrower sense) compositions transmitted beyond the middle of the second millennium BCE, surviving in a couple of bilingual manuscripts from the seventh century BCE, that is, 1000 years after the bulk of the manuscripts was copied. Looking at Lines 58 and following, you can see that he enters the much later manuscripts $B_1$ and $B_2$ with their Akkadian translation separately, which is the right editorial choice to make. Whereas one can justify using contemporary manuscripts from different sites—usually Nippur and Ur—when editing a composition, manuscripts written after the passage of a millennium must be kept apart and utilized not to reconstruct the literary

[20] For example, Hallo and Van Dijk (1968). However, Wilcke (1976: 89) preferred those authors' method to the 'sonst üblichen, auf meist subjektiver Auswahl beruhenden Komposittext' that Kramer had pioneered, but the real problem was not the 'subjective' quality of an editor's choices, but the inability of a reader to easily see which manuscripts had which readings, a problem solved by the 'score' method introduced by Edzard (1974) (see below).

nu-èš-bi šita$_x$(= REC 316)-kù-ge du$_7$-a-me-eš[1]
engar-maḫ-bi sipa-zi-kalam-ma
u$_4$[1]-du$_{10}$-ga zi-dè-eš tu[2]-da-a[3]

engar[1] gána[2]-dagal-la[3] ḫé[4]-du$_7$-àm
ši[1]-im-da-DU nidba[2]-gal-gal-la-kam
giš-lá[1]-bi nu-mu-un-du é-kur-za-gìn-šè

$^{d}$en-líl á-dam-kù[1] ki-a ḪAR-ra-a-za
nibru$^{ki}$ uru ní-za ši-im-mu-e[1]-dù-dù-àm
ki-ùr[1] kur ki-sikil-zu a-bí-du$_{10}$-ga
ub-da-limmú-ba[1] murub$_4$-ba dur-an-ki-ka[2] ki ba-ni[3]-tag-ge

saḫar-bi zi-kalam-ma zi-kur-kur-ra-ka[1]
sig$_4$-bi kù-ḫuš-a uru$_4$[1]-$^{na_4}$za[2]-gìn-na[3]-ka

am-gim ke-en-ge-ra si-mul[1] ba-ni-ib-íla[2]
kur-kur-re sag im-da[1]-ab[2]-sìg-ge
ezem-gal-gal-ba[1] ukù-e nam-ḫé[2]-a u$_4$-bi[3] im[4]-di-ni-ib-zal-e

$^{d}$en-líl[1] $^{d}$uraš[2]-kù ḫi-li du$_8$-du$_8$-a-zu[3]
abzu bára[1]-kù-ga[2] gal-bi[3] túm-ma[4]-zu

59. 1 T du$_7$-me-en, fehlerhaft, s. Z. 58 und 61.
61. 1 So nach R und T, darnach B itu- zu korrigieren. 2 R und T + -ud-. 3 T -da-me-en, fehlerhaft, s. Z. 58—59.
62. 1 A$_1$ engar!. 2 So nach T; B é. 3 In A$_1$ fehlt -la. 4 A$_1$ ḪAR × ×[ ] statt -la ḫé-.
63. 1 B anscheinend ḫi-. 2 A$_1$ wohl nidba!-.
64. 1 So nach R und T; B AB-, wohl in giš-lá- zu korrigieren.
65. 1 A$_1$ anscheinend -igi, vielleicht unvollständig erhaltenes -kù.
66. 1 R -mi- für -mu-e-. 67. 1 A$_1$ -ùr!.
68. 1 A$_1$ -⟨ba⟩. 2 In R und T fehlt -ka. 3 B -e-, R -e-ni-.
69. 1 T -ke$_4$. 70. 1 T uš-. 2 In R und T fehlt $^{na}$$_4$. 3 R -DI-, fehlerhaft.
71. 1 B AN.AN, fehlerhaft. 2 R und T -bé statt -íla in D.
72. 1 E wohl auch -da-. 2 In D, R und T fehlt -ab-; B im-ma-da-.
73. 1 B, R und T -bi. 2 D -ḫi-. 3 In R fehlt -bi. 4 So nach D; B, R und T mu-un-; E ist unklar.
74. 1 E + -le. 2 So nach E; B, D, R und T nur uraš. 3 E -ba, Kopiefehler, auf Tf. XLI für Z. 74—76 korrigiert.
75. 1 E AD, Kopiefehler, auf Tf. XLI korrigiert; B engur, ebenfalls Kopiefehler. 2 E und T -ge. 3 E -bi!. 4 D + -a-.

**Fig. 10.2** Falkenstein 1959: 14

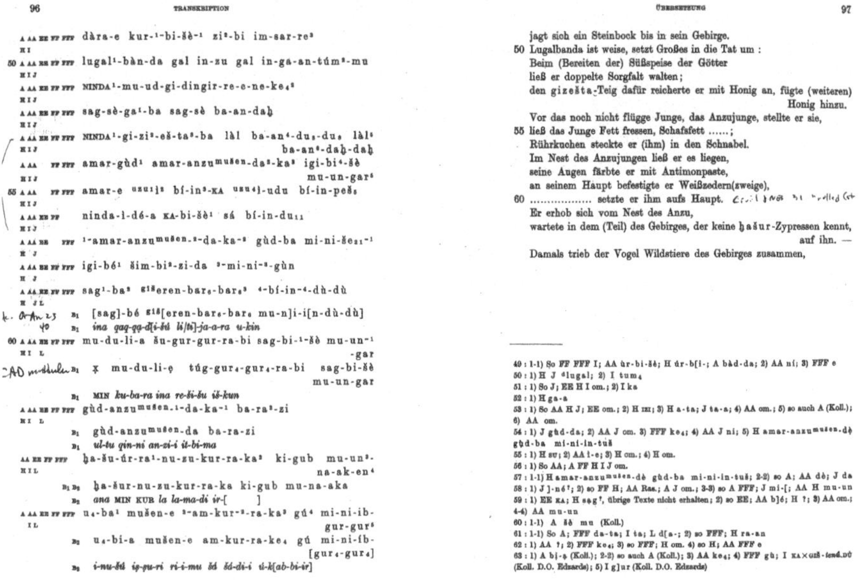

96 TRANSKRIPTION

A AA EE FF FFF dàra-e kur-1-bi-šè-1 zi2-bi im-sar-re3
H I
A AA EE FF FFF lugal1-bàn-da gal in-zu gal in-ga-an-túm2-mu
H I J
A AA EE FF FFF NINDA1-mu-ud-gi-dingir-re-e-ne-ke4 2
H I J
A AA EE FF FFF sag-sè-ga1-ba sag-sè ba-an-daḫ
H I J
A AA EE FF FFF NINDA1-gi-zi2-eš-ta3-ba làl ba-an4-du8-du8 làl5 ba-an6-daḫ-daḫ
H I J
A AA FF FFF amar-gùd1 amar-anzumušen-da2-ka3 igi-bi4-šè mu-un-gar5
H I J
A AA FF FFF amar-e uzu1ì2 bí-in3-KA uzu4ì-udu bí-in-peš5
H I J
A AA EE FF ninda-ì-dé-a KA-bi-šè1 sá bí-in-du11
H I J
A AA EE FFF 1-amar-anzumušen-2-da-ka-2 gùd-ba mi-ni-še21-1
H J
A AA EE FF FFF igi-bé1 šim-bi2-zi-da 3-mi-ni-3-gùn
H J
A AA EE FF FFF sag1-ba2 gišeren-bar6-bar6 3 4-bí-in-4-dù-dù
H J L
B1 [sag]-bé giš[eren-bar6-bar6 mu-n]i-i[n-dù-dù]
B1 *ina qaq-qa-d[i-šú li/ti]-ja-a-ra u-kin*
A AA EE FF FFF mu-du-li-a šu-gur-gur-ra-bi sag-bi-1-šè mu-un-1-gar
H I L
B1 x mu-du-li-e túg-gur4-gur4-ra-bi sag-bi-šè mu-un-gar
B1 MIN *ku-ba-ra ina re-ši-šu iš-kun*
A AA EE FF FFF gùd-anzumušen-1-da-ka-1 ba-ra2-zi
H I L
B1 gùd-anzumušen-da ba-ra-zi
B1 *ul-tu qin-ni an-zi-i it-bi-ma*
AA EE FF FFF ḫa-šu-úr-ra1-nu-zu-kur-ra-ka2 ki-gub mu-un3-na-ak-en4
H I L
B1 B2 ḫa-šur-nu-zu-kur-ra-ka ki-gub mu-na-aka
B2 *ana* MIN KUR *la la-ma-di ir-*[ ]
A AA EE FF FFF u4-ba1 mušen-e 2-am-kur-2-ra-ka3 gú4 mi-ni-ib-gur-gur5
I L
B2 u4-bi-a mušen-e am-kur-ra-ke4 gú mi-ni-íb-[gur4-gur4]
B2 *i-nu-šú iṣ-ṣu-ri ri-i-mu šá šá-di-i ú-k[ab-bi-ir]*

ÜBERSETZUNG 97

jagt sich ein Steinbock bis in sein Gebirge.
Lugalbanda ist weise, setzt Großes in die Tat um:
Beim (Bereiten der) Süßspeise der Götter
ließ er doppelte Sorgfalt walten;
den gizešta-Teig dafür reicherte er mit Honig an, fügte (weiteren) Honig hinzu.
Vor das noch nicht flügge Junge, das Anzujunge, stellte er sie,
ließ das Junge Fett fressen, Schafsfett ......;
Rührkuchen steckte er (ihm) in den Schnabel.
Im Nest des Anzujungen ließ er es liegen,
seine Augen färbte er mit Antimonpaste,
an seinem Haupt befestigte er Weißzedern(zweige),
................. setzte er ihm aufs Haupt.
Er erhob sich vom Nest des Anzu,
wartete in dem (Teil) des Gebirges, der keine ḫašur-Zypressen kennt, auf ihn. —
Damals trieb der Vogel Wildstiere des Gebirges zusammen,

49 : 1-1) So FF FFF I; AA ùr-bi-šè; H úr-b[i-; A bàd-da; 2) AA ní; 3) FFF e
50 : 1) H J dlugal; 2) I tum4
51 : 1) So J; EE H I om.; 2) I ka
52 : 1) H ga-a
53 : 1) So AA H J; EE om.; 2) H IZI; 3) H a-ta; J ta-a; 4) AA om.; 5) so auch A (Koll.); 6) AA om.
54 : 1) J gùd-da; 2) AA J om. 3) FFF ke4; 4) AA J ni; 5) H amar-anzumušen-dè gùd-ba mi-ni-in-tuš
55 : 1) H SU; 2) AA ì-e; 3) H om.; 4) H om.
56 : 1) So AA; A FF H I J om.
57 : 1-1) H amar-anzumušen-dè gùd-ba mi-ni-in-tuš; 2-2) so A; AA dè; J da
58 : 1) J ]-né?; 2) so FF H; AA Ras.; A J om.; 3-3) so A FFF; J mi-[; AA H mu-un
59 : 1) EE KA; H sag?, übrige Texte nicht erhalten; 2) so EE; AA b]é; H ?; 3) AA om.; 4-4) AA mu-un
60 : 1-1) A šè mu (Koll.)
61 : 1-1) So A; FFF da-ta; I ta; L d[a-; 2) so FFF; H ra-an
62 : 1) AA ?; 2) FFF ke4; 3) so FFF; H om. 4) so H; AA FFF e
63 : 1) A bi-ṣ (Koll.); 2-2) so auch A (Koll.); 3) AA ke4; 4) FFF gù; I KA×GIŠ-tenû.DÚ (Koll. D.O. Edzards); 5) I g]ur (Koll. D.O. Edzards)

**Fig. 10.3** Wilcke 1969: 96–97

composition as it was known in 1800 BCE, but rather to understand how Sumerian language and literature were transmitted and received in 800 BCE.[21]

Despite Wilcke's very thorough apparatus, which enables the user to imagine the state of the manuscripts for a given variant, Sumerian compositions were still presented as a succession of lines of reconstructed text, with a forest of footnotes constituting the critical apparatus. The characteristics of individual manuscripts could only be ascertained by assembling the various copies and photographs of tablets used by the editor. This changed in the mid-1970s when another Falkenstein student, D. Edzard, challenged the accepted method (Edzard 1974). He proposed that the entire text of each witness for each line be presented, one after the other (Fig. 10.4), in an arrangement he likened, mistakenly I think, to a musical score, and the name 'score', or '*Partitur*' in German, has stuck. His solution threatened to bring on what Pollock (2009: 952) has called 'Text Panic', because, as is evident from his example, there is no editorial decision, no reconstructed 'best text', data but no interpretation. This is quite user unfriendly when there are as many individual manuscripts as in his example here, but not quite as bad in the case of the first complete composition to be so edited (Fig. 10.5), a Sumerian hymn with only a few extant manuscripts that Edzard published together with Wilcke in 1976.

[21] Of course, in the very rare cases when all of the early manuscripts for a line or a portion of a composition are broken, but that text portion is preserved in a later manuscript, the later manuscript must serve as the basis for reconstruction, with all due caveats.

106 D. O. Edzard

Das Arbeiten mit "Partituren" zeitigt verblüffende Ergebnisse. Es treten auf einen Blick Manuskriptzusammenhänge zutage, die sonst im Dickicht des Variantenapparates verloren gehen. Es wird leichter möglich, Manuskriptstemmata herzustellen, wie sie in der klassischen Philologie eine Selbstverständlichkeit sind [13]. Nur mit Hilfe von 'Partituren' lassen sich auch Vergleiche von einem Literaturwerk zum anderen ziehen, womöglich der Einfluss bestimmter Schreiber erkennen und vieles mehr. Sumerische Philologie und Literaturwissenschaft wird auf die Dauer an einer Änderung der Editionsweise nicht vorbeikommen, wenn wirkliche Fortschritte erzielt werden sollen.

Das hier Gesagte möchte ich an Z. 58-59 der Keš-Hymne verdeutlichen. Die 'Partitur' sieht so aus [14]:

A uru$^{ki}$ ga-àm uru$^{ki}$ ga-àm šà-bi a-ba mu-zu
é-kèš$^{ki}$ uru$^{ki}$ ga-àm šà-bi a-ba mu-zu
B [×] in-ga-àm uru in-ga-àm šà-bi a-ba-a mu-un-[z]u
é-kèš$^{ki}$ uru in-ga-àm šà-bi a-ba-a mu-[u]n-[z]u
C uru in-ga-àm uru in-ga-[...]
é-kèš$^{ki}$ uru in-ga-àm [...]
I [...-à]m uru in-ga-àm šà-bi a-ba-a mu-un-zu
[... ur]u in-ga-àm šà-bi a-ba-a mu-un-zu
J uru$^{ki}$ in-ga-àm uru$^{ki}$ in-ga-àm ... a-ba-a mu-un-zu x̣
(Ende des Textes)
O uru$^{ki}$ ga-àm uru$^{ki}$ ga-àm ... a-ba-a mu-un-zu
(Ende des Textes)
BB ...
(abgebrochen)
DD ...
(abgebrochen)
FF uru in-ga uru in-ga ... a-ba-a mu-zu
... uru in-ga ... a-ba-a mu-un-zu
GG ...]-àm ... in-ga-àm ...
... uru in-ga-àm ...
KK uru in-g[a ...]
é-kè[š ...]
RR [... i]n-ga-àm uru i[n-...]
[...] x̣ x̣ [...]

**Fig. 10.4** From Edzard 1974: 106

My own effort in 1983 to devise a method of presenting and editing Sumerian manuscripts strove to preserve the advantages of Edzard's 'score' method but to streamline the representation of individual manuscripts in order to enable the differences and similarities of the manuscripts to be grasped more quickly (Fig. 10.6). The manuscripts were arrayed beneath a reconstructed eclectic 'best text' for each line: when a manuscript was identical to the reconstructed text, a dash was placed beneath each preserved sign; an omitted sign was indicated by zero and only

| | | |
|---|---|---|
| 241' | A | lú udu [sa$_{10}$-s]a$_{10}$-dè igi-g̃ál-la-ni nu-mu-na-ab-bé |
| | E$_2$ | [lú] udu sa$_{10}$-sa$_{10}$-dè igi-g̃ál-la-ni nu-mu-n[a- . . .] |
| | F | lú udu$^{!}$ šám-šám igi-g̃ál-[l]a-ni nam-mu-na-ab-bé |
| 242' | A | $^{d}$ḫen[dur-s]ag̃-g̃á-ra m[u-n]a-ab-bé |
| | E$_2$ | [t]ukumbi $^{d}$ḫendur-sag̃-g̃á-k[e$_4$. . .] / [udu] sa$_{10}$-sa$_{10}$-dè mu-na-ab-b[é] |
| | F | tukumbi $^{d}$ḫendur-sag̃-g̃á-ke$_4$ / udu šám-šám-da mu- na-ab-bé |
| 243' | A | $^{d}$utu [amaš ni]g̃ín-na-bi nam im-ma-ab-tar-re |
| | E$_2$ | [$^{d}$]utu amaš nig̃ín-na-bi nam mu-ni-ib-[tar-re] |
| | F | $^{d}$utu <amaš> nig̃ín-n[a]-bi-a$^{!}$ nam im-<mi>-ib-tar-re |
| 244' | A | tu[kum]bi lú sag̃ in-sa$_{10}$-sa$_{10}$ |
| | E$_2$ | [tukum]bi [lú] sag̃ ib$^{!}$-sa$_{10}$-s[a$_{10}$] |
| | F | tukumbi lú-u$_x$ sag̃ in-šám-šám |
| 245' | A | $^{d}$utu $^{d}$ḫendur-sag̃-g̃á-ke$_4$ èn mu-na-tar-re |
| | E$_2$ | [. . . $^{d}$]ḫendur-sag̃-ra èn mu-na-tar-r[e] |
| | F | $^{d}$utu $^{d}$ḫendur-sag̃-g̃á-ke$_4$ èn mu-na-tar-re |
| 246' | A | lú sag̃ sa$_{10}$-sa$_{10}$-dè ig[i-g̃]ál-la-ni nu-mu-na-ab-bé |
| | E$_2$ | [lú] sag̃ sa$_{10}$-sa$_{10}$-dè igi-g̃ál-la-ni nu-mu-n[a-ab-bé] |
| | F | lú sag̃ šám-šám igi-g̃ál-la-ni nam-{MU}-mu-na-ab-bé |
| 247' | A | $^{d}$ḫendur-sag̃-g̃á-ra mu-na-ab-bé |
| | E$_2$ | [tuku]mbi $^{d}$ḫendur-sag̃-g̃á-[x] / [sag̃] sa$_{10}$-sa$_{10}$ (Rasur) mu-na-ab-bé |
| | F | tuk[umbi $^{d}$ḫendur]-sag̃-g̃á-ke$_4$ / [sag̃ šám-šám-da mu-na-a]b-bé |
| 248' | A | $^{d}$utu á-g̃iš$^{!}$-rín-na-ka$^{!}$ mu-un-na-gi-né$^{!}$ |
| | E$_2$ | [. . .] ki-g̃iš-rín-na-ka mu-un-[. . .](!) |
| | F | [$^{d}$utu . . . -n]é |
| 249' | A | tukumbi lú dam in-TUK.TUK |
| | E$_2$ | [tuku]mbi lú dam in-TUK.TUK |
| 250' | A | $^{d}$utu $^{d}$ḫundur-sag̃-[g̃]á-ke$_4$ èn mu-na-tar-re |
| | E$_2$ | [. . . $^{d}$ḫ]endur-sag̃-ra èn mu-na-ta[r-re] |
| | F | [$^{d}$ut]u $^{d}$ḫendur-sag̃-g̃á-ke$_4$ [. . .] |
| 251' | A | lú dam TUK.TUK-dè igi-g̃ál-la-ni nu-mu-na-ab-bé |
| | E$_2$ | [. . .] igi-g̃á[l- . . .] |
| | E$_1$ | lú dam TUK.TUK [. . .] |
| | F | lú dam TUK.TUK igi-g̃ál-la-ni [. . .] |
| 252' | A | <$^{d}$ḫendur-sag̃-g̃á-ra mu-na-ab-bé> |
| | E$_1$ | tukumbi [. . .] / dam TUK.TUK-dè [. . .] |
| | F | tukumbi {$^{d}$utu} $^{d}$ḫendur$^{!}$-sag̃-ke$_4$ / [dam] TUK.TUK-da mu-na-ab-bé |
| 253' | A | lú-u$_x$-bi dam in-ga-an-TUK.TUK dumu in-ga-an-ù-tu |
| | E$_1$ | lú-u$_x$-bi dam in-ga-[. . .] |
| | F | [lú]-u$_x$-bi dam in-ga-TUK.TUK du[mu in]-ga-ù-tu |
| 254' | A | é in-ga-dù-e GÁ in-ga-NIGÍN.NIGÍN |
| | E$_1$ | é in-ga-dù-e G[Á . . .] |
| | F | [. . . -g]a-dù-e GÁ [i]n-ga-NIGÍN.NIGÍN |
| 255' | A | nita-bi dam-a$^{!}$-n[i$^{?}$] du$_{10}$-bi mu-da-ab-bé |
| | E$_1$ | nita-bi dam-a-ni du$_{10}$-b[i . . .] |
| | F | [. . .]-a-ni du$_{10}$-bi m[u]-un-da-bé |

**Fig. 10.5** Edzard and Wilcke 1976: 158

```
127.               a-gàr mah a-eštub^ku6 dagal-la-gim
C                  - -   - ⌈-?  -?  -?⌉[            ]
J                  - ⌈-   -⌉[               ] ⌈-?⌉
P        a-šà      - -   - -  ⌈-⌉  ø   -   ŧ    ]
T                  - -   - -  ⌈-⌉  ø   ŧ        ]
V        a-šà      - -   - -  -    ø   -   -  ke4
G1       a-šà      - -  ⌈-⌉ -  ŧ                ]
S1       [     ] ⌈-  -   -⌉[                    ]
M2                 - -  ⌈-⌉ -  -   ŧ            ]
T2                [                            ŧ?
W2                 - -   - -  -    ŧ            ]
Y2       a-sa      - ŧ]  - -  -    ø   -   ŧ    ]
B4                 - -  ⌈-⌉[                    ]
D4                 - -   -* -  ŧ                ]**

R3                 ugur2  - - -    ø  dugud a  -***
S3                 ugur2  - - -    ø   -   ø / -****
                           *adds gim?
                          **skips to 173
                         ***127f. between 113 and 115
                        ****127f. follow 114
```

**Fig. 10.6** Cooper 1983: 150

variants were written out.[22] Note here the two manuscripts at bottom ($R_3$, $S_3$), spaced apart from the others. This composition is one of the few that has manuscript witnesses from ca. 2000 BCE—three centuries earlier than the others—and it is these

[22] I no longer remember whether this was all of my own devising or whether, more probably, I had been influenced by unpublished manuscripts of M. Civil, which I had seen (see below). I am certain that part of my motivation was that inserting the diacritics needed for transliteration was so cumbersome in those days.

**COMPOSITE TEXT AND TRANSLATION**

***Composite Text*[1]**

1. sag-ki gíd-da $^{d}$en-líl-lá-ke$_4$
2. kiš$^{ki}$ gu$_4$ an-na-gim im-ug$_5$-ga-ta
3. é ki unu$^{ki}$-ga gu$_4$-maḫ-gim saḫar-ra mi-ni-ib-gaz-a-ta
4. KI.UD-ba šar-ru-GI lugal a-ga-dè$^{ki}$-ra
5. sig-ta igi-nim-šè $^{d}$en-líl-le
6. nam-en nam-lugal-la mu-un-na-an-sum-ma-ta
7. u$_4$-ba èš a-ga-dè$^{ki}$ kù $^{d}$inanna-ke$_4$
8. ama$_5$ maḫ-a-ni-šè im-ma-an-dù-dù
9. ul-maš$^{ki}$-a $^{giš}$gu-za ba-ni-in-gub
10. lú-tur gibil-bi é dù-ù-gim
11. dumu-bàn-da ama$_5$ gá-gá-gim
12. é níg-ga-ra níg sá-di-dè
13. uru$^{ki}$-bé dúr ki-gar sum-mu-dè
14. un-bé ú nir-gál kú-ù-dè
15. un-bé a nir-gál nag-nag-dè
16. sag a-tu$_5$-a kisal ḫúl-le-dè
17. ki ezem-ma un sig$_7$-ge-dè
18. lu zu-ù-ne téš-bi kú-ù-dè
19. lú bar-ra mušen nu-zu-gim an-na nigin-dè
20. mar-ḫa-ši$^{ki}$ le-um-ma gur-ru-dè
21. $^{ugu}$ugu$_4$-bi am-si maḫ áb-za-za ú-ma-am ki bad-rá
22. šà sila-dagal-la-ke$_4$ téš-bi tag-tag-ge-dè
23. ur-gi$_7$ ur-nim dara$_3$[2] kur-ra udu-a-lum SÍG.BU-si
24. kù $^{d}$inanna-ke$_4$ ù nu-um-ši-ku-ku
25. u$_4$-ba a-ga-dè$^{ki}$ é-ÁŠ-a-ba kù-GI mi-ni-in-si
26. é-ÁŠ babbar-ra-ba kù-babbar mi-ni-in-si
27. araḫ$_4$ še-ba urudu an-na na$_4$-lagab za-gìn-na sá im-mi-in-du$_{11}$-du$_{11}$[3]
28. gur$_7$-bi bar-ta im[4] ba-an-ùr

50

***Translation*[5]**

*I. The Rise of Agade*

After Enlil's frown
Had slain Kish like the Bull of Heaven,
Had slaughtered the house of the land of Uruk in the dust like a mighty bull,
And then, to Sargon, king of Agade,
Enlil, from south to north,
Had given sovereignty and kingship—
At that time, holy Inanna built
The sanctuary Agade as her grand woman's domain,
Set up her throne in Ulmaš.
Like a youngster building a house for the first time,
Like a girl establishing a woman's domain,
So that the warehouses would be provisioned,
That dwellings would be founded in that city,
That its people would eat splendid food,
That its people would drink splendid beverages,
That those bathed (for holidays) would rejoice in the courtyards,
That the people would throng the places of celebration,
That acquaintances would dine together,
That foreigners would cruise about like unusual birds in the sky,
That (even) Marḫaši would be reentered on the (tribute) rolls,
That monkeys, mighty elephants, *water buffalo*, exotic animals,
Would jostle each other in the public squares—
*Thoroughbred dogs, lions, mountain ibexes,*[6] *alu*-sheep with long wool—
(So that all this might happen), Holy Inanna did not sleep.
At that time, she filled Agade's . . . with gold,
She filled its shining . . . with silver,
Delivered copper, tin, and blocks of lapis to its granaries,
Sealed them away in its silos.

**Fig. 10.7** Cooper 1983: 50–51

that I have kept apart. For the user's convenience, as well to emphasize that what we have here is a literary text and not just a collection of Sumerian words and sentences, I separately laid out the eclectic 'best text' facing an English translation (Fig. 10.7), with only significant variants indicated in footnotes.

Thus, by the mid-1980s, Sumerian text editions looked quite a bit different from those of Kramer or Falkenstein, but one thing had not changed: manuscript sigla were alphabetic, with double letters or letters with numerical subscripts used when the number of manuscripts exceeded twenty-six. A 1996 publication of S. Tinney changed that (Fig. 10.8): Manuscripts are identified by place of origin: N for Nippur manuscripts, U for Ur manuscripts and X for unknown. Within geographical groups, manuscripts are distinguished only by subscript. But, as can be seen, most manuscripts are from Nippur (N), and it is rather difficult, especially for older colleagues, to distinguish the subscript numbers. Even if you can read the numbers easily, the manuscripts are harder to get to know than when they each have their own upper-case siglum. Nevertheless, Tinney made an important point, and other text editions, beginning in the mid-1990s (see below), have managed to present manuscript provenance while preserving the distinctiveness of the upper-case sigla.

Just before Tinney's publication, M. Civil (1994) published an edition that altered somewhat the representation of manuscripts, adhering to a method he told me he had employed since the late 1950s in the hundreds of unpublished editions he has made of Sumerian literary compositions (Fig. 10.9). Civil considered each line of text with its witnesses a 'textual matrix', and the representation of the composition as a whole 'textual matrices', a far better term than the misleading 'score' (though I fear that the latter's conciseness ensures that it will continue to be the preferred

Score 225

**188.** ama$_5$-mu dù-ma-ab é-mu si-ma-ab in-na-an-dug$_4$-ga-àm
$N_1$ ⌜⌝ - - - - - - ⌜⌝ [ ] / - - - - - ⌜⌝
$N_7$ [ ] ⌜⌝ - ⌜⌝ [ ] / - ⌜⌝ - ⌜⌝ g[a ]
$N_{13}$ - - - - - - [ ]
$N_{14}$ é! - - - - ⌜ama$_5$⌝- ⌜⌝ - - - - - - - a
$X_1$ é - - - - GÁ - - - - - - - - - *
In $N_1$ ⌜àm⌝ is A.[AN].

**189.** šà-ba-ni-a ad i-ni-in-gi$_4$-eš-àm e-ne-èm-bi bí-in-dug$_4$-ga
$N_1$ - - - - - - - ⌜⌝ [ ] // - - - * - [ ]
$N_7$ - - - - - - - ⌜⌝ [ ] / - - - ⌜⌝ [ ]
$N_{13}$ - - - - ⌜⌝ - - - ⌜⌝ [ ]
$N_{14}$ - - - - - - - - - - a - - - * - - eš a
$X_1$ - - - - - - - - - - - - - - - / - - - -

**190.** $^d$mu-ul-líl e-ne-èm zi dug$_4$-ga-ni in-na-a-gin-na
$N_1$ - - ⌜- - ⌝[ ]
$N_7$ ⌜⌝ - - - - - ⌜⌝ [ ] / - - - - [ ]
$N_{13}$ - - - -e - - [ ]
$N_{14}$ - - - - - - - - - - -a - ⌜⌝ * - -
$X_1$ - - - - - - - - - ba - dè - - -

**191.** é zi ág-ḫa-lam-ma ba-ab-aka sa$_6$-ga-aš i-ni-in-ku$_4$-ra
$N_7$ ⌜- ⌝ - ⌜- - ⌝ [ ] / ⌜⌝ - - ⌜⌝[ ]
$N_{13}$ - - - - - x[ ]
$N_{14}$ - - - - - - ba-e-a-aka-a - - - - - - - -
$X_1$ - - - - - - - - - - - - - - - ⌜-/ ⌝

**192.** ág ur$_5$-ra-ra e-ra-ni-ib-gi$_4$-a e-ne íb-ta-an-e$_{11}$-dè
$N_7$ - - - * - - - - - ⌜- ⌝ - ⌜-? - ⌝ x[ ]
$N_{13}$ ⌜⌝ - - ⌜- - - ⌝ [ ]
$N_{14}$ - - - - e - - íb - - - - - - - - -
$X_1$ - - - - ér - - - - - - - - / ⌜- - ⌝ zi-ga
In $N_7$ e-ra is written over erased A.IGI, obviously as a result of confusing e-ra with ér-ra. Note the similar variant in $X_1$. In $N_{14}$ there is an erasure after the second ra.

**Fig. 10.8** Tinney 1996: 225

*The Farmer's Instructions*

```
004   a-šà a-dé-a-zu-  dè a-bi šà-ba nu-   íl
 A    + o  o .  + +  - NI + +  + ba nu     o
 B    o .  + +  + +  ù x  o o  o  o  o     o
 C    + +  + .  . +  - dè + +  + ba ga-ab x
 E    o o  o o  o o  o o  o .  + ba nu     +
 A2   + +  + +  + +  - dè + +  + bi nu     o
 A3   . +  +.+  + +  - dè + +  + bi nu     +

005   u4 a-ta   im-mu-e-a-e11-dè-a
 A    +  + +  - +  mu + + e11 .  +
 B    .  + +  - +  me + o o   o  o
 C    +  - +  - +  x  - - e11 .  +
 A2   .  + +  i +  me - - e11 +  o
 A3   +  + +  - +  mu + - è   +  +

006   a-šà ki-duru5-bi en-nu-un ak- ab ki-še-er a-ra-ab-tuku
 A    + +  + +      + + + +  ak  x  o  .  +  + +  +  +
 B    + +  + +      + + + +  a[k o  o  o  o  o o  o  o
 C    + +  + +      + + + +  k[a ab +  +  .  + .  +  +
 E    . .  . o      . o o .  ak  a  +  +  o  o o  +  +
 A2   + +  + +      + + + +  ak  ab +  +  +  + +  .  o
 A3   + +  + +      + + + +  nag ab +  +  +  + +  +  +
*D edge: ki-du-ru-bi (eras.) e-ja

007   gud suhúb  gìr na-ra-ab-zukum-e
 A    +   suhúb  +   + + +  .    +
 B    +   suhúb  +   + + +  .    o
 C    +   suh[úb o   o o o  .    +
 A2   .   suhub  +   + + +  +    o
 A3   +   suhub  +   + + +  +    +

008   ú-sag11- bi ù-bí-  zé a-šà ki-gar ù-bí-dug4
 A    + +      - + + bi   + + +  .  .   + + +
 B    + +      - + + ba-e + + +  .  o   o o o
 C    + +      a + o o    o o o  o  o   o o o
 E    o +      - + o o    + o o  .  +   + + +
 A2   + +      - + + bi   + + +  +  +   + + +
 A3   + +      - + + bi   + + +  +  +   + + .

009   10-àm uruda.gín-sal 2/3 ma-na-ta-   sá-a-ab
 A    + - +       .  +    +   + + +   -   + + +
 B    + - +       +  +    +   + + +   -   + . o
 E    + + -       +  +    +   + + +   -   + + +
 A2   + - +       +  +    +   + + +   -   + + +
 A3   + - +       +  +    +   + + .   àm  + + +

010   giš.al-šub-bé umbin gud  ha-ra-   šú-šú sa ha-ra-ab-lá-lá
 A    +   + +   + +       gud? + .  -   o  o  .  +  +  +  +  +
 B    +   + +   + +       bi?  + +  -   +! +! +  +  o  o  +  +
 E    +   + .   - .       gud  + +  -   +  +  +  .  .  .  +  +
 A2   +   + +   + +       bi?  + +  ab  +  +  +  +  +  +  +  +
 A3   +   + -   + +       gu[d o .  .   o  x  +  +  +  +  o  o
```

**Fig. 10.9** Civil 1994: 208

term). A plus-mark is used when a sign is the same as that in the eclectic text, a minus when it is omitted, a dot indicates a partially preserved sign and a zero is used where the tablet is broken. Civil pointed out to me an important but as yet unnoticed difference in the way he treated variants: when variants to a sign in the eclectic text are present, the sign in every manuscript is spelled out, even when it agrees with the eclectic text. Should an editor at some time before publication decide that another of the variants is to be preferred to the one he already chose for his eclectic text, the editor need only change the eclectic text and not the individual manuscript lines.

Meanwhile, our Germanophone colleagues were uneasy about the dots, dashes, plusses and zeros. For reasons that were at first difficult to comprehend, they wanted to see every sign of every manuscript spelled out.[23] Seemingly, this is a conflict between the maximal presentation of data and an economical one,[24] yet I remain doubtful that anything is gained by the former, whereas the advantages of the latter are clear. K. Volk (1995) (Fig. 10.10) uses upper-case sigla, adding provenience in lower case (n = Nippur, u = Ur), but his maximalist presentation makes it difficult to tell at a glance that in Line 129, the only variant is in manuscript Fu, which reads /ni/ where other manuscripts read /na/. Volk conveniently adds the column and line number for each line in each manuscript, also done by Zgoll (1997) (Fig. 10.11), who, however, introduces a very complicated system for her sigla: Each siglum is composed of a two-letter geographic abbreviation—Ni for Nippur, Ur for Ur, La for Larsa and Un for Unbekannt—and a capital letter, whereby each geographical category has its own sequence of letters. So, for example, there is a NiB as well as an UrB. Neither Volk nor Zgoll precede the individual readings for each line with their eclectic 'best' text, given elsewhere facing a connected translation, so a reader going through Zgoll's 'score', for example (Fig. 10.11), will not immediately see that in Line 151, the reading of NiA is not the reading the editor accepts as correct. Mittermayer (2009) (Fig. 10.12) follows Volk's construction of sigla (upper-case plus lower-case indication of provenience) and the addition of manuscript column and line numbers of both Volk and Zgoll, but precedes the manuscript readings with each line's eclectic text. To my mind, this is the most successful of the maximalist formats.

More recently, a pioneering monograph on editing Sumerian texts has been published by P. Delnero (2012b). His primary concern is the motivation and evaluation of textual variants, based on a painstaking analysis of 740 manuscripts of ten frequently copied Sumerian compositions, thought to be among the first advanced literary compositions copied by students at early second millennium BC Nippur. Although Delnero is American, he studied for many years in Germany and hence presents the entire text of each manuscript, highlighting variants from his eclectic text in boldface (Fig. 10.13). His sigla are also maximalist in the information they provide and revert to Tinney's construction of sigla using upper-case indicators of provenience: Nippur manuscripts are labeled N (Ur stands for a manuscript from Ur and X for unknown provenience), followed by a subscript identifying (only for the

[23] This was, of course, becoming much easier with electronic text production.

[24] As Glenn Most explained to me at the Paris conference.

*K. Volk, Inanna und Šukaletuda* **105**

| | | | | |
|---|---|---|---|---|
| **123** | Cn$_1$ | iii | 4’ | g̃iš im-ma-n[i-in-du$_{11}$ ] |
| | Dn | iii | 20’ | g̃iš im-ma-ni-in-du$_{11}$ ne im-ma-[ni-in-su-ub] |
| | Fu | iii | 5 | g̃iš im-ma-ni-in-d[u$_{11}$ ] |
| **124** | Cn$_1$ | iii | om. | |
| | Dn | iii | om. | |
| | Fu | iii | 6 | g̃iš ba-ni-in-du$_{11}$-g[a ne ba-ni-in-su-ub-ba] |
| **125** | Cn$_1$ | iii | 5’ | zà-sar-ra-k[a-ni ] |
| | Dn | iii | 21’ | ⸢zà-sar-ra-ka⸣-ni im-ma-ši-i[n-gi$_4$] |
| | Fu | iii | 7 | zà-sar-ra-ka-ni i[m-ma-ši-in-gi$_4$] |
| **126** | Cn$_1$ | iii | 6’ | u$_4$ im-zal [ ] |
| | Dn | iii | 22’ | [u$_4$ i]m-⸢zal⸣ $^{d}$utu im-ta-⸢è-a⸣-[ra] |
| | Fu | iii | 8 | u$_4$ im-zal $^{d}$⸢utu⸣ i[m-ta-è-a-ra] |
| **127** | Cn$_1$ | iii | 7’ | munus-e ní t[e-a-ni ] |
| | Dn | iii | 23’ | *m[unus]-⸢e⸣ ní te-a-ni igi im-kár-ká[r] |
| | Fu | iii | 9 | ⸢munus⸣-e ní te-na ⸢igi⸣ im-k[ár-kár] |
| **128** | Cn$_1$ | iii | 8’ | kù-$^{d}$inanna-ke$_4$ [ ] |
| | Dn | iii | 24’ | *⸢$^{d}$⸣[inann]a-ke$_4$ ní te-a-ni igi im-kár-kár |
| | Fu | iii | 10 | ⸢kù⸣-$^{d}$inanna-ke$_4$ ní te-na igi im-kár-[kár] |
| **129** | Cn$_1$ | iii | 9’ | u$_4$-ba munus-⸢e⸣ [*na]m-⸢gal$_4$⸣-la-⸢na⸣-*š[è ] |
| | Dn | iii | 25’ | *⸢u$_4$⸣-[ba] ⸢munus⸣-e nam-gal$_4$-la-na-šè a-na im-gu-lu-u$_8$-a-bi |
| | Fu | iii | 11 | u$_4$-ba munus-e nam-gal$_4$-la-ni-[šè] a-na im-gu-lu-⸢u$_8$⸣-[a-bi] |
| | Hn | Vs. | 1’ | [ na]m-gal$_4$-⸢la⸣-na-⸢šè a-na⸣ *i[m-gu-lu-u$_8$-a-bi] |
| **130** | Cn$_1$ | iii | 10’ | kù-$^{d}$inanna-ke$_4$ nam-gal$_4$-la-*ni(über na?)-š[è ] |
| | Dn | iii | 26’ | *⸢$^{d}$⸣[inanna]-⸢ke$_4$⸣ nam-gal$_4$-la-na-šè a-na im-kè-{lu-u$_8$}-a-bi |
| | Fu | iii | 12 | kù-⸢$^{d}$inanna-ke$_4$⸣ nam-gal$_4$-la-ni-šè a-na ak-k[a?-bi] |
| | Hn | Vs. | 2’ | [kù-$^{d}$inanna-*k]e$_4$ nam-gal$_4$-la-ni-šè a-na [ ] |
| **131** | Cn$_1$ | iii | 11’ | pú-*kalam-ma-ka úš bí-[ib-si-si] |
| | Dn | iii | 27’ | pú-[kala]m-ma-ka úš (Rasur) bí-ib-si-si |
| | Fu | iii | 13 | pú-kalam-ma-ka úš bí-íb-s[i-si] |
| | Hn | Vs. | 3’ | [pú-kalam-ma]-ka úš bí-ib-[si-si] |
| **132** | Cn$_1$ | iii | 12’ | pú-$^{g̃iš}$ki]ri$_6$-kalam-ma-ka úš-àm ⸢ì⸣-[tùm-tùm] |
| | Dn | iii | 28’ | pú- $^{g̃[iš}$ki]ri$_6$-kalam-ma-ka úš-⸢àm⸣ ì-tùm-tùm |
| | Fu | iii | 14 | pú-$^{g̃iš}$kiri$_6$-kalam-ma-ka úš-àm ì-[tùm-tùm] |
| | Hn | Vs. | 4’ | [pú-$^{g̃iš}$kiri$_6$-kal]am-ma-ka úš-àm ⸢ì⸣-[tùm-tùm] |
| **133** | Cn$_1$ | iii | 13’ | úrdu l[ú]-⸢ù⸣ ú íl-i-dè g̃en-n[a ] |
| | Dn | iii | 29’ | úrdu lú-⸢ù⸣ ú íl-i-dè g̃e[n]-⸢na *úš⸣-àm ì-na$_8$-na$_8$ |
| | Fu | iii | 15 | [úr]du ú íl-i-dè g̃en-na ⸢úš⸣-à[m ] |
| | Hn | Vs. | 5’ | [úrdu lú]-*⸢ù⸣ú íl-i-dè g̃en-na *ú[š-àm ] |

**Fig. 10.10** Volk 1995: 105

**Fig. 10.11** Zgoll 1997: 292

292 11. PARTITUR

| | |
|---|---|
| UnJ iv 1’ | [ m]u-na-⌜ab⌝-[ ] |

**ninmešara 151**

| | |
|---|---|
| NiA iv 22 | nu-gig-ra du$_{11}$-ga-bi maḫ--a |
| NiB iv 12’ | nu-gig-ra du$_{11}$-ga-[ ] |
| NiC iv 27’ | [ ] du$_{11}$-ga-ni m[aḫ- ] |
| NiH Rs 14’ | [ ] ⌜du$_{11}$-ga-*n[i ] |
| NiM Rs 1’ | n[u]-gig [ ] |
| NiU iv 40 | nu-gig-r[a ] maḫ-àm |
| NiW iv 30 | [ -r]a du$_{11}$-<ga>-ni maḫ-àm |
| NiAA iv 18’ | [ -gi]g-ra du$_{11}$-ga-ni maḫ-àm |
| NiDD iv 12’ | [ ]-gig-ra [ ] ⌜maḫ⌝-[à]m |
| NiHH Rs 3 | [ -gi]g-ra ⌜du$_{11}$⌝-ga-ni maḫ-a |
| NiYY Rs 5 | ⌜nu-gig⌝-[ ] maḫ-[ ] |
| UrB iv 18’ | nu-gig-ra -- du$_{11}$-ga-ni maḫ--àm |
| UrE Rs 6 | nu-gig-ra ⌜du$_{11}$⌝-ga-a-ni maḫ$^!$-[ ] |
| UrG$^?$ iii 11’ | ⌜nu-gig⌝-r[a ] ⌜x x x⌝ |
| LaC$_3^?$ li.Rd. ii 2 | nu-gig-ra du$_{11}$-ga-ni maḫ-àm |
| UnJ iv 2’ | [ -g]a-ni maḫ-àm |

**ninmešara 152**

| | |
|---|---|
| NiA iv 23 | kur gul-gul An-da me -- ba--a |
| NiB iv 13’ | kur gul-gul An-d[a ] |
| NiC iv 28’ | [ *A]n-da me ba-[ ] |
| NiM Rs 2’ | kur gul-gul [ ] |
| NiU iv 41 | kur gul-g[ul ] me ba---⌜a⌝ |
| NiW iv 31 | [ -gu]l An-da me ba-àm |
| NiAA iv 19’ | [ -g]ul An-da me ba-⌜a⌝ |
| NiDD iv 13’ | [ -g]ul-⌜gul x⌝ [ ] ⌜ba-a⌝ |
| NiHH Rs 4 | [ ]-⌜gul An⌝ [ ] |
| NiYY Rs 6 | [ ] ⌜x x x⌝ |
| UrB iv 19’ | kur gul-gul -- An-da me -- ba-a [ ] [883] |
| UrE Rs 7 | kur gul-gul An-da me-da [ ] |
| LaC$_3^?$ li.Rd. ii 3 | -- *kur-kur gul-gul An-da me-da -- ba [884] |
| UnJ iv 3’ | [ ] An-da me ba----a |

[883] Rasur oder Abbruch

[884] Eingerückt, am Beginn der Zeile, Rasur von nin.

182 Textedition

| | |
|---|---|
| 225 | arata$^{ki}$ unu$^{ki}$-še$_3$ gu$_2$ a-gen$_7$ i$_3$-ĝa$_2$-ĝa$_2$ |
| An iv 48 | arata$^{ki}$ unu$^{ki}$-še$_3$ ⸢gu$_2$ a-gen$_7$ i$_3$⸣-ĝa$_2$-ĝa$_2$ |
| En v 1 | arata$^{ki}$ unu$^{ki}$-še$_3$ g[u$_2$ ] |
| Mn iv 11' | [*arat]a$^{ki}$ unu$^{ki}$-še$_3$ gu$_2$ a-gen$_7$ i$_3$-ĝa$_2$-ĝa$_2$ |
| Nn ii 16' | arata$^{ki}$ unu$^{ki}$-še$_3$ gu$_2$ a-ge[n$_7$ ] |
| Vu Rs. 12' | arata$^{ki}$ unu$^{ki}$-še$_3$ gu$_2$ a-gen$_7$ i$_3$-ĝa$_2$-ĝa$_2$ |
| Xunb ii' 1' | arata$^{ki}$ unu$^{ki}$-⸢še$_3$ gu$_2$⸣ [ ] |
| 226 | arata$^{ki}$ unu$^{ki}$-še$_3$ gu$_2$ ĝa$_2$-ĝa$_2$ nu-ĝal$_2$ e-ne-ra du$_{11}$-mu-na-ab |
| An iv 49 | arata$^{ki}$ unu⸢$^{ki}$-še$_3$ gu$_2$⸣ ĝ[a$_2$-ĝa$_2$] nu-ĝal$_2$ / e-ne-ra du$_{11}$-mu-na-[a]b |
| En v 2 | arata$^{ki}$ unu$^{ki}$-š[e$_3$ ] / e-ne-ra [ ] |
| Mn iv 12' | [arat]a$^{ki}$ unu$^{ki}$-še$_3$ gu$_2$ ĝa$_2$-ĝa$_2$ nu-ĝal$_2$ / e-ne-ra du$_{11}$-mu-na-ab |
| Nn ii 17' | arata$^{ki}$ ⸢unu$^{ki}$⸣-še$_3$ gu$_2$ ĝ[a$_2$-ĝa$_2$ ( )][440] |
| Vu Rs. 13'f. | arata$^{ki}$ unu$^{ki}$-še$_3$ gu$_2$ ĝa$_2$-ĝa$_2$ nu-ĝal$_2$ // e-ne-ra du$_{11}$-mu-na-ab |
| Xunb ii' 2' | arata$^{ki}$ unu$^{ki}$-še$_3$ gu$_2$ ĝa$_2$-ĝ[a$_2$ ] |
| 227 | ur$_5$-gen$_7$ ḫu-mu-na-ab-be$_2$-a-ka |
| An iv 50 | ur$_5$-gen$_7$ ḫu-mu-na-ab-be$_2$-⸢a-ka⸣ |
| En v 3 | ⸢ur$_5$-gen$_7$⸣ [x-x]-⸢un⸣-n[a$^?$- ] |
| Mn iv 13' | (ur$_5$-g)en$_7$ ḫu-mu-na-ab-be$_2$-a-ka |
| Nn ii 18' | ur$_5$-gen$_7$ ḫ[u]-mu-na-ab-be[$_2$- ] |
| Vu Rs. 15' | ur$_5$-gen$_7$ ḫu-mu-un-na-ab-e-a-ka |
| Xunb ii' 3' | ur$_5$-gen$_7$ ḫu-mu-u[n- ] |
| 228 | kiĝ$_2$-ge$_4$-a en arata$^{ki}$-ra mu-na-ni-ib-ge$_4$-ge$_4$ |
| An iv 51 | kiĝ$_2$-ge$_4$-a en arata$^{ki}$-[r]a / mu-na-ni-ib-ge$_4$-[g]e$_4$ |
| En v 4 | kiĝ$_2$-ge$_4$-a e[n ] |
| Mn iv 14' | (kiĝ$_2$-g)e$_4$-a en arata$^{ki}$-ra mu-na-ni-ib-ge$_4$-ge$_4$ |
| Nn ii 19'-20' | ⸢kiĝ$_2$-ge$_4$-a⸣ [x] arata⸢$^{ki}$⸣-[x] / [m]u-ni-ib-⸢ge$_4$⸣-[x][441] |
| Vu Rs. 16' | ⸢kiĝ$_2$⸣-ge$_4$-a en arata$^{ki}$-ke$_4$ mu-na-ni-ib-ge$_4$-ge$_4$ |
| Xunb ii' 4' | kiĝ$_2$-ge$_4$-a en arata$^{ki}$-[ ] |
| 229 | nin gal an-na me ḫuš-a u$_5$-a |
| An iv 52 | nin gal an-na me ḫuš-a ⸢u$_5$-a⸣ |
| En v 5 | nin gal an-n[a ] |
| Mn iv 15' | (n)in gal $^d$innana me *ḫuš-a u$_5$-a |
| Nn ii 21' | [*n]in$^!$ ⸢*gal⸣ [ ] ⸢ḫuš⸣-a [ ] |
| Vu Rs. 17' | nin gal $^d$innana ⸢me ḫuš⸣-a u$_5$-a |
| Xunb ii' 5' | nin gal an-na [ ] |

[440] e-ne-ra du$_{11}$-mu-na-ab von Text Nn wurde entweder ausgelassen oder am Rand geschrieben.

[441] Die Spuren in Kramers Kopie am Zeilenanfang gehören nach dem Foto zur folgenden Zeile 229; die 2. Hälfte von 228 ist eingerückt und ohne Zeilentrennstrich.

Partitur

| | |
|---|---|
| 230 | ḫur-saĝ kur subi$_2$-ka dur$_2$ ĝar-ra |
| An iv 53 | ḫur-saĝ kur subi$_2$-ka dur$_2$ ĝar-[x] |
| En v 6 | ḫur-saĝ kur sub[i- ] ⸢dur$_2$$^?$⸣ [ ] |
| Mn iv 16' | ḫur-saĝ kur subi$_x$-ka dur$_2$ ĝar-ra |
| Nn ii 22' | [ ] ⸢dur$_2$ x⸣ [ ] |
| Pn Vs. 1' | [ MU]Š$_{2/3}$-ka d[ur$_2$ x x (x)] |
| Vu Rs. 18' | ḫur-saĝ kur subi$_3$-ka dur$_2$ ĝar-ra |
| Xunb ii' 6' | ḫur-saĝ kur subi$_3$-ka [ ] |
| 231 | para$_{10}$ kur subi$_2$-ka še-er-ka-an du$_{11}$-ga |
| An iv 54 | para$_{10}$ kur subi$_2$-ka še-er-ka-an du$_{11}$-[x] |
| En v 7 | para$_{10}$ kur subi-ka [š]e-er-ka-an du$_{11}$-⸢ga⸣ |
| Mn iv 17' | para$_{10}$ kur subi$_x$-ka še-er-ka-an du$_{11}$-ga |
| Pn Vs. 2' | [ -k]a še-er-ka-a[n x x] |
| Vu Rs. 19' | para$_{10}$ kur subi$_3$-ka še-er-ka-an du$_{11}$-ga |
| Xunb ii' 7' | ⸢para$_{10}$$^?$ kur⸣ [ -k]a$^?$ [ ] |
| 232 | en lugal-ĝu$_{10}$ subur-a-ni(-im) |
| An iv 55 | en lugal-ĝu$_{10}$ subur-a-[x] |
| En v 8 | en lugal-ĝu$_{10}$ subur-a-ni |
| Mn iv 18' | en lugal-ĝu$_{10}$ subur-a-ni-im |
| Pn Vs. 3' | [ ] ⸢subur-ra$^?$⸣-[x x] |
| Qn Vs. 1 | en *l[ugal ] |
| Vu Rs. 20' | en lugal-ĝu$_{10}$-u$_8$$^?$ subur-ra-na-še$_3$ |
| 233 | $^d$nin-e$_2$-an-na-ka mu-na-ni-in-ku$_4$-re(-eš) |
| An v 1 | $^d$nin-e$_2$-an-na-ka mu-un-di-ni-ib-ku$_4$-re-eš |
| En v 9 | $^d$nin-e$_2$-an-[x]-ka mu-na-ni-in-ku$_4$-re-eš |
| Mn iv 19' | nin[442]-e$_2$-an-na-ra mu-na-ni-in-ku$_4$-re |
| Pn Vs. 4' | [ ]-⸢ke$_4$⸣ mu-⸢na⸣-ni-in-k[u$_4$-x] |
| Qn Vs. 2 | $^d$nin-e$_2$-[ ] |
| Vu Rs. 21' | nin-e$_2$-an-na-ka mu-na-ni-in-ku$_4$-re |
| 234 | en arata$^{ki}$ gu$_2$ ki i$_3$-ĝal$_2$ |
| An v 2 | en arata$^{ki}$ gu$_2$ ki-še$_3$ ba-ni-in-ĝal$_2$ |
| En v 10 | en arat[a]$^{ki}$ gu$_2$ ki i-in-ĝal$_2$ |
| Mn iv 20' | en arata$^{ki}$ gu$_2$ ki i$_3$-ĝal$_2$ |
| Nn iii 1' | [ ] ⸢I⸣-[x] |
| Pn Vs. 5' | [ ]$^{ki}$ gu$_2$ ⸢ki i$_3$⸣-[x] |
| Qn Vs. 3 | en arata[$^{ki}$ ] |
| Vu Rs. 22' | en arata$^{ki}$ gu$_2$ ki bi$_2$-in-⸢ĝal$_2$⸣ → |

[442] Nach NIN ist ein zweites, fehlerhaftes NIN radiert.

**Fig. 10.12** Mittermayer 2009: 182

Nippur manuscripts) whether the manuscript is a multi-column tablet (Roman I), single-column (Roman III) or prism (P).[25] Each manuscript within those three groups is assigned a subscript Arabic numeral which appears after the format identification. While the information provided is useful for some purposes, the subscripts are not only difficult to read, but note how similar the subscript of the third manuscript in this example—N-Roman I-numeral 4—is to the that of the last, X-numeral 14. Also, as stated earlier, it is much harder to become acquainted with individual manuscripts when they are distinguished only by subscripts. When working through an edition, one begins to recognize the idiosyncrasies of Manuscript M and the frequent errors of Manuscript R; when both are instead identified by N plus a series of subscripts, their distinctive characteristics are more difficult to spot and keep in mind. But, as Delnero explains (2012b: 183), this system was devised as a

[25] The use of Roman I and III derives from Civil's classification of the manuscripts of lexical texts (e.g. Civil et al. 1979: 5; 1995: 2308); for the problems in the application of Civil's manuscripts types to literary texts, which only rarely appear on Civil's Type II and IV tablets, see Veldhuis (1997: 65–66).

| | |
|---|---|
| *ŠA* 66: | murgu-ĝa$_2$ dub-dab$_5$ he$_2$-em-mi-ib-za |
| N$_{I1}$ | [...] ĝa$_2$ dub-dab$_5$ h***e***$_2$-***m***i-ib [...] |
| N$_{I2}$ | murgu-ĝa$_2$ dub-dab$_5$ he$_2$-em-mi-ib-za |
| N$_{I4}$ | murgu-ĝa$_2$ dub-dab$_5$ he$_2$-em-mi [...] |
| N$_{III6}$ | murgu-***ĝu***$_{10}$ [...] |
| N$_{III21}$ | murgu-ĝa$_2$ dub [...] |
| N$_{III23}$ | murgu-ĝa$_2$ dub-dab$_5$ he$_2$-em-mi-ib-za |
| N$_{III24}$ | ⸢murgu⸣ [...] |
| N$_{III25}$ | [...] he$_2$-em-mi-ib-za |
| N$_{P1}$ | [...] za |
| N$_{P2}$ | [...] ⸢murgu⸣-ĝa$_2$ ***bu*** / ⸢***ud***⸣-***ba-ad he*** / [...] ***ib***$_2$-za-***am***$_3$ |
| Ur$_1$ | ⸢murgu⸣-ĝa$_2$ dub-***dab*** he$_2$-em-mi-⸢***ib***$_2$⸣ [...] |
| X$_1$ | murgu-ĝa$_2$ dub-dab$_5$? h***e***$_2$-***m***i-***ib***$_2$-za |
| X$_3$ | murgu-ĝa$_2$ dub-***dab***$_2$(TAB) he$_2$-em-⸢mi⸣-***ib***$_2$-za |
| X$_{12}$ | murgu-⸢***ĝu***$_{10}$⸣? ⸢dub⸣?-***be***$_2$ / he$_2$-em-mi-⸢ib⸣?-⸢za⸣ |
| X$_{13}$ | [...] ⸢he$_2$⸣-em-mi-ib-za |
| X$_{14}$ | murgu-ĝa$_2$ ⸢dub⸣-***dab***$_4$(DUB) h***e***$_2$-***mi***?-***z***a |

NOTE: N$_{P2}$, X$_1$, X$_{12}$, and X$_{14}$ - This line was written on the same line as the previous line.

NOTE: X$_1$ - It is unclear whether the sign read as "dab$_5$" is "dab$_5$" or "ba."

**Fig. 10.13** From Delnero 2012b: 182–183

research tool to enable him to quickly identify the provenience of each manuscript and its format for the purposes his study of variants across his corpus, and he is not advocating it as a standard for text editions.

The real importance of Delnero's work is that he is able to establish that when there are four or more manuscript witnesses, the variants of the majority are nearly always correct, and thus to be favored when reconstructing a 'best' text (2012: 185). There appears to be some circularity to his argument, but if we remember that what we are reconstructing is simply a text that would be recognized by a Nippur schoolmaster as correct, and that the manuscripts of individual compositions are 90–95% identical (Delnero 2012b: 10), then there is a compelling logic to his contention, all the more so because variation from the majority can almost always be explained as errors or as simple orthographic variation. Delnero, I believe, has defined what Pollock might call the proper 'editorial strokes' for studying the 'historical folks' learning Sumerian in eighteenth-century BCE Babylonia. Nearly all the textual variants can be considered, in Pollock's words, 'textually transmitted disease' (Pollock 2009: 952–953), keeping in mind that the symptoms of disease can teach us much about the patient.

What we have in the end is a reconstructed corpus of literature whose only certain context is the schoolroom. Not only is authorship and original context left to our educated guesses, but we have no idea if, in eighteenth-century BCE Babylonia, this literature had any other context at all outside the school, or any function other than the construction of an educated elite identity, an identity whose existence is detectable only in these same school texts (Cooper 2015). We can point to nothing resembling scriptoria with scribes laboring a lifetime to preserve ancient sacred traditions, no odd sectarians copying manuscripts in the Judean desert, no mandarins collating texts in a peony pavillion—just schoolboys learning a mostly irrelevant literature in a dead language. If ever Said's statement (2004: 57) that 'philology is just about the … least sexy … of any of the branches of learning' were true, it would be true for the all-consuming, ceaselessly fascinating task of editing the Sumerian literary manuscripts produced by Babylonian students nearly 4000 years ago.

## References

Black, J., G. Cunningham, E. Robson, and G. Zólyomi. 2004. *The literature of ancient Sumer*. Oxford: Oxford University Press.

Civil, M. 1994. *The farmer's instructions. A Sumerian agricultural manual*, Aula Orientalis Supplementa 5. Sabadell-Barcelona: Editorial AUSA.

———. 1995. Ancient Mesopotamian lexicography. In *Civilizations of the ancient Near East*, ed. J. Sasson, vol. IV, 2305–2314. New York: Charles Scribner's Sons.

———. 2013. Remarks on AD-GI$_4$ (A.K.A. 'Archaic Word List C' or 'Tribute'). *Journal of Cuneiform Studies* 65: 13–67.

Civil, M., M.W. Green, and W.G. Lambert. 1979. *Ea A = nâqu, Aa A = nâqu, with their forerunners and related texts*, Materials for the Sumerian Lexicon 14. Rome: Pontificium Institutum Biblicum.

Cooper, J. 1983. *The curse of Agade*. Baltimore: The Johns Hopkins University Press.

———. 1992. From Mosul to Manila: early approaches to funding Ancient Near Eastern studies research in the United States. *Culture and History* 11: 133–164.

———. 2015. Sumerian literature and Sumerian identity. In *Problems of canonicity and identity formation in Ancient Egypt and Mesopotamia*, CNI publications 43, ed. K. Ryholt and G. Barjamovic, 1–18. Copenhagen: Museum Tusculanum Press.

Delnero, P. 2010. Sumerian extract tablets and scribal education. *Journal of Cuneiform Studies* 62: 53–69.

———. 2012a. Memorization and the transmission of Sumerian literary compositions. *Journal of Near Eastern Studies* 71: 189–208.

———. 2012b. *The textual criticism of Sumerian literature*, Journal of Cuneiform Studies, supplementary series 3. Boston: American Schools of Oriental Research.

———. 2015. Text and performance. In *Texts and contexts*, Studies in ancient near eastern records 9, ed. P. Delnero and J. Lauinger, 87–118. Boston: De Gruyter.

Edzard, D. 1974. Zur sumerischen Hymne auf das Heiligtum Keš. *Orientalia* 43: 103–113.

Edzard, D., and C. Wilcke. 1976. Die Ḫendursanga Hymne. In *Kramer anniversary volume*, Alter Orient und Altes Testament 25, ed. B. Eichler et al., 139–176. Kevelaer: Butzon & Bercker.

Falkenstein, A. 1959. *Sumerische Götterlieder. I. Teil*, Abhandlungen der Heidelberger Akademie der Wissenschaften, Philosophisch-Historische Klasse 1. Heidelberg: C. Winter Universitätsverlag.

George, A. 2005. In search of the É.DUB.BA.A: the ancient Mesopotamian school in literature and reality. In *'An experienced scribe who neglects nothing.' Ancient Near Eastern studies in honor of Jacob Klein*, ed. Y. Sefati et al., 127–137. Bethesda MD: CDL Press.

Gibson, M., D. Hansen, and R. Zettler. 2001. Nippur B. Archäologisch. *Reallexikon der Assyriologie* 9: 546–565.
Hallo, W., and J. Van Dijk. 1968. *The exaltation of Inanna*, Yale Near Eastern Researches 3. New Haven: Yale University Press.
Houston, S., ed. 2004. *The first writing: Script invention as history and process*. Cambridge: Cambridge University Press.
Houston, S., J. Baines, and J. Cooper. 2003. Last writing: Script obsolescence in Egypt, Mesopotamia and Mesoamerica. *Comparative Studies in Society and History* 45: 430–479.
Kramer, S. 1937. Inana's descent to the Nether World. *Revue d'Assyriologie* 34: 93–134.
———. 1986. *In the world of Sumer: An autobiography*. Detroit: Wayne State University Press.
Liverani, M. 2014. *The ancient Near East: history, society and economy*. London: Routledge.
Michalowski, P. 2008. Sumerian. In *The ancient languages of Mesopotamia, Egypt, and Aksum*, ed. R. Woodard, 6–46. Cambridge: Cambridge University Press.
Mittermayer, C. 2009. *Enmerkara und der Herr von Arata*, Orbis Biblicus et Orientalis 239. Fribourg: Academic.
Oelsner, J. 2014. Überlegungen zu den 'Graeco-Babyloniaca'. In *He has opened Nisaba's house of learning. Studies in honor of Å. W. Sjöberg*, Cuneiform monographs 46, ed. L. Sassmannshausen, 147–164. Leiden: Brill.
Ong, W. 1982. *Orality and literacy*. London: Methuen.
Pollock, S. 2009. Future philology? The fate of a soft science in a hard world. *Critical Inquiry* 35: 931–961.
Reiner, E. 2002. An adventure of great dimension: the launching of the *Chicago Assyrian Dictionary*. *Transactions of the American Philosophical Society* 92: 3.
Robson, E. 2001. The tablet house: a scribal school in Old Babylonian Nippur. *Révue d'Assyriologie* 95: 39–66.
Rubio, G. 2009. Sumerian literature. In *From an antique land: an introduction to ancient Near Eastern literature*, ed. C. Ehrlich, 11–75. Plymouth: Rowman & Littlefield.
Said, E. 2004. The return to philology. In *Humanism and democratic criticism*, ed. E. Said, 57–84. New York: Columbia University Press.
Taylor, J. 2011. Tablets as artefacts, scribes as artisans. In *The Oxford handbook of cuneiform culture*, ed. K. Radner and E. Robson, 5–31. Oxford: Oxford University Press.
Tinney, S. 1996. *The Nippur lament. Royal rhetoric and divine legitimation in the reign of Išme-Dagan of Isin (1953–1935 B.C.)*, Occasional publications of the Samuel Noah Kramer Fund 16. Philadelphia: Samuel Noah Kramer Fund.
———. 2011. Tablets of schools and scholars: a portrait of the Old Babylonian corpus. In *The Oxford handbook of cuneiform culture*, ed. K. Radner and E. Robson, 577–596. Oxford: Oxford University Press.
Van de Mieroop, M. 2015. *A history of the ancient Near East*. 3rd ed. Malden: Wiley-Blackwell.
Vanstiphout, H. 2003a. *Epics of Sumerian kings. The matter of Aratta*, Writings from the ancient world 20. Atlanta: Society of Biblical Literature.
———. 2003b. The Old Babylonian literary canon. In *Cultural repertoires. Structure, function and dynamics*, Groningen studies in cultural change 3, ed. G. Dorleijnand and H. Vanstiphout, 1–28. Leuven: Peeters.
Veldhuis, N. 1997. *Elementary education at Nippur*. Ph. D. dissertation, Rijksuniversiteit Groningen.
———. 2003. Sumerian literature. In *Cultural repertoires. Structure, function and dynamics*, Groningen studies in cultural change 3, ed. G. Dorleijnand and H. Vanstiphout, 29–43. Leuven: Peeters.
———. 2004. *Religion, literature, and scholarship: the Sumerian composition 'Nanše and the Birds'*, Cuneiform monographs 22. Leiden: Brill/Styx.
———. 2014. *History of the cuneiform lexical tradition*, Guides to the Mesopotamian textual record 6. Münster: Ugarit-Verlag.

Volk, K. 1995. *Inanna und Šukaletuda. Zur historisch-politischen Deutung eines sumerischen Literaturwerkes*, *Santag* 3. Wiesbaden: Harrassowitz.

———. 2000. Edubba'a und Edubba'a-Literature: Rätsel und Lösungen. *Zeitschrift für Assyriologie* 90: 1–30.

Waetzoldt, H., and A. Cavigneaux. 2009. Schule. *Reallexikon der Assyriologie* 12: 294–309.

Wilcke, C. 1969. *Das Lugalbandaepos*. Wiesbaden: Harrassowitz.

———. 1976. Nin-me-šár-ra—Probleme der Interpretation. *Wiener Zeitschrift für die Kunde des Morgenlandes* 68: 79–92.

———. 2012. *The Sumerian poem Enmerkar and En-suḫkeš-ana: epic, play, or? Stage craft at the turn from the Third to the Second Millennium B.C*, *American oriental series essay* 12. New Haven: American Oriental Society.

Woods, C. 2010. *Visible language: Inventions of writing in the ancient Middle East and beyond*, Oriental Institute Museum publications 32. Chicago: Oriental Institute.

Zgoll, A. 1997. *Der Rechtsfall der En-ḫedu-Ana im Lied nin-me-šara*, Alter Orient und Altes Testament 246. Münster: Ugarit-Verlag.

# Part IV
# Publishing Ancient Mathematical and Astronomical texts: Comparative Perspectives

# Chapter 11
# The Critical Edition of the Mathematical Texts of Greek Antiquity: Challenges and Methods

**Micheline Decorps-Foulquier**

**Abstract** The present study aims to analyze the evolution of methods used in the publication of mathematical works from Greek antiquity. The progress made in the history of texts offers the instruments necessary for measuring what can separate the author's text in its original form from the various editions produced in the course of time. The evolution of book production techniques, the influence exerted by the circles that took over the transmission of scientific knowledge, and the conceptual bases that have influenced the intervention of publishers are all elements to be taken into account by the historian of science in judging the exact nature of the transmitted text and evaluating what its present form owes to the path taken to reach us.

The authorial text is one of the most important sources for our understanding of the knowledge acquired by the Greeks in the field of the mathematical sciences. However, a great number of works which were known and used in Antiquity have either completely disappeared[1] or survive only in fragments; furthermore, the realities relating to writing and copying books at that time[2] and the fact that the circulation of mathematical treatises was under the sway of the scholastic tradition can often take us a long way from the original form of the transmitted works. Access to

The research leading to these results has received funding from the European Research Council under the European Union's Seventh Framework Programme (FP7/2007–2013)/ERC Grant Agreement No. 269804.

[1] Other losses include a large number of collections, excerpta and compilations drawn up by the scholastic tradition and used by commentators and historians of science in Antiquity.

[2] For these historical questions, I refer the reader to works which remain very useful: Devreesse (1954), Dain (1975) and Reynolds and Wilson (1986). See also Dorandi (2000).

M. Decorps-Foulquier (✉)
Université d'Auvergne LSH-PHIER, Maison des Sciences de l'Homme,
Clermont-Ferrand Cedex 1, France
e-mail: micheline.decorps@uca.fr



A. Keller, K. Chemla (eds.), *Shaping the Sciences of the Ancient and Medieval World*, Archimedes 69, https://doi.org/10.1007/978-3-031-49617-2_11

the author's text can therefore only be obtained through a series of relatively complex operations.

The aim of the present study is to analyze the various editorial methods that were used from Antiquity up to the first critical editions. Such a perspective enables us to understand the reasons for the distance that has gradually appeared between the original text and the transmitted work. It also throws a clearer light on the concepts underlying the edition of an ancient Greek scientific text. A historian of science can only benefit from evaluating the nature of the mathematical document he has at his disposal and being perfectly familiar with the textual realities behind it.

I will first review how the needs of scientific circles influenced the very nature of transmitted mathematical literature; I will then study how editorial practice evolved and led up to the first critical editions at the end of the nineteenth century. Finally, I will use concrete examples to present the latest issues in modern editions.

## 11.1 The Nature of Conserved Mathematical Writings and Their Means of Transmission

### *11.1.1 Factual Data from the History of Texts*

Despite all the losses endured, Greek Antiquity has handed down a certain number of texts in the field of science and technology which give us a very good idea of the level attained in scientific research.[3] Their transmission was by no means haphazard. With the exception of papyri discovered by archeological methods, the paths taken by conserved manuscripts until they reached us show that they were distributed throughout Antiquity through clearly identified scientific circles and schools.[4] What enabled the transmission of most of the works which did not purely and simply disappear was the fact that as soon as they were put into circulation, they were revised, commented on, collected in thematic works and used as teaching aids in specific disciplines.[5] A number of elements back this up.

[3]Amongst the numerous works on the history of Greek mathematics, one of the leading reference works is Heath (1921). For an overview, see Knorr (1996) and Cuomo (2001). For present-day lines of thought and research, see Sidoli and Van Brummelen (2014).

[4]The textual history of the transmitted mathematical treatises, knowledge of which is increasingly accurate, is highly instructive in this respect. All information on this has been grouped together in the introductions to the critical editions and translations. On the role played by schools in the transmission of texts, see Goulet (2007).

[5]A certain number of classifications made it possible to identify the disciplines which fell to a variable extent within the field of mathematics. To a greater or lesser degree, they inspired the organization of texts used for teaching. The best known are the following: (1) the four-part division of mathematics inherited from the Pythagoreans; it distinguished arithmetic, music, geometry and spherics; (2) an Aristotelian-inspired division which the Neoplatonic philosopher Proclus (fifth century CE) attributed to Geminus in his commentary on Book I of Euclid's *Elements*; this classification made it possible to take into account the development of numerous applied sciences by

First, the analysis of the information provided by the history of the texts can be considered (identifying the sources of medieval manuscripts, putting evidence from the indirect tradition[6] into perspective, classifying manuscripts, identifying copyists, annotators and owners). Alongside this is the study of evidence from authors' and editors' prefaces.[7] All these elements make it possible to determine with a relative degree of accuracy where works written by ancient Greek mathematicians first appeared and which scientific circles undertook to circulate them.

A great amount of additional information confirms the activity of these circles. The most relevant comes from accounts given by historians of science in Antiquity[8]; additionally, there is information drawn from transmitted technical corpuses. The internal organization and mode of transmission of numerous corpuses, whether mathematical, astronomical, musical, mechanical, tactical, geographic, gromatic, botanical, astrological, alchemical or medical, in fact show which circles used them and the pedagogical requirements they satisfied.[9] All these facts make it possible to identify the context in which a technical work in Antiquity was written, and also to study its reception. This work of reconstitution is an integral part, and indeed a foundation, of critical editions.

In the specific field of the mathematical sciences, the indications given by the very nature of the conserved writings and of their textual traditions reveal the great influence of centers of higher education on the transmission of texts. It suffices to note the essentially vertical character of the tradition of the founding treatises and to remember that within Euclid's tradition, amongst the ancient manuscripts of the *Elements*,[10] only one (the *Vaticanus graecus*. 190 from between 830 and 850)[11]

contrasting the two sciences concerned with intelligible objects—arithmetic and geometry (with its two parts, plane geometry and stereometry)—with six other disciplines concerned with perceptible objects (geodesy, logistic, optics, canonics, mechanics and astronomy), each of which is further subdivided into several branches.

[6] The indirect tradition of a Greek mathematical treatise essentially consists of ancient commentaries, quotations by later mathematicians (or historians of science) as well as Arabic and Latin medieval translations. The great variety of forms in which mathematicians used the works of their predecessors may also offer very precious indirect testimonies.

[7] On the great interest of these texts, which range from the personal introductory letter written by a Greek mathematician to a friend or correspondent to didactic introductions to a work, see Aujac (Aujac 1979–1980), Mansfeld (1994, 1998), Vitrac (2008), and Taub and Doody (2009).

[8] The most eminent figures are: (1) Eudemus of Rhodes, a disciple of Aristotle, whose works have been preserved only in fragments; see Schneider (2000). His history of geometry was largely used by later historians and provided a precious testimony on mathematics before Euclid; (2) Geminus; see Tood (2000); (3) Pappus of Alexandria (fourth century CE); see Goulet (2012); (4) Proclus; see Endress (2012); (5) Archimedes' commentator Eutocius (sixth century CE) and the Neoplatonist philosopher Simplicius (sixth century CE). For an overview of all testimonies, see Michel (1950).

[9] They are at present the subject of vast publishing projects, in particular in the *Collection des Universités de France* (Les Belles Lettres).

[10] See Stamatis post Heiberg (1969–1977); Heath (1926); Vitrac (1990–2001).

[11] See Irigoin (2003: 215).

gives us a text that is independent of Theon of Alexandria's version (fourth century CE).[12]

## *11.1.2 The Nature of the Transmitted Texts*

It is important for an editor to determine which genre a mathematician's text belongs to, the context in which it was written, and how it was read and circulated. This historical approach, which was already used by commentators and editors in Antiquity, underlies any attempt to correct and restore the transmitted texts.

Except for the two treatises on mathematical astronomy by Autolycus of Pitane (between 360 and 290 BCE), *On the Moving Sphere* and *On Risings and Settings*,[13] no complete mathematical work from the Archaic and Classical periods has come down to us. Our textual documentation essentially consists of the *loci mathematici* found in Plato's work and in the Aristotelian corpus, and also of the fragmentary accounts found in works by later historians of science and commentators (Neoplatonists or Neopythagoreans). These testimonies illustrate the intensity of mathematical activity at the time[14] and show the extent of what has been lost.[15]

It should be pointed out that the two treatises by Autolycus were only conserved because they were included in an Alexandrian collection of minor writings on astronomy, whose first core was possibly constituted by the Alexandrian librarian and scholar Eratosthenes of Cyrene in the third century BCE.[16]

Indeed, it was at that time that a single event entirely changed the situation. One cannot help being struck by the correlation between the number of mathematical treatises that have been conserved since then and the creation of the Museum of Alexandria and its library. During the Hellenistic Period, mathematicians not only found learned circles who were disposed to accept their work but they also

---

[12] See Heiberg (1903). For a summary of Heiberg's works on the Euclidean tradition, see Heath (1926: 1–151) and Caveing (1990: 45–55). See also Vitrac (2012).

[13] See Aujac (1979).

[14] The circles in which the important theoretical knowledge was developed were philosophical communities centered around the master's oral teachings; they were to be found in places as far apart as the rich cities of Asia Minor, such as Miletus, where Thales lived, and Magna Graecia with the teachings of the Pythagoreans. It goes without saying that this primitive configuration evolved in the course of the fifth century, in particular with the Sophists. As they travelled from one city to another offering high-level teachings in all fields, places of learning appeared in private houses, gymnasiums and public places; at the same time, the Sophists fostered the book trade. This legacy was handed down to the first communities who were members of the Academy and the Lyceum. All these contributions paved the way for the Euclidean synthesis and provided Greek mathematics with the epistemological tools required for their development. See Netz (1999: 271–312) and Levy (2007).

[15] In the case of technical literature, it is necessary to add to all the reasons which explain the considerable losses incurred during this period and the following periods, the fact that the circulation of a new treatise or a new commentary rendered the previous books on the subject obsolete in the eyes of contemporaries.

[16] This is the hypothesis defended by Aujac (1977).

benefitted from the book-centered activity of the Library's scholars[17] and their counterparts in Pergamon or Athens. Through this activity, they acquired full awareness of the reality of text edition and an interest in preserving their works.

During this period, the intense activity in the field of books and libraries in the great Greek and Roman centers of learning was also reflected in the multiple book forms which mathematicians were to adopt. While the expository model used for geometric demonstration[18] scarcely varied and even spread to the applied sciences, a great variety can be observed in the style and nature of the mathematical works circulated.[19] Amongst these figure works of synthesis, such as Ptolemy's *Almagest* (second century CE)[20] and Heron of Alexandria's *Mechanics*[21] (first century CE); these were written to group together in one place everything that was necessary for complete knowledge of a given science and the author completed the work by adding the results of his own research. One also finds other types of works: the exploration of a precise question such as *The Quadrature of the Parabola*[22] by Archimedes; the methodical exposition of fundamental properties of a mathematical object, such as the *Conics* by Apollonius of Perga[23]; a structured collection of 'elementary' propositions to be used for establishing other results, as in the work by Euclid.[24] Authors can, following the Euclidean model, build a logical chain made of a succession of propositions or adopt a continuous discourse by inserting demonstrative reasoning at the appropriate moment.

In the late scientific tradition fostered in philosophical schools at the end of Antiquity, other genres appeared such as introductions to a science[25] or to a spe-

---

[17] Scholars at the Musaeum Library, which received all the books written in Ancient, Classical and Hellenistic Greece, had not only to classify all these documents (the first classifications had certain repercussions on the transmission of texts), but also to take charge of reproducing and circulating them. The editorial activity undertaken to obtain new vulgates, to standardize the language and spelling, and to reorganize text divisions by adapting them to the form of books at the time, enabled critical and exegetical methods to be perfected and to be exceptionally long-lived. On the activity of Alexandrian grammarians and their role in the transmission of texts, see in particular Irigoin (2003: 133–173).

[18] On the formal model of mathematical propositions, see Caveing (1990: 114–148) and Netz (1999: 240–270).

[19] See Decorps-Foulquier (2011) and Taub (2013).

[20] See Heiberg (1898–1903).

[21] Transmitted in Arabic only. See Carra de Vaux (1988).

[22] See Mugler (1971).

[23] Greek text: Decorps-Foulquier (2008–2010); Heiberg (1891–1893). Arabic text: Rashed (2008–2010); Toomer (1990).

[24] See Caveing (1990: 84–88).

[25] These works examined the epistemological status of the science in question, its relations to other disciplines, its fundamental principles, and also its history and its main results. One can quote the *Introduction to Arithmetic* by the Neoplatonic Nicomachus of Gerasa (second century CE) and Book VIII of the *Collection* by Pappus of Alexandria (Hultsch 1876–1878), which is an introduction to mechanics; it was in this form that Pappus' contribution was known to the mathematician Eutocius as attested in his commentary on the treatise *On the Sphere and Cylinder* by Archimedes (Mugler 1972: 54.1).

cific treatise[26] using as a model the kind of exegesis of Aristotle or Plato that was developed in neo-Platonic schools.[27] The same scholastic tradition produced manuals summarizing elementary knowledge imparted in the teaching of the four sciences. The work of Theon of Smyrna (second century CE), whose title *Expositio rerum mathematicarum ad legendum Platonem utilium* speaks for itself, offers the best example. To this should be added the collection and rewriting of texts from the previous tradition[28] to discuss a given question such as, for example Books IV to V of Pappus' *Mathematical Collection* on geometric constructions. Commentaries in due form also stem from this scientific activity. In both content and subject matter, they belong to the corpus of mathematical works, but in the method they apply, they pertain to exegetical literature and come under the same tradition as commentaries on poets, rhetoricians and philosophers. Greek Antiquity has conserved continuous commentaries on major works such as the commentary by Proclus on Book I of Euclid's *Elements,*[29] the commentaries by Theon of Alexandria on Ptolemy's *Almagest* and *Handy Tables* or those by Eutocius on Archimedes' treatises and the first four books of *Conics* by Apollonius of Perga.

All the books quoted above were transmitted as works in due form with a title and the author's name,[30] rarely individually, and more often grouped together in thematic corpuses.[31] The presence of a title and a preface by no means guarantees the homogeneity of the transmitted text. In a certain number of texts, a list of textual inconsistencies reveals that different versions of the text have been superimposed but without the key to elucidating them.[32]

There is another category of textual documents which require specific investigation. This is all the text fragments, more often than not anonymous, with which our manuscripts teem. In this mass, one finds demonstrations of auxiliary theorems used without proof by mathematicians (lemmas), variants of demonstrations, figures, collections of tables, collections of scholia, geometric definitions and extracts of historical works,[33] all of which are learned documents connected with our manuscripts in various forms.

[26] For example, the 'Prolegomena' to Euclid's *Data* written by Proclus' pupil Marinus of Neapolis (Menge 1896); see Mansfeld (1998: 61–65).

[27] See Hadot (1987) and Hoffmann (2000).

[28] See Cuomo (1998).

[29] See Friedlein (1873).

[30] For the particular case of Pappus' works, see below.

[31] The mathematical corpuses transmitted in our medieval manuscripts frequently have a guiding principle which shows that they were designed to satisfy specific needs and that they are not a heap of fragments which just happened to be saved when they were discovered. An ancient core can be identified in some of them. The replacement of *volumen* (a scroll) by the *codex* (a book with pages) favored the establishment of antique corpuses from the second century CE onwards.

[32] Theon's *Great Commentary on Ptolemy's Handy Tables* is a good illustration of this; see Tihon and Mogenet (1985: 72).

[33] For the Euclidean tradition, see Vitrac (2003); for astronomical texts, see Tihon (1992); for the Heronian corpus, see Vitrac and Acerbi (2014: 429–556).

Finally, there are the 'books' created from scratch by assembling previous works or the teachings of scholars. Cases in point are Pappus' *Mathematical Collection*[34] and the anonymous compilation *Introduction to Almagest*[35] (sixth century CE). Sometimes, the 'books' are works which have been reorganized and completed to clarify a topic of research, such as Diocles' *On Burning Mirrors*.[36] In other cases, on the contrary, they are treatises which have been rewritten to supply a simplified version such as the transmitted version of Archimedes' *Measurement of a Circle*.[37]

The dynamism of this editorial activity sometimes had most unfortunate consequences for the textual history of mathematical treatises. Once the simplified or amended versions were circulated, they either co-existed with older versions with the risk of their being combined with them and giving arise to a new tradition, or they simply led to their disappearance.

In the case of the tradition of the *Conics* treatise,[38] the initiative taken by Eutocius when he collated several sources to produce an edition of Books I to IV, accompanied in addition by a marginal commentary, had two major effects: firstly it introduced a new screen which led us even further away from the original work; secondly its success certainly contributed to making copies of the whole treatise scarcer, as Books V to VII (Book VIII was lost) are only available in Arabic thanks to the translation of Books I to VII made in Bagdad in the ninth century under the aegis of the Banû Mûsâ brothers.

---

[34] It brings together in a single work—of which Books II to VII have been preserved—various writings by mathematician. Some of these writings must have originally been separate treatises, as shown by remaining fragments of their prefaces. The way in which the works were grouped together leaves no doubt as to the pedagogical intentions of the editors (see Decorps-Foulquier 2000: 43–59).

[35] This anonymous compilation (still partly unpublished) was written according to the rules in use amongst Neoplatonic commentators in the sixth century. It groups together under one heading three separate expositions which were clearly assembled to complement the reading of Book I of the *Almagest*. The first of these three expositions reproduces the lost treatise by the Hellenistic mathematician Zenodorus on isoperimetric figures. See Acerbi et al. (2010), Decorps-Foulquier and Federspiel (2014: XVI–XVII).

[36] The treatise was transmitted in Arabic only; see Toomer (1976) and Rashed (2000). The preface and the title concern burning mirrors only, which are the subject of the first group of propositions (1–5). They do not mention the other two groups (7–8 and 10–16), which are not connected with burning mirrors as they deal respectively (1) with the problem to which Archimedes reduced the division of a sphere into segments having a given ratio in Proposition II.4 of the treatise *On the Sphere and Cylinder* and (2) with the doubling the cube problem. Assembling different contributions to the study of the applications of conic sections in catoptrics and geometry only makes sense if it is done by the editor and not by the author. Eutocius who used the geometric Propositions 78 in his commentary on Proposition II.4 of Archimedes' treatise was already familiar with the work in this form and with this title (Mugler 1972: 105.1).

[37] See Mugler (1970).

[38] See below.

## 11.2 The Edition of Mathematical Texts Up to Critical Editions

Generally speaking, the editorial activity deployed around Greek texts took many forms before the concept of critical edition appeared. Editors mainly wanted to make texts from Antiquity available and readable; however, with the appearance of the critical edition, it became necessary to be as faithful as possible to the original form of the written work and to direct philological research towards this ultimate aim.

### *11.2.1 Editorial Issues in Antiquity*

As shown by the differences from one tradition to another, mathematical treatises were far more open to amendment than literary texts. The need to find tools which could be used for the work partly explains the creation of these new texts, even if editors in Antiquity did not totally relinquish all critical ambition.

#### 11.2.1.1 The Need for Revision

The genre of transmitted mathematical works and the scarcity of our papyrological documents[39] show that the mathematical treatises were read and circulated in fairly closed circles even if the mathematicians themselves came from very different horizons and countries. Their activity was centered in cultural metropolises where the very circles which received their work were responsible for circulating the texts and organizing teaching based on their exegesis. Mathematical treatises were consequently work tools which were useful for the scientific community and teaching. They were naturally subject to revision and this was occasionally undertaken by the author.

As soon as they were first circulated, the works of mathematicians became a subject of discussion between scholars. This is clearly shown in Archimedes' prefaces and the preface to Book XIV of *Elements*, written by Hypsicles of Alexandria.[40] It is clear that the intention was to participate in scientific advance even if it meant writing in haste. This explains why temporary versions were circulated to impatient friends and disciples. The existence of these copies complicated the textual history of these treatises when they were circulated at the same time as the final version.[41]

Once the mathematical text had become a subject of study in schools, it was actively read. In turn, this gave rise to alternative demonstrations, new figures, new

[39] Extant papyrological documentation mostly contains school exercises.

[40] See Stamatis post Heiberg (1977: 1.1–2.6).

[41] See Reynolds and Wilson (1986: 148).

cases and lemmas intended to compensate for elliptical reasoning. All these materials fostered commentaries which have come down to us as well as the numerous scholia which were copied into our medieval manuscripts. Parts of these materials were occasionally introduced directly into the text when new revisions were made.

A comparative analysis of direct and indirect witnesses shows, in a certain number of traditions, that editorial modifications took the same form as the explanatory material which accumulated in the commentaries and scholia: the original demonstrations were replaced by alternative proofs, which were pure products of the scholastic tradition, lemmas were interpolated, propositions and the demonstration of other cases added, the corpus of figures modified, the order of theorems inverted and internal references and explanations introduced.[42]

That the mathematical treatise was so easily rewritten in the course of later recensions, is also because the language of demonstration is partly coded,[43] as are the structure of the proposition, the formulae used for cross-references[44] and the writing of specific procedures, such as the application of reasoning by exhaustion[45] in the demonstration. On the other hand, personal prefaces are rarely altered when the author writes in the language of his day; likewise, complex enunciations such as those in *Conics* were on the whole spared, as were all sorts of explanatory passages which did not fall within the category of 'text parts' subjected to writing conventions.

#### 11.2.1.2 Corruption and Restoration of Sources

An editor, whoever he may be, is necessarily confronted with the problem of the corruption of the sources. When reproducing a mathematical text, there are numerous opportunities for making mistakes. Above all, one copyist after another may have made any number of mistakes when reproducing the letters designating geometrical objects or numbers, thereby making the text incomprehensible in places. Repetition of the same sequences in syllogisms in the demonstrative parts caused numerous leaps from the same to the same. The use of abbreviations, which was a common practice in copying technical texts, gave rise to much misreading.

A certain number of literary testimonies show that in general ancient writers were fully aware not only of the errors which could slip into the work at any stage before it was circulated,[46] but also of the mistakes stemming from copyists' shortcomings. In the field of science and technology, two learned bibliophiles, the physician Galen in *De indolentia*[47] and the mathematician Eutocius in his commentary on

[42] The traditions of Euclid, Apollonius and Archimedes contain numerous examples of this type.

[43] See Mugler (1958: 7–32), Federspiel (1995, 2003).

[44] See for example Federspiel (1994: 213–214).

[45] On the formulae expressions used in reasoning by exhaustion, see Vitrac (1997).

[46] See (Dorandi 2000: Chaps. 5 and 6).

[47] Boudon-Millot and Jouanna (2010: 5.18–8.11).

Propositions II.I and II.4 of Archimedes' treatise *On the Sphere and Cylinder*,[48] give relatively precise accounts of the corruption of the ancient texts they discovered in the course of their research and of their own attempts to correct them. Galen wanted to set up his own library of Aristotelian authors and ancient physicians; Eutocius wanted to make available for the public the ancient solutions to two problems which Archimedes took to be solved without giving a demonstration.

A comparison between Eutocius' versions of solutions attributed in his commentary on Archimedes treatise to Heron (Commentary on II.I),[49] Pappus (Commentary on II.I)[50] and Diocles (Commentary on II.4) and the conserved treatises of these three mathematicians show that they had been rewritten and extensively completed. In the solution proposed by Diocles,[51] Eutocius restored all the intermediary steps which had been omitted in the demonstration, added the synthesis and the lemmas concerning the construction of the hyperbola with given asymptotes. He also altered the vocabulary: the version attributed to Menaechmus (fourth century BCE)[52] contains the terminology used by Apollonius when writing about conics. I have discussed elsewhere[53] his intention of producing as far as possible a mathematical text that is 'clear and written in everyday language' (κοινοτέρᾳ καὶ σαφεστέρᾳ κατὰ τὸ δυνατὸν λέξει γράφομεν)[54] and pointed out his faithfulness to his sources when he expressly claims to quote them.[55]

#### 11.2.1.3 Issues of Attribution and Authenticity

The issue of authenticity is at the heart of the concept of the critical edition. This was already the concern of the Alexandrian editors of Homer.[56] It also held an important place in the history of literary criticism throughout Antiquity.[57] The prefaces to two of Galen's works, *De ordine librorum suorum* and *De libris propriis*,[58] clearly show the fears of an ancient author for the fate of his works: uncontrolled

[48] Mugler (1972: 45.2–45.17; 88.13–88.89.15; 101.4–101.9).

[49] Mugler (1972: 47.6–48.20).

[50] Mugler (1972: 54.1–56.5).

[51] Mugler (1972: 106.22–114.17). The solution reported by Eutocius corresponds to Proposition 8 of *On Burning Mirrors* (Rashed 2000: 122–125).

[52] Mugler (1972: 58.17–60.11).

[53] Decorps-Foulquier (1998).

[54] Mugler (1972: 89.11).

[55] This is the case for Proposition 7 of *On Burning Mirrors* (Rashed 2000: 119–121) which introduces the solution itself given in Proposition 8. Eutocius faithfully reports this introductory passage (Mugler 1972: 105.1–106.21).

[56] See Irigoin (2003: 175–187); see also Reynolds and Wilson (1986: 8–10).

[57] In the late Neoplatonic tradition, questions of authenticity constitute one of the points to be examined (τὸ γνήσιον) in the introductions which specifically concern Aristotle's treatises; on the type of questions raised, see the article by Moraux (1974).

[58] See Boudon-Millot (2007).

circulation, plagiarism, misappropriation, additions and deletions. When Eutocius used old sources in his commentary on Proposition II.1 of the treatise *On the Sphere and Cylinder*[59] to discover ancient solutions to the problem of finding two mean proportional lines between two given lines, to which Archimedes merely assumed the solution without giving a proof, he was directly confronted with problems of attribution. He informs the reader of the criteria which led him to consider some solutions unauthentic, such as the solution attributed to Eudoxus of Cnidus[60]; on the other hand, in his commentary on Proposition II.4, he believes he recognizes in 'an old book', despite the copying mistakes in the text and the figures, the original text of the problem that Archimedes promised to deal with at the end of the proposition and which is missing.[61] The criteria he used, which were those of his edition of *Conics*, concern vocabulary and style, and also as well as the choice of demonstrative modes and the accuracy of reasoning. Eutocius offers in the field of mathematical sciences a unique insight into how an ancient editor evaluated the quality of his sources and selected them.

#### 11.2.1.4 Combining Traditions

In his commentary on Conics,[62] Eutocius once again gives a precious account of the use of sources in the edition of mathematical texts. In a very important passage in the prologue, he gives indications on the editorial method he used for his edition of the treatise:

> Since several editions exist, as stated by Apollonius himself in his dedication, I thought that it would be convenient to gather them into a single one, placing in the text the most evident parts among those that I had, to help with initiation, and noting in the margins, as is customary, in the commentaries I made, the different kinds of demonstrations that could be found.[63]

Further on in his commentary, Eutocius does in fact mention how the discrepancies between manuscripts troubled him and occasionally gives *in extenso* the text of the demonstrations he decided not to include in his edition of Apollonius' text. However, contrary to his commentary on the treatise *On the Sphere and Cylinder*, he does not enter into a reflection on authenticity either in the prologue or in the commentary on the propositions (the only explicitly formulated criteria for choosing between

---

[59] Mugler (1972: 45.1–75.13).

[60] Mugler (1972: 45.6–15).

[61] Mugler (1972: 88.25–89.7). For a mathematical analysis of this demonstration, see Knorr (1986: 171–173).

[62] See Decorps-Foulquier and Federspiel (2014).

[63] Πλειόνων δὲ οὐσῶν ἐκδόσεων, ὡς καὶ αὐτός φησιν ἐν τῇ ἐπιστολῇ, ἄμεινον ἡγησάμην συναγαγεῖν αὐτὰς ἐκ τῶν ἐμπιπτόντων τὰ σαφέστερα παρατιθέμενος ἐν τῷ ῥητῷ διὰ τὴν τῶν εἰσαγομένων εὐμάρειαν, ἔξωθεν δὲ ἐν τοῖς συντεταγμένοις σχολίοις ἐπισημαίνεσθαι τοὺς διαφόρους ὡς εἰκὸς τρόπους τῶν ἀποδείξεων; Decorps-Foulquier and Federspiel (2014: 11.26–11.31).

divergent versions are clarity and simplicity)[64]; nor does he make any explicit allusion to predecessors who might have added their own remarks or modified a traditional text; he formulates no separating line between a basic text that he would follow and one or several other recensions he might have at his disposal. Eutocius seems to attribute equal authority to his manuscripts since Apollonius himself refers to the existence of different versions that had been circulated; hence the obligation Eutocius imposes on himself to combine them. It is easy to understand that his method of work sets up a screen which prevents us from going further back in the tradition.

A late recension has therefore the same repercussions on textual history as the Alexandrian editions.[65] The end of Antiquity is a crucial period for the transmission of texts. Subsequent editions, whether Greek or Arabic, largely depend on the form the texts took between the fourth and fifth centuries.

### *11.2.2 From the Middle Ages to Eighteenth-Century Editions*

The traditions of late Antiquity were collected by Medieval translators in Western Europe (through southern Italy and Arabic cultural zones), on the one hand, and by Byzantine scholars on the other hand. In both cases, the transmission gave rise to numerous philological works and saved the major texts.

Two fundamental operations for the edition of scientific texts appeared during the Byzantine Middle Ages: (1) in the ninth and tenth centuries, ancient majuscule manuscripts which had been inherited from the end of Antiquity were transliterated into minuscule[66] These new manuscripts are at the origin of the subsequent tradition. (2) In the thirteenth and fourteenth centuries, higher scientific education started to be organized, bringing with it wide-scale copying of treatises which were then collected in vast corpuses of science and philosophy.[67] We are indebted to scholars of the Palaeologan Revival for discovering and saving large numbers of forgotten texts, one of which was Ptolemy's *Geography*.[68] Numerous mathematical texts were

[64] See my article cited above (Decorps-Foulquier 1998).

[65] Works edited by the Musaeum of Alexandria form a screen for works which preceded the Hellenistic period and the best a philologist can hope to do is to attain this form of the text. Works from the Hellenistic and Roman periods or from the end of Antiquity, which is the case for most Greek mathematical texts, have a tradition which can theoretically go directly back to the edition which the author himself circulated. But the picture is far from being as simple for all the works written before the codex was progressively adopted (from the first century CE). The discontinuation of the papyrus scroll and the adoption of the codex form for the transmission of writings caused an enormous upheaval in book-related activities which led to new editorial choices; this forms a second screen which appeared in the course of the Roman period and which takes us even further away from the original form of extant treatises.

[66] See for example Wilson (1983: Chaps. 3–7) and Irigoin (2003: 197–232).

[67] For an overview see Wilson (1983: Chaps. 12 and 13).

[68] See Gautier Dalché (2009: 82–84).

the object of erudite activity. Given the mediocre mathematical skills possessed by men of letters, the recension of these texts centered on the form; these Byzantine recensions can be easily recognized in the textual traditions collected by Renaissance scholars.[69]

The frenzy which seized scholars in Western Europe during the Renaissance when Greek manuscripts arrived from Byzantium[70] and the increase in the number of copying centers resulted in a significant growth in the number of copies of mathematical texts available for reading and studying. Even though native Greek copyists were used, some of whom were specialized in copying scientific texts, libraries and private collections accumulated manuscripts that had been severely distorted by all sorts of carelessness. Mathematical scholars and amateurs then used the margins of their copies to undertake genuine editorial work and their influence can be traced in derived manuscripts. In these marginal annotations, one finds corrections of copying mistakes, personal conjectures, large numbers of textual variants drawn from the consultation of other copies, new figures, glosses and lemmas to make up for elliptical reasoning. These elements were sometimes directly integrated into the text by subsequent copyists. This is how new texts were created; they may be considered to be new recensions.[71] Starting with one single copy from Byzantium, this scholarly activity contributed to multiplying the branches of the manuscript tradition.

The results of these works fed the Renaissance editions.

#### 11.2.2.1 The *Editio Princeps* of the Renaissance

In the early days of printing,[72] there was an increase in the number of editions of the first Latin and Greek documents, which gave partial access to mathematical texts.[73] But the corruptions were such that they gave rise to numerous reproaches from informed users. It was mainly in the second half of the sixteenth century that complete editions of Greek texts started to appear along with systematic reflection on the nature of the mistakes made by copyists and on the practice of textual criticism.[74] We are indebted to these scholars[75] for the 'vulgates' on which seventeenth and eighteenth-century editions were based and also for the Latin translations which

[69] See Decorps-Foulquier (1987, 2000: 199–226, 2014: LXIV–LXV); see also Tihon (1992: 135–136).

[70] On the role of an eminent figure such as Bessarion in this transmission to the West, see Märtl et al. (2013).

[71] For the tradition of *Conics*, see below.

[72] On first editions and the mechanical reproduction of contemporary manuscripts, see Dain (1975: 160–163).

[73] For an overview of this editorial activity related to mathematical texts, see Rose (1975).

[74] For an overview of this progress, see Reynolds and Wilson (1986: Chaps. 5 and 6).

[75] For an updated bibliography of works on great figures such as Giorgio Valla, Regiomontanus, Maurolycus, Commandino and Dasypodius, see D'Alessandro and Napolitani (2012: 15–43).

were to become the work tools of subsequent mathematicians. Despite the great competence of these scholars, the scope of their publishing programs and their use of the learned work of their predecessors, they continued to remain extremely dependent on the manuscripts they had at hand.[76] Seventeenth century editors carried on working in the same way as their predecessors had.[77]

#### 11.2.2.2 Later Editions

A more systematic investigation of collections and libraries together with increasingly numerous discoveries of papyri and palimpsests led to a renewal of the sources of all texts from Antiquity from which eighteenth- and early nineteenth-century editions were to benefit. In the field of mathematics, high-level scientists with encyclopedic ambitions took control of the edition of major works with the intention of grouping explanatory notes and accounts from the indirect tradition. From then on, the prefaces contain a clear formulation of the editor's aims, the presentation of his scientific collaboration and a detailed account of the discovery of the manuscripts.

These great scholars, who were at the head of genuine editorial enterprises, remained faithful to the methods of their predecessors. Like David Gregory in his edition of the *Elements* (1703), they continued to use the humanistic *editio princeps* or, when it did not exist, the text of readily available manuscripts, as did Edmund Halley in his *editio princeps* of Apollonius' *Conics* (1710). And even when they had at their disposal an ancient manuscript whose quality they had accurately gauged against the Renaissance copies, they continued to base their work on the text of the humanistic edition; this was the case for François Peyrard's edition of Euclid (1814–1818).[78] They trusted their own judgement to produce an accurate and readable text. The prefaces contain no sign of any critical reflection on the nature of the sources that had been used. Textual problems were posed only in terms of *emendatio*.[79]

The astronomer Edmund Halley produced an imposing piece of work on Apollonius' *Conics* treatise (Halley 1710). For the first time, his Graeco-Latin edition gave access to the Greek text of Books I–IV, to their commentators,

[76] Jean Pena's preface to his Latin translation in his edition of *Spherics* by Theodosius (1558) clearly shows how well aware editors were of the shortcomings of their manuscript sources and what liberties they had to take to propose an acceptable text.

[77] The recent critical edition of *In Nicomachi Arithmeticam* by Jamblique (see Vinel 2014: 54) bears this out once again for the late *editio princeps* produced by Tennulius in 1668.

[78] See Caveing (1990: 80).

[79] Hence the importance that eighteenth-century editions attached to collecting variants from manuscripts or old editions and presenting them in an appendix or in endnotes. This also holds for Peyrard's edition and for many contemporary editions such as Friedrich Ast's edition of the *Theologumena arithmeticae* and of the *Introductio arithmetica* of Nicomachus (Ast 1817: 157–202, 205–329).

Eutocius and Pappus,[80] as well as to the Arabic text of Books V–VII given in Latin translation. The design behind the whole organization of the edition centers on giving the reader a better understanding of the text of *Conics*; this goes as far as page layout, which is designed to enable Eutocius' commentary on the relevant propositions to be read simultaneously. On the other hand, he gives no information about the modifications he introduced into the text of manuscripts which he used as a basis for the edition of Books I–IV of *Conics* and for Eutocius' commentary. It is the study of the text and the consultation of the manuscripts that were copied and annotated for edition,[81] which show that Halley took full advantage of Commandino's excellent translation (1566) and of his own consultation of Arabic sources; it also shows that he did not hesitate to rewrite, on his own initiative, distorted passages without using other manuscript sources. When the text is elliptical and lends itself to several interpretations, he likewise does not hesitate in distancing himself from the source; he modifies the text and the figure according to his own understanding of the passage. In his eyes, the main thing was to produce a clear text which reflects the intrinsic logic of the reasoning.

In any case, and despite his efforts, he did not have the means to accomplish a truly critical piece of work.[82] The Oxford manuscripts that he used for *Conics* and Eutocius' commentary[83] were defective secondary copies, and a great many copying mistakes added at each step remained uncorrected.

### *11.2.3 Towards a Critical Edition*

New working methods gradually appeared in the course of the nineteenth century with the unprecedented arrival of new sources and the discovery of exceptional manuscripts such as the *Marcianus graecus* 454[84] (tenth century), which Jean-Baptiste Caspar d'Ansse de Villoison used as a basis for his edition of the *Iliad* in 1788. The scientific foundations of modern editions are heavily indebted to New Testament philology[85] which was confronted with the problem of multiple sources.

---

[80] At the beginning of the Books of *Conics* in question, Halley presents the Greek text and the Latin translation of lemmas to *Conics* collected by Pappus in Book VII of the *Collection*. In Book VII, Pappus brings together and demonstrates the auxiliary theorems which were used in the treatises linked with what he calls the 'domain of analysis' (ἀναλυόμενος τόπος) and which included Apollonius' works; see Jones (1986) and Decorps-Foulquier (2000: 43–59, 237–270).

[81] See Decorps-Foulquier (2001: 113–115, 2008: LXIII–LXIV).

[82] If we define textual criticism as the art of evaluating the differences between the transmitted text and the original text and the most appropriate means for correcting it.

[83] See his preface to *Conics* (Halley 1710).

[84] This is the famous *Venetus A*. It contains excerpts from the commentaries of Alexandrian scholars, which are copied into margins or between lines of the text itself.

[85] The Berlin editor of the New Testament in Greek, Karl Lachmann (1793–1851), is the emblematic figure.

The nature of editors' ambitions changed. Their aim was to produce a text which was as close as possible to the original form. To achieve this, they used any tools which would help them go further back in the history of the text. Given the quantity of documentation which was then available and the tools from historical, paleographic and codicological research that had been perfected to hierarchize it, it had henceforward become completely unjustified to produce an author's text based on one or several sources, even after careful scrutiny by the editor, without an examination of their authority.

The second half of the nineteenth century marked the beginning of the period of critical editions. Scientific texts arrived on the scene fairly late owing to the specific skills they called for.[86] The Danish philologist Johan Ludvig Heiberg (1854–1928) is the emblematic figure to whom we are indebted for the establishment in due form of so many mathematical texts.

The most fundamental theoretical contribution of the critical edition resides in its collection and classification of the entire existing manuscript tradition.[87] The scientific basis of a critical edition centers on determining the nature of the witnesses and the greater or lesser antiquity of the different forms of the text to which they give access. Critical edition cannot exist without an evaluation of the respective authority of all the witnesses. This evaluation enlightens the editor on the choice of variants and bestows rigor to editorial decisions. At the same time, it gives the reader the means to verify the validity of the choices that have been made.

After all the manuscripts have been collected, the operation of *recensio* reduces the number that will be used as a basis for edition; only the manuscript or manuscripts that make it possible to go back to the most ancient form of the text[88] will be retained. The result of this operation is recorded in the critical apparatus; this informs the reader of the direct and indirect witnesses that attest the reading that has been retained or discarded. It is written according to strict rules so as to avoid the brevity required giving rise to uncertainty or ambiguity.

To satisfy all these demands, editions of mathematical texts were progressively enriched with extremely precious para-texts: glossaries giving exhaustive lists of technical terms and linguistic characteristics; the edition of scholia; the edition of testimonia; and above all notes on the analysis of the information used as a basis for the critical work: codological description of manuscripts, classification of sources, use of the indirect tradition to go back to earlier stages in the history of the text, overviews of the paths of transmission, and editorial principles adapted to the specific case of the text being edited.

---

[86] For example, in 1852, the philological work carried out by Ernest Nizze on *Sphaerics* by Theodosius gave rise to an edition based on earlier editions without reference to manuscripts; it was not until Heiberg's work (1927) that a soundly established text was produced. The traditions of mathematical texts supply numerous other examples.

[87] The methods of *recensio* may vary with the nature of the tradition. On this question, see Irigoin's fundamental article (2003: 19–36).

[88] Irrespective of the age of the manuscript or the quality of the copy.

The collection of this information radically changed the operation of *emendatio*. Corrections and attempts at restoration were based on prior analysis of the type of mistake or alteration, on the knowledge of its origin and on the mechanism which led to the corruption.

Late nineteenth-century editors considerably increased the confidence which could be placed in the texts they produced, as the precise description of sources and the publication of the critical apparatus gave readers the means of checking their working method and editorial choices.[89] But they were sometimes over-conservative. The systematic reticence about making corrections as seen in Heiberg's editions can sometimes lead to an extremely mediocre rendering being accepted or to tolerating distortions of syntax, without equivalents elsewhere, thereby creating totally artificial *hapax*.

Furthermore, editors like Heiberg were confronted with the necessity of producing editions of large numbers of hitherto unexplored mathematical texts; at the same time, they had to solve extremely complex questions raised by the tradition of the major corpuses of Euclid, Archimedes and Heron of Alexandria. They sometimes worked too quickly[90] and were too systematic in applying principles which corresponded to their vision of an ancient mathematical text.[91] Consequently, it is necessary to beware of what they eliminated but also to reassess a certain number of their historical reconstructions as these had a direct impact on their editorial decisions.

## 11.3 Modern Editorial Issues

The ultimate aim of modern editions has not changed. It is still to obtain a text that is as close as possible to the one the author wrote. It is our vision of the nature of the ancient text that has changed. Increasingly detailed research into the primitive forms of writing a Greek text, the conditions in which copyists wrote and the diversity of the paths of transmission have revealed complex realities. Our representations have changed and led to different types of questions.

Renewals in paleography, codicology and textual history have largely contributed to a change in perspective. Advances in paleography have, in particular, made it possible to identify copying mistakes more easily and above all to situate a certain

[89] On the other hand, they were not very forthcoming on the principles they applied to edit the figures.

[90] In a good many of his editions, Heiberg contented himself with a very superficial collation of the manuscripts he discarded when establishing the text.

[91] The first pitfall is preconceived ideas. An example of this is the attitude of Friedrich Hultsch towards the transmitted text of the *Collection* by Pappus of Alexandria: in his edition of Books II–VIII (1876–1878), he athetizes without any justification a certain number of passages, on the grounds of an ideal representation of what a Greek mathematical treatise should be. These passages supply important historical information, and specific analytical tools are required to judge whether they are late interpolations by the editors who produced the collection or whether they were written by Pappus himself and drawn from works he personally consulted.

number of them in determined periods. In some traditions, researchers have succeeded in going far enough back into history to postulate several transliterations in regions other than the Byzantine capital. A much closer scrutiny of the material details of layout and codicological signs has made it possible to go as far back as when a corpus was first formed, to the transition from *volumen* to *codex*. The systematic exploration of Latin translations from Greek and translations from the Arabic-Latin tradition have given us the means to go far beyond the archetype[92] of the Greek manuscript tradition.

At the same time, constant technical progress in processing images, scientific analysis of different types of paper and ink, the exploration of library collections and codicological descriptions have provided a large amount of historical information about where our sources came from; we can add to these the digital processing of data in lexicological studies and the analysis of variant readings in the classification of manuscripts.[93] All these means have resulted in editors renewing the forms of textual criticism.

Editors of Greek mathematical texts are all the more concerned by this progress as the transmission of treatises conserved in educational circles largely contributed to the evolution of their primitive form, which was itself often difficult to delimit. These remarks clearly show that the scientific edition of an ancient mathematical treatise cannot be undertaken without prior analysis of information on its textual history. Indeed, an answer must be found to these two fundamental questions: given the conditions in which mathematical literature was transmitted, what state of text can an edition aim to reach and which text should be edited?

## 11.4 *Conics*, a Case in Point

The tradition of the *Conics* treatise offers a good example of how the multiple states of a text intermingle during transmission. Its study enables one to understand in concrete terms what becomes of a treatise which is considered essential reading in a disciplinary field. The disciplinary field in question here is that of the geometry of conic sections, a topic developed if not started by the Greeks in Plato's Academy. When Apollonius wrote a voluminous work of synthesis on the theory of conic sections, his name permanently entered the history of mathematics he consequently became an enduring name in the history of mathematics. In this work he faithfully represented the lines of research of his predecessors and his contemporaries while also opening new, extremely promising avenues.

---

[92] In other words, the common ancestor which has either been conserved or reconstituted and which is closest to all witnesses of the text in question. On this fundamental notion, see Irigoin (2003: 1–18, 37–53).

[93] See Irigoin (2003: 56–66).

### 11.4.1 The Vicissitudes of Transmission

The transmission of Apollonius' work suffered a great many vicissitudes.[94] None of his other geometric and astronomical treatises quoted in the indirect tradition have come down to us, with the exception of the treatise entitled *Section des droites selon des rapports*, transmitted in Arabic.[95] As seen above, *Conics*, transmitted in seven books with approximately sixty propositions in each (Book VIII is lost), did not escape this fate.

On reading the introductory letter to Book I sent to his friend Eudemus, which acts as a general preface to the whole work, we can identify three problematic points: (1) first of all, sending rough drafts to his disciples[96] was the surest way of having several versions in circulation and to pave the way for traditions to diverge.[97] (2) The second weak point is that the writing of the final version was staggered. Apollonius had the work delivered book by book (or, strictly speaking, roll by roll). It was consequently circulated progressively. Furthermore, before Book IV was sent,[98] Apollonius had to cope with the death of his friend Eudemus and change correspondents. (3) The third weak point was the internal organization of the work, which easily lent itself to partial editions or to the risk that the last parts might be separated from the rest. In Apollonius' presentation of the content of each book, one can immediately detect the dividing line which was to fix the fate of the treatise. The last book, Book VIII, which was lost, did not contain theorems, as did the preceding books, but problems; Books I to IV are described as books of 'elements', and Books V to VIII are considered to be the richest. It is consequently not surprising that the first four books were taken as the subject of Eutocius' recension at the end of Antiquity, with the dramatic consequences that we know for the survival of the whole treatise.

The Greek tradition only transmitted the text of Eutocius' recension,[99] as is attested by the head title and the final title in Book IV in our medieval tradition. In his commentary on *Conics*, Eutocius states that the original form of his work was an edition of the first four books[100] in which the reader could find the commentary on the edited propositions in the margin.[101] The terms used in the passage show that he

---

[94] For a summary of the text history, see the introductory parts in Decorps-Foulquier and Federspiel (2008, 2014).

[95] This is the title under which the French *editio princeps* appeared; see Rashed and Bellosta (2010).

[96] See Decorps-Foulquier and Federspiel (2008: 3.8–3.18).

[97] Apollonius himself points out to Eudemus that different versions of Books I and II are already in circulation.

[98] See his introductory letter to Book IV, sent to a second correspondent in Pergamon, a certain Attalus (Decorps-Foulquier and Federspiel 2010: 342.2–342.6).

[99] On Eutocius' recension of *Conics*, see Decorps-Foulquier and Federspiel (2014: IX–XCVII).

[100] Decorps-Foulquier and Federspiel (2014: 212.12–212.16).

[101] See the passage cited above. The position of the commentary in the margins of the edition is totally confirmed by the way in which the references to the edited text function and by the material damage the text suffered.

undertook real editorial work and that his commentary, so to speak, had the same function as a critical apparatus.

Yet another mishap for the *Conics* treatise: the Greek tradition did not transmit Eutocius' recension in its original form. The edition of Apollonius' text was separated from the commentary. The commentary is to be found transcribed as a continuous text in a ninth-century manuscript, *Vaticanus graecus* 204 in the collection known as the 'Little Astronomy'. Followed by two treatises by the mathematician Serenus *On the Section of a Cone* and *On the Section of a Cylinder*,[102] the edition itself was transmitted in a relatively late manuscript *Vaticanus graecus* 206, which can be dated to the late twelfth century or early thirteenth century. It can be seen that the end of the original form of Eutocius' 'commented edition', certainly at the very end of Antiquity, gave rise to other textual units and groupings which corresponded to new trends in Byzantine intellectual curiosity.

The Arabic translation which has come down to us is not strictly speaking a literal translation. A reading of the preface written by the Banû Mûsâ[103] shows that it is also a real piece of editorial work, based on two copies—an old corrupted manuscript of Books I to VII and a copy of Eutocius' edition found in Syria.[104] It informs us that Apollonius' treatise was the subject of a collective undertaking for its translation and commentary, bringing together the elite mathematicians of Baghdad, which included Thâbit ibn Qurra.

To conclude, the Greek text and the Arabic translation were edited very late in the day. The *editio princeps* of the Greek text (together with Eutocius' commentary) dates from no earlier than 1710. The first critical edition of *Conics* and Eutocius' commentary by Heiberg was published in 1891–1893. A second critical edition by Michel Federspiel and myself only appeared in 2008–2010 and 2014. It contains the first French translation of Eutocius' commentary. The *editio princeps* of the Arabic text of Books V to VII, which is also a critical edition, only dates from 1990 (by the historian of science Gerald J. Toomer). The complete critical edition of the Arabic text of Books I to VII was published by the historian of science, Roshdi Rashed in 2008–2010. It is also the *editio princeps* of Books I to IV.

### *11.4.2 Issues Editors Face*

After reading these facts, one immediately understands how important it is for an editor to know the real nature of the witnesses of a treatise and to identify what states of the text they represent, as it is necessary to know what one should do with the material one has.

[102] See Decorps-Foulquier and Federspiel (2019).

[103] Rashed (2008: 500–506).

[104] The preface is silent on the exact layout of Eutocius' edition, nor does it mention whether or not there was a commentary.

#### 11.4.2.1 Answers from Heiberg's Edition

Heiberg's work on *Conics* meets these requirements despite its shortcomings. We are indebted to Heiberg for proving that *Vaticanus gr.* 206 is the ancestor of the whole manuscript tradition of *Conics*. But he was content with a very superficial study of the rest of the tradition. An exhaustive study of the manuscript tradition reveals the existence of a Byzantine recension[105] and a humanistic recension which both Heiberg and Halley accessed through manuscripts[106]; however, they failed to recognize the nature and the origin of these two recensions and thereby distorted their representations of the lessons they took from them.

We are also indebted to Heiberg for showing that it is necessary to compare the testimony of our manuscripts with testimonies from the ancient indirect tradition, namely Eutocius' commentary on the *Conics* and Pappus' lemmas. In a fundamental article 'Ueber Eutokios' on the scientific environment of the mathematician, his work and his library,[107] Heiberg broaches the question of the nature of Eutocius' intervention in the text of *Conics*. He rightly refers the reader to the commentary but without giving himself any other means of checking Eutocius' statements. In the *prolegomena* of his edition, he uses Pappus' indirect testimony as follows: on the principle that Pappus' lemmas were exclusively used to compensate for Apollonius' implicit postulations, he dogmatically deduced that all the passages in *Conics* where the auxiliary theorems which Pappus demonstrated are used and explicitly justified had been interpolated after the fourth century. We will see below the results when such hypotheses are re-examined.

But neither Halley nor Heiberg, who nevertheless had these sources at their disposal, used the testimony of the Arabic tradition when editing Books I to IV. The explanation is simple. At the end of their preface, the Banû Mûsâ state that the translation of the first four Books was based on the books—to quote the text—'as Eutocius corrected them'[108] and that the translation of Books V to VII was based on the books—to quote the text again—'as Apollonius had composed them'.[109]

The editors, and likewise the historians of science who came after them, did not consider it was necessary to translate and study this part of the Arabic translation. But in places, the Arabic text of Books I to IV differs considerably from the Greek text in Eutocius' recension. This discrepancy, which varies in scale, can give rise to very different and debatable interpretations; in any case, it requires a study in its own right and a case-by-case analysis. This was missing from Heiberg's edition.

---

[105] See above.

[106] See Decorps-Foulquier (2001: 83–86).

[107] See Heiberg (1880).

[108] Rashed (2008: 506).

[109] Rashed (2008: 506).

#### 11.4.2.2 Recent Directions of Investigation

The major research issue for the new critical edition of the treatise was to determine which editorial principles should be adopted when the text history revealed so many superimposed screens, two of which concerned far-reaching editorial projects, one in Alexandrian circles at the end of Antiquity, the other in Baghdad in the ninth century.

The fundamental issue for the Greek tradition of Books I to IV was whether it was possible to determine the nature of the sources which Eutocius had at his disposal and the real extent of his interventions. As regards the Arabic tradition of the corresponding books, it was theoretically necessary to answer three questions: (1) what was the state of the text in Eutocius' edition found in Syria? (2) How had the Arabic translators used the two sources they had at their disposal, that is, Eutocius' edition and their copy of the whole treatise? (3) What modifications did they introduce into the transmitted text, given the excellent scientific level of the scientists recruited by the Banû Mûsâ?

Below is a four-point summary of the approaches used to tackle the issue.[110]

(1) The first line of investigation was to determine as accurately as possible which ecdotic and exegetical tradition Eutocius belonged to. On the one hand, recent historical testimonies have made it possible to confirm that Eutocius frequented Alexandrian philosophical circles, while the study of his commentary on Archimedes' treatise *On the Sphere and Cylinder* has supplied a large number of concrete examples of the work method he used for editing texts.

As already mentioned, it also made possible the verification that Eutocius kept to the text of his sources faithfully when he said expressly that he was quoting them. It was important to check this point to avoid overestimating his editorial interventions.

Furthermore, much could be expected from Eutocius' references to manuscripts in the commentary of *Conics*. Unfortunately, he remains very allusive. Nonetheless, an analysis of the demonstrations which he found in the text of certain sources, but which he rejected on the grounds of their being unauthentic, makes it possible to advance safely. The lack of generality and the clumsy, verbose style in which these demonstrations were written betray their scholastic origin. They reveal the existence of contaminated sources in Eutocius' documentation; they also reveal that Eutocius gives priority to his mathematical judgement when a choice has to be made.

There is no trace of these demonstrations in the Arabic translation. Consequently, this attests to the fact that relatively late scholarly works contributed to diversifying the manuscript tradition Eutocius had at his disposal.

(2) The second line of research concerned the testimonies of the indirect tradition prior to Eutocius. The most important testimony are Pappus of Alexandria's lemmas.

[110]All the results of this study are found in the introductory parts and the additional notes of the Greek text in our edition of *Conics* (Decorps-Foulquier and Federspiel 2008, 2010).

A great majority of them complete an ellipsis in Apollonius' reasoning[111] and amongst them we find equivalents in the lemmas Eutocius recorded in his commentary.[112] The two authors propose different versions of theorems from the same exegetical tradition of the *Conics* treatise; they show that they have taken them from the same sources. Other lemmas are clearly linked with certain propositions, but cannot complete an ellipsis since the auxiliary theorem is justified in Apollonius' text. When the justification bears signs of having been rewritten, Pappus' testimony supports doubts on the authenticity of the passage. As there is no particular reason for attributing these interventions to Eutocius, one must suppose that his own sources had integrated the modifications. Finally, there are other lemmas, which without ever losing their link with certain propositions in the *Conics*, do not find a precise point of application in the transmitted text. In some parts of the Apollonius treatise, Pappus was clearly not reading the same text as the one that has come down to us, but in the absence of any precise means of comparison, we cannot suggest a chronology for the observed discrepancies: certain lemmas may refer to a period prior to the fourth century and if that is the case, Pappus would seem to attest the existence of different branches of the tradition from those Eutocius used; others may illustrate an evolution of Apollonius' text after Pappus (this may be accounted for by recensions possibly prior to Eutocius' recension).

(3) The third line of research was the comparison with the Arabic sources. This reveals both a very close correspondence with the Greek text and some considerable discrepancies. In some propositions in the Arabic text, operations on proportions are developed in the greatest detail and each case is the subject of a complete exposition. In the first part of Book IV, the differences concern the organization of the propositions themselves and the nature of the demonstrative procedures.

A case-by-case examination makes it possible to put a certain number of the most glaring discrepancies into perspective: in the first part of Book IV, the difference from the Arabic translation was accentuated by the intervention of a scholar who either preceded or followed Eutocius[113] and who obviously rewrote and shortened some propositions; in certain passages, the translators seem to be trying to avoid an elliptic rendering of the Greek text.

On the other hand, when compared with Pappus, the two Greek and Arabic traditions show no differences,[114] which leads us to believe that the differences between the two traditions were accentuated after the fourth century.

---

[111] In accordance with the definition of the lemmas given by Proclus in his commentary on Book I of Euclid's *Elements*; see Caveing (1990: 141).

[112] See Decorps-Foulquier (2000: 267–270).

[113] As Eutocius did not comment on this part of Book IV, there is no means of establishing a relative chronology.

[114] With regard to Pappus' lemmas which complete an ellipsis in Apollonius' reasoning, it can in fact be observed that the ellipsis is in the Greek and Arabic texts; when the lemmas give a demonstration for which the content is in the Greek text, we have the same text in Arabic. This also holds for the lemmas which have no direct point of application in Apollonius' text.

(4) The most reliable results were provided by internal and linguistic analyses. The anomalies and breaks noted in the structure of the text and in the terminology attest to the relative lack of homogeneity, thereby confirming information from the textual history.

The study of the language used[115] reveals that while the scientific vocabulary in the treatise is the same from Book I onwards, this does not always hold for the style of writing and the use of linguistic structures. In Books III and IV and in the problems,[116] there is an accumulation of unusual turns of phrase which are closer to the language used by Archimedes than to that found in Book I. In places, these expressions bring to light the language used by Apollonius' sources, which shows that he did not lessen the resulting terminological divergence.[117]

The internal analysis shows that the style is far from being homogeneous. Clear signs of rewriting are found, but it is not possible to attribute them to Eutocius or to later scholars. Instances are also found of passages which escaped the standard writing codes and seem to bear traces of the author's unrevised first drafts.

The analysis of the text also reveals a lack of internal coherence, which might suggest interventions prior to Eutocius; indeed, there is a faint echo of them in Pappus' lemmas, in the exegetical material collected by Eutocius and less frequently in the Arabic translation.

Finally, this analysis shows a number of occasional but marked discrepancies with the text of *Conics* to which Eutocius refers to in his commentary. In these passages, the testimony of Eutocius' commentary and that of the Arabic text agree but differ from the text transmitted by our medieval manuscripts, proving without doubt that the treatise, as edited by Eutocius in the sixth century, was altered after him at the end of Antiquity.

## 11.5 From Text History to Text Criticism: Methodological Issues

The *Conics* treatise is not an isolated case. The study of text history and philological analyses converge to draw a relatively complex picture of the traditions that have been transmitted to us. A number of questions about the nature of witnesses and the real status of the texts they represent remain open. Therefore, it is clear that the corrections and the choice of variants directly depend on the outcomes resulting from the study of the text's history.

---

[115] For each of the books from Book I to IV, a study of the language used was the subject of critical and exegetical notes by M. Federspiel in *Revue des Études Grecques*; see Federspiel (1994) (1999) (2000) (2002) (2008a, b) (2009). These notes complete a certain number of the entries in Mugler's dictionary (1958).

[116] See Federspiel (2008a, b).

[117] See Federspiel (2012).

Which state of the text should one restore? It is out of the question to distort the testimony of a text by altering it with readings from independent or more ancient traditions. Equally valid as testimonies are the modifications the text has undergone in the course of its history and whose traces it bears. It is the function of the body of notes accompanying the text to collect these data which inform us about how the texts were received and read. In the case of the first four books of Apollonius' work, it is clear that one cannot go beyond the last known recension, that of Eutocius. The existence of the commentary supplies precious points of reference for going beyond the testimonies of the manuscript witnesses of the recension and for reaching with relative certainty the text of *Conics* as edited by Eutocius. The philological tradition to which he seems to belong and his mathematical competences make us hope that he transmitted without any major modification all the propositions for which the manuscript copies he had at his disposal showed no difference and that satisfied him scientifically.

However, in fact, mathematical manuscripts accumulate a large number of errors which cannot be left uncorrected. This goes without saying concerning scribal errors. But even then, the error mechanism must be taken into account when correcting the mistakes; this is the surest means of finding the original text. When the mistakes concern the use of language, mathematics or the deductive line of reasoning, correction is essential but this requirement nonetheless rests on a theoretical representation of the process of writing a text, which stipulates that in principle, an error of this type should not be attributed to the author of the treatise or, in the case of Apollonius's *Conics*, to his editor Eutocius.

Furthermore, ignorance of the real modes of operating of mathematical language was the source of many misinterpretations. Greek mathematicians used everyday language to develop particular turns of phrase, introduce short forms and make specific syntactical use of the article as well as of moods and voices. Ignorance of these points led editors to make detailed corrections that were pointless but there continue to be repercussions today on translations into the vernacular which frequently fail to avoid two major stumbling blocks: a too modernized translation or an excessively literal translation which respects neither the source language nor the target language.

## 11.6 Which Figures Should Be Edited?

The edition of a mathematical, geometric or technical treatise must also address the problems posed by the representation of diagrams.[118] Given the poor condition in which the manuscripts had transmitted them, they were frequently redrawn in editions which preceded critical editions. The figures have suffered in our manuscripts. Being more often than not drawn in the margins, they were either cut off by binders or deleted. Furthermore, the errors made in copying the letters designating the points, and the blatant errors in their drawing resulted in their losing all credibility

[118] See Saito (2006).

in the eyes of editors. But to misrepresent the choice of the cases that they illustrate, to misrepresent the position of the points on a straight line or a curve, to misrepresent their orientation, as can be seen in the editions produced by the great eighteenth and nineteenth century scientists, and even sometimes in Heiberg's editions without the slightest warning, all this amounts to depriving them of their status as testimony. The figures had every chance of being more carefully copied than the text; given their lack of scientific competence, medieval copyists had in fact no alternative but to faithfully reproduce the drawings they saw in their model.

On the other hand, one cannot be certain that all the figures reproduced were authentic (anomalies are easily detected in a mathematical proposition by the lack of correspondence between the figure in the manuscript and the text of the *ekthesis* which names the objects) as, in the course of the successive recensions throughout Antiquity, a certain number of them may have been added to the original corpus to illustrate other cases.

Observance of the manuscript testimony does not imply that the edition should reproduce a photograph so to speak of the figures drawn by the copyists. It is obvious that the drawings, on papyrus in the first Hellenistic editions and on codex thereafter, had to fit into the spaces they were allocated. Local practices appeared. Some figures were drawn with a ruler and compass, others freehand; in these conditions, parallel lines and right angles could only be approximate.

For geometrical figures in space, the part of convention was inevitable.[119] In the case of plane sections of cylindrical or conic bodies, the lack of appropriate instruments made it impossible to draw the curves in one continuous movement and the copyists did not have the competence to draw the curves point by point. Ellipses, for example, were represented by two circular arcs. One cannot expect such figures to keep the metrical relations assumed by the text. They only enable points to be identified. However, it should be noted that in the case of *Conics*, the figures of conic sections reproduced in the *Vaticanus graecus* 206 seem to have kept a a faint echo of what originally was a correct proportion between the measurement of a diameter and its 'erect side' (the *latus rectum* or 'parameter'—the length of the rectangle constructed on the abscissa).

Unlike the copyists, editors who wished to find the exact proportions between the geometrical elements obviously had the technical means to improve the accuracy of the figures considerably. There are two situations in which the representation depends on a personal editorial decision. What should one do when the transmitted figure illustrates a special case while the proposition is general? And how should one represent a wrong figure in the numerous cases of *reductio ad absurdum* used by Greek mathematicians?

A certain number of transmitted figures represent special cases, such as for example, the representation of an isosceles triangle when the triangle referred to in the text is unspecified, a right cone when the cone in the proposition is scalene, etc.[120]

---

[119] For treatises on mechanics and pneumatics, see Bogen (2013).

[120] On this question, see Netz's remarks (2011: 280–285).

This possibly bears the stamp of mere scribal practices but one cannot discard the hypothesis of a deliberate choice made by ancient mathematicians to represent a generic object through the form which is visually the simplest one.

The testimony of the editor Eutocius[121] is very interesting in this respect. In his commentary on Proposition I.16 of *Conics*, showing his concern not to reproduce a figure which is contradictory to the hypotheses, he speaks up against observed habits and gives strict rules for copyist drawers to respect. He remarks that the *erect side* is always to be drawn perpendicularly to the corresponding diameter. He requires that the ordinates and the *second diameter* should form an acute angle with the diameter so that the ordinates can be clearly distinguished from the parallel lines to the *erect side*. Here Eutocius observes the mode of producing the conic sections adopted in the work by Apollonius, which means that the diameter of the curve is not necessarily its axis.

These material realities confront editors with a choice: to remain faithful to the manuscript tradition or to redraw the figure. Commandino and Halley were dependent on sources in which diagrams had already been revised by humanistic scholars. Furthermore, they did not hesitate to produce other diagrams. This was not so for Heiberg, who had decided to remain faithful to his source, the *Vaticanus graecus* 206, but there are numerous cases in which he can be seen to have behaved in a somewhat offhand manner by being inconsistent or editing erroneous figures as a result of careless collations.

In conclusion, I wish to remark that while intellectual requirements and the goals assigned to editions have seen significant variations, the fragmentary nature of our Greek sources, their corruption and the complexity of the states of the texts that they represent have led all editors to address the same issues. Highlighting them and analyzing the solutions aims to enrich the critical reflection of historians of science on the nature and origin of the texts they quote and use.

## References

### *Primary Sources*

Acerbi, F., N. Vinel, and B. Vitrac. 2010. Les Prolégomènes à l'Almageste. Une édition à partir des manuscrits les plus anciens. *SCIAMUS* 11: 53–110.

Ast, F. 1817. *Theologumena arithmeticae ad rarissimum exemplum Parisiense emendatius descripta. Accedit Nicomachi Gerasini Institutio arithmetica ad fidem codicum Monacensium emendata*. Leipzig: Libraria Weidmannia.

Aujac, G. 1979. *Autolycos de Pitane*. Paris: Les Belles Lettres.

Boudon-Millot, V. 2007. *Galien. Introduction générale. Sur l'ordre de ses propres livres. Sur ses propres livres. Que l'excellent médecin est aussi philosophe*. Paris: Les Belles Lettres.

Boudon-Millot, V., and J. Jouanna. 2010. *Galien. Ne pas se chagriner*. Paris: Les Belles Lettres.

[121] See Decorps-Foulquier (1999).

Carra de Vaux, B. 1988. *Héron d'Alexandrie. Les Mécaniques ou l'élévateur des corps lourds*. Paris: Les Belles Lettres.

Decorps-Foulquier, M., and M. Federspiel. 2008. *Apollonius de Perge, Coniques. Tome 1.2: Livre I (Greek text)*. Berlin: Walter de Gruyter.

———. 2010. *Apollonius de Perge, Coniques. Tome 2.3: Livres II–IV (Greek text)*. Berlin: Walter de Gruyter.

———. 2014. *Eutocius d'Ascalon. Commentaire sur le traité des Coniques d'Apollonius de Perge (Livres I–IV)*. Berlin: Walter de Gruyter.

———. 2019. *Sérénus, La section du cylindre. La section du cône*. Paris : Les Belles Lettres.

Friedlein, G. 1873. *Procli Diadochi In primum Euclidis Elementorum librum commentarii*. Leipzig: Weidmann.

Gregory, D. 1703. *Euclidis quae supersunt omnia*. Oxford: Theatro Sheldoniano.

Halley, E. 1710. *Apollonii Pergaei Conicorum libri octo et Sereni Antissensis De sectione cylindri et coni libri duo*. Oxford: Theatro Sheldoniano.

Heath, T.L. 1926. *The thirteen books of Euclid's elements*. 3 vols. 2nd ed. Cambridge UK: University Press.

Heiberg, J.L. 1891–1893. *Apollonii Pergaei quae graece exstant cum commentariis antiquis*. Leipzig: Teubner.

———. 1898–1903. Claudii Ptolemaei Opera quae exstant. In *Syntaxis Mathematica*, vol. 1.1–2. Leipzig: Teubner.

———. 1927. *Theodosius Tripolites Sphaerica*. Berlin: Teubner.

Hultsch, F. 1876–1878. *Pappi Alexandrini Collectio quae supersunt*. 3 vols. Berlin: Weidmann.

Jones, A. 1986. *Pappus of Alexandria book 7 of the collection*. New York: Springer–Verlag.

Menge, H. 1896. *Euclidis Data cum commentario Marini et scholiis antiquis*. Leipzig: Teubner.

Mugler, Ch. 1970. *Archimède*. Vol. 1. Paris: Les Belles Lettres.

———. 1971. *Archimède*. Vol. 2. Paris: Les Belles Lettres.

———. 1972. *Archimède*. Vol. 4. Paris: Les Belles Lettres.

Nizze, E. 1852. *Theodosii Tripolitae Sphaericorum libros tres Ernestus Nizze recognovit...*. Berlin: Georg Reimer.

Pena, J. 1558. *Theodosii Tripolitae Sphaericorum libri tres nunquam antehac graece excusi*. Paris: André Wechel.

Peyrard, F. 1814–1818. *Les œuvres d'Euclide, en grec, en latin et en français, d'après un manuscrit très ancien, qui était resté inconnu jusqu'à nos jours*. Paris: M. Patris.

Rashed, R. 2000. Livre de Dioclès sur les miroirs ardents. In *Les Catoptriciens*, ed. R. Rashed, 3–151. Paris: Les Belles Lettres.

———. 2008. *Apollonius de Perge, Coniques. Tome 1.1: Livre I (Arabic text)*. Berlin: Walter de Gruyter.

———. 2008–2010. *Apollonius de Perge, Coniques (5 vols.). Books I-VII (Arabic text)*. Berlin: Walter de Gruyter.

Rashed, R., and H. Bellosta. 2010. *Apollonius de Perge. La section des droites selon des rapports*. Berlin: Walter de Gruyter.

Stamatis, E.S., and I.L. post Heiberg. 1977. *Euclidis Elementa. Prolegomena critica, Libri XIV–XV, Scolia in Libros I–V. Euclidis Elementa*. Leipzig: Teubner.

Tennulius, S. 1668. *Jamblichi Chalcidensis ex Cœle–Syria In Nicomachi Geraseni Arithmeticam introductionem*. Arnhem: J.F. Hagium.

Tihon, A., and J. Mogenet. 1985. *Le 'Grand Commentaire' de Théon d'Alexandrie aux Tables Faciles de Ptolémée. Livre I. Studi e Testi 315*. Rome: Biblioteca Apostolica Vaticana.

Toomer, G.J. 1976. *Diocles. On burning mirrors*. New York: Springer–Verlag.

———. 1990. *Apollonius. Conics books V to VII*. New York: Springer–Verlag.

Vinel, N. 2014. Jamblique. In *Nicomachi Arithmeticam. Introduction, texte critique, traduction française et notes de commentaire*. Pisa: Fabrizio Serra Editore.

Vitrac, B. 1990–2001. *Euclide d'Alexandrie. Les Eléments*. Paris: Presses Universitaires de France.

Vitrac, B., and F. Acerbi. 2014. *Héron d'Alexandrie. Metrica. Introduction, texte critique, traduction française et notes de commentaire*. Pisa: Fabrizio Serra Editore.

## *Secondary Sources*


Aujac, G. 1977. Eratosthène, premier éditeur de textes scientifiques? *Pallas* 13: 3–24.

———. 1979–1980. La lettre à teneur scientifique à l'époque alexandrine, Bulletin de la Société toulousaine d'études classiques de l'Université de Toulouse Le Mirail 179–180: 79–102.

Bogen, S. 2013. Diagrammatic reasoning: The foundations of mechanics. In *Writing science. Medical and mathematical authorship in Ancient Greece*, ed. M. Asper, 279–298. Berlin: Walter de Gruyter.

Caveing, M. 1990. Introduction générale. In *Euclide d'Alexandrie. Les Eléments*, ed. M. Caveing and B. Vitrac, vol. I, 13–148. Paris: Presses Universitaires de France.

Cuomo, S. 1998. Collecting authorities, construction authority in Pappus of Alexandria's Συναγωγή. In *Gattungen wissenschlaftlicher Literatur in der Antike*, ed. W. Kullmann, J. Althoff, and M. Asper, 219–237. Tübingen: Gunter Narr Verlag.

———. 2001. *Ancient mathematics*. London: Routledge.

D'Alessandro, P., and P.D. Napolitani. 2012. *Archimede Latino*. Paris: Les Belles Lettres.

Dain, A. 1975. *Les manuscrits*. Paris: Les Belles Lettres.

Decorps-Foulquier, M. 1987. Un corpus astronomico-mathématique au temps des Paléologues. Essai de reconstitution d'une recension. *Revue d'Histoire des Textes* 17: 15–54.

———. 1998. Eutocius d'Ascalon éditeur du traité des *Coniques* d'Apollonios de Pergé et l'exigence de 'clarté': Un exemple des pratiques exégétiques et critiques des héritiers de la science alexandrine. In *Sciences exactes et sciences appliquées à Alexandrie*, ed. G. Argoud and J.-Y. Guillaumin, 49–60. Saint-Etienne: Université de Saint-Etienne.

———. 1999. Sur les figures du traité des *Coniques* d'Apollonios de Pergé édité par Eutocius d'Ascalon. *Revue d'Histoire des Mathématiques* 5: 61–82.

———. 2000. *Recherches sur les Coniques d'Apollonios de Pergé et leurs commentateurs grecs*. Paris: Klincksiek.

———. 2001. La traduction manuscrite du texte grec des *Coniques* d'Apollonios de Pergé (Livres I–IV). *Revue d'Histoire des Textes* 31: 61–116.

———. 2011. Le livre mathématique dans l'Antiquité grecque. In *L'Homme et la Science*, ed. J. Jouanna, M. Fartzoff, and B. Bakhouche, 207–220. Paris: Les Belles Lettres.

Devreesse, R. 1954. *Introduction à l'étude des manuscrits grecs*. Paris: Klincksieck.

Dorandi, T. 2000. *Le stylet et la tablette. Dans le secret des auteurs antiques*. Paris: Les Belles Lettres.

Endress, G. 2012. Proclus de Lycie. In *Dictionnaire des Philosophes antiques*, ed. R. Goulet, vol. Vb, 1546–1674. Paris: CNRS Editions.

Federspiel, M. 1994. Notes critiques sur le Livre I des *Coniques* d'Apollonius de Pergè. *Revue des Études Grecques* 107: 203–218.

———. 1995. Sur l'opposition *défini/indéfini* dans la langue des mathématiques grecques. *Les Études classiques* 63: 249–293.

———. 1999. Notes linguistiques et critiques sur le Livre II des *Coniques* d'Apollonius de Perge (Première partie). *Revue des Études Grecques* 112: 409–443.

———. 2000. Notes linguistiques et critiques sur le Livre II des *Coniques* d'Apollonius de Perge (Deuxième partie). *Revue des Études Grecques* 113: 359–391.

———. 2002. Notes linguistiques et critiques sur le Livre III des *Coniques* d'Apollonius de Pergè (Première partie). *Revue des Études Grecques* 115: 110–148.

———. 2003. Sur quelques effets du 'principe d'abréviation' chez Euclide. *Les Études classiques* 71: 321–352.

———. 2008a. Notes linguistiques et critiques sur le Livre III des *Coniques* d'Apollonius de Pergè (Seconde partie). *Revue des Études Grecques* 121: 515–545.

———. 2008b. Les problèmes des Livres grecs des *Coniques* d'Apollonius de Perge. Des propositions mathématiques en quête d'auteur. *Les Études classiques* 76: 321–360.

———. 2009. Notes linguistiques et critiques sur le Livre IV des *Coniques* d'Apollonius de Pergè. *Revue des Études Grecques* 122: 293–317.

———. 2012. Quelques traits de la modernisation de la langue mathématique par Apollonius. *Les Études classiques* 80: 319–342.
Gautier Dalché, P. 2009. *La Géographie de Ptolémée en Occident (IV$^e$–XVI$^e$)*. Turnhout: Brepols.
Goulet, R. 2007. La conservation et la transmission des textes philosophiques grecs. In *The libraries of the Neoplatonists*, ed. C. D'Ancona, 29–61. Leiden: Brill.
———. 2012. Pappus d'Alexandrie. In *Dictionnaire des Philosophes antiques*, ed. R. Goulet, vol. Va, 147–149. Paris: CNRS Editions.
Hadot, I. 1987. Les introductions aux commentaires exégétiques chez les auteurs néoplatoniciens et les auteurs chrétiens. In *Les règles de l'interprétation*, ed. M. Tardieu, 99–122. Paris: Editions du Cerf.
Heath, T.L. 1921. *A history of Greek mathematics*. Oxford: Clarendon Press.
Heiberg, J.L. 1880. Ueber Eutokios. *Philologische Studien zu griechischen Mathematikern, Jahrbücher für classische Philologie* Suppl. XI: 357–384.
———. 1903. Paralipomena zu Euklid. *Hermes* 38: 46–74. 161–201, 321–356.
Hoffmann, P. 2000. Bibliothèques et formes du livre à la fin de l'Antiquité. Le témoignage de la littérature néoplatonicienne des V$^e$ et VI$^e$ siècles. In *I manoscritti greci tra riflessione e dibattito*, ed. G. Prato, 601–632. Florence: Gonnelli.
Irigoin, J. 2003. *La tradition des textes grecs*. Paris: Les Belles Lettres.
Knorr, W. 1986. *The ancient tradition of geometric problems*. Boston: Birkhäuser.
———. 1996. Mathématiques. In *Le Savoir grec*, ed. J. Brunschwig, G. Lloyd, and P. Pellegrin, 409–438. Paris: Flammarion.
Levy, C. 2007. De la Grèce à Rome: l'espace-temps des philosophes antiques. In *Lieux de savoir. Espaces et communautés*, ed. C. Jacob, 1019–1048. Paris: Albin Michel.
Mansfeld, J. 1994. *Prolegomena. Questions to be settled before the study of an author or a text*. Leiden: Brill.
———. 1998. *Prolegomena Mathematica. From Apollonius of Perga to the late Neoplatonists*. Leiden: Brill.
Märtl, C., C. Kaiser, and T. Ricklin, eds. 2013. *Inter Graecos Latinissimus, inter Latinos Graecissimus. Bessarion Zwischen den Kulturen*. Berlin: Walter de Gruyter.
Michel, P.-H. 1950. *De Pythagore à Euclide. Contribution à l'histoire des mathématiques préeuclidiennes*. Paris: Les Belles Lettres.
Moraux, P. 1974. La critique d'authenticité chez les commentateurs grecs d'Aristote. In *Mélanges Mansel*, ed. E. Akurgal and U. Bahadir Alkim, vol. I, 265–288. Ankara: TTK.
Mugler, C. 1958. *Dictionnaire historique de la terminologie géométrique des Grecs*. Paris: Klincksieck.
Netz, R. 1999. *The shaping of deduction in Greek mathematics*. Cambridge UK: Cambridge University Press.
———. 2011. The place of codex C in Archimedes scholarship. In *The Archimedes palimpsest*, ed. R. Netz, W. Noel, N. Tchernetska, and N. Wilson, vol. I, 266–320. Cambridge UK: Cambridge University Press.
Reynolds, L.D., and N.G. Wilson. 1986. *D'Homère à Erasme. La transmission des classiques grecs et latins*. Paris: Editions du CNRS.
Rose, P.L. 1975. *The Italian renaissance of mathematics*. Genève: Droz.
Saito, K. 2006. A preliminary study in the critical assessment of diagrams in Greek mathematical works. *SCIAMUS* 7: 81–144.
Schneider, J.-P. 2000. Eudème de Rhodes. In *Dictionnaire des Philosophes antiques*, ed. R. Goulet, vol. III, 285–289. Paris: CNRS Editions.
Sidoli, N., and G. Van Brummelen, eds. 2014. *From Alexandria through Bagdad*. Berlin: Springer–Verlag.
Taub, L. 2013. On the variety of 'genres' of Greek mathematical writing: Thinking about mathematical texts and modes of mathematical discourse. In *Writing science. Medical and mathematical authorship in ancient Greece*, ed. M. Asper, 333–365. Berlin: Walter de Gruyter.

Taub, L., and A. Doody, eds. 2009. *Authorial voices in Greco-Roman technical writing*. Trier: Wissenschaftlicher Verlag Trier.

Tihon, A. 1992. Propos sur l'édition de textes astronomiques grecs des IVe et Ve siècles de notre ère. In *Les Problèmes posés par l'édition critique des textes anciens et médiévaux*, ed. J. Hamesse, 113–137. Louvain la Neuve: Université Catholique de Louvain.

Tood, R.B. 2000. Géminos. In *Dictionnaire des Philosophes antiques*, ed. R. Goulet, vol. III, 472–477. Paris: CNRS Editions.

Vitrac, B. 1997. Théon d'Alexandrie et la *Mesure du cercle* d'Archimède. *Oriens-Occidens* 1: 48–81.

———. 2003. Les scholies grecques aux *Eléments* d'Euclide. *Revue d'Histoire des Sciences* 56 (2): 275–292.

———. 2008. Les préfaces des textes mathématiques grecs anciens. In *Liber amicorum Jean Dhombres*, ed. P. Radelet de Grave, 518–556. Turnhout: Brepols.

———. 2012. The Euclidean ideal of proof in *the elements* and philological uncertainties of Heiberg's edition of the text. In *The history of mathematical proof in ancient traditions*, ed. K. Chemla, 69–134. Cambridge UK: Cambridge University Press.

Wilson, N.G. 1983. *Scholars of Byzantium*. London: Duckworth.

# Chapter 12
# Shaping a Mathematical Text in Sanskrit: H. T. Colebrooke, Sudhākara Dvivedin, and Pṛthūdaka's Commentary on the Twelfth Chapter of the *Brāhmasphuṭasiddhānta*

**Agathe Keller**

**Abstract** In 1817, Henry Thomas Colebrooke (1765–1837) published English translations of different Sanskrit mathematical texts, including a translation of the twelfth chapter of Brahmagupta's *Brāhmasphuṭasiddhānta* (*'Theoretical Treatise on the true Brahmā [school of astral science]'* 628 CE), with the help of Pṛthūdaka's *Vāsanābhāṣya* (*'Commentary with explanations'* fl. 960). Almost ninety years later, in 1902, Sudhākara Dvivedin (1855–1910/11) published an edition of the *Brāhmasphuṭasiddhānta* with his own commentary, which refers also to Pṛthūdaka's commentary. Since both texts were crucial to the access one could have to Brahmagupta's mathematics, their point of view durably marked the historiography. Dvivedin's and Colebrooke's publications are striking for the different ways they deal with the structure of Brahmagupta's mathematical chapter, their relation to Pṛthūdaka's commentary, or numerical practices.

I will argue that these authors' views on commentaries, structure and number went together with a specific idea of what a Sanskrit mathematical text was to be, informing their editorial choices, and shaping the views we have of the text today.

The research leading to these results has received funding from the European Research Council under the European Union's Seventh Framework Programme (FP7/2007–2013)/ERC Grant Agreement No. 269804. I would like to thank A. Graheli, K. Preisendanz, M. Schneider, X. Zhou, C. Montelle, K. Margolis, K. Chemla and D. Bayuk for their helpful comments on previous drafts of this chapter.

A. Keller (✉)
Faculté Sociétés et Humanités / UFR IHSS / Laboratoire SPHère,
CNRS - Université Paris Cité, Paris, France
e-mail: agathe.keller@u-paris.fr



A. Keller, K. Chemla (eds.), *Shaping the Sciences of the Ancient and Medieval World*, Archimedes 69, https://doi.org/10.1007/978-3-031-49617-2_12

## Abbreviations

| | |
|---|---|
| BSS | *Brāhmasphuṭasiddhānta* of Brahmagupta |
| PBSS | Pṛthūdaka's *Vāsanabhāṣya,* a commentary on the BSS |
| KK | *Khaṇḍakhādyaka* of Brahmagupta |
| PKK | Pṛthūdaka's commentary on the *Khaṇḍakhādyaka* of Brahmagupta |

## 12.1 Introduction

In 1817, Henry Thomas Colebrooke (1765–1837) published English translations of different Sanskrit mathematical texts, including a translation of the twelfth chapter of Brahmagupta's *Brāhmasphuṭasiddhānta* ('a theoretical treatise on the true Brahmā [school of astral science]', 628 CE), with the help of the *Vāsanābhāṣya* ('commentary with explanations' by Pṛthūdaka (fl. 860)).[1] Eighty-five years later, in 1902, Sudhākara Dvivedin (1855–1910/11) published the first printed edition of the *Brāhmasphuṭasiddhānta* with his own commentary based on that of Pṛthūdaka.[2] Since both texts were crucial for the access they provided to Brahmagupta's mathematics, these publications have enduringly marked the way he has been perceived. Brahmagupta is, indeed, one of the most famous ancient Indian mathematicians, celebrated more or less faithfully for his rules of computation on zero, negative quantities, and sometimes for his work on cyclic quadrilaterals or indeterminate analysis.

Brahmagupta's and Pṛthūdaka's works raise a certain number of issues that are as open to research today as they were at the time of Colebrooke or Dvivedin. In the following, we will look at three such problems:

1. The organisation of the mathematical chapter of the *Brāhmasphuṭasiddhānta*,
2. The integration into the published text of the crucial importance of Pṛthūdaka's commentary to understand Brahmagupta's rules, and
3. Finally, some elements concerning numbers and quantities as revealed in the treatise and the commentary.

The editorial choices of both Colebrooke and Dvivedin concerning these questions will be described and analyzed. Such answers were not necessarily argued for; they were, in most cases, acts that molded the printed texts they produced, and thus influenced the way such texts were subsequently perceived in a silent but pervasive manner. The aim of this article is to describe the context in which the publications took place and highlight which ideas about mathematics, about mathematical texts, and

[1] Colebrooke (1817), which also includes translations of Chap. 18 of the *Brāhmasphuṭasiddhānta*, and of the *Līlavatī* and *Bījagaṇita* by Bhāskara (b. 1114).

[2] Dvivedin (1902).

which of the editors' motivations for publishing the texts had an impact on their editorial choices. We shall also, where possible, briefly point out the impact such choices had on later historiography.

## 12.2 Some Unresolved Problems

### *12.2.1 Sources: Brahmagupta and Pṛthūdaka's Commentary*

Two works of Brahmagupta (b. 598)[3] are known. The *Brāhmasphuṭasiddhānta* (henceforth BSS) and the *Khaṇḍakhādyaka* (665, 'sweet/edible pieces', henceforth KK). The first is a *siddhānta*, that is, a theoretical text of the branch of astral science (*jyotiṣa*) concerned with mathematical astronomy. The second is a *karaṇa*, that is, a work of mathematical astronomy with a focus on the ease and rapidity of procedures. It is noteworthy that it was the most theoretical text that attracted both Colebrooke's translation and then S. Dvivedin's edition, while historically, the other text seems to have been favored if we take into account the number of circulated manuscript copies.[4] Brahmagupta's work, as we shall see, will be constantly compared with the twelfth-century *Siddhāntaśiromāṇi* of Bhāskara, a landmark in Sanskrit mathematics and astral science. Bhāskara was familiar with Brahmagupta's work.[5] In fact, Colebrooke translated the two mathematical chapters of Bhāskara's *Siddhāntaśiromāṇi* together with those of Brahmagupta.

The twelfth chapter of the BSS is devoted to mathematics, *gaṇita*. This will be our focus here, together with Pṛthūdaka's commentary on it. The eighteenth chapter of the BSS is devoted to algebra, and, more specifically, to indeterminate analysis. It contains rules for computing with zero, additive and subtractive quantities. We will also discuss it here, but to a lesser extent.

The oldest preserved commentaries on the BSS and the KK were authored by Caturveda Pṛthūdakasvāmin (henceforth Pṛthūdaka, fl. 864).[6] His commentary on the BSS, the *Vāsanābhāṣya* ('commentary with explanations', henceforth PBSS) examines many topics in detail: he describes the construction of an armillary sphere, how to perform a multiplication of integers on a working surface, etc.; in addition,

[3] In verses 7 and 8 of Chap. 24 of the *Brāhmasphuṭasiddhānta* (BSS.XXIV.7–8), Brahmagupta explained that he composed this text at the age of 30 in Śakā 550 (=628 CE) during the reign of King Vyāghramukha. From this we can gather that he was born in 598, CESS (1970–1995:A4: 254) (Ikeyama 2003: S2).

[4] As evidenced by CESS (A4:256–257). The first published edition of the *Khaṇḍakhādyaka* was Babuā (1925). Dvivedin evokes this text and quotes from some of its commentaries, which shows that he had access to this work, probably through manuscript copies.

[5] As noted by Colebrooke (1817: xxxii) and Dvivedin (1902: 1).

[6] The oldest known commentary of both works, by Balabhadra, has not been preserved. See Pingree (1983).

as indicated in the title of his commentary, he gives 'explanations' (*vāsanā*) of Brahmagupta's rules.

In preparing their translations and editions, both Colebrooke and Dvivedin had only scarce manuscript material in a fairly bad state. Colebrooke, for his translation of the mathematical chapter of the BSS, seems to have mostly used the single manuscript available to him that contained Pṛthūdaka's commentary. The three manuscripts Colebrooke had for the BSS, with or without Pṛthūdaka's commentary, reached him in the form of disorganized folios.[7] Dvivedin explicitly used Colebrooke's manuscripts for his edition.[8] It is clear from Colebrooke's introduction that he had the feeling that he had obtained fragments of an old, very badly preserved text.[9] The same complaint is voiced in Dvivedin's preface, as we will see.

### *12.2.2 Problem 1: A Text Understood with a Badly Preserved Commentary*

Colebrooke and Dvivedin relied heavily on Pṛthūdaka's commentary: as noted above, the manuscripts they worked with included parts of it.[10] Bhrahmagupta's rules are not always cryptic but, like most *sūtras*, they are composed for interpretation, and can thus give rise to several contradictory readings. Commentaries then testify to how a treatise was read. They give a partial point of view, but one from within the Sanskrit scholarly world itself. How should one deal with the fact that commentaries are essential keys for understanding and interpreting such treatises? How should one deal with the multiple possibilities offered when reading Brahmagupta's rules? Such questions have given rise to many options in the editing and translating of Brahmagupta's rules, and no standard exists to this day.[11]

[7] See Eggeling (1896: 993–995), as also pointed out by Dvivedin (1902:3–4). The correct order of the BSS had thus to be reconstructed, taking into account that Pṛthūdaka's commentary did not follow the order of the root text.

[8] See below for a discussion of Dvivedin (1902: 3–4). It seems that he mainly relied on a corrected copy of Colebrooke's manuscript. As an editor, I noticed that Dvivedin's copy often corrected some mistakes of Colebrooke's manuscripts but added others.

[9] (Colebrooke 1817: iv–vi; xxviii–xxxiii).

[10] Their manuscript copies are now the basis for today's editions of PBSS. Setsuro Ikeyama edited Pṛthūdaka's commentary on Chap. 21 of the BSS (Ikeyama 2003). I have started an edition of Pṛthūdaka's commentary on Chap. 12 of the BSS using the only three known manuscripts of this part of the commentary: a manuscript that belonged to Colebrooke and was annotated by him, a copy of this manuscript and a copy (Pingree's actual copy) of the manuscript used by Sudhākara Dvivedin to edit Brahmagupta's text with detailed annotated readings and his own handwritten corrections.

[11] See, for instance, the contemporary attempt by Kichenassamy (2010, 2012) to give a new interpretation of Brahmagupta's rules on cyclic quadrilaterals, arguing that the Sanskrit tradition actually misunderstood them. By contrast, B. Chatterjee gave a very literal translation of the text of the KK, peppered with Sanskrit words, as interpreted from Utpalla's point of view; see Chatterjee (1970). S. Pollock's chapter in this volume touches on the wider implications of these questions for the Sanskrit philological tradition.

### 12.2.3 Problem 2: Which Table of Contents?

The material problem of transmission of the manuscripts added another layer of difficulty: the BSS seems to have been made of two contradictory parts. Indeed, the BSS contains 24 chapters[12] (ten *adhikāra* (sections) and fourteen *adhyāya* (lessons)) on topics ranging from determining eclipses and gnomonic shadows to the movement of planets but also the combination of long and short syllables in a verse. As it is understood today, the first part of the BSS is an exposition; the second is a review of the first exposition, with emendations and alternative rules, to which all sorts of supplementary topics are adjoined, from mathematics (*gaṇita*) to measuring units (*māna*) and from metrics (*chandasciti*) to astronomical instruments (*yantra*). This structure of the BSS seems from the beginning to have been a matter of discussion and rearrangement. Thus, Pṛthūdaka's commentary regroups and dismembers the whole treatise, and starts with a commentary of Chap. 21. This rearrangement can be interpreted as a way of attempting to show the coherence of the treatise's different parts.[13]

More specifically, the structure of the mathematical chapter is puzzling and, as such, was a challenge to both Colebrooke and Dvivedin. This chapter, generally considered to be composed of sixty-five *āryā* verses, starts with what can be understood as a somewhat elliptical definition of mathematics[14]:

> Whoever distinctly and severally knows the twenty operations (*parikarman*) beginning with addition, and the eight practices (*vyavahāra*) ending with shadows, is a mathematician (*gaṇaka*). (BSS.12.1)[15]

Should this delimitation of a mathematician's knowledge be understood as a table of contents of the chapter itself?

Both Colebrooke and Dvivedin would indeed understand this definition of the knowledge to be mastered by a mathematican as a list of topics that are treated in their enumerated order in the mathematical chapter. However, such a superimposition between the definition of mathematics and the table of contents of the chapter does not account for the last nine verses of the chapter. Indeed, as seen in Table 12.1, verses 2–9 of the mathematical chapter deal with 'operations'. Verses 10–54 deal

---

[12] See Ikeyama (2003:2–3); Colebrooke (1817: Appendix B, xxviii–xxix); Dvivedin (1902: 2). It contains in total 25 chapters, if you count the versified tables (*dhyanagrahopadeṣādhyāya*) that Dvivedin attached to his edition and that have progressively been considered as belonging to the treatise as well.

[13] See Ikeyama (2003: S7 and S9). Here, Pṛthūdhaka may also be writing a specific kind of commentary which could include both adding explanations and not following the order of the treatise. It is possible that he may be following Balabhadra in this respect. Pṛthūdaka does the same with the KK. Both Utpalla and Āmarāja would also rearrange the verses of the KK in their commentaries, which quote and are inspired by PBSS and PKK. Rearrangement for coherence and the genre of commentary probably go hand in hand here.

[14] *parikarmaviṃśatiṃ saṅkalitādyāṃ pṛthag vijānāti| aṣṭau ca vyavahārān chāyāntaṃ bhavati gaṇakas saḥ||*

[15] Dvivedin (1902:172).

**Table 12.1** The structure of the mathematical chapter of the BSS

| Verses | Contents |
|---|---|
| 1 | Definition of a mathematician |
| 2–9 | 20 'operations' (*parikarman*) |
| 10–54 | 8 'practices' (*vyavahāra*) |
| 55–64 | Rules on multiplications, squares, computations with sexagesimal quantities. |
| 65 | End verse |

with topics that are identified by all commentators as 'practices'. Verses 55–64, however, seem to reinstate 'operations'. This latter treatment, a sort of complement on operations, evokes comments in Pṛthūdaka's commentary, showing that the arrangement of the mathematical chapter already needed to be justified in his times.

The inner logic of this mathematical chapter still needs to be unraveled today. Further, understanding this classification of mathematics, and situating it in history, is still also a question open to research. In particular, it needs to be specified how one should understand Brahmagupta's and Pṛthūdaka's use of the terms *parikarman*, (operation), and *vyavahāra* (practices).

### *12.2.4 Problem 3: Practices with Numbers and Quantities*

Sanskrit astral and mathematical texts famously use (and even define) the decimal place value notation. They further use different devices to state values and numbers with words. How should an edition and a translation take into account the various ways in which numbers are arranged and stated? The norms defined in this very book, or the stories told in this volume (such as those evoked by C. Proust and C. Michel), show that these questions are still alive today for the history of mathematics in general, and thus indeed for the history of mathematics written in Sanskrit. What assumptions about Brahmagupta's conception of quantities were at work in this case? This is also an open question. Indeed, the exploration of a given author's conception of quantity has rarely been undertaken because the presence of a decimal place value notation often encouraged the projection of our own conceptions of quantities into the text, making such undertakings seem obvious and transparent, not requiring study or commentary. Further, Brahmagupta's text is famous for the rules he gives on zero. As stated above, these rules do not appear in the chapter on which we focus here, but in Chap. 18, which is devoted to algebra and indeterminate analysis. Nonetheless, we will briefly contemplate how these rules were presented. Was their algebraic nature tackled? Can we understand why these rules were considered as belonging to algebra?

Both Colebrooke and Dvivedin dealt with these questions concerning the genesis of the work, its original structure, the way it was handed down and understood by Pṛthūdaka, and its use of numbers. As we shall see, none of these topics were their direct subject of inquiry; their treatment of these questions was molded by the assumptions concerning them, and not necessarily explained and justified. The description of the editorial choices of Colebrooke and Dvivedin is the aim of the following sections.

## 12.3 Colebrooke's Translation

Colebrooke (1765–1837) is often described as embodying eighteenth-century eclecticism and nineteenth-century specialization, both in the service of a colonial power. He is usually referred to as one of the founders of Indology. His works range from studies of the Veda and reflections on traditions of logic and grammar and on ancient Indian chronology, to astronomy and, of course, mathematics. He turned to Indian languages in the ambivalent colonial setting as a revenue collector in the service of the East India Company, but also as the young heir of William Jones. By the end of his time in India, he had become the head of the Asiatic Society of Bengal.[16]

In the biography written by one of his sons, Sir Thomas Edward Colebrooke (1813–1890), we learn that from the beginning of his study of Indian languages, he acquired manuscripts on mathematics and astral science. But it was only after retiring as the head of the Asiatic society of Bengal in 1815, at the age of 50, and according to Thomas Edward Colebrooke, on the boat journey back to London, that he would turn to these texts. He would do so, having matured the topic for many years and at the moment when he was at the peak of his mastery of Sanskrit.[17] Colebrooke's works testify to the fact that from the outset, questions relating to what he saw as a unified Indian tradition in astronomy, astrology and mathematics were among the important and powerful incentives to studying the language and the texts that came from the Indian subcontinent.[18]

---

[16] See Rocher and Rocher (2012) for a detailed biography.

[17] Colebrooke (1839: 16–17): 'The original object for which he sought to acquire a knowledge of Sanscrit (sic) language had a reference to the treatises on Algebra, which he afterwards translated and gave to the world. However, when he had mastered the language, his attention was diverted to other subjects, and that design was not executed until 20 years afterwards.'Colebrooke (1839:50–51): 'His earliest labour, after his arrival in England, was to prepare for publication a work on which he had been engaged during his homeward voyage. It consisted of a translation of the most celebrated treatises on Indian Algebra, accompanied by a dissertation on the state of the science as cultivated by the Hindus. The subject is interesting in the history of his writings, as being (as I have already mentioned) that which first led him to the study of the Sanscrit language, and which was laid aside for about 20 years'

[18] Along the way (sometime in the first half of the twentieth century), this topic seems to have 'fallen off the Indological wagon' and it became a separate technical part of the history of science in India, undertaken by trained mathematicians rather than by Indologists.

### 12.3.1 Singling Out Mathematics

Colebrooke understood both the *Siddhāntaśiromāṇi* and the BSS as 'a course of astronomy'.[19] Consequently, he considered the chapters he selected as autonomous 'treatises'.[20] Was this the prevalent point of view on these texts, transmitted to him by the *paṇḍits* who helped him collect these manuscripts and probably decipher them?

Colebrooke did not comment on the fact that he operated such a selection. However, the choice of extracting chapters from greater works can be understood in many different ways. It can be justified, first, by the encyclopedic quality of both works: the scope was perhaps too large for a first attempt. Furthermore, the two mathematical chapters of Bhāskara's *Siddhāntaśiromāṇi*, the *Līlāvatī* and the *Bijagaṇita*, that were published by Colebrooke with chapters taken from Brahmagupta's treatise, had been copied as autonomous tracts within the Sanskrit manuscript tradition itself. When Colebrooke selected these texts, he was continuing a tradition of independent copy. This was not the case for the two chapters of the BSS. When Colebrooke extracted the mathematical parts of this astronomical text, he may also have done so as a strategy to deal with only a small part of a text which reached him in a rather corrupt state and concerned many different topics he could not master all at once.[21]

This extraction can further be understood in the light of Colebrooke's interest in mathematics, which concerned a separate set of questions from those sparked by his interest in astronomy. Indeed, ultimately, the aim of Colebrooke's work was to achieve a comparative history of algebra.[22] Brahmagupta's chapters are translated after Bhāskara's: they are at the end of the book. Such an order, reversely chronological, encouraged comparison of its content with Bhāskara's. The book's organization may also have contributed to developing an ahistorical view of Sanskrit mathematics.[23] For Colebrooke, the essential work was the one most famous in

---

[19] The expression is used in Colebrooke (1817: ii).

[20] Colebrooke (1817: ii): (a quotation respecting his yet non standardised orthography of Sansrkit names) 'The treatises in question, which occupy the present volume, are the *Víja-gańita* and *Lílávatí* of BHÁSCARA ÁCHÁRYA and the *Gańitád'haya* and *Cuťťacád'hyaya* of BRAHMEGUPTA. The two first mentioned constitute the preliminary portion of BHÁSCARA's Course of Astronomy, entitled Sidd'hánta-śirómańi. The two last are the twelfth and eighteenth chapters of a similar course of astronomy, by BRAHMEGUPTA, entitled Brahma-sidd'hánta.'

[21] Colebrooke (1817: Appendix D, xxxi–xxxiii) confirmed the fact that he had read other parts of the BSS as well.

[22] Since this is well known and documented, we will not dwell on it. See Raina (2012:231–239; 242–246), Rocher and Rocher (2012:133–135) and Smadja (2016:19). Indeed, at the outset, Colebrooke (1817: i): writes 'observations will be added (to the translations), tending to a comparison of the Indian, with the Arabian, the Grecian and the modern Algebra.'

[23] See, for instance, the consequence in Peacock's understanding of Indian mathematics, as analyzed in Raina (2011: 243): 'It is interesting that in tracing the linkages between Āryabhaṭa and Bhāskara, Peacock is gesturing towards not only a genealogy of the arithmetical tradition in India but also towards a continuity of the tradition spanning several centuries.' Peacock, we know, read Colebrooke carefully.

his time, Bhāskara's. Brahmagupta's treatise was seen as an older version, which had inspired the newer treatise.[24]

This choice of extracting mathematical chapters solidified a disciplinary difference between the history of astronomy and the history of mathematics. Although such a divide may have already existed beforehand, both in the Sanskrit traditions and in the difference in professionalization between mathematicians and astronomers at the end of the eighteenth century in Europe, Colebrooke's publication was to increase the divide between the study of the history of astronomy in Sanskrit and of the history of mathematics.

### *12.3.2 A Translation: Establishing an Original Text*

Colebrooke made the content of ancient Sanskrit texts available to others through translations. Making editions was probably impossible because of the lack of an adequate printing press (see the chapter by Graheli in the present volume).[25] However, it is probable that Colebrooke planned to publish editions of the texts he had at hand.[26] Translating and editing are two different—but related—tasks. For Colebrooke, it seems that a translation had to be made from an established 'exact text'.[27] In the dissertation that opened his book, Colebrooke specified how he had worked at establishing an original text for Bhāskara's work by collation of the text alone and of its commentaries[28]:

> A careful collation of several commentaries, and of three copies of the original work, has been made, and it will be seen in the notes to the translation how unimportant are the discrepancies.

A similar endeavor was undertaken for the text of Brahmagupta's treatise and for Pṛthūdaka's commentary, extending the search to indirect references[29]:

> … the translator proceeded to collate text and commentary, numerous quotations from both, which he found in Bháscara's writing or in those of his expositors.

---

[24] Colebrooke (1817: i).

[25] But we know he was intent on finding a way of having Devanagari printed, as evidenced by his correspondence with Schlegel and his proximity to William Carrey. Rocher (2013: 76–77).

[26] The British Library has a copy of Colebrooke's manuscript of Pṛthūdaka's commentary in book form and clear reading, partly corrected, which, as suggested to me by A. Graheli, might have been a preprint copy of the manuscript. However, I do not know when or by whom such a copy was made.

[27] Colebrooke (1817: i) (my emphasis): 'The object of the present publication is to exhibit the science in the state in which the Hindus possessed it, by an exact version of the most approved treatise on it in the ancient language of India, with one of the earlier treatises (the only extant one) from which it was compiled...' He annotated the manuscript he used. Further study might reveal something of how he worked in relation to the text.

[28] Colebrooke (1817: ii).

[29] Colebrooke (1817: v–vi).

In Colebrooke's work, then, the translation is implicitly based on an unpublished edition of the text. However, it is difficult to know whether Colebrooke made a sharp distinction between an original text and a best text: both criteria seem to be operated together when Colebrooke evokes the established text.[30]

Commentaries are then used to establish the text and also, obviously, to help in translating because, for Colebrooke, they helped in recovering the meaning of the text.[31]

### 12.3.3 *The Commentary in a Footnote*

Colebrooke relied on Pṛthūdaka to read Brahmagupta and, in most cases, did not differ from him in his interpretation of the text of the mathematical chapter. Indeed, different parts of the commentary are given in footnotes, as seen in Fig. 12.1, and as stated explicitly by Colebrooke.[32]

More specifically, for his translation of the mathematical chapter of the BSS, the parts of the commentary not dealing with the syntax of the root text, and not corrupted in his version of the text to the point of not being emendable, are systematically translated. Colebrooke very thoroughly and precisely translates:

- Every explication of technical vocabulary provided by Pṛthūdaka;
- The problems stated in the commentaries and their resolutions (it is in the resolution of problems that he sometimes paraphrases rather than translates literally, but even this is exceptional)[33];
- The *vāsanā*s (i.e., the explanations or proofs) given by Pṛthūdaka.

Colebrooke's translations are to be distinguished from his own remarks on the text, which are quite rare in this part of his text. Translations are marked by an abbreviated signature at the end of the footnotes: CH, sometimes followed by 'Ib' (for *Ibidem*) in the next note. Thus, Colebrooke's printed translation gives a shape to the text, a shape that does not imitate the landscape-format pages (evoked in the chapters by Graheli and Preisendanz in the present volume) of the manuscript he

[30] Thus at the beginning of the Dissertation the authenticity of the texts is discussed. The collation establishes 'the original' text. In the appendixes, however, commentaries help additionally to provide a correct interpretation.

[31] Colebrooke is quite explicit about both aspects. See Colebrooke, (1817: xxv–xxvi) and also Keller (2010a: 228–231), Raina (2012:n240).

[32] Colebrooke (1817: 278).

[33] Note that Colebrooke hesitated concerning the authorship of the versified examples found in the manuscript of the BSS—after all, the *Līlavatī* and the *Bījagaṇita* included such examples in the base text: 'It is not quite clear whether the examples are the author's or the commentator's. The metre of them is different from that of the rules; and they are not comprehended, either in this or in the chapter on Algebra, in the summed contents at the close of each. They are probably the commentator's; and consigned therefore to notes.' (Colebrooke 1817:278).

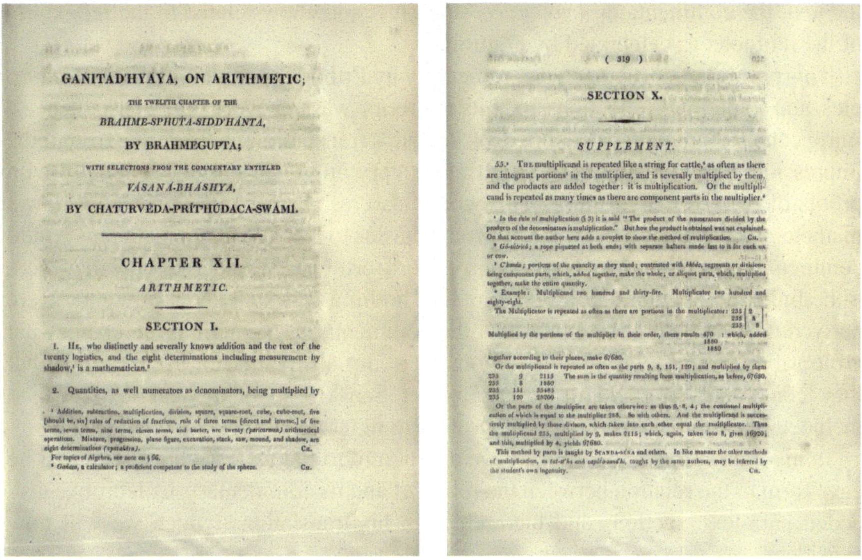

GANITÁDHYÁYA, ON ARITHMETIC;

THE TWELFTH CHAPTER OF THE

*BRAHME-SPHUTA-SIDD'HÁNTA,*

BY BRAHMEGUPTA;

WITH SELECTIONS FROM THE COMMENTARY ENTITLED

*VÁSANÁ-BHÁSHYA,*

BY CHATURVÉDA-PRÍT'HÚDACA-SWÁMI.

CHAPTER XII.

*ARITHMETIC.*

SECTION I.

1. He, who distinctly and severally knows addition and the rest of the twenty logistics, and the eight determinations including measurement by shadow,[1] is a mathematician.[2]

2. Quantities, as well numerators as denominators, being multiplied by

[1] Addition, subtraction, multiplication, division, square, square-root, cube, cube-root, five [should be, six] rules of reduction of fractions, rule of three terms [direct and inverse,] of five terms, seven terms, nine terms, eleven terms, and barter, are twenty *(paricarman)* arithmetical operations. Mixture, progression, plane figure, excavation, stack, saw, mound, and shadow, are eight determinations *(vyavahára)*. Ch.

For topics of Algebra, see note on § 66.

[2] *Ganaca*, a calculator; a proficient competent to the study of the sphere. Ch.

( 319 )

SECTION X.

*SUPPLEMENT.*

*55.*[1] The multiplicand is repeated like a string for cattle,[2] as often as there are integrant portions[3] in the multiplier, and is severally multiplied by them, and the products are added together: it is multiplication. Or the multiplicand is repeated as many times as there are component parts in the multiplier.[4]

[1] In the rule of multiplication (§ 3) it is said "The product of the numerators divided by the products of the denominators is multiplication." But how the product is obtained was not explained. On that account the author here adds a couplet to show the method of multiplication. Ch.

[2] *Gó-sútricá*; a rope piqueted at both ends; with separate halters made fast to it for each ox or cow.

[3] *C'handa*; portions of the quantity as they stand; contrasted with *bhéda*, segments or divisions; being component parts, which, added together, make the whole; or aliquot parts, which, multiplied together, make the entire quantity.

[4] Example: Multiplicand two hundred and thirty-five. Multiplicator two hundred and eighty-eight.

The Multiplicator is repeated as often as there are portions in the multiplicator: 235 | 2, 235 | 8, 235 | 8

Multiplied by the portions of the multiplier in their order, there results 470, 1880, 1880 : which, added together according to their places, make 67680.

Or the multiplicand is repeated as often as the parts 9, 8, 151, 120; and multiplied by them

| | | |
|---|---|---|
| 235 | 9 | 2115 |
| 235 | 8 | 1880 |
| 235 | 151 | 35485 |
| 235 | 120 | 28200 |

The sum is the quantity resulting from multiplication, as before, 67680.

Or the parts of the multiplier are taken otherwise: as thus 9, 8, 4; the continued multiplication of which is equal to the multiplier 288. So with others. And the multiplicand is successively multiplied by those divisors, which taken into each other equal the multiplicator. Thus the multiplicand 235, multiplied by 9, makes 2115; which, again, taken into 8, gives 16920; and this, multiplied by 4, yields 67680.

This method by parts is taught by Scanda-séna and others. In like manner the other methods of multiplication, as *tat-st'ha* and *capáta-sand'hi*, taught by the same authors, may be inferred by the student's own ingenuity. Ch.

**Fig. 12.1** Colebrooke's translation of the mathematical chapter of the BSS, incorporating PBSS (noted with Ch) in the footnotes

was dealing with.[34] In this case, the shape given to the text and the commentary seems to imitate the Medieval marginal manuscript gloss, the footnote being the book's way of rendering it. The commentary, however, does not come to Colebrooke in manuscripts as bits and fragments. He is clear about this in his dissertation, noting in relation to the commentary on the *Brāhmasphuṭasiddhānta*[35]:

> The commentary is a perpetual one; successively quoting at length each verse of the text; proceeding to the interpretation of it, word by word; and subjoining elucidations and remarks: and its colophon, at the close of each chapter, gives the title of the work and name of the author

Why does he break up the published translation?

It is possible that Colebrooke conceived of commentaries as interesting in bits and pieces. Colebrooke may also have consciously chosen not to deal with the syntactical glosses, which are difficult to translate. In all cases, Colebrooke's footnotes in his translation (contrary to those in his dissertation) are not yet either the editor's critical apparatus, making the reader aware of alternative possible readings and emendations—although this happens at times—or the historian's footnote 'making clear the limitations of [the historian's] own theses even as they try to back them up'.[36] The footnote here is only a partial window onto another voice concerning

[34] See also Raina (2011: 240).

[35] Colebrooke (1817: v).

[36] Grafton (1997: 23).

the text, the commentator's voice. As such, it is somehow related to the reflexivity of the footnote, as celebrated by Grafton.[37]

This portrait of a fragmented commentary by Pṛthūdaka containing solved examples, and sometimes reflections and bits of proofs, was to have lasting consequences: European mathematicians reading Colebrooke's translation would derive from it the impression that commentaries were fragmentary and dealt only occasionally with proofs of the given algorithms. This would also resonate in questions about how a mathematical text has to be written. From Chasles to Hankel, mathematicians would comment on Brahmagupta's 'propositions' or 'formulas', wondering how they were to be linked to one another and noting that nothing was said about how they were derived. They would also remark that the commentaries, sometimes and somewhat arbitrarily, offer proofs. Such assertions were not always meant to be critical, but they constructed a standard description of Sanskrit mathematical texts, which included a rather negative assessment of them in terms of apparent coherence.[38]

In his translation, Colebrooke gave a structure to the text: we have seen that he gave form to the relation between the root text and its commentary. Colebrooke also added para-text, sections and hierarchies in his translations, which we will now consider.

### 12.3.4 A Table of Contents of Antiquated Topics: Logistics and Determinations

The question of mapping a definition of mathematics to the contents of a chapter raises two issues. The first one concerns how Colebrooke understood the organization of the chapter, while the second relates to how Colebrooke understood the definition of mathematics he had at hand. The table of contents of Colebrooke's translation lists the contents of Bhāskara's two mathematical tracts, followed by Brahmagupta's chapters. As seen in Figs. 12.2 and 12.3, it shows how Colebrooke understood both texts as defending similar conceptions of arithmetic and algebra.

Although the tables of contents of Bhāskara's texts are more detailed, in both cases the arithmetical part consists of 'algorithms' or 'operations' followed by 'Mixtures, Progression, Plane Figure, Excavations, Stacks, Saw, Mounds of Grain,

[37] Grafton (1997: 26–29). In the book itself, Brahmagupta's text appears after the translations of Bhāskara's mathematical texts. For Bhāskara, several commentators are used, in relation to the main text. In the case of the commentary to Brahmagupta's text, Colebrooke, at just one crucial point concerning cyclic quadrilaterals, in a part of the manuscript he considers to be corrupt, supplies not only a translation of Pṛthūdaka's explanations but also a part of Gaṇeśa's commentary on the *Līlāvatī*, that he actually understands as dealing with this part of Brahmagupta's text. See also Smadja (2015).

[38] Smadja (2015).

BHÁSCARA.

*ARITHMETIC (Lílávati.)*

| | | | |
|---|---|---|---|
| CHAPTER I. | Introduction. Axioms. Weights and Measures | | 1 |
| CHAPTER II. | *Sect. I.* | Invocation. Numeration | 4 |
| | *Sect. II.* | Eight Operations of Arithmetic: Addition, &c. | 5 |
| | *Sect. III.* | Fractions | 13 |
| | *Sect. IV.* | Cipher | 19 |

CONTENTS.

| | | | |
|---|---|---|---|
| CHAPTER III. | Miscellaneous. | | |
| | *Sect. I.* | Inversion | 21 |
| | *Sect. II.* | Supposition | 23 |
| | *Sect. III.* | Concurrence | 26 |
| | *Sect. IV.* | Problem concerning Squares | 27 |
| | *Sect. V.* | Assimilation | 29 |
| | *Sect. VI.* | Rule of Proportion | 33 |
| CHAPTER IV. | Mixture. | | |
| | *Sect. I.* | Interest | 39 |
| | *Sect. II.* | Fractions | 42 |
| | *Sect. III.* | Purchase and Sale | 43 |
| | *Sect. IV.* | A Problem | 45 |
| | *Sect. V.* | Alligation | 46 |
| | *Sect. VI.* | Permutation and Combination | 49 |
| CHAPTER V. | Progression. | | |
| | *Sect. I.* | Arithmetical | 51 |
| | *Sect. II.* | Geometrical | 55 |
| CHAPTER VI. | Plane Figure | | 58 |
| CHAPTER VII. | Excavations and Content of Solids | | 97 |
| CHAPTER VIII. | Stacks | | 100 |
| CHAPTER IX. | Saw | | 101 |
| CHAPTER X. | Mound of Grain | | 103 |
| CHAPTER XI. | Shadow of a Gnomon | | 106 |
| CHAPTER XII. | Pulverizer (*Cuttaca*) | | 112 |
| CHAPTER XIII. | Combination | | 123 |

**Fig. 12.2** Colebrooke's table of contents for Bhāskara's texts

Measure by Shadow'.[39] This list of eight topics is, indeed, a translation of the standard subjects of *vyavahāra* ('practices' or 'determinations'). The table of contents of Bhāskara's text has a 'miscellaneous' section which occurs between operations

[39] Colebrooke (1817:1) specifies that the 'axioms' found at the beginning of Bhāskara's text are 'consisting in definitions of technical terms'.

*ALGEBRA* (*Vïja-gańita.*)



CONTENTS.



**Fig. 12.2** (continued)

and mixtures. The table of contents of Brahmagupta's arithmetical chapter ends with a 'Supplement' containing the nine verses that do not seem to fit into the simple mapping of Brahmagupta's definition of mathematics onto the contents of his 'arithmetical' chapter. In both cases, the similarities and differences in the tables of contents simultaneously underline the aim of highlighting what could appear as a standard definition of mathematics and the fact that the organization of the chapters do not exactly follow such an order.[40]

[40]Looking more closely at Colebrooke's sub-sectionings further highlights such difficulties. Bhāskara's texts are subdivided into chapters and sections, giving the feeling that each is an individual treatise, while Brahmagupta's chapters are each subdivided in sections. For instance, in the *Līlāvatī*, Section III (p. 122) and Sect. V (p. 125) of the chapter 'Miscellaneous rules', which precisely includes unclassified procedures, have no title. The first sections of Chaps. 12 and 18 of the BSS have no title either. Some chapters or sections are long, and others very short; thus in Chap. 12, the section on plane figures has 16 pages (p. 295–311), while those covering the topics of stacks (p. 314) and saws (p. 315) have but one page; the sectioning thus makes sense here in relation to the definition of mathematics but not in relation to the chapter's scope of investigation. In addition, the specifications of titles are sometimes given in headers, which appears to be another level of sub-sectioning in Colebrooke's published text.

BRAHMEGUPTA.

CHAPTER XII. *ARITHMETIC* (*Gańita.*)



CHAPTER XVIII. *ALGEBRA* (*Cuttaca.*)



**Fig. 12.3** Colebrooke's table of contents for Brahmagupta's texts

Further, the table of contents presents both authors as articulating a relation between arithmetic and algebra. In both cases, arithmetic and algebra define similar operations before looking at specialized problems. Thus Bhāskara's arithmetical 'eight operations, fractions' and 'cipher' in arithmetic correspond to the 'algorithms' of 'negative and affirmative quantities', 'cipher', 'unknown quantity' and 'surds' in his algebra.[41] In the case of Brahmagupta, the 20 operations of the arithmetical chapter, and the thirty-six ones of the algebraical part are both simply summarized as 'algorithms'.

[41] Bhāskara's algebraical text is subtitled 'Elemental Arithmetic or Algebra' (Colebrooke 1817: 129).

Here, we can surprise Colebrooke departing from his usual philological cautiousness, as he attempts to underline a continuity between arithmetic and algebra, as well as between Bhāskara and Brahmagupta.

First, about arithmetic and algebra, indeed, we have seen that Colebrooke states in a footnote that *parikarman*.s are considered in Bhāskara's algebraical text, the *Bījagaṇita*. However, this word only appears once in Bhāskara's commentary on the *Bījagaṇita*, in what is maybe a spurious colophon, which counts thirty-six *parikarman*.s.[42] In the *Bījagaṇita*, the equivalent of the *parikarman* of arithmetic seems to be rather *vidha*, which appears systematically in colophons of subsections in algebra, counting the number of 'methods' treated, on different kinds of quantities.[43]

Second, about Bhāskara and Brahmagupta, Colebrooke goes further, since in his 'Sect. II' of Brahmagupta's Chap. 18, he inserts a title 'Algorithm' and, in a footnote appended to this title, quotes the colophon which belongs to the commentary on the *Bījagaṇita*, the one mentioning thirty-six *parikarman*.s, as if it was part of Brahmagupta's text. However, in the Sanskrit, no general term referring to operations/methods can be found in this section, each being treated by its proper name of 'addition', 'subtraction', etc.[44]

The parallel between Bhāskara's and Brahmagupta's texts is further underlined—again, rather elliptically—in the way Colebrooke translates the chapter titles of the BSS. As seen in Fig. 12.4, he understands the term *gaṇita* ('mathematics') to mean 'arithmetic' here, while the term *kuṭṭaka* (transliterated as *cut'taca*), usually translated by Colebrooke himself as 'pulverizer', is understood here as 'algebra'.

Thus, Colebrooke, in partitioning his translations and giving titles to the sections, is clearly attempting to mold the text and its underlying conception of the relations of arithmetic and algebra into a specific structure—a structure that provides a standard definition of arithmetic and algebra, coherent in relation to the two texts by different authors. Thus, while translating both texts and dividing them into sections, he is shaping a uniform, coherent structure for Indian/Hindu mathematics.

It is noteworthy that Colebrooke hesitated precisely in the translation of *parikarman*. Thus, in Bhāskara's arithmetical text, it is rendered as 'operations', but a footnote specifies 'modes of process': logistics or algorism.[45] When, in relation to the operations of algebra in Bhāskara's text, the *parikarman* is announced as 'Algorithm or Logistics', a footnote specifies 'operations or modes of process', referring back to the arithmetical text, the *Līlāvatī*.[46] The *parikarman* of 'arithmetic'

[42] (Hayashi 2009: 15, note 97)

[43] (Hayashi 2009: 3–4, 7–8, 15). As underlined by Hayashi, six types of quantities each have six *vidha*s, which amounts to thirty-sixty.

[44] Colebrooke (1817: 339).

[45] Colebrooke (1817: 5).

[46] Colebrooke (1817: 129).

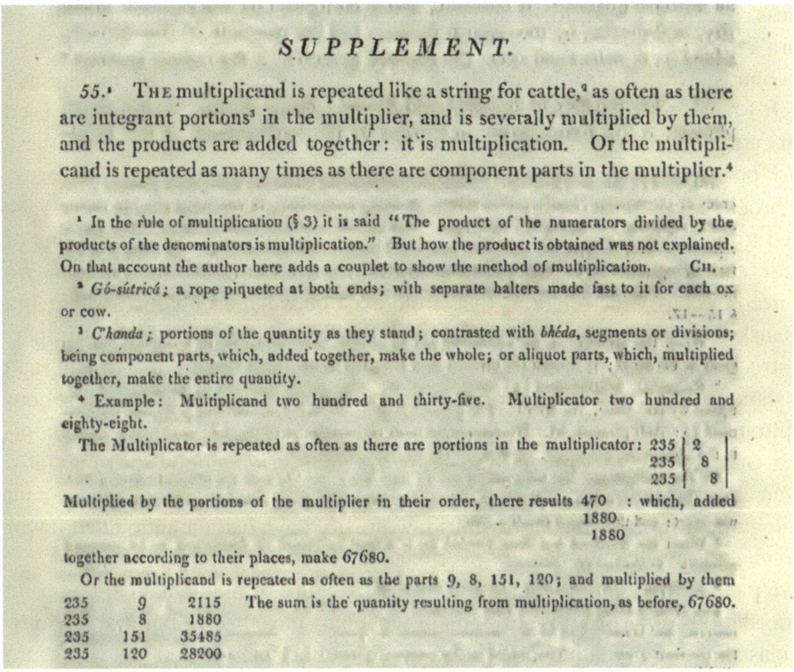

*SUPPLEMENT.*

*55.*[1] THE multiplicand is repeated like a string for cattle,[2] as often as there are integrant portions[3] in the multiplier, and is severally multiplied by them, and the products are added together: it is multiplication. Or the multiplicand is repeated as many times as there are component parts in the multiplier.[4]

[1] In the rule of multiplication (§ 3) it is said "The product of the numerators divided by the products of the denominators is multiplication." But how the product is obtained was not explained. On that account the author here adds a couplet to show the method of multiplication. CH.

[2] *Gó-sútricá;* a rope piqueted at both ends; with separate halters made fast to it for each ox or cow.

[3] *C'handa;* portions of the quantity as they stand; contrasted with *bhéda*, segments or divisions; being component parts, which, added together, make the whole; or aliquot parts, which, multiplied together, make the entire quantity.

[4] Example: Multiplicand two hundred and thirty-five. Multiplicator two hundred and eighty-eight.

The Multiplicator is repeated as often as there are portions in the multiplicator: 235 | 2, 235 | 8, 235 | 8

Multiplied by the portions of the multiplier in their order, there results 470, 1880, 1880 : which, added together according to their places, make 67680.

Or the multiplicand is repeated as often as the parts 9, 8, 151, 120; and multiplied by them

| | | |
|---|---|---|
| 235 | 9 | 2115 |
| 235 | 8 | 1880 |
| 235 | 151 | 35485 |
| 235 | 120 | 28200 |

The sum is the quantity resulting from multiplication, as before, 67680.

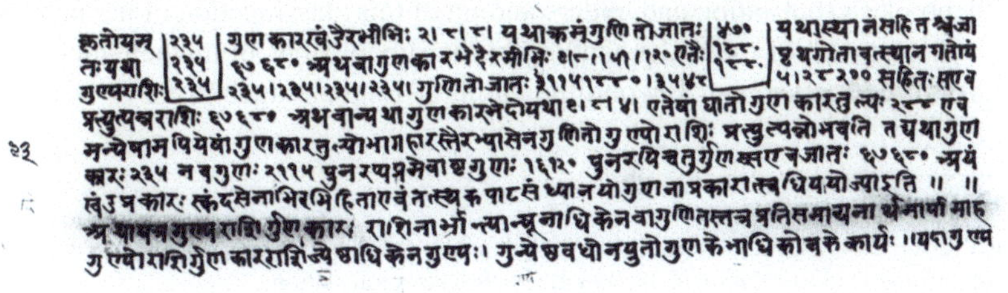

**Fig. 12.4** Colebrooke's presentation of the multiplication described in PBSS.12.55 in tabular form, in contrast with what appears in his manuscript. (Colebrooke (1817: 319) and BL-IOSAN1304, folio 73 verso)

(*gaṇita*) in the BSS is translated as 'logistics' in the root text but as 'operations' in Pṛthūdhaka's commentary.[47] The ghost *parikarman* of Brahmagupta's algebraical chapter is translated as 'algorithm'. In the footnote, Colebrooke specifies 'or modes

[47] Colebrooke (1817: 277). The reader can compare the translation of verse 1 with the translation in Colebrooke's Footnote 1. The header on page 279 is 'Logistics: Fractions'.

of process'. He refers back to Brahmagupta's chapter 'On Arithmetic', but also to Bhāskara's *Bījagaṇita.*[48]

By contrast, *vyavahāra* is translated with consistency as 'determinations'. Although I have not found the origin of this translation, which is often still used today, it certainly underlines how Colebrooke's realm of understanding links these topics to 'practical' arithmetic and the applications of procedures to specific cases.

Colebrooke's popularity with English algebraists is well known. However, we still need to understand why he tended to homogenize not only the definitions of arithmetic and algebra, but also the way they are articulated to one another. We further need to clarify the relations that could exist between arithmetic and algebra in the Sanskrit texts he published. We know that his reading of Playfair was important in structuring his approach to these texts, while Peacock and de Morgan read his translations closely and incorporated them into their own works.[49] In this context, Hutton's *Philosophical and Mathematical Dictionary* (1795) and his *School-Master's Guide: Or, A Complete System of Practical Arithmetic* (1766) may also have been an important source of inspiration for Colebrooke.[50] Perhaps the hesitant terminology of 'logistics' or 'algorism' may already have had an antiquated flavor with Greek and seventeenth-century overtones in Colebrooke's time.

In every case, in the homogenization of the structure of mathematical texts and in the translations of terms, this adoption of what may have been a familiar but antiquated classification of mathematics had a twofold effect: it rendered the works into intellectually manageable pieces, suitable for comparison, while introducing familiar categories. In other words, this shaping of a homogenized form of arithmetic and algebra may have been a way of preparing the texts for comparison among themselves and with others.

Colebrooke's translation and understanding of this classification of mathematics was to shape much of the understanding of the subdivisions of mathematics found in Sanskrit texts after him. Beyond the circle of English algebraists, in Datta and

---

[48] Colebrooke (1817: 339).

[49] Raina (2011: 244–245) underlines how Peacock paid particularly careful attention to the definitions of arithmetic in its relation to algebra in Colebrooke's translation.

[50] Hutton (1795) seems to have been regularly re-edited and augmented. Sections on algebra and arithmetic in Vol.II of the 1815 edition are abbreviated from Hutton (1812). Colebrooke may have read both. Indeed, E.T. Colebrooke's biography transcribes many letters from Colebrooke to his family; in one dating from 1798–1799, we see him asking, among many other books, for Charles Hutton's *Dictionary of Mathematics*; Colebrooke (1839: 30). Hutton's dictionary is quoted twice in Colebrooke's dissertation; Colebrooke (1817: x, lxxiii). The section on algebra in Hutton's mathematical dictionary contains a detailed historical exposition, including a synthesis of Davis and Stratchey's publications in which he claims that algebra probably originated in India (Hutton 1812: 65, Column 2). Colebrooke may have borrowed Hutton's definition of algebra as analysis Hutton (1795: Vol. I, 60); of logistics as related to practical arithmetic, sexagesimal astronomical computation and occasionally algebra (Hutton 1795: Vol. I, 300). He also defined algorithms in relation to algebra (Hutton 1795: Vol. I, 103). Further, Colebrooke's table of contents seems to resonate in Hutton's *Practical Arithmetic*.

Singh's reference manual *A History of Hindu Mathematics* published in 1937 in Lahore, 'logistics' is retained with 'operations' to translate the term *parikarman*, and 'determinations' is used for *vyavahāra*.[51] Standard textbooks and articles borrowing from this manual were to propagate not only the vocabulary, but also the idea that such a classification was part of all Sanskrit mathematical texts.[52] The persisting use of the term 'logistics' in relation to Brahmagupta can be seen in M.S. Sriram's 2013 online class on Brahmagupta.[53]

### 12.3.5 Care with Decimal Place Value Notation

Colebrooke was curious, as we are today, about mathematical practices: he noted diagrams and tabular displays of numbers, and the different uses of the tabular resources of decimal place value notation. Commentaries, for Colebrooke, not only explained the root text and not only furnished explanations of the rules at times, but also opened doors onto such practices. Such tabulations are, for instance, obvious when dealing with mathematical operations. As shown in Fig. 12.4, Colebrooke actually displayed the whole multiplication operation in a table, contrary to what could be seen in the manuscript available to him.[54]

Here again, Hutton's influence can perhaps be traced, as his course on practical arithmetic emphasized the uses of tables for computing and making conversions.[55] But it is possible this was just due to the wider way in which 'practical arithmetic' was understood to be practiced.

The section in Brahmagupta's text on computations with zero, and positive and negative quantities is strikingly devoid of footnotes, as if the text was self-evident and understandable to the reader of those times.[56] Such a reading, however, neglects that it appears in an algebraic context that was not accounted for, and would seldom be considered in what followed. In Datta & Singh's textbook, for instance, whereas Brahmagupta's zero is described under the section 'zero in algebra', the chapter on zero belongs to Vol. I, which is devoted to arithmetic.[57]

[51] Datta and Singh (1935: Vol. I, 124).

[52] See, for instance, Bag (1979: 76). See Keller (2010b) on Plofker (2009: 296) for difficulties linked to such assumptions.

[53] Sriram 2013.

[54] This is examined in Keller and Morice-Singh (2022).

[55] Hutton (1775: 23–25), for example.

[56] Colebrooke (1817: 339–340).

[57] Datta & Singh (1935: Vol I, p. 241).

### 12.3.6 A Fragmented, Hierarchized and Organized Text as a Tool for Comparison and as a Grid to Mold and Evaluate

Colebrooke's translation is very much that of an editor who wants to publish a composition which is as close to the original text and as accurate as possible.

But in the way he integrated the role of commentaries in the understanding of the root text, in the paratextual segments he provided of the chapters, he created a fragmented whole which presented many details of mathematical practice molded into categories that may have seemed familiar but slightly antiquated to the European scholars who would read him. Colebrooke did not conflate Bhāskara with Brahmagupta, or mix up Pṛthūdaka the commentator with the author of the root text. They were presented with a view to comparison and their similarities were highlighted. Overall, the form of Colebrooke's publication encourages comparisons, which was, no doubt, the aim of his text in the first place. But it also provided a grid in which all of Sanskrit mathematics could be approached and possibly evaluated. Indeed, as we have seen, the hierarchy and segments derived by Colebrooke are understood as an implicit definition of mathematics and its structure. All mathematical texts can then be approached and interrogated in relation to this grid. This probably helped in forging the idea of a homogenized Sanskrit mathematical tradition, with a more or less standard shape.

## 12.4 Dvivedin

We have seen how Colebrooke's assumptions and questions modeled his translation and had a lasting influence on the way the text was perceived and described by those who did not have access to the original Sanskrit and, surprisingly, also to those who had. Did Dvivedin's 1902 edition of the *Brahmasphuṭasiddhānta*, which is still the authoritative edition today, have a similar impact? Did it show traces of Colebrooke's work? What were his criteria for establishing the text? In which way have his choices enduringly marked our understanding and our access to the text today? In the following, we shall look at how Dvivedin responded to problems of the structure of the text and outline the underlying notion of mathematical text, commentary and quantity his edition relied on.

We know little of the biography of Sudhākara Dvivedin, a member of the constellation of Indian scholars who made editions of Sanskrit texts in the nineteenth and at the beginning of the twentieth century, as described in the chapters by Karin Preisendanz and Alessandro Graheli in the present volume.[58] Dvivedin was born in 1855 in Khajuri, a village near Varanasi where he studied mathematics with Paṇḍit

[58] His last name is sometimes transcribed Dvivedī, or Dvivedi. I have adopted Dvivedin, sticking to a strict transliteration from the Devanagari. The scarce bibliographical information comes from Gupta (1990) and Gupta in Dauben and Scriba (2002: 312–313).

Devakṛṣṇa. In 1883, he was appointed librarian at the Government Sanskrit College in Benares, and became professor of mathematics and astronomy when Bāpūdevā Śastrī retired in 1889. He was also a lecturer at Allahabad University. Dvivedin wrote a number of textbooks in indigenous languages on 'new' topics such as conic sections, calculus and logarithms, as well as creating Sanskrit editions of mathematical texts. He also wrote a work in Sanskrit on the lives and works of Indian astronomers and mathematicians, and a history of mathematics in Hindi.[59] He is typical—and maybe the last, —of the *paṇḍits* fostered in the milieu created by Lancelot Wilkinson in Sehore in the 1830s. To quote Michael Dodson, this milieu developed 'a constructive orientalist pedagogy through a comparative methodology, stressing the shared components, but relative advancement, of European over Indian sciences.[60] Like his predecessor who held the *jyotiṣa* chair in Benares, Bāpū Deva Śāstrī, Sudhākara Dvivedin was an ardent 'modernist', while also establishing the basis for what would soon be a nationalist claim for the existence of an important and relevant tradition of exact sciences in Sanskrit.

### *12.4.1 Establishing an Author and a Text*

Before his edition of the BSS, Dvivedin had edited both a purely astral text and a purely mathematical one. Indeed, he had previously published (the first known) edition of the *Śiṣyadhīvṛddhidatāntra* of Lālla (fl. c.a. 650–750) in 1867.[61] Footnotes were used in this edition to add some commentaries here and there, but there was no discussion of the edition itself. In Dvivedin's edition of the *Triśatika* of Śridhāra (c.a. 750–900), published in 1899, however, the preface discusses the date and time of Śridhāra, and the metrical consistency of the text. The edition is published with a brief commentary pursued in footnotes, which occasionally notes differences between the two manuscripts used to compile the edition.[62] In between the two texts, Dvivedin had authored the Sanskrit commentary that went with the edition of the *Pañcasiddhāntika* written by Varāhamihira (fl. c.a. 550) by the German Indologist G.F.W Thibaut (1848–1914), which was published in 1889.[63] In this edition, the versified root text is contrasted with a 'traditional' (possibly oral) version of the text, in a presentation in two columns. This is followed by a Sanskrit commentary on the one hand, and an English translation on the other, each placed in a separate part of the book. In Dvivedin's edition of 1902, more than 10 years later, there is no English translation, and the commentary is given with the editing. In appearance, then, the co-publication with Thibaut looks like a scholarly critical edition, with an

[59] Dvivedin (1892), (1910b).

[60] Dodson (2007a: 162).

[61] Dvivedin (1867).

[62] Dvivedin (1899). See, for instance, the footnote on page 21.

[63] Thibaut and Dvivedin (1889).

apparatus of translation and analysis, while the publication of the BSS seems to be the printed version of a Sanskrit manuscript tradition in which texts are published with new commentaries.

However, the situation in both cases is more complex. First, the principles on which Thibaut (and, as we shall see, Dvivedin with him) edited the text are not exactly those of a standard critical edition. Indeed, the preface, signed by Thibaut, states[64]:

> … texts of purely mathematical or astronomical content may, without great disadvantages, be submitted to a much rougher and bolder treatment than texts of other kinds. What interests us in these works, is almost exclusively their matter, not either their general style or the particular words employed, and the peculiar nature of the subject often enables us to restore with nearly absolute certainty the general meaning of passages the single words of which are past trustworthy emendation.

Two arguments are evoked here: first, that what counts in these texts is their content, not their wording; second, that because you can grasp the contents independently from the text, the scientific content can help in restoring the text. Thibaut expresses a point of view here that seems to have been endorsed by Dvivedin. Indeed, he also specifies[65]:

> What, in the attempt to reconstitute the text of an astronomical or mathematical work, has chiefly to be kept in view, is of course to arrive at rules which are capable of being proved mathematically. This consideration has, in more than one place, led us to introduce changes even where such appeared hardly to be required by the external form of the traditional text (…). In a few cases, notably in the last chapter, where we were unable to emend the readings of the Manuscripts in any satisfactory way, Pandit Sudhâkara has substituted for the traditional text, rules founded on the general principles of Hindû astronomy (…).

First, then, the edition is critical only in the sense that it edits a mathematically correct text. Further, it edits a text which is correct in relation to 'Hindû' astronomy perceived as an ahistorical system with its specific 'principles', a sort of subsystem within mathematical astronomy.[66] The person who embodied the authority of such a system was no other than *Paṇḍit* 'Sudhâkara' (i.e., Dvivedin). The result is that Dvivedin's and Thibaut's edition of the *Pañcasiddhāntika* was shaped by what appears to be their shared idea of what a mathematical and Sanskrit astronomical text should be.[67] It cannot be considered as a standard 'critical' edition, although their emendations are visible and justified.

[64] Thibaut & Dvivedin (1889: v).

[65] Thibaut & Dvivedin (1889: lxi).

[66] Thibaut & Dvivedin (1889: lxi) also note in the case of Agatsya's theory of heliacal rising: 'The considerable deviation from the text of the manuscripts which our emended text exhibits in that place appears to be absolutely called for by the general principles on which such calculations have to be performed according to all scientific Hindû treatises.'

[67] Although we shall return below to the similar principles on which Dvivedin edited Brahmagupta's text, let us note here that Thibaut insisted in his preface and introduction how closely he worked with Dvivedin, whom he referred to several times as his 'collaborator'. See especially Thibaut & Dvivedin (1889: vi).

Second, the paratext of Dvivedin's edition of the BSS in 1902 reveals that he did not merely take a version of the BSS and add his commentary to it but, that he also conducted extensive research of existing manuscripts before undertaking a minute classification and reading of the folios of Colebrooke's manuscript. In a way, although not 'critical' (Dvivedin was not looking for Brahmagupta's original text), he actually carried out editorial work, influenced by his acquaintance with Thibaut.

Indeed, Dvivedin's preface (*bhūmikā*) to the 1902 publication indicates a network in which G. Thibaut (*ḍā. tibo sāhib/mahāśaya*), and a certain tradition of German philology exported to Great Britain, seem to play an important role. Thus, together with local sources for his manuscripts, the main reference in this preface is to Eggeling's 1896 catalog of the Indian Office manuscripts. Julius Eggeling, like Max Müller, was a German Sanskrit scholar working in Britain. Both Eggeling and Thibaut were in personal contact with Müller, the first as a colleague; the second as a former student.

Dvivedin, however, before establishing a text, established an author. He started by fixing the name, place, date and origin of the author of the text he was going to edit. This groundwork was done by accumulating quotations of ancient scholars, evoking Brahmagupta's origins. He thus quotes, first, Bhāskara and his scholiasts, and then Pṛthūdaka, before mentioning the medieval traveler Hieun Tsang (Xuanzang, fl. 650) and the Persian astronomer and poet al-Bīrūnī (973–1048).[68] Afterwards, Dvivedin turned to the manuscript material (*grantha, pustaka*) he had at hand. He noted that three were full of mistakes (*atīvāśuddhaṃ*) and often incomplete (*bahuc skhalitaṃ*), but that the one used by Colebrooke, and presently in the India Office, was very reliable. He mentioned that the copies of the former three texts had been given to him by respectively Thibaut, the chief astrologer of the king of Ayodhyā, Vicchrī Yajñadata Śarman, and the Benares Rājakīyapāṭha.[69] However, he does not explain to us how he gained access to the manuscript(s), which belonged

[68] Dvivedin (1902:2–3).

[69] Dvivedin (1902: 3): *etatkṛtasyāsya siddhāntagranthasyaikā pratiḥ kāśikarājakīyapāṭhālayato dvitīyā ḍā. thibo sāhiba-ahmahāśayatas tṛtīyā cāyodhyānareśapradhānajyotir vicchrīyajñadata śamaṇo mayā labdhā| (...) idaṃ pustakatrayam atīvāśuddhaṃ bahucc skhalitaṃ cāsti yatpustakānusāreṇa vyaktāvyaktādhyāyayor dvādaśāṣṭādaśasaṃkhyayor aṅgulabhāṣāyām anuvādaḥ kolabrūkasāhibena kṛtas tat pustakam etat trayato bhinnam ity asaṃśayaṃ vibhāti paṭhavibhedāt| tat pustakasya kuṭṭakādhyāyaḥ saṃprati iṇḍīyā-āphis-sarasvatībhavane vartate.* (See the *Catalogue of the Sanskrit Manuscripts in the Library of India Office*, Part V, p. 995).

I have obtained manuscripts (*grantha*) of this *siddhānta,* a first from the Rājakīyapāṭha in Kāśikā (Benares), a second from the Mahāśaya by Mister (*sāhib*) G. Thibaut and a third from Vicchrī Yajñadata Śarman, the head astrologer of the king of Ayodhyā, <each manuscript> had been produced for them. (...) These three manuscripts (*pustaka*) are very faulty indeed and are incomplete. A translation in English was made by Mister Colebrooke with an authoritative manuscript having the chapters on arithmetic and algebra numbered 12 and 18 which is different from the three others since it is devoid of difficulties <or > dissent (reading *vimati* for *vibhāti*) which would originate from divergent readings. The Chapter of this manuscript on Pulverizers can be found at present in the Indian Office Library. (See *Catalogue of the Sanskrit Manuscripts in the Library of India Office*, Part V, p. 995).'

to the Colebrooke collection and that were kept in London.[70] A footnote gives the impression that Dvivedin had a look at Colebrooke's PBSS manuscript himself. He describes the document's script and notes its specificities in terms of *sandhi* (phonetic transformations) and caesuras.[71] He describes the manuscript's bindings, and the loose sheets that contributed to making Colebrooke's manuscript a disordered set of folios.[72] Eggeling's catalogue, which Dvivedin refers to, does not contain the details described here, but it does include the three cataloged manuscripts of Brahmagupta's text belonging to Colebrooke's collection. No known biography of Dvivedin mentions a journey to London. The traces we have of Dvivedin's editing are provided by his annotated copy of Colebrooke's manuscript, which is still kept in Varanasi to this day and serves as one of the bases for today's editions.

To establish the text, Dvivedin painstakingly rearranged the slips of Colebrooke's main manuscript, identifying folios mistakenly glued together, focusing on the parts of the manuscript which included a commentary, aiming at a newly ordered text to enable its correct reading (*pratipāṭhakrama*).[73] He established a best text by homogenizing verses in accordance with meaning and metrics. This was realized in practice in the edition rather than being proclaimed in the preface. Footnotes at the heart

[70]At times, he seems to assume that there were two manuscripts: one with commentary, another without. But mostly, he evokes the manuscript with which Colebrooke made his translation, in the singular.

[71]Dvivedin (1902: 3, footnote) *idaṃ pustakaṃ kasyacin maithilākṣair likhitasya pustakasya pratyantaram iti niḥsaṃśayaṃ pratibhāti| 11 patre 9 ity asya sthāne 7 iti lekhāt| evam anyatrāpi | visargāt parataḥ 'ka' sthāne 'kva' iti lekhāt| yathā 12 patreḥ 6 paṅkto | tathā visargasthāne 'v' iti lekhāt, yathā 12 patreḥ 9 paṅkte| vākyavirāme 'pi sandhikaraṇāt | yathā golādhyāyasya 21 patre 2 paṅkte sarvam upapapannam uktam akhaṇḍena| kutracit 'nta' sthāne 'nū' iti lekhāt| ekasmin patre methilākṣare 'oṃgaṇeśāya namaḥ' iti lekhāc ca|*

This manuscript, which belongs to a codex written in Maithili script, appears devoid of error. On Sheet 11, 9 is written in the place of 7. It is like this elsewhere as well. When a '*ka*' stands behind a *visarga* it is written '*kva*'. For instance, for Sheet 12, Line 6 in place of a *visarga*, there is a *v*, it is likewise for Sheet 12, on line 9. At the end of a sentence also *sandhi* is performed. For instance in the Chapter on the Sphere, in Leaf 21, Line 2, the whole explanation (*upapapanna*) is stated without any caesura. Sometimes '*nū*' is written instead of '*nta*'. On one page, '*Om*, homage to Gaṇeśa' is written in Maithili script.

[72]Dvivedin (1902: 3–4): *aho iṇḍiyāphis sarasvatībhavane asya pustakasya puṭaka-bandhanakāle 'navadhāna-tayā pattrāṇy asaṇgatāni jātāni, bahuc khaṇḍitāni ca santi| tāni kadācid an-upayuktapattrāṇāṃ madhye syur iti teṣāṃ samyag-anveṣaṇaṃ samucitam. (See Catalogue of the Sanskrit Manuscripts in the Library of India Office, Part V, p. 993–995).*

Now, in the Indian Office library, at the time of the bundling (*bandhana*) of this manuscript, leaves were doubled over as a pocket (*puṭaka*), independent (*asaṅgata*) sheets (*pattra*) were produced by inattention and many were cut off (*khaṇḍita*). Some of these unattached (*anupayukta*) leaves were in the middle. (See *Catalogue of the Sanskrit Manuscripts in the Library of India Office*, Part V, p. 993–995)'.

This statement of Dvivedin's preface is followed by a list of loose and cut out sheets, their number and the parts of the BSS they belonged to.

[73]Dvivedin (1902: 4–5).

of the edition note his emendations to the text of the manuscript.[74] Although not systematic, they read like the fragments of a critical apparatus, providing alternate possible readings of a text as well as implicit arguments for the editor's choice. Dvivedin verified that versification was metrically consistent. He checked that the announced number of verses per chapter actually corresponded to the number he had at hand.[75] The preface ends with a huge addition.[76]

In this sense, both form and content are taken into account to establish the text. As underlined in S. Pollock's article in the present volume, as well as in Thibaut's preface and introduction to the PS, the text seems to have been established with the implicit idea that form and content should be true and coherent, rather than authentically ancient.[77] It is not clear whether this true and coherent text is considered to be (an approximation of) the ancient text.

Sudhākara Dvivedin would continue to edit texts in this manner, providing an important number of source texts and thus shaping an enduring corpus for the historians of mathematics and astral science that would come after him. Subsequent editors, such as K. S. Shukla and K. V. Sarma, would sometimes make stemmas and allow for mathematical mistakes, insofar as they made sense for the 'system' of astronomy and mathematics they were describing. They would, however, implicitly always edit 'best texts', and not attempt to recover an author's original text.

### *12.4.2 Integrating the Commentary Into a Commentary*

We have seen that, in appearance, Dvivedin's edition seems to be the printed in continuation of scholarly Sanskrit traditions. We argue that Dvivedin's choice in editing 'best texts' belongs to a common ground in which the *śāstric* culture of editing 'true texts' met with Thibaut's conception of the edition of texts in the exact sciences. Dvivedin's Sanskrit commentary can further be seen as a common ground in which elements of Sanskrit commentaries harmonize with a 'modernist' aim.

Indeed, by publishing an edition with a commentary *in Sanskrit* (and not in English or, for example, in Hindi), Dvivedin positioned himself in the continuity of a lineage of scholarly commentary production, of a 'tradition' also conferred on him by his title as *paṇḍit*. Dvivedin's commentary refers to previous Sanskrit authors. In the mathematical chapter, as announced in his preface, he systematically quoted

[74] The manuscript's alternative readings (*pāṭhāntaram*) are given in the footnotes. See, for instance, Dvivedin (1902: 38).

[75] Dvivedin (1902: 4–6).

[76] Dvivedin (1902: 7).

[77] The published edition marks some variants, but given the fact that Dvivedin noted the existence of very corrupt manuscripts, he does not seem to have adopted 'a maximally inclusive edition' of the kind that could be found for Sanskrit literature, as described by Pollock in the present volume.

the versified problems which Pṛthūdaka gave in his commentary.[78] I am not sure for what purpose Dvivedin did this, but maybe it was a way of reflecting that such versified problems belonged to the intellectual realm and erudition of the rule. This may also be a legacy of Colebrooke's translation, since this is what Colebrooke did. But then again, it is difficult to decide, because perhaps Colebrooke was reproducing a way of treating examples that had been transmitted to him by his *paṇḍits*. More generally, the mathematical achievements of the texts are not isolated but presented as belonging to a coherent—and thus quite homogenous—Sanskrit tradition. Indeed, the continuity and coherence with Bhāskara II's later *Siddhāntaśiromāṇi* is underlined. Thus rules that share common ideas are always referred to in the commentary. In other words, the treatise by Bhāskara II is canonized in the text. Then again, it is difficult to decide whether such a point of view is a Colebrooke legacy or something that existed beforehand within the *jyotiṣa* milieus of the late eighteenth century.

Further, the purpose of Dvivedin's commentary is to make Brahmagupta's rules mathematically and astronomically clear, to ground them with modern mathematics when they prove difficult. Such functions of a mathematical commentary can be found in the requirements for a commentary within both European and *śāstric* textual cultures. A strikingly 'modernist' streak, however, can be found in the introduction of algebraic symbolism into the commentary. Dvivedin had already introduced such symbolism in his Sanskrit commentary on the *Pañcasiddhānta*. In doing so, Dvivedin established himself as a double authority in mathematics and astral science on the one hand, and in Sanskrit lore on the other. The more general aim of providing a view of 'the principles of Hindu astronomy' through clarifying one of its texts both in terms of language and mathematics provides a basis for understanding Dvivedin's different editorial choices.

### 12.4.3 Concealing Problems of Uncertainty Related to Sectioning

Although his preface articulates questions of the coherence of the general structure of the treatise, Dvivedin did not tackle the problems of structure of chapters themselves head on. The verses which feature as 'supplements' in Colebrooke's translation appear here as part of the section 'Practices on Shadows' (*chāya-vyavahāra*), as seen in the heading reproduced in Fig. 12.5. Nonetheless, as shown in the table of contents (see Fig. 12.6), the two chapters translated by Colebrooke receive special treatment: their contents are enumerated in parenthesis.

[78] Dvivedin (1902:6): *asya siddhāntasya dvādaśādhyāye gaṇitādhyāye pāṭīgaṇitarūpe mayā pāṭhakānāṃ vinodāya nijaṭīkāyāṃ caturvedācāryapṛthūdakaṭīkāt udāharaṇāni saṃhṛtya likhitāni|*

For this treatise, in the twelfth chapter —which is the mathematical chapter, consisting of *Pāṭīgaṇita* (arithmetic)—after collecting examples from Master Caturveda, that is, Pṛthūdaka's commentary, I have written a commentary for students, for pleasure. (Dvivedin 1902:6)

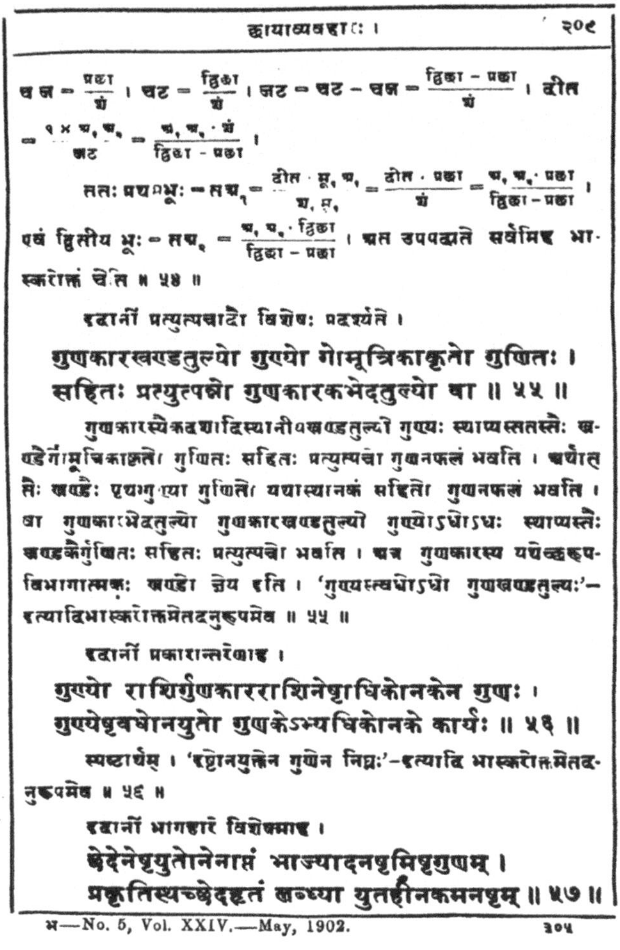
छायाव्यवहारः । २०९

चज = प्रछा / शं । चट = द्विछा / शं । जट = चट − चज = (द्विछा − प्रछा) / शं । दीप = (१ × भ१, भ२) / जट = (भ१, भ२ · शं) / (द्विछा − प्रछा) ।

ततः प्रथमभूः = तभ१ = (दीप · भू१, भ१) / भ१, प्र१ = (दीप · प्रछा) / शं = (भ१, भ२ · प्रछा) / (द्विछा − प्रछा) ।

एवं द्वितीय भूः = तभ२ = (भ१, भ२ · द्विछा) / (द्विछा − प्रछा) । अत उपपद्यते सर्वमिदं भास्करोक्तं चेति ॥ ५४ ॥

इदानीं प्रत्युत्पन्नादौ विशेषः प्रदर्श्यते ।

गुणकारखण्डतुल्यो गुण्यो गोमूत्रिकाकृतो गुणितः ।
सहितः प्रत्युत्पन्नो गुणकारकभेदतुल्यो वा ॥ ५५ ॥

गुणकारस्यैकदशादिस्थानीयखण्डतुल्यो गुण्यः स्थाप्यस्ततस्तैः खण्डैर्गोमूत्रिकाकृतो गुणितः सहितः प्रत्युत्पन्नो गुणनफलं भवति । अर्थात्तैः खण्डैः पृथग्गुण्या गुणितो यथास्थानकं सहितो गुणनफलं भवति । वा गुणकारभेदतुल्यो गुणकारखण्डतुल्यो गुण्योऽधोऽधः स्थाप्यस्तैः खण्डैर्गुणितः सहितः प्रत्युत्पन्नो भवति । अत्र गुणकारस्य यथेच्छरूपविभागात्मकः खण्डो ज्ञेय इति । 'गुण्यस्त्वधोऽधो गुणखण्डतुल्यः'—इत्यादिभास्करोक्तमेतदनुरूपमेव ॥ ५५ ॥

इदानीं प्रकारान्तरेणाह ।

गुण्यो राशिर्गुणकारराशिनेष्टाधिकोनकेन गुणः ।
गुण्येष्टवधोनयुतो गुणकेऽभ्यधिकोनके कार्यः ॥ ५६ ॥

स्पष्टार्थम् । 'इष्टोनयुक्तेन गुणेन निघ्नः'—इत्यादि भास्करोक्तमेतदनुरूपमेव ॥ ५६ ॥

इदानीं भागहारे विशेषमाह ।

छेदेनेष्टयुतोनेनाप्तं भाज्यादनष्टमिष्टगुणम् ।
प्रकृतिस्थच्छेदहृतं लब्ध्या युतहीनकमनष्टम् ॥ ५७ ॥

भ—No. 5, Vol. XXIV.—May, 1902. ३०५

**Fig. 12.5** Dvivedin's edition and commentary of BSS.12.55. The heading reads: 'Practices on Shadows' (*chāyavyavahāra*)

Like Colebrooke, Dvivedin considers Chaps. 12 and 18 as some kind of older version or variation of Bhāskara's works on arithmetic and algebra. Consequently, in his preface, he refers to them as evoking, respectively, determined *(vyakta)* and undetermined *(avyakta)* quantities.[79] Concerning the mathematical chapter, Dvivedin specifies in the table of contents 'it is *pāṭīgaṇita*; there are twenty operations and eight practices.' The use of the category *pāṭīgaṇita*, which is not employed in Brahmagupta's text, is revealing here, since this is the name adopted by Bhāskara to designate arithmetic.[80] Thus, Chap. 12 of the BSS is equated with categories

[79] Dvivedin (1902: 3).

[80] This title is repeated in his preface. See Dvivedin (1902: 5).

विषयानुक्रमः ।



७५५

**Fig. 12.6** Table of contents of Dvivedin's edition

belonging to the *Līlavatī*. Note that operations are mentioned in Dvivedin's enumeration of the topics treated in the chapter on algebra as 'additives and subtractives, and so forth' (*dhanarṇādi*). No specific remark on computations with zero is made, while the different forms of equations and problems of indeterminate analysis are detailed. Dvivedin, however, departs from Colebrooke's attempt to mirror algebraic operations in arithmetical ones. Even in his comments, he carefully avoids giving a generic name to the different additions, subtractions etc., that quantities are subject to.

Thus, the table of contents in Dvivedin's edition first emphasizes Chaps. 12 and 18, demonstrating a probable influence of Colebrooke's translation and further, illustrates Dvivedin's own aim of providing a coherent homogenous tradition of arithmetic and algebra, to which Brahmagupta and Bhāskarācārya both testify.

### 12.4.4 What Is Interesting About Numbers?

Dvivedin's reflections in his commentaries are not textual; he is interested in ideas, not in practices. The multiplication we briefly saw Colebrooke reconstruct, using the tabular resources of decimal place value notation, is notable in Dvivedin's commentary for its absence of numbers or tables, as can be seen in Fig. 12.5. On this page, Dvivedin comments on the rule BSS.12.55. If we compare this commentary with that made by Colebrooke as illustrated in Fig. 12.4, it is striking that no tabular configuration is provided in Dvivedin's commentary. He did not care for such reconstructions at all. His awareness of debates on the origins of the decimal place value notation, and the discussions on the kind of numerals to be found in words can, perhaps, be seen in the care he took to transcribe the many numbers given in words in Brahmagupta's verses with the decimal place value notation. Some of the versified numerical tables provided by Brahmagupta thus appear visually as lists of numbers, while there is no reason to believe that the root text used the decimal place value notation to write numbers.[81] Yet again, in this case, Dvivedin might be following manuscript traditions in which word numerals transcribed into decimal place value notation are well attested. Sometimes expressions such as the 'subdivisions of a circle' (*cakrāṃśa*) are readily transcribed as the numbers they refer to (here, 360). This makes Brahmagupta's text appear visually as a repository of numbers, parameters that are to be used in computation.[82]

We had noted that the table of contents did not display the presence of computations on zero in the 'algebraic' Chap. 18. However, Dvivedin's editorial subsections provide a title before verse 30/31 of Chap. 18 which includes zero: 'Computations with 'capital' (e.g. positives), 'debt' (subtractives) and zero' (*dhanarṇaśūnyānāṃ saṅkalanam*). His commentary on this part of the text dwells at length, textually, on these computations. In BSS.18.34/35,[83] Brahmagupta writes that 'zero divided by zero is zero' (*khaṃ khabhaktaṃ svam*). Dvivedin does not comment on the validity of the rule but recalls that Bhāskara states a rule for writing zero divided by zero without computing the division. He adds that such an understanding had caused 'commotion in the world' (*etad arthaṃ valanakalanaṃ vilokyam*).[84] We can recognize here the standard treatment of this moment in the history of zero computations in India, which often follows the same narrative path: from the first, Brahmagupta, to the correct one, Bhāskara. It would need to be further established, but I think that this narrative of zero may have its origin in Dvivedin's commentary on Brahmagupta's text. However, Dvivedin's remarks are carefully devoid of judgment in this case.

[81] See, for instance, the tables of BSS.1.15–24 in Dvivedin (1902: 6–7 in the main text).

[82] See, for instance, BSS.2.49 in Dvivedin (1902: 42).

[83] Colebrooke's numbering is also given as an alternative BSS.18.35–36. See Colebrooke (1817: 339–340).

[84] Dvivedin (1902: 310).

### 12.4.5 A Publication for Whom?

We have seen that Dvivedin's work on the BSS was infused with Colebrooke's previous publication, and influenced by George Thibaut. We have, however, noted that Dvivedin's 1902 edition, contrary to Colebrooke's translation, does not carve a mathematical text out of an astronomical text, and that he did not include an English translation in it. Dvivedin seems to have shared with Thibaut several aims, notably exhibiting the 'principles of Hindu astronomy' and making the authors of Sanskrit mathematical and astral texts known to a readership of Sanskrit scholars in India and perhaps beyond its borders. These features can also be attributed on the part of Dvivedin to a growing nationalist pride, to his (servile?) adherence to the viewpoints of Thibaut (and other colonial civil servants-*cum*-Orientalists) on Sanskrit scholarship, or to a mixture of both. How much can we recover at this point of the specificity of Dvivedin's point of view? Dvivedin's association with Thibaut in the late 1880s could be interpreted as partiality not only to 'Western' science but also to the orientalism, or philology, of Westerners. Indeed, in 1884, Thibaut started a reform of the teaching of Sanskrit in Benares, requiring both students and *paṇḍits* to read the publications of Western Orientalists such as Max Müller and H.T. Colebrooke. This reform met with a degree of resistance from the Benares *paṇḍits*, and in the larger nationalist milieus of the Ārya Samāj.[85] Dvivedin seems to have sided with Thibaut, as testified by their collaborative publications and Dvivedin's many references to Thibaut. As we have seen, he did not adopt the techniques of critical edition, but in this, he was in agreement with Thibaut.

Nonetheless, with his solo publications, Dvivedin inaugurated a new type of work on the astral sciences in India, that of texts edited by natives of the Indian subcontinent in Sanskrit. This leads us to ask who the intended readership of Dvivedin's publication was. The BSS was first published in *The Paṇḍit,* a journal which was initially edited by Ralph T. H. Griffith. In Michael Dodson's words[86]:

> [The publication of The Paṇḍit, which began in July of 1866] was intended not principally as a tool for the spread of useful knowledge, however, but rather as a medium for bringing the expertise and authority of the college's paṇḍits to bear upon the field of Sanskrit scholarship. The journal's mandate was described thus:
>
> 'to publish rare Sanskrit works which appear worthy of careful editing hereafter; to offer a field of discussion of controverted points in old Indian philosophy, philology, history, and literature; to communicate ideas between the Aryan scholars of the East and the West; between the Pandits of Benares and Calcutta and the Sanskritists of the Universities of Europe.'

This program may have planned to leave 'the careful editing' to the Sanskritists of the universities of Europe. However, Dodson underlines that the Sanskrit publications of *The Paṇḍit* 'also demonstrate the potentially disputative uses which

[85] Dodson (2007b: 53–54).

[86] Dodson (2007a: 174).

expertise in Sanskrit literature could be put to at this time.'[87] Although not polemical, Dvivedin's knowledge of Sanskrit and his erudition in incorporating known Orientalist themes places the text within a larger landscape of Sanskrit lore.

Further, this publication should be situated within the context of a very rich landscape of journals in Benares and in Calcutta, where editions could be made and histories could be written. Twenty or thirty years after *The Paṇḍit*, the early history of mathematics and astronomy written by Indian scholars used an intricate network where publication series such as the *Chowkhamba Sanskrit Series* co-existed with journals such as the *Indian Historical Quarterly* and the *Bulletin of the Calcutta Mathematical Society*, in which articles, translations and editions of Sanskrit mathematical and astral texts were published.[88]

We have underlined how Dvivedin's edition worked in the realm of a common ground on which *paṇḍits* and certain European philologists could meet. This could have contributed to anchoring such editorial practices over time. Indeed, one of the consequences of this kind of perspective on the editing of scientific texts is that there has been, until recently, relatively little self-reflection on how *gaṇita* (mathematics) and *jyotiṣa* (astral science) texts should be edited. Further, Dvivedin's work, seen as located in a middle ground in which collaborations with colonial institutions went along with a revalorization of one's own tradition, may perhaps be seen as a publication intended just as much for the awareness of European 'sahibs' as for the Indian students he was teaching in Benares.[89]

## 12.5 Conclusion

Colebrooke perceived the mathematical texts he had at hand as providing elements of practical mathematics together with interesting theoretical considerations. In a way, they could seem familiar, not unlike Hutton's manual of arithmetic, and in the case of Brahmagupta, they also had a rather antiquated flavor. This can be seen in Colebrooke's choices of sections for his translation. His interest in algebraic and arithmetical practice, and his faithful engagement with Pṛthūdaka's commentary, also shaped the printed text. On the other hand, for Dvivedin, the BSS was an all-encompassing text of astral science, and its mathematics was to be clarified when establishing the Sanskrit text.

---

[87] Dodson (2007a: 176).

[88] Keller (2011). Milieus dealing with astral science in Varanasi seem to have been located in a middle ground between the practitioners of ayurveda encountered in K. Preisendanz in the present volume (Chap. 6), and A. Graheli's (Chap. 5) conservative epistemologists based on the outskirts of Calcutta. Indeed, a specialist in *jyotiṣa* could equally be a scholar studying ancient texts and a practicing astrologer, or sometimes both at the same time.

[89] Thus, in the sentence of his preface quoted above, the examples of PBSS seem to have been included in his commentary for the enjoyment of students.

Translation and edition are tools for comparing, analyzing and understanding. Colebrooke's translations can be seen as operating a great variety of cuts and sectioning, with a comparative aim. Perhaps his translation can also be seen as reducing each problem to more sizeable ones. In a history of colonial translations, they often are understood as a mode of appropriation and transformation into something that can literally be possessed.[90] More than 80 years later, Colebrooke's legacy can be found in Dvivedin's edition, which adopted some of his parallels and distinctions. But while, on the one hand, Dvivedin's edition can be seen as a product of a colonial institution, on the other hand, it can also be viewed as that of a *paṇḍit* reclaiming and proclaiming his own scientific tradition.

However nuanced Colebrooke's approach may be, he is narrating a history of algebra in which 'Hindu' mathematicians' works are compared to the achievements of the 'Greeks' and the 'Arabs'.[91] Even if Dvivedin's work enabled subsequent nation-building histories of science, it is striking that he did not endorse such comparisons. If he spoke as an authority on 'the principles of Hindu astronomy', this authority was implicit, and did not involve an explicit positioning in relation to the world outside the world he was describing.

If translations, on the one hand, and editions, on the other, do not imply necessarily the same readership, there is nonetheless a striking continuity in relation to questions of sectioning, of editorial reflections on 'original' or 'best texts' and, more generally, on the questions one might want to raise in relation to the mathematical texts.

### 12.5.1 Making a Corpus of Great Texts

Both Colebrooke and Dvivedin read Brahmagupta in the light of Bhāskara. The story of these editions and translations is thus very much that of the bricks that contributed in making a canonical work out of Bhāskara's *Siddhāntaśiromaṇi*. How much was this canonization already at work within the Sanskrit traditions that Colebrooke received?

The legacy of both editions still needs to be precisely described. We might enumerate a few:

The canonization of Bhāskara's text, putting Brahmagupta's in the shadow, lasted until at least the middle of the twentieth century. The reification of the topics of mathematics as something in the realm of practical arithmetic with an antiquated tinge revived by the use of the term 'logistics' and the separation of questions concerning the history of mathematics from those on the history of astronomy are still alive today. Reflections on the possibility of critically editing Sanskrit mathematical texts are being re-examined.

---

[90] Dodson (2007b).

[91] Raina (2012: 245–246).

These are some of the legacies that these two texts have contributed to developing, each in its own way.

### *12.5.2 Questions for the History of Mathematics in India*

Studies in the historiography of mathematics open new arenas for the history of mathematics itself. I would like to end this article by underlining some of the questions raised in the introduction to this chapter that can specified by this study.

The existence of reflections on the relationship between arithmetic and algebra, whether the notion of quantity is coherent for both disciplines, is salient in Bhāskara's text. This is particularly striking as Bhāskara qualifies quantities (*rāśi*) in arithmetic as *vyakta* ('determined') and in algebra as *avyakta* ('undetermined'). As we have seen, this has been implicitly interpreted in the historiography as being a general point of view for all Sanskrit mathematical texts. Further research could try to restore their specific points of views, if they exist, to each author. How these lessons on arithmetic or algebra were organized, how Brahmagupta on the one hand and Bhāskara on the other conceived of the relation between both topics, and the nature of their theories of operations and quantities, are domains that still need to be investigated.

The questions of Brahmagupta's definition of a mathematician's knowledge, and the organization of his mathematical chapter, superimpose themselves onto a larger question, which may have been occasioned by Colebrooke's extraction of mathematical texts from those of texts on astral science. Why was the mathematical chapter of the BSS included in this astronomical text? What is the function of the mathematical tools given in that chapter?

This is a question that can be raised for many mathematical writings found in Sanskrit astral texts: indeed, they cannot simply be understood as providing tools for mathematical astronomy. On the one hand, there are tools provided in Brahmagupta's mathematical chapter that seem unrelated to astral science (like rules concerning sums of interest or mounds of grains). On the other, other chapters may include procedures of interpolation, for instance, that are not given in the mathematical chapter. In other words, what is meant by *gaṇaka* in Brahmagupta's verse is not only an issue internal to mathematics, it concerns also the relation of mathematics to astral science. In some cases, mathematical tools (the Rule of Three, the Pythagorean procedure, and perhaps algebra) could be used to provide proofs of astral procedures. But this only seems to be true for certain texts and procedures. Moreover, how mathematics is perceived is actually linked to the more general question of understanding the nature of the functions of 'lessons'/chapters in the BSS. These unresolved questions remain with us today, and continue to shape the critical editions we produce.

# References[92]

## Critical Editions and Some Other Publications by S. Dvivedin

Dvivedin, Sudhākara. 1867. *The S'ishyadhi'vṛddhida, the treatise on astronomy by Lall'ach'arya edited by Pandit Sudh'akara Dvivedi'*. Benares: Medical Hall.

———. 1885. *A catalogue of Sanskrit manuscripts in the North-Western provinces, compiled by order of government. Part IX*. Allahabad: Printed at the M. -W. F. and Oudh Goverment Press.

———. 1892 [1933]. Gaṇakataraṅgiṇī. Benares: Hemant Sanskrit Series Granthamala.

———. 1899. *Triśatikà by Śridharàchàrya*. Benares: Jagannath Mehta.

———. 1902 *Brāhmasphuṭasiddhānta and Dhyanagrahopadeśādhyāya by Brahmagupta edited with his own commentary*. Reprint from *The Pandit*, no. 12, Vol. 24. Benares: Printed at the Medical Hall Press.

———. 1908. *Yājusha-Jyautisha, with the Bhāshyas of Somākara Śesha & Sudh'akara Dvivedin and Ārcha-Jyautisha with the Bhāshya of Sudh'akara Dvivedin and Professor Mural'idhar Jh'a's explanatory*. Benares.

———. 1910a. *The Mahāsiddhānta of Āryabhaṭa; Mah'asiddh'anta, a treatise on astronomy by 'Aryabhaṭ, edited with his own commentary by Mah'amahop'adhy'aya Sudh'akara Dvivedi, first professor, Govt. Sanskrit college, Benares, and fellow of the Allahabad University &c.* Benares Sanskrit series 148, 149, 150. Benares: Printed at the Medical Hall Press.

Dvivedin, Sudhākra. 1910b. *Gaṇita Kā Itihāsa, a history of mathematics, Phalā Bhāga, Pāṭīgaṇita, first part, Arithmetics*. Benares: Prabhakar Printing Works.

Dvivedin, Sudhākara & Śrī Muralīdhara Jhā. 1927. Bījagaṇita (Elements of algebra) of S'rī Bhāskarāchārya with expository notes and illustrative examples by M. *M. Pandit S'rī Sudhākara Dvivedi. Edited with further notes by Mahāmahopādhyāya Pandit S'rī Muralīdhara Jhā, first professor, Sanskrit college, Benares*. Benares Sanskrit series 159. Benares: Krishna Das Gupta.

Thibaut, Georges F., and S. Dvivedin. 1889. *Pañchasiddhântikâ*. Benares: EJ Lazarus and Company.

## Primary Sources

Babuā, Miśra. 1925. *Khaṇḍa khādyakaṁ of Brahmagupta with the commentary called Vasana-Bhasya by Āmarâja*. Calcutta: University of Calcutta.

Chatterjee, Bina. 1970. *The Khaṇḍakhādyaka (an astronomical treatise) of Brahmagupta: With the commentary of Bhaṭṭotpala, edited and translated*. Calcutta: World Press.

Colebrooke, H.T. 1817. *Algebra, with arithmetic and mensuration, from the Sanscrit of Brahmegupta and Bhàscara*. London: J. Murray.

Colebrooke, T.E. 1839. Notices of the life of Henry Thomas Colebrooke, Esq., by his son. *Journal of the Royal Asiatic Society of Great Britain and Ireland* 5: 1–60.

Datta, Bibhutibhushan, and Narayan Avadesh Singh. 1935. *History of Hindu mathematics*. Lahore: Motilal Banarsidass.

Hutton, Charles. 1775. *A Complete Treatise on Practical Arithmetic and Book-Keeping Both by Single and Double Entry Adapted to the Use of Schools*. Seventh Edition. London: R. Baldwin.

———. 1795 [1815] *A philosophical and mathematical dictionary, containing an explanation of the terms, and an account of the several subjects, comprised under the heads mathematics, astronomy, and philosophy both natural and experimental ... Also memoirs of the lives and writings of the most eminent authors, both ancient and modern*. London: J. Johnson; and GG and J. Robinson.

[92] In the following the non-standard transliteration of Sanskrit words have been preserved when giving titles of nineteenth century works.

———. 1812. *Tracts on mathematical and philosophical subjects. Vol. II*. London: Printed for F. C and J. Rivington; G. Wilkik and J. Robinson; J. Walker. Lackington, Allen and Co.; Cadell and Dayies & I. J. Cuthill; D. and R. Crosby and Co. & J. J Richardson; J. M. Richardson & J. H. Baldwin and A. Robinson.

Sharma, R.S., M. Mishra, V. Jha, D.S. Dikshita, and O.D.S. Shastra, eds. 1966. *Brāhma-Sphuṭa-Siddhāṇta with Vāsanā, Vijñāna and Hindi Commentaries*. New Delhi: Indian Institute of Astronomical and Sanskrit Research.

## *Secondary Sources*

Bag, A.K. 1979. *Mathematics in ancient and medieval India*. New-Dehli: Chaukhambha oriental research studies.

CESS (see Pingree (1970–1995)).

Dauben, J.C., and Joseph W. Scriba, eds. 2002. *Writing the history of mathematics: Historical development*. Basel: Birkhäuser.

Dodson, M. 2007a. *Orientalism, Empire, and National Culture; India, 1770–1880*, Cambridge Imperial and post-colonial studies. Basingstoke: Palgrave Macmillan.

———. 2007b. Contesting translations: Orientalism and the interpretation of the Vedas. *Modern Intellectual History* 4 (01): 43–59.

Eggeling, Julius. 1896. *Catalogue of the Sanskrit manuscripts in the library of the India office. Vol. Part V*, Sanskrit literature: A. scientific and technical literature IX Medicine X. Astronomy. London: Secretary of State for India in Council.

Grafton, Anthony. 1997. *The footnote: A curious history*. Cambridge MA: Harvard University Press.

Gupta, R.C. 1990. Sudhākara Dvivedī (1855–1910), historian of Indian astronomy and mathematics. *Gaṇita Bhāratī* 12: 17–26.

———. 2002. India. In *Writing the history of mathematics: Its historical development*, ed. Joseph Dauben and C.W. Scriba, 306–315. Historical Studies. Science Networks. Basel: Birkhäuser.

Hayashi, Takao. 2009. Bījagaṇita of Bhāskara. *SCIAMVS* 10.

Ikeyama, Setsuro. 2003. *Brāhmasphuṭasiddhānta (Ch. 21) of Brahmagupta with commentary of Pṛthudaka, Critically Ed. with Eng. Tr. and Notes. Vol. XXXVIII*. New Delhi: Indian Journal of History of Science.

Keller, Agathe. 2010a. On Sanskrit commentaries dealing with mathematics (VIIth–XIIth Century). In *Looking at it from Asia: The processes that shaped the sources of history of science*, ed. Florence Bretelle-Establet, 211–244. Boston: Boston Studies in the Philosophy of Science.

———. 2010b. Review of Kim Plofker, mathematics in India. *Isis* 101 (1): 199–200.

———. 2011. George Peacock's arithmetic in the changing landscape of the history of mathematics in India. *Indian Journal of History of Science* 46 (2): 205–234.

Keller, Agathe, and Catherine Morice-Singh. 2022. Multiplying integers: On the diverse practices of medieval Sanskrit authors. In *Cultures of computation and quantification in the ancient world: Numbers, measurements, and operations in documents from Mesopotamia, China and South Asia*, Why the sciences of the ancient world matter, ed. Karine Chemla, Agathe Keller, and Christine Proust, 495–552. Cham: Springer International Publishing.

Kichenassamy, Satyanad. 2010. Brahmagupta's derivation of the area of a cyclic quadrilateral. *Historia Mathematica* 37: 28–61.

———. 2012. Brahmagupta's propositions on the perpendiculars of cyclic quadrilaterals. *Historia Mathematica* 39 (4): 387–404.

Pingree, David. 1970–1995. *Census of the exact sciences in Sanskrit (CESS)*. Vol. 5 Volumes. American Philosophical Society.

———. 1983. Brahmagupta, Balabhadra, Pṛthūdaka and Al-Bīrūnī. *Journal of the American Oriental Society* 103 (2): 353–360.

Plofker, Kim. 2009. *Mathematics in India*. Princeton: Princeton University Press.

Raina, Dhruv. 2011. Situating the history of Indian arithmetical knowledge in Peacock's arithmetic. *Indian Journal of History of Science* 46 (2): 235–250.

———. 2012. Contextualizing Playfair and Colebrooke on proof and demonstration in the Indian mathematical tradition (1780–1820). In *The history of mathematical proof in ancient traditions*, ed. Karine Chemla, 228–259. Cambridge: Cambridge University Press.

Rocher, L., and R. Rocher. 2012. *The making of Western Indology, Henry Thomas Colebrooke and the East India company*. London: Routledge.

———. 2013. *Founders of Western Indology: August Wilhelm Von Schlegel and Henry Thomas Colebrooke in correspondence 1820–1837*. Wiesbaden: Harrassowitz.

Smadja, Ivahn. 2015. Sanskrit versus Greek 'proofs': History of mathematics at the crossroads of philology and mathematics in nineteenth-century Germany. *Revue d'Histoire Des Mathématiques* 21 (2): 217–349.

———. 2016. On two conjectures that shaped the historiography of indeterminate analysis: Strachey and Chasles on Sanskrit sources. *Historia Mathematica* 43: 241–287.

Sriram, M.S. 2013. Brahmasphutasiddhanta of Brahmagupta – Part 1 in *Mathematics in India – From Vedic period to modern times*. Indian Institute of Technology Madras. https://www.youtube.com/watch?v=ez60V_IdITU. Accessed 9 Aug 2017.

# Chapter 13
# On the First Printed Edition of *Mathematical Book in Nine Chapters* (1842)

**Yiwen Zhu and Cheng Zheng**

**Abstract** *Mathematical Book in Nine Chapters* (*Shushu Jiuzhang*, hereafter *Mathematical Book*), completed by Qin Jiushao (1208–1268) in 1247, was one of the representative mathematical works in thirteenth-century China. Previous scholarly studies heavily rely on the first printed edition of *Mathematical Book* (*Yijiatang* edition) that was published in *Yijiatang* Collectanea in 1842. However, our recent research shows that a seventeenth century manuscript of *Mathematical Book* (Zhao Qimei's manuscript, 1616) contains very important pieces of information on the use of counting rods, which were not reproduced in the first printed edition. This chapter aims to revisit the formation of the 1842 edition, and evaluate its influence on the subsequent historiography of mathematics. Firstly, we introduce the textual history of *Mathematical Book* before 1800, the scholarly background in the Qian-Jia Period (1736–1820), and Yu Songnian's and Song Jingchang's contributions to the *Yijiatang* Collectanea and the first printed edition. Secondly, we analyze the writing system and three representations of the Chinese Remainder Theorem in the 1616 manuscript, highlighting the difference between the 1616 manuscript and the 1842 printed edition: the key type of representation using lines and diagrams for the operational procedures of counting rods was lost in the printed edition. Finally, the reception of *Mathematical Book* in the nineteenth century is discussed. We show the influence of the printed edition, we also point out that, in this context, scholars were specifically interested in interpreting ancient mathematics using modern mathematics.

In conclusion, we argue that the first printed edition of *Mathematical Book* exerted a strong influence in shaping the image of the book in the historiography.

---

Y. Zhu (✉)
Institute of Logic and Cognition, Department of Philosophy, Sun Yat-sen University, Guangzhou, China
e-mail: zhuyiwen@ymail.com; zhuyiw3@mail.sysu.edu.cn

C. Zheng
Institute for the History of Natural Sciences, Chinese Academy of Sciences, Beijing, China
e-mail: zhengcheng@ihns.ac.cn



A. Keller, K. Chemla (eds.), *Shaping the Sciences of the Ancient and Medieval World*, Archimedes 69, https://doi.org/10.1007/978-3-031-49617-2_13

## 13.1 Introduction[1]

Qin Jiushao (秦九韶, 1208–1268) was a scholar-official during the Song dynasty (960–1279) who is regarded as one of China's most brilliant mathematicians.[2] His only extant work, titled *Mathematical Book in Nine Chapters* (*Shushu Jiuzhang* 數書九章, hereafter *Mathematical Book*), includes the author's preface of 1247 and eighty-one problems in nine categories relating to calendar calculations, surveying, tax, construction, military engineering and logistics, commerce, and so forth. *Mathematical Book* is famous especially for providing the Chinese Remainder Theorem (*dayan zongshu shu* 大衍總數術, literally the 'Great inference procedure for all numbers')—which is the general algorithm for solving simultaneous linear congruences—and a method for solving numerical equations of higher degree (*zhengfu kaifang shu* 正負開方術, literally the 'Procedure for extracting roots with positive and negative [numbers]'). This work is also a valuable source for social and economic history. In 1842, the Yijiatang edition of *Mathematical Book* appeared. This edition was the first printed edition, which was published in *Yijiatang Congshu* (宜稼堂叢書 *Yijiatang Collectanea*) in Shanghai with a detailed critical apparatus. The editor and publisher of *Mathematical Book* were, respectively, Song Jingchang 宋景昌 and Yu Songnian 郁松年.

Modern scholars mainly rely on the 1842 edition to research Qin's work. Qian Baocong 錢寶琮 (1966) systematically studied *Mathematical Book*. Then Libbrecht (1973) completed a monograph on *Mathematical Book* titled *Chinese Mathematics in the Thirteenth Century*. In the collected papers on *Mathematical Book* (Wu 1987), Li Di 李迪 (1987b) provided research on the textual history of *Mathematical Book*, and Shen Kangsheng 沈康身 (1987) corrected some errors in the 1842 edition.

[1]The research leading to these results has received funding from the European Research Council under the European Union's Seventh Framework Programme (FP7/2007–2013)/ERC Grant Agreement No. 269804. We also thank Chinese National Planning Office of Philosophy and Social Science [the National Social Science Fund of China, 21VJXG022] for supporting funding on this research. The results presented in this chapter emerged from the two authors' cooperation on the textual history of *Mathematical Book in Nine Chapters* in 2009, when both of us were pursuing our doctoral degrees at the Institute for the History of Natural Sciences, Chinese Academy of Sciences. In 2013, when Zhu Yiwen took up the postdoctoral fellowship in Paris, following discussions with the European project SAW team, we decided to complete a paper in English. These results were initially presented on 17 June 2013, and improved on 3 December 2014 within the framework of the seminars and conferences of the European Project 'SAW'. We are very grateful to the audiences for their questions and comments, which contributed to improving our study. We continued working on this chapter after the 2014 conference, and would like to thank Karine Chemla and Agathe Keller for their careful reading of this chapter, which helped us greatly. We deeply appreciate their comments and advice. We would also like to thank Sun Yat-sen University and the Chinese Academy of Sciences for their support on this research. We are also grateful to Julia Legh-Smith, Jens Høyrup, Karen Margolis and Christine Mousset for the English editing of this chapter.

[2]For a brief introduction to Qin Jiushao and his work, see Martzloff (2006: 149–152).

Wang Shouyi 王守義 (1992) provided a *New Annotated* Mathematical Book in Nine Chapters. There are also several studies on Qin Jiushao's life.[3]

In 2010, we examined a manuscript of *Mathematical Book* transcribed in 1616 and presented new research on the textual history of *Mathematical Book*, arguing that the 1616 manuscript is the most complete edition to be found today. This manuscript retains very important features of Qin Jiushao's mathematics, notably the counting-diagrams with lines for writing procedures of counting rods and numbers with colours representing two opposite numbers (Zheng and Zhu 2010). These features are missing in the 1842 edition. Because of the importance of the 1616 edition, it is thus necessary to revisit the formation of the 1842 edition and evaluate its influence on the study of the history of mathematics in China. This is the aim of the present chapter.

## 13.2 Textual History of *Mathematical Book in Nine Chapters* Before 1800

Some thirteenth-century authors referred to the title of *Mathematical Book* as *Shushu* (數術, *Mathematical Methods*), *Shushu Dalüe* (數術大略, *Outline of Mathematical Methods*) or *Shuxue Dalüe* (數學大略, *Outline of Mathematics*). From the fifteenth to the seventeenth century, the standard title of *Mathematical Book* was *Shuxue Jiuzhang* (數學九章, *Mathematics in Nine Chapters*), not *Shushu Jiuzhang* (數書九章, *Mathematical Book in Nine Chapters*). Although a little different, the meaning is similar. In 1441, the first catalogue of the Imperial Library (*Wenyuange Shumu*, 文淵閣書目) of the Ming dynasty (1368–1644), which recorded approximatively 5800 titles—notably 24 mathematical works—kept in Beijing, included an entry '*Shuxue Jiuzhang*, one copy, three volumes, complete.'[4] In 1605, the revised catalogue of the same library only contained 2447 titles. One of the three extant mathematical works was '*Shuxue Jiuzhang*, three volumes, complete, manuscript, written by Qin Jiushao of Lu county, during the Chunyou period [淳祐 1241–1252] of the Song dynasty.'[5] This manuscript is the master copy for all the editions that still exist today, but it was lost, probably in the turmoil after the fall of Beijing in 1644.[6]

In 1616, a famous bibliophile, Zhao Qimei (趙琦美, 1563–1624), acquired his own copy of *Mathematical Book* in Beijing and wrote postscripts in this manuscript (1616 [2010]). He called the book *Shushu Jiuzhang* for the first time and recorded

[3] For Qin Jiushao's life, see Qian (1966: 60–62), Peng-Yoke (1971), Libbrecht (1973: 23–30), Yan (1987), Li (1987a) and Guo (2010: 359–362).

[4] The original Chinese text: 數學九章，一部三冊，完全。

[5] The original Chinese text: 數學九章，三冊全，鈔本，宋淳祐間魯郡秦九韶撰。

[6] For the bibliographical records, see Libbrecht (1973: 35–41), and Zheng and Zhu (2010).

that Wang Yingling had made a manuscript transcribed from the copy kept in the Imperial Library. Zhao was able to borrow Wang's copy to make this new duplicate.[7]

Another bibliophile, Wang Yingling (王應遴, ?–1644) was especially interested in astronomy (the first two chapters of *Mathematical Book* are about calendar calculations). In the 1630s, as a minor official, Wang was involved in the calendar reform in Beijing (Zheng and Zhu 2010). His name appears as a reviser on the title page of *On Ancient and Modern Eclipses* (*Gujin Jiaoshi Kao* 古今交食考) (Schall von Bell, ca. 1635), a Chinese pamphlet written by a famous Jesuit, Johann Adam Schall von Bell (1592–1666).[8] But there is no trace of the Jesuits in Beijing ever having had access to any Chinese mathematical work of the thirteenth century. Wang compiled a series of writings which he called *Ce Shu* (測書, *Book of Surveying*), *Shu Shu* (數書, *Book of Mathematics*), *Bei Shu* (備書, *Book of Defence*), etc. (Shi and Song 2006). It is reasonable to assume that Wang changed the title of *Mathematical Book* from ***Shuxue*** *Jiuzhang* to ***Shushu*** *Jiuzhang* out of personal taste to give all his works similar names (with the form 'X *Shu*').

Wang's manuscript was lost. Luckily, Zhao Qimei brought his own copy back to his hometown of Changshu 常熟, a city located 80 kilometres northwest of Shanghai. Both Changshu and Shanghai are situated in the Yangtze River Delta, which was the most prosperous region, the centre of literary culture and publishing in the Ming-Qing periods. From the mid-seventeenth to the early twentieth century, Zhao's manuscript of *Mathematical Book* was preserved by several local bibliophiles in Changshu. Some duplicates transcribed from Zhao's manuscript were shared by scholars from the late eighteenth century onward. Since the 1950s, Zhao's manuscript (or a fairly accurate copy) has been preserved in the National Library of China (Zhao Qimei 1616 [2010]; Zheng and Zhu 2010).[9]

Another lineage of *Mathematical Book* is more complicated, somewhat like a jigsaw puzzle. At the beginning of the fifteenth century, Zhu Di (朱棣, 1360–1424), the third emperor of the Ming Dynasty, ordered his officials to compile *Great Compendium of Yongle Era* (*Yongle Dadian* 永樂大典, 1405–1408 [1986]). This huge encyclopedia, which had 22,937 sections in 11,095 manuscript volumes, drew its material mainly from the collection of the Imperial Library in Nanjing 南京. *Mathematical Book* was copied into it. In 1421, when Beijing replaced Nanjing as the new capital of the Ming dynasty, the main part of the library's collection in Nanjing was moved to Beijing and was later housed in the new Imperial Library there. It is thus highly probable that the particular copy of *Mathematical Book* that

[7] For Zhao Qimei's postscripts in the 1617 manuscript of *Mathematical Book*, see the facsimile edition (1616 [2010]: 236, 325).

[8] *Chongzhen Lishu* 崇禎曆書 Vol. 24; woodblock edition printed in 1630s. Kyujanggak Institute for Korean Studies, Seoul National University.

[9] The publication date of the *Siku tiyao zhulu congshu* is not very clear. The copyright page states December 2010, but the editor-in-chief signed his preface in March 2014. We saw this facsimile edition in December 2014. Before that, we used a microfilm of Zhao Qimei's manuscript (call number 3485) in the Rare Books & Special Collections Reading Room of the National Library of China.

Wang Yinglin transcribed in around 1616 is the same as the one transcribed into *Great Compendium* in around 1408 (Zheng and Zhu 2010).

According to the catalogue of *Great Compendium*, there were 37 sections (from Sect. 16 329 to Sect. 16 365) listed under 'Mathematics' (*suan* 筭). However, in accordance with the structure of the encyclopedia, the preface and the 81 problems of *Mathematical Book* were subdivided into various entries and arranged in different sections, which corresponded to different mathematical subjects. The original version of *Great Compendium* was totally lost in the war in the mid-seventeenth century. Thanks to the Ming government, which made a faithful manuscript copy of *Great Compendium* around the year 1565, today, there are approximately 408 volumes extant, less than 4% of the total 1565 edition. Three problems of *Mathematical Book* survive in Sect. 16 343 of the only extant volume under 'Mathematics' (Sect. 16 343 and 16 344), which is now held in the library of Cambridge University.[10]

In the turbulent era of the mid-seventeenth century, the 1565 edition of *Great Compendium* survived almost intact in Beijing and entered the collection of the new Manchu emperors of the Qing dynasty (1644–1911). In 1773, Qianlong Emperor (乾隆帝, 1711–1799) launched another cultural project, compiling *Complete Works in the Four Treasuries* (*Siku Quanshu* 四庫全書, hereafter abbreviated to *Complete Works*). More than ten thousand titles were assembled in Beijing for investigation and censorship.[11] In the end, the incomparably huge manuscript collection constituted around 3500 titles (approx. 36,000 volumes). All of the books were transcribed according to the same layout. Each of them was prefaced by a summary by the editors.[12] Around 10% of all the titles in *Complete Works* were restored from *Great Compendium*. Many of them were rare or unique editions.

The first manuscript copy of *Complete Works* (*Wenyuange Siku Quanshu* 文淵閣四庫全書) was completed in 1781. It comprised 3458 titles, with 386 titles—including *Mathematical Book*—derived from *Great Compendium*. There were 25 mathematical works in this edition, including traditional ones like Qin's *Mathematical Book* and European ones like the first six books of Euclid's *Elements*, which had been translated into Chinese in 1607 by the Italian Jesuit Matteo Ricci (1552–1610) and Xu Guangqi (徐光啟, 1562–1633). By 1787, all seven manuscript copies of *Complete Works* had been finished. Four of them were housed in different Imperial palaces in the North (Beijing, Rehe, Shenyang), while another three were delivered to the South for the imperial libraries in Hangzhou 杭州, Yangzhou 揚州 and Zhenjiang 鎮江, three prosperous cities in the Yangtze River Delta. Scholars were invited to consult *Complete Works* in these three libraries. In fact, it was a privilege for high-ranking officials.

[10] The extant volume, which contains two sections, survived by chance to the present day. See the facsimile edition of *Yongle Dadian* (1565 [1986]).

[11] In this project, more than 3000 titles were ordered to be banned or destroyed by the Qing government.

[12] The 'summary' (*tiyao* 提要) was always a short book review essay with less than a thousand Chinese characters. The essays were unsigned. The general chief editor Ji Yun (紀昀, 1724–1805) and other senior editors were responsible for them.

All 3500 summaries and a further 6700 summaries of 'less important but useful books' were collected together as the *Comprehensive Catalogue of Complete Works in the Four Treasuries* (*Siku Quanshu Zongmu* 四庫全書總目) and published in 1795. This *Comprehensive Catalogue* has been the most influential guide for Chinese classical studies since the nineteenth century. The summary of *Mathematical Book* criticizes Qin Jiushao's algorithm for calendar calculation as obsolete but highly praises one of his mathematical methods, which was later named the Chinese Remainder Theorem. The summary describes it as exquisite and ingenious, and even hails it as the origin of the European algebraic technique *Jiegenfang* (借根方, literally 'the borrowing of roots and powers [of the unknown]'), the algebraic system the Jesuit missionaries introduced to the Qing court in the late seventeenth century.

*Complete Works* provided the first critical edition of *Mathematical Book*. We shall call it the *Siku* edition. The title page lists the proof readers Gu Zhixiong 古之雄 and Ni Tingmei 倪廷梅 as two subordinate officials from the Imperial Bureau of Astronomy. We know little about them. Meanwhile, according to the extant archive of the Qing dynasty, in 1781 Chen Jixin 陳際新, who was the Imperial astronomer and chief editor for the astronomical and mathematical books in *Complete Works*, was punished for the clerical errors in the manuscript of *Mathematical Book* (Zheng and Zhu 2010).[13] Chen was born into a family that had worked for the Imperial Bureau of Astronomy for several generations. Perhaps Chen was in charge of compiling the new edition of *Mathematical Book* and wrote the summary.

To retrieve *Mathematical Book* from *Great Compendium* is not a simple task. Given that *Great Compendium* quotes materials without the title or number of the chapters, the editors (who were unaware of the existence of Zhao Qimei's manuscript) could only rely on Qin Jiushao's preface, which gave a brief description of each chapter, and reorganized the 81 problems derived from *Great Compendium* into nine chapters. The sequence of the problems in the *Siku* edition is therefore different from Zhao Qimei's manuscript.

The summary of *Mathematical Book* tells us that in the *Siku* edition 'errors are corrected' and 'commentaries are added'.[14] Libbrecht (1973: 43) already remarked on this, saying: 'The author of the descriptive notes tries to explain some of Ch'in's [i.e., Qin Jiushao's] methods, with the result that he proves himself incapable of understanding much of the text; and the greater part of his note is useless talk.' What is more important is that in the *Siku* edition the text and the illustrations of the copy-text were always modified (Zhu 2023: 235–241). In fact, the editors reorganized the structure of the first problem of *Mathematical Book* so as to emphasize the Great inference procedure for all numbers, which was not Qin's originally meaning. They also mistakenly intrepreted *ji* 奇 (odd number) and *ou* 偶 (even number), and

[13] For the archive material, see Zhongguo Diyi Lishi Dang'anguan (1997: 1451, 2694).

[14] The original Chinese text: 今即《永樂大典》所載，於其誤者正之，疎者辨之，顛倒者次序之，各加案語於下，庶得失不掩，俾算家有所稽考焉。'Here, on the basis of the [edition] in *Great Compendium of the Yongle Era*, we correct errors, examine defects, and put [back in the right] order what had been reversed; for each of [the modifications], we add commentaries under [it], to uncover the achievements and mistakes [of Qin Jiushao's *Mathematical Book in Nine Chapters*], and to help mathematicians pursue their research.' (JI Yun 紀昀 ed. 1795 [2019]: 233).

thus misguided later scholars' studies on the reduction of the modules in the Great inference procedure for all numbers. They further believed Qin's counting-diagrams were usually wrong, hence they changed them and omitted lines within the counting diagrams. Red-colored signs are missing and lines in counting-diagrams are omitted or distorted at random. In a word, it was not an accurate transcription of *Mathematical Book* from *Great Compendium*. In fact, casual changes and arbitrary alterations were very common in *Complete Works*, because of the censorship (which purged any possible anti-Manchu words), as well as carelessness and oversight.[15]

## 13.3 Scholarship and Textual Criticism in the Qian-Jia Period

The reigns of Emperors Qianlong and Jiaqing 嘉慶, from 1736–1820, often called the Qian-Jia period, are viewed as the golden age of 'evidential scholarship' (*kaozhengxue* 考證學) during the Qing dynasty.[16] The project of compiling *Complete Works* (1773–1781), which brought together rare books and brilliant scholars, was one factor that stimulated the development of scholarship. The late eighteenth and early nineteenth century also witnessed the flourishing of publishing. Classical works were often published in collections of writings. The powerful officials or wealthy gentry were willing to provide financial support to these endeavors and they invited scholars as editors. The principles with which the scholarly editions were carried out in the context of evidential scholarship demanded serious textual criticism.

In this period, the landmark of textual criticism would be the *Critical Apparatus for Commentary and Sub-commentary on the Thirteen Classics* (*Shisanjing Zhushu Jiaokanji*, 十三經注疏校勘記, 1808), in 217 sections. The *Thirteen Classics* are the core literature of Confucianism. They include *Book of Changes* (*Yi Jing* 易經), *Book of Poetry* (*Shi Jing* 詩經), *Analects of Confucius* (*Lun Yu* 論語) and other titles, all compiled before the second century BCE. The commentaries and subcommentaries were written by scholars from the second century CE to the eleventh century. Ruan Yuan (阮元, 1764–1849), a high-ranking official and an excellent scholar, initiated an ambitious project for preparing a new edition of the *Commentary and Sub-commentary on the Thirteen Classics* (1816) and its *Critical Apparatus*.[17] His editorial team included outstanding scholars like Gu Guangqi (顧廣圻, 1770–1839)[18] and the mathematician Li Rui (李銳, 1769–1817).[19] Gu Guangqi was

---

[15] In this case, we think the omissions and distortions in the *Siku* edition of Qiu Jiushao's work were due to carelessness and oversight. For the eighteenth-century book censorship of Qing dynasty, see Guy (1987: 157–200).

[16] For 'evidential scholarship', see Elman (1984: 39–85).

[17] For the project of Ruan Yuan's new edition of the *Thirteen Classics*, see Wang (1963: 25–60).

[18] For Gu Guangqi's life and works, see Hummel (1944: 417–419).

[19] For a very brief introduction to Li Rui, see Hummel (1944: 144).

perhaps the most renowned textual critic of the Qing period. He had a famous saying: 'editing without modification' (*bu jiao jiao zhi* 不校校之). The editorial policy followed this dictum. As copy-text, Ruan Yuan tried his best to collect early printed editions which had often been printed in the thirteenth and fourteenth centuries. He republished them without modification of the texts. The *Critical Apparatus* often quotes more than ten printed and manuscript editions of each commentary and subcommentary, and lists and discusses the variants and readings. However, compared with this strict model, publishing editions of ancient works that were thoroughly revised editions was much more popular. Nevertheless, at that time, it was generally accepted that a reliable new edition of ancient books should be equipped with a critical apparatus.

Partly owing to the project of *Complete Works in the Four Treasuries*, in the second half of the eighteenth century, a series of ancient mathematical masterpieces were rediscovered and studied by competent scholars. Some critical editions were published subsequently. The leading evidential scholar, Dai Zhen (戴震, 1724–1777), collated *Ten Mathematical Classics* (*Suanjing Shishu*, 算經十書).[20] Dai's new edition was used as the master copy in *Complete Works in the Four Treasuries* and it was finally published in *Weiboxie Collectanea* (微波榭叢書) in 1773. Moreover, the works of a talented mathematician, Li Ye (李冶, 1192–1279), were edited by Li Rui and published in *Zhubuzuzhai Collectanea* (知不足齋叢書) in 1778. In addition, Gu Guangqi mentioned in a preface (written ca. 1810) for *Mathematical Book* that a bibliophile, Qin Enfu (秦恩復, 1760–1843), provided his own copy for the republication of Qin Jiushao's work. But this plan seems to have remained unfulfilled.

## 13.4 The First Printed Edition in 1842 of *Mathematical Book in Nine Chapters*

### *13.4.1* Yu Songnian and the *Yijiatang Collectanea*

At the beginning of the nineteenth century, manuscript duplicates of Qin Jiushao's *Mathematical Book*, transcribed from Zhao Qimei's copy and the *Siku* edition, circulated among a small community of evidential scholars interested in mathematics. Li Rui, Gu Guangqi, Qian Daxin (錢大昕, 1728–1804), Li Huang (李潢, 1746–1812), Jiao Xun (焦循, 1763–1820) and others acquired their own copies (Li 1987b). They studied the work and shared ideas privately. The publication of *Mathematical Book* in *Yijiatang Collectanea* at an auspicious juncture is no surprise.

[20] For Dai Zhen's life and work, see Hummel (1944: 695–700). *Ten Mathematical Classics* refer to ten ancient mathematical books, written from the Han to Tang periods, including the famous *Nine Chapters on Mathematical Procedures* (*Jiuzhang suanshu,* 九章筭術). For Dai's work on the *Ten Mathematical Classics*, see Guo (2010: 678–684).

Yijiatang (宜稼堂, literally 'favorable farming hall') was the study room of Yu Songnian (郁松年, 1799–1865), the second son of a very rich family who owned more than seventy cargo ships in Shanghai. Yu was a bibliomaniac rather than a businessman and used his wealth to become one of the most important bibliophiles in the nineteenth century (Matsuura 2012: 201–210).

Between 1840 and 1843, Yu Songnian published the *Yijiatang Collectanea*, which consisted of critical editions of seven works by six authors, all of whom had lived in the thirteenth and fourteenth centuries (during the Song dynasty and the Yuan dynasty). These included two historical works on the Three Kingdoms period (third century), two anthologies of prose and poetry, and three mathematical works, one by Qin Jiushao and two by Yang Hui (楊輝, fl. 1260). The master copies were rare manuscripts owned by Yu himself or borrowed from his friends. Yu published the revisions with independent volumes of *Reading Notes* (*Zhaji* 札記) as critical apparatus. He compiled *Reading Notes* of the historical works and anthologies by himself and invited Song Jingchang 宋景昌 to edit the three mathematical works. It is noteworthy that editions of the first four titles and of Qin's *Mathematical Book* had already been copied into *Complete Works*. Yu might have felt proud that he could provide a reliable text, which was better than the Emperor's *Complete Works*.

This was the period of the first Sino-British war (known as the Opium War, 1839–1842). At the end of 1842, Yu Songnian wrote as follows: 'When I was preparing *Reading Notes* for publication, the English barbarians triggered a crisis. I was terribly worried that the manuscript would be destroyed. Then peace returned and I can complete the printing now.' (Yu Songnian 1842: 1)[21] The Qing dynasty was defeated. The British army occupied Shanghai county town in June 1842. The year after the Treaty of Nanjing (August 1842), Shanghai became a treaty port officially open to the Western world. It was the beginning of a new era which would have a great impact on the intellectual world. In any case, Yu's publishing project was suspended after 1843.[22]

### *13.4.2* Song Jingchang and His *Reading Notes*

In March 1842, Yu Songnian wrote the preface for *Reading Notes of* Mathematical Book in Nine Chapters. In it, he wrote that Mao Yuesheng (毛嶽生, 1791–1841) suggested he should publish *Mathematical Book* because it contained marvelous mathematical methods unknown for hundreds of years and the extant copies had accumulated a great many errors and omissions due to continuous transcriptions. Mao Yuesheng provided two copies. One was borrowed from a famous scholar Li

[21] Yu's preface of *Reading Notes* for *A Sequel to the History of Later Han Dynasty* (續後漢書, i.e., History of the Three Kingdoms) by Hao Jing (郝經, 1223–1275). The original Chinese text: 札記刊未畢，暎夷不靖，余大懼是書之散亡。夷氛息，鳩工卒事，至是告竣。

[22] We only know that, in 1846, he published one more title, *Yinglian Conghua* (楹聯叢話) by Liang Zhangju 梁章鉅.

Zhaoluo (李兆洛, 1769–1841), a duplicate transcribed from Zhao Qimei's manuscript. Another, owned by Mao himself, was a manuscript transcribed from the *Siku* edition with the critical notes of the mathematician Li Rui. Mao also recommended Song Jingchang, who specialized in mathematics, as the editor. Song acquired a posthumous manuscript from his teacher Shen Qinpei (沈欽裴, ? –before 1840). It was Shen's reading notes of *Mathematical Book* based on Zhao Qimei's manuscript (probably not the original, but a duplicate).

Song used the duplicate of Zhao Qimei's manuscript as a copy-text and the duplicate of the *Siku* edition as a reference text. In *Reading Notes*, as Libbrecht (1973: 49) summarizes, Song 'point[ed] out the textual errors' and where he introduced modifications, he quoted the comments from the *Siku* edition as well as from his teacher Shen Qinpei, Li Rui and Mao Yuesheng: 'For some of the problems, he gives a mathematical reconstruction or correction, but as a whole, this work is a textual collation' (Libbrecht 1973: 49). *Reading Notes* is a very useful tool for the study of *Mathematical Book.*

At the beginning of *Reading Notes*, Song gives a brief description: 'In the copy-text, every folio has 20 columns and each column has 20 characters.'[23] The format is the same as that of Zhao Qimei's manuscript in the National Library of China. Comparing the readings, we discover that some textual errors which Song pointed out in the copy-text are actually correct in Zhao's manuscript. This shows that Song Jingchang did not actually see Zhao Qimei's manuscript. More importantly, *Reading Notes* does not mention the lines in the counting-diagrams, which are a special feature of Zhao's manuscript. The 1842 edition of *Mathematical Book* contains only the counting-diagrams without the lines.

We find a small piece of paper with handwritten notes in Chinese ink inserted in Chap. 1 of Zhao Qimei's manuscript. It reads, 'It is not necessary to write down the lines [between the numbers]; but it is possible to only write down the signs on both sides.'[24] It seems to be an instruction for scribes, written in the eighteenth or nineteenth century. This note might have been written by someone who borrowed the manuscript from its owner and wanted to make a new copy for himself. It reminds us that some copies derived from Zhao's manuscript were produced without lines. We know that Song Jingchang mainly referred to a copy transcribed from Zhao Qimei's manuscript. It is possible that Song did not find any lines in this copy.

The first printed edition of *Mathematical Book* and its *Reading Notes* became the most influential edition. It was republished several times in the twentieth century. It also resulted in *Shushu Jiuzhang* replacing *Shuxue Jiuzhang* as the standard title of *Mathematical Book*. Nearly all the modern studies of Qin Jiushao's work are based on the 1842 edition. This edition is quite different from Zhao Qimei's manuscript. We can say that it loses some important information about Qiu Jiushao's mathematical practice.

---

[23] The original Chinese text reads: 原本每葉二十行，每行二十字。 (Song 1842: 1)

[24] The original Chinese text reads: 不必寫線，但寫兩邊頭號可也。 (Qin 1616 [2010]: 114)

## 13.5 Divergences: Counting-Diagrams and Text

### *13.5.1* The Writing Systems in *Mathematical Book in Nine Chapters*

A significant feature of *Mathematical Book* is that Qin Jiushao uses different layers of writing systems to describe the same mathematical procedure. To be precise, in the problems (*ti* 題), Qin formulated questions (*wen* 問), which were followed by answers (*da* 答), procedures (*shu* 術), detailed solutions (*cao* 草) and counting-diagrams (*suantu* 筭圖), in that order (e.g., Fig. 13.1). According to Qin's preface, the latter three are given in succession, and the counting-diagrams are used to explain the detailed solutions (Zhao Qimei's manuscript 1616 [2010]: 98).[25] However, if we check the whole book, the relationships between them differ from what Qin states. Only 45 of all the 81 problems have counting-diagrams. Of these 45 problems, there are 23 in which procedures, detailed solutions and

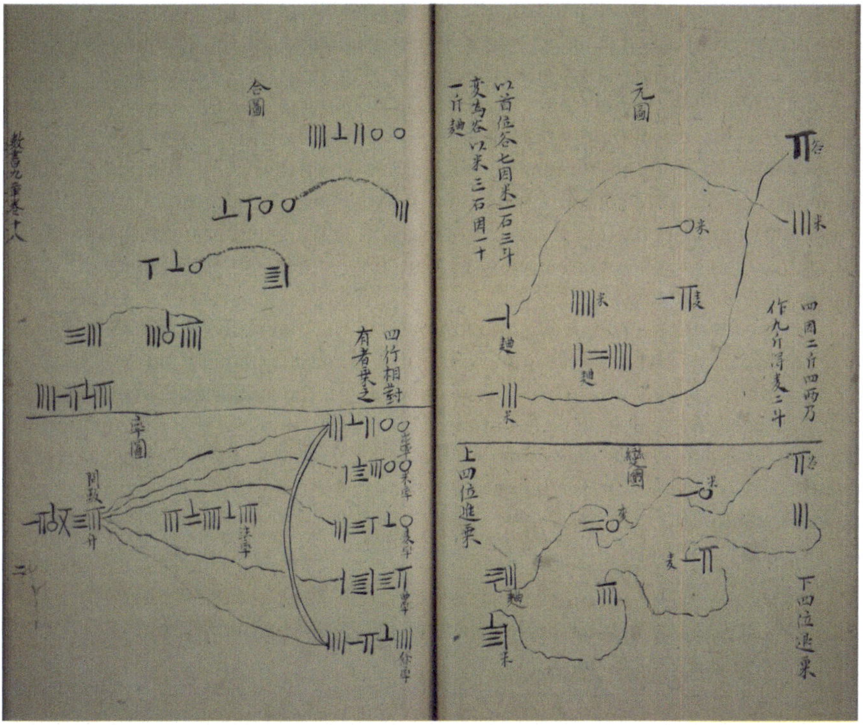

**Fig. 13.1** A typical counting-diagram for the rule of three in Zhao Qimei's manuscript of *Mathematical Book*. Courtesy of the National Library of China

[25] 'I set up procedures with detailed solutions, and sometimes elucidate them by counting-diagrams.' The original Chinese text: 立術具草，閒以圖發之。

**Table 13.1** Writing numeral system in Zhao Qimei's manuscript of *Mathematical Book*

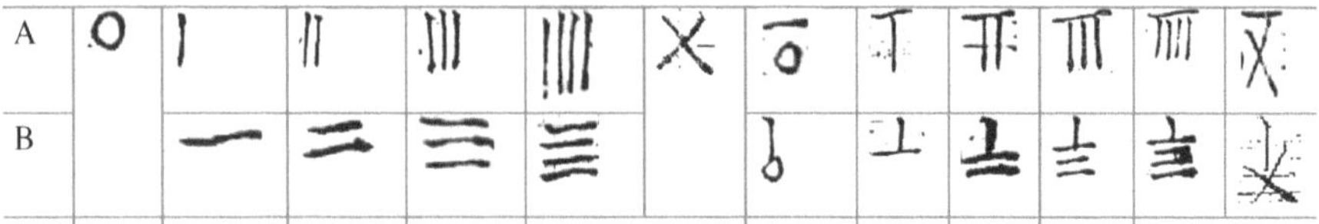

| | | | | | | | | | | | | |
|---|---|---|---|---|---|---|---|---|---|---|---|---|
| A | | | | | | | | | | | | |
| B | | | | | | | | | | | | |
| | 0 | 1 | 2 | 3 | 4 | | 5 | 6 | 7 | 8 | 9 | |

counting-diagrams are separate, while the remaining 22 problems have counting-diagrams presented within the text of the procedures or detailed solutions. This means the relationships are not uniform, and are more complex than Qin says. We have to examine them one by one in order to understand them.

Li Di (1991) primarily studied counting-diagrams and the writing system used in *Mathematical Book*. Karine Chemla (2001) studied the change of diagrams [*tu* 圖], especially geometrical diagrams, during the Song and Yuan periods. In the present chapter, we aim to show the divergences between different writing layers.

It is important to mention that, as a mathematical book in thirteenth century China, this book was written in Chinese characters except for the counting-diagrams (i.e., the procedures and detailed solutions are in Chinese, but the counting-diagrams include so-called rod-signs; see Table 13.1). Figure 13.1 contains the typical counting-diagrams copied from Zhao Qimei's manuscript of *Mathematical Book*. It is clear that the counting-diagram has three writing systems: (1) written numeral systems used to represent numbers (Table 13.1), which derived from counting rods; moreover, colors were sometimes used for positive and negative numbers; (2) lines connecting numbers used to represent operations (Zheng and Zhu 2010; Zhu 2017, 2020); (3) Chinese characters used to explain procedures and give names and measuring units to numbers.

The three writing systems are all employed in the diagrams. In the 45 problems that include such diagrams, the function of the diagrams depends on which of the three writing systems are used. It also depends on the relationship between diagrams, procedures and detailed solutions. This remark suggests that the functions of diagrams could also vary depending on the cases. Counting-diagrams are a characteristic part of *Mathematical Book* and tell us something that the procedures and detailed solutions do not reveal.

### *13.5.2 Three Representations: The Case of the Chinese Remainder Theorem*[26]

The case of the Chinese Remainder Theorem is discussed below to illustrate the use of counting-diagrams and the divergence between counting-diagrams and texts. As mentioned above, the procedure called the '*dayan zongshu shu*' (大衍總數術,

[26] For the case study of Qin's writing on the Chinese Remainder Theorem, see Zhu (2011, 2017).

literally the 'Great inference procedure for all numbers') occupies the most important position in *Mathematical Book* and brought fame to Qin Jiushao. In fact, there are ten problems in *Mathematical Book* relating to the Chinese Remainder Theorem.[27] A sub-procedure of *dayan zongshu shu*, called *dayan qiuyi shu* (大衍求一術, literally the 'Great inference procedure looking for one'), is at the heart of the Chinese Remainder Theorem. Qin Jiushao used different writing systems to represent this procedure. We will first begin by showing its various representations and analyze them for differences.

Representation A only uses textual description. Qin gives the procedure in Chinese characters in the answer to Problem 1[28] and a simplified description in the procedure for Problem 1.[29] These two text descriptions are slightly different, but the key point is that both are used for describing the procedure, which should actually be carried out with counting rods outside the text. In other words, thirteenth-century readers would resort to use of counting rods to understand the text.

Representation B uses counting-diagrams involved in detailed solutions. For instance, in the detailed solution of Problem 1, Qin presents, between the sentences of the text, the counting-diagrams representing the computations with counting rods referred to in the text (Fig. 13.2). By this method, counting-diagrams and text descriptions are written alternately. The whole diagram certainly refers to the same

---

[27] Nine problems in Chap. 1 and one problem called 'making calendars and computing eras' (*Zhili yanji* 治曆演紀) in Chap. 2.

[28] "Great inference procedure looking for One' reads: One places the remainder [*ji* 奇] on the upper [position] of the right [column]; one places the determined [divisor] [*ding* 定] on the lower [position] of the right [column]; one sets the Celestial Source [*tianyuan* 天元], one, on the upper [position] of the left [column]. One first divides the lower right number by the upper right number. The quotient obtained multiplies the upper left one and one places the result on the lower [position] of the left [column]. After that, one uses numbers on the upper and lower [positions] of the right [column], dividing the larger number by the smaller number, successively mutually dividing one by another. The quotients obtained immediately successively and mutually repeatedly multiply [the numbers on the left column inside], adding the results to the upper and lower [positions] of the left [column]. One must make the remaining upper right number one, and cease the procedure. Thus, one checks what has been obtained on the upper left. That is LÜ OF THE MULTIPLIER. Sometimes the remainder is already a single one [at the start], so it is LÜ OF THE MULTIPLIER [*chenglü* 乘率]. 大衍求一術云: 置奇右上，定居右下，立天元一於左上。先以右上除右下，所得商數與左上一相生，入左下。然後乃以右行上下，以少除多，遞互除之。所得商數隨即遞互累乘，歸左行上下。須使右上末後奇一，而止。乃驗左上所得，以爲乘率。或奇數已見單一者，便爲乘率。'(Zhao Qimei's manuscript 1616 [2010]: 103)

[29] "Great Inference Procedure to Find One' reads: One places the remainder [*ji* 奇] on the upper [position] of the right [column]; one places the determined divisor [*dingmu* 定母] on the lower [position] of the right [column]; one sets the Celestial Source [*tianyuan* 天元], one, on the upper [position] of the left [column]. One first uses two numbers on the upper and lower [positions] of the right [column], dividing the larger number by the smaller number. Thus, the obtained quotients successively mutually multiply numbers on the left [column] inside. One stops, when one has made the upper [position] on the right yield one. The number on the upper left is LÜ OF THE MULTIPLIER [*chenglü* 乘率]. 大衍求一術云: 以奇於右上，定母於右下，立天元一於左上。先以右行上下兩位以少除多，所得商數乃遞互乘內左行，使右上得一而止。左上爲乘率。'(Zhao Qimei's manuscript 1616 [2010]: 103)

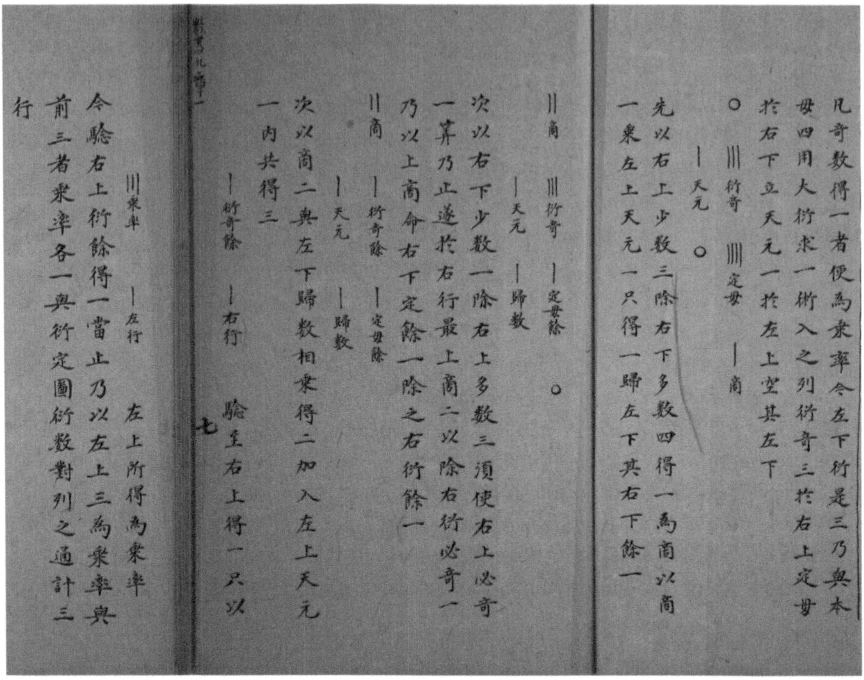

**Fig. 13.2** Counting-diagrams involved in detailed solutions in Zhao Qimei's manuscript of *Mathematical Book* (to be read from right to left). Courtesy of the National Library of China

procedure carried out with counting rods, as the one written only in Chinese characters (Case A). However, readers have no need to use counting rods as in Case A, because a part of the procedure is depicted by counting-diagrams.

Representation C uses independent counting-diagrams. This can be illustrated using the other representation of the procedure that Qin gives in the problem called 'making calendars and computing eras' in Chap. 2 (Fig. 13.3).[30] Here, Chinese characters are only used to associate names with numbers. Lines within counting-diagrams are used to indicate an operation applied to the numbers that they connect (e.g., addition, subtraction, multiplication, division). They thus yield a dynamic inscription of the procedure. Therefore, although the counting-diagrams refer to the same use of counting rods as in Cases A and B, readers do not need to be able to operate counting rods to understand the diagrams.

We can conclude that Qin Jiushao used different writing systems for writing down the so-called Chinese Remainder Theorem. The different writing systems comprise different representations of the same procedure with counting rods. In other words, the same procedure is described using different representations. This means that although there is only one procedure, different representations offer

[30] There are two cases in which independent counting-diagrams are used. As they are similar, we only show one here.

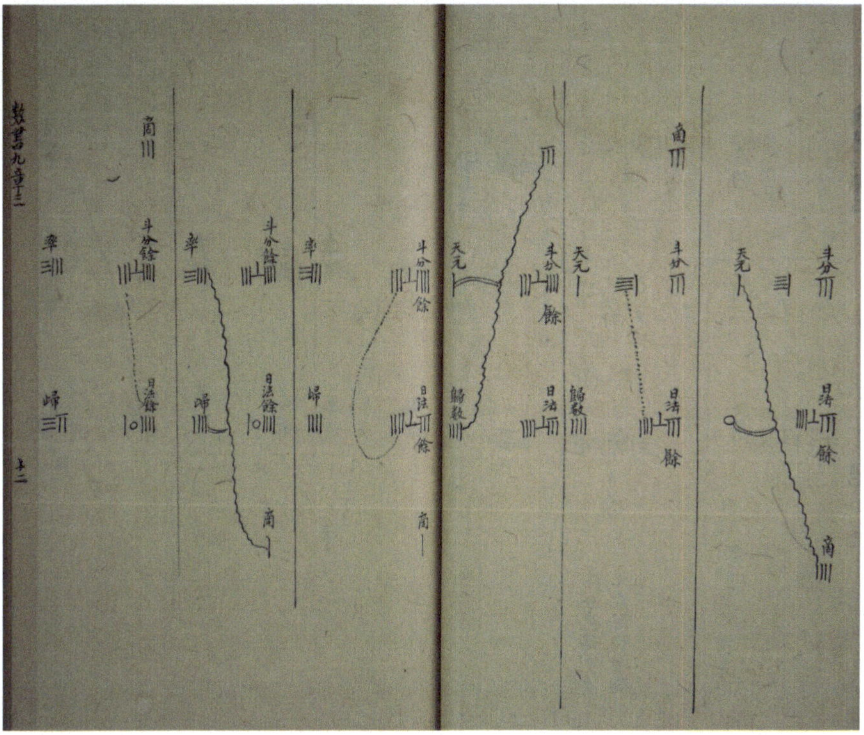

**Fig. 13.3** Independent counting-diagrams in Zhao Qimei's manuscript of *Mathematical Book*. Courtesy of the National Library of China

readers different ways to perceive the procedure.[31] In other words, if only text descriptions were given, readers' understanding in the thirteenth century would have relied on the operation of counting rods. When counting-diagrams were involved in describing detailed solutions, readers were less dependent on counting rods. When independent counting-diagrams were given, they naturally became a replacement for counting rods for readers.

### *13.5.3 Differences Between Zhao Qimei's Manuscript and the First Printed Edition of 1842*

The characteristic of Zhao Qimei's manuscript is that it preserves lines within counting-diagrams, which offers an opportunity to see a crucial original feature of *Mathematical Book*. At the same time, lines within counting-diagrams are missing

[31] We became aware of the issue of representations for mathematics through the work of Brown (2008, 93–97).

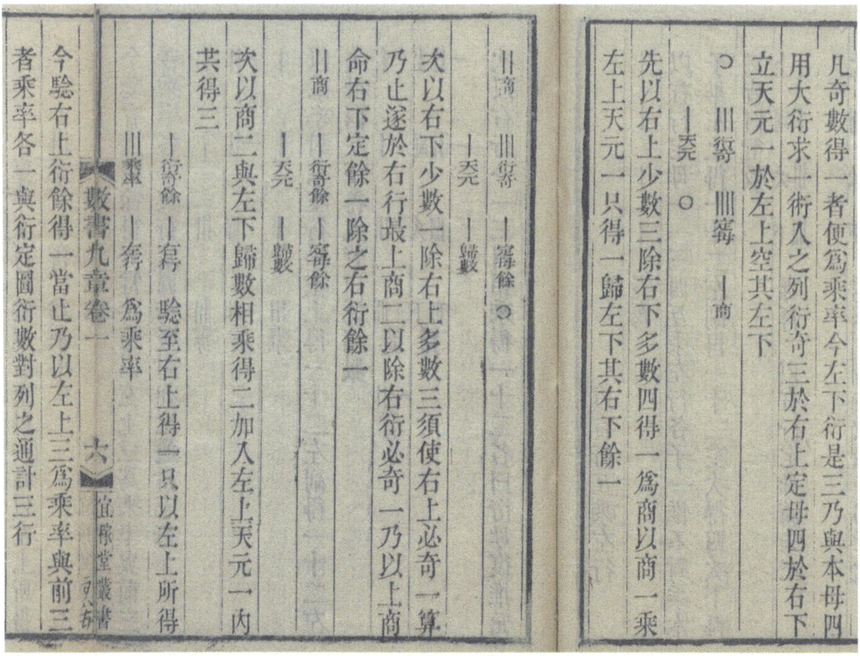

**Fig. 13.4** Counting-diagrams involved in detailed solutions in the 1842 edition corresponding to Fig. 13.2. Courtesy of the Institute for the History of Natural Sciences, Chinese Academy of Sciences

in the first printed edition (1842). Despite the fact that they seem to have been present in the Yongle edition (1565 [1986]), the omission of lines also happened in the *Siku* edition. Figures 13.4 and 13.5 show the parts in the 1842 edition that correspond to those we have seen in Zhao Qimei's manuscript in Figs. 13.2 and 13.3. Figure 13.4 is almost the same as Fig. 13.2, because lines are not involved in these diagrams. The absence of lines in Fig. 13.5 makes it look very different from Fig. 13.3.

As mentioned previously, Qin used different representations to write the same procedure. These representations actually show different aspects of a procedure. The independent counting-diagrams with lines tell us more about how to use counting rods than text descriptions. However, since lines were omitted in the first printed edition, counting-diagrams became static and lost the function of writing procedures. Thus, the differences between Zhao Qimei's manuscript and the first printed edition lie not only in the omission of lines in the counting-diagrams, but also in the inclusion of representations by independent counting-diagrams, which helped readers to understand the operations of counting rods in the same procedure.

Li and Feng (1998) suggested that there were two systems used for written calculations in the Qing dynasty. One was based on the Western system introduced in the seventeenth century. The other one was the new rod-signs system derived from the thirteenth-century mathematical writings. A key piece of evidence in favor of the

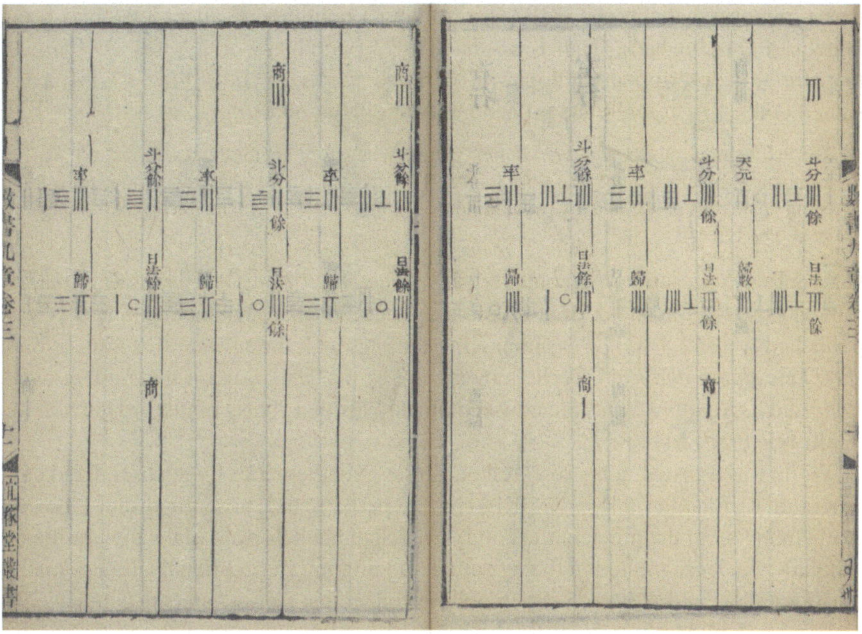

**Fig. 13.5** Independent counting-diagrams in the 1842 edition corresponding to Fig. 13.3. Courtesy of the Institute for the History of Natural Sciences, Chinese Academy of Sciences

above argument is found in the first printed edition of *Mathematical Book.* However, the omission of lines in the first printed edition results in the loss of key features of the representation used in *Mathematical Book* to refer to the operation of counting rods, and it makes the counting diagrams useless and irrelevant asa representation of procedures. In China after the fifteenth century, counting rods were gradually completely replaced by the abacus and other tools for calculation. Therefore, it was difficult for readers of the first printed edition of *Mathematical Book* to understand the operational procedure of counting rods. Instead, in the nineteenth century, *Mathematical Book* would be understood in terms of the new mathematical writing system.

## 13.6 The Reception of *Mathematical Book* in the Nineteenth Century

In 1847, a British Protestant missionary, Alexander Wylie (1815–1887), arrived in Shanghai. He supervised the London Missionary Society's press in the city and engaged in studying Chinese culture. In 1852, Wylie published a series of articles under the title *Jottings on the Science of the Chinese*. In one article, he translated a passage from *Mathematical Book* and pointed out that William George Horner's newly published theorem (1819) 'for solving Equations of all orders' had been known to Qin Jiushao ('Tsin Kew-chaou').[32] In another work, *Notes on Chinese Literature*, a descriptive catalogue of Chinese works, Wylie (1867: 93–94) connected Qin's 'Great inference procedure' to the 'Hindoo *Cuttaca*' as follows:

> The first section (of *Mathematical Book*) contains a new formula for the resolution of indeterminate problems, called 大衍 *Ta yen* (ZYW & ZC: i.e. 'Great inference procedure'), being analogous to the better known Hindoo process *Cuttaca*, which Colebrooke[33] translates 'Pulverizer'.

He went on to describe the contents of *Mathematical Book* briefly, noting that 'A critical examination and correction of the typographical and other errors in this was published in 1842, by 宋景昌 Sung King-ch'ang, with the title 數書九章札記 *Soo shoo kew chang cha ke*.'[34] Wylie's own copy of the 1842 edition of *Mathematical Book* and its *Reading Notes* are now kept in the Weston Library of Oxford University (Helliwell 1985: 51).

Wylie was assisted in his research on Chinese mathematics by his friend and collaborator Li Shanlan (李善蘭, 1810–1882), perhaps the most important Chinese mathematician in the nineteenth century.[35] Li Shanlan belonged to the generation of scholars who were trained in their youth in the mixture of traditional Chinese mathematics and the old Western mathematics introduced in the seventeenth century. The 1842 edition of *Mathematical Book* and the special method 'for solving Equations of all orders' were also new to him. It is quite likely that Li introduced *Mathematical Book* to Wylie.

Qin's reputation in mathematics rests mainly on two phenomenal achievements: the method for solving high-order equations, originally named 'Procedure for extracting roots with positive and negative [numbers]' (*zhengfu kaifang shu* 正負開方術), and the method for solving congruence equations, originally named 'Great inference procedure for all numbers' (*dayan zongshu shu* 大衍總數術). *Mathematical Book* interested Chinese and Western scholars in the nineteenth century not only because of its great achievements in mathematics but also because they could compare these two methods to analogous procedures in Western and Hindu mathematics. The first method was usually compared to Horner's or Newton's

[32] 'It appears some have thought proper to dispute the right of Horner to the invention, and it will perhaps be an unexpected occurrence to our European friends, to find a third competitor coming forward from the celestial empire, with a very fair chance of being able to establish his claim to priority.' (Wylie 1852, 1897: 184–185).

[33] Henry Thomas Colebrooke (1765–1837) was an English scholar who specialized in the mathematics in Sanskrit sources (note of the authors).

[34] Accent signs are omitted in the quotation.

[35] Li met Wylie in Shanghai in 1852. They cooperated on translating a series of mathematical and scientific works into Chinese, including the last nine books of *Euclid's Elements* (1857), William Herschel's *Outlines of Astronomy* (1859) and Elias Loomis' *Elements of Analytical Geometry and of the Differential and Integral Calculus* (1859). They also prepared an unfinished translation of Newton's *Mathematical Principles of Natural Philosophy*. See Han (1998).

method, while the second one was given the name Chinese Remainder Theorem, and compared to the Hindu *Cuttaca* and later to Gauss's method.[36]

Qin's mathematical practice was based on counting rods, and the counting-diagrams in *Mathematical Book* had guided thirteenth-century readers to follow the operation of counting rods representing the procedures. After the first printed edition of *Mathematical Book* was published in 1842 and became the main edition, the key representation of procedures using counting-diagrams and the operation of counting rods were lost. In a sense, the first printed edition misled readers and prevented them from understanding the mathematical procedures which Qin carried out with counting rods in *Mathematical Book*.

However, Chinese and Western scholars in the nineteenth century focused on interpreting ancient mathematics by means of new mathematics, and also compared Chinese mathematics to Western and Hindu mathematics. Although they did not fully understand Qin's methods of writing mathematics, they successfully applied the new representation for the procedures in *Mathematical Book*, and evaluated the merits of *Mathematical Book* by comparison with modern Western mathematics.

## 13.7 Conclusion

*Mathematical Book in Nine Chapters* was written in the context of a practice of mathematics in which practitioners used counting rods, and Qin Jiushao transcribed the counting-diagrams as the new mathematical representation of the operations with counting rods (Zhu 2020). However, from the fifteenth century onward, the abacus gradually replaced counting rods in China. Scholars slowly lost the knowledge of how to use counting rods. In the early years of the seventeenth century, with the compilation of *Mathematical Guidance in the Common Language* (*Tongwen suanzhi* 同文算指, 1613) by Matteo Ricci and Li Zhizao (李之藻, 1565–1630), Western written calculations were introduced into China (Chemla 1996; Chemla 1997; Martzloff 2006: 375–376; Zhu 2018). In the nineteenth century, most Chinese scholars understood ancient mathematics in the context of using the abacus and written calculation. Therefore, the transformation from counting rods to the abacus and written calculations changed the context of reading mathematical texts in which counting rods were the main mathematical instrument. These facts had a strong impact not only on the comprehension of the counting-diagrams in *Mathematical Book* but also on the compilation of its critical edition in 1842.

In the same way as in the nineteenth century, Western and Chinese scholars who studied the history of mathematics in China in the twentieth century focused on applying the new representation with modern signs to traditional mathematics.

[36] In 1874, a German scholar, Ludwig Matthiessen (1830–1906), proved that the 'Great inference procedure for all numbers' is equivalent to Gauss's method. See Libbrecht (1973: 310–327).

Generally speaking, the reconstruction of ancient mathematical practices had not received enough emphasis. Although there are a few lines in the fragments of *Yongle Dadian* (1565 [1986]) and the *Siku* edition (1781 [1986]), and although Zhao Qimei's manuscript has been preserved in the National Library of China for half a century, the dominance of the first printed edition of *Mathematical Book* encouraged the neglect of counting-diagrams. In other words, when scholars created the modern representation for Qin Jiushao's *Mathematical Book*, they were shaping a new historical image, which has a strong connection with the first printed edition.

# References

## *Primary Sources*

Qin Jiushao 秦九韶. 1247. *Mathematical Book in Nine Chapters* (1 & 3 *Shuxue Jiuzhang* 數學九章; 2 & 4 *Shushu Jiuzhang* 數書九章).

## *Main Editions*

1. ca. 1565 [1986] Three problems in *Yongle Dadian* 永樂大典 (*great compendium of the Yongle Era*), Sect. 16 343. Ming manuscript copy of *great compendium* [1405–1408]. Facsimile edition. Beijing: Zhonghua Book Company.
2. 1616 [2010]. Zhao Qimei's manuscript 趙琦美抄本. Facsimile edition. *Siku tiyao zhulu congshu. Zibu* 四庫提要著録叢書.子部(*Selected Facsimile Edition of the Books recorded in the Comprehensive Catalogue of Complete Works in the Four Treasuries. Part 3*) Vol. XX. Beijing: Beijing Publishing House.
3. 1781 [1986]. *Siku* edition 四庫全書本, manuscript copy. 1781. Facsimile edition. *Yingyin Wenyuange Siku Quanshu* 景印文淵閣四庫全書 (*Facsimile Edition of Complete Works in the Four Treasuries in Wenyuange*). Vol. DCCXCVII. Taipei: Taiwan Commercial Press.
4. 1842. Yijiatang edition (the first printed edition) 宜稼堂本. ed. Song Jingchang 宋景昌. In Yu Songnian 郁松年 ed. (1840–1843).

JI Yun 紀昀 ed. 1795 [2019]. *Siku Quanshu Zongmu* 四庫全書總目 (Comprehensive catalogue of *complete works in the four treasuries*). Sect. 107. Facsimile edition. Vol. XXIX. Beijing: National Library of China Publishing House.

Schall von Bell, J.A., ca. 1635. *Gujin Jiaoshi Kao* 古今交食考 (*On ancient and modern eclipses*). In *Chongzhen Lishu* 崇禎曆書 Vol. XXIV. Kyujanggak Institute for Korean Studies, Seoul National University.; woodblock edition.

Song Jingchang 宋景昌. 1842. Shushu Jiuzhang Zhaji 數書九章札記 (*Reading notes of* mathematical book in nine chapters). In *Yijiatang Congshu,* ed. Yu Songnian. Shanghai: Yu Songnian 郁松年; woodblock edition.

Wylie, A. 1852. Jottings on the science of the Chinese: Arithmetic. The North China Herald, November 6.

———. 1867. *Notes on Chinese literature*. Shanghai: American Presbyterian Mission Press.

———. 1897. *Chinese researches*. Shanghai.

Yu Songnian 郁松年 ed. 1840–1843. *Yijiatang Congshu* 宜稼堂叢書 (*Yijiatang Collectanea*). Woodblock edition. Shanghai: Yijiatang.

Yu Songnian 郁松年. 1842. Xu Houhanshu Zhaji 續後漢書札記 (*Reading Notes for a Sequel to the History of Later Han Dynasty*). In Yu Songnian 郁松年 ed. (1840–1843).

Zhongguo Diyi Lishi Dang'anguan 中國第一歷史檔案館. 1997. *Zhuanxiu Sikuquanshu Dang'an* 纂修四庫全書檔案 *(Archives of the project for compiling complete works in the four treasuries)*. Shanghai: Shanghai Guji Press.

## *Secondary Literature*

Brown, J.R. 2008. *Philosophy of mathematics*. New York: Routledge.

Chemla, K. 1996. Que signifie l'expression 'mathématiques européennes' vue de Chine? In *L'Europe mathématique. Histoires, Mythes, Identités. Mathematical Europe. History, Myth, Identity*, eds. C. Goldstein, J. Gray, and J. Ritter, 220–245. Paris: Editions de la Maison des Sciences de l'Homme.

———. 1997. Reflections on the world-wide history of the rule of false double position, or how a loop was closed. *Centaurus* 39 (2): 97–120.

———. 2001. Variété des modes d'utilisation des *tu* dans les textes mathématiques des Song et des Yuan. Presentation for the conference: From Image to Action: The Function of Tu-Representations in East Asian Intellectual Culture. Paris. https://halshs.archives-ouvertes.fr/halshs-00000103/document. Accessed 5 July 2017.

Elman, B.A. 1984. *From philosophy to philology: Intellectual and social aspects of change in late Imperial China*. Cambridge, MA: Council on East Asian Studies, Harvard University.

Guo, Shuchun 郭書春, ed. 2010. *Zhongguo Kexue Jishu shi: Shuxue Juan* 中國科學技術史·數學卷 (*A history of science and Technology in China: Mathematics*). Beijing: Science Press.

Guy, R.K. 1987. *The Emperor's four treasuries: Scholars and the state in the late Ch'ien-lung Era*. Cambridge, MA: Harvard University Press.

Han, Qi 韓琦. 1998. Chuanjiaoshi Weilieyali zaihua de kexue huodong 傳教士偉烈亞力在華的科學活動 (Alexander Wylie's scientific activities in China), *Ziran bianzhengfa tongxun* 自然辯證法通訊 (*Journal of Dialectics of Nature*) 20(2): 57–70.

Helliwell, D. 1985. *A catalogue of the old Chinese books in the Bodleian library. Vol. II. Alexander Wylie's Books*. Oxford: The Bodleian Library.

HO, Peng-Yoke. 1971. Ch'in Chiu-shao. In *Dictionary of scientific biography*, ed. C.C. Gillispie, vol. III, 249–256. New York: Charles Scribner's Sons.

Hummel, A.W. 1944. *Eminent Chinese of the Ch'ing period (1644–1912)*. Washington: U.S. Government Printing Office.

Li, Di 李迪 1987a. Qin Jiushao zhuan lue 秦九韶傳略 (A brief biography of Qin Jiushao). In *Qin Jiushao yu Shushu Jiuzhang*, ed. Wu Wenjun, 25–42. Beijing: Beijing Normal University Press.

Li, Di 李迪 1987b. *Shushu Jiuzhang* liuchuan kao 數書九章流傳考 (A textual history of *mathematical book in nine chapters*). In *Qin Jiushao yu Shushu Jiuzhang*, ed. Wu Wenjun, 43–58. Beijing: Beijing Normal University Press.

Li, Di 李迪 1991. Song Yuan shiqi shuxuexingshi de zhuanbian 宋元時期數學形式的轉變 (Changes of mathematical forms in Song and Yuan periods). In *Zhongguo Kexuejishushi Lunwenji* 中國科學技術史論文集 (*Collected papers on the history of Chinese science and technology*), ed. Li Di, 219–233. Hohhot: Inner Mongolian Education Press.

Li, Di 李迪 and Lisheng Feng 馮立昇. 1998. Qingdai shuxuejia shiyong bisuan luelun 清代數學家使用筆算略論 (*A study on mathematicians' calculations with pen and paper in the Qing dynasty*). *Xibei daxue xuebao* (*Ziran kexue ban*) 西北大學學報 (自然科學版) (*Journal of Northwest University, Natural Science Edition*) 28(6): 461–466.

Libbrecht, U. 1973. *Chinese mathematics in the thirteenth century: The Shu-shu chiu-chang of Ch'in Chiu-shao*. Cambridge, MA: MIT Press.

Martzloff, J. 2006. *A history of Chinese mathematics*. New York: Springer Verlag.

Matsuura, A. 松浦章. 2012. *Qingdai Shanghai Shachuan Hangyunyeshi Yanjiu* 清代上海沙船航運業史研究 (*A study on the history of Shanghai junk freight business in Qing Era*), Trans. Yang Lei 杨蕾 et al. Nanjing: Jiangsu People's Publishing House.
Qian, Baocong 錢寶琮. 1966. Qin Jiushao *Shushu Jiuzhang* yanjiu 秦九韶數書九章研究 (A study of Qin Jiushao's *mathematical book in nine chapters*). In *Song Yuan Shuxueshi Lunwenji* 宋元數學史論文集 (*Collected papers on the history of mathematics in Song and yuan dynasties*), ed. Qian Baocong, 60–103. Beijing: Science Press.
Shen, Kangshen 沈康身. 1987. Yijiatang ben *Shushu Jiuzhang* zhengwu 宜稼堂本數書九章正誤 (Some corrections of *mathematical book in nine chapters* [*Yijiatang* edition]). In Qin Jiushao yu Shushu Jiuzhang, ed. Wu Wenjun, 59–72. Beijing: Beijing Normal University Press.
Shi, Yunli 石雲里 and Bing Song 宋兵. 2006. Wang Yinglin yu *Jingtiangai* guanxi de xinxiansuo 王應遴與《經天該》關係的新線索 (New evidence concerning Wang Yinglin's possible authorship of the *Jingtian Gai*), *Zongguo kejishi zazhi* 中國科技史雜志 (*The Chinese Journal for the History of Science and Technology*) 23 (3): 189–196.
Wang, Shaoying 汪紹楹. 1963. *Ruanshi Chongke Songben Shisanjing Zhushu Kao* 阮氏重刻宋本十三經注疏考 (On Ruan Yuan's new edition of the thirteen classics). *Wenshi* 文史 3: 25–60.
Wang, Shouyi 王守義. 1992. *Shushu Jiuzhang Xinshi* 數書九章新釋 (*A new annotated* mathematical book in nine chapters). Hefei: Anhui Science & Technology Publishing House.
Wu, Wenjun 吴文俊, ed. 1987. *Qin Jiushao yu Shushu Jiuzhang* 秦九韶與數書九章 (*Qin Jiushao and his* mathematical book in nine chapters). Beijing: Beijing Normal University Press.
Yan, Dunjie 嚴敦傑. 1987. Qin Jiushao nianpu chugao 秦九韶年譜初稿 (First draft of Qin Jiushao's chronicle biography). In Qin Jiushao yu Shushu Jiuzhang, ed. Wu Wenjun, 12–24. Beijing: Beijing Normal University Press.
Zheng, Cheng 鄭誠 and Yiwen Zhu 朱一文. 2010. *Shushu Jiuzhang* liuchuan xin kao: Zhao Qimei chaoben chutan 數書九章流傳新考：趙琦美鈔本初探 (A new study on the textual transmission of *Shushu Jiuzhang*), *Ziran kexueshi yanjiu* 自然科學史研究 (*Studies in the History of Natural Sciences*) 29 (3): 319–328.
Zhu, Yiwen 朱一文. 2011. Qin Jiushao lijia suiyong yongerbuzhi jie 秦九韶曆家雖用用而不知解 (A history of *Dayan* ('Great inference procedure'): From *Fangcheng* to *Dayan* ('Great inference procedure looking for one'), *Ziran kexueshi yanjiu* 自然科學史研究 (*Studies in the History of Natural Sciences*) 30 (2): 193–206.
———. 2017. Qin Jiushao dui dayanshu de suantu biaoda: jiyu *Shushu Jiuzhang* Zhao Qimei chaoben (1616) de fenxi 秦九韶對大衍術的筭圖表達：基於《數書九章》趙琦美鈔本 (1616) 的分析 (Qin Jiushao's writing of the *Dayan* algorithm 'great inference procedure': An analysis based on Zhao Qimei's manuscript of *mathematical book in nine chapters* (1616)), *Ziran kexueshi yanjiu* 自然科學史研究 (*Studies in the History of Natural Sciences*) 36 (2): 244–257.
———. 2018. How were Western written calculations introduced into China? — An analysis of the *Tongwen suanzhi* (*Arithmetic Guidance in the Common Language 1613*), *Centaurus* 60 (1–2): 69–86.
———. 2020. On Qin Jiushao's writing system. *Archive for History of Exact Sciences* 74 (4): 345–379.
———. 2023. *Suanxue yu jing xue: Zhongguo shuxue xinshi* 算學與經學：中國數學新史 *(Mathematics and the Study of Confucian Classics: A New History of Chinese Mathematics).* Beijing: The Commercial Press.

# Chapter 14
# Babylonian Astronomy: Editing and Interpreting an Ancient Science

**Mathieu Ossendrijver**

**Abstract** The year 1881 marks the birth of Babylonian astronomy as a modern area of scholarship with the first publication about this topic. From the outset, its scholarly practices were based on a variety of methods rooted in Assyriology, astronomy, mathematics and the history of science. In this chapter, the editing practices that were applied to Babylonian astronomical texts and their relationship to research methods in the abovementioned disciplines are traced through the works of the main scholars who were active in the field of Babylonian astronomy between 1881 and 1955. In this period, two phases can be distinguished: a pioneering phase (1881–1935) in which mainly individual tablets were edited while editing practices remained fluid, and a subsequent phase characterized by the appearance of standard editions of complete and coherent corpora of texts. In both phases, the translations and interpretations were strongly shaped by modern mathematics and astronomy.

For the benefit of the reader, this contribution begins with a very short introduction to Babylonian astral science, its main genres, its chronology and some important concepts that will be mentioned repeatedly. Between 750 BCE and 250 BCE, a host of new genres of astral science emerged in Babylonia. They include astronomical texts, which are mainly concerned with the observation and prediction of celestial phenomena, and astrological texts, which primarily deal with the terrestrial consequences of these phenomena. The largest astronomical corpus, comprising over one

The research leading to these results has received FUNDING from the European Research Council under the European Union's Seventh Framework Programme (FP7/2007-2013) / ERC Grant Agreement No. 269804. This chapter was partly written with support from the project "ZODIAC – Ancient Astral Science in Transformation" which is funded by the European Research Council (ERC) under the Horizon 2020 framework (Grant Agreement No. 885478).

M. Ossendrijver (✉)
Institut für Wissensgeschichte des Altertums, Freie Universität Berlin, Berlin, Germany
e-mail: mathieu.ossendrijver@fu-berlin.de



A. Keller, K. Chemla (eds.), *Shaping the Sciences of the Ancient and Medieval World*, Archimedes 69, https://doi.org/10.1007/978-3-031-49617-2_14

thousand tablets, consists of astronomical diaries and related tablets, all written between 650 BCE and 90 CE. Diaries, the most common type, are six-monthly observational reports of astronomical, meteorological, economic and historical data.[1] Also part of this corpus are goal-year texts, almanacs and normal star almanacs, which are compilations of data extracted from diaries for the purpose of prediction.[2] The underlying predictive method is nowadays referred to as the goal-year method.[3] In the fifth century BCE, Babylonians constructed the zodiac by dividing the path of the sun into twelve sections of thirty degrees. This coordinate system was subsequently used for predicting lunar and planetary phenomena in the corpus of mathematical astronomy, which comprises about 450 tablets written between 400 and 50 BCE. Most of these tablets are computed tables, the remainder being procedure texts with instructions on how to compute the tables. The zodiac also underlies various forms of astrology including Babylonian horoscopes, which contain computed planetary positions for the date of birth. For a proper understanding of editing practices in Babylonian astronomy, some knowledge of the sexagesimal place-value notation is also required. In this notation, numbers are represented as sequences of digits, each having a value between zero and fifty-nine. Every digit is associated with a power of sixty that decreases by one in the rightward direction. A special sign is used for indicating vanishing intermediate digits (0). As opposed to our decimal system, which is an absolute notation, the Babylonian notation is relative or floating, because vanishing initials were rarely written and because there is no equivalent of our decimal point. Hence the power of sixty corresponding to each digit can only be inferred from the context.

By limiting the scope of this investigation to astronomical texts, other tablets with astral science edited during or before the same period are disregarded.[4] This can be justified by the distinct nature of Babylonian astronomy as a scholarly discipline, a consequence of the high mathematical and astronomical complexity of these texts. Since they abound with technical expressions and mathematical computations, they can hardly be translated and interpreted using purely Assyriological methods. This is particularly true for the tabular texts of mathematical astronomy, which contain little more than sequences of sexagesimal numbers. In fact, these tablets can, to a large extent, be translated and analyzed without any knowledge of the Akkadian language. Not surprisingly, Babylonian astronomy became the domain

[1] For a complete edition of the astronomical diaries, see Sachs and Hunger (1988–1996, Vols. I–III), Hunger (2022, Vol. IV). For related astronomical reports, see Hunger and Sachs (2001); for goal-year texts, see Hunger (2006); for almanacs and normal star almanacs, see Hunger (2014).

[2] For the goal-year method see Brack-Bernsen and Hunger (2002); Steele (2011).

[3] This term was introduced by Abraham Sachs, whose classification of the astronomical diaries and related texts (Sachs 1948) is still in use today.

[4] This particularly concerns editions of tablets with celestial divination (omens), astrology and star lists produced by the British Assyriologist Archibald Sayce (1845–1933), the German orientalists Carl Bezold (1859–1922) and Ernst Weidner (1891–1976), and the French orientalists Jules Oppert (1825–1905), François Thureau-Dangin (1872–1944) and Charles Virolleaud (1879–1968).

of scholars who were relative outsiders to Oriental studies, being primarily trained in modern mathematics and astronomy. Their distinct educational and professional background shaped the editions and interpretations of the astronomical texts. For reasons of space, the investigation is also limited to the most important scholars, which leaves out several scholars who occasionally worked on this topic.[5] Furthermore, this chapter is not concerned with Babylonian astronomy itself, the history of its reconstruction or the biographies of the involved scholars. For the latter two aspects, the reader is referred to de Jong (2016), who covers the same time period and, by and large, the same scholars. A brief outlook on developments after 1955 concludes this chapter.

Editing practices are traced here through a selection of exemplary astronomical tablets, labeled T1–T4. Most of them were edited multiple times, usually because a new fragment was found or progress was made with the astronomical interpretation. All four are kept in the British Museum and belong to what is nowadays referred to as the Babylon collection. The approximately 20,000 tablets of this collection were excavated unscientifically between 1876 and the 1880s and acquired through explorers and antiquities dealers. Only decades later, it became clear that the astronomical and many other tablets in this collection largely originated from Babylon. In order to aid the reader, some guidance to the confusing nomenclature of the tablets seems warranted.[6] Originally, each group of tablets within the Babylon collection was identified by a date (e.g. 81–7–6 = 7 June 1881) or the name of the individual through whom it was acquired (e.g. Rm = Rassam, SH = Shemtob, Sp = Spartoli). For some groups of tablets, both types of labels were in use (e.g. 76–11–17 and S+, or 81–7–6 and SH).[7] Individual tablets within each group are identified by a running number (e.g. 81–7–6, 277). At a later stage, a single running British Museum (BM) number was assigned to each fragment, which is how they are usually referred to in modern publications. Text T1 is a tabular text from the corpus of mathematical astronomy. In modern terms it is classified as a synodic table of lunar System B.[8] The editions of its fragments cover the entire history of research on Babylonian astronomy. Text T2 is a normal star almanac, a predictive compilation of data drawn from astronomical diaries. Text T3 is a procedure text for lunar System A. Text T4 is another procedure text concerned with Jupiter System A'. The following table lists all the major publications on these tablets, including the latest editions.

---

[5]The Dutch astronomer Anton Pannekoek (1873–1960), the German astronomer Paul Viktor Neugebauer (1878–1940), the German orientalists Paul Schnabel (1887–1947) and Ernst Weidner (1891–1976), and the Dutch mathematician Bartel van der Waerden (1903–1996).

[6]Leichty (1986, xii–xxxvi)

[7]The significance of the abbreviation S+ appears to be unknown. Perhaps it refers to the Assyriologist George Smith, who acquired the 76–11–17 collection in Baghdad (Leichty 1986: xiv).

[8]For the lunar and planetary 'systems' (families of algorithms) of Babylonian mathematical astronomy, see Neugebauer (1955) and Ossendrijver (2012).

| | Publication | (Partial) edition of |
|---|---|---|
| T1 | Epping and Strassmaier (1881: 283–286) | 81–7–6, 277 |
| | Epping (1889: 8–16, Appendix) | 81–7–6, 277 ('A') and 81–7–6, 272 ('C') |
| | Epping (1890: 225–240, Appendix) | 81–7–6, 277 ('A') |
| | Kugler (1900: 9–16) | Sp 2, 52 + 75[9] + 81–7–6, 272 + 277 + 331 + 333 + 386 + 589 ('no. 272') |
| | Schaumberger (1935: 375–394, Plates XV–XVII) | BM 34580 + BM 42690 = Sp 2, 52 + 75 + 81–7–6, 272 + 277 + 331 + 333 + 386 + 589 ('BM 34580') |
| | Neugebauer (1936) | review of Schaumberger (1935) |
| | Pinches and Sachs (1955: No. 66) | Copy of BM 34580 + BM 42690 |
| | Neugebauer (1955: No. 122) | BM 34580 + BM 42690 |
| | Steele (2010: Text G) | BM 34580 + BM 42690 + BM 42869 + BM 42902 + BM 43000 + BM 43030 |
| T2 | Epping and Strassmaier (1881: 287–289) | Discussion of Sp 129 |
| | Epping (1889: 152–159, Plates 1–3) | Sp 129, including copy |
| | Hunger (2014: No. 87) | BM 34033 (Sp 129) |
| T3 | Kugler (1900: 59, 66–67, 72–73, 76, 78–79, 116, 142–147, 160–165, 169–173, 180–181, 191–192) | BM 32651 ('S+ 2418') |
| | Neugebauer (1955: Nos. 200 and 200aa) | BM 32651 and BM 32172 |
| | Ossendrijver (2012: No. 53) | BM 32167 + BM 32172 + BM 32451 + BM 32651 + BM 32663 + BM 32744 + BM 32752 |
| T4 | Kugler (1907: 136–140, Plate XVI) | BM 33869 (Rm 4, 431) |
| | Neugebauer (1955: No. 810) | BM 33869 |
| | Ossendrijver (2012: No. 32) | BM 33869 |

Until 1881, little more was known about Babylonian astronomy than what was compiled from scattered passages about Chaldeans in the works of Greek and Roman scholars.[10] However, they reveal almost nothing substantial about Babylonian methods and practices. The appearance in 1881 of a pioneering paper by Epping and

[9] Erroneously referred to as Sp I 75 by Kugler.

[10] For references to Chaldeans in Greek and Latin sources see Rochberg (1988: 2–5).

Strassmaier about newly discovered Babylonian astronomical tablets kept in the British Museum marks the beginning of modern research on Babylonian astronomy. The history of this field may be divided into three phases.[11] The first phase (1881–1935) began with the pioneering works of Epping and Strassmaier, culminating in the monumental volumes of Franz Xaver Kugler. Throughout this phase, individual tablets were edited, often only partially, while editing practices remained fluid. The main interest of these scholars was to reconstruct the Babylonian astronomical methods. The following phase (1935–1986) is characterized by the appearance of standard editions of complete and coherent corpora of texts. Otto Neugebauer was the dominant scholar, with significant contributions by Abraham Sachs, Bartel van der Waerden and several other scholars. The translations and interpretations were strongly shaped by modern mathematics and astronomy. The beginning of a third phase characterized by a more holistic, contextualizing approach to Babylonian astral science can be dated to about 1986.

## 14.1 Epping and Strassmaier

Babylonian astronomy was rediscovered in 1880 by the orientalist Johann Nepomuk Strassmaier (1846–1920) and the astronomer Joseph Epping (1835–1894), German priests of the Jesuit order (*Societas Jesu*).[12] Strassmaier knew Epping from the time when he studied theology and philology at the Jesuit *Collegium Magnum* in Maria Laach, where Epping taught mathematics and physics. Like many Jesuits, Strassmaier left Germany in 1872 when the order was banned by Bismarck. From 1878 onwards, he spent much of his time in the British Museum, studying cuneiform tablets as they were arriving.[13] In 1880, while visiting the philosophical seminary of the German Jesuits in castle Bleijenbeek (the Netherlands), Strassmaier mentioned the existence of astronomical tablets to Epping, suggesting a collaborative project in order to decipher them. In 1881, this resulted in a publication titled *Zur Entzifferung der astronomischen Tafeln der Chaldäer* (On the Decipherment of the Astronomical Tablets of the Chaldeans),[14] the first of several papers on this topic to appear in *Stimmen aus Maria-Laach* (Voices from Maria Laach). This scholarly journal of the German Jesuits was founded essentially as a platform to counteract the increasing attempts by liberals, nationalists and socialists to reduce the role of

[11] This chronology is inspired by a similar one developed by Høyrup (1996) for Babylonian mathematics. For a brief characterization, see also Ossendrijver (2012: 3–4) and Rochberg (2014).

[12] For a detailed account including biographical information, see de Jong (2016).

[13] At about the same time, the German orientalist Max Müller (1823–1900) was studying Sanskrit and other manuscripts in the British Library, then located in the British Museum. It is unclear whether Strassmaier or any of the other German pioneers of Babylonian astronomy were in contact with him (see also the contribution by A. Keller in this volume).

[14] In the contents of Vol. XXI of *Stimmen aus Maria-Laach*, only Epping appears as the author of this paper.

Catholicism in German public and intellectual life.[15] However, not all contributions have an apologetic tone and they cover a wide range of topics including ancient and modern history, church history, literature, theology, philosophy, education, contemporary social, political and ethical issues and, more rarely, the natural sciences. History of astronomy was an exceptional topic in the journal,[16] but astronomy was a major focus of Jesuit scholarship during the nineteenth century and early twentieth century. Jesuits founded numerous observatories across the globe in order to promote astronomy and, by implication, a new societal role for the Catholic church as a sponsor of modern science and education (Udías 2003, 2015). Epping was actively involved in these endeavors, having taught mathematics and astronomy along with other Jesuits at the newly founded polytechnic school and observatory in Quito, Ecuador, from 1872 to 1876 (de Jong 2016; Udías 2015: 149–150, 218). At the same time, Assyriology was gaining prominence as a matter of public and scholarly debate in Germany (see below). Against this background, Jesuits could justifiably consider Babylonian astronomy to be an appropriate and promising area of scholarship. However, the writings of Strassmaier and Epping reveal almost nothing about such considerations. For a recent discussion of their 1881 paper and a selection of translated passages the reader is referred to de Jong (2016: 269–272). Here it suffice to summarize some important features of their approach to the problem of editing and interpreting the newly discovered Babylonian astronomical tablets. The paper consists of a five-page historical introduction in which Strassmaier reports on the current state of knowledge of 'Chaldean astronomy'—the term commonly used in this era for Mesopotamian astral science. For him, these texts are mainly of interest because he believes that they will provide a solid basis for understanding the Babylonian calendar and other aspects of ancient chronology. In that view, the recently found Assyrian tablets from Kuyuncik (ancient Nineveh) published by the French-German orientalist Jules Oppert (1825–1905) and the British Assyriologist Archibald Sayce (1845–1933) add only little to the sparse information drawn from classical sources, since they mainly deal with celestial omens ('portents')[17]:

> The largest part of the texts which the mentioned scholars [Oppert and Sayce] used for their works are so-called tablets of portents, which on many occasions give the astrological significance of constellations, usually in a rather incomprehensible manner and, as far as we can judge, without any interest (Epping and Strassmaier 1881: 279).

[15] Founded in 1865, the *Stimmen aus Maria-Laach. Katholische Blätter* (Voices from Maria Laach. Catholic leaflets), was published by the Catholic publishing house Herder in Freiburg im Breisgau until 1872. Between 1872 and 1914, it was printed in Belgium and the Netherlands; since 1914, again by Herder under the name *Stimmen der Zeit* (Voices of Time).

[16] The only other contribution about the history of astronomy published in the *Stimmen aus Maria-Laach* before Epping and Strassmaier's pioneering paper was by the Jesuit and church historian Gerhard Schneemann (1829–1885), editor of the journal, who published a long serialized paper about Galileo Galilei (Schneemann 1878).

[17] 'Der größte Theil der Texte, den die obengenannten Gelehrten für ihre Arbeiten benützten, sind die sogen. Portenttafeln, welche an vielen Stellen die astrologische Bedeutung von Konstellationen geben, meistens ziemlich unverständlich und, soweit wir urtheilen können, ohne alles Interesse.' (Epping and Strassmaier 1881, 279). Here Strassmaier certainly has in mind Sayce (1874).

Apart from mentioning classical sources about Chaldeans, neither Strassmaier nor Epping makes an effort to position their research in the wider history of science. When Strassmaier turns to the astronomical tablets from Babylonia,[18] he stresses how difficult it is to read these often fragmentary and cursively written pieces of unburnt clay, and how important it is to copy them, before they eventually disintegrate (Epping and Strassmaier 1881, 280–281; de Jong 2016, 269). The remaining ten pages, entitled 'Astronomical Revelations' [*Astronomische Enthüllungen*], were written by Epping. After humbly conveying his initial doubts about his own competence, claiming, for instance,[19]

> I did not believe I was a computer sufficiently skilled to solve an equation with so many unknowns and so few knowns…,

Epping reports that he was thrilled to contribute to the project by using his astronomical knowledge. It is interesting to note that Epping here uses the mathematical metaphor of solving an equation with many unknowns in order to characterize the problem of deciphering and interpreting the astronomical tablets. He was also the first scholar to use modern equations for analyzing Babylonian astronomical tables (see below). Epping continues with a classification of the tablets into a computational, an observational and an intermediate category[20]:

> Among the tablets that were available to me, a twofold kind could immediately be distinguished: computational and observational tablets; a third kind probably constituted the key to both. If the latter were deciphered one would, of course, have a great tool for clarifying the remaining ones; but that was a matter of the learned gentlemen Assyriologists. However, as long as they were unable to dispel the darkness, the other road had to be taken on the astronomical side; first the other tablets were to be understood and then it might be possible to draw conclusions for the explanatory tablets. Easier things first is a recognized practice; those tablets that contained almost only numbers had to constitute the first object of attack. In some of them one could discover a constant difference between the subsequent numbers;

[18] Strassmaier believed that the astronomical tablets originate mainly from Birs Nimrud, later equated to ancient Borsippa, and Abu Habba, already then equated to ancient Sippar. Babylon, nowadays considered to be the site where most of these tablets were excavated, is not mentioned at all by Strassmaier.

[19] 'Ich glaubte doch nicht ein solcher Rechenkünstler zu sein, daß ich eine Gleichung lösen könnte, die der unbekannten Größen so viele und er bekannten so wenig hatte.' (Epping and Strassmaier 1881, 282–283; de Jong 2016, 270).

[20] 'Bei den Tafeln nun, die mir vorlagen, ließ sich sofort eine doppelte Art unterscheiden: Rechnungstafeln und Beobachtugstafeln; eine dritte Art bildetet wahrscheinlich den Schlüssel für beide. Wäre diese letztere entziffert, so hätte man natürlich ein großes Hilfsmittel für die Aufklärung der übrigen gehabt; doch das war ja Sache der gelehrten Herren Assyriologen. So lange aber diese das Dunkel nicht zu heben vermochten, mußte astronomischerseits der entgegengesetzte Weg eingeschlagen werden; erst hieß es die andern Tafeln verstehen, und dann ließen sich vielleicht Schlüsse ziehen auf die Erklärungstafeln. Das Leichtere zuerst, ist eine anerkannte Praktik; diejenigen Tafeln, welche fast lauter Zahlen enthielten, mußten das erste Angriffsobject bilden. Bei manchen von diesen konnte man eine constante Differenz in den aufeinanderfolgenden Zahlen entdecken; sie bilden also eine sogenannte arithmetische Progression, analog der gewöhnlichen Zahlenreihe; darauf wird dann diese so gewonnene Reihe dazu benützt, um aus ihr eine andere abzuleiten und aus letzterer wieder eine dritte. Die mir vorliegende Tafel geht bis zur Bildung einer Differenzreihen von dritter Ordnung voran.' (Epping and Strassmaier 1881: 283)

> hence they formed a so-called arithmetical progression, analogous to the usual number sequence. From that, this sequence of numbers thus obtained is used for deriving from it another one, and from the latter a third one. The tablet available to me proceeds until the formation of a third-order difference sequence.

The first category concerns tablets now classified as mathematical astronomy; the second one concerns a normal star almanac. It is not fully clear what kind of tablets gave rise to the third category. Epping's background as an astronomer is obvious from his preference for the numerical tables. He then presents results of his investigations of two tablets, including a partial edition of the reverse of fragment 81–7–6, 277 (T1). Of the seven partly preserved columns of numbers, nowadays referred to as Ψ'', F, G, H, J, K and L of lunar System B, Epping edits and discusses only K and L, which he calls A and B. Strassmaier's copy is not included, nor are any details provided about it, but there is no doubt that Epping used the copy shown in Fig. 14.1.[21]

By choosing capital letters as labels for the columns, Epping initiated a convention used by all subsequent scholars. With Epping, they still have a double role, namely to indicate the position of the column on the fragment and, secondly, to serve as a purely conventional modern symbol for the quantity tabulated in the column.[22] As we shall see, Epping also used these symbols for expressing

| | A. | | | | B. | | | | | |
|---|---|---|---|---|---|---|---|---|---|---|
| 1) | 2 | 51 | 40 | 10 | Arahsamna | 28 | 2 | 47 | 23 | — |
| 2) | 2 | 29 | 34 | 10 | Kislev | 28 | 5* | 16 | 57 | 10 |
| 3) | 2 | 33 | 7 | 30 | Tebet | 29 | 1 | 50 | 4 | 40 |
| 4) | 2 | 40 | 17 | 30 | Sebat | 28 | 4 | 30 | 22 | 10 |
| 5) | 2 | 42 | 55 | — | Adar | 29 | 1 | 13 | 17 | 10 |
| 6) | 2 | 52 | 20 | — | Nisan | 28 | 4 | 5 | 37 | 10 |
| 7) | 3 | 8 | 32 | 30 | Airu | 29 | 1 | 14 | 9 | 40 |
| 8) | 3 | 30 | 32 | 30 | Sivan | 28 | 4 | 44 | 42 | 10 |
| 9) | 3 | 59 | 4 | — | Tammuz | 28 | 2 | 43 | 46 | 10 |
| 10) | 3 | 59 | 48* | 10 | Abu | 29 | 0 | 43 | 34 | 20 |
| 11) | 3 | 58 | 10 | 40 | Elul | 28 | 4 | 41 | 45 | — |
| 12) | 3 | 40* | 0 | 40 | Tisri | 29 | 2 | 21 | 45 | 40 |
| 13) | 3 | 26* | 54* | 20 | Arahsamna | 29 | 5* | 48 | 40 | — |

**Fig. 14.1** Left: Strassmaier's unpublished partial copy of the reverse of 81–7–6, 277 (T1) dated 3 June 1879. Right: translation by Epping and Strassmaier (1881)

[21] This copy by Strassmaier, dated 3 June 1879, is preserved in the Nachlass of A. Aaboe and O.N. Neugebauer that is kept in the Institute for the Study of the Ancient World (New York University). Epping also mentions that Strassmaier copied the fragment in 1879 (Epping 1889: Appendix 'Tablet A'). Hence the collection number 81–7–6 does not correspond to the date of arrival in this case. Indeed Leichty (1986: xv–xvi) reports that this collection received its registration number much later than the date of arrival, which had by then been forgotten, and that other tablets in this collection are known to have been in the museum by 1879.

[22] That is, the symbol is arbitrary and has no intrinsic connection to the astronomical quantity. The same can be said for nearly all symbols for lunar columns introduced by subsequent scholars such as Kugler and Neugebauer.

mathematical relations between the columns in modern formulas.[23] Sexagesimal numbers are represented by Epping in tabular form, with each digit vertically aligned in its own subcolumn separated from the next one by a vertical ruling. This formatting of the digits reflects Epping's essentially correct understanding of the underlying algorithm, but no strict alignment is apparent in Strassmaier's copy or in the cuneiform original. The dash is used as a distinct symbol for vanishing final digits, but these digits are not written, as can be seen for instance in Column A, lines 5–6, 9. Sexagesimal numbers are translated using a relative notation as in the original (i.e. they do not include any markers of the power of sixty that corresponds to each digit; e.g. the number 2 41 40 10 in Column A). Only in the commentary does Epping explain that the first digit of A and B represents the number of 'hours', where one 'hour' corresponds to one-sixth of a day.[24] The translation is followed by a critical apparatus to which the reader is directed by asterisks (*). Epping's reconstruction of the algorithmic connection between A and B prompted him to correct some of the numbers copied by Strassmaier. In his words: 'the numbers designated with stars are improvements of the copy; perhaps the original is damaged there.' (Epping and Strassmaier 1881, 284, footnote 1).[25] The following numbers are provided with an asterisk:

| | |
|---|---|
| (A10), 48 | Epping changes Strassmaier's 28 to 48, which is the correct reading (the 20 is in fact a crammed 40). However, Strassmaier's copy offers no clue that the 20 might be read differently. |
| (A12), 40 | Strassmaier copied a 50, which is the correct reading. Epping changed it to 40 on the basis of the entry in B12, using the algorithmic connection between A and B, but the number in B12, 2 21 45 40, was wrongly copied by Strassmaier. The actual number on the tablet is 2 31 45 40, consistent with the 50 in A12. |
| (A13), 26 54 | Strassmaier correctly copied 35 3, which Epping changed to 26 54 on the basis of the entries in B12–13, again using the algorithmic connection between A and B, but these numbers contain errors. |
| (B2), 5 | Epping changed Strassmaier's 2 to a 5. Again, Epping inferred this correction from the algorithmic connection between A and B. Indeed the tablet has a 5 that was wrongly copied as a 2 by Strassmaier. |

Each asterisk correctly identifies a problem in the text as copied by Strassmaier. However, nothing in Strassmaier's copy indicates that any of these readings are doubtful. Epping expresses the relation between columns A and B in the modern

[23] The modern convention, whereby a given quantity is always referred to by the same capital letter, while the column's position on a particular fragment is indicated by a Roman numeral, was introduced by Schaumberger (1935) on a suggestion by Neugebauer (see below).

[24] This unit, called UŠ in cuneiform, corresponds to 1/360 of a day, i.e. 1 day = 6,0 UŠ in sexagesimal notation. The first digit of A and B pertains to 60, unless this digit vanishes, in which case there is an open space and the first written digit pertains to 1. Epping's statement that the first digit of A and B corresponds to 1/6 of a day is valid if it pertains to 60. The unit UŠ is sometimes referred to as time degree in the literature.

[25] 'Die mit Sternchen bezeichneten Zahlen sind Verbesserungen der Copie; vielleicht ist das Original dort beschädigt'.

formula $B_{n-1} + A_n = B_n$, where $n$ labels the month, corresponding to row $n$ in the table (i.e. the value of B in row $n$ is obtained by adding the value of A in that row to the value of B in the previous row ($n - 1$). Already with Epping, modern mathematical equations had become a tool for representing and analyzing Babylonian astronomical algorithms.[26] In this, he was followed by all subsequent researchers. Epping also provides an essentially correct astronomical interpretation of A and B.

The paper continues with a discussion of Sp 129 (T2), a tablet now classified as a normal star almanac. Epping provides a lively account of his efforts to understand its content by comparing the reported data with his own astronomical computations for the year 123 BCE, but he refrains from editing the text[27]:

> In the meantime, we now turn to the other category of tablets, that is, those on which observations are reported. This is where the real difficulty begins... In the running text, some expressions could be identified with more or less certainty; thus, to list the most important ones: *dil-bat* Venus, *gut-tu* Jupiter, *ū* (*ammatu*) cubit or degree, *attalu* eclipse, *an* (*kakkabu*) star, same heaven, *si* (*namir*) visible, *bir* (*nûru*) glow and some others of lesser significance... The fruitless computations lasted until the third Sunday after Easter, the Feast of St. Joseph. Then I wanted to have a last try. And having begun the work, suddenly the thought came to me: why are you so possessed with the fatal *guttu* = Jupiter? Try *guttu* = Mars. Thought, done... In the meantime, my honored brother P. Strassmaier, whom I had immediately informed about my finding, surprised me with an extremely valuable present. As he saw that the investigation had been directed into better waters, he immediately set out to copy the entire tablet of the year 123 once more with the utmost accuracy and render it pronounceable for me in Latin letters.[1] ... The new copy was indeed Classical; everything was much clearer now. [Footnote 1:] Copying is no futility: apart from a great aptitude, it requires unusual combinatorial sharpness of mind. If a certain philological tact is lacking then one will separate signs that belong together and, conversely, join the wrong ones.

---

[26] On the dangers of using modern mathematical tools for representing Babylonian algorithms, see Ossendrijver (2012: 38–39).

[27] 'Wir wenden uns unterdessen der andern Kategorie von Tafeln zu, d. h. denjenigen, auf welchen die Beobachtungen verzeichnet stehen. Hier beginnt die eigentliche Schwierigkeit... In dem laufenden Text konnten einige Ausdrücke als mehr oder weniger gesichert angegeben werden; so, um die wichtigsten anzuführen: dil-bat Venus, gut-tu Jupiter, ū (ammatu) Elle oder Grad, attalu Finsternis, an (kakkabu) Stern, same Himmel, si (namir) sichtbar, bir (nûru) Glanz und noch einige andere von weniger Bedeutung... Das vergebliche Rechnen dauerte bis zum dritten Sonntag nach Ostern, dem Schutzfest des hl. Joseph. Da wollte ich noch einen letzten Versuch machen. Und kaum hatte ich die Arbeit begonnen, kommt mir plötzlich der Gedanke: Warum bist du so versessen auf das fatale guttu = Jupiter? probire guttu = Mars. Gedacht, gethan... Unterdessen überraschte mich mein hochverehrter Mitbruder P. Straßmaier, dem ich gleich von meinem Funde Nachricht gegeben, mit einem äußerst werthvollen Geschenk. Da er sah, daß die Untersuchung in ein besseres Fahrwasser geleitet war, so machte er sich sofort daran, die ganze Tafel des Jahres 123 nochmals mit der möglichsten Genauigkeit zu copiren[1] und eigens für mich durch lateinische Lettern mundgerecht zu machen... Die neue Abschrift war in der That classisch; Alles trat viel bestimmter und klarer hervor... [Fussnote 1: Das Copiren ist keine Kleinigkeit: es erfordert neben großer Geschicklichkeit einen nicht gewöhnlichen combinatorischen Scharfsinn. Wenn ein gewisser philologischer Takt fehlt, so wird man beim Copiren zusammengehörige Zeichen auseinanderreißen, dafür falsche miteinander vereinen.' (Epping and Strassmaier 1881: 287–289).

As becomes clear from this passage, even basic astronomical terms like the names of planets were not yet clarified,[28] which explains why Epping and Strassmaier chose not to offer a translation. Epping concludes by stating that 'Assyriology has opened up a fruitful field of activity for astronomy.' (Epping and Strassmaier 1881: 291).[29] In other words, he envisions Babylonian astronomy as a field of scholarship not only for Assyriologists, but also for astronomers.

In the following years, Epping and Strassmaier collaborated on their project in Bleijenbeek Castle (the Netherlands), until Strassmaier returned to London in 1884. In 1889, their results appeared as a 200-page supplement to the *Stimmen aus Maria Laach* entitled *Astronomisches aus Babylon oder das Wissen der Chaldäer über den gestirnten Himmel* (Astronomy from Babylon or the Chaldean Knowledge of the Starry Sky). In this impressive monograph, an enormous step forward is made in reconstructing the Babylonian astronomical methods and, to some extent, also in editing practices. While Epping wrote the astronomical chapters, Strassmaier contributed an introduction and several Assyriological appendices with transliterations, copies and a glossary. Epping's style is deductive and remarkably informal. Rather than systematically arranging their final results, he presents a lively chronological narrative with insights, mistakes and anecdotes. Apart from reflecting his personality, he may have considered this style to be appropriate for drawing the attention of his fellow Jesuits to an unusual topic. As in 1881, mathematical astronomy has his priority. The reverse of the lunar table 81–7–6, 277 (T1) is now edited more completely. An additional fragment, 81–7–6, 272 ('Tablet C'), physically joins it, but Epping does not report this until we get to page 101. Both are rendered in a tabular format, with some logograms given in transliteration, while other signs are translated. In particular, the month names are translated with Hebrew equivalents – a practice that remained common well into the twentieth century. Logograms of which the Akkadian reading was unknown are transliterated, using the conventions of his time (*bar*, *sik*, *num*, *tab*, *lal*, *šu*).[30] If a logogram is repeated in subsequent rows, Epping uses an abbreviation for this („). Several explanatory columns with modern data are inserted between the Babylonian columns. In Epping's words[31]:

[28] Corresponding modern readings and interpretations of these planets: *dil-bat* = idem, Venus; *gut-tu* = $GU_4$.UD, Mercury.

[29] 'Die Assyriologie hat der Astronomie ein fruchtbares Feld der Thätigkeit eröffnet.' .

[30] Modern transliterations and translations: BAR = marker of eclipse possibility, SIG = 'depth' (distance below ecliptic), NIM = 'height' (distance above ecliptic), TAB = 'add', $LA_2$ = 'subtract', $ŠU_2$ = 'setting' (time until sunset).

[31] 'Die Columnen a, b, $c_1$, $c_2$, d, &c sind in der Reihenfolge abgeschrieben, wie sie sich in den Keilschrifttafeln vorfanden; nur sind die mit x, y, v bezeichneten der Uebersichtlichkeit und Vergleichung wegen beigefügt, von denen x die jedesmalige Horizontalreihe bestimmt, y die Anzahl der Tage angibt, welche nach dem Rechnungsresultat der danebenstehende Monat enhalten mußte, und v endlich die Reihe e nach unserer Tageseintheilung von 24 Stunden wiedergibt.' (Epping 1889: 8–9)

> The Columns a, b, $c_1$, $c_2$, d, etc. were copied in the order in which they were found on the cuneiform tablets, but the ones marked x, y, v were added for clarity and comparison, of which *x* determines the horizontal row, *y* the number of days which the month written next to it should contain according to the computation and, finally, v represents the row e in accordance with our division of the day into 24 hours.

Epping now edits all but one of the columns, independent of whether or not he understood their computation.[32] He introduced a new set of symbols for them, such that d and e correspond to A and B in the 1881 paper. The explanatory columns (x, y, v) are typographically indistinguishable from the other columns. On the other hand, Epping has become more hesitant to correct Strassmaier's readings if they do not match his expectations. For instance, numbers which he had modified in 1881 (Fig. 14.1) are now maintained and qualified by the remark *undeutl.* (uncl.) (Fig. 14.2). Epping does not explain these changes, but they seem to reflect his developing understanding of the transliteration as a more or less objective representation of what is legible on the tablet as opposed to what should have been there according to his reconstructions. On Page 101 Epping returns to the tablet, remarking that the fragment 81–7–6, 272 ('Tablet C') must join it on the right side. The

**Tablet A.**

| x | a | b | $c_1$ | $c_2$ | d | e | y | v | m |
|---|---|---|---|---|---|---|---|---|---|
| — | — | — | — | — | — | Thischri 28 5 55 42 50 | 30ᵈ | — | — |
| I | — | — | — | — | 2 51 40 10 | Marches. 28 2 47 23 | 29 | 28ᵈ 11ʰ 10ᵐ | šu 28 1 .... |
| II | bar 15 2 | 1 57 46 40 | 1 45 | 31 47 30 tab | 2 29 34 10 | Kislev 28 5 16 57 10 | 29 | 28 21 8 | " 29 1 .... |
| III | sik 14 26 | 2 9 52 30 | 8 32 30 | 23 15 " | 2 33 7 30 | Tebeth 29 1 50 4 40 | 30 | 29 7 20 | " 29 0 .... |
| IV | sik 13 50 | 2 32 22 30 | 15 20 | 7 55 " | 2 40 17 30 | Schebat 28 4 30 22 10 | 29 | 28 18 1 | " 29 0 9 |
| V | sik 13 14 | 2 54[1] 52 30 | 19 52[1] 30 | 11 57 30 lal | 2 42 55 | Adar 29 1 13 17 10 | 30 | 29 4 53 | " 29 15 |
| VI | sik 12 38 | 3 17 22 30 | 13 5 | 25 2 30? " | 2 52 20 | Nisan 28 4 5 37 10 | 29 | 28 16 22 | " 28 34 |
| VII | bar 12 2 | 3 39 52 30 | 6 17 30 | 31 20 " | 3 8 32 30 | Ijar 29 1 14 9 40 | 30 | 29 4 57 | " 29 0 1(2) |
| VIII | num 11 26 | 4 2 22 30 | 0 30 | 31 50 " | 3 30 32 30 | Sivan 28 4 44 42 10 | 30 | 28 18 59 | " 28 0 3 |
| IX | num 11 20 10 | 4 24 52 30 | 7 17 30 | 25 48 30 " | 3 59 4 | Thammuz 28 2 43 46 10 | 29 | 28 10 55 | " 28 1 29 |
| X | num 11 56 10 | 4 11 31 40 | 14 5 | 11 43 30 " | 3 59 48[1] 10 | Ab 29 0 43 34 20 | 30 | 29 2 54 | " 29 30? |
| XI | num 12 32 10 | 3 49 1 40 | 20 52 30 | 9 9 tab | 3 58 10 40 | Elul 28 4 41 45 | 29 | 28 18 47 | " 29 9 |
| XII | num 13 8 10 | 3 26 31 40 | 14 20 | 23 29 " | 3 50 0 40 | Thischri 29 2 31 45 40 | 30 | 29 10 7 | " 29 30 |
| XIII | — | — | — | 3 1 30 " | 3 unbeutl 20 | Marcheschvan unbeutlich. | — | 28 23 15 | " 29 |
| | | [1] Nicht 52! | [1] Nicht 51! | | [1] Nicht 28! | | | | |

Diese drei Tafeln befinden sich im Britischen Museum und zwar A und C in der Sammlung von Shemtob, B in der Sammlung von Spartoli, waren aber nicht numerirt, als sie 1879 copirt wurden, und konnten nicht weiter collationirt werden, da der Assistent Th. G. Pinches dieselben nicht wieder auffinden konnte.

Die höchste Zeile in e A ist ein gesetzmäßig gebildeter Zusatz zum spätern Gebrauch.

In e B sind I und II mangelhaft und unleserlich im Original, sie sind gesetzmäßig reconstruirt.

**Tablet C.**

| g | h | i | k |
|---|---|---|---|
| qur 30 10 25 | bat 20 10 | Marchesch. 1 17 20 | 27 17 |
| " 29 7 27 | " 17 50 | Kislev 30 14 | 27 20 10 |
| " 29 4 36 | " 10 10 | Tebeth 30 10 20 | 28 9 50 |
| " 30 8 25 | " 15 30 | Schebat 1 16 20 | 27 11 10 |
| " 29 5 51 | " 11 30 | Adar 30 12 | 27 12 |
| " 30 9 17 | " 16 30 | Nisan 1 17 40 | 27 11 20 |
| " 29 6 34 | " 10 10 | Ijar 30 10 50 | 27 20 10 |
| " 30 9 40 | " 16 | Sivan 1 15 40 | 27 15 10 |
| " 30 12 3 | " 18 30 | Thammuz 1 18 50 | 27 13 20 |
| " 29 8 2 | " 11 50 | Ab 30 11 10 | 28 11 |
| " 30 9 57 | " 16 50 | Elul 1 14 20 | 27 20 40 |
| " 29 5 51 | " 10 us | Thischri 30 7 30 | — |
| " 30 7[1] 52 | " 1' 30 | — | — |
| [1] Nicht 8! | | | |

**Fig. 14.2** Epping (1889): translation of the reverse of 81–7–6, 277 ('Tablet A') and 81–7–6, 272 ('Tablet C'). Modern equivalents of Epping's designations: a = F, b = G, $c_1$ = H, $c_2$ = J, d = K, e = L, m = M, g = N, h = O, i = $NA_1$, k = KUR

[32] Comparison with Strassmaier's unpublished copy of 'Tablet C' reveals that a partly preserved column to the right side of Column k is omitted.

following remarks, placed next to the translation of 'Tablet A' (Fig. 14.2), also suggest an increased sensitivity to the original sources[33]:

> These three tablets are located in the British Museum, in particular A and C in the collection of Shemtob, B in the collection of Spartoli, but they were not numbered when they were copied in 1879 and could not be collated anymore, since the assistant Th. G. Pinches could no longer find the same. The upper line in eA is an addition for later use, computed from rules. In eB, I and II are erroneous and illegible in the original, they have been reconstructed from rules.

Here, eA stands for Column e of Tablet A, eB for Column e of Tablet B, and the Roman numerals refer to line numbers.

Chapters 3 and 4 are dedicated to 'ephemerides', Epping's term for almanacs, normal star almanacs and other compilations related to the astronomical diaries. In a paragraph titled *Realübersetzung mit Text in Transcription* (Factual translation with text in transcription), we find an edition of Sp 129 (T2) (Epping 1889: 152–159, Chap. 4, §10). Epping explains the purpose of his 'factual translation' as follows[34]:

> The aim of the following factual [*sachlich*] translation is to represent the concrete meaning of the cuneiform signs insofar as they signify identical phenomena. However, meanings that are already known from Assyriology or made probable in the present investigation have also been put to use. For the month Nisan the lunar dates have also been added in the translation.

The edition includes, in this order and on facing pages, a translation and a transliteration (called 'transcription'), while the copy is placed in an appendix. This is the earliest edition of a Babylonian astronomical text to include all three representations. Note that the format commonly used in other branches of Assyriology in the nineteenth and early twentieth centuries, with alternating lines of printed cuneiform text, transliteration and translation,[35] was never adopted for editions of Babylonian astronomical texts, most likely because of the mentioned problems of transliteration and translation. The transcription by Epping and Strassmaier consists mainly of normalized Akkadian words with some transliterated signs, sometimes followed by a normalization in brackets, all in the same font. Signs for which the Akkadian reading is uncertain are transliterated in a cursive font, damaged or missing signs are rendered by dots, as can be seen in the first section for the month Nisannu[36]:

---

[33] 'Diese drei Tafeln befinden sich im Britischen Museum und zwar A und C in der Sammlung von Shemtob, B in der Sammlung von Spartoli, waren aber nicht numerirt, als sie 1879 copirt wurden, und konnten nicht weiter collationirt werden, da der Assistent Th. G. Pinches dieselben nicht wieder auffinden konnte. Die höchste Zeile in eA ist ein gesetzmäßig gebildeter Zusatz zum spätern Gebrauch. In eB sind I und II mangelhaft und unleserlich im Original, sie sind gesetzmäßig reconstruirt'. (Epping 1889: 191)

[34] 'Bei der folgenden sachlichen Uebersetzung handelt es sich darum, die concrete Bedeutung der Keilzeichen, insofern sie auf identische Erscheinungen hinweisen, wiederzugeben; dabei sind allerdings die assyriologisch schon bekannten oder durch vorstehende Arbeit wahrscheinlich gemachten Bedeutungen verwerthet. Für den Monat Nisan sind auch die Monddaten in Uebersetzung beigegeben'. (Epping 1889, 152)

[35] See, for instance, the editions of Assyrian astrological tablets by Sayce (1874).

[36] For a modern translation of this passage, see Hunger (2014: 198–199).

**Ina a-mat Bel u Bilit -ia purussû.**

(Auf Geheiß von Bel und Beltis, meiner Herrin, eine Entscheidung.)

**(2) Nisannu** [2].

| | | | | |
|---|---|---|---|---|
| 1 | 20 | 30 | *tab* | . . . . mahrû, 4 ammat . . . . ⅓ ammat; |
| 12 | 1 | 10 | *shu* | mushu 12: lilâtu . . . . *sik* (= elish) . . . . 1 ammat; |
| 13 | 8 | 40 | *lal* | mushu 20: lilâtu *Gut-tu* . . . . *is-da* (= pidnu) . . . .; |
| 13 | 9 | | *na* | mushu 27: lilâtu *Te-ut e shur-gar* (= shûru narkabti?) sha |
| 14 | 8 | 30 | *mi;* | shûtu ⅓ ammat. |
| 26 | 15 | | *mat.* | |

**Nisannu** (1. N. = 25. März — 122).

Am 1. des Abends ist die Mondsichel 20° 30′ (= $44^{m}$) sichtbar vollständig.

Am 12. geht der Mond 1° 10′ (= $4^{m}$ $40^{s}$) vor Sonnenaufgang unter; aber am 13. 8° 40′ (= $34^{m}$ $40^{s}$) vor Sonnenuntergang auf als Vollmond, um am andern Morgen 9° (= $36^{m}$) vor Sonnenaufgang zu verschwinden; ferner wird er am 14. 8° 30′ (= $34^{m}$) nach Sonnenuntergang, also des Nachts aufgehen.

Am 26. des Morgens ist die Mondsichel noch 15° (= $1^{h}$) sichtbar.

Des Nachts am 27. (erscheint) am Abendhimmel Jupiter, darunter shûru narkabti gen Süden (= ζ im Stier oder das südliche Horn) (auf eine Entfernung von) ⅓ Ellen.

(At the command of Bel and Beltis, my lady, a decision.)

**Nisannu** (1. N. = 25 March –122)

On the first in the evening, the lunar crescent was visible 20° 30' (=$44^{m}$) completely.
On the twelfth, the moon sets 1° 10' (=$4^{m}$ $40^{s}$) before sunrise; but on the thirteenth, 8° 40' (=$34^{m}$ $40^{s}$) before sunset as the full moon, disappearing the next morning 9° (=$36^{m}$) before sunrise; furthermore, it will rise on the fourteenth 8° 30' (=$34^{m}$) after sunset, i.e. at night.
On the twenty-sixth in the morning, the lunar crescent is still 15° (=$1^{h}$) visible.
In the night on the twenty-seventh in the evening sky Jupiter (appears), below it *shûru narkabti* to the south (= ζ in Taurus or the southern horn) (at a distance of) ⅓ cubits.

**Fig. 14.3** Transliteration and 'factual translation' of the first section of T2 in Epping (1889)

Compared to the terse, technical formulation of the Babylonian original, the 'factual translation' is rather free and verbose. Signs and phrases whose meaning could not be established are omitted from the translation without any typographical markers (Fig. 14.3). Epping's remarks on the 'factual translation' can therefore be taken to mean that he primarily aims to render the effective, technical meaning of the cuneiform signs inferred through his astronomical and mathematical interpretations, with a lesser role for meanings established by philological means or established in other domains of Assyriology.

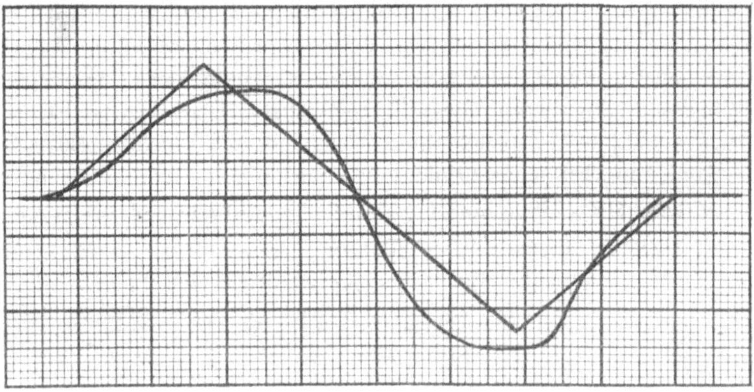

**Fig. 14.4** Graph from Epping (1890) with the zigzag curve for 'Column b' (nowadays: G) and a corresponding modern curve for the duration of the mean synodic month

The monograph was followed in 1890 by a short paper with new results on text T1 accompanied by an edition in tabular format similar to the previous one, except that the rightmost column (m) is omitted and one more explanatory column is added. The transliterated signs *tab* ('add') and *lal* ('subtract') in Column $c_2$ (nowadays: J) are replaced by the symbols + and –, and the sexagesimal numbers in Columns a–e are expressed in an absolute notation with modern symbols for degrees, minutes and seconds of arc attached to the first number in each column.[37] This paper is the first publication on Babylonian astronomy to contain a graphical representation of the zigzag algorithm (Fig. 14.4), referred to by Epping as '*gebrochene Differenzreihe erster Ordnung*' (broken first-order difference sequence).[38] Epping included the figure '*um ein Bild von der babylonischen Annäherung zu gewinnen* (in order to gain an impression of the Babylonian approximation) (i.e. in order to assess visually how well b approximates the empirical value of the quantity it aims to describe, which he identified, essentially correctly, as the mean synodic month).[39] The objective of assessing how accurately an ancient algorithm reproduces empirical behavior is clearly rooted in Epping's professional background as an astronomer and was later also pursued by other scholars of Babylonian astronomy with a similar background. As we shall see, Kugler refrained from including modern graphs of Babylonian algorithms, but with Neugebauer, they became a common feature in any work on Babylonian astronomy.

---

[37] A similar notation was adopted by the French Assyriologist F. Thureau-Dangin in his editions of Babylonian mathematical tablets (e.g. Thureau-Dangin 1932).

[38] Epping (1890: 236)

[39] To be precise, b (G) represents an approximation of the synodic month minus 29 days that accounts only for the lunar variation (i.e. it ignores the solar (annual) variation), which is modeled by the correction $c_2$ (J), as was already understood by Epping.

In Epping's time, Assyriology was a young discipline lacking firmly established conventions and channels of publication. Editing practices were fluid to the extent that transliteration and translation were sometimes combined in a single representation and restored text was not always distinguishable from preserved text. Little attention was paid to a tablet's physical features and archaeological context. However, in the late 1880s, Assyriology was consolidating its position in the academic world with the founding of several disciplinary journals (e.g. *Revue d'Assyriologie* (1884) and *Zeitschrift für Assyriologie* (1886)). Epping's 1890 paper was his last one on Babylonian astronomy to appear in the *Stimmen aus Maria Laach*.[40] Between 1890 and 1893, Epping and Strassmaier published several further papers on Babylonian astronomy in the *Zeitschrift für Assyriologie*. In the latter papers, their editing practices have advanced considerably, which presumably reflects the different standards and readership of this journal. For instance, their edition of a Goal-Year text (Epping and Strassmaier 1891) is arranged in two columns, of which the left one contains a transliteration and the right one a translation.[41]

Ever since the pioneering years of Strassmaier and Epping, the integration of Assyriological and astronomical competences has been a central and often problematic issue for scholars of Babylonian astronomy. As we shall see, Franz Xaver Kugler strongly felt that one should master both, clearly in response to the strict division of these competences that existed between Strassmaier and Epping. Even though he and later scholars such as Otto Neugebauer did acquire significant Assyriological competences, the traces of their original profession in the natural sciences and mathematics can hardly be overlooked in their works.

## 14.2 Franz Xaver Kugler

Franz Xaver Kugler (1862–1929) studied chemistry in Munich and received a PhD in that field from the University of Basel in 1885. After entering the Jesuit order in 1886, he studied philosophy and theology at Jesuit colleges in the Netherlands and England. From 1894 until his death in 1929, he taught mathematics at the German Jesuit college then located in Valkenburg (the Netherlands). By 1897, he followed in the footsteps of Epping. Before exploring Kugler's editing practices, it is worth noting that he emerged as a scholar of Babylonian astronomy exactly when Near Eastern studies were undergoing a spectacular development in Germany. In 1899, the Deutsche Orientgesellschaft was established, under whose auspices Robert Koldewey began his excavations in Babylon in 1900. That same year Friedrich Delitzsch was installed as professor of Assyriology at the Friedrich-Wilhem University (now Humboldt University) in Berlin. Three widely disseminated public

[40] Note that Kugler, who was much more of a public figure than Epping, continued to publish in the *Stimmen aus Maria Laach* and other Catholic journals until the 1920s (see below).

[41] For a list of publications on Babylonian astronomy written by Epping and Strassmaier, see de Jong (2016).

lectures held by Delitzsch in 1899–1901 mark the beginning of the era of Pan-Babylonism, a school of thought in which Judaism and Christianity were viewed as deeply dependent on Mesopotamian precursors.[42] Some Panbabylonists construed their theories around the notion of an ancient Mesopotamian astral cult.[43] It is primarily the efforts of these scholars to prove a very old age of Babylonian astronomy that were attacked by Kugler. In several publications, he used his profound knowledge of the astronomical texts to demonstrate conclusively, often with cynical disdain for his opponents, that Babylonian astronomy did not emerge before the first millennium BCE .[44]

Already in 1900, Kugler published an impressive monograph titled *Babylonische Mondrechnung. Zwei Systeme der Chaldäer über den Lauf des Mondes und der Sonne* (*BMR*) (Babylonian Lunar Computus. Two Systems of the Chaldaeans Concerning the Course of the Moon and the Sun). A novel feature of this work is that it deals with a single, coherent group of texts from the corpus of mathematical astronomy. The observational tablets were to be dealt with in a separate publication. Kugler is the first scholar of Babylonian astronomy to use the term 'critical edition'. In this connection, he issues a disclaimer in the introduction, because he considers himself to be incapable of producing one[45]:

> That the following investigations are primarily of a mathematical-astronomical nature is obvious; however, the hope is not unfounded that also the Assyriologist will find something in them not unworthy of his attention – even though the form in which it is presented may contain several flaws... The cuneiform tablets included in the book were reproduced from the copies of Strassmaier. Even if my less practiced hand could not mimic the clear, recog-

---

[42] On Delitzsch, Panbabylonism and the *Babel–Bibel-Streit* (Babel–Bible dispute), see Lehmann (1990).

[43] Primarily the German orientalists Hugo Winkler (1863–1913) and Alfred Jeremias (1864–1935).

[44] Kugler (1909, 1910).

[45] 'Dass der Charakter der folgenden Untersuchungen vorwiegend ein mathematisch-astronomischer ist, leuchtet ein; allein es ist die Hoffnung nicht unbegründet, dass auch der Assyriologe einiges darin finden wird, was seiner Beachtung nicht ganz unwürdig ist — wenn auch die Form, in der es dargeboten wird, mancherlei Mängel aufweisen mag... Die dem Buche beigegebenen keilinschriftlichen Tafeln sind eine Abschrift der Copien Strassmaiers; wenn es aber auch meiner wenig geübten Hand nicht gelingen wollte, die klaren, markanten Züge des erfahrenen Meisters nachzuahmen, so sind die einzelnen Keilzeichen doch alle richtig und erfüllen ganz ihrem Zweck. Dieser ist ein doppelter. Vor allem bilden jene Keilinschriften die einzige Basis für die vorliegenden Untersuchungen, und darum hat der Leser der letztern auch ein Recht auf einen Einblick in die erstern. Das ist aber nicht nur schicklich, sondern sogar nothwendig, da ohne den vorliegenden Keiltext gewisse Argumentationen des Verfassers gar nicht verständlich sind, und weil obendrein mehrere Keilzeichen nur provisorisch transscribirt werden konnten, die späterhin von competenter Seite richtig gestellt werden müssen. Eine kritische Textausgabe ist also keineswegs beabsichtigt; eine solche wird durch die vorliegende Arbeit erst möglich und verlangt ausserdem eine sorgfältige Collation mit dem Original im Britischen Museum. Aber auch so bieten die Copien Strassmaiers alles, was für eine gedeihliche Bearbeitung nothwendig ist. An mehreren Stellen mussten bei der Transscription allerdings Aenderungen vorgenommen werden; aber abgesehen davon, dass es sich hier in der Regel um schadhafte Partien handelt, kann der Fehler ebenso gut auf Seite des chaldäischen Abschreibers wie auf derjenigen des Paläographen zu suchen sein.' (Kugler 1900: viii–ix)

> nizable features of the experienced master, all cuneiform signs are correct and they wholly serve their purpose, which is twofold. First of all, these cuneiform texts are the only basis of the present investigations and therefore the reader of the latter has the right to view the former. That is not only appropriate but even necessary, since without the cuneiform text being available, certain arguments of the author are not comprehensible at all and, moreover, several cuneiform signs could only be provisionally transcribed and will need to be rectified later by someone competent. Therefore, a critical text edition [*kritische Textausgabe*] is not intended at all; only by the present work will such become possible, and it also requires a careful collation with the original in the British Museum. But even in this form the copies of Strassmaier yield all that is required for a profitable investigation [*gedeihliche Bearbeitung*]. On several locations changes had nevertheless to be made in the transcription; but, apart from the fact that this usually concerns damaged parts, the error may be sought on the side of the Chaldean copyist or that of the palaeographer.

These reflections on editing methods reveal that Kugler's Assyriological scholarship was more developed than that of Epping, to the extent that he was rather critical of Strassmaier's achievements.[46] Nevertheless, Kugler continued to rely on his copies and, as far as known, he never copied or collated a cuneiform tablet himself. He also displayed the same tendency as Epping to 'correct' Strassmaier's copies if his reconstructions of the astronomical algorithms appeared to require it, blaming the error on the 'Chaldean copyist or that of the palaeographer' (i.e. Strassmaier). In *BMR*, which comprises 215 pages and thirteen appendices, Kugler discusses a dozen tablets, including the ones previously analyzed by Epping and Strassmaier. Instead of Epping's informal style, Kugler's writing is more strongly guided by systematic considerations. For instance, he begins (Kugler 1900: 1–8) by introducing the term 'system' —the set of algorithms underlying a group of tables, a central concept in all subsequent historiography on Babylonian mathematical astronomy. He identifies two of them, which he labels I and II, corresponding to B and A in the modern terminology. The systematic approach permeates the entire monograph which is divided into three parts[47] further subdivided into sections, each of which typically deals with one column and the algorithm reconstructed from it. Like Epping, Kugler uses modern formulas for representing Babylonian algorithms but, unlike Epping, he refrains from including modern graphs of them: the few graphs that are contained in *BMR* are concerned with modern astronomical computations.

Kugler's new edition of T1 begins with a description of the physical characteristics of the tablet and a discussion of previous research (Kugler (1900: 9–16).[48] He

[46] On the problematic relationship between Kugler and Strassmaier, see de Jong (2016).

[47] Part I (9–53): '*Der Mondlauf und die Berechnung des Neu- und Vollmondes nach System I*' (The lunar motion and the computation of new and full moon according to System I); Part II (54–114) '*Der Sonnenlauf nach System II und I*' (The solar motion according to Systems II and I); Part III (115–206) '*Der Mondlauf, die Syzygien und Finsternisse nach System II*' (The lunar motion, the syzygies and eclipses according to System II).

[48] Kugler refers to the tablet as 81–7–6, 272, a shorthand for eight joined fragments (see Table 1). His nomenclature is all the more confusing, since he actually omits the fragment 81–7–6, 272 from the edition.

**Neulicht-TafelNr. 272** 81—7—6

(linke Seite = ½ der Tafel)

von 208 bis 210 S. Ä. = 144 bis 146 A. Ä.

| | Zeile | Monat | A | B | C | D | E | F | G | H | I | K | L |
|---|---|---|---|---|---|---|---|---|---|---|---|---|---|
| Obv. | 1. | Adâru | 29 6 39 18 | 2 2 6 20 Arietis | 2 56 | 1 32 | 6 5 30 sik[1] | 11 30 | 3 59 52 30 | 20 20 | 7 19 lal[1] | 3 52 33 30 | Adâru 29 1 2 43 50 |
| | 2. | Nisannu | 28 50 39 18 | 0 52 45 38 Tauri | 3 14 | 1 23 | 9 46 30 sik | 11 16 10 | 4 22 22 30 | 14 52 30 | 22 11 30 lal | 4 0 11 0 | Nisannu 28 5 2 54 50 |
| | 3. | Airu | 28 32 39 18 | 29 25 24 56 „ | 3 26 | 1 17 | 5 54 sik | 11 52 10 | 4 14 1 40 | 8 5 | 30 16 30 lal | 3 43 45 10 | Airu 28 2 46 40 0 |
| | 4. | Simannu | 28 14 39 18 | 27 40 4 14 Gemin. | 3 34 | 1 13 | 2 1 30 sik | 12 28 10 | 3 51 31 40 | 1 17 30 | 31 34 lal | 3 19 57 40 | Simannu 29 0 6 37 40 |
| | 5. | Dûzu | 28 24 40 2 | 26 4 44 16 Cancri | 3 32 | 1 14 | 1 51 bar[2] | 13 4 10 | 3 29 1 40 | 5 30 | 27 52 lal | 3 1 9 40 | Dûzu 28 3 7 47 20 |
| | 6. | Âbu | 28 42 40 2 | 24 47 24 18 Leonis | 3 24 | 1 18 | 2 43 30 num[3] | 13 40 10 | 3 6 31 40 | -12 17 30 | 15 34 30 lal | 2 50 57 10 | Âbu 28 5 58 44 30 |
| | 7. | Ulûlu I | 29 0 40 2 | 23 48 4 20 Virginis | 3 9 | 1 25 | 6 36 num | 14 16 10 | 2 44 1 40 | 19 5 | 3 30 30 tab[2] | 2 47 32 10 | Ulûlu I 28 2 46 16 40 |
| | 8. | Ulûlu II | 29 18 40 2 | 23 6 44 22 Librae | 2 51 | 1 34 | 9 16 num | 14 52 10 | 2 21 31 40 | 16 7 30 | 19 38 tab | 2 41 9 40 | Ulûlu II 28 5 27 26 20 |
| | 9. | Tišrîtu | 29 36 40 2 | 22 43 24 24 Scorpii | 2 36 | 1 42 | 5 33 30 num | 15 4 | 1 59 1 40 | 9 20 | 28 58 tab | 2 27 59 40 | Tišrîtu 29 1 55 26 0 |
| | 10. | Araḫ-s. | 29 54 40 2 | 22 38 4 26 Arcit. | 2 27 | 1 46 | 1 31 num | 14 28 | 2 8 37 30 | 2 12 30 | 31 30 30 tab | 2 40 8 0 | Araḫ-s. 28 4 35 34 0 |
| | 11. | Kislimu | 29 51 17 58 | 22 29 22 24 Capri | 2 27 | 1 46 | 2 21 30 bar | 13 52 | 2 31 7 30 | 4 15 | 29 10 30 tab | 3 0 18 0 | Kislimu 29 1 35 52 0 |
| | 12. | Tebitu | 29 33 17 58 | 22 2 40 22 Amph. | 2 36 | 1 42 | 3 14 0 sik | 13 16 | 2 53 37 30 | 11 2 30 | 18 8 tab | 3 11 45 30 | Tebitu 28 4 47 37 30 |
| | 13. | Šabâtu | 29 15 17 58 | 21 17 58 20 Piscium | 2 50 | 1 35 | 7 6 30 sik | 12 40 | 3 16 7 30 | 17 50 | 0 18 tab | 3 16 25 30 | Šabâtu 29 2 4 3 0 |
| | 14. | Adâru | 28 57 17 58 | 20 15 16 18 Arietis | 3 7 | 1 27 | 8 45 30 sik | 12 4 | 3 38 37 30 | 17 22 30 | 17 4 30 lal | 3 21 33 0 | Adâru 28 5 25 36 0 |
| | 15. | Nisannu | 28 39 17 58 | 18 54 34 16 Tauri | 3 22 | 1 19 | 4 53 sik | 11 28 | 4 1 7 30 | 10 35 | 27 39 30 lal | 3 33 28 0 | Nisannu 28 2 59 4 0 |
| | 16. | Airu | 28 21 17 58 | 17 15 52 14 Gemin. | 3 32 | 1 14 | 1 0 30 sik | 11 18 10 | 4 23 37 30 | 3 47 30 | 31 27 lal | 3 52 10 30 | Airu 29 0 51 14 30 |
| | 17. | Simannu | 28 18 1 22 | 15 33 53 36 Cancri | 3 35 | 1 12 | 2 52 bar | 11 54 10 | 4 12 46 40 | 3 | 30 29 lal | 3 49 17 40 | Simannu 28 4 33 32 10 |
| | 18. | Dûzu | 28 36 1 22 | 14 9 54 58 Leonis | 3 28 | 1 16 | 3 44 30 num | 12 30 10 | 3 50 16 40 | 9 47 30 | 20 41 30 lal | 3 29 35 10 | Dûzu 28 2 3 7 20 |
| | 19. | Âbu | 28 54 1 22 | 13 8 56 20 Virginis | 3 15 | 1 22 | 7 37 0 num | 13 6 10 | 3 27 46 40 | 16 35 | 4 6 30 lal | 3 23 40 10 | Âbu 28 5 26 47 30 |
| | 20. | Ulûlu | 29 12 1 22 | 12 15 57 42 Librae | 2 58 | 1 31 | 8 15 num | 13 42 10 | 3 5 16 40 | 18 37 30 | 14 31 tab | 3 19 47 40 | Ulûlu 29 2 46 35 10 |
| Rev. | 21. | Tišrîtu | 29 30 1 22 | 11 45 59 4 Scorpii | 2 41 | 1 40 | 4 22 30 num | 14 18 10 | 2 42 46 40 | 11 50 | 26 21 tab | 3 9 7 40 | Tišrîtu 28 5 55 42 50 |
| | 22. | Araḫ-s. | 29 48 1 22 | 11 34 0 26 Arcit. | 2 29 | 1 45 | 0 30 0 num | 14 54 10 | 2 20 16 40 | 5 2 30 | 31 28 30 tab | 2 51 40 10 | Araḫ-s. 28 2 47 23 0 |
| | 23. | Kislimu | 29 57 56 38 | 11 31 57 4 Capri | 2 25 | 1 47 | 3 22 30 bar | 15 2 | 1 57 46 40 | 1 45 | 31 47 30 tab | 2 29 34 10 | Kislimu 28 5 16 57 10 |
| | 24. | Tebitu | 29 39 56 38 | 11 11 53 42 Amph. | 2 31 | 1 44 | 4 15 0 sik | 14 26 | 2 9 52 30 | 8 32 30 | 23 15 tab | 2 33 7 30 | Tebitu 29 1 50 4 40 |
| | 25. | Šabâtu | 29 21 56 38 | 10 33 50 20 Piscium | 2 43 | 1 38 | 8 7 30 sik | 13 50 | 2 32 22 30 | -15 20 | 7 35 tab | 2 40 17 30 | Šabâtu 28 4 30 22 10 |
| | 26. | Adâru | 29 3 56 38 | 9 37 46 58 Arietis | 3 1 | 1 29 | 7 44 30 sik | 13 14 | 3 54 52 30 | -19 52 30 | 11 57 30 lal | 2 42 55 0 | Adâru 29 1 13 17 10 |
| | 27. | Nisannu | 28 45 56 38 | 8 23 43 36 Tauri | 3 18 | 1 21 | 3 52 0 sik | 12 38 | 3 17 22 30 | -13 5 | 25 2 30 lal | 2 52 20 0 | Nisannu 28 4 5 37 10 |
| | 28. | Airu | 28 27 56 38 | 6 51 40 14 Gemin. | 3 29 | 1 15 | 0 0 30 bar | 12 2 | 3 39 52 30 | -6 17 30 | 31 20 lal | 3 8 32 30 | Airu 29 1 14 9 40 |
| | 29. | Simannu | 28 11 22 42 | 5 3 2 56 Cancri | 3 35 | 1 12 | 0 53 num | 11 26 | 4 2 22 30 | 0 30 | 31 50 lal | 3 30 32 30 | Simannu 28 4 44 42 10 |
| | 30. | Dûzu | 28 29 22 42 | 3 32 25 38 Leonis | 3 31 | 1 14 | 4 45 30 num | 11 20 10 | 4 24 52 30 | 7 17 30 | 25 48 30 lal | 3 59 4 0 | Dûzu 28 2 43 46 10 |
| | 31. | Âbu | 28 47 22 42 | 2 19 48 20 Virginis | 3 20 | 1 20 | 8 38 num | 11 56 10 | 4 11 31 40 | -14 5 | 11 43 30 lal | 3 59 48 10 | Âbu 29 0 43 34 20 |
| | 32. | Ulûlu | 29 5 22 42 | 1 25 11 2 Librae | 3 4 | 1 28 | 7 14 num | 12 32 10 | 3 49 1 40 | -20 52 30 | 9 9 tab | 3 58 10 40 | Ulûlu 28 4 41 45 0 |
| | 33. | Tišrîtu | 29 23 22 42 | 0 48 33 44 Scorpii | 2 46 | 1 37 | 3 21 30 num | 13 8 10 | 3 26 31 40 | 14 20 | 23 29 tab | 3 50 0 40 | Tišrîtu 29 2 31 45 40 |
| | 34. | Araḫ-s. | 29 41 22 42 | 0 29 56 26 Arcit. | 2 33 | 1 43 | 0 31 0 bar | 13 44 10 | 3 4 1 40 | 7 32 30 | 31 1 30 tab | 3 35 3 10 | Araḫ-s. 29 0 6 15 40 |
| | 35. | Kislimu | 29 59 22 42 | 0 29 19 8 Capri | 2 26 | 1 47 | 1 23 sik | 14 20 10 | 2 41 31 40 | 0 45 | 31 46 30 tab | 3 18 18 10 | Kislimu 28 3 20 7 0 |
| | 36. | Tebitu | 29 46 35 18 | 0 15 54 26 Amph. | 2 28 | 1 46 | 5 16 sik | 14 56 10 | 2 19 1 40 | 6 2 30 | 27 7 tab | 2 46 8 40 | Tebitu 29 0 6 48 50 |
| | 37. | Šabâtu | 29 28 35 18 | 29 44 29 44 Amph. | 2 39 | 1 40 | 9 8 30 sik | 15 | 1 56 31 40 | 12 50 | 14 17 tab | 2 10 48 40 | Šabâtu 28 2 17 4 20 |
| | 38. | Adâru I | 29 10 35 18 | 28 55 5 2 Piscium | 2 54 | 1 33 | 6 43 30 sik | 14 24 | 2 11 7 30 | 19 37 30 | 5 20 30 lal | 2 5 47 0 | Adâru I 28 4 22 51 20 |
| | 39. | Adâru II | 28 52 35 18 | 27 47 40 20 Arietis | 3 12 | 1 24 | 2 51 sik | 13 48 | 2 33 37 30 | 15 35 | 20 55 30 lal | 2 12 42 0 | Adâru II 29 0 35 33 20 |

[1] *sik* = unten, südlich. [2] *bar* = Knoten. [3] *num* = oben, nördlich.

[1] *lal* = minus. [2] *tab* = plus.

Anmerk. 1. In dieser und allen folgenden Tabellen sind die Ergänzungen der Fragmente durch Kleincursivdruck wiedergegeben.

2. Die Sternbilder der Ekliptik sind nach dem bekannten Gedächtnissvers des Ausonius benannt; doch möchten dem Nichtassyriologen auch die nachstehenden chaldäischen Namen erwünscht sein:

1. Ku(sarikku) . . . . . aries (♈).
2. Te(mennu, pidnu) . . . taurus (♉).
3. Mâšu . . . . . . . . gemini (♊).
4. Pulukku . . . . . . cancer (♋).
5. Arû . . . . . . . . leo (♌).
6. Serû . . . . . . . . virgo (♍).
7. Zibânîtu . . . . . . libra (♎).
8. Aqrabu . . . . . . scorpius (♏).
9. PA . . . . . . . . arcitenens (♐).
10. Enzu . . . . . . . . caper (♑).
11. GU . . . . . . . . amphora (♒).
12. Nûnu (ZIB) . . . . pisces (♓).

Die in Klammer beigefügten astronomischen Zeichen kommen der Kürze halber später fast ausschliesslich zur Anwendung.

**Fig. 14.5** Partial edition of T1 in Kugler (1900: 12–13)

correctly identifies the word *tersitu*, which appears in the colophon as the Babylonian term for this kind of astronomical table. Although it is the first edition of T1 to cover both sides of the tablet (Fig. 14.5), it is nevertheless incomplete, because Kugler omits the columns beyond L, which are contained on the fragment 81–7–6, 272 (Epping's 'Tablet C'). He also omits Strassmaier's copy of the tablet, which was published much later by Schaumberger (1935). Kugler's understanding of the algorithms for A–L enables him to restore any missing number in these columns. These restorations are printed in a cursive letter of smaller size, as explained in a footnote (Fig. 14.5), but otherwise not marked. His editing practice also differs from the modern one in that signs of which the Akkadian reading was unclear are rendered in transliteration (the logograms *tab*, *lal*, *sik*, *bar*, *num*),[49] other signs are rendered in normalized Akkadian (the month names) and still others in translation (the zodiacal signs). Like Epping, Kugler separates the digits of the sexagesimal numbers by small spaces and he strictly aligns them; the latter feature is not apparent in Strassmaier's copy or on the original tablets. Moreover, he adds horizontal rulings in order to mark the extrema in each column, such that a single ruling indicates a minimum and a double one a maximum (Fig. 14.5). This convention, which was

[49] These are the same logograms TAB, $LA_2$, SIG, BAR, NIM mentioned above in connection with Epping (1889).

later adopted by O. Neugebauer (1955) in *Astronomical Cuneiform Texts* (*ACT*) (see below), has no counterpart on the tablets either. The edition is followed by a detailed analysis of the columns (Kugler 1900:16–40, 88–114). Kugler introduced his own nomenclature for the columns based on capital letters, which was later slighthly modified by Neugebauer.

In Part III, which deals with lunar 'System II' (nowadays: A), Kugler includes a nearly complete edition of the *Lehrtablet* (instructional tablet) S+ 2418 (T3), the first ever edition of a procedure text of mathematical astronomy. It is not edited as a coherent text but carved up into individual procedures, each edited in a different paragraph. This approach is a consequence of the rigorous systematic structure that Kugler imposed on his monograph. Note that he was later followed in this by Neugebauer, who split up the procedure texts along similar lines in *ACT* (see below). Kugler considered this text to be of great importance for reconstructing 'System II', as is apparent from the following statement in the introduction to Part III[50]:

> The instructional tablet S+ 2418 which was mentioned and utilized earlier offers a comforting confirmation of our conjectures and computations, while, on the other hand, the number columns of the lunar tables help to uncover the meaning of the technical expressions of that tablet, which would otherwise remain wrapped in eternal darkness (Kugler 1900: 116).

A revealing example of Kugler's approach is his edition of the procedure for 'Column G' (nowadays: F), which he correctly interprets as the moon's daily displacement along the ecliptic.[51] The edition consists of a transliteration (Fig. 14.6), a 'factual translation' [*Realübersetzung*] (Fig. 14.7) and a 'justification of the preceding translation'.[52]

**(85) S + 2418, Obvers, Col. I [Zeile 14 bis 19].**

[Z. 14.] ***Epiš UŠ ša Zi Sin araḫ ana araḫ 42 tab u lal lib-bu-u ša 15 56 54 22 30 mat gir.***

[Z. 15.] ***ša al 15 56 54 22 30 tir ultu 15 56 54 22 30 lal lib-bu-u ša 11 4 4 41 15***

[Z. 16.] ***sik gir-tab . ša al 11 4 4 41 15 lal u itti 11 4 4 41 15 tab .***

**Fig. 14.6** Selected lines from Kugler's transliteration of the procedure for 'G' (=F) in T3 (Kugler 1900)

[50] 'Eine beruhigende Bestätigung für unsere Conjecturen und Rechnungen bietet glücklicherweise die schon früher erwähnte und benutzte Lehrtafel S+ 2418, während andererseits die Zahlencolumnen der Mondrechnungstafeln den Sinn der technischen Ausdrücke jenes Tablets erschliessen helfen, die sonst in immerwährendes Dunkel gehüllt blieben.'

[51] For the most recent edition of this procedure, see Ossendrijver (2012: 358–372, P7').

**Realübersetzung.**

**[Z. 14.] Um zu machen die Aenderung der (täglichen) Bewegung des Mondes von Monat zu Monat, werden 42′ addirt oder subtrahirt. Als Grenzwerth wird 15° 56′ 54″ 22‴ 30⁗ oben erhalten.**

**[Z. 15.] Ueber 15° 56′ 54″ 22‴ 30⁗ der Ueberschuss (d. h. von 42′) wird von 15° 56′ 54″ 22‴ 30⁗ subtrahirt (und so) als Grenzwerth 11° 4′ 4″ 41‴ 15⁗**

**[Z. 16.] unten erhalten. Ueber 11° 4′ 4″ 41‴ 15⁗ hinab der Ueberschuss (d. h. von 42′) wird zu 11° 4′ 4″ 41‴ 15⁗ addirt.**

**Factual translation.**

[l. 14.] In order to make the change of the (daily) motion of the moon from month to month 42' are added or subtracted. As the boundary value, 15° 56' 54" 22''' 30'''' is obtained above.

[l. 15] Above 15° 56' 54" 22''' 30'''' the excess (i.e. of 42') is subtracted from 15° 56' 54" 22''' 30'''' (and thus) as the boundary value 11° 4' 4" 41''' 15''''

[l. 16] is obtained below. Beyond 11° 4' 4" 41''' 15'''' downwards the excess (i.e. of 42') is added to 11° 4' 4" 41''' 15''''.

**Fig. 14.7** 'Factual translation' of the transliterated text from Fig. 14.6 (Kugler 1900: 160).

Whenever Kugler feels confident about the Akkadian reading of an expression, he puts this into the transliteration. Another difference with modern editing practices concerns the transliteration of logograms, called *Wortzeichen* (word signs) by Kugler. Even though the Sumerian origin of the logograms was quite well understood by 1900, Kugler avoids this issue and he does not make a typographical distinction between 'word signs' and Akkadian phonetic signs. By 1907, when Book I of *Sternkunde und Sterndienst in Babel* (*SSB*) (Kugler 1907) appeared, Kugler had sorted out these things (see below).

The term 'factual translation' was also used by Epping (1889) in connection with T2. Similar to Epping's approach, Kugler appears to mean by this an effective translation that is primarily based on mathematical and astronomical interpretation. Inspection of his translation reveals that it is literal in the case of signs with a well-established meaning and also in terms of word order, but pragmatic and speculative for signs for which Assyriology could not yet offer a suitable meaning. Indeed, although Kugler's Assyriological competence was more advanced than that of Epping, this did not help him very much in the case of the procedure texts, due to their stenographic and idiosyncratic terminology. Thus 'boundary value' is Kugler's translation of '*lib-bu-u*' (modern reading: *lib*$_3$*-bu-u*, 'whereby') and 'obtained above' is his translation of '*mat gir*', which is nowadays read KUR-*ad*$_2$, 'it reaches [an extremum]'. Other procedures from the same tablet are presented only in transliteration, because their terminology was still too inaccessible. Even though many signs are nowadays read or translated differently, Kugler's algorithmic reconstructions are largely correct, but some of his astronomical interpretations later turned out to be wrong. This is not surprising, because the algorithms could be reconstructed by analyzing procedures and tables irrespective of their astronomical significance, which is rarely mentioned explicitly in the texts. There are numerous examples of algorithms that were reconstructed essentially correctly in this manner by Kugler and Epping before him, while their astronomical interpretation continued to be debated, sometimes until recently.[53]

With *BMR*, Kugler established himself as the most authoritative and prolific scholar in the field of Babylonian astronomy. This position was solidified completely when, between 1907 and 1924, his magnum opus, *Sternkunde und Sterndienst in Babel* (*SSB*) (Astral Science and Astral Worship in Babylon) was published. This vast encyclopedic work of scholarship has a much wider scope than *BMR*, covering all corpora of Mesopotamian astral science of interest to Kugler. It consists of two rather different books. Book I, subtitled 'The Development of Babylonian Planetary Science from its Beginnings until Christ. After Mostly Unpublished Sources of the British Museum',[54] appeared in 1907. In spite of the title, this coherent monograph deals with Babylonian astronomical texts from the first millennium BCE, i.e. diaries and related texts and mathematical astronomy. Book II, which appeared in several installments between 1909 and 1924, deals with a wider range of texts about astronomy, astrology, divination and chronology from the second and the first millennia BCE. In the Preface to Book I Kugler justifies the wide scope of *SSB* as follows[55]:

> Certainly we must regret both aberrations of the human mind [polytheism and astrology], but a comforting thought can serve to soften our judgement. Astral religion was the noblest form of polytheism, and astrology even yielded large scientific benefits. Astral religion lifted the human heart from the bleakness of everyday life and purely material pleasure to a higher conception of life and it instructed him to recognize the majesty and the powerful rule of divinity in the radiance and world-encompassing movement of the stars. Moreover,

---

[52] Kugler (1900: 160–165).

[53] For instance, Kugler's interpretation of Column B as the diameter of the Moon (Kugler 1900: 116). Nowadays, this quantity is named Φ and interpreted as the duration of 223 months (the so-called saros period) in excess of the nearest whole number (6585) of days. For other examples, see Ossendrijver (2012).

[54] Entwicklung der babylonischen Planetenkunde von ihren Anfängen bis auf Christus. Nach zumeist ungedruckten Quellen des Britischen Museums.

[55] 'Gewiß müssen wir beide Verirrungen des Menschengeistes beklagen; ein trostreicher Gedanke ist aber geeignet, unser Urteil zu mildern. Die Sternreligion war die edelste Form des Polytheismus, und die Astrologie hat sogar großen wissenschaftlichen Nutzen gestiftet. Die Astralreligion erhob das Menschenherz aus der Öde des Alltags und dem rein materiellen Genuß zu einer höheren Lebensauffassung und wies ihn an, in dem Glanz und der weltumspannenden Bewegung der Gestirne die Majestät und das mächtige Walten der Gottheit zu erkennen. Die Astrologie war außerdem die Mutter der Astronomie, d. h. ohne die mächtige Überzeugung von dem notwendigen Zusammenhang zwischen Sternkonstellation und Menschengeschick wäre es im Altertum — wenn wir etwa von Sonne und Mond als den Bestimmern der Zeit absehen — kaum jemals zur Pflege einer wissenschaftlichen Sternkunde gekommen... Astralreligion, Astrologie, Astronomie (und Chronologie) bilden in der Himmelserforschung der Babylonier ein Ganzes, wie denn auch die Ämter des Priesters und des Astrologen bezw. Kalenderbeamten in einer Person vereint waren. Was aber historisch verbunden ist, darf auch in einem Werk vereint dargestellt werden, und in unserem Falle um so mehr, als ein tieferes Verständnis der babylonischen Religion oft genug erst durch eingehende astronomische Untersuchungen - natürlich im Bunde mit der Sprach- und der Religionswissenschaft - erschlossen werden kann.' (Kugler 1907: vii–viii)

> astrology was the mother of astronomy, i.e. without the powerful conviction of a necessary connection between stellar constellation and human fate—if we momentarily disregard the sun and the moon as indicators of time—a scientific astronomy would hardly ever have been fostered in antiquity... Astral religion, astrology, astronomy (and chronology) form one whole in the celestial sciences of the Babylonians and, accordingly, the offices of priests and astrologers or calendar officials were united in a single person. What is connected historically can also be presented united in a single work, and in our case even more so, since a deeper understanding of the Babylonian religion can, often, be revealed only through astronomical investigations, naturally in conjunction with scholarship in language and religion.

While dutifully adding a rejection of polytheism and astrology, Kugler reveals that his approach to Babylonian astronomy was, in hindsight, quite modern. For him, all corpora of Babylonian astral science were interconnected, not only in terms of their subject matter but also because they were produced by the same actors. Note that Kugler made ample use of the concept of astral religion, even though he bitterly fought its Panbabylonian interpretation. Neither in *BMR* nor in *SSB* does Kugler consider it necessary to present a detailed justification of his research or define its place in the wider history of science, apart from repeating, like his predecessors, how much the tablets add to what is reported on Babylonian astronomy in the classical sources. It is rather clear that he considered Mesopotamian astral science in general and Babylonian astronomy in particular to be of sufficient interest on its own terms. However, from scattered remarks, such as the passage quoted above, one can infer that he was also of the opinion that his work on these topics would contribute to a better understanding of Mesopotamian religion and chronology. The preface continues with some considerations on the appropriate research methodology and editing practices for the astronomical texts[56]:

> However, a truly fruitful collaboration of such dissimilar sciences cannot at all be achieved through the principle of division of labor, or it can only with great difficulty. Surely the astronomer, supported only by the numbers in the astronomical cuneiform inscriptions, may succeed in clarifying certain technical expressions and thus provide the Assyriologist here und there with a handle for further investigations. Surely the Assyriologist may likewise succeed in transcribing and translating more or less correctly the most primitive astronomical observational reports and stereotypical omen formulas, so that they can be analysed

[56] 'Ein wahrhaft gedeihliches Zusammenwirken so ganz ungleichartiger Wissenschaften läßt sich aber nach dem Prinzip der Arbeitsteilung entweder gar nicht oder nur sehr schwer erreichen. Freilich vermag schon der Astronom als solcher, einzig und allein auf die Zahlenangaben astronomischer Keilschriftinschriften gestützt, den Sinn mancher technischer Ausdrücke klarzustellen und so dem Assyriologen hie und da eine Handhabe für weitere Untersuchungen zu bieten; freilich vermag auch der Assyriologe als solcher die primitivsten Beobachtungsberichte und stereotype Ominaformeln im allgemeinen sinngerecht zu transkribieren und zu übersetzen, so daß sie sich auch astronomisch weiter verwerten lassen. Allein darüber hinaus reicht die Wirksamkeit der vereinten Kräfte im besten Falle nicht. Es fehlt eben der gemeinsame Boden, das wechselseitige Verständnis. So bleibt nur ein Ausweg: die Vereinigung des sprachlichen und des mathematisch-astronomischen Wissens in ein und demselben Kopf, d. h. es muß entweder der Assyriologe sich entschließen, ein paar Jahre der Mathematik und Astronomie zu widmen oder der Astronom, mit Sprache und Geist der Babylonier und Assyrer sich vertraut zu machen. Hierauf waren denn auch seit mehreren jahren meine Bemühungen gerichtet.' (Kugler 1907: viii)

> astronomically. But even in the best case the results of joined forces cannot exceed that. Hence there is only one way out: the unification of linguistic and mathematical-astronomical knowledge in one and the same head, that is, either the Assyriologist decides to devote himself to mathematics and astronomy for a couple of years, or the astronomer to acquaint himself with language and mind of the Babylonians and the Assyrians. This is then what my efforts were aimed at for several years.

For Kugler, a scholar of Babylonian astronomy can be only be successful by mastering all the relevant disciplines like himself. In spite of this claim, Kugler continued to rely on Strassmaier and, as far as known, never copied or collated a cuneiform tablet in the British Museum. *SSB* Book I is nevertheless a mature piece of interdisciplinary scholarship with editions of tablets; commentaries; philological and astronomical analysis; a fourteen-page glossary; an index with the names of months, stars, planets, gods, kings, cities and persons; an astronomical index; a list of corrigenda; a list of tablets and, finally, copies of tablets.[57] Kugler's editing practices have evolved in several ways. With the exception of mathematical astronomy (see below), most tablets are edited completely and without interruption, using a standardized arrangement with transliteration and translation placed on facing pages. Second, the rough and pragmatic 'factual translations' (*Realübersetzungen*) of *BMR* have given way to more accurate 'translations', which reflects Kugler's increasing grasp of the technical terminology. Thirdly, Kugler adopts the typographical distinction between sumerograms and Akkadian phonetic signs also used by other Assyriologists: the former are written in capitals, the latter in small cursive letters. Fourthly, restorations are marked by square brackets rather than by a different font size. In his transliterations, Kugler continues to render as accurately as possible the Akkadian pronunciation of a text. Logograms are generally replaced by Akkadian normalized words, but they are now occasionally followed by the logogram in round brackets. Kugler still makes an exception for certain procedure texts, which he presents only in transliteration, followed by a commentary. His edition of the Jupiter procedure Rm 4, 431 (T4) is introduced as follows[58]:

> We first want to transcribe the text in a provisional manner by introducing phonetic writing instead of ideographic writing only when the correct reading is immediately obvious from the present investigation... Only after a complete decipherment and explanation has the time come for a phonetic representation (in as far as it is possible at all) and translation of the entire text. In order not to print the tablet twice, the restorations of parts destroyed or

[57] Two supplements to Books I and II published by Kugler in 1913 and 1914 are mainly concerned with star names, astrology and chronology. Since they do not contain any editions of Babylonian astronomical texts from the first millennium BCE, they are ignored here.

[58] 'Zunächst wollen wir denselben provisorisch transskribieren, indem wir an Stelle der ideographischen Schreibung nur da die phonetische einführen, wo die richtige Lesung schon auf Grund der bisherigen Untersuchung sich sofort ergibt... Erst nach völlig abgeschlossener Entzifferung und Erklärung ist die Zeit für die phonetische Wiedergabe (insoweit sie überhaupt möglich ist) und Übersetzung des ganzen Textes gekommen. Um jedoch die Tafel nicht zweimal drucken zu müssen, werden die sich als notwendig ergebenden Ergänzungen von zerstörten oder vom Schreiber ausgelassenen Partien der Umschrift des Originals sofort in [ ] beigefügt und die einzelnen Sätze durch ** getrennt (während im keilinschriftlichen Text kein Satztrenner — sei es ein leerer Raum oder das zuweilen gebrauchte Trennungszeichen — zu bemerken ist).' (Kugler 1907: 138)

**3.** ***ultu 2 Enzu adi 17 Mulmullu 36 TAB * ŠA AL.LA 17 Mulmullu TIR A.DU 56 15 DU****

**4.** ***ultu 17 Mulmullu adi 9 Pulukku 33 4[5] TAB * ŠA AL.LA 9 Pulukku***
////β/////
***TIR A.DU 53 20 DU****

**Fig. 14.8** Transliterated lines from a procedure for Jupiter (Kugler 1907)

> omitted by the scribe that turn out to be necessary are inserted in the transcription of the source between [ ] and the individual sentences are separated by ** (while no separator is noticable between sentences in the cuneiform text—be it an empty space or the occasionally used separator sign).

Kugler's new convention with the asterisk (*) as a semantic separator, which has no counterpart in the cuneiform text, can be observed in the following lines of his transliteration of T4. On this occasion, he also introduces a new, ad hoc notation for damaged signs by placing vertical lines above them (as he explains in a footnote labeled β)[59] (Fig. 14.8).

Book II of *SSB* appeared between 1909 and 1924 in three parts comprising a total of 630 pages. This last major work by Kugler is less coherent than the previous one and much more difficult to use, since it lacks any kind of index. The first two installments (1909, 1912) are devoted to celestial omens, astrology, meteorology and chronology, and therefore do not concern us here. However, note that the former begins with a section called 'Retrospection to the First Book' in which Kugler declares his definitive victory over the Panbabylonists by having demonstrated that Babylonian astronomy is a late development.[60] The third installment (1924)[61] does include several editions of tablets with mathematical astronomy and astronomical diaries and related texts, but no copies, which were published much later by Schaumberger (1935). Also noteworthy are editions of two Babylonian horoscopes—a testimony to Kugler's holistic view on Babylonian astral science.

That Kugler's editing practices remained eclectic becomes especially clear in his edition of the normal star almanac Sp 1, 175 + Sp 2, 777 + 782 + 61 (Kugler 1912 = *SSB* II: 532–534).[62] For reasons that remain unclear, this tablet is not edited in the usual format with a transliteration followed by a translation. Instead, Kugler produced a hybrid edition in which the planetary sections are translated, while the invocation and the lunar data are transliterated. Over the years, Kugler kept modifying his editing practices, but his transliterations of numbers remained consistent in that he always separated the digits by empty spaces. Furthermore, Kugler never took up Epping's initiative to use modern graphs for representing Babylonian algorithms.

---

[59] Kugler (1907: 137).

[60] For an account of Kugler's engagement with the Panbabylonists, see de Jong (2016).

[61] Kugler (1924 = *SSB* II: 464–562).

[62] For a modern edition of this tablet see Hunger (2014), Text 86.

## 14.3 Johann Baptist Schaumberger

After Kugler's death in 1929, Johann Baptist Schaumberger (1885–1955), a German Catholic priest of the order of the Redemptorists,[63] continued his unfinished work by writing a third supplement to *SSB*, which appeared in 1935. Schaumberger presents numerous new results of his own investigations of the tablets edited by Kugler in *SSB* Book II, and he includes Strassmaier's hand copies of these tablets, as originally planned by Kugler. Schaumberger's most significant accomplishment is his investigation of 'SH 272' (T1), which appears thirty-five years after the previous edition by Kugler (1900)[64]:

> Finally, I dared to investigate and present Kidinnu's large lunar table SH 272 in its full extent, including the final columns (computation of first crescent and last crescent) that were never dealt with before, Kidinnu's lunar computation is one of the biggest achievements of ancient astronomy.

The twenty-page edition[65] begins with an introduction in which Schaumberger summarizes the history of research on this tablet[66]:

> Since decades, work has been done on the decipherment of this famous tablet. Strassmaier copied the first fragments in 1879. Later he succeeded in reconstituting almost the entire tablet from many pieces. In the meantime, Epping had investigated the first pieces [Epping 1889]. That he was successful in explaining the text, in spite of its fragmentary state, is one of his most splendid accomplishments... The final columns, namely XIV and XVII, still resist interpretation. They will be investigated here. Kugler (1900: p. 12) published the largest part of the text (Cols. A until L = I until XI) in transcription. Appendices XV and XVI of this booklet offer the entire preserved cuneiform text for the first time. The autograph is

[63] Official designation: *Congregatio Sanctissimi Redemptoris* (CSsR) (Congregation of the Most Holy Redeemer). Like the Jesuits, the Redemptorists were banned from Germany between 1872 and 1894. For the biography of Schaumberger, see de Jong (2016).

[64] 'Endlich habe ich es gewagt, KIDINNUS große Mondrechnungstafel SH 272 in ihrem ganzen Umfang zu untersuchen und vorzulegen, auch die noch nie bearbeiteten letzten Kolumnen (Neulicht- und Altlichtrechnung), KIDINNUS Mondrechnung ist eine der größten Leistungen der antiken Astronomie.' (Schaumberger 1935, iv).

[65] Schaumberger (1935: 375–394).

[66] 'An der Erschließung dieser berühmten Tafel ist seit Jahrzehnten gearbeitet worden. Die ersten Bruchstücke hat STRASSMAIER im Jahre 1879 kopiert. Später gelang es ihm, aus vielen Stücken fast die ganze Tafel zusammenzustellen. Inzwischen hatte EPPING die ersten Stücke untersucht (Astronomisches aus Babylon = 44. Ergänzungsheft zu den Stimmen aus Maria Laach, Freiburg 1889, 8–108). Daß ihm trotz des fragmentarischen Zustandes seines Textes die Erklärung größtenteils gelang, ist eine seiner glänzendsten Leistungen... Die letzten Kolumnen, namentlich XIV und XVII harren noch der Bearbeitung. Sie sollen hier untersucht werden. Den größeren Teil des Textes (Kol. A bis L = I bis XI) hat KUGLER (Mondrechnung 12 f.) in Transkription mitgeteilt. Den ganzen erhaltenen Keilschrifttext gibt zum ersten Male die Beilage XV und XVI dieses Heftes. Die Autographie beruht auf der zweiten Kopie STRASSMAIERS und auf einer Photographie. Letztere verdanke ich Herrn O. NEUGEBAUER, der selbst die Güte hatte, die ersten zehn Kolumnen mit ihr zu kollationieren, sowie eine Zeichnung herzustellen, die den Zusammenhang der verschiedenen Bruchstücke der Tafel veranschaulicht (Beilage XVII). Für all das sei ihm auch hier herzlichst gedankt.' Schaumberger (1935: 375–376).

> based on the second copy of Strassmaier and on a photograph. The latter I owe to Mr. O. Neugebauer, who himself had the grace to collate the first ten columns with the photograph and produce a drawing demonstrating the connection between the different fragments of the tablet. For all that he is cordially thanked.

With Schaumberger, photographs of the tablets play an increasingly important role in research on Babylonian astronomy, but they are not yet included in his editions. Instead, Schaumberger offers a transcription and Strassmaier's copy of T1. His edition is the first one to include all columns, but in terms of editing practices, it offers little change to the one by Kugler (1900). Some logograms (e.g. TAB, 'add', LAL, 'subtract' and those for the zodiacal signs) are replaced by modern symbols and restored text is printed in a different font but not enclosed in brackets. Remarkably, Schaumberger's commentary includes a much more modern transliteration of a lunar table from the Louvre, AO (= Antiquités Orientales) 6491 (Schaumberger 1935: 387).[67] An enduring innovation concerns the nomenclature of the lunar columns. Schaumberger takes up a suggestion by Neugebauer to introduce a double designation for each column: first, a Roman numeral indicating its position on a given fragment; second, a capital letter that uniquely identifies its content.[68] The final appendix contains Neugebauer's drawing of the arrangement of the fragments of T1, a reflection of Neugebauer's awareness that a proper edition should include a full documentation of the material aspects of tablets (Fig. 14.9).

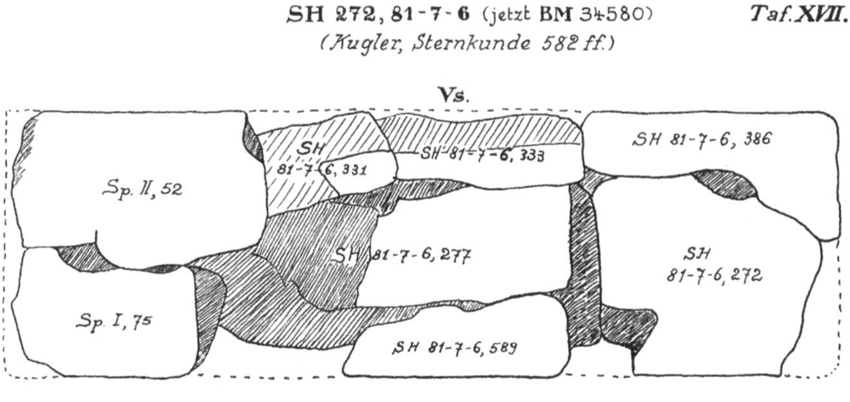


**Fig. 14.9** Neugebauer's drawing of T1 (obverse) in Schaumberger (1935)

[67] For this tablet, see Neugebauer (1955: No. 102).

[68] Neugebauer introduced analogous symbolic notations for astronomical quantities in other areas of ancient astronomy (e.g. Claudius Ptolemy).

Schaumberger's supplement marks the end of a pioneering era that began with Epping and culminated in Kugler's *SSB*, a monumental, Byzantine work covering all aspects of Mesopotamian astral science. After 1935, editing practices developed in a direction away from Kugler's encyclopedic and eclectic style. Henceforth, different corpora of Mesopotamian astral science were investigated by different scholars and edited in separate monographs. Only in the 1990s, when the work of editing the various different corpora was nearing completion, have these divisions been softened due to a growing awareness of the connections between the different forms of astral science and a renewed interest in its social and institutional context—a development that Kugler would certainly approve of. Schaumberger continued to publish several papers on Mesopotamian astral science, but he never again worked on Babylonian mathematical astronomy, which had by then become the realm of Otto Neugebauer, who emerged as the translator and interpreter of this corpus by 1937.

## 14.4 Otto Neugebauer

Otto Neugebauer (1899–1990) ushered in a new phase in the historiography of Babylonian astronomy. Having recently published a complete edition of the Babylonian mathematical corpus (Neugebauer 1935–1937), Neugebauer switched his attention to Babylonian mathematical astronomy. Henceforth, the astronomical texts were edited as coherent text corpora. After making a brief appearance in Schaumberger (1935), his first publication on this topic was an admiring review of Schaumberger's achievements (Neugebauer 1936). Since the methodology of Neugebauer's research on Babylonian astronomy has been discussed elsewhere (Ossendrijver 2016), it suffices to recall some important features. One pillar of his research comprised various methods and tools for dating and reconstructing astronomical tables; a second one included his methods for reconstructing the underlying astronomical theories and empirical data. The production of critical editions was the third pillar of his research. The former two have in common that they largely consist of mathematical analysis. Especially in the early stages of his research, his focus lay squarely on the tabular texts with their sequences of numbers. A universalist mathematical perspective, which can be traced back to his mathematical education at Göttingen, underlies all of his investigations on ancient mathematical astronomy.[69] In a similar vein, Neugebauer frequently uses modern formulas and graphs in order to represent the Babylonian algorithms, unlike Kugler. Only later, when Neugebauer began to tackle the procedure texts, did the problem of translation attract his attention (see below). This contrasts with the more eclectic approach of Kugler, who, from the very beginning, combined the analysis of tables with efforts to translate procedure texts.

[69] See especially Neugebauer (1975).

In a short programmatic paper (Neugebauer 1937) published in the newly founded journal *Quellen and Studien* (Sources and Studies), Neugebauer portrays the unsatisfactory state of publication of the astronomical texts. After decades of pioneering research, the modern reader is facing a forest of badly accessible monographs and papers, comprising layer upon layer of corrections on each tablet, with ever changing notations. He then announces the following ambitious plan[70]:

> I have decided … to make available the complete source material in a special text edition. I hope that, in this manner, approximately the same state of scientific methodology is reached in the field of pre-Greek mathematics and mathematical astronomy that has long ago become self-evident in the field of classical antiquity through the untiring efforts of Hultsch, Heiberg, Manitius, Tannery and many others, namely that one considers complete text editions to be the indispensable foundation of all further work.[2] [Footnote 2: For Egyptian mathematics there are the editions by Peet, Chace and Struve; for Babylonian, my edition.] This edition of the astronomical cuneiform texts of a mathematical character is part of a more comprehensive plan. The gentlemen L. Hartmann, J. Schaumberger and A. Schott and possibly other collaborators will also edit and publish the totality of the other classes of astronomical cuneiform texts, observational texts and astrological-astronomical texts, so that we finally hope to be able to present a truly complete collection of source materials on Babylonian astronomy. In the first part ['*Texts*'] of this edition of the mathematical-astronomical cuneiform texts, which is in progress (I will henceforth simply cite it as MACT), I shall publish the entire textual material as completely as possible.

In a paper entitled 'History of Ancient Astronomy: Methods and Problems', published in 1946, Neugebauer reiterates some of the same points,

> The most essential task is that of making the original sources accessible as easily as possible in their best available form. By the indefatigable work of Heiberg, Hultsch, Tannery and many others, we possess today a great part of the extant writings of the Greek scientists in excellent editions… To make Greek and Oriental source material more generally accessible, supplemented, of course, by modern translations and commentaries, will be the foremost problem of the future (Neugebauer 1946: 131).

[70] '… so habe ich mich entschlossen, … das gesamte Tatsachenmaterial der Originaltexte in einer besonderen Textedition allgemein zugänglich zu machen. Ich hoffe, daß damit auf dem Gebiete der vorgriechischen Mathematik und der mathematischen Astronomie ungefähr derselbe Zustand wissenschaftlicher Methodik erreicht ist, wie er auf klassisch-antikem Gebiet durch die unermüdliche Arbeit von Hultsch, Heiberg, Manitius, Tannery und vielen Andern schon längst selbstverständlich geworden ist, daß man nämlich vollständige Textausgaben als das unumgängliche Fundament aller weiteren Arbeit ansieht.[2] [Fussnote 2 Für die aegyptische Mathematik liegen die Textausgaben von Peet [1], Chace [1] und Struve [1] vor, für die babylonische meine Ausgabe [2].] Diese Edition der astronomischen Keilschrifttexte mathematischen Charakters ordnet sich einem umfassenderen Plane unter. Die Herren L. Hartmann, J. Schaumberger und A. Schott und gegebenenfalls auch noch andere Mitarbeiter werden nämlich auch die Gesamtheit der übrigen Klassen astronomischer Keilschrifttexte, Beobachtungstexte wie astrologisch-astronomischer Texte, bearbeiten und herausgeben, so daß wir zusammen schließlich eine wirklich vollständige Sammlung des Quellenmaterials der babylonischen Astronomie vorlegen zu können hoffen. Ich werde im ersten Teil („*Texte*") dieser in Arbeit befindlichen Edition der mathematisch-astronomischen Keilschrifttexte (ich zitiere sie in Hinkunft einfach als MAKT) das Gesamtmaterial der Texte so vollständig wie mir möglich herausgeben.' (Neugebauer 1937: 29–30) For the quoted editions of Egyptian and Babylonian mathematical texts, see Peet (1923), Chace et al. (1927), Struve (1930) and Neugebauer (1935–1937).

For Neugebauer, it was clear that systematic research on Babylonian astronomy must proceed from complete editions of entire corpora produced in accordance with the same high standards that had been established in classical philology. The works of renowned historians of ancient Greek and Roman astronomy, mathematics and metrology[71] serve as models for the edition that he was aiming for. Of the editions announced by Neugebauer, only the one that he intended to produce himself was actually realized within the decades after 1937. This became *Astronomical Cuneiform Texts* (*ACT*), which appeared in 1955. The other editions of the astronomical diaries and related texts and the astrological texts were not realized until after 1985 by a new generation of scholars. Neugebauer was a competent Assyriologist, but he relied on other Assyriologists to copy and collate tablets, not unlike Kugler. While continuing to work with the unpublished copies of Strassmaier, he also used the partly overlapping collection of copies produced by Theophilus Goldridge Pinches (1856–1934), who was subsequently assistant and curator of the cuneiform collection of the British Museum. These copies, produced between 1878 and 1900—at about the same time when Strassmaier was a regular visitor of the museum—were eventually published as Pinches and Sachs (1955). The latter author, the Assyriologist Abraham Sachs (1915–1983), became Neugebauer's main collaborator in philological matters by 1941. Neugebauer was the first scholar of Babylonian astronomy to systematically use photographs of tablets along with the copies. This reflects the careful attention he gave to all physical aspects of the tablets. Virtually all of the editions by Neugebauer were based on copies by Strassmaier and Pinches, photographs that were sent to him by the British Museum, and various other collections and collations by Sachs and other fellow Assyriologists. Only few of these photographs were published in *ACT* or elsewhere.[72]

Appearing some 18 years after Neugebauer's announcement, *ACT* is the first ever edition of a complete corpus of astronomical texts and a milestone in editing practices. *ACT* is structured in a systematic manner, Vol. I being devoted to the moon, Vol. II to the planets, while Vol. III contains transliterations of the tabular texts as well as photographs and copies of selected tablets. Vol. I begins with an introduction about the history of research on Babylonian mathematical astronomy, documentary aspects of the tablets, colophons, mathematical methods, calendar issues, units and other general topics. The critical editions are preceded by introductory chapters about the algorithms that were reconstructed from the texts. In these chapters, Neugebauer typically combines a discussion of the Babylonian algorithms with explanations about the modern mathematical tools developed by him for

[71] Friedrich Otto Hultsch (1833–1906), a German philologist and historian of Greek and Roman mathematics and metrology; Johan Ludvig Heiberg (1854–1928), a Danish philologist and historian of Greek mathematics and astronomy; Karl Heinrich August Manitius (1848–1921), a German philologist and historian of Greek astronomy; Paul Tannery (1843–1904), a French philologist and historian of mathematics.

[72] The original photographs, notes and manuscripts on which *ACT* was based are preserved in the Nachlass of O. Neugebauer, which is kept in the Institute for the Study of the Ancient World (New York University).

reconstructing and analyzing these algorithms. By contrast, Neugebauer barely addresses editing practices, nor does he discuss his research methodology. His strongly mathematical approach is nevertheless immediately obvious in the introductory chapters, and various traces of it can be identified in the editions themselves. Within each section devoted to the moon or a planet, the tabular texts are always given priority, the procedure texts being edited in second position. All tablets are edited in their entirety, but not necessarily in the form of a single uninterrupted edition. If a tablet contains both tables and procedures, these different textual components are edited in separate chapters. Furthermore, both types of texts are edited differently. Each edition of a tabular text begins with basic documentary information[73] including a reference to the hand copy in Pinches and Sachs (1955), the transcription in Vol. III, followed by a critical apparatus and a commentary. The latter is usually divided into paragraphs, each devoted to a different column. For reasons of economy, no translations, hand copies or photographs of the tabular texts are included in *ACT*. Compared with the editions published by Kugler, who often mixed translation and transliteration, Neugebauer's editions have a standardized and more modern appearance. However, they do contain mathematically motivated features that have no counterpart on the tablets. For instance, in columns with periodically varying number sequences, dashed, drawn and double horizontal rulings indicate where a minimum, maximum or zero crossing is reached, a notation introduced by Kugler (1900). Vertical rulings are drawn between all columns, even if they are not present on the tablet. Hence the tabular texts are transliterated in an idealized layout with added rulings that reflect the mathematical structure of the columns.

The editions of the procedure texts are arranged rather differently. Each one includes a transcription, a critical apparatus, a translation, a commentary and a reference to a hand copy, either in Pinches and Sachs (1955) or in Vol. III, and, usually, a photograph in Vol. III. Unlike the tabular texts, the tablets with procedures are not edited as coherent texts. Instead, each procedure is edited separately in the form of a transcription, critical apparatus, translation and commentary. By splitting up the texts in this manner, Neugebauer follows a practice initiated by Kugler (1900). Neugebauer's editions of the procedure texts represent a tremendous step forward in our understanding of the highly technical astronomical and mathematical terminology. However, the translations are, in hindsight, quite strongly shaped by Neugebauer's background in modern mathematics and astronomy. Arithmetical and astronomical terms are replaced by overly modern supposed equivalents, and semantic variations in the Babylonian terminology are not preserved in the translations (Ossendrijver 2016). A similar tendency underlies Neugebauer's translations of Old Babylonian mathematical problem texts (Høyrup 1996). When editing procedure texts, Neugebauer did not aim for optimal semantic accuracy but for adequacy in a pragmatic, mathematical sense, in accordance with his universalist perspective on ancient astronomy.

[73] Usually comprising Contents, Arrangement of the sides, Provenance, and previous publications, but no information about physical dimensions.

Since Neugebauer, research on Babylonian astronomy has shifted its aims towards interpreting the Babylonian methods and practices and the underlying concepts in Babylonian terms rather than in modern terms. Moreover, the focus on reconstructing astronomical algorithms and the empirical data from which they were constructed is making way for a more holistic approach, whereby Babylonian astronomy is interpreted in its diverse institutional, political, religious and social contexts, while fully acknowledging the connections between the different forms of Babylonian astral science, including celestial divination and astrology.

## References

### *Primary Sources*

Brack-Bernsen, L., and H. Hunger. 2002. TU 11. *A collection of rules for the prediction of lunar phases and of month lengths*. *SCIAMVS* 3: 3–90.

Chace, A.B., H.P. Manning, and R.C. Archibald. 1927. *The rind mathematical papyrus British museum 10057 and 10058*. Vol. I and II. Oberlin: Mathematical Association of America.

Epping, J. 1889. *Astronomisches aus Babylon oder das Wissen der Chaldäer über den gestirnten Himmel. Stimmen aus Maria-Laach*, Ergänzungsheft 44. Freiburg im Breisgau: Herder Verlag.

———. 1890. Die babylonische Berechnung des Neumondes. *Stimmen aus Maria-Laach* 39: 225–240.

Epping, J., and J.N. Strassmaier. 1891. Neue babylonische Planeten-Tafeln II. *Zeitschrift für Assyriologie* 6: 89–102.

Hunger, H. 2006. *Astronomical diaries and related texts from Babylonia. Volume VI. Goal-Year Texts*. Vienna: Verlag der Österreichischen Akademie der Wissenschaften.

———. 2014. *Astronomical diaries and related texts from Babylonia. Volume VII. Almanacs and Normal-Star Almanacs*. Vienna: Verlag der Österreichischen Akademie der Wissenschaften.

———. 2022. *Astronomical diaries and related texts from Babylonia. Vol. IV: Undated diaries and addenda*. Vienna: Verlag der Österreichischen Akademie der Wissenschaften.

Hunger, H., and A. Sachs. 2001. *Astronomical diaries and related texts from Babylonia V. Lunar and planetary texts*. Vienna: Verlag der Österreichischen Akademie der Wissenschaften.

Kugler, F.X. 1900. *Die babylonische Mondrechnung. Zwei Systeme der Chaldäer über den Lauf des Mondes und der Sonne*. Freiburg im Breisgau: Herder Verlag.

———. 1907. *Sternkunde und Sterndienst in Babel, Buch I: Entwicklung der babylonischen Planetenkunde*. Münster: Aschendorffsche Verlagsbuchhandlung.

———. 1909. Sternkunde und Sterndienst in Babel, Buch II, Teil I. In *Natur, Mythus und Geschichte als Grundlagen babylonischer Zeitordnung nebst eingehenden Untersuchungen der älteren Sternkunde und Meteorologie*. Münster: Aschendorffsche Verlagsbuchhandlung.

———. 1912. *Sternkunde und Sterndienst in Babel, Buch II, Teil II, Heft 1*. Münster: Aschendorffsche Verlagsbuchhandlung.

———. 1913. *Ergänzungen zum I. und II. Buch*. Münster: Aschendorffsche Verlagsbuchhandlung.

———. 1924. *Sternkunde und Sterndienst in Babel, Buch II, Teil II, Heft 2*. Münster: Aschendorffsche Verlagsbuchhandlung.

Neugebauer, O.N. 1935–1937. *Mathematische Keilschrifttexte I–III*. Berlin: Springer.

———. 1955. *Astronomical cuneiform texts*. London: Lund Humphries.

Ossendrijver, M. 2012. *Babylonian mathematical astronomy: Procedure texts*. New York: Springer.

Peet, T.E. 1923. *The Rhind mathematical papyrus British museum 10058 and 10058*. London: University Press of Liverpool.
Pinches, T.G., and A. Sachs. 1955. *Late Babylonian astronomical and related texts*. Providence, R.I: Brown University.
Rochberg, F. 1988. *Aspects of Babylonian celestial divination: The lunar eclipse tablets of Enūma Anu Enlil*, Archiv für Orientforschung, Beihefte 22. Horn (Austria): Verlag Ferdinand Berger & Söhne.
Sachs, A.J., and H. Hunger. 1988–1996. *Astronomical Diaries and related texts from Babylonia, I–III*. Vienna: Verlag der Österreichischen Akademie der Wissenschaften.
Sayce, A. 1874. The astronomy and astrology of the Babylonians. With translations of the tablets relating to these subjects. *Transactions of the Society of Biblical Archaeology* 3: 145–339.
Schaumberger, J. 1935. *Sternkunde und Sterndienst in Babel. Assyriologische, astronomische und astralmythologische Untersuchungen*, 3. Ergänzungsheft. Münster: Verlag der Aschendorffschen Verlagsbuchhandlung.
Steele, J. 2010. Newly identified lunar and planetary tables from Babylon in the British museum. *SCIAMVS* 11: 211–239.
Struve, W.W. 1930. *Mathematischer Papyrus des Staatlichen Museums der schönen Künste in Moskau*. Berlin: Springer.

## Secondary Sources

de Jong, T. 2016. Babylonian astronomy: The players and the field. In *A Mathematician's journeys: Otto Neugebauer and modern transformations of ancient science*, eds. A. Jones, C. Proust, and J. Steele, 265–302. New York: Springer.
Epping, J., and J.N. Strassmaier. 1881. Zur Entzifferung der astronomischen Tafeln der Chaldäer. *Stimmen aus Maria-Laach* 21: 277–292.
Høyrup, J. 1996. Changing trends in the historiography of Mesopotamian mathematics: An insider's view. *History of Science* 34: 1–32.
Kugler, F.X. 1910. *Im Bannkreis Babels. Panbabylonistische Konstruktionen und religionsgeschichtliche Tatsachen*. Münster: Aschendorffsche Verlagsbuchhandlung.
Lehmann, R. 1990. *Friedrich Delitzsch und der Babel–Bibel-Streit. OBO 133*. Freiburg (Switzerland): Universitätsverlag Freiburg Schweiz.
Leichty, E. 1986. *Catalogue of the Babylonian tablets in the British museum, Vol. VI: Tablets from Sippar 1*. London: Trustees of the British Museum.
Neugebauer, O.N. 1936. Review of J. Schaumberger, Drittes Ergänzungsheft zu F. X. Kugler, S.J., *Sternkunde und Sterndienst in Babel*. *Quellen und Studien B* 3: 271–286.
———. 1937. Untersuchungen zur antiken Astronomie I. *Quellen und Studien B* 4: 29–33.
———. 1946. History of ancient astronomy: Problems and methods. *Publications of the Astronomical Society of the Pacific* 58 (17–43): 104–142.
———. 1975. *History of ancient mathematical astronomy*. New York: Springer.
Ossendrijver, M. 2016. Translating Babylonian mathematical astronomy: Neugebauer and beyond. In *A mathematician's journeys: Otto Neugebauer and modern transformations of ancient science*, eds. A. Jones, C. Proust, and J. Steele, 333–342. New York: Springer.
Rochberg, F. 2014. The history of science and ancient Mesopotamia. *Journal of Ancient Near Eastern History* 1: 37–60.
Sachs, A.J. 1948. A classification of the Babylonian astronomical tablets of the Seleucid period. *Journal of Cuneiform Studies* 2: 271–290.
Schneemann, G. 1878. Galileo und der römische Stuhl. *Stimmen aus Maria-Laach* 14:113–130, 254–270, 389–403.

Steele, J. 2011. Goal-year periods and their use in predicting planetary phenomena. In *The empirical dimensions of ancient near eastern studies: Die empirische dimension altorientalischer Forschungen*, Wiener Offene Orientalistik 6, ed. G. Selz, 101–110. Vienna/Berlin: Lit-Verlag.
Thureau-Dangin, F. 1932. *Esquisse d'une histoire du système sexagésimal*. Paris: Geuthner.
Udías, A. 2003. *Searching the heavens and the earth: The history of Jesuit observatories*. Dordrecht: Kluwer Academic Publishers.
———. 2015. *Jesuit contribution to science. A history*. New York: Springer.

# Chapter 15
# Postface

**Glenn Most**

**Abstract** The relationship between science and philology is far more complex, and far more interesting, than would be suggested by the popular imagination of a banal antithesis between them: for scientists have usually also been philologists, no science is possible without textualization, and the institutions and practices involved are fundamentally deeply similar. Yet much still needs to be done to help the conversation between philology and the sciences to develop all its potential. This pioneering book is an important step in that direction.

In the popular imagination, scientists stand in brightly lit laboratories surrounded by gleaming arrays of metal equipment, while teams of assistants bustle about to study the instruments and report their findings to them; philologists, by contrast, sit in solitude in darkened libraries, studying dusty tomes, transcribing their contents on foolscap, and breaking their silence only to shush the pedants at other tables whose whispers interfere with their lucubrations. Scientists turn their backs on the past, which they consider as little more than an unbroken succession of errors, and stride resolutely into a future of their own making; philologists immerse themselves in a historically remote culture that would be lost but for their efforts and are dragged, unaware or reluctant, into a future they cannot or will not imagine.

As is usually the case, such images tell us far more about the transient fashions of popular culture than about the real nature of their apparent objects. In the present case, indeed, they are simply false. For in at least three essential regards, the relationship between science and philology is far more complex, and far more

The research leading to these results has received funding from the European Research Council under the European Union's Seventh Framework Programme (FP7/2007–2013)/ERC Grant Agreement No. 269804.

G. Most (✉)
The University of Chicago, Committee on Social Thought, Chicago, IL, USA

The Max-Planck Institut für Wissenschaftsgeschichte, Berlin, Germany
e-mail: most@sns.it



A. Keller, K. Chemla (eds.), *Shaping the Sciences of the Ancient and Medieval World*, Archimedes 69, https://doi.org/10.1007/978-3-031-49617-2_15

interesting, than would be suggested by this kind of banal antithesis. First, with regard to the past, during most of the history of science in most parts of the world, scientists were fundamentally also philologists, in the sense not only that they were trained in their fields by the careful study of canonical texts authorized by long-established traditions of edition and commentary, but also that they often conceived of their own work of research as taking the form of explication and correction of those very same texts—that is, research and commentary were often indiscernible. Second, with regard to the present, no scientist of any kind is trained except by the institutionalized and sustained oral and written communication of preceding results and current problems, and no reputable scientists of any kind fail to make their findings known by the oral and written communication of them to their disciplinary community; that is, there is no science of any kind without textualization, and where there are texts, there must also be philologists to care for them. And third, theoretically, if we abstract from the kinds of objects which tend to be studied by the natural sciences and the humanities and ask instead how these fields of knowledge production are structured in terms of their institutions and practices, it turns out that any definition of a science that applies to natural scientists in their laboratories will also be applicable to philologists in their libraries. For if a science can in general be defined as 'a rational, disciplined and institutionalized form of interpersonal research, testing and communication,' then philology in particular is nothing other than 'the rational, disciplined and institutionalized form of interpersonal research, testing and communication, directed to (above all written) texts'[1]; that is, philologists are themselves a kind of scientist.

So there are good reasons for studying the various kinds of intersections between philology and other kinds of scientific disciplines that have obtained, that should obtain, and that might obtain, especially within the terms of a broad inter-cultural and inter-disciplinary context that makes it easier to avoid hasty pseudo-naturalizations of merely local phenomena. And precisely this is the ambition and the achievement of the present collection of essays. Yet despite the evident interest and significance of this topic, this volume is innovative—indeed, there is nothing quite like it, and it explores issues that have rarely, if ever, been investigated at this level of depth and with this breadth of range. This is in itself a matter not only for reflection but also for a certain disquiet. Why have there been so few comprehensive studies examining the textual criticism, critical editions and translations of ancient and medieval scientific texts (to say nothing of modern ones)? The fact that no single scholar could possibly possess both all the requisite first-hand knowledge of the disparate scientific disciplines involved and all the necessary linguistic and historical expertise presupposed by this type of comparative research is not a satisfactory explanation; a topic like this can only be handled by a group of experts working closely together, with mutual respect and stimulation, as is the case here. Instead, factors on both sides seem hitherto to have hampered the conversation between philology and the sciences. On the side of philology, the paradigm for investigations

[1] Daston and Most (2015).

into the history and methodology of textual studies has, for the most part, been their application to central literary and especially poetic texts, as it is these that have often, especially in the West, enjoyed the greatest prestige, the widest audiences and the longest traditions of scholarship; by contrast, research into the philological study of scientific texts—often (no doubt self-protectively) dismissed as laborious in content and derivative in method—has tended to lag behind. To take an example from Ancient Greek (in other cultures, matters may well be different), Euclid, too, was certainly an important school text, but he was taught to far fewer pupils than Homer was; correspondingly, the history of Homeric philology has been studied in minute detail (which does not mean that there is no longer any point in investigating it—quite the contrary), while the history of Euclidean philology has only recently begun to attract the attention it deserves. And on the side of the sciences, the long-lasting focus on the part of the history of science above all upon scientific revolutions, inventions and progress, within the context of a broad commitment to defending the superiority of the West against other cultures, has led it to downplay the enduring structures of traditional instruction and stabilization in favor of heroic individuals and their disruptions of inherited configurations. This very volume is, by its existence, evidence for the gradual breakdown of such inherited barriers to the understanding of so widespread and significant a phenomenon as the philology of scientific texts—and, it is to be hoped, it will itself make a forceful further contribution to this development.

In fact, we may well suspect that in many regards, the philology of scientific texts derives in general from that of other canonical or scholarly literary texts (poetic, religious, legal, etc.): historically, in that often (though not always, such as in Mesopotamia) the poems were codified earlier; institutionally, in that these generally (though with exceptions, such as in India) played a more central role in education, especially in practices of language refinement and character formation, and were inculcated into a larger number of students; and methodologically, in that central categories of scientific philology (edition, commentary, scholia, translation, etc.) were established first in the case of non-scientific texts and then came to be applied to scientific ones as the need arose. And so it is not surprising that a number of the issues raised by the essays in this volume are not limited in their applicability to the philology of scientific texts but recur in virtually any philological activity whatsoever: for example, the fundamental ideological link between the sustained effort to stabilize texts and the reassuring hope that the world as a whole can thereby be stabilized too; the anchoring of editions and commentaries in schools and libraries and the role of patronage, especially royal patronage, in the foundation, legitimation, and protection of such institutions and of the scholars who flourished among them; the immeasurable and, on principle, unknowable penumbra of oral commentary and explanation surrounding the written documents which are our only available source of knowledge; the tension between center and periphery, and the effects of regionalization, dialects and different writing systems; the genres of scholarly activity (edition, commentary, translation); and the layout of the page (text, glosses, marks of separation and reference).

But the differences specific to scientific discourse inevitably pose at least some kinds of philological problems that differ from those raised by literary texts. For one thing, numbers and diagrams play a far more important role in scientific texts, especially in mathematical ones, than they do in other kinds of texts, and they pose particularly delicate problems, addressed by several of the authors in this volume, both for their pre-modern manuscript transmission and for their modern scholarly edition. For another, the agents involved in the philology of scientific texts—not only the authors and scholars but also the readers and students—must be presumed to have a higher degree of specialized knowledge of the field than is usually the case in other kinds of philology (and this leads to specific differences in the textual phenomena, such as abbreviations, brachylogy, revision, etc.). Again, scientific texts tend to belong to the genre of *Gebrauchstexte*, ones valued not so much for their specific linguistic form as rather for the useful information they conveyed, and hence are much more exposed to all kinds of modifications (additions, substitutions, deletions, updatings, corrections) than literary texts were. To give another example from Ancient Greek, Euclid was certainly a classic, valued not only for the accuracy of the information that could be detached from his text but also for the beauty and rigor of his text itself—and yet, with only a single exception, every one of our manuscripts of his *Elements* derives not from Euclid's own text but from Theon of Alexandria's revision of it, composed seven centuries later.

But more generally and, more significantly, the philology of scientific texts must always negotiate a double-edged set of loyalties: on the one hand, it must retain allegiance to the text as a linguistic document, with all the advantages and disadvantages of any such document; but on the other hand, it must also remain loyal to the objects to which that text refers and which provide its justification and meaning. A treatise of astronomy is not only a treatise but also one that makes certain verifiable claims about celestial phenomena, and a philological treatment of this treatise must do its best to do justice to both aspects. The difficulties involved are evident, and are made all the more complex by the fact that we must distinguish, with regard to the objects designated by such a treatise, between the way they are (i.e. are thought to be by ourselves) and the way they were (i.e. were thought to be by the authors). One expression of this dilemma is the tension, in the goals of the edition of such texts, between authenticity (following back the textual transmission to the earliest state to which the surviving witnesses give us access) and veridicity (presenting the text in such a way that its statements are in accordance with the phenomena to which it is referring). In terms of kinds of editorial interventions, this is the tension between conjectural emendations intended to correct errors introduced into scientific texts, like any other kind of texts, by scribal corruptions during the centuries of their transmission, and conjectural emendations intended to produce a text that is correct or accurate from the scientific point of view (these two kinds of emendations can, of course, coincide but they need not do so, and they often seem to be conflated in the minds of modern editors). Tacit correction of scribal errors in order to produce a scientifically correct text, transformation of non-decimal numeric systems into decimal ones, reduction of the absurdity of intentionally absurd geometric diagrams, and much more: there is a vast gamut of modern (and sometimes pre-modern) scholarly practices, all tending to standardize the ancient scientific texts and make them seem more like the modern ones with which most readers may be thought to be more familiar. It is in general only at a comparatively late stage in the history of a discipline that the value of authenticity can come to be thought to take general precedence over that of veridicity, so that later scholars can disdainfully criticize as unhistorical the standardizations performed by their earlier colleagues—ones which, in fact, had the effect (if not necessarily the intention) of making the texts in question seem more attractive and accessible to readers, in an early phase when too rebarbative an authenticity might well have put the very survival of the discipline at risk.

Already in pre-modern scholarship, there is a widespread preference to present texts, especially scientific ones, in such a way as to make them seem not textually authentic but instead scientifically correct—quite understandably, given the application in practical use for which such texts were usually intended and employed, and given the absence of a generally accepted historicization of textual transmission and the evolution of scientific disciplines. The complicity between this pre-modern state of affairs and the history of modern philological treatment of these texts, especially in the nineteenth century, raises perhaps the most delicate of all the questions posed by the essays in this volume: whether it is really quite safe for us to use our modern concept of the critical edition as the standard by which to judge all earlier philological activities and to consider that concept as the teleological goal towards which all philological endeavors were always aiming—or should have been. It is not accidental that so much of the rich historical documentation provided in this volume focuses on the nineteenth century: not only because, in the case of Mesopotamian studies, the field itself did not exist before that century and had to be invented then from scratch without the advantage of having any ancient models to follow, but also because it was during that same century that for the first time, both foreign and indigenous scholars in such cultures as India and China prepared for local traditional texts critical editions and scholarly translations that could be compared with, and were often modeled upon, those that were prevalent in the most up-to-date European philological disciplines. This is, of course, undeniably progress of a sort: but it must not be forgotten that it came not only with considerable advantages but also at the price of the transformation or indeed disruption of traditional cultural patterns and the loss of various kinds of knowledge and expertise, and that this price has not ceased to haunt the politics of philology until and including our own times. We might distinguish two extreme attitudes towards the question of the relationship between modern philological techniques and pre-modern ones—let us call them a *hermeneutical* view, which approaches both sides of the opposition with a cautious sense of their relative merits and defects, and an *empirical* view, which sees only progress from pre-modern scholarship to its modern successor—and if so, we might note that most of the authors in this volume tend to aggregate to one pole or the other of this antinomy and that this tendency is often linked to their disciplinary identity. Some of the essays here stress, hermeneutically, how little we know about the questions that we would most like to be able to answer and suggest that this is not only a matter of incomplete documentation but of the fundamental impossibility of finding satisfactory answers to them, given the impossibility of finding a stable starting point for assessing the evidence impartially; others suggest, empirically, that relatively simple remedies provided by the best modern philology could be applied to solve the nagging difficulties posed by pre-modern or modern philology.

Both of these tendencies are understandable—indeed, are inevitable. We are, all of us, modern scholars and we cannot pretend to be able to reproduce the techniques and consciousness of the ancient scholars we study; nor would we want to, even if we could, for it is only by understanding them in their difference from us that we can understand them at all. But it is one thing to recognize that there has been progress in certain regards from pre-modern philology to modern philology, and quite another thing to claim that this progress was unidirectional, necessary and free of any drawbacks. The goal of the study of the history of the philology of scientific texts, as of any other kind of philology, must surely be not only to prepare the best possible editions of those texts, but also to make us better able to assess with caution apparent similarities in the techniques attested in different periods or places, to treat with sympathy what might seem to be errors and defects of methods and results, and, more generally, to understand not only the differences but also the continuities between pre-modern and modern scholarship and to recognize not only what was gained but also what was lost in the passage from the one to the other.

The editors and authors of this volume are all established scholars, ones who have been contributing to deepening our understanding of the intersection between philology and science for many years. And the new insights and discoveries that mark so many of its pages are breathtaking: far beyond any reasonable expectation, they expand our horizons about the history of science, the history of philology and the history of the philology of science. Yet the impression that this book—itself a notable intersection of science and philology—leaves is that we are still only at the very beginning. This is a pioneering book: it points the way forward.

## Reference

Daston, Lorraine, and Glenn W. Most. 2015. History of science and history of philologies. *Isis* 106 (2): 378–390.

# Correction to: Editing the Sumerians, How and Why?

**Jerrold S. Cooper**

## Correction to:
## Chapter 10 in: A. Keller, K. Chemla (eds.), *Shaping the Sciences of the Ancient and Medieval World*, Archimedes 69, https://doi.org/10.1007/978-3-031-49617-2_10

In page 412, in Chapter 10 – Editing the Sumerians, How and Why?, the following corrections were updated:

1) The sentence in 2nd paragraph has been updated as follows:

"known today (many of these would be identified by Kramer himself over the following four decades)."

2) Note 1 has been replaced with Note 2, and Note 3 has been replaced with Note 4.

The updated version of this chapter can be found at
https://doi.org/10.1007/978-3-031-49617-2_10



A. Keller, K. Chemla (eds.), *Shaping the Sciences of the Ancient and Medieval World*, Archimedes 69, https://doi.org/10.1007/978-3-031-49617-2_16

# Annex A

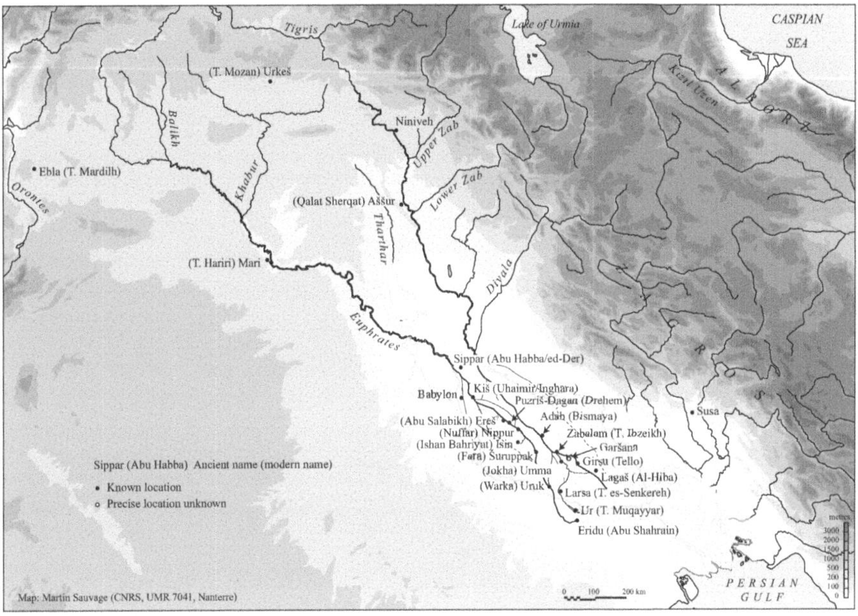


Map of Mesopotamia. (by Martin Sauvage CNRS, UMR 7041 ArScAn, Nanterre)


A. Keller, K. Chemla (eds.), *Shaping the Sciences of the Ancient and Medieval World*, Archimedes 69, https://doi.org/10.1007/978-3-031-49617-2

# Index





A. Keller, K. Chemla (eds.), *Shaping the Sciences of the Ancient and Medieval World*, Archimedes 69, https://doi.org/10.1007/978-3-031-49617-2

**F**

## O



## P

The manufacturer's authorised representative in the EU is Springer Nature Customer Service Centre GmbH, Europaplatz 3, 69115 Heidelberg, Germany. If you have any concerns regarding our products, please contact ProductSafety@springernature.com

Printed and bound by CPI Group (UK) Ltd, Croydon, CR0 4YY
01/10/2026
02225899-0017